THE BIOLOGY OF CYANOBACTERIA

BOTANICAL MONOGRAPHS

BOTANICAL MONOGRAPHS · VOLUME 19

THE BIOLOGY OF CYANOBACTERIA

EDITED BY

N. G. CARR
BSc, DPhil
Reader in Biochemistry
University of Liverpool
England

AND

B. A. WHITTON
MA, DSc
Reader in Botany
University of Durham
England

UNIVERSITY OF CALIFORNIA PRESS
BERKELEY AND LOS ANGELES 1982

UNIVERSITY OF
CALIFORNIA PRESS
Berkeley and Los Angeles

Library of Congress
Cataloging in Publication Data

The biology of cyanobacteria.

(Botanical monographs; v. 19)
1. Cyanophyta. I. Carr, N.G.
II. Whitton, B. A. III. Title: Cyanobacteria.
IV. Series.
QK569.C96B56 1982 589.4′6 82-1906
ISBN 0-520-04717-6 AACR2

Printed in Great Britain

CONTENTS

CONTRIBUTORS

D. G. ADAMS *Department of Biochemistry, University of Liverpool, P.O. Box 147, Liverpool L69 3BX, England*

H. BOTHE *Botanisches Institut der Universität Köln, II Lehrstuhl, Gyrhofstrasse 15, D-5000 Köln 41, Germany*

D. A. BRYANT *Microbiology Programme, Pennsylvania State University, S 101 Frear Building, University Park, Pennsylvania 16802, USA*

N. G. CARR *Department of Biochemistry, University of Liverpool, P.O. Box 147, Liverpool L69 3BX, England*

R. W. CASTENHOLZ *Department of Biology, University of Oregon, Eugene, Oregon 97403, USA*

Y. COHEN *Heinz Steinitz Marine Laboratory, P.O.B. 469, Elat, Israel*

G. COHEN-BAZIRE *Institute Pasteur, 28 Rue du Dr Roux, 75724 Paris Cedex 15, France*

W. F. DOOLITTLE *Department of Biochemistry, Dalhousie University, Halifax, Nova Scotia, B3H 4H7, Canada*

G. DREWS *Institut für Biologie II Mikrobiologie, Albert-Ludwig Universität, D-7800 Freiburg i. Br., Schänzlestrasse 1, Germany*

G. E. FOGG *Department of Marine Biology, Marine Science Laboratories, Menai Bridge, Anglesey LL59 5EH, Wales*

C. E. GIBSON *Freshwater Biological Investigation Unit, Greenmount Road, Muckamore, Antrim, BT41 4PX, Northern Ireland*

J. R. GOLECKI *Institut für Biologie II Mikrobiologie, Albert-Ludwig Universität, D-7800 Freiburg i. Br., Schänzlestrasse 1, Germany*

F. D. HEALEY *Freshwater Institute, 501 University Crescent, Winnipeg, Manitoba R3T 2N6, Canada*

M. HERDMAN *Laboratoire de Photosynthèse, Centre National de la Recherche Scientifique, 91190 Gif-sur-Yvette, France*

K. K. HO *Department of Biochemistry, Purdue University, West Lafayette, Indiana 47907, USA* (Present address: *Department of Botany, National University of Singapore, Singapore 0511, Republic of Singapore*)

D. W. KROGMANN *Department of Biochemistry, Purdue University, West Lafayette, Indiana 47907, USA*

J. M. NICHOLS *Department of Biochemistry, University of Liverpool, P.O. Box 147, Liverpool L69 3BX, England*

E. PADAN *Department of Microbiological Chemistry, Hebrew University–Hadassah Medical School, Jerusalem, Israel*

H. W. PAERL *Institute of Marine Sciences, University of North Carolina, Morehead City, North Carolina 28557, USA*

M. POTTS *Department of Biological Science, Florida State University, Tallahassee, Florida 32306, USA*

J. W. SCHOPF *Earth and Space Sciences Department, 3806 Geology Building, Los Angeles, California 90024, USA*

A. J. SMITH *Department of Biochemistry, University College of Wales, Aberystwyth, Wales*

R. V. SMITH *Freshwater Biological Investigation Unit, Greenmount Road, Muckamore, Antrim, BT41 4PX, Northern Ireland*

L. VAN LIERE *Laboratorium voor Microbiologie, Universiteit van Amsterdam, Nieuwe Achtergracht 127, 1018 WS Amsterdam, Netherlands*

A. E. WALSBY *Department of Botany, University of Bristol, Bristol BS8 1UG, England*

M. R. WALTER *Bass Becking Geological Laboratory, P.O. Box 378, Canberra, Australia*

J. WECKESSER *Institut für Biologie II Mikrobiologie, Albert-Ludwig Universität, D-7800 Freiburg i. Br., Schänzlestrasse 1, Germany.*

B. A. WHITTON *Department of Botany, University of Durham, Durham DH1 3LE, England*

C. P. WOLK *MSU-DOE Plant Research Laboratory, Michigan State University, East Lansing, Michigan 48824, USA*

FOREWORD

R. Y. STANIER

The reader will note that this treatise bears the title of *The Biology of Cyanobacteria*, not *The Biology of Blue–Green Algae*, the title of its predecessor. This is a significant departure from tradition. For some 180 years, the cyanobacteria have been classified as *algae*, by virtue of their typical oxygenic phototrophy.

Many properties of the cell set the cyanobacteria apart. Even studying the cyanobacteria without the aid of electron microscopy, Fritsch (1945) gives a perceptive summary:

> The Myxophyceae or Blue–green Algae are more clearly circumscribed than any other class of the pigmented Protophyta. The low state of cell differentiation, the usual blue–green pigmentation, the abundant production of compounds containing glycogen . . . the absence of advanced states of thallus organization, the often copious production of mucilage . . . and, lastly, the customary absence of clearly differentiated reproductive organs, all tend to give a very definite stamp to members of this class. Both in cell structure and in the lack of specialization of the plant body, the Myxophyceae exhibit a relatively low state of differentiation. . . . One is left with the impression of an archaic group that has possibly persisted with little alteration during long epochs of the earth's history.

What to do about this problem, taxonomically speaking? For well over a century, it has remained an open question. Ferdinand Cohn (1875) showed a clear awareness of it, in his discussion of the relationships between bacteria and blue–green algae:

> Perhaps the designation of *Schizophytae* may recommend itself for this first and simplest division of living beings, which appears to me to be naturally segregated from the higher plant groups, *even though its distinguishing characters are negative rather than positive* [author's italics].

Indeed, that was the catch about it. As recently as 40 years ago, Stanier and van Niel (1941) could do little better, in an attempt to define collectively these two groups. The issue was at last resolved (at least, to the author's satisfaction) by the discovery of a major evolutionary discontinuity, at the cellular level, amongst all biological systems. I allude to the distinction of two super-kingdoms, eukaryotes and prokaryotes. I think it is profoundly significant that the fundamental differences between eukaryotes and prokaryotes could not be rigorously formulated prior to approximately 1960 (Stanier & van Niel, 1941; Stanier, 1961; Stanier & van Niel, 1962). It was André Lwoff who proposed these two names during a discussion that I had with him in 1961. He revived these historically appropriate names from oblivion, citing as his authority an equally obscure and rare publication from the great French protozoologist Edouard Chatton (1938). Nine years later, I failed to mention Chatton's proposal (Stanier, 1970) in the course of preparing a chapter in 'Organization and Control in Prokaryotic and Eukaryotic Cells', for the 20th annual SGM symposium: a sure sign that these terms have entered history. This induced the editors, H.P. Charles and B.C.J.G. Knight, to make a false attribution. Quite independently (and about 20 years later than Chatton) Dougherty (1957) had proposed the prokaryotic–eukaryotic dichotomy.

In addition to the organisms long recognized as 'bacteria' (viz., the chemo- and photolithotrophs and chemo- and photoorganotrophs), the super-kingdom of prokaryotes includes a large and widely distributed group of prokaryotes, among which predominate the oxygen-evolving cyanobacteria. They can be defined succinctly as microorganisms that harbour, within a typically prokaryotic cell, a photosynthetic apparatus closely similar in structure and function to that located within chloroplasts of phototrophic eukaryotes. Speaking with hindsight, their traditional classification as algae was based on resemblances between the cyanobacterial cell as a whole, and one constituent of the (eukaryotic) algal cell: the chloroplast.

ROGER Y. STANIER
died at his home near Paris on
29 January 1982

PREFACE

During the last decade cyanobacteria have become generally recognized as a major group which provides experimental material for all aspects of microbiology—with the commendable exception of invasive animal pathogenicity! The pace of accumulation of knowledge has demanded a new volume to present a broad view of their biology; this is very evident from the large proportion of recent publications in our bibliography. We did not attempt, in 1973, to bring together within *The Biology of Blue–Green Algae* a totally comprehensive account of these organisms, nor is this the aim of the present volume. Instead we have aimed for an eclectic account in which the currently most interesting aspects of cyanobacterial biology are seen within the context of the most recent descriptions of their structure, biochemistry and ecology. Inevitably some interesting topics will, given the exigences of publishing life, escape full coverage: perhaps cyanobacterial soil ecology and their symbiotic association are examples. The use of the name Cyanobacteria, rather than its fondly associated precursor, reflects their position in biology; as cyanobacteria increasingly contribute to microbiology, it is important that an appropriate name be used, a change which has occurred in many major journals. One of our contributors has retained the use of 'blue–green algae'—and his long-term contributions to our subject have stayed the editorial hand! We thank warmly all our contributors, and others who have aided with advice and provided photographs; Robert Campbell and Mike Bodinham at Blackwell Scientific Publications have again combined enthusiasm with patience and Derek Phillips gave invaluable help at the proof stage. It is necessary to close with a recognition of loss; Roger Stanier, who contributed our Foreword, died earlier this year. This is not the place to record his many contributions to microbiology; suffice to say here that our understanding of cyanobacteria has been enlightened by his perception, and our volume distinguished by probably his last written work.

N.G. Carr
B.A. Whitton

CHAPTER 1
CYANOBACTERIA: CURRENT PERSPECTIVES

B. A. WHITTON & N. G. CARR

During the past two decades cyanobacteria have become generally recognized as a major group in the prokaryote kingdom—certainly one of the most diverse, judged by variety and number of species, and perhaps the largest. As discussed in the Foreword, popular recognition of their true place in the biological hierarchy was preceded by increasingly precise electron microscopy, which helped to undermine previous views about their links with other algae. This grouping with other algae had been based largely on the fact that all these organisms possess essentially common photosynthetic structures and mechanisms. The explanation for these similarities, of course, lies behind the resurgence of the endosymbiotic hypothesis for the evolutionary origin of chloroplasts, a topic that has become extensively studied and reviewed (see Fredrick, 1981). The expansion of variety caused by the inclusion of cyanobacteria in the prokaryotes has been accompanied by the ever-increasing recognition of the variety in form, metabolism and ecological roles of bacteria themselves.

Although there are predominant traits in their metabolism and ultrastructure, cyanobacteria do exhibit a wider range than was previously thought. As many people without specialist knowledge are likely to use this book, it may be helpful both to summarize these traits and to comment on some of the exceptions. Cyanobacteria include all, or all but one, known prokaryotes capable of oxygen-evolving photosynthesis. Just which definition is used depends on whether or not the genus *Prochloron* is accepted as a cyanobacterium (see below). Cyanobacteria can form not only chlorophyll *a*, but (again with the exception of *Prochloron*) the phycobilin–protein pigment complexes based upon phycocyanin, phycoerythrin and allophycocyanin (see Chapter 7) which contribute to the organism's characteristic colour. Certain genera of colourless filamentous gliding bacteria are morphologically quite similar to some cyanobacteria (*Beggiatoa* to *Oscillatoria* for instance) but, in spite of several claims in the literature, nobody has yet produced a viable colourless mutant of a photosynthetic form. There is a widespread dependence upon carbon dioxide as a carbon source, with the attendant importance of the reductive pentose phosphate pathway in biosynthetic mechanisms. The majority of cyanobacteria are easy to recognize by electron microscopy, with thylakoid and polyhedral bodies being the structures most obvious in a typical thin section. The thylakoids are considered to be sites for both photosynthesis and respiration, although the location of the latter activity requires much further work. In the case of a remarkable isolate by Rippka *et al.* (1974), *Gloeobacter violaceus*, the cell lacks thylakoids altogether: photosynthesis and respiration are apparently associated with the plasma membrane (Guglielmi *et al.*, 1981). It is not known whether the plasma membrane also has this role in some of the other cyanobacteria (Peschek & Schmetterer, 1980). The polyhedral bodies, named originally for their shape (Pankratz & Bowen, 1963) are now recognized as carboxysomes (Chapter 3).

There are several other attributes of cyanobacteria which, though restricted to a minority of species, can still be considered as important features of the group. These include complementary chromatic adaptation, the presence of gas vacuoles, morphological changes, particularly the formation of heterocysts associated with the onset of nitrogen deficiency and formation of tapered hairs in whole sections of filament associated with the onset of phosphate deficiency. Only the changes associated with nitrogen fixation are unique to the group. Gas vacuoles occur in other types of planktonic prokaryotes, while complementary chromatic adaptation and morphological changes in filaments associated with phosphate deficiency occur in certain groups of eukaryotic algae. Nevertheless these are all cyanobacterial phenomena which increasingly attract interest at various levels of study and for which the group is particularly well suited for experimental analysis. If we may paraphrase a sentence of Doolittle's (1979) 'cyanobacteriologists would do well to study those things that cyanobacteria do well'.

At the metabolic level the very widespread and possibly universal distribution of the reserve material cyanophycin, a copolymer of asparate and arginine (Chapter 3) would appear to be unique to cyanobacteria—there are no reports of its detection outside the group. The presence of cyanophycin and the now established role of phycocyanin as a protein reserve material as well as a photopigment, endow the cyanobacteria with nitrogen reserve capacity. This is an important feature of metabolism with the potentially great selective advantage that it is absent from other prokaryotes although, of course, well established within certain plant tissues. It would be instructive to know the extent to which bacteria may mobilize reserves of nitrogen, presumably from sources other than specific reserve materials and to compare the efficacy of the process with that found in cyanobacteria.

The two most important examples of changed views on the understanding of the metabolic limits of the cyanobacteria concern photosynthesis and nitrogen fixation: in both cases it was the study of anaerobic or microaerobic environments which helped lead to the relevant discoveries. The frequent association of cyanobacteria with waters rich in hydrogen sulphide and the fact that addition of small amounts of sulphide may promote the growth of some cyanobacteria in culture led several authors to speculate that sulphide may be directly involved in the metabolism of these organisms. The realization that a true anoxygenic photosynthesis may take place in certain cyanobacteria comes, however, from studies on *Oscillatoria limnetica* from Solar Lake near the mouth of the Gulf of Elat (Chapter 9). The metabolic strategies of this organism in exploiting aerobic and anaerobic photosynthesis and its dissimilatory activities under either condition results in its ability to grow in widely fluctuating environments. Perhaps many readers will agree that there is an intellectual satisfaction to be derived from the recognition that a cyanobacterium is capable of switching from a 'bacterial' photosynthesis dependent on one photosystem to the oxygen-evolving photosystem of 'chloroplast' type involving two photosystems. This demonstrates the essential unity of the photosynthetic process and ease of transition between the two with apparent sharing of light capture and electron transport molecules. It illustrates beautifully, over 40 years after its enunciation, the famous van Niel unifying hypothesis, in which all photosynthetic reactions are viewed as having a common basic mechanism. Thus we have the situation whereby some cyanobacteria have been described as having close structural and phylogenetic affinities to chloroplasts of particular higher algae, whilst others become, under specific environments, physiologically very similar to certain photosynthetic bacteria.

As more and more examples of cyanobacteria with heterocysts were shown during the 1940s and 1950s to be capable of fixing atmospheric nitrogen, the hypothesis (Fay *et al.*, 1968) that it was the heterocyst itself which is the site of fixation became accepted very rapidly when positive evidence was first put forward to support it (Chapter 14). Within a short period of time the International Biological Programme was encouraging a survey of heterocyst frequency in cyanobacteria-dominated communities in order to get some indication of the importance of cyanobacterial fixation in these communities. It is perhaps not suprising therefore that when a brief paper reporting fixation by the unicellular cyanobacterium *Gloeocapsa* was published (Wyatt & Silvey, 1969), it was at first treated sceptically by some researchers. However, when a strain of the non-heterocystous *Plectonema boryanum* was also shown to exhibit nitrogenase activity (Stewart & Lex, 1970), though only under anaerobic or microaerobic conditions, the theory that the presence of a heterocyst was always essential for nitrogen fixation had to be discarded. However, the heterocyst is typically needed for aerobic conditions, providing a microaerobic site within the filament. A relatively large number of non-heterocystous strains have now been found in the laboratory to show nitrogenase activity, usually under near-anaerobic conditions, but it is far from clear just how important non-heterocystous nitrogen fixation is in nature. While it seems likely that the process is important in many environments with anaerobic conditions near the surface, such as salt marshes and paddy fields, no extensive studies have yet been carried out to document this. Further, we have little idea how far it will prove possible to predict the ability to carry out non-heterocystous nitrogen fixation by observing morphology as seen with the light microscope. It is possible that all strains of the marine *Trichodesmium* exhibit nitrogenase activity (see Chapter 19). In the case of *Microcoleus chthonoplastes* it may be that some strains possess the ability, while others lack it (see Chapter 20). Almost certainly it will eventually become clear that the ability is entirely lacking in some species, but it is at present impossible to do more than guess which these might be.

The development of heterocysts and akinetes, cells which are structurally modified and functionally specialized, illustrates what is perhaps the most important characteristic of cyanobacteria relative to other prokaryotes. There is no equivalent degree of differentiation and development elsewhere. Other important examples of cellular differentiation among the cyanobacteria are hormogonia formation, development of hair cells and cell cycle-associated changes. Full accounts of heterocysts and akinetes appear in later chapters; what may be emphasized here are the constraints and controls over their formation. Like all other types of prokaryote differentiation, the onset of their formation is triggered environmentally; in this way it contrasts with differentiation in higher organisms, which is frequently inate or obligately integrated into the cell sequences. Within the imposition of an ordered pattern of heterocysts in response to nitrogen limitation, there lies a response of the whole filament to the environmental change. The ordered degree of cell differentiation leading to the maintenance, as growth proceeds, of a tightly controlled pattern of heterocysts throughout the whole filament produces what may justly be thought of as a multicellular structure. Thus a filament of vegetative cells may change from several hundred essentially independent cells into two population types which are virtually interdependent and which have quite different roles and features. The exact degree of independence of the constitutive cells of undifferentiated filaments should be assessed with some caution, as there is now clear evidence of interaction between the components of such filaments. Interaction between individuals of cyanobacterial populations may in fact be general, especially when the individuals are held together in a matrix as in *Aphanothece stagnina*, *Nostoc* and *Rivularia*. Cyanobacteria which exhibit some of the attributes of multicellular organisms may prove to be valuable experimental systems for the study of intercellular activities.

It is generally assumed that the presence of heterocystous cyanobacteria in a natural population reflects a relative deficiency of combined nitrogen within that population. There is in fact good evidence to support this from the study of changes taking place within particular lakes, but the evidence so far is more ambiguous when comparisons are made of widely differing types of environment. An ability to interpret physical and chemical features of the environment by observing the dominant cyanobacteria present is clearly of considerable importance for the management of lakes, fish ponds and rice fields. It is a pity, therefore, that we still know remarkably little about the physiological differences between the main genera of cyanobacteria—between, for instance, *Anabaena*, *Anabaenopsis* and *Aphanizomenon* in lakes and between *Anabaena*, *Aulosira*, *Cylindrospermum* and *Gloeotrichia* in rice fields. Part of the difficulty lies with the time and trouble required to get most strains into axenic clonal culture. The application of good microbiological practice has resulted in a greatly increased number of such cultures and there are now several hundred strains in the culture collections. Nevertheless, from an ecological viewpoint these still do not represent anything approaching a cross-section of the variety of cyanobacteria seen in nature. Further, many isolates have been obtained with scant regard for the original environment from which they were obtained. Enrichment cultures have frequently been made with levels of, for instance, potassium and phosphate, well above anything ever likely to be encountered at the site from which the material was taken. Presumably this has often led to the selection of strains which would stand little chance of survival in the field. The bias toward the study of only a few selected species, with the consequent restrictions on generalization, is especially pronounced with the more biochemical types of study. The papers of Kratz and Myers (1955a,b), which described conditions that permitted rapid growth and characterized the basic physiological behaviour of three species, were crucial in the establishment of '*Anacystis nidulans*' as a major experimental organism. We need a similar approach today to bring many more cyanobacteria from nature into the laboratory culture vessel. This time, though, full details of the field environment should be recorded, considered during the various stages of laboratory isolation and kept at the back of the mind when subsequently designing experiments. Further, a more critical approach is needed towards the maintenance of genetic stability in key research strains.

Not only have many of the forms described in classical botanical floras (e.g. Geitler, 1932) yet to become the subject of modern study, but further strange forms are still being found. For instance, there are several widely scattered records of planktonic Oscillatoriaceae with trichomes well over 100 μm wide. In contrast, *Starria zimbabweënsis*, whose trichomes are triradiate in section and were illustrated with such elegance by Lang (1977), has so far been recorded from only one site. Many types of tropical environment likely to be rich in cyanobacteria, such as limestone rocks in regions of high rainfall, have still received only cursory study, so possibly further quite different forms remain to be found. In view of their present diversity, it is intriguing to speculate on the morphological range of Precambrian cyanobacteria, at a time when the majority of eukaryotic algae had not yet evolved to compete with them.

A feature of the cyanobacteria is the ease with which they appear to have entered into symbiotic associations. On the world scale other types of symbiotic association such as the corals and green alga-containing lichens are probably of greater quantitative importance, but no other group of microorganisms has entered into such a wide range of symbiotic associations. Examples of symbiotic cyanobacterial associations exist with one or more representatives of the following groups: green alga, diatom, liverwort, moss, fern, cycad, angiosperm, phycomycete, ascomycete, protozoa, sponge, bryozoa, echiuroid worm, opisthobranch, didemnid ascidian, shrimp, mammal. There are also several descriptions, none as yet characterized unequivocally, of loose, but specific, associations between cyanobacteria and other prokaryotes. Cyanobacteria may be extra- or intracellular symbionts and may supply to, or receive from, the other organism dissolved organic carbon; wherever the cyanobacterium is capable of nitrogen fixation, then fixed nitrogen is apparently passed to the other organism. It is from former intracellular symbionts that it is assumed that modern cyanelles have arisen, such as those inside *Cyanophora paradoxa* and *Glaucocystis nostochinearum*, and perhaps also, much earlier in evolution, true chloroplasts. At least two modern intracellular cyanobacterial symbionts can fix nitrogen, so it is strange that this ability appears to be lacking in all chloroplasts (unless two recent reports for *Chlorella* become confirmed).

One group of symbiotic organisms is in particular need of critical study, if the limits of the cyanobacteria are to be clearly understood. These are the photosynthetic organisms associated with many didemnid ascidians. Some of these are apparently fairly typical cyanobacteria (Kott, 1980 and pers. comm.), whereas others are chlorophyll *b*-containing, phycobilin-lacking prokaryotes for which the genus *Prochloron* has been erected (Lewin, 1977). Other features are also known to separate *Prochloron* from the rest of the cyanobacteria, such as the stacking of the thylakoids and the central area free of all thylakoids (Whatley, 1977): nevertheless, the separation of this genus as a separate phylum, the Prochlorophyta (Lewin, 1976), seems premature (Antia, 1977) in view of the lack of research on the blue–green coloured prokaryotes associated with other didemnid ascidians. As the Didemnidae are relatively young in evolutionary terms (perhaps not evolving until the Tertiary (P. Kott, pers. comm.), *Prochloron* must presumably either have existed previously as a free-living organism or have evolved since then. In view of the hypotheses already erected around this organism, we hope the reader will forgive our suggesting a still further one. This is that *Prochloron* evolved from a typical cyanobacterium, acquiring the ability to synthesize chlorophyll *b* either by molecular evolution from chlorophyll *a* or its precursors or by a chance transfer of the relevant genetic information from some green alga. As a consequence of the possession of chlorophyll *b*, further evolutionary changes such as the loss of phycobilins and the changed thylakoid arrangement might have occurred rapidly. If there is a strong selective advantage for some prokaryotes to possess chlorophyll *b* rather than phycobilin pigments on the tropical reef-ridges where these organisms occur, then it would seem possible that further examples of chlorophyll-*b*-containing prokaryotes may be found in shallow marine waters. If *Prochloron* is eventually accepted to belong with the cyanobacteria rather than being separated into a phylum of its own, then this clearly affects the descriptive limits of the cyanobacteria. (See the recent study of ribosomal RNA sequences by Seewaldt and Stackebrandt, 1982.)

Cyanobacteria have not been neglected in the resurgence of interest in applied microbiology, which is hardly surprising in view of their photosynthetic and nitrogen-fixing capabilities. The mass culture of algae, fashionable 30 years ago, has re-emerged as a legitimate goal. Cyanobacteria growing in natural situations are used by several human cultures as food and several commercial undertakings successfully market *Spirulina* from Mexican and Taiwanese lagoons. This has encouraged the belief that *Spirulina* may have a major role as cattle or human food substitutes. Whether the extreme claims for the health-giving properties of this organism currently being advocated by several Japanese and USA organizations help or hinder this belief remains to be seen. However, much as we are attracted to the idea of *Spirulina* or other cyanobacteria being used widely as food, we should be cautious in advocating the extension from harvesting semi-natural populations to their mass culture on an industrial scale. There is no evidence that the energy balance of cultures which become light-limited at relatively low densities and which require active harvesting, will ever become favourable. Cyanobacteria may have a more realistic role to play in the immediate future as a source for the extraction of high-value individual components, such as phycobilin pigments for colouring foodstuffs.

Time alone will tell whether the more glamorous studies concerned with cyanobacterial nitrogen fixation eventually prove to be of practical value: whether, for instance, entirely new symbiotic associations can be developed or whether genes for nitrogenase activity can be transferred successfully to higher plant crops. At present the lines of research which appear to offer the most immediate return are those which concern the nitrogen economy of certain intensively farmed areas of the tropics and subtropics, such as fish ponds and soils which carry rice for all or part of the year. Although early studies on attempts to increase the annual contribution of cyanobacterial nitrogen fixation in rice fields were often rather naïve, some modern research projects appear very promising. These include the distribution of dried cyanobacteria with bags of fertilizer, the growth in village ponds of inocula suitable for local fields (Agarwal, 1979) and the development of the right conditions in any one area for maximizing nitrogen fixation by *Azolla* with its symbiotic *Anabaena* (Lumpkin & Plucknett, 1980).

Some of the most exciting discoveries about cyanobacteria could not have been carried out without extensive research funds, and future studies on their molecular biology are likely to prove even more expensive. Nevertheless, some important discoveries of the past decade were relatively simple to carry out and depended mainly on the common sense or originality of the researchers concerned. One of the most striking is the finding (Johnson & Sieburth 1979; Waterbury *et al.*, 1979; Morris & Glover, 1981) that small-celled cyanobacteria can form an important component of the oceanic phytoplankton crop (Chapter 19). Prior to this, most oceanographers collected plankton samples on membrane filters too coarse to hold these cyanobacteria and it had been assumed that the chlorophyll passing through these filters was mainly degradation product. Millions of dollars must have been spent on oceanographic surveys of chlorophyll content, with the intent of measuring biological productivity, which must now be considered to have provided results of doubtful quality; certainly it will require many further surveys before the value of the earlier work can be assessed properly.

The final comments in this chapter will deal with a problem that has troubled many people researching on cyanobacteria, the difficulties that can occur in giving an organism a name. This is of most obvious concern in descriptive field surveys (Chapter 20), and the production of regional floras such as those currently being planned for China, Mexico and no doubt many other countries. It can, however, also prove a serious hindrance to laboratory studies. Here problems of synonymy can be a nuisance (see Chapter 11), but the most serious difficulties arise because descriptive taxonomy has developed over a period of one and a half centuries and descriptions have been made according to the nomenclatural rules of the Botanical Code. There appears at present no way of avoiding these 'botanical' species for the provision of floristic lists from different communities, but the lack of a living holotype makes the system of little value for many types of modern critical study. For this and other reasons, Stanier *et al.* (1978) have proposed that cyanobacteria should be treated under the conventions of the Bacteriological Code. Although the acceptance of one or other of the codes has led to heated argument among their protagonists, it seems clear that at present neither is fully adequate to deal with all the types of situation for which a name is needed. Whatever anyone argues, it is likely that both codes will continue to be used in parallel for some years to come. In view of this, it is important that the approaches used should be made as compatible as possible with each other and whenever standardization is feasible, such as is the use of descriptive terms and size categories, then it should be adopted.

CHAPTER 2
INTERACTIONS OF CYANOBACTERIA WITH LIGHT

L. VAN LIERE & A. E. WALSBY

2.1 Introduction

All cyanobacteria are photoautotrophic organisms capable of oxygenic photosynthesis. While some of them are able to live photoheterotrophically in dim light and a few are able to live heterotrophically in the dark, photoautotrophy is their predominant mode of nutrition and the availability of light is one of the principal factors which determine their growth.

This chapter deals with the ways in which the growth and physiology of cyanobacteria are affected by light and the consequences this has for their biology and ecology. Central to these considerations is first the dependence of these organisms on light for a supply of radiant energy, and that the photosynthetic system which makes the energy conversion will be limited at low irradiance, saturated at high irradiance and inhibited at very high irradiance. Secondly, the light-dependent activities of cyanobacteria are strongly wavelength-dependent because they are based on photochemical reactions by pigments which absorb in discrete regions of the spectrum. There are genotypic differences in the pigments of different species of cyanobacteria that determine the scope of the responses of these organisms to light. Thirdly, like other photosynthetic organisms, cyanobacteria are able to vary their pigment content and other aspects of their physiology in such a way as to optimize their performance in light of different irradiance and quality. Fourthly, some cyanobacteria have the ability to move along light gradients in a way that may take them to positions which are suitable, if not optimal, for their growth.

There are many light-based reactions which determine the success of particular cyanobacteria under certain conditions, and some are beyond the scope of this review—in which we have restricted our considerations to the main responses that determine photosynthetic growth in relation to the light field. Thus we have omitted the specialist topics of photomorphogenesis, covered in detail by Lazaroff (1973), and the mechanisms of damage by ultraviolet light and light-mediated nucleic acid repair. The subjects of photorespiration, photooxidation and photoheterotrophy are given brief mention here, mainly to put them into context. Photosynthesis itself is reviewed in Chapter 8 and anoxygenic photosynthesis in Chapter 9. Other intractions with light are covered in Sections 16.4, 18.5.5, 19.3.1, 19.3.2 and 20.3.4.

We have entitled our chapter *interactions* of cyanobacteria with light because we wish to emphasize that while light affects cyanobacterial growth, the growth of cyanobacteria results in the attenuation of light, a phenomenon which is often ignored in laboratory cultures and one which ultimately determines the success of cyanobacteria in competition with other organisms in nature.

2.2 Measurement and attenuation of light

2.2.1 Systems of light measurement

In studying the effect of light on the growth of photoautotrophic organisms it is important to know what the distribution is of light in the culture vessel. This is also true when studying the effects of other factors since the light levels should then be neither inhibiting nor limiting. In this section we briefly discuss some of the problems of measuring light in cultures because we feel that they have not been fully appreciated by cyanobacteriologists in general and that this has generated difficulties in comparing results from different laboratories. For a good introduction to light measurement and absorption we recommend the book of Clayton (1977).

Light is electromagnetic radiation which, strictly speaking, falls within the visible part of the wavelength spectrum. However, for the purposes of photobiology this definition is broadened to include the ultraviolet and infrared wavebands that generate photochemical reactions in living organisms. Three units of measurement are used in the SI system to describe the rate at which light falls on a surface: lumen m^{-2} or lux, W m^{-2}, μEinstein s^{-1} m^{-2}.

Until recently, the first of these was the most widely used in biology. The lux measures light intensity with reference to a standard source and judged according to the sensitivity of the human eye. The measurement integrates, albeit unevenly, light over the visible spectrum, from about 400 to 700 nm and as such provides a rough-and-ready estimate for photochemical reactions, like photosynthesis, which occur within the same region. If the same light source is used (e.g. a given type of incandescent or fluorescent light), lux measurements can be used for comparative purposes, but comparisons between readings with different light sources may give large errors, as the human eye, and photometers calibrated in lux, are much less sensitive to blue and red light than to green. Hence, the sensitivity of the lux meter, usually a photoelectric cell with a correcting filter in front of it, does not correspond to that of photosynthesis over the visible spectrum. It gives rather a poor indication of the amount of photosynthetic energy or quanta that a culture experiences, especially when there are changes in spectral distribution of light by selective absorption (Fig. 2.1, Table 2.1).

Absolute values of light flux can be measured or expressed as either a quantum flux or an energy flux, related in the following way.

Absorption of one quantum can cause a photochemical reaction by one molecule; 1 mol of a substance requires N quanta, where N is the Avogadro number. The SI units of quantum flux, i.e. the number of quanta which arrive at a unit of surface per unit time are Einstein s^{-1} m^{-2}, where

$$1 \text{ Einstein} \equiv N \text{ quanta} \qquad (2.1)$$

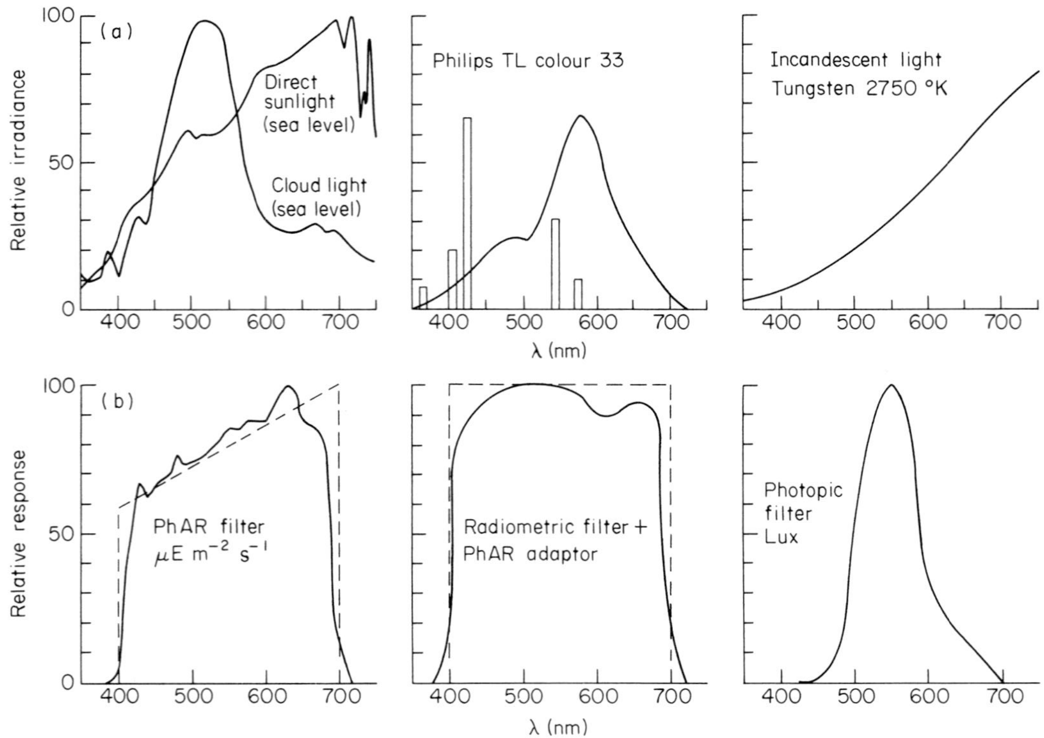

Fig. 2.1. (a) Relative irradiance of three light-sources and (b) relative response with filters used in a MACAM Q 301 head unit. Broken lines denote ideal responses; values calculated from data given by the manufacturers. The total irradiance of the Philips TL colour 33 is obtained by summing the histogram and the relative irradiance curve.

Table 2.1. Conversion of units used in light measurements for some light sources. (A more complete survey is given by Wetzel, 1975)

Nature of light-source	Units		
	$W\ m^{-2}$	μEinstein $m^{-2}\ s^{-1}$	lux‡
Monochromatic light			
$\lambda = 400$ nm	1	3.3	
$\lambda = 500$ nm	1	4.2	
$\lambda = 600$ nm	1	5.0	
$\lambda = 700$ nm	1	5.9	
Philips TL 33 (white)	1	4.6*	280
Philips TL 55 (cool white)	1	4.5*	250
Incandescent light (Tungsten 2750°K)	1	5.1*	250
Sunlight sea level	1	4.6†	250

* Calculated from data given by the manufacturer.
† Independent of weather conditions (Morel & Smith, 1974).
‡ Gaastra (1959).

Because of the small numerical values which otherwise result, photobiologists usually use μEinstein $s^{-1}\ m^{-2}$ as the units for quantum flux.

If the wavelength of the light is known, the equivalent energy flux can be calculated using the relationship

$$E = h \cdot c / \lambda \qquad (2.2)$$

where E is the energy per quantum (in Joules), λ is the wavelength (in metres), h is Planck's constant (6.6×10^{-34} J·s) and c is the velocity of light (3×10^{8} m s^{-1}). Using the relationship given in eqns (2.1) and (2.2) the equivalence between quantum flux density and energy flux density can be calculated. For any light other than monochromatic light it is necessary to integrate over the region of the spectrum considered. The choice of 'quantum sensor' or 'energy sensor' depends on the aim of the researcher but for knowledge of both, either both instruments or a spectroradiometer are needed. In practice, the commercially available sensors are usually photodiodes (with linear response in current per time) screened by the appropriate correction filters that determine the slope of the wavelength response curve and exclude ultraviolet and infrared radiation (see Fig. 2.1). The photodiode-filter sensors come calibrated against standard light sources or thermopiles. Measuring devices (photocells, photodiodes or photoresistors) without correcting filters can be used to measure irradiance or quantum flux for light of a given spectral distribution by calibrating them against standard instruments. However, if the spectral composition is changed, by change in the type of light source or by selective absorption within a culture, the calibration will no longer hold, and recalibration is needed (van Liere, 1979; Loogman & van Liere, 1978).

The commercial instruments are also cosine-corrected. This means that if the sensor is rotated in parallel light, the response is linearly proportional to the cosine of the angle of incidence. As a result one can measure the average irradiance inside a culture by measuring in two opposite directions at each point.

2.2.2 *Measurement of mean light irradiance in a culture*

Light scattering and selective absorption complicate the measurement of the mean light irradiance in a culture. Considerable differences may exist between the irradiance at the surface and centre of the vessel, particularly with dense cultures. The concentration of organisms, their pigmentation, the presence of gas vacuoles and other refractile organelles, the geometry of the culture vessel and the shape and nature of the light source may all markedly affect the distribution of light in a culture. There are several approaches to determining the mean light irradiance in a culture and the proportion of the light absorbed.

Light penetration is diminished by both absorbance and scattering. In dense cultures as little as 5% of the incident light may re-emerge (Loogman & van Liere, 1978). Such cultures, which appear virtually black, are useful for determining the efficiency of phototrophic growth (see below). In the case of gas-vacuolate cyanobacteria, however, the gas vacuoles back-scatter a proportion of the incident light with the result that less penetrates the culture (Fig. 2.2), which has a lighter appearance. The effect of collapsing the gas vacuoles on light penetration is discussed below (2.8.2).

At the other extreme, very thin layers of dilute cultures (e.g. Shear & Walsby, 1975) can be used when it is necessary to know the light irradiance at the cell surface accurately, as under these conditions it is virtually the same as the irradiance incident at the culture surface. This incident irradiance is the one usually reported even when there is substantial absorption in the culture. Under these conditions an improved estimate of the mean irradiance (E_m) in the culture would be obtained by integrating the irradiance measured at the front (E_o) and back (E_d) faces of the vessel, with respect to the light path. Then

$$E_m = \frac{E_o - E_d}{\ln E_o - \ln E_d}. \qquad (2.3)$$

This formula assumes that Beer's law holds, and this is strictly true only when the light enters in parallel rays and when there is no change in the spectral composition of the light. Thus, even application of this correction

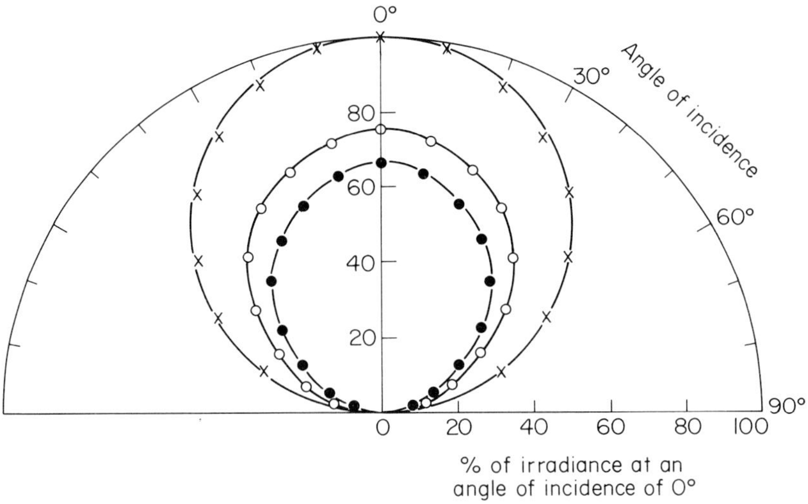

Fig. 2.2. The percentage angular distribution of light irradiance in a circular vessel illuminated from one side in an otherwise darkened room with a black background. The measurements were made at 10° intervals with a small photocell (surface area 28 mm^2) at the end of a vertical glass rod placed in the centre of the vessel and rotated through 360°. The readings in the backward-facing semicircle (not shown) were approximately zero. Vessel containing water (×); vessel containing a culture of *Anabaena flos-aquae* with gas vacuoles intact (●), and with gas vacuoles collapsed (○). The gas vacuoles cause back-scattering of light, thereby reducing light penetration in the culture. The total amount of light received from all directions by a cyanobacterium at the centre of the vessel will be proportional to the area of the polar plot. (A.E. Walsby & M.J. Booker, unpubl.)

leads to inaccuracies with increasing culture density where light usually becomes enriched in the green region. It then becomes necessary to measure the light-irradiance profile through the culture and from this to calculate the mean value. Loogman and van Liere (1978) and van Liere (1979) have described a means of doing this. They designed a small light sensor, which could be placed at any point inside a vessel, and they developed a method of integrating the measurements made at all the different positions to give a mean irradiance for the culture as a whole. This is equivalent to the average irradiance a cell would experience as it is stirred around inside the vessel. This method can be used with vessels of various shapes and sizes as long as the sensor can be manipulated to all parts of the culture, and it makes no assumptions about the arrangement of the light source. Van Liere (1979) used a cylindrical culture vessel surrounded by one or more circular fluorescent lamps (Fig. 2.6) for culturing *Oscillatoria agardhii*. The vessel can be used in continuous cultures (see 2.5) and sufficiently large quantities of cyanobacteria or algae may be grown under a defined light regime for many physiological investigations.

Another approach was used by Aiba and Ogawa (1977) who grew *Spirulina platensis* in both photostat and turbidostat continuous cultures in square cross-section cuvettes which enabled them to measure the incoming and outgoing light irradiance. In this way they could determine the production of cells (g h^{-1}) per unit of light energy (J h^{-1}). If the energy content of the cells had also been measured it would have been possible to calculate the efficiency of the light energy conversion, i.e. the amount of energy incorporated into the cells per light energy supplied (see 2.5.1).

Ogawa *et al.* (1979) carried out continuous culture experiments with their cuvette inside a spherical spectrophotometer sensor whitened on the inside and equipped with a small hole to admit a light beam of known irradiance to the culture. All of the light energy that escaped the culture system through non-absorbance could thus be measured and, by difference, the amount absorbed could be calculated. This is a novel approach but is rather limited in usefulness by the small size of the culture vessel which produces only small amounts of cell material.

2.3 Effect of light irradiance on rates of photosynthesis and growth

Photosynthesis is limited by light at low irradiance. As the radiant intensity is raised the photosynthetic rate will approach the maximum or light-saturating level determined by the intrinsic rate of enzyme catalysed reactions. Such intrinsic rates rise with temperature. At very high light irradiance the rate of photosynthesis may actually decrease owing to photorespiration and photo-oxidative damage. These inhibiting reactions increase with time of exposure to high irradiance because they are brought about by time-dependent activities such as the build-up of oxidants.

If the cell carbon content remains constant then the relative rate of carbon increase μ_c (mg C increase per mg cell C), will be identical to the instantaneous growth rate, μ. Comparison of this rate with the relative rate of carbon fixation (mg C fixed per mg cell C) will reveal the rate of C loss from the system by processes such as respiration and extracellular production. To date there has been no simultaneous measurement of growth rate and photosynthetic rate at different light irradiances. However, W. Zevenboom (pers. comm.) has determined the growth rate of a strain of *Anabaena flos-aquae* whose photosynthesis has been investigated by Shear and Walsby (1975). The comparison is shown in Fig. 2.3, where it is seen that at low irradiance the rate of exponential carbon increase estimated by the two methods is very similar. Photosynthesis saturates at a higher radiant intensity than growth, however, indicating that the maximum growth rate is limited by some process other than carbon assimilation; this may be the uptake of a mineral nutrient, or perhaps the assimilation of nitrogen. The comparisons made in Fig. 2.3 make

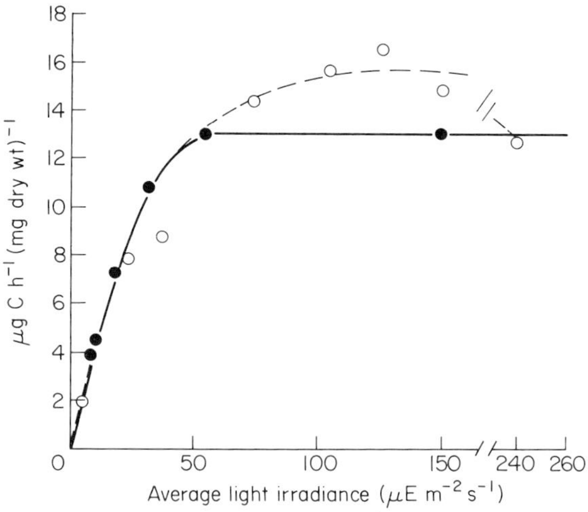

Fig. 2.3. Growth rate (●) and rate of photosynthesis (○) of *Anabaena flos-aquae* as a function of light irradiance. The rate of photosynthesis was calculated from the data of Shear and Walsby (1975) and the growth rate from the data of W. Zevenboom (pers. comm.)

certain assumptions (that cell C is 46% of the dry weight—the figure determined for *Oscillatoria agardhii* by L. van Liere) and the exercise should now be repeated using the same cultures for both sorts of measurement. A particularly important experiment to do is the measurement of the photosynthetic rate (mg C fixed per mg cell C) at different radiant intensities of cells grown at different irradiance, as this will reveal whether the organism is able to adapt its photosynthetic mechanism to the prevailing light conditions.

The growth of another, apparently related strain (see Booker & Walsby, 1979, for strain histories) of *Anabaena flos-aquae* over a range of light irradiances has been investigated by Foy *et al.* (1976). The maximum growth rate of this organism (equivalent to a doubling time of 21 h) is a little faster than that described above (minimum doubling time 24 h) and was reached at a much lower irradiance. This emphasizes the importance of fully describing the organism used in such studies and demonstrates that generalizations cannot be made about the growth responses of cyanobacteria to light.

There have been a number of other studies on the rates of photosynthesis by cyanobacteria at different radiant intensities but almost all of these are in relative units, e.g. ct/min ^{14}C fixed per ml culture, which may be useful for demonstrating the relative sensitivity of photosynthesis to different conditions but cannot be used for predicting the growth rate.

There are a number of studies on growth rates of planktonic cyanobacteria expressed under a range of different radiant intensities (Table 2.2). The value at which half-maximum growth rate generally is reached (K_s) is low as

Table 2.2. Data on growth constants for light-energy (Monod kinetics) recalculated to appropriate units from growth curves presented in the literature (see also Table 18.4)

Organism and source of isolate	Illumination	T (°C)	pH	N source	μ_m (h^{-1})	μ_m (d^{-1})	K_s ($W\ m^{-2}$)	Inhibition irradiance ($W\ m^{-2}$)	Reference
Anabaena flos-aquae (Windermere)	Continuous	20	8.0	NO_3^-	0.029	0.70	3.0	>70	W. Zevenboom (pers. comm.)
				N_2	0.022	0.53	2.0		
Anabaena flos-aquae (Windermere)	Continuous	10	7.6	NO_3^-	0.014	0.33	1.0	<8	Foy *et al.* (1976)
	Continuous	20	7.6	NO_3^-	0.033	0.78	0.7		
	Light/dark 8/16	10	7.6	NO_3^-	—	0.19	1.7	>8	
	Light/dark 8/16	20	7.6	NO_3^-	—	0.35	1.0		
Aphanizomenon flos-aquae (Brielse Meer)	Continuous	20	8.0	NO_3^-	0.033	0.60	3.0	>80	W. Zevenboom (pers. comm.)
				N_2	0.024	0.55	3.0		
Aphanizomenon flos-aquae fa *gracile* (Lough Neagh)	Continuous	10	7.6	NO_3^-	0.014	0.33	0.3	>8	Foy *et al.* (1976)
	Continuous	20	7.6	NO_3^-	0.041	0.98	1.0		
	Light/dark 8/16	10	7.6	NO_3^-	—	0.18	0.7	>8	
	Light/dark 8/16	20	7.6	NO_3^-	—	0.27	1.3		
Oscillatoria agardhii (Loughgall)	Continuous	10	7.6	NO_3^-	0.010	0.24	0.3	>8	Foy *et al.* (1976)
	Continuous	20	7.6	NO_3^-	0.034	0.82	0.7		
	Light/dark 8/16	10	7.6	NO_3^-	—	0.17	1.0		
	Light/dark 8/16	20	7.6	NO_3^-	—	0.31	1.7		
Oscillatoria agardhii (Veluwemeer)	Continuous	20	8.0	NO_3^-	0.036	0.86	1.0	>40	van Liere & Mur (1980)
	Continuous	28	8.0		0.048	1.15	1.0	>40	
Oscillatoria redekei (Lough Neagh)	Continuous	10	7.6	NO_3^-	0.015	0.37	<0.3	>8	Foy *et al.* (1976)
	Continuous	20	7.6	NO_3^-	0.046	1.09	<1.0		
	Light/dark 8/16	10	7.6	NO_3^-	—	0.14	0.7	>2	
	Light/dark 8/16	20	7.6	NO_3^-	—	0.30	0.7	>3	
Oscillatoria rubescens (Lac de Nantua)	Light/dark 8/16	20	7.3	NO_3^-	—	0.25	2.0		Feuillade (1972)

compared with green algae (van Liere & Mur, 1980). Both photosynthesis and growth rate as a function of the light irradiance need to be studied at limiting, saturating and inhibiting radiant intensities (see also 2.4.2).

2.3.1 *Inhibition of photosynthesis at high irradiance*

At high irradiance photosynthetic rates are depressed by photooxidation and photorespiration. The principal agent of photooxidation is the superoxide ion, O_2^-, This is generated through reduction of the oxygen molecule by electrons generated in the photosynthetic system. The production of O_2^- occurs particularly in high concentrations of oxygen and low concentrations of CO_2, which provides an alternative sink for electrons; both of these conditions are promoted by high light irradiance. The superoxide ion may cause damage to cytochromes, pigments and other components of the photosynthetic apparatus, thereby depressing subsequent photosynthetic rates. The ion has a fleeting existence, disappearing by the dismutase reaction

$$O_2^- + O_2^- + 2H^+ \rightarrow H_2O_2 + O_2$$

Although this takes place very rapidly without the enzymic catalysis, the O_2^- ion is so toxic that an enzyme, superoxide dismutase (SOD), is necessary to rid the cell of it (McCord & Fridovich, 1977). Cyanobacteria, like all other aerobic organisms, contain active SOD enzymes. The enzymes isolated from this group are of the ferri type and manganese type.

Abeliovich *et al.* (1974) found that SOD was an inducible enzyme system in *Anacystis nidulans*. It was present in cells grown under aerobic conditions, in cultures bubbled with air, but was not detectable when the cultures were bubbled with nitrogen gas containing 5% CO_2, even though oxygen was present at reduced concentrations owing to the photosynthetic activity of the cyanobacterium. They found that the SOD level dropped under conditions which Abeliovich and Shilo (1972) showed to cause photooxidative death, i.e. in the presence of 100% O_2, absence of CO_2 and under a radiant intensity of 10 W m^{-2}. The onset of photooxidative death varied according to the pre-treatment conditions but in each case it was observed to occur after the SOD level in the cells had fallen to about 10% of its normal value. For example, pretreating the cells in air with 5% CO_2 prolonged to 40 h both their subsequent survival under photooxidative conditions and the maintenance of SOD levels. The activity of SOD was negligible in cells grown in cultures bubbled with N_2 gas containing 5% CO_2 (giving microaerobic rather than anaerobic conditions because the cells generated oxygen by photosynthesis) and such cells died after only a short period if transferred to photooxidative conditions.

Eloff *et al.* (1976) showed that photooxidative conditions existed, in Israel, at the surfaces of fish ponds at which cyanobacterial waterblooms were observed to form. Samples of several laboratory strains of cyanobacteria, including species of *Nostoc*, *Microcystis*, *Aphanocapsa*, *Anabaena* and *Plectonema*, all became photooxidized when suspended in dialysis tubing at the pond surface and analysis showed that their demise was again correlated with a sharp fall in their SOD level. However, when samples of the natural *Microcystis* population were similarly exposed, they survived and this survival was correlated with cell SOD activity being maintained at a high level. Either the SOD was not degraded in these natural strains or it was replaced as fast as it was destroyed. The latter is more likely because it was observed that treatment with a protein synthesis inhibitor, chloramphenicol, resulted in a decreased SOD activity. In the absence of axenic strains of *Microcystis*, Steinitz *et al.* (1979) isolated a mutant of *Plectonema boryanum* resistant to photooxidation which possessed similar properties, i.e. it survived and maintained high SOD levels under photooxidative conditions. The wild type of *Plectonema* was found to have two SOD isoenzymes, a ferri-SOD, which is inactivated in the presence of hydrogen peroxide, and a manganese-SOD which is stable under these conditions. While the peroxide-sensitive isoenzyme accounts for 90% of the activity in the wild type, they found the other form to be more abundant in the strains resistant to photooxidation.

The conditions which promote photorespiration are essentially the same as those which give rise to photooxidation. Photorespiration is the light-stimulated uptake of oxygen which is brought about by the oxidation of ribulose 1,5 bisphosphate (RuBP). The reaction is catalysed by the enzyme RuBP carboxylase, which is responsible for photosynthetic carbon fixation. Oxygen and carbon dioxide compete for the RuBP substrate on the active sites of the enzyme. Instead of being carboxylated, RuBP is oxygenated to form phosphoglycerate and phosphoglycolate; the latter is hydrolysed to glycolate. The oxygenase reaction is favoured by high pH and, of course, low CO_2 and high O_2 availability (Tolbert, 1974). In contrast to dark respiration, photorespiration does not conserve, but consumes energy. However, due to this energy requirement plants are protected from oxygen-dependent photoinduced damage (Heber & Krause, 1980).

2.4 Adaptation to light

2.4.1 *Alternating light–dark regimes*

In natural conditions cyanobacteria will normally experience alternating light and dark periods. The amount of growth sustained over such alternating periods will depend on the relative length of the light period, the irradiance during the light period, and the respiratory activity. Loogman *et al.* (1980) have investigated the effect of the length of the light period on the growth of *Oscillatoria agardhii* using saturating light irradiance. As shown in Fig. 2.4, the overall growth rate increased with the length of the light period but not in direct proportion. It appears that as the length of the light period increases the efficiency of light-energy incorporation decreases. There are two possible explanations for this. First, the faster growing cells under the longer photoperiods will have a higher respiratory rate. Second, there may be a

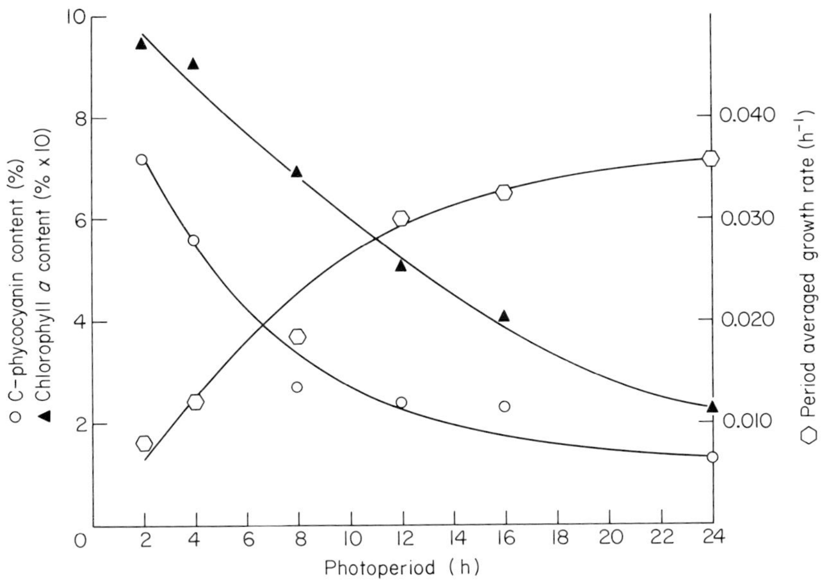

Fig. 2.4. Growth rate and pigmentation of *Oscillatoria agardhii* as a function of the length of the photoperiod. The total length of the light–dark cycle was uniformly 24 h. Growth rate was period averaged. Irradiance was saturating. (From Loogman *et al.*, 1980 and pers. comm.)

time-dependent inhibition of photosynthesis which progressively increases during the light period. The light-energy harvesting pigments, chlorophyll *a* and C-phycocyanin, increase in content as the photoperiod is shortened (Fig. 2.4). This adaptation to a light–dark regime gives *O. agardhii* the possibility of maintaining a high photosynthetic activity during short photoperiods, so that it may survive prolonged periods in the dark.

Foy *et al.* (1976) found a similar increase in growth rate relative to light period on transferring cells from continuous light to a 6:18 light–dark cycle. They showed that the increase was different in *O. agardhii* and *O. redekei* and that this difference could be used to explain the relative success of the two organisms in Lough Neagh, Northern Ireland, where they form an important component of the phytoplankton.

2.4.2 *Adaptation to light irradiance*

Undoubtedly, different species of cyanobacteria show marked differences in their tolerance and photosynthetic efficiency to irradiance. The cyanobacteria of hot springs, desert varnishes and tropical pools evidently flourish in radiant intensities of full sunlight (exceeding 300 W m^{-2}) which are far in excess of irradiances which kill other species. *Oscillatoria agardhii*, a planktonic species of temperate lakes, is killed by prolonged exposure to irradiance exceeding 70 W m^{-2}. This is evidence of genotypic variation, parallel to the sun-and-shade species of higher plants.

Like the eukaryotic algae investigated by Steemann Nielsen (1976) and Beardall & Morris (1976), the prokaryotic cyanobacteria also show phenotypic adaptation to light irradiance. Jones and Myers (1965) found that there was a fourfold decrease in the chlorophyll-*a* content of *Anacystis nidulans* when grown at high irradiance. This pigment reduction can be viewed as likely to increase the efficiency of the photosynthetic system since, with an increased photon flux, less pigment is required to sustain the energy requirement of cells at their maximum division rate. However, it has yet to be proved that the changes which occur in the photosynthetic system actually produce an increase in the instantaneous growth rate under a given set of conditions. Experiments with continuous cultures of the sort described by Beardall and Morris (1976) with the diatom *Phaeodactylum tricornutum*, should be carried out, though to calculate instantaneous growth rates the rate of carbon fixation should be related to the cell carbon content, rather than cell number or chlorophyll content which may give misleading comparisons.

2.4.3 *Adaptation to light quality*

2.4.3.1 Pigment composition

Jones and Myers (1965) found that the chlorophyll content of *Anacystis nidulans* changed with the spectral composition of light used. On increasing the wavelength through the series 619, 659, 666 and 680 nm the cell chlorophyll content dropped while the phycobilin content was little changed. As the wavelength is increased an increasing proportion of the light is absorbed by chlorophyll, supporting photosystem I, and a decreasing proportion by the phycobilins, supporting photosystem II. An imbalance between the two photosystems is redressed by reducing the chlorophyll content, as in the light irradiance response discussed above.

2.4.3.2 Complementary chromatic adaptation

Certain cyanobacteria which possess both the blue pigment C-phycocyanin and the red pigment C-phycoerythrin show marked changes in the proportions of these phycobiliproteins when grown under light of different spectral composition (Chapter 7). This phenomenon is termed complementary chromatic adaptation, because of the contrasting colours of the organism and the light in which it is grown. The significance of this adaptation is that it permits cyanobacteria to exploit light filtered by other overlaying pigmented organisms. In particular, the C-phycoerythrin can be used to scavenge the green light unused by most other algal groups.

2.5 Role of continuous culture

2.5.1 Introduction

For studying kinetics of microbial growth a choice of culture system must be made between a closed batch culture and open continuous culture. A closed system has limited volume and limited life. It is often started with an artificially high nutrient concentration in order to obtain measurable amounts of cell material. The growth cycle in this system passes through a series of phases from the inoculation to an eventual cessation of growth. In these phases there are significant changes in the physiological and biochemical properties of the cells. The supposition that these phases are fundamental to growth of microorganisms is, of course, fallacious; they are imposed by the system (Tempest, 1970; Droop, 1974a; Rhee, 1979). With phototrophic organisms an additional complication is the change in light irradiance caused by mutual shading during the increase in biomass. Thus, to study growth kinetics of phototrophic organisms the batch culture is of value only in its early stages when the cell concentration does not result in significant light attenuation. This requires cultures which are shallow in a plane normal to the incident light and which do not reach high cell densities, and it imposes considerable restriction on the uses to which the culture can be put (e.g. Guillard *et al.*, 1973; Steemann Nielsen, 1978). These theoretical and technical difficulties can be avoided by studying growth kinetics in the continuous culture. The great advantage of this system is that by virtue of constant dilution with fresh medium the environment is of constant and reproducible composition and the cells are maintained in a fixed physiological state, which makes studies of physiological properties much more reliable than with closed systems. There are several types of continuous culture system. In the turbidostat the culture density is kept constant by means of an external system, which regulates the rate of medium inflow. In the chemostat one nutrient is present in a growth-limiting concentration, so that the medium flow rate determines the growth rate. The dilution rate (flow rate/culture volume) is the independent variable. If the light-energy supply is the growth-limiting factor it is better to refer to the apparatus as a photostat, i.e. the biomass is determined by the constant influx of light energy. When in steady state, which means that growth rate equals dilution rate, all of these continuous culture systems are comparable. However, the turbidostat operates most effectively near or at the maximum growth rate of the organism under study while the chemostat and photostat must be operated at sub-maximum growth rates in order to maintain them in equilibrium conditions. A simple scheme for a continuous culture is shown in Fig. 2.5, while a system suitable for cultivating fragile filamentous cyanobacteria is shown in Fig. 2.6.

Normally, the continuous culture is used for investigating one limiting factor at a time. For example, van Liere and Mur (1979b) have used the photostat to investigate the effect of light irradiance on the growth efficiency of *Oscillatoria agardhii*. Van Liere (1979) determined maximum growth rates by means of a turbidostat and van Liere *et al.* (1979) studied the pigment variation as a function of growth rate and irradiance using a chemostat culture. Recently, continuous cultures have been used to study interactions between limiting factors; for example, light and nutrients, temperature and nutrients (Rhee & Gotham, 1981a,b). Continuous cultures can also be used to study the physiology of cyanobacteria and other algae during light–dark cycles (Rhee *et al.*, 1981; van Liere, Mur, Gibson & Herdman, 1979).

For detailed information on the theory and applications of the continuous culture see Tempest (1970) and a recent review by Rhee (1979). The kinetics of

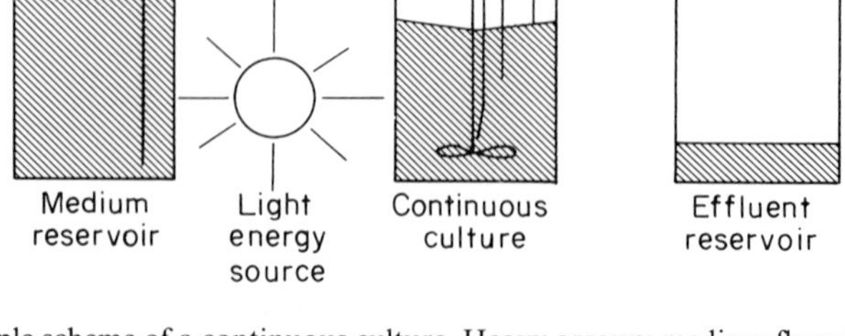

Fig. 2.5. A simple scheme of a continuous culture. Heavy arrows; medium flow: light arrows; gas flow.

Fig. 2.6. Continuous culture system (van Liere, 1979). (1) Light-source Philips TLEM 40W/33, (2) medium and pump, (3) thermostat, (4) pH control, (5) flow of moistened air, (6) cotton wool filters, (7) sampling system, (8) effluent, (9) filter of sintered glass.

light-energy limited growth are dealt with by Pipes and Koutsoyannis (1962) and van Liere and Mur (1979b), but will briefly be considered in this chapter. Although continuous cultures cannot reproduce natural environments, they have tremendous capacity to reveal basic mechanisms of what is happening in natural waters (Rhee, 1979).

2.5.2 *Light-energy limited growth*

Suppose a phototrophic organism is cultured in a closed system at constant and optimum irradiance with all nutrients in excess. The growth rate will depend on the light energy absorbed by the biomass. This will initially be high when a small inoculum is used. The organisms will express their maximum growth rate potential, and an exponential phase will ensue. By their own growth, however, mutual shading increases, which will decrease the specific (per unit of biomass) light-energy uptake rate. As a result, the growth rate begins to fall.

As the biomass continues to increase, so the growth rate continues to fall; an interesting feature of this phase is that the total biomass of the culture increases linearly with time because the production per unit of time is proportional to the absorption of light by the culture (Tamiya *et al.*, 1953; van Liere & Mur, 1979b). Ultimately a stationary phase will be reached, in which all the available energy is used for population maintenance.

The question arises as to whether it is possible to maintain steady state in a continuous culture in light-energy limited conditions. As the dilution rate is exponential, production over time is constant when light energy limits growth (see above). Thus, $\mu X = K$ (mg l^{-1} h^{-1}). The rate of biomass change in a continuous culture, which generally is depicted as $dX/dt = \mu X - DX$ (X = biomass concentration; $\mu = (dX/dt)(1/X)$ specific growth rate and D = dilution rate), will change into $dX/dt = K - DX$. On integration this linear differential equation will give

$$X = \frac{K}{D} + \left(X_s - \frac{K}{D}\right) e^{-Dt} \tag{2.4}$$

with boundary conditions $t = 0$ and $X = X_s$ (light-saturated biomass). After the culture has been running for a certain time the last term becomes negligible and a steady state is reached. The steady-state biomass concentration will be inversely proportional to the dilution rate (Pipes & Koutsoyannis, 1962; van Liere & Mur, 1979b).

Van Liere and Mur (1979b) have described the relationship between the growth rate and the specific light-energy uptake rate of photoautotrophs with the following formula

$$\mu = q_E \cdot c - \mu_e \tag{2.5}$$

in which μ is the specific growth rate $(dX/dt)(1/X)$, q_E the specific light-energy uptake rate $(dE/dt)(1/X)$, μ_e is the specific maintenance rate constant and c is the growth efficiency factor.* Without any direct knowledge of the mechanism one can say that $q_E \cdot c$ is the potential maximum growth rate that would be achieved if no maintenance energy was required, and μ_e is the amount by which growth rate is reduced below this potential maximum because of diversion of energy for maintenance. It is convenient to express the biomass X in terms of the energy it contains (J) and to determine it by measuring the heat of combustion; c then becomes dimensionless. Four examples of the

* Equation (2.5) is essentially the same as the formula derived by Pirt (1965) for heterotrophic bacteria.

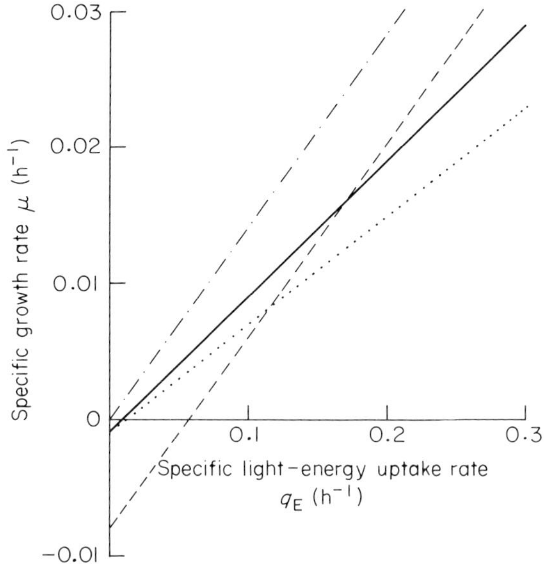

Fig. 2.7. Energy balance calculated after extrapolation to comparable growth conditions: pH 8.0; temperature 20°C; incident irradiance 12 W m^{-2}. –·–, Theoretical line when there is no maintenance energy requirement; – – – – – –, *Scenedesmus protuberans*, calculated from Gons and Mur (1980); and J.G. Loogman (pers. comm.); ————, *Oscillatoria agardhii* (van Liere & Mur, 1979b); · · · · · ·, *Aphanizomenon flos-aquae* (Zevenboom & Mur, 1980). (The first two organisms were grown on nitrate; *A. flos-aquae* was cultivated under nitrogen-fixing conditions.)

relationship expressed in eqn. (2.5) are shown in Fig. 2.7. Theoretically, if no maintenance was required a linear function with slope c passing the origin would be found. However, a maintenance requirement causes the line to be shifted to the right as is observed, for example, with *Scenedesmus*: μ_e is then given by the intercept with the ordinate. If c changes, the function would have a different slope (compare the results obtained with *Oscillatoria* and *Aphanizomenon*). Theoretically, it is debatable whether μ_e is constant or a function of growth rate (Tempest & Neijssel, 1976). However, eqn. (2.5) is a valid description of growth both in light-energy limited and nutrient-limited conditions (van Liere & Mur, 1979b). At high light irradiance and at high growth rates the model deviates from linear.

The growth efficiency factor c is influenced by any process that leads either to a change in efficiency with which light energy is converted into chemically stored energy (such as changes in efficiency of light trapping and transfer by the photosystem, fluorescence and uncoupling of growth and energy generation), or to losses of the stored energy (e.g. by photorespiration or excretion). Another factor influencing c is the nitrogen source in the growth medium. It requires more energy per mol N to reduce and assimilate nitrate than it does for ammonium.

Differences in μ_e and c of different organisms will result in selection of species in light-energy limited ecosystems. A comparison of three strains is given in Fig. 2.7. Although *Scenedesmus protuberans* has a higher growth efficiency factor than *Oscillatoria agardhii*, its μ_e value is also an order of magnitude higher. This means that when the specific light-energy uptake rate is low, *O. agardhii* will express a higher growth rate. This will be so at low light irradiance in dilute culture (Fig. 2.8). Van Liere and Mur (1979b) have

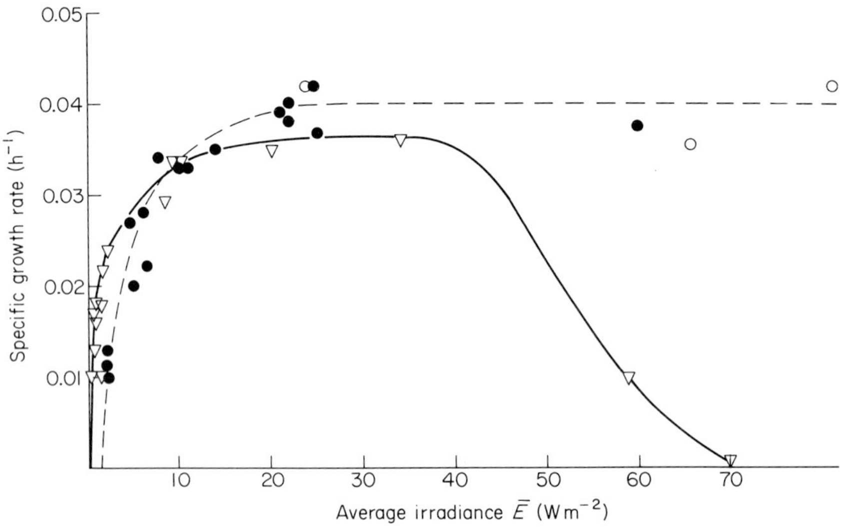

Fig. 2.8. Specific growth rate at 20°C as a function of average irradiance. ▽, *Oscillatoria agardhii* Gomont, pH 8.0 and continuous illumination; ●, *Scenedesmus protuberans* with saturating photoperiods; ○, continuous illumination. pH varied between 7.5 and 8.2. (Data from van Liere *et al.* (1978) and J.G. Loogman pers. comm.)

compared several strains of cyanobacteria and green algae and concluded that μ_e of the cyanobacteria considered varied within the range 0.001–0.004 h^{-1}, while μ_e of green algae varied from 0.008 to 0.015 h^{-1}.

A similar argument was used to explain why the N_2-fixing *Aphanizomenon flos-aquae* does not become dominant in well-mixed, light-energy limited systems (Zevenboom & Mur, 1980). Although the μ_e value was comparable with that of such species as *Oscillatoria agardhii*, the growth efficiency factor was considerably lower, probably as a result of heterocyst formation which results in energy being diverted into non-dividing cells.

2.6 Competition and metabolic differences

2.6.1 Experiments involving two organisms

Differences in μ_e and c (2.5.2) will determine the outcome of competition in systems where light-energy absorption is low, that is at low average light irradiance, but also at high light irradiance with high biomass concentration.

Competition experiments in continuous cultures have been performed by Mur *et al.* (1978) with *Oscillatoria agardhii* and *Scenedesmus protuberans*. Both strains were adapted before the experiment to their future environment and then inoculated into the 'competition' vessel in which subsequent growth was followed. The first experiment was done at low irradiance (Fig. 2.9a). At this irradiance the cyanobacterium can express a higher growth rate (see Fig.

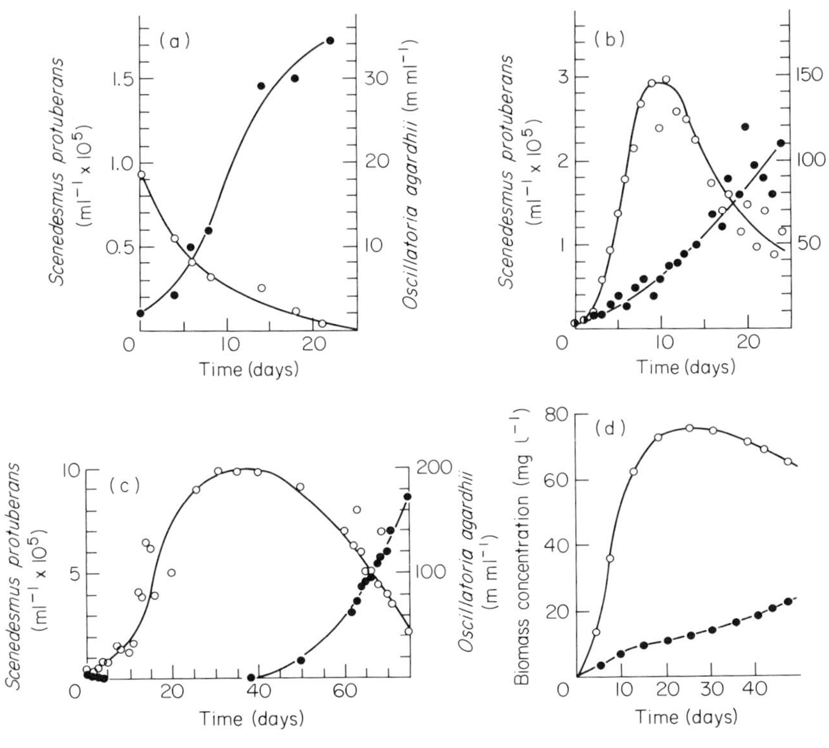

Fig. 2.9. Competition for light energy between the green alga *Scenedesmus protuberans* (○) and the cyanobacterium *Oscillatoria agardhii* (●). (a), (b), and (c) in continuous culture, temperature 20°C, pH 8.0, continuously illuminated and with nitrate as nitrogen source (Mur *et al.*, 1978). (a) Incident irradiance 1.0 W m^{-2}; dilution rate 0.01 h^{-1}. (b) Incident irradiance 25 W m^{-2}; dilution rate 0.03 h^{-1}. (c) Incident irradiance 39 W m^{-2}; dilution rate 0.03 h^{-1}. (d) Computer simulation of biomass increase in a shallow lake (Mur & Beijdorff, 1978).

2.8) and it increased in biomass. *Scenedesmus protuberans* also expressed positive growth but the rate was lower than the dilution rate, and hence it was progressively washed out and eventually eliminated from the culture.

The second experiment (Fig. 2.9b) was carried out at optimum irradiance for both organisms and both rapidly increased in biomass concentration. By virtue of the increase in concentration of both organisms the mean light irradiance in the culture decreased and their specific light-energy uptake rate therefore decreased. This had an adverse effect on *S. protuberans* before *Oscillatoria agardhii* with the result that the growth rate of *Scenedesmus protuberans* after 10 days fell below the dilution rate and subsequent washout was observed. The cyanobacterium, however, remained in conditions which enabled it to increase in biomass.

The third experiment was performed at 39 Wm^{-2} (Fig. 2.9c). In the culture system used (see Fig. 2.6) this implied an irradiance still optimum for *Scenedesmus* growth, but inhibitory to *Oscillatoria*. Subsequently the latter was washed out completely, *Scenedesmus protuberans* increased at maximum rate and finally reached steady state. At this point the biomass concentration of the green alga was very high, so a low specific light-energy uptake rate was established. At this low q_E the cyanobacterium was reinoculated. The shading in the culture was then sufficient to permit *Oscillatoria* to grow. The increase in biomass of *O. agardhii* next disturbed the steady state growth of *Scenedesmus protuberans*. The energy absorbed by *S. protuberans* decreased below the equilibrium value, which resulted in its washout.

Mur and Beijdorff (1978) have used μ_e values and growth rate responses to light irradiance (see Fig. 2.8) of *Oscillatoria agardhii* and *Scenedesmus protuberans* in a computer simulation of phytoplankton growth. They considered the case of a hypothetical well-mixed lake divided in distinct layers of light irradiance and calculated growth rates in these layers, which, averaged over the water column, gave the biomass increase (Fig. 2.9d).

The succession, from green algae or diatoms to 'shade-type' cyanobacteria (observed after the first groups had reached high biomass concentration as a result of nutrient enrichment) is well described (Berger, 1975; Skulberg, 1978; Jones, 1977b; Ahlgren, 1971) although simultaneous measurements of radiant intensity performed during the succession are scarce (but see Fig. 2.10).

In light-energy limited competition experiments, as described above, Zevenboom *et al.* (1981) isolated a heterocystless mutant of *Aphanizomenon flos-aquae*. The higher energy demand needed in the wild type for heterocyst production, which was not fully suppressed by the nitrate in the growth medium, led to an increase in μ_e, and a decrease in c under the conditions of the competition experiment. The mutant was also able to maintain a higher growth rate at lower light-energy absorption. The outcome of the competition is shown in Fig. 2.11.

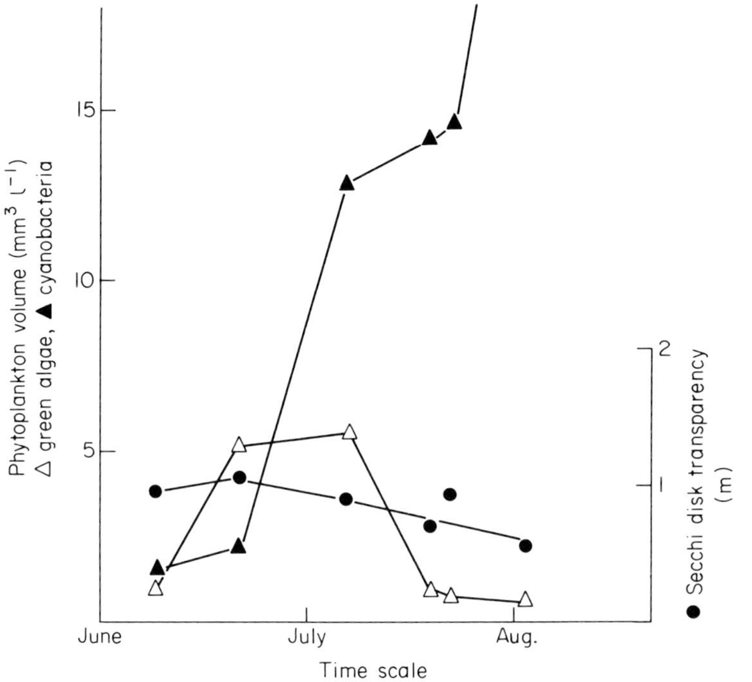

Fig. 2.10. Succession from green algae to cyanobacteria (dominated by *Aphanizomenon flos-aquae* and *Oscillatoria agardhii*) in Norrviken (Sweden). (From Ahlgren, 1971.)

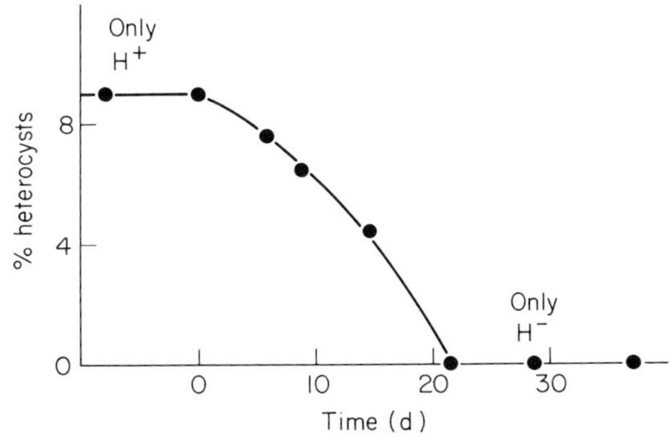

Fig. 2.11. Competition experiment with *Aphanizomenon flos-aquae*. H^+, heterocyst producing wild type, H^-, mutant that did not produce heterocysts. The latter was inoculated at $t = 0$ into a steady state culture of the wild type organism. (Zevenboom *et al.*, 1981.)

Walsby and Booker (in prep.) have carried out competition experiments in semi-continuous culture, diluted manually at intervals, between the wild type and a gas-vacuoleless mutant of *Anabaena flos-aquae* (described by Walsby, 1977). Over seventeen generations there was no significant difference in the growth rate of the two forms.

2.6.2 *Time-dependence of growth kinetics*

The eutrophic light-energy limited environment in which several cyanobacteria become dominant is of increasing ecological importance. However, most planktonic species probably evolved in nutrient-limited environments, where various species coexist. Another important factor is that they usually experience alternating light–dark regimes, which have a profound influence on their growth-kinetics (2.3). The time-dependence of growth in the light–dark cycle is reflected in various physiological properties.

Nutrient consumption rates seem, by and large, to be time-dependent, i.e. when organisms are cultured in a light–dark cycle, different species reach their maximum uptake potential at different times during the cycle. This has been shown already for several diatoms and green algae (Rhee *et al.*, 1981), and with two species of phototrophic bacteria (van Gemerden, 1974), but to date there seem to be no accounts of time-dependent nutrient uptake in cyanobacteria.

The time-dependent nutrient consumption, however, can explain the coexistence of species (van Gemerden, 1974). It is, however, not only the time-dependence of nutrient uptake which determines the success of one species in the presence of competitors. The ratio of the time spent in the light to that spent in the dark may also be important (2.4.1).

2.6.3 *Decay in the dark*

When photoautotrophic organisms are deprived of light, their energy source, they will be expected to decrease in biomass because of respiratory activity and the energy consumed in maintenance processes. In natural conditions this deprivation occurs every night. At the biochemical level, survival in the dark can be considered as a problem of maintaining the internal ATP concentration. To do this cyanobacteria degrade stored glycogen (Lehmann & Wöber, 1976) mainly by the oxidative pentose phosphate cycle. The evidence for this and possible alternative pathways is reviewed by Doolittle (in press). Foy and Smith (1980) and van Liere, Mur, Gibson and Herdman (1979) have shown that in cultures of *Oscillatoria* species exposed to alternating light–dark cycles, carbohydrates were metabolized in the dark to support the continued synthesis of proteins and other macromolecules. When cultures are darkened for prolonged periods the overall biomass decreases until a certain 'maintenance rate' of energy consumption is achieved. This rate should be comparable with μ_e (see 2.6). Fallon and Brock (1979a) found that in samples of a natural waterbloom in Lake Mendota, Wisconsin, containing *Aphanizomenon*, *Anabaena* and *Phormidium*, the cell volume decreased by more than 99% in 10 days (average decay rate $> 0.02\ h^{-1}$). Van Liere and Mur (1979b) placed photostat cultures of *Oscillatoria agardhii* with contaminant bacteria in the dark and also observed rapid decay. After a few days the cultures collapsed almost completely and did not grow again when exposed to light. This is consistent with the finding that cyanobacteria lost their recognizable cellular structure rapidly during dark incubation (Gunnison & Alexander, 1975; Fallon & Brock, 1979a). However, when *O. agardhii*, taken from exponentially growing batch cultures or from nitrogen-limited chemostats, was exposed to darkness, decay was equal to μ_e ($0.001\ h^{-1}$) for the first 15–20 days, but it then increased tenfold. The carbohydrate content of nitrogen-limited cells was found to be much higher than that of photostat-grown cells. This suggests that stored carbohydrates are metabolized to maintain the cells' functions and that it is not until carbohydrates are exhausted that other macromolecules and structural carbohydrates are broken down, resulting eventually in cell death. Conditions giving rise to high carbohydrate content enable the cells to survive for prolonged periods in darkness. Another possible mechanism for dark survival is heterotrophic growth demonstrated in a number of species, e.g. *Nostoc muscorum* (Allison *et al.*, 1937), *Tolypothrix tenuis* (Kiyohara *et al.*, 1960), and *Chlorogloea fritschii* (Fay, 1965) but so far little investigated in planktonic species (Saunders, 1972). Apparently, *Microcystis* has the ability to maintain its cell morphology in the dark (Fallon & Brock, 1979a) and this suggests dark metabolism which may explain its survival throughout the winter months in virtual darkness on the bottom mud of lakes (Fallon & Brock, 1979a; Reynolds, 1973).

2.7 Light shielding

2.7.1 Introduction

There are two strategies an organism can adopt to avoid exposure to supra-optimal light irradiance; one is to move to a position where the intensity is lower (dealt with in the next section, 2.8) and the other is to produce light-shielding structures. Shielding may be effected by structures which scatter light away from the light-sensitive regions or by pigments which absorb the light. Whichever method is used, two general principles apply. First, the structure or pigment which performs the shielding should surround the light-sensitive region that is liable to be damaged by excessive light.

Secondly, the light shield should give protection against the potentially damaging wavelengths, perhaps selectively. Shorter wavelengths are generally the most harmful. Nucleic acids and proteins may be damaged directly by ultraviolet light, and of the visible wavelengths, blue light appears to be most active in causing photooxidations (Jagger, 1967).

2.7.2 *Light shielding by pigments*

Pigments which absorb preferentially in the shorter visible wavelengths appear red, yellow or brown in colour. Pigments of these colours that may have light-shielding properties in cyanobacteria include the lipophilic carotenoids and xanthophylls and a number of diverse extracellular pigments.

2.7.2.1 Extracellular pigments

The filtrates of aged cultures of cyanobacteria are often the colour of weak tea (Fogg, 1952), indicating that these organisms release pigments or substances that become pigmented. The brown pigment released by *Anabaena cylindrica*, studied by Whitton (1965), was shown to be at least partly formed from amino acid-containing, u.v.-fluorescent compounds which became pigmented on polymerization after they were released into the culture medium (Walsby, 1974a,b). However, the coloration of these compounds, which complex iron strongly, may be incidental to their biological role. For example, other iron-complexing pigmented substances, termed siderochromes, have been shown to have a role in the uptake of iron in *Anabaena* (Simpson & Neilands, 1976) and *Agmenellum quadruplicatum* (Armstrong & Van Baalen, 1979). No experiments have been conducted to test the possibility that their light absorption may also bring benefits to organisms producing them.

If extracellular pigments are of value in light shielding they would be more effective if retained in a layer or sheath surrounding the cells. (Pigments released into free solution would shield other organisms as well and would be of less selective advantage.) A number of cyanobacteria are known to produce pigmented sheaths (20.3.4).

2.7.2.2 Intracellular pigments

Cyanobacteria produce a wide variety of carotenoids and xanthophylls, which have been painstakingly isolated and characterized (see Nichols, 1973 and Wolk, 1973 for reviews). These pigments are particularly in evidence in cyanobacteria cultured under a high light irradiance when these organisms invariably become coloured bright yellow or orange. In other microorganisms and in higher plants there is ample evidence that carotenoids protect cells from light by quenching triplet sensitizers which otherwise initiate potentially damaging photochemical reactions (Krinsky, 1979). Evidence for the protection provided by carotenoids was first provided by Griffiths and Stanier (1956) who showed that a carotenoid-lacking mutant of the purple photosynthetic bacterium *Rhodopseudomonas sphaeroides* was killed by light in the presence of oxygen whereas the carotenoid-containing wild-type survived. Parallel experiments have been carried out with other microorganisms (e.g. Kunisawa & Stanier, 1958) but to date there has been no similar investigation with cyanobacteria. In view of the knowledge amassed on the chemistry and distribution of these pigments in this group, an investigation of their light-shielding function is overdue.

Steinitz *et al.* (1979) found that their mutant stains of *Plectonema boryanum*, which were resistant to photooxidation (see 2.3.1), also differed from the sensitive wild-type in possessing water-soluble pigments absorbing in the region of 459–490 nm. It was speculated that these pigments may have a similar quenching role to that demonstrated for carotenoids.

2.7.3 *Light shielding by refractile structures*

2.7.3.1 Gas vacuoles

Structures which scatter light may redirect part of the incident light away from underlying light-sensitive regions. It has been suggested that gas vacuoles and calcified sheaths may provide light shielding in this way. Tyndall scattering is proportional to λ^{-4} and is therefore much more effective in dispersing the shorter, more damaging, wavelengths.

The gas vesicles which make up the gas vacuoles of cyanobacteria scatter light strongly because they are very small and because the refractive index of their contents, gas, is very different from that of the surrounding aqueous medium (Walsby, 1970). When they are collapsed by pressure the gas-filled spaces disappear and nearly all of the light scattering is lost. There is controversy, however, over whether the gas vesicles provide useful light shielding. This topic has been reviewed previously (Walsby 1972, 1978b), but the essential points will be summarized here and new evidence discussed.

Shear and Walsby (1975) showed that when the gas vacuoles of *Anabaena flos-aquae* were collapsed by pressure there was little change in the photosynthetic rate over a range of light irradiance values from limiting through saturating to inhibiting irradiance (Fig. 2.12). This showed that the intact gas vesicles could not be shielding the photosynthetic lamellae, since if they had been the photosynthetic rate would have increased at limiting light irradiance and decreased at inhibiting values, on collapsing the gas vacuoles. Further-

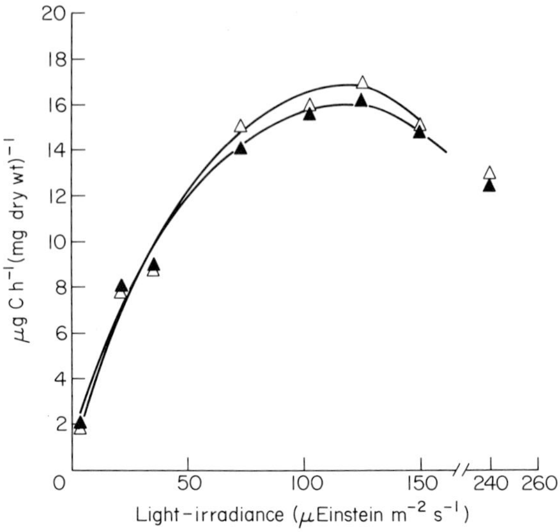

Fig. 2.12. Photosynthetic rates of *Anabaena flos-aquae* with intact (△) and collapsed (▲) gas vacuoles grown at different light irradiance. (Shear & Walsby, 1975.)

more, the survival-duration curves of cells with intact and collapsed gas vacuoles, exposed to ultraviolet light were identical. This demonstrated that the gas vacuoles did not provide any protection from u.v. irradiation. These experiments were performed with shallow layers of dilute suspension and show that in this cyanobacterium gas vesicles provide no shielding of structures within the cells in which they occur.

Waaland *et al.* (1971) provided evidence from spectrophotometry that in *Nostoc muscorum* the intact gas vacuoles decrease the light absorbed by the photosynthetic lamellae. Shear and Walsby (1975) pointed out that part of the change in the absorption spectrum which occurs on collapsing the gas vacuoles is an artefact due to the optical sieve effect. Thus while there seems to be some shading by gas vesicles in *Nostoc* it is not possible to quantify it by the method used; the shielding should be reinvestigated by the method of Shear and Walsby (1975).

Any object like a gas vacuole which scatters light will increase the back scattering from a suspension and thereby decrease the amount of light which penetrates the suspension. (This explains why dense suspensions of cyanobacteria look much darker when their gas vacuoles are destroyed by pressure (see fig. 1 of Walsby, 1972), when they are illuminated from the same side as the observer.) Walsby (1972) described this sort of shielding in suspensions *en masse* as 'sacrificial' shading because the cells with gas vacuoles at the surface of the culture will shade cells (with or without gas vacuoles) deeper in the culture. This certainly explains the results of Porter and Jost (1976) who found a shielding effect in cultures of *Microcystis aeruginosa* en masse which increased progressively with cell concentration. At low concentrations their results indicate no difference between cells with intact or collapsed gas vesicles. Ogawa *et al.* (1979) investigated the absorption characteristics of dilute suspensions of *Microcystis aeruginosa* with intact and collapsed gas vacuoles using an integrating sphere photometer in which all of the light not absorbed by a sample (i.e. all the light both scattered and transmitted from the sample) was measured by the photocell. They found little difference in the absorption on gas-vacuole collapse and concluded that (at low concentrations) there was no light shielding by gas vacuoles.

The biological importance of sacrificial shading by gas-vacuolate cells remains to be assessed. First, in uniform suspensions this will have no selective advantage because the cell with gas vacuoles will receive no shielding from the gas vacuoles it contains; while it is true that the light penetrating the suspension will be more rapidly attenuated, this will benefit cells of other species present. An exception to this is the case of colonial cyanobacteria such as *Microcystis*, *Aphanizomenon* and *Trichodesmium*. Here, the gas vacuoles in cells at the surface of the colony will decrease the light reaching those in the centre, a sacrifice for genetically identical cells. Light shielding in suspensions en masse also has a particular significance in surface waterblooms. The gas vacuoles which buoy the bloom-forming cyanobacteria to the surface will also decrease light penetration. The light levels may be very low under dense surface blooms.

Finally, it can be stated that if gas vesicles do provide any useful light shielding it can hardly pay for their keep or they would be found in terrestrial cyanobacteria (Walsby 1978b).

2.7.3.2 Calcified sheaths

Cyanobacteria are associated with several forms of calcium carbonate deposits, described in detail by Golubić (1973). It was shown by Pentecost (1978), who measured the attenuation of light from a laser by slabs of tufa ground to different thicknesses, that in this form of calcite deposit only 1% of the incident light penetrated per millimetre. This rapid attenuation may be attributed to scattering by the calcite microcrystals. Pentecost (1978) determined the compensation point of the cyanobacteria present in the tufa (mainly *Schizothrix calcicola*) and calculated that its photoautotrophic growth would be possible down to a depth of about 2 mm. Thin sections normally revealed a layer of cyanobacteria 1–2 mm thick, though the organisms could be found at greater depths. Pentecost found that the equilibrium change by photosynthetic CO_2 uptake could account for only a small proportion of the calcium carbonate precipitated in the sheaths of the tufa cyanobacteria. The formation of tufa must therefore be due to precipitation from $CaCO_3$ supersaturated ground water, though the cyanobacterial sheaths may provide favourable nucleation sites for the calcite crystals.

Cyanobacteria which bore into established limestone deposits (Golubić, 1973; Section 20.4.3) will similarly obtain shielding from sunlight. This may be of great value in regions of high irradiance (see Krumbein *et al.*, 1978, 1979).

2.8 Positioning with respect to light field

Microorganisms are usually dispersed by random agents such as wind and water currents. Where they develop a substantial population the conditions must be suitable for their growth in competition with other organisms. If the conditions then change to become inimical to growth the population may collapse unless it is able to move to a new suitable position. Clearly, microorganisms able to make such movements will be at a considerable advantage. Changes in the average light irradiance may have an overriding effect on growth of cyanobacteria. In this section we consider two types of movement they make to maintain themselves in an optimum light field. The first of these is the phototactic gliding movement of cyanobacteria which attach to surfaces. The second is brought about by light-regulated buoyancy changes of planktonic cyanobacteria suspended in stable bodies of water. In each case the response to light should enable the organism to move along a light irradiance gradient to the optimum position.

2.8.1 Positioning by gliding movement

Many cyanobacteria are capable of moving over surfaces by gliding movement. It is thought that this type of movement is effected by movements of microfibrils which are present just external to the peptidoglycan layer of the cell wall and run parallel to the direction of movement (16.2.2.2). A thin layer of mucilage usually covers the surface of the gliding cell or filament and serves to make a good contact between the gliding mechanism and surface of the substrate (Walsby, 1968). Filaments suspended in water are not able to swim by this mechanism, but they can move with respect to other filaments in suspended colonies and, as for example in *Aphanizomenon flos-aquae*, this can result in constant changes in the relative position of the filaments and shape of the colony. The rate of gliding movement varies from 1 to 11 $\mu m\ s^{-1}$ in filamentous cyanobacteria and from 0.03 to 0.1 $\mu m\ s^{-1}$ in unicellular forms.

Motile cyanobacteria may respond to light in three different ways, by photokinesis, phototopotaxis and photophobotaxis, the mechanisms of which are described briefly below and in more detail in Chapter 16. The phototactic responses enable these organisms to move up a light-irradiance gradient from dim light of limiting irradiance and down a light gradient away from inhibiting irradiance. From this it can be concluded that they would position themselves at the optimum irradiance on the gradient, though this has not yet been demonstrated directly.

Photokinesis is the change in gliding speed in response to light irradiance. If the speed increases with an increase in light irradiance the response is said to be positive; if the speed decreases with an irradiance increase the response is negative. Nultsch (1962b, 1965; see also 16.4.2) has made detailed studies of photokinesis in *Phormidium autumnale* which shows positive photokinesis up to 30 klx and negative at higher irradiance. The maximum positive response was at 2 klx, however. By measuring the response in light of different wavelengths he obtained an action spectrum, which in various species of *Phormidium* showed peaks in the blue and red which suggest that chlorophyll *a* is the principal pigment involved. This supports the idea that photokinesis is dependent on ATP produced by photophosphorylation (Nultsch, 1965). The action spectrum indicates that this is driven mainly by photosystem I (PS I) cyclic electron flow, though inhibition of photokinesis by DCMU suggests photosystem II (PS II) must also be involved.

Not all cyanobacteria behave identically. In *Anabaena variabilis* the action spectrum for photokinesis indicates phycocyanin participates in the response (Nultsch & Hellman, 1972) while only red light absorbed by chlorophyll seems to be active. Halfen and Castenholz (1971a) reported that *Oscillatoria princeps* continues to glide until it dies in the dark at an undiminished rate, apparently mobilizing carbohydrate food reserves to provide an energy source. In this organism DNP (dinitrophenol), which uncouples oxidative phosphorylation, completely suppresses movement in the dark and inhibits movements by 75% in the light, again indicating that gliding is dependent on ATP.

The majority of those motile cyanobacteria investigated also show phototactic movements, usually responding positively (moving towards the light) at low irradiance and negatively at high irradiance. Two sorts of phototaxis are distinguished. The first is phototopotaxis in which the filament glides actively toward (or away from) the light source. The mechanism is said to be different in the Nostocaceae and the Oscillatoriaceae (Nultsch *et al.*, 1979). In the Oscillatoriaceae the filaments are not able to change their direction and steer towards the light, but filaments which lie parallel to the light irradiance gradient will reverse their direction of movement sooner when gliding away from the light than when gliding towards it (Drews, 1959). By this technique of three steps forward and one step backward they gradually make their way towards the light. Filaments not positioned directly towards the light may eventually have their path redirected by collisions with obstacles or other gliding trichomes. Burkholder (1934) described how curved filaments make their way toward the light in a series of arcs; the filament reverses when the trajectory of the arc begins to point away from the light source.

In the Nostocaceae the filaments are made up of chains that are constricted where the cells join. They are more flexible and are able to steer actively towards the light source (Drews 1957, 1959). The filaments may either turn end on to the light or glide as a U-shape with its centre directed towards the light; the mechanism of steering is unknown (see Nultsch *et al.*, 1979). However, there are two features of the gliding movement of this group which may explain 'steering'. First, the filaments do not rotate as they glide forward. This permits the looped filament to move forward without twisting on itself. Secondly, the gliding mechanism may operate in different directions in different sections of the same filament (Walsby, 1968). This would explain how the U-shaped bend forms in the filament originally oriented orthogonal to the light direction. This is also the necessary explanation for how the bend is maintained in the looped filament gliding forward. Hence, the mechanisms of phototactic movement may not be so fundamentally different in these two groups of organisms after all.

Photophobotaxis describes the shock reaction of a filament or cell moving across a boundary from a region of high to low light irradiance; if the filament responds positively it will reverse back into the high-irradiance region. The response can be assessed quantitatively by observing filaments accumulate in a light trap (see Häder, 1974). Using white light, filaments of *Phormidium uncinatum* will usually respond positively, even up to intensities as high as 50 klx (Drews, 1957; Nultsch, 1962b). In the photophobotactic response the filament behaves in a polarized fashion. It is much more likely to reverse its direction of movement if the front, rather than the back end is shaded (Drews, 1959). Studies by Nultsch (1967, 1968) showed that photosynthetic electron transport was involved in this response, rather than photophosphorylation as in the phototopotactic response. The role of electron transport in the photophobotactic response has been elucidated in a series of experiments performed by Häder (1974, 1975, 1977, 1978a,b). His hypothesis is that a positive photophobotactic response occurs when electrons are drained from the electron pool held in the electron transport chain connecting the two photosynthetic photosystems. Hence, if a filament moves from a light trap illuminated by light (of 561 nm) supporting PS II this will decrease the supply of electrons to the pool and elicit a positive response and the filament will reverse back into the trap (Häder, 1974). The effect is enhanced if the area surrounding the trap is illuminated in light of longer wavelength, supporting PS I, as this will drain electrons from the pool. This explains why longer wavelengths elicit a negative photophobotactic response. Häder's scheme also neatly explains the previously puzzling results obtained with DCMU as this blocks the flow of electrons from PS II into the pool; as the concentration of the inhibitor is increased the positive photophobotactic response of filaments will decrease and the response will then become negative (Häder, 1974, 1975).

Häder (1977) next proposed that the sensory transduction of the photophobic reaction is mediated by a change in the endogenous membrane potential. He showed that externally applied electric fields exceeding 3 V would eliminate the phobic response if the field was roughly parallel to the long axis of the filament. It is calculated that these applied voltages would induce an electrical potential exceeding 10 mV along the length of the filament. He then demonstrated that shining light on trichomes of *Phormidium uncinatum* arranged so as to bridge the gap between two platinum electrodes induced a change in potential, which increased up to a saturation value of 12 mV at 10 klx (Häder, 1978a). The change in potential across the length of a single filament, which was threaded through the end of a close-fitting glass micropipette, was also measured and changes of up to 10 mV were demonstrated on illumination (Häder, 1978b). By penetrating individual cells with a micro-electrode Häder (1978b) also demonstrated a membrane potential of about −32 mV which changed by about 2 mV (i.e. to −34 mV) in the light and back again when returned to the dark.

These studies point to the gliding movement on the surfaces of different cells in the filament being coordinated by electrical phenomena controlled by the photosystems. There are feedback mechanisms in the phototactic responses described above which suggest that the filaments will move to a position on a light irradiance gradient which is optimum for their photosynthetic requirements. These may be expected to change according to limits imposed by temperature and availability of nutrients. To date there has been little investigation of gliding movements in relation to changing light conditions in the environment. Castenholz (1968) showed that the hot spring *Oscillatoria terebriformis* formed dispersed mats under shaded conditions, but that under full sunlight these mats contracted into dense clumps where the filaments presumably obtain mutual benefit from self-shading. There are many other ecological situations, such as cyanobacterial mats in the upper-littoral mud flats, where phototactic movements should be investigated.

2.8.2 *Positioning by light-regulated buoyancy changes*

2.8.2.1 Introduction

There have been many observations of planktonic cyanobacteria forming vertically stratified populations in stable lakes (see Reynolds & Walsby, 1975 for references). In at least some of these instances the population remains stratified for long periods (e.g. Baker & Brook, 1971) and from this it can be concluded that the cells or filaments of the population must maintain a density roughly equal to that of the water in which they are suspended. All of these stratifying planktonic cyanobacteria possess gas vacuoles, organelles made up of assemblages of hollow cylindrical structures called gas vesicles (Bowen & Jensen, 1965). The density of the cyanobacterial gas vesicle is about 0·12 g cm^{-2} (Walsby & Armstrong, 1979), about an eighth of that of water, and if sufficient gas vesicles are present in a cell it can become positively buoyant (see Walsby, 1972). By regulating their gas vacuole content these organisms are able to adjust their buoyancy. Studies on planktonic cyanobacteria in culture have shown that gas-vacuolation is regulated in response to light irradiance in a way which permits them to select a position of optimum irradiance on a vertically-decreasing light irradiance gradient.

The mechanism of buoyancy control is described below. Further details of gas vacuole structures and processes which regulate them are given in Chapter 10.

2.8.2.2 Gas vacuole regulation by *Anabaena flos-aquae*

When the planktonic cyanobacterium *Anabaena flos-aquae* is grown in low light irradiance it produces enough gas vacuoles to make the cells buoyant (Walsby, 1969), but when transferred to a high irradiance the buoyancy is lost. This loss of buoyancy is correlated with a decrease in gas vacuolation and can be explained by the irreversible collapse of a proportion of the constituent gas vesicles by rising cell turgor pressure. Dinsdale and Walsby (1972) showed that the cell turgor pressure can rise by 1 bar or more when the light irradiance is increased. The turgor rise is clearly driven by photosynthesis since it is inhibited by DCMU and does not take place in the absence of CO_2. Grant and Walsby (1977) found that up to half of the rise could be explained by the increase in osmotic pressure that would be generated by the extra photosynthetic products of low molecular weight which accumulate in the cell water. The balance may be generated by K^+ ion uptake (Allison & Walsby, 1981) though a quantitative balance sheet has yet to be drawn up.

In *A. flos-aquae* there are several thousand gas vesicles per cell and the new vesicles are made with a critical collapse pressure which varies from 4.5 to 7.5 bars normally distributed about a mean of 6 bars (A.E. Walsby, unpubl.). The weaker gas vesicles are lost progressively when the turgor pressure rises above 4.5 bars. It may be that this mechanism accounts entirely for the variation in gas vacuolation of the cells. It is not known if the protein of the collapsed gas vesicle membranes is recycled but, if it is wasted, it is unlikely to account for more than 1% of the dry weight of the cell. It is possible that as well as losing gas vesicles at high irradiance *Anabaena flos-aquae* makes them less quickly under these conditions, so that the gas vesicle content is diluted by cell growth (Walsby, 1970). However, no measurement of gas vesicle formation *per se*, independent of collapse, has yet been made.

Walsby and Booker (1980) showed that in dilute batch cultures of *A. flos-aquae* all of the filaments became buoyant at light irradiance below 5 μE m^{-2} s^{-1} and all sank above 35 μE m^{-2} s^{-1}; approximately equal numbers of sinking and floating filaments occur at between 13 and 22 μE m^{-2} s^{-1}, equivalent to about 0.65–1.1 klx. This response enables *A. flos-aquae* to stratify on a vertically decreasing light irradiance gradient. In a 2.2 m deep experimental water column it was found to accumulate in a layer where the intensity was about 1 klx, as long as the water did not become depleted of dissolved inorganic carbon. Foy *et al.* (1976), investigating another strain of *A. flos-aquae*, found that 1 klx just about saturates growth. It would seem likely, therefore, that the cyanobacterium selects a position of optimal light irradiance in the column.

The mechanism of light-regulated buoyancy regulation in *A. flos-aquae* is summarized in Fig. 2.13. Clearly, the relationship between light irradiance and buoyancy is not one of direct cause and effect but an indirect one involving a chain of events, each of which may be perturbed by exogenous factors.

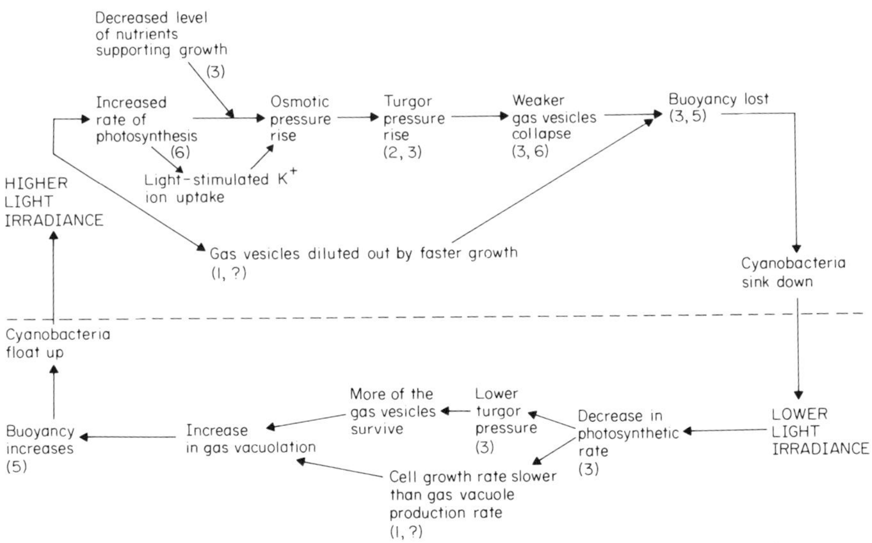

Fig. 2.13. Diagram summarizing the possible sequence of events in buoyancy regulation by planktonic cyanobacteria (from Reynolds & Walsby, 1975). (1) Walsby, 1970; (2) Walsby, 1971; (3) Dinsdale & Walsby, 1972; (4) Grant & Walsby, 1977; (5) Walsby & Booker, 1980; (6) Allison & Walsby (1981); (?) no evidence as yet.

2.8.2.3 Gas vacuole regulation by other cyanobacteria

The rise in turgor pressure with light irradiance has been found in a number of other gas-vacuolate cyanobacteria including *Oscillatoria agardhii* var. *isothrix* (Walsby & Klemer, 1974), *Microcystis aeruginosa* (Reynolds, 1973), *Aphanizomenon flos-aquae* (Konopka *et al.*, 1978), *Anabaena spiroides* and *Gloeotrichia echinulata* (E.M. Allison & A.E. Walsby, unpubl.). It probably occurs in cyanobacteria as an inevitable consequence of the increased photosynthetic activity. However, for this turgor pressure rise to result in gas vacuole regulation the pressure reached must exceed the lowest gas vesicle pressure. This definitely occurs in the species of *Aphanizomenon*, *Anabaena* and *Gloeotrichia* listed above and in these organisms the low critical pressure range should probably be considered to be the basis of the evolved buoyancy-regulation response.

Although *Microcystis aeruginosa* has been observed to stratify in lakes (Reynolds, 1973) it is not certain that it is able to collapse its gas vesicles by turgor pressure, as they have a higher critical pressure. *Microcystis* spp. are often found at the water surface in calm conditions and may be adapted to withstanding high light irradiance (2.7.3.1). Very strong gas vesicles are found in the oceanic *Trichodesmium thiebautii*, where the mean critical pressure is 37 bars. There is no possibility of collapse by turgor pressure in this case. The strength of the gas vesicles in this organism is evidently an adaptation to surviving exposure to hydrostatic pressure at the considerable depths to which it may be carried by water currents (Walsby, 1978a).

Some non-planktonic cyanobacteria produce gas-vacuolate hormogonia, perhaps as an aid to dispersal in conditions of flooding (see Walsby, 1972). In one case, *Nostoc muscorum*, the gas vacuoles are induced by increased light irradiance (Waaland *et al.*, 1971).

2.8.2.4 Buoyancy regulation in natural waters

Measurements were carried out on *Oscillatoria agardhii* var. *isothrix*, in Deming Lake (Minnesota, USA), which showed that the gas vacuolation of filaments that formed a stable layer at a depth of 5 m was within 5% of the level required to give neutral buoyancy. Samples of the population incubated in bottles nearer the surface lost their buoyancy while samples incubated at greater depths increased in buoyancy (Walsby & Klemer, 1974). Probably, the position of the layer is determined by the maximum light irradiance during the day. It is not known if the layer moves up and down significantly according to the diurnal light variation but it is unlikely that the filaments would move up very far at night as they are individually dispersed and, by Stokes's law, as a consequence of their small size they would have a low rate of flotation.

Cyanobacteria which form large colonies, on the other hand, will float or sink rapidly for a small change in excess density and such colonial forms may show diurnal migration. Konopka *et al.* (1978) showed that *Aphanizomenon flos-aquae*, which forms large fusiform colonies, became buoyant during the night and floated to the surface of Lake Mendota (Wisconsin, USA) if the lake was calm. During the morning a rise in cell turgor pressure occurred causing gas vesicle collapse and buoyancy loss. By noon the population peak had moved from the surface to a depth of 2 m, and by 1800 h it had moved down to 4 m.

CHAPTER 3
MODES OF CYANOBACTERIAL CARBON METABOLISM

A. J. SMITH

3.1 Introduction

The dominant growth physiology of the diverse cyanobacteria, now available for laboratory investigation, is photoautotrophic. However, these organisms also display other metabolic capabilities; one of these is of particular importance to cyanobacteria in their natural environment while the significance of the others is less certain. In this chapter the various modes of carbon metabolism that cyanobacteria can adopt will be reviewed against the background of the chapter on the biosynthesis of metabolic intermediates (Smith, 1973) in *The Biology of Blue Green Algae*. The main emphasis will be on areas where major advances have been made in the past decade and where progress is likely to be made in the future.

Several of the cyanobacteria which have been used extensively in laboratory studies have been cited in the literature by different names. In order to avoid confusion regarding the identity of organisms mentioned in this chapter, strain nomenclature has been standardized as far as possible. The generic designations recommended by Rippka *et al.* (1979) are used in conjunction with the Paris culture collection numbers (Table 3.1). In those cases where organisms were not among those surveyed by Rippka and her colleagues the name given in the original paper is used.

Table 3.1. Nomenclature of cyanobacteria

Strain notation used in this chapter	ATCC* strain number	Alternative names for strains
Anabaena 7118	27892	*Anabaena variabilis* Kütz.
Anabaena 7122	27899	*Anabaena cylindrica* Lemm.
Anabaena 29413†	29413	*Anabaena variabilis, Anabaena flos-aquae*
Chlorogloeopsis 6912	27913	*Chlorogloeopsis fritschii, Chlorogloea fritschii*
Nostoc 6720	27895	*Anabaenopsis circularis*
Plectonema 73110	29407	*Plectonema boryanum*
Synechococcus 6301	27144	*Anacystis nidulans*
Synechococcus 7418	29534	*Aphanothece halophytica*
Synechocystis 6308	27150	*Aphanocapsa* 6308, *Gloeocapsa alpicola*
Synechocystis 6714	27178	*Aphanocapsa* 6714

* ATCC: American Type Culture Collection.

† This *Anabaena* strain was not included in the survey of Rippka *et al.* (1979) but is included in this table because it has been referred to in the literature by various different names.

3.2 Photoautotrophic growth and metabolism

3.2.1 Assimilation of CO_2

As a group, cyanobacteria are capable of extended growth under photoautotrophic conditions with light energy driving the formation of ATP and NADPH needed for the net assimilation of CO_2 and for various biosynthetic processes in the cell. Uncertainties about the route for the fixation of CO_2 have been largely resolved by detailed kinetic analyses of the products of $^{14}CO_2$ assimilation by several unicellular strains (Pelroy & Bassham, 1972; Ihlenfeldt & Gibson, 1975a). Most of the isotope incorporated during the shortest time of exposure was recovered in 3-phosphoglyceric acid (Fig. 3.1). With longer exposure times, label appeared in the sugar phosphates, ribulose-1,5-bisphosphate and glucose-6-phosphate followed by fructose-6-phosphate, sedoheptulose-7-phosphate and carboxylic and amino acids including phosphoenolpyruvate, aspartate and glutamate. In addition, a shift from light to dark had the expected reciprocal effect on the size of the pools of 3-phosphoglyceric acid and of ribulose-1,5-bisphosphate, the former increasing and the latter decreasing (Fig. 3.1). These patterns, obtained at light intensities and CO_2 concentrations optimal for growth (Ihlenfeldt & Gibson, 1975a) are wholly consistent with the net fixation of CO_2 via the reductive pentose phosphate cycle (Fig. 3.2a). Under conditions of active growth,

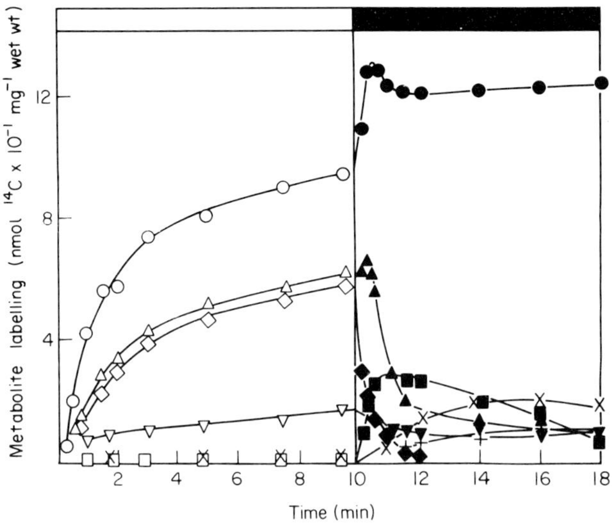

Fig. 3.1. Metabolite labelling by $^{14}CO_2$ in *Synechocystis* 6308 during photosynthesis and a subsequent dark period. The organisms were incubated in the light for 10 min before being transferred to the dark for 8 min. The metabolites investigated were 3-phosphoglyceric acid (○, ●), glucose-6-phosphate (△, ▲), ribulose-1,5-bisphosphate (◇, ◆), 6-phosphogluconic acid (□, ■), fructose-1,6-bisphosphate (×), sedoheptulose-1,7-bisphosphate (+) and sedoheptulose-7-phosphate (▽, ▼). (From Pelroy & Bassham, 1972.)

intermediates will be removed from the cycle at various points for the synthesis of cell constituents in a balanced manner. However, most of the carbon assimilated in suspension experiments was not incorporated equally into all of the growth polymers but primarily into a polymer containing glucose (Pelroy & Bassham, 1972).

In these short-term labelling experiments, C_4 compounds such as malate and aspartate were also labelled. The isotope in aspartate was present in both carboxyl groups which can be attributed to the synthesis of C_3 precursors from intermediates of the reductive pentose phosphate cycle and to their subsequent carboxylation to give the C_4 precursor of this amino acid (Fig. 3.3). The latter does not constitute a primary carboxylation mechanism because it is not possible to regenerate the C_3 acceptor without loss of carbon as CO_2. In the labelling experiments already described, the amount of isotope in C_4 compounds after short exposures to $^{14}CO_2$ was very small compared with that in 3-phosphoglyceric acid. This is in marked contrast to the patterns of metabolite labelling obtained by Döhler (1974) and by Weathers and Allen (1978) who found a relatively large amount of radioactivity in malate and aspartate after brief exposure to $^{14}CO_2$. The carboxylation of a C_3 metabolite is an essential step in the synthesis of precursors of amino acids belonging to

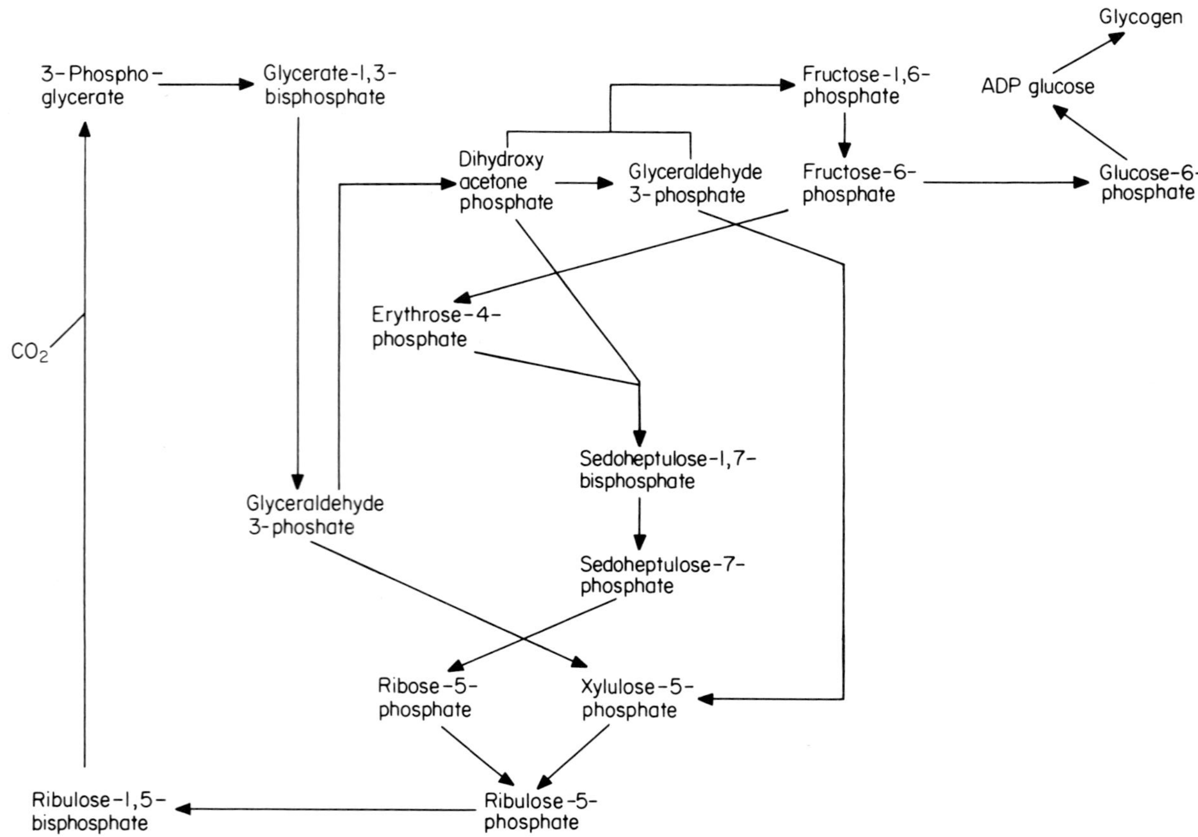

Fig. 3.2a. Primary routes of intermediary metabolism in cyanobacteria. Relationships between the pentose-phosphate cycles. CO_2 assimilation via the reductive pentose-phosphate cycle.

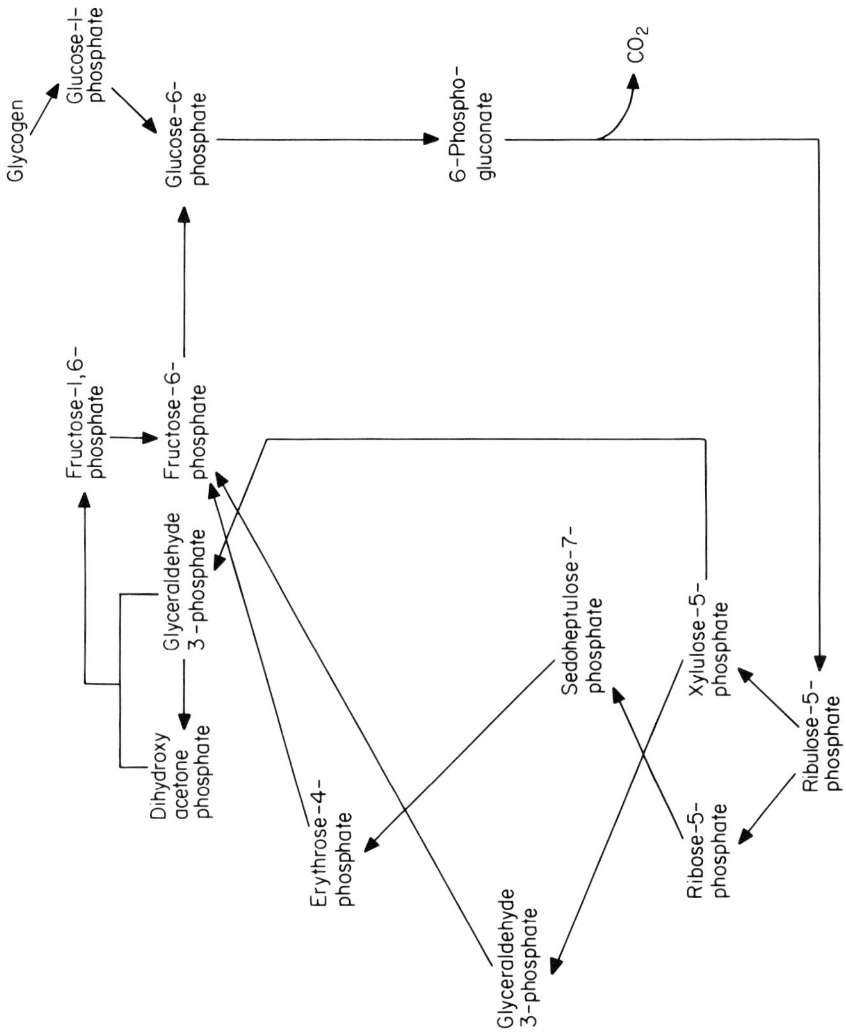

Fig. 3.2b. Carbohydrate catabolism via the oxidative pentose-phosphate cycle.

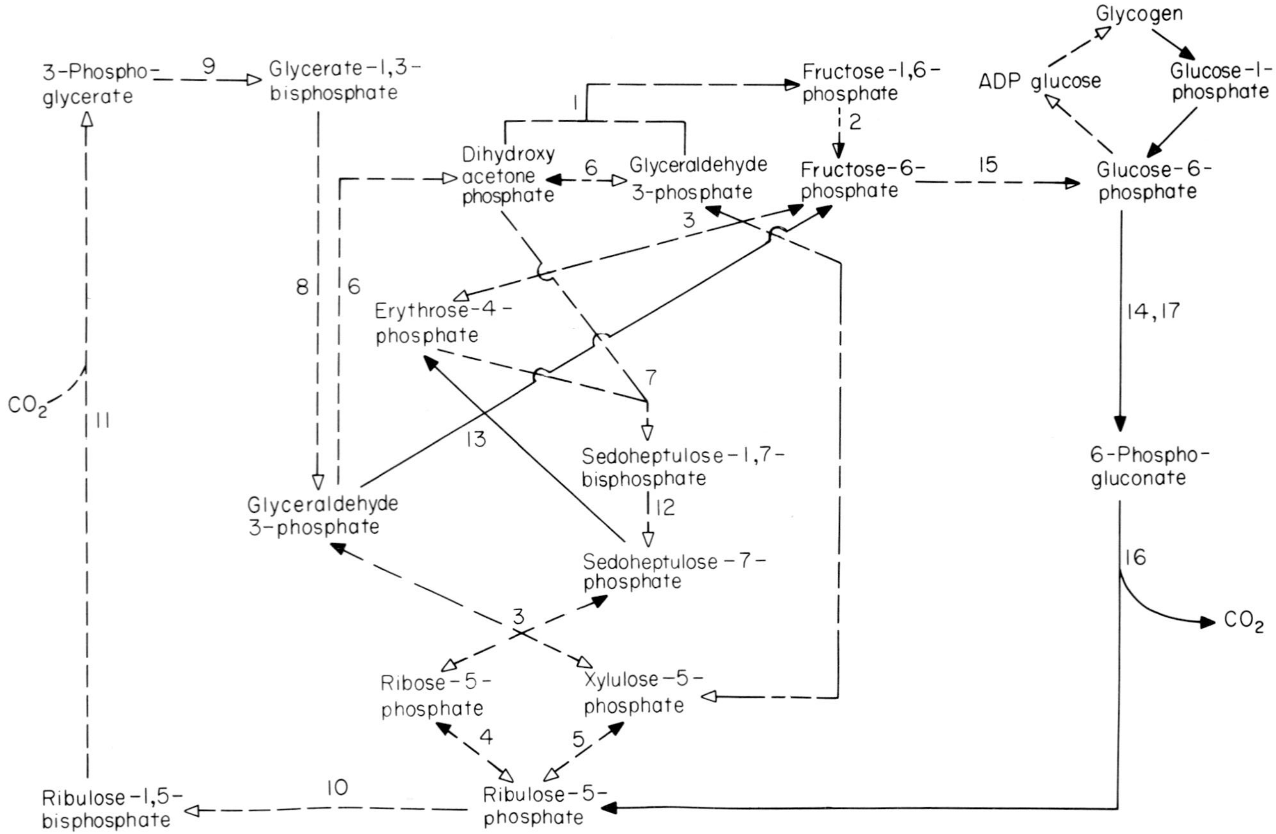

Fig. 3.2c. Reactions specific to CO_2 fixation (— — —), to carbohydrate catabolism (——) and common to both (– – – –). To identify the enzymes of the pentose-phosphate cycles refer to Table 3.3.

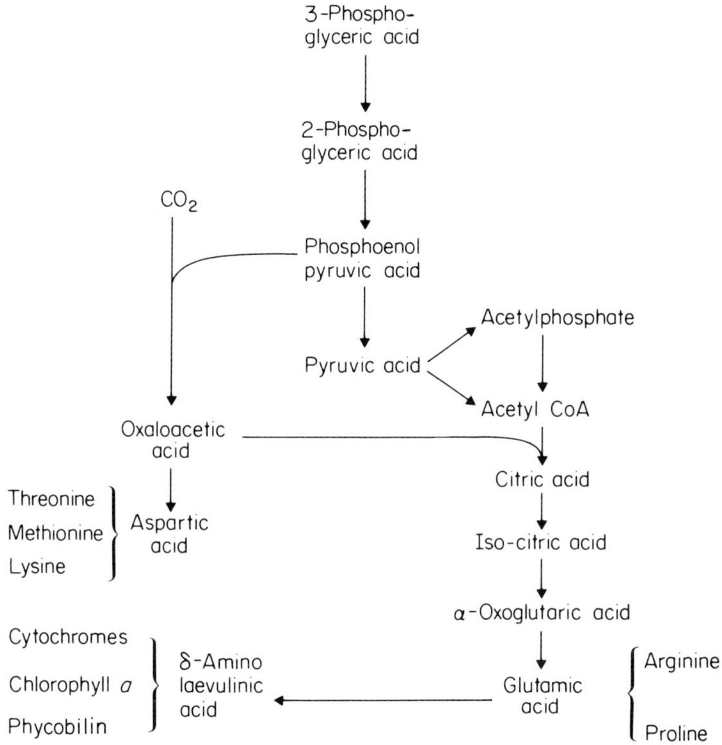

Fig. 3.3. Metabolism of C_3 compounds derived from the reductive pentose-phosphate cycle.

the aspartic and glutamic acid families and of tetrapyrroles (Fig. 3.3). Aspartate and arginine are not only constituents of cell protein but they are also the sole constituents of a polymer unique to cyanobacteria:multi-L-arginine poly-L-aspartic acid known by the trivial name cyanophycin (see 3.2.2.2). Under conditions of balanced growth and cell synthesis, C_4 labelling by brief exposure to $^{14}CO_2$ is expected but should be less important on quantitative grounds than the labelling of 3-phosphoglyceric acid. The preponderance of label in C_4 compounds under other conditions may be due to an increased flow of CO_2 carbon into intermediary metabolism via C_3 carboxylation as a result of a shift in the relative activities of the two carboxylation reactions. This might serve to channel intermediary metabolites into cyanophycin synthesis rather than into polysaccharide which is the main product of carbon assimilation in other short-term suspension experiments. However, Allen *et al.* (1980) have reported that relatively little carbon from CO_2 is incorporated into cyanophycin; this polypeptide is apparently produced largely at the expense of cell protein.

Under conditions optimal for growth, the greater part of the CO_2 assimilated by cyanobacteria enters intermediary metabolism via the reaction catalysed by ribulose-1,5-bisphosphate (RuBP) carboxylase which has been detected in all cyanobacteria investigated. In morphologically complex strains that form heterocysts for the purpose of fixing nitrogen the distribution of this enzyme in the filament is not uniform. Using an autoradiographic technique Wolk (1968) showed that isotope assimilated by filaments of *Anabaena cylindrica* exposed to $^{14}CO_2$ was present initially in vegetative cells and immature heterocysts but not in mature heterocysts. In agreement with this, assays for RuBP carboxylase have shown that heterocysts contain very little enzyme relative to vegetative cells (Winkenbach & Wolk, 1973; Stewart & Codd, 1975; Codd & Stewart, 1977c). The distribution of the ability to fix CO_2 and of RuBP carboxylase in filaments correlate with the occurrence of polyhedral bodies in heterocystous cyanobacteria (Stewart & Codd, 1975); they are present in vegetative cells but not in heterocysts. These intracellular inclusions, first described by Pankratz and Bowen (1963), appear in thin sections under the electron microscope as polygonal structures. Similar structures have been seen in thin sections of thiobacilli (Wang & Lundgren, 1969) and nitrifying bacteria (Murray & Watson, 1965). Purified preparations of these inclusions have been shown to contain RuBP carboxylase (Shively *et al.*, 1973) and, in recognition of this, they have been termed carboxysomes. Studies with the cyanobacteria *Anabaena* 7122 and *Chlorogloeopsis* 6912 suggest that the polyhedral bodies of these organisms also contain RuBP carboxylase (Codd & Stewart, 1976; Lanaras & Codd, 1981a,b). While it is virtually certain that polyhedral bodies will prove to contain this enzyme, the function of such structures has yet to be clearly established in any autotrophic microorganism. It has been suggested that the carboxysome may be a storehouse of RuBP carboxylase (Shively, 1974) which is consistent with the observation that *Anabaena* 7122 contains few carboxysomes when growing rapidly, whereas stationary and lag-phase organisms contain larger numbers of these inclusions (Codd & Stewart, 1976). An alternative explanation which has yet to be investigated is that such a compartmentation of the enzyme might serve to protect it from excessive turnover in the cell.

The variety of molecular types of carboxylase in photosynthetic bacteria (see McFadden & Tabita, 1974) has stimulated interest in the molecular characteristics of the enzyme in cyanobacteria. A preliminary investigation of the molecular properties of RuBP carboxylase from cyanobacteria suggested that it was of a molecular weight comparable to the enzyme of eukaryotic algae and green plants (Anderson *et al.*, 1968). This high molecular weight form of the enzyme (approximately 500 000) containing large (L) and small (S) subunits has been isolated from various strains of cyanobacteria and from the cyanelles of *Cyanophora paradoxa* (Table 3.2). In contrast, F.R. Tabita's

Table 3.2. Molecular properties of D-ribulose-1,5-bisphosphate carboxylase in cyanobacteria and cyanelles

Organism	Native enzyme mol wt	Subunit composition	Subunit Mol wt		Reference
(1) Unicellular species					
Microcystis aeruginosa	518 000	L_8S_8	50 000	14 000	Stewart *et al.* (1977)
Synechococcus 73109	456 000	L_8	56 000	—	Tabita *et al.* (1974)
Synechococcus 7418	237 000	L_4	57 000	—	Codd *et al.* (1979)
Synechococcus sp	530 000	L_8S_8	57 000	12 000	Andrews *et al.* (1981)
Synechocystis 6308	525 000	L_8S_8	51 000	15 000	Codd & Stewart (1977a)
(2) Filamentous species					
Anabaena 7122	449 000	L_8	54 000	—	Tabita *et al.* (1976)
	500 000	L_8S_8	52 000	12 000	Takabe (1977)
Anabaena variabilis	[18S]	L_8S_8	52 000	13 000	Takabe *et al.* (1976)
Chlorogloeopsis 6912	520 000	L_8S_8	53 000	13 000	Lanaras & Codd (1981b)
Plectonema boryanum	[18S]	L_8S_8	54 000	13 000	Takabe *et al.* (1976)
(3) Cyanelle					
Cyanophora paradoxa	525 000	L_8S_8	51 000	15 000	Codd & Stewart (1977b)

group have consistently isolated a high molecular weight carboxylase containing only the large (L) type of subunit (Table 3.2). Despite differences in apparent subunit composition, both forms of the carboxylase are reported to have high specific enzyme activity. There appear to be two alternative explanations for these contradictory data. Firstly, the small (S) type of subunit is a normal component of the enzyme *in vivo* but it is lost either during purification by procedures involving acid precipitation (Codd & Stewart, 1977a; Andrews & Abel, 1981) or in the dialysis of preparations of the enzyme dissociated with sodium dodecylsulphate (see McFadden *et al.*, 1975). Repeated acid precipitation of the purified enzyme from a marine *Synechococcus* has been shown to result in the loss of the small subunit (Andrews & Abel, 1981). The enzyme activity lost as a result of this treatment was recovered by incubating the acid treated enzyme with a preparation of the small subunit. Alternatively, the small subunit is not a normal component of the carboxylase *in vivo* but it is acquired during the preparation of extracts and the isolation of the enzyme (McFadden and Tabita, 1974). This question might be settled by directly comparing the two forms of the enzyme from either *Anabaena* 7122 or *Synechocystis* 6308. Such a study would have to take account of the procedures that are now used to activate RuBP carboxylase. These data could then be compared with the properties of the enzyme in cells made permeable to small charged molecules by treatment with toluene (Tabita *et al.*, 1978). This method which has been used to compare the catalytic properties of citrate synthase isolated from *Escherichia coli* with the enzyme in the intact cell (Weitzman, 1973) has recently been applied to the study of ribulose-1,5-bisphosphate carboxylase in several cyanobacteria (Tabita & Colletti, 1979). Such an approach offers the possibility of resolving the uncertainty that remains regarding the form of the enzyme in the intact cyanobacterial cell. Recent work indicates that the carboxylase in purified carboxysomes contains large and small subunits (Lanaras & Codd, 1981a).

Codd *et al.* (1979) have recently isolated and characterized the carboxylase of *Synechococcus* 7418, a cyanobacterium capable of anoxygenic photosynthesis. Using procedures which have consistently yielded the high molecular weight form of the enzyme containing large and small subunits from other cyanobacteria, they obtained a pure enzyme containing only the large subunit and with an aggregate molecular weight consistent with the subunit constitution L_4 (Table 3.2). It remains to be established whether this organism is unique among cyanobacteria or whether various molecular forms of RuBP carboxylase will be a common feature of cyanobacteria as is the case among photosynthetic bacteria (see McFadden & Tabita, 1974).

Besides comparisons of molecular size and subunit composition, carboxylases isolated from cyanobacteria have been compared with those of other photosynthetic organisms using immunochemical techniques. Antiserum raised against the intact enzyme of *Euglena gracilis* gave single precipitation bands with extracts of four cyanobacteria (Lord *et al.*, 1975). The partial fusion of the bands for cyanobacterial and eukaryotic enzymes was attributed to partial immunological identity and the complete fusion of the precipitin bands given by all the cyanobacterial enzymes was taken to indicate the presence of identical antigenic determinants. Similar conclusions have been reached by Takabe (1977) using antiserum raised against spinach enzyme and pure RuBP carboxylase isolated from *Anabaena cylindrica.* Two groups of workers (Lord *et al.*, 1975; Takabe *et al.*, 1976) also found that the catalytic activity of the carboxylases of cyanobacteria was inhibited by antiserum raised against the native enzyme and against the large, catalytic subunit.

Few of the cyanobacterial RuBP carboxylases have been fully characterized in terms of their catalytic properties. Badger (1980) has characterized the carboxylase of *A. variabilis* in terms of various kinetic parameters and compared it with the higher plant enzyme. While both enzymes have similar K_m values for ribulose-1,5-bisphosphate, the K_m for CO_2, the combined equilibrium constant for activation and the maximum turnover rate of the cyanobacterial carboxylase were all greater than the corresponding values for the plant enzyme. The enzyme from *Synechocystis* 6308 has been shown, like other large two-subunit types of carboxylase, to be inhibited by 6-phosphogluconate (Codd & Stewart, 1977a). In addition, the activity of RuBP carboxylase isolated from this unicellular cyanobacterium was inhibited competitively by oxygen (Codd & Stewart, 1977a). This feature, which is common to carboxylases from diverse sources, has been attributed to the RuBP oxygenase activity of RuBP carboxylase preparations; the protein catalyses not only a CO_2-dependent conversion of RuBP to two molecules of 3-phosphoglycerate but also an oxygen-dependent conversion of the same substrate to 3-phosphoglycerate and phosphoglycollate (see Andrews and Lorimer, 1978; Lorimer, 1981). The purified RuBP carboxylase of *Synechocystis* 6308 has been shown to have oxygenase activity (Codd & Stewart, 1977a) comparable to preparations of carboxylase from other photosynthetic organisms.

The formation of glycollate by oxygenic photosynthetic organisms including cyanobacteria has been known for many years (Norris *et al.*, 1955; Wilson & Calvin, 1955) and it is now accepted that the oxygenase activity of RuBP carboxylase preparations is the principal source of this compound (see Andrews & Lorimer, 1978; Lorimer, 1981). However, cyanobacteria may differ from other oxygenic photosynthetic organisms in the route they use to recycle glycollate formed by the oxygenase reaction. Codd and Stewart (1973) have confirmed the observations of Miller *et al.* (1971) that glycollate metabolism in cyanobacteria is sensitive to α-hydroxysulphonates but not to iso-nicotinylhydrazide. They attributed this to the operation of a pathway (Fig. 3.4) in which glycollate is converted to glycerate by way of tartronic semialdehyde rather than the more common route via glycine and serine

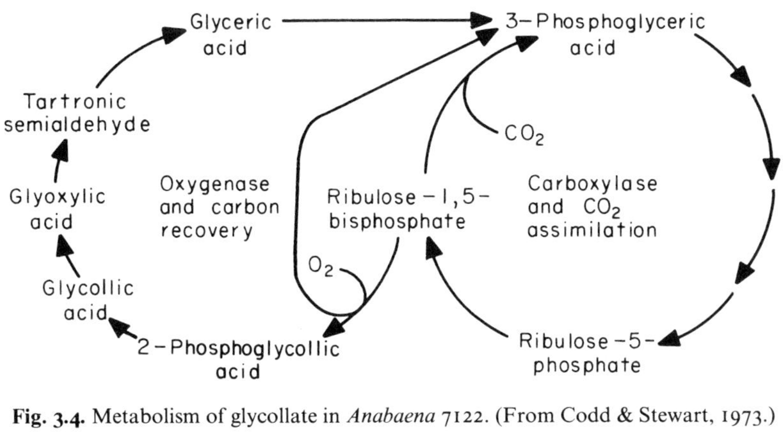

Fig. 3.4. Metabolism of glycollate in *Anabaena* 7122. (From Codd & Stewart, 1973.)

(Codd & Stewart, 1973). In the heterocystous organism *Anabaena* 7122 at least some of the enzymes of this pathway are, like ribulose-1,5-bisphosphate carboxylase, only present in vegetative cells (Codd *et al.*, 1980).

3.2.2 *Biosynthesis of selected cell constituents*

In autotrophic organisms, intermediates of the reductive pentose-phosphate cycle are withdrawn at various points for the biosynthesis of cell constituents. The routes to most of the latter in cyanobacteria are likely to be the same or very similar to those that operate in other organisms. The synthesis of only two particular categories of cell constituent will be considered here in detail; these are tetrapyrroles and reserve polymers.

3.2.2.1 Tetrapyrroles

In recent years, evidence has accumulated that δ-aminolaevulinic acid (δALA) is produced in plants and eukaryotic algae by a route which is distinct from that in bacteria and in animals. In the latter groups, δALA is formed from succinate and glycine (Jacobs, 1974; Shemin & Russell, 1953). On the basis of patterns of precursor labelling by various compounds (Beale *et al.*, 1975; Jurgenson *et al.*, 1976) and, more recently of cell-free studies (Ford & Friedman, 1979, 1981), it has been proposed that oxygenic photosynthetic organisms produce δALA from C_5 precursors related to glutamate (Fig. 3.3). As oxygenic photosynthetic organisms cyanobacteria might be expected to use the C_5 pathway, although on the basis of their prokaryotic structure they would be more likely to produce δALA by the succinate–glycine route. Kipe-Nolt and Stevens (1980) and Avissar (1980) have investigated the biosynthesis of this tetrapyrrole precursor in cyanobacteria by determining the distribution of isotope from specifically labelled organic compounds in δALA accumulated in the presence of laevulinic acid. Glutamate carbon was incorporated into δALA in appreciable amounts whereas very little carbon from glycine or succinate was found in this compound. Although the recoveries of isotope in the products of partial degradation of the labelled δALA were less than 100% and variable, it was apparent that C_1 and C_5 of glutamate had contributed significantly to C_5 and C_1 of δALA. This distribution of isotope is consistent with the conversion of the carbon skeleton of glutamate to the tetrapyrrole precursor. Because of uncertainties about the ability of cyanobacteria to assimilate glycine and succinate these labelling studies do not eliminate the possibility of a contribution from the succinate–glycine route to the synthesis of δALA in these organisms.

The route to δALA in these organisms can also be investigated by exploiting the incomplete tricarboxylic-acid cycle characteristic of all cyanobacteria (Fig. 3.3). Because of the block in the cycle, isotope from [^{14}C] acetate is incorporated into C_5 compounds leading to glutamate but not into C_4 compounds en route to asparatate. From this it follows that δALA made via the C_5 route will yield tetrapyrroles labelled to a degree similar to glutamate whereas that made from succinate and glycine, the former being derived from oxaloacetate, will lead to tetrapyrroles containing very little isotope. Applying this approach to *Synechococcus* 6301 it was found that the specific radioactivities of phaeophorbide *a*, glutamate and aspartate isolated from organisms grown in the presence of [^{14}C] acetates were in the ratio 500:200:1 for the [1-^{14}C] substrate and 8600:850:1 for [2-^{14}C] acetate (McKie *et al.*, 1981). If δALA is produced by the C_5 route the labelling of phaeophorbide *a* and glutamate by [1-^{14}C] and [2-^{14}C] acetate should be in the ratio 2:1 and 8:1 respectively. The observed values were 2.5:1 and 10.2:1 which are wholly consistent with the operation of the C_5 pathway as the sole route to δALA in cyanobacteria.

3.2.2.2 Reserve polymers

As a group, cyanobacteria are able to accumulate a variety of putative reserve materials. These include polysaccharide, polypeptide, polyphosphate and, in one strain, poly-β-hydroxybutyrate. Polysaccharide and poly-β-hydroxybutyrate are usually considered to be carbon and energy reserves, polyphosphate is a potential energy or phosphate reserve and the polypeptide cyanophycin (multi-L-arginine poly-L-aspartic acid) is, in principle, a nitrogen reserve with a possible additional minor role as a source of limited amounts of carbon and energy.

Cyanophycin is the trivial name given to the major component of the structured granules which are visible in thin sections of cyanobacteria (Drews & Niklowitz, 1956). They are particularly evident in cells from older cultures and in spores (Miller & Lang, 1968) and they retain many of their characteristic morphological features when isolated in a purified form (Lang *et al.* 1972). Cyanophycin ranges in molecular weight from 25 000 to 100 000 (Simon, 1971) and consists of L-arginyl residues attached to a poly-L-aspartic acid core via peptide links between the α-amino groups of the former and the free carboxyl groups of the latter (Simon & Weathers, 1976). An enzyme has been detected in *Anabaena cylindrica* which elongates an existing polymer chain adding arginyl and aspartyl residues in the presence of ATP, Mg^{2+} and a sulphydryl reagent (Simon, 1976). Polymer degradation is likely to be mediated by peptidases like that detected in *Anabaena* species (Gupta & Carr, 1981); aspartic acid-arginine was the product of this reaction.

Synthesis of this polypeptide is independent of ribosomes which accounts for the failure of chloramphenicol to block cyanophycin formation (Simon,

1973b). Although the identity of the immediate precursors of cyanophycin is clear their origin in the cell is less certain. In *Synechocystis* 6308 there is evidence that cyanophycin is formed from the products of catabolism of cell protein and not from the immediate products of CO_2 fixation (Allen *et al.*, 1980). Polymer accumulation in the presence of chloramphenicol is accompanied by a decrease in cell protein, and only a small amount of the isotope in $^{14}CO_2$ is incorporated into cyanophycin by light-limited stationary-phase organisms.

Cyanophycin commonly accumulates in the range 8–16% of dry weight after the end of exponential growth (Simon, 1973a; Allen *et al.*, 1980). Dilution of a culture in the stationary phase to re-establish growth results in a marked decrease in cyanophycin content (Simon, 1973a). In addition, the amount of polymer in *Synechocystis* 6308 decreases after transfer to a medium free of combined nitrogen (Allen *et al.*, 1980). When chloramphenicol is added to cultures to block protein synthesis, cyanophycin accumulates at levels up to 7% of dry weight and the removal of the antibiotic results in the restoration of protein synthesis and a marked drop in cyanophycin content (Simon, 1973b). The features of cyanophycin production and utilization give no clear indication of its role in the metabolism of the cyanobacterial cell; although its high content of nitrogen (approximately 24% by weight) suggests that it may function as nitrogen reserve. Studies of the role of cyanophycin are complicated by the fact that changes in the availability of nitrogen are also accompanied by marked shifts in the phycobiliprotein content of cyanobacteria (Allen & Smith, 1969). During nitrogen starvation the complement of biliprotein decreases markedly and normal levels are only restored on the provision of a source of combined nitrogen. Such changes in accessory photosynthetic pigment content have been taken to indicate an additional role for phycobiliproteins as an intracellular nitrogen reserve (Boussiba & Richmond, 1980). Cyanophycin is probably a reserve in the classical sense being present in relatively small amounts in rapidly growing cyanobacteria and accumulating in light-limited stationary-phase organisms. In contrast, phycobiliprotein may be an additional nitrogen reserve which is called upon in more extreme circumstances such as nitrogen starvation. In natural environments, combined nitrogen is likely to be a limiting nutrient for the growth of cyanobacteria which cannot fix molecular nitrogen (van Liere *et al.*, 1977). For these organisms, and for strains adapting to nitrogen fixation, cyanophycin reserves built up while combined nitrogen was available, together with phycobiliprotein, may prove to be significant nitrogen reserves. Marked variations have been reported in the amount of cyanophycin in a natural population of *Microcystis flos-aquae* (see Stewart, 1977). During the period of peak cell density relatively little polymer was present in the cyanobacteria while organisms before and after the appearance of the bloom contained relatively large amounts. More detailed investigations of the conditions promoting cyanophycin accumulation and utilization are needed to confirm the role of this compound in the nitrogen economy of the cell.

Cyanophycin may also provide cyanobacteria with a limited source of energy and carbon. Certain microorganisms can produce ATP by substrate-level phosphorylation during the catabolism of arginine by the dihydrolase route (Shimke *et al.*, 1966; Sperry & Wilkins, 1976; Blakemore & Canale-Parola, 1976).

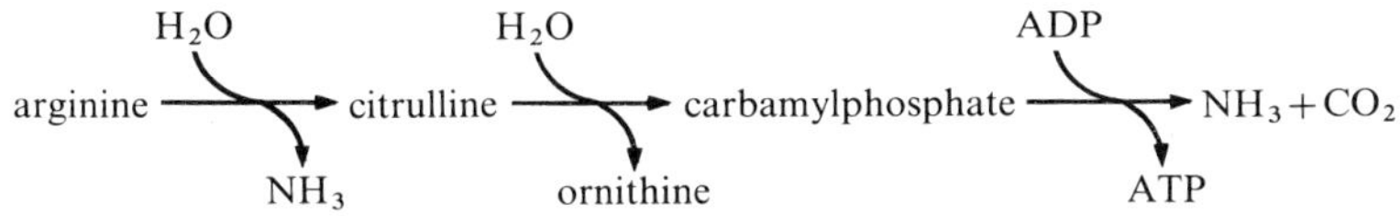

The enzymes involved in this pathway have been demonstrated in extracts of *Synechocystis* 6308 (Weathers & Allen, 1978). If they are also present in *Synechococcus* 6301 they may account for the survival in the dark of the double mutant of this organism which does not possess a functional oxidative pentose-phosphate cycle because it lacks the specific dehydrogenases for glucose-6-phosphate and 6-phosphogluconate (Doolittle & Singer, 1974). Arginine has been reported to support the slow growth of *Synechocystis* 6308 when CO_2 is excluded from cultures (Weathers *et al.*, 1978). Growth under these conditions may depend on CO_2 produced from the guanido-carbon of arginine catabolized by the pathway given above. These limited contributions to the energy and carbon metabolism of the cell which may benefit organisms under extreme conditions are likely, however, to be an incidental bonus that is a consequence of the use of an organic polymer to store nitrogen within the cell.

Many microorganisms, including cyanobacteria, contain granules of polyphosphate which may function as a source of phosphate for the synthesis of nucleic acids and phospholipids or as a source of energy for ATP formation (Merrick, 1979). Polyphosphate like cyanophycin commonly accumulates in cyanobacteria in the stationary phase or when phosphate is added to a culture of starved organisms (Stewart, 1977; Lawry & Jensen, 1979). Such reserves disappear during phosphate starvation and are able to support a limited amount of growth (Stewart & Alexander, 1971). These effects need to be defined more precisely in order to identify the function of polyphosphate in cyanobacteria. If it is involved in the production of ATP, it may provide an alternative explanation for the survival in the dark of the Doolittle and Singer (1974) double mutant of *Synechococcus* 6301.

Poly-β-hydroxybutyrate is a polymer unique to bacteria which serves as a carbon and energy reserve (see Dawes & Senior, 1973). Although it has been seen in thin sections of *Chlorogloeopsis* 6912 (Jensen & Sicko, 1971a) and has been isolated from it by conventional techniques (Carr, 1966), its role in this organism, the only cyanobacterium known to produce it, is obscure. It can play only a very minor role in the economy of the cell because acetate which is produced from it can neither be further catabolized for the generation of energy nor used in the synthesis of more than a few cell constituents.

Like many other microorganisms, cyanobacteria produce an intracellular polysaccharide with many of the features of a carbon and energy reserve material. In thin sections of these organisms polysaccharide takes the form of discrete granules or rods located between the thylakoids (Ris & Singh, 1961; Jost, 1965). This cell constituent has been isolated from several cyanobacteria and characterized in reasonable detail (Chao & Bowen, 1971; Weber & Wöber, 1975; Dudkin *et al.*, 1976). Material isolated from lysed cells by centrifugation is similar to the granules present in sections of intact organisms. These preparations, which were contaminated with a small amount of protein, gave only D-glucose when hydrolysed with acid, while α-amylase treatment yielded maltose and an α-macrodextrin accounting for 11–13% of the polysaccharide. The average chain length of α (1-4) linked units was 13 which is within the range (10–14) quoted for bacterial glycogens (Merrick, 1979). In contrast the size range of α (1-4) polyglucose molecules released by treatment with iso-amylase was comparable to that for phytoglycogen but distinct from that for other microbial glycogens (Weber & Wöber, 1975).

Glycogen synthesis involves three enzymes: a nucleoside-diphosphate glucose pyrophosphorylase, glycogen synthetase and a branching enzyme. These enzymes mediate, respectively, the synthesis of the monosaccharide donor, the polymerization of these monosaccharide units to form an α(1-4) polymer of glucose, and the rearrangement of this polymer to give α(1-6) branches in the chain. Levi and Preiss (1976) have characterized the ADP-glucose pyrophosphorylase of *Synechococcus* 6301; this activity was inhibited by phosphate and stimulated 8–25-fold by 3-phosphoglycerate, which is an intermediate of the reductive pentose-phosphate cycle. They have proposed that glycogen synthesis in cyanobacteria is regulated by the balance of these two metabolites in the cell as in other oxygenic photosynthetic organisms. In contrast the activity of the pyrophosphorylases of other prokaryotes is modulated by different metabolites including pyruvate, and fructose phosphates (Shen & Preiss, 1964; Furlong & Preiss, 1969).

Cyanobacteria in the exponential phase of growth in batch photoautotrophic culture contain glycogen amounting to 10–20% of dry weight (Allen & Smith, 1969; Lehmann & Wöber, 1976; van Liere, 1979) whereas it accounts for up to 60% of dry weight in cultures starved of nitrogen (Allen & Smith, 1969; Lehmann & Wöber, 1976). The addition of combined nitrogen to a starved culture results in a marked decrease in glycogen content before cell division commences (Allen & Smith, 1969) and in the absence of CO_2, endogenous polymer supports a doubling in chlorophyll-*a* content (Lehmann

& Wöber, 1976). In these experiments, polymer accumulated under conditions where the supply of carbon and energy was plentiful and nitrogen was the limiting nutrient. The decrease in glycogen content following the addition of combined nitrogen suggests it has a role in supplementing or even replacing the normal supply of carbon during the period before restoration of exponential growth. Glycogen also has an important role in the dark metabolism of cyanobacteria and this will be considered in detail in the next section.

3.3 Dark-endogenous metabolism

The natural environment of most cyanobacteria is one in which there is a regular diurnal light–dark cycle. During the period of illumination energy sufficient for CO_2 assimilation and the synthesis of cell constituents leading to net growth is derived from light and conserved by the photosynthetic apparatus (see Chapter 8). In the dark period cyanobacteria have to resort to alternative energy-yielding mechanisms so that, at the very least, they can maintain cell integrity, function and viability until the next light period. Doolittle and Singer (1974) have shown that *Synechococcus* 6301 can maintain its viability over a period of 20 h under aerobic conditions in the dark. In the light cyanobacteria have an intracellular ATP concentration of 2.5–3.5 μmol ATP/g dry wt (Bornefeld & Simonis, 1974; Ihlenfeldt & Gibson,

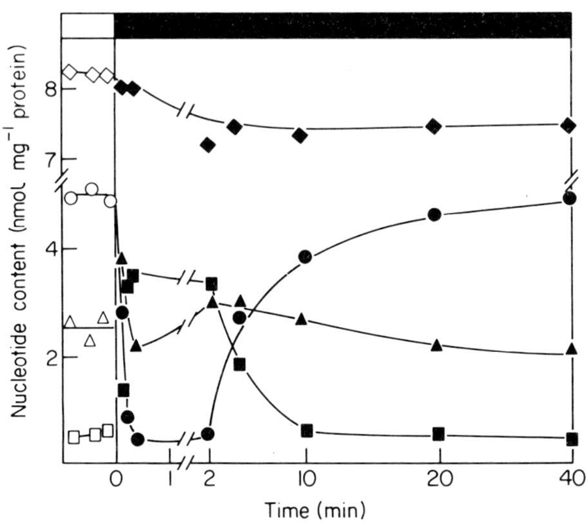

Fig. 3.5. Effect of a light-to-dark transfer on the adenylate pools of *Synechococcus* 6301. After incubation in the light the organisms were transferred to the dark at time zero. ATP (○, ●), ADP (△, ▲), AMP (□, ■) and total adenine nucleotides (◊, ♦) were measured. (From Ihlenfeldt & Gibson, 1975a.)

1975a; Bottomley & Stewart, 1976a). Transfer of organisms from light to darkness results in an immediate decrease in the concentration of ATP in the cell (Fig. 3.5) and a reciprocal effect on the ADP plus AMP pool (Pelroy & Bassham, 1972; Ihlenfeldt & Gibson, 1975a; Imafuku & Katoh, 1976; Bottomley & Stewart, 1976a). This effect is transient lasting approximately 1–2 min and within 10–12 min of transfer, the intracellular concentration of ATP is restored to that of illuminated organisms. However, the rate of ATP synthesis in the dark, 0.33 μmol mg^{-1} dry wt h^{-1} *Anabaena* 7122, is only 10% of the rate supported by photophosphorylation (Bottomley & Stewart, 1976a). Although the ability of cyanobacteria to produce ATP is much lower in the dark than in the light, the maintenance of a high intracellular concentration of this nucleotide indicates that the demand for ATP in the dark is much lower than in growing organisms. Presumably it is required largely for the maintenance of processes essential to the survival of the cell. When cyanobacteria are transferred in the dark from aerobic to anaerobic conditions or incubated in the dark with cyanide, dinitrophenol or carbonyl cyanide m-chlorophenyl hydrazone (CCCP) to block respiration there is a marked decline in ATP content. All cyanobacteria exhibit a low rate of dark-endogenous respiration (Webster & Frenkel, 1952; Kratz & Myers, 1955b; Pearce & Carr, 1967) which is likely to be involved in the maintenance of the intracellular ATP pool. When *Synechococcus* 6301 is transferred from aerobic to anaerobic conditions in the dark the level of NADPH in the cell increases suggesting that this nucleotide rather than NADH is the substrate for endogenous respiration (Biggins, 1969). The oxidative synthesis of ATP using NADPH as electron donor has been described in extracts of *Anabaena* 7118 (Leach & Carr, 1970)

Polysaccharide is the major intracellular reserve of oxidizable substrate in cyanobacteria; incubation in the dark under aerobic conditions results in a gradual decrease in glycogen content (Lehmann & Wöber, 1976; Oren & Shilo, 1979). This endogenous reserve role of cyanobacterial glycogen has been confirmed in the elegant studies of van Liere, Mur, Gibson & Herdman (1979). The continuous culture of *Oscillatoria agardhii* with a light–dark cycle resulted in the accumulation of glycogen during the light phase and the utilization of this reserve material during the subsequent aerobic-dark period (Fig. 3.6). Nothing is known of the mechanisms that regulate the mobilization of endogenous polysaccharide reserves in cyanobacteria.

Various approaches have been used to identify the principal metabolic route involved in the maintenance of energy levels of cyanobacteria in the dark; these include rapid labelling experiments (Pelroy & Bassham, 1972), assay of enzymes in cell-free extracts (Pelroy & Bassham, 1972) and mutant studies (Doolittle & Singer, 1974). Transfer to the dark of *Synechocystis* 6308 exposed to $^{14}CO_2$ in the light resulted in the appearance of a readily detectable

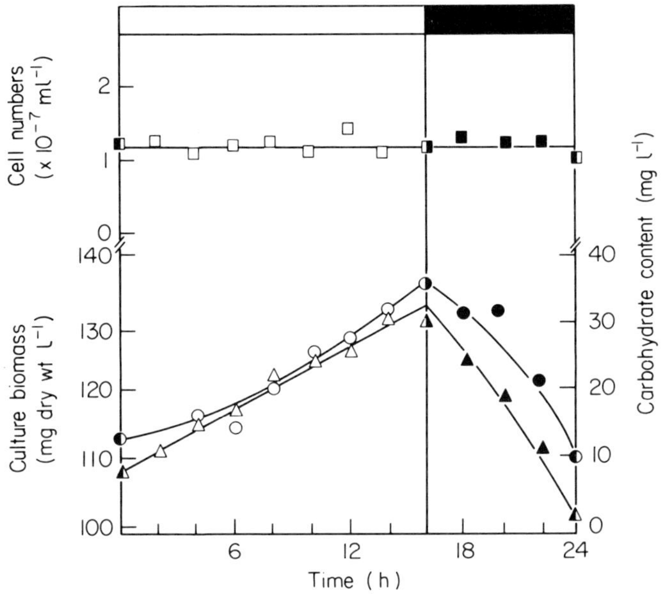

Fig. 3.6. Growth of *Oscillatoria agardhii* grown in continuous culture with a light–dark cycle. The organism was grown with a diurnal cycle of 16 h light and 8 h dark at a dilution rate of 0.022 h^{-1}. The parameters measured included cell numbers (□, ■), carbohydrate (○, ●) content, and dry weight (△, ▲). (From van Liere, Mur, Gibson & Herdman, 1979.)

pool of 6-phosphogluconate and an increase in the pool sizes of fructose-1,6-bisphosphate and sedoheptulose-1,7-bisphosphate (Fig. 3.1). The amount of isotope in sugar monophosphates (glucose-6-phosphate, sedoheptulose-7-phosphate and ultimately 6-phosphogluconate as well) decreased although the pool sizes of these metabolites did not change (Pelroy & Bassham, 1972). This was attributed to a relatively rapid turnover of these compounds in the dark. In contrast, the turnover of intermediates of glycolysis and the biosynthetic routes to glutamate and aspartate was much slower. These data are consistent with the complete catabolism of endogenous carbohydrate in the dark by way of the oxidative pentose-phosphate cycle (Fig. 3.2b). All cyanobacteria contain NADP-specific dehydrogenases for glucose-6-phosphate and 6-phosphogluconate (Fewson *et al.*, 1962; Pearce & Carr, 1969; Pelroy *et al.*, 1972) which are involved in this catabolic process. They catalyse the formation of NADPH which is then oxidized using the respiratory chain with electron transport to oxygen coupled to ATP production. The drastic reduction in the respiratory activity of *Synechococcus* 6301 following the loss of both of these enzymes by mutation clearly implicates them in the endogenous dark metabolism of this organism (Doolittle & Singer, 1974).

When cyanobacteria are transferred from light to dark their metabolism changes from the net fixation of CO_2 via the reductive pentose-phosphate cycle (Fig. 3.2a) to the catabolism of endogenous polysaccharide through the oxidative pentose-phosphate cycle (Fig. 3.2b). Although these two metabolic routes have several enzyme-catalysed reactions in common they are distinguished by the reactions involved in the reductive trapping of CO_2 on the one hand and in the oxidative decarboxylation of hexose phosphate molecules on the other (Fig. 3.2c; Table 3.3). This major change in metabolism will depend on control mechanisms which prevent CO_2 fixation in the dark and, under normal circumstances, block the catabolism of endogenous polysaccharide in the light. On the basis of labelling studies with intact organisms (Pelroy & Bassham, 1973b; Pelroy, Levine & Bassham, 1976) the cessation of CO_2 fixation that is the immediate consequence of transfer from light to dark has been attributed to an inhibition of phosphoribulokinase, and of sedoheptulose-1,7- and fructose-1,6-bisphosphatases. Although the mechanism of inhibition has yet to be identified, it is likely to be activated by the loss of non-cyclic electron flow (Pelroy, Kirk & Bassham, 1976). Under these conditions control at the level of glyceraldehyde-3-phosphate dehydrogenase may also be of importance (Ihlenfeldt & Gibson, 1975a). 6-Phosphogluconate inhibits the RuBP carboxylase activity of permeabilized cells (Tabita &

Table 3.3. Enzymes of the pentose phosphate cycles

	Enzymes unique to:	
Common enzymes	Reductive pentose phosphate cycle	Oxidative pentose phosphate cycle
Fructose-1,6-bisphosphate aldolase (1)	Aldolase (7)	Dihydroxyacetone transferase (transaldolase) (13)
Fructose-1,6-bisphosphate phosphatase (2)	Glyceraldehyde 3-phosphate dehydrogenase (8)	Glucose 6-phosphate dehydrogenase (14)
Glycoaldehyde transferase (transketolase) (3)	3-Phosphoglycerate kinase (9)	Hexose-6-phosphate isomerase* (15)
	5-Phosphoribulokinase (10)	6-Phosphogluconate dehydrogenase (16)
Ribose-5-phosphate isomerase (4)	Ribulose-1,5-bisphosphate carboxylase (11)	6-Phosphogluconolactonase (17)
Ribulose-5-phosphate 3-epimerase (5)		
Triose phosphate isomerase (6)	Sedoheptulose-1,7-bisphosphate phosphatase (12)	

* Although hexosephosphate isomerase is not an enzyme of the reductive pentose phosphate cycle, it is involved in the further metabolism of cycle intermediates (see Fig. 3.2a).
The numbers in parentheses identify the reactions in Fig. 3.2c that are catalysed by these enzymes.

Colletti, 1979) as well as the purified carboxylase (Codd & Stewart, 1977a). Although a detectable pool of this metabolite accumulates immediately after the transfer of cyanobacteria from light to dark, the concomitant decline in the concentration of RuBP (Fig. 3.1) suggests that the carboxylase is still functioning, albeit for the relatively short time that its substrate is still available (Pelroy, Kirk & Bassham, 1976). It is therefore unlikely that 6-phosphogluconate is of major importance in the shut-down of CO_2 fixation resulting from the shift to darkness.

The appearance of a pool of 6-phosphogluconate in cyanobacteria after the transfer of organisms from light to dark is consistent with a mechanism for the control of the oxidative pentose-phosphate cycle which blocks this route in the light at the level of glucose-6-phosphate dehydrogenase. There are however differences of opinion about the properties of this enzyme which are relevant to its control in cyanobacteria. Glucose-6-phosphate dehydrogenase has been partially purified from *Anabaena* 7120 (Schaeffer & Stanier, 1978) and from *Synechococcus* 6301 (Grossman & McGowan, 1975; Cséke *et al.*, 1981). These preparations of the enzyme have been characterized kinetically and some of their molecular properties determined. The catalytic activity of both purified enzyme preparations was inhibited by NADPH and ATP (Grossman & McGowan, 1975; Schaeffer & Stanier, 1978; Cséke *et al.*, 1981). In addition the enzyme isolated from *Anabaena* 7120 dissociated reversibly resulting in a marked decrease in activity; the balance between the various molecular forms of the enzyme was influenced by several factors including substrate concentration and pH (Schaeffer & Stanier, 1978). While the activity of crude preparations of the *Anabaena* enzyme was inhibited by RuBP, an effect attributed to the formation of the low molecular weight, hypoactive enzyme, the purified enzyme from this organism was not affected by this metabolite (Schaeffer & Stanier, 1978). Enzyme activity in crude extracts of several unicellular cyanobacteria is also sensitive to RuBP (Pelroy & Bassham, 1972; Pelroy, Kirk & Bassham, 1976) but the activity in *Plectonema* 73110 is not inhibited by RuBP, NADPH or ATP (Raboy *et al.*, 1976). In the latter organism, glucose-6-phosphate dehydrogenase is markedly inhibited by preincubation with the reducing agent dithiothreitol. Although some of these effects on enzyme activity can be fitted into a logical scheme for the regulation of this enzyme and of the oxidative pentose-phosphate cycle (Stanier, 1973), it would be premature to generalize about the properties of this enzyme and intracellular mechanisms for the regulation of its activity because of inconsistencies in the data available.

The catabolism of polysaccharide reserve material is a characteristic feature of the dark aerobic endogenous metabolism of cyanobacteria. The coupling of this process to an aerobic respiratory phosphorylating electron transport chain serves to maintain the level of ATP in the cell. Unlike other cyanobacteria, *Oscillatoria limnetica* is exposed to marked fluctuations in conditions, the most notable of which are frequent shifts between aerobic and anaerobic states (Cohen, Jørgensen, Padan & Shilo, 1975). This is accommodated in the light by an ability for anoxygenic as well as oxygenic photosynthesis (Oren & Padan, 1978; see Chapter 9). In the dark, this organism catabolizes endogenous polysaccharide both aerobically and anaerobically. In the absence of oxygen, two different metabolic strategies can be adopted, namely anaerobic respiration and fermentation (Oren & Shilo, 1979). In the presence of sulphur, carbohydrate catabolism is coupled to the reduction of sulphur to sulphide, 5–9 mol of the latter being formed for every mol of glucose degraded to CO_2, the other major end product. The catabolic pathway involved has not been characterized but it is likely to be the oxidative pentose-phosphate cycle. In the absence of sulphur, the breakdown of intracellular polysaccharide is accompanied by a decrease in pH due to the formation of lactic acid. The yield of the latter is in the range 1.4–1.8 mol for each mol of glucose catabolized. In the catabolism of glucose via the oxidative pentose-phosphate cycle (Fig. 3.2b) 2 mol of 3-phosphoglyceraldehyde are produced for every 6 mol of glucose-6-phosphate entering the cycle. These could be diverted to lactate but this would not dispose of the pyridine nucleotide reduced during the catabolism of glucose by this route. It is more likely that glucose is converted to lactate via glycolysis as in certain lactic acid bacteria. Although nothing is known of the relative activities of enzymes involved in anaerobic carbohydrate breakdown in *Oscillatoria limnetica*, it is known that *Synechocystis* 7616 possesses all the enzymes necessary for the homolactic fermentation of glucose (Sanchez *et al.*, 1975). The inability of several cyanobacteria to maintain a high intracellular concentration of ATP under dark, anaerobic conditions (Biggins, 1969; Bottomley & Stewart, 1976a) suggests that an anaerobic energy-yielding metabolism based on the lactic fermentation of endogenous carbohydrate may not be a general feature of all strains.

Since the endogenous reserves of polysaccharide in cyanobacteria are limited in amount, they would be incapable of supporting more than a small net increase in the growth polymers of the cell. No such increase has been demonstrated in suspensions of batch-grown organism (Oren & Shilo, 1979). In contrast, an economical use of these limited reserves would ensure that the cell is able to survive for an extended period in the dark. It is therefore of great interest that an apparent synthesis of growth polymers in *Oscillatoria agardhii* maintained in continuous culture with a 24 h light–dark cycle (16 h light/8 h dark) continued largely uninterrupted despite regular shifts between light and dark (van Liere, Mur, Gibson & Herdman, 1979). In such cultures, polysaccharide accumulated during a particular light period was used as a source of carbon and energy for the biosynthesis of cell constituents during the succeeding dark period. The amount of growth occurring during the dark period was, however, rather small, corresponding to between 8 and 18% of the culture density. In cultures exposed to diurnal cycles in which the dark period was greater than 12 h, growth polymer synthesis was not maintained beyond the first few hours in the dark.

3.4 Heterotrophic growth and metabolism

Some cyanobacteria will grow not only in the light on CO_2 but also under conditions where growth is dependent on exogenous organic compounds. Such heterotrophic growth can take two different forms: photoheterotrophy, which is growth in the light on an organic compound in the absence of net CO_2 fixation, and chemoheterotrophy, which is growth on an organic substrate in complete darkness. In the latter case, the organic compound provides the organism with a source of carbon and energy while in photoheterotrophy the organic compound is used only as a source of carbon, light supplying the energy needs of the cell. Photoheterotrophy has been demonstrated among cyanobacteria using several different procedures for reducing or eliminating net CO_2 fixation. This has been achieved by the exclusion of CO_2 from the growth medium (Ingram *et al.*, 1973) and by blocking its assimilation indirectly with the herbicide DCMU (3′-(3,4-dichlorophenyl)1′,1′-dimethylurea) (Rippka, 1972) which inhibits photosystem II, non-cyclic electron flow and the production of NADPH. Alternatively, net CO_2 fixation can be greatly reduced by the more physiological device of lowering the incident light intensity (Van Baalen *et al.*, 1971). These two modes of heterotrophic growth in which an increase in culture density is totally dependent on exogenous organic compounds are distinct from mixotrophic growth in which CO_2 and organic carbon are assimilated simultaneously in amounts which can vary with the culture conditions (see Smith, 1973).

3.4.1 Distribution and features of heterotrophy among cyanobacteria

Isolated reports of the growth of individual strains on organic media (Fay, 1965; Watanabe & Yamamoto, 1967; Hoare *et al.*, 1971; White & Shilo, 1975; Wolk & Shaffer, 1976) together with the results of surveys of the heterotrophic potential of large collections of cyanobacteria (Khoja & Whitton, 1971; Rippka *et al.*, 1979) have provided ample proof that some cyanobacteria are capable of growth on exogenous organic compounds. As part of a detailed taxonomic study of cyanobacteria, Rippka *et al.* (1979) screened a wide variety of strains for photoheterotrophic growth using the DCMU procedure (Table 3.4). About half the strains tested grew with an exogenous organic compound as carbon source. This ability was not, however, distributed

Table 3.4. Distribution of photoheterotrophy among cyanobacteria

Taxonomic group	Species screened	Photo-hetero-trophic species	Species that grew on: Glucose	Fructose	Ribose	Sucrose	Glycerol
Synechococcus	17	4	3	2	0	1	2
Gloeothece	5	0	0	0	0	0	0
Gloeobacter	1	0	0	0	0	0	0
Synechocystis	11	3	3	0	0	1	1
Gloeocapsa	4	2	2	1	1	1	0
Chamaesiphon	2	1	1	1	0	1	0
Dermocarpa	6	4	4	2	—	3	—
Dermocarpella	1	1	1	1	—	1	—
Xenococcus	3	1	1	0	—	1	—
Myxosarcina	2	2	1	0	—	2	—
Chroococcidiopsis	3	3	2	3	—	2	—
Pleurocapsa	12	10	4	4	—	8	—
Spirulina	2	0	0	0	0	0	0
Oscillatoria	5	3	3	3	0	0	0
Pseudanabaena	8	0	0	0	0	0	0
Lyngbya-Plectonema-Phormidium	20	11	10	8	3	8	1
Anabaena	3	0	0	0	0	0	0
Nodularia	1	1	1	1	0	1	0
Cylindrospermum	3	1	0	1	0	1	0
Nostoc	12	9	4	6	3	6	0
Scytonema	1	1	1	1	0	1	0
Calothrix	10	10	8	9	3	7	0
Fischerella	7	7	7	7	3	6	0
Chlorogloeopsis	1	1	1	1	1	1	0

These data have been taken from the survey of Rippka *et al.* (1979). In those cases in which strains were considered to be independent isolates of the same species, they have been counted as one species.

uniformly among all taxonomic groups. At one extreme all of the isolates of *Dermocarpella* (1), *Myxosarcina* (2), *Chroococcidiopsis* (3), *Nodularia* (1), *Scytonema* (1), *Calothrix* (10), *Fischerella* (7) and *Chlorogloeopsis* (1) grew on at least one organic compound. In contrast none of the strains of *Gloeothece* (5), *Gloeobacter* (1), *Spirulina* (2), *Pseudanabaena* (8) or *Anabaena* (3) grew photoheterotrophically and it was rare among *Synechococcus* (17) and *Synechocystis* (11) isolates. Although a relatively small range of organic compounds was tested, it emerged that D-glucose supported the growth of more strains than any of the other compounds tested; sucrose and D-fructose were somewhat less effective and relatively few isolates grew on D-ribose or glycerol. A similar pattern emerged from a survey of nineteen organic compounds for the ability to support the photoheterotrophic growth of nine different cyanobacteria (More *et al.*, 1979); D-glucose, D-fructose, D-ribose and sucrose supported significant heterotrophic growth of several strains and marginal growth was observed with D-galactose, maltose, cellobiose, mannitol and D-xylose. One of the most distinctive features of heterotrophic growth among cyanobacteria is the low rate relative to growth under photoautotrophic conditions (Table 3.5). While the growth of *Synechocystis* 6714 on D-glucose was not affected by the addition of other organic compounds to the medium (Rippka, 1972), a mixture of amino acids enhanced significantly the growth of *Nostoc* strain MAC (Hoare *et al.*, 1971) and *Plectonema* 73110 (White & Shilo, 1975). Whether this is due to a few amino acids or to all the components of the mixture has yet to be established. It is worth noting in passing that eukaryotic algae like cyanobacteria commonly grow more slowly on organic compounds than in the light on CO_2 (see Droop, 1974b). In most cases, cyanobacteria capable of heterotrophic growth are apparently able to grow without a significant lag on transfer between autotrophic and heterotrophic conditions (Fig. 3.7). It is not therefore surprising that dark-grown organisms retain high levels of photosynthetic pigments (Hoare *et al.*, 1971). In contrast, *Plectonema* 73110 requires an extended period of training before it is capable of growth on glucose at a rate comparable to other cyanobacteria (Raboy *et al.*, 1976). On transfer to the dark, growth on this substrate took

Table 3.5. Growth rates of cyanobacteria under various conditions

	Growth rate (doubling time in hours)			
Organism	Photo-autotrophic	Photohetero-trophic	Chemohetero-trophic‡	Reference
Anabaena 29413	14	25(F)*	40(F)	Author (unpubl.)
Chlorogloeopsis 6912	30	—	80(S)	Joset-Espardellier *et al.* (1978)
Synechocystis 6714	15	30(G)*	50(G)	Rippka (1972)
Nostoc strain MAC	6.5	20(G)†	48(G)	Ingram *et al.* (1973)
Plectonema 73110	24	—	40(G)	Raboy *et al.* (1976)

* Growth in the light in the presence of DCMU (10^{-5} M)
† Growth at low light intensity (850 lx)
‡ The letters in parentheses refer to the organic substrate: D-glucose (G), D-fructose (F) and sucrose (S).
The figures in this table have been taken from those sources quoting rates for all of the specified conditions.

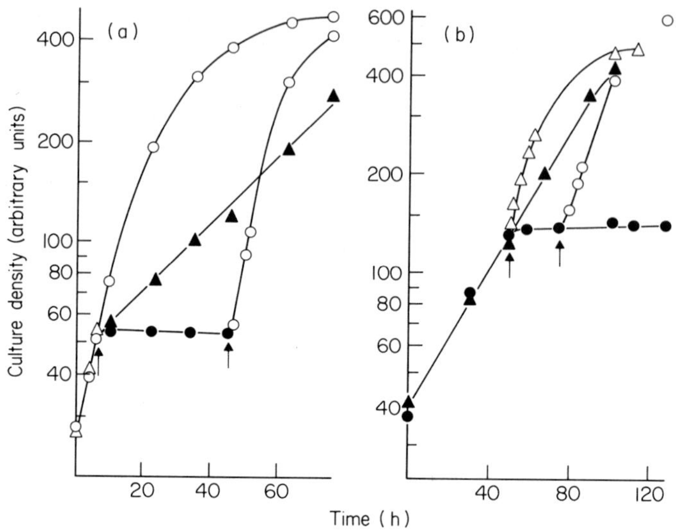

Fig. 3.7. Growth of *Synechocystis* 6714 on transfer between light and dark. (a) Organisms growing in the light on CO_2 (○) and on CO_2 pluse D-glucose (△) were transferred to the dark after 6 h (●, ▲) incubation. The dark culture lacking D-glucose was returned to the light after 40 h in the dark. (b) Organisms growing in the dark on D-glucose at a final concentration of 50 mmol (▲) and 10 mmol (●) were transferred to the light (△, ○) after 52 and 80 h incubation. The arrows indicate the point at which cultures were transferred from one set of conditions to another. (From Lucas, 1974.)

place at a very low rate, equivalent to a doubling time of 200 h. The maintenance of this organism on glucose in the dark for 4 years resulted in a five-fold improvement in heterotrophic growth rate. When fully adapted organisms were grown for several generations under photoautotrophic conditions and then returned to glucose media in the dark, they required a period of approximately 12 days adaptation before the maximum heterotrophic growth rate was restored. *Chlorogloeopsis* 6912 also appears to require a period of adaptation before a reproducible rate of growth in the dark on organic media is established (Fay, 1965). In addition, after prolonged cultivation of this organism in the dark on sucrose there is an appreciable lag before certain photoautotrophic processes are re-established (Evans & Carr, 1975).

3.4.2 Heterotrophic metabolism

All of the organic compounds that support the heterotrophic growth of cyanobacteria can be metabolized via the oxidative pentose-phosphate cycle. The exceptions to this generalization are the unconfirmed reports of the growth of *Chlorogloeopsis* 6912 (Miller & Allen, 1972) and of *Nostoc* strain MAC (Ingram *et al.*, 1973) on acetate. Studies with *Synechocystis* 6714 (Pelroy *et al.*, 1972; Pelroy & Bassham, 1973a,b; Pelroy, Kirk & Bassham, 1976) and to a lesser extent with other heterotrophic cyanobacteria (Raboy *et al.*, 1976; Joset-Espardellier *et al.*, 1978; Bottomley & Van Baalen, 1978; Lucas, 1974) have firmly established the oxidative pentose-phosphate cycle as the route for the aerobic metabolism of exogenous glucose. The carbon atoms of this carbohydrate are converted to CO_2 in a highly asymmetric fashion by *Synechocystis* 6714; the initial rates of CO_2 formation from C_1, C_3+C_4 and C_6 were in the ratio 13:5:1 whereas all four carbon atoms were incorporated into glycogen at essentially the same rate. A similar asymmetric catabolism of exogenous glucose has been observed with *Plectonema* 73110 (Raboy *et al.*, 1976), *Nostoc* strain MAC and *Chlorogloeopsis* 6912 (Lucas, 1974) and at a much lower net rate with cyanobacteria of doubtful heterotrophic status (Cheung & Gibbs, 1966; Wildon & ap Rees, 1965; Pearce & Carr, 1969). The size of intermediary metabolite pools in *Synechocystis* 6714 incubated with glucose are also consistent with the operation of the oxidative pentose-phosphate cycle; these include large pools of glucose-6-phosphate, 6-phosphogluconate and sedoheptulose-7-phosphate (Pelroy *et al.*, 1972; Pelroy & Bassham, 1973b). Under photoautotrophic, photomixotrophic, photo-heterotrophic and chemoheterotrophic conditions, the intracellular pools of glucose-6-phosphate were in the ratio 1:18:37:140. Cell-free preparations of *Synechocystis* 6714 contain readily detectable levels of glucose-6-phosphate and 6-phosphogluconate dehydrogenases, the key enzymes of the oxidative pentose-phosphate cycle, and little or no phosphofructokinase, ketodeoxy-phosphogluconate aldolase or 6-phosphogluconate dehydrase (Pelroy *et al.*, 1972). Both dehydrogenases have been demonstrated in other heterotrophic cyanobacteria (Lucas, 1974; Raboy *et al.*, 1976; Joset-Espardellier *et al.*, 1978) and, in addition, the activity of glucose-6-phosphate dehydrogenase is several-fold higher in organisms grown under chemoheterotrophic conditions (Pelroy *et al.*, 1972; Raboy *et al.*, 1976; Joset-Espardellier *et al.*, 1978). These enzymological data not only confirm the involvement of the oxidative pentose-phosphate cycle, and little or no phosphofructokinase, ketodeoxy-phosphogluconate aldolase or 6-phosphogluconate dehydrase (Pelroy *et al.*, route operate at a significant level. The role of D-glucose dehydrogenase which has been detected in several heterotrophic cyanobacteria including *Chlorogloeopsis* 6912 and *Nostoc* strain MAC (Pulich & Van Baalen, 1973) is uncertain. This enzyme, which is distinct from D-glucose-6-phosphate dehydrogenase (Pritchard *et al.*, 1975) has a very high K_m for D-glucose ($1-2 \times 10^{-2}$M) and relatively low activity with other carbohydrates, apart from the substrate analogue 2-deoxy-D-glucose (Pritchard *et al.*, 1975; Pulich *et al.*, 1976). Since the affinity of the enzyme for its substrate is very low and, in addition, this enzyme is not present in the heterotrophic cyanobacterium *Synechocystis* 6714 (Joset-Espardellier *et al.*, 1978) it is unlikely to have an important role in cyanobacterial heterotrophy.

The relatively high rate of CO_2 production from exogenous glucose by *Synechocystis* 6714 in comparison with cyanobacteria that do not grow on this substrate and the saturation kinetics of this catabolic process have been attributed to the presence of a mechanism for the transport of exogenous glucose into the cell (Pelroy *et al.*, 1972). The uptake of D-glucose by three heterotrophic cyanobacteria, *Synechocystis* 6714, *Nostoc* strain MAC and *Plectonema* 73110, which are all capable of growth on D-glucose, has been investigated using structural analogues of D-glucose which are not metabolized by the cell (Beauclerk & Smith, 1978; Raboy & Padan, 1978). *Synechocystis* 6714 took up 3-O-methyl-D-glucose in relatively small amounts when compared with D-glucose but the uptake of each substrate was inhibited by the other (Fig. 3.8). As these inhibitory effects were competitive, it was concluded that the analogue, 3-O-methyl-D-glucose, was taken into the cell by the mechanism which mediated the uptake of the physiological substrate (Beauclerk & Smith, 1978). Similar conclusions have been drawn from studies of the uptake of D-glucose and α-methyl-D-glucoside by *Plectonema* 73110 (Raboy & Padan, 1978). While the transport mechanisms in *Synechocystis* 6714 and *Nostoc* strain MAC were of similar activity in organisms grown in the light or in the dark, glucose incorporation in *Plectonema* 73110 was

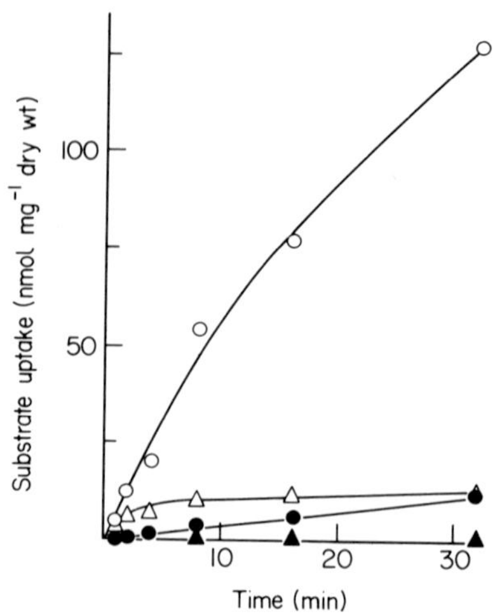

Fig. 3.8. Uptake of D-glucose and 3-O-methyl-D-glucose by *Synechocystis* 6714. Organisms were incubated in the dark in the presence of D-[U-^{14}C]glucose (○, ●) and 3-O-methyl-D-[U-^{14}C]glucose (△, ▲) alone (○, △) and in the presence of unlabelled 3-O-methyl-D-glucose (●) and D-glucose (▲). (From Beauclerk & Smith, 1978.)

induced seven-fold on transfer from photoautotrophic to chemoheterotrophic conditions (Raboy *et al.*, 1976). The transport mechanisms in these organisms were further distinguished by their analogue specificity; α-methyl-D-glucoside was a substrate for the *Plectonema* 73110 transport process whereas 3-O-methyl-D-glucose and 6-deoxy-D-glucose were both taken up by *Synechocystis* 6714 and *Nostoc* strain MAC, α-methyl-D-glucoside being inactive with these latter organisms. The uptake of 3-O-methyl-D-glucose by *Synechocystis* 6714 and of α-methyl-D-glucoside by *Plectonema* 73110 both resulted in the accumulation of unmodified analogue within the cell at concentrations 10–100-fold higher than the extracellular concentration. From these and other features of analogue uptake it has been concluded that these heterotrophic cyanobacteria possess mechanisms for the active transport of the organic growth substrate into the cell. Estimates of the rate of transport of the physiological substrate made from measurements of the initial rates of uptake are consistent with the observed rates of growth under heterotrophic conditions: *Plectonema* 73110 grows chemoheterotrophically with a mean generation time of 40 h and took up D-glucose at a rate of 3.5 nmol min^{-1} mg^{-1} dry wt (Raboy & Padan, 1978) while *Synechocystis* 6714 which grows chemoheterotrophically with a doubling time of 35 h took up this carbohydrate at a rate of 8 nmol min^{-1} mg^{-1} dry wt (Beauclerk & Smith, 1978).

It has yet to be established that all heterotrophic cyanobacteria possess structurally specific mechanisms for the transport of their organic growth substrates into the cell. The rates at which carbohydrates are taken up by other

Table 3.6. Rates of utilization of carbohydrates by illuminated cell suspensions of cyanobacteria grown photoautotrophically

Organism	D-glucose	D-fructose	D-ribose	sucrose	maltose
	(rate of substrate uptake: nmol mg^{-1} dry wt 10 min^{-1})				
Anabaena 29413	0.9	35.0*	8.4	1.4	0.1
Nostoc strain MAC	17.6*	2.0*	2.6	0.7	0.1
Nostoc 6720	18.3*	15.7*	0.8	0.5	—
Synechocystis 6714	44.6*	0.5	0.2	0.8	—
Chlorogloeopsis 6912	4.4*	0.6	0.4	1.5*	1.7
Plectonema 73110	2.6*	5.1*	6.1*	3.6*	2.3*

* These substrates support the heterotrophic growth of the specified strains (see Fay, 1965; Watanabe & Yamamoto, 1967; Hoare *et al.*, 1971; Rippka, 1972; White & Shilo, 1975; Wolk & Shaffer, 1976).
Rates of substrate uptake are from unpublished results (J.E. More, M.M. Williams & A.J. Smith).

heterotrophic cyanobacteria have been shown to vary (Table 3.6). Those compounds that supported growth were taken up at relatively high rates by suspensions of several strains grown under photoautotrophic conditions. In contrast, other heterotrophic cyanobacteria took up organic compounds including those that supported heterotrophic growth at relatively low rates suggesting that growth on the latter may involve adaptive effects influencing substrate uptake. The utilization of organic substrates by several heterotrophic strains including *Plectonema* 73110 (sucrose), *Anabaena* 29413 (D-fructose) and *Chlorogloeopsis* 6912 (D-glucose) exhibited saturation kinetics (More *et al.*, 1979). In addition, sucrose uptake by two strains was inhibited by trehalose and maltose but not by other disaccharides or by D-fructose or D-glucose. These observations imply that all heterotrophic cyanobacteria possess mechanisms for the transport of organic growth substrates into the cell.

The accumulation of exogenous organic substrates within prokaryotic or eukaryotic cells at concentrations greater than in the medium is necessarily dependent on a source of energy. Under photoheterotrophic conditions, energy for the uptake and assimilation of organic growth substrates and the synthesis of cell constituents is provided by cyclic photophosphorylation whereas in the dark substrate utilization is coupled to the catabolism of carbohydrate obtained from the medium. The relation between the energy metabolism of the cell and the uptake of D-glucose and its analogues has been investigated to a limited extent in *Synechocystis* 6714 (Beauclerk & Smith, 1976) and in greater detail in *Plectonema* 73110 (Raboy & Padan, 1978). Agents such as carbonyl cyanide chlorophenylhydrazone (CCCP), tetrachlorosalicylanilide (TCS) (Beauclerk & Smith, 1976), carbonyl cyanide p-trifluoromethoxy phenylhydrazone (FCCP) (Raboy & Padan, 1978) and cyanide which interfere with various stages in energy conservation and anaerobiosis (Raboy & Padan, 1978) blocked the uptake of the physiological substrate and its analogues. This has been attributed to an obligatory coupling of substrate transport to the ion gradient maintained across the cytoplasmic membrane (Raboy & Padan, 1978). Because a combination of valinomycin and nigericin was required to block the accumulation of α-methyl-D-glucoside, it is likely that the uptake of exogenous D-glucose and its analogues can be coupled to the chemical and to the electrical component of the proton electrochemical gradient across the cytoplasmic membrane (Raboy & Padan, 1978).

Accumulation of glucose analogue (Raboy & Padan, 1978), CO_2 production from glucose (Pelroy *et al.*, 1972), glycogen synthesis (Pelroy & Bassham, 1973a) and metabolite labelling by [$^{14}CO_2$] glucose (Pelroy & Bassham, 1973a) were all much lower under anaerobic conditions. In addition, the concentration of ATP within the cell declined when heterotrophic organisms were transferred from aerobic to anaerobic conditions in the dark in the presence of D-glucose (Pelroy & Bassham, 1973a). On the basis of these observations, fermentation and anaerobic respiration can be discounted as contributing to the generation of energy in the dark in heterotrophic cyanobacteria. Instead, aerobic respiration coupled to the catabolism of carbohydrate obtained from the medium provides energy for the growth of these organisms in the dark. The chemoheterotrophic energy metabolism of these organisms is therefore similar in qualitative terms to that in cyanobacteria catabolizing endogenous carbohydrate. It has already been mentioned that all cyanobacteria exhibit a low rate of endogenous respiration. With strains capable of chemoheterotrophic growth, organic growth substrates cause little more than a two- or three-fold increase in the rate of oxygen uptake (Lucas, 1974). In this respect, heterotrophic cyanobacteria differ from other heterotrophic microorganisms which commonly show a marked respiratory response to exogenous oxidizable substrates. Nevertheless, this process is able to support the growth of heterotrophic cyanobacteria in the dark although the rates are typically much lower than under photoautotrophic conditions (Table 3.5). An approximate estimate of the yield of ATP from respiration has been made from a comparison of glycogen production with either O_2 consumption or CO_2 production. For *Synechocystis* 6714, the estimated P/2e values were in the range 1.7–3.1 (Pelroy & Bassham, 1973a). Taking into account the likely errors in this calculation, it appears that the coupling efficiency of respiration in cyanobacteria is not significantly lower than in many heterotrophic microorganisms. Attempts have been made to estimate the magnitude of ATP turnover in cyanobacteria under different conditions. Pelroy, Kirk & Bassham (1976) have derived rates of ATP utilization from data on the synthesis of glycogen by cell suspensions incubated under photoautotrophic, photoheterotrophic and chemoheterotrophic conditions and found these to be approximately 1.7, 0.9 and 0.1 μmol ATP mg^{-1} dry wt h^{-1} respectively. Although these calculations did not take into account the energy needed to drive the transport of exogenous glucose into the cell under either of the heterotrophic conditions, it is apparent that energy consumption is markedly different under the three sets of conditions. These figures should be regarded as minimum values for ATP production because they are estimates of rates of ATP consumption. The only data available on rates of ATP synthesis are for *Anabaena* 7122 (Bottomley & Stewart, 1976a) which cannot be grown heterotrophically; in this organism ATP production in the light and in the dark has been estimated to be 3.8 and 0.3 μmol ATP mg^{-1} dry wt h^{-1}. It is likely that the capacity of a heterotrophic cyanobacterium to generate energy in the dark by the catabolism of an exogenous organic substrate is also relatively low. As yet there is no direct information about the spatial relationship between the respiratory and photosynthetic electron transport chains of cyanobacteria. The features of energy conservation reactions in the photosynthetic membranes of *Plectonema* 73110 have been taken to indicate that there are two distinct types of energy transducing membranes in this organism: the intracellular photosynthetic (thylakoid) membranes involved in the conservation of light energy and the peripheral cytoplasmic membrane which is presumably the site of the respiratory process (Padan & Schuldiner, 1978). However, there is indirect evidence that both processes are located in the thylakoids of *Anacystis nidulans* (Paschinger, 1977) and *Nostoc* strain MAC (Smith, 1981).

Under photoautotrophic conditions an appreciable amount of assimilated carbon is channelled into the synthesis of protein while in the dark, in the presence of glucose, polysaccharide is the predominant polymer formed (Pelroy & Bassham,1973b; Pelroy, Kirk & Bassham, 1976). An explanation for the marked change in the relative rates of synthesis of these cell constituents has been sought in changes in metabolite pool sizes and turnover in *Synechocystis* 6714. Although the rate of protein synthesis was greatly reduced in organic media in the dark, the usual metabolic routes were operating though at a much lower rate. A rapid flow of carbon from exogenous carbohydrate into monophosphates in the dark was apparently not balanced by the removal of metabolites at a comparable rate. This has been attributed to a partial block early in glucose metabolism possibly at the level of glucose-6-phosphate oxidation; this could, in turn, have the effect of greatly reducing growth rate without necessarily impeding the flow of carbon from an exogenous organic compound into polysaccharide (Pelroy, Kirk & Bassham, 1976). This partial block might be due to the increased pool of NADPH and to the higher NADPH/NADP ratio in *Synechocystis* 6714 in the dark on organic media (Pelroy, Kirk & Bassham, 1976). Such an effect could serve to balance the generation of NADPH by the oxidative pentose-phosphate cycle to the capacity of respiratory electron transport to reoxidize this cofactor.

The studies of Pelroy and his colleagues have been particularly influential in formulating our present view of the intermediary metabolism of heterotrophic cyanobacteria. It is, however, worth noting that photoautotrophically grown organisms were used in many incubations not only under photoautotrophic but also under photoheterotrophic and chemoheterotrophic conditions. As a result, some of the effects observed may have been exaggerated because the organisms were adjusting to new conditions of incubation. This can be the only interpretation of the marked discrepancy in the exponential rates of protein and polysaccharide synthesis in photoautotrophically grown *Synechocystis* 6714 incubated aerobically in the dark in the presence of glucose (Pelroy, Kirk & Bassham, 1976). Such a gross imbalance in the synthesis of cell constituents on organic media in the dark would be incompatible with the ability of *Synechocystis* 6714 for sustained growth under these conditions. Nevertheless, differences between organisms grown autotrophically and heterotrophically are likely to be qualitatively similar to those reported by Pelroy and his colleagues with organisms grown on organic compounds containing a large amount of polysaccharide relative to other cell constituents.

3.4.3 Significance of cyanobacterial heterotrophy under natural conditions

Because the heterotrophic modes of metabolism of cyanobacteria have been characterized under what are for the most part unnatural, laboratory conditions, their significance in the natural environment is uncertain. Are they irrelevant or are they further examples of 'physiological processes acquired in the course of evolution to cope with the viscissitudes of life outside the laboratory' (Tempest & Neijssel, 1976)? This problem is not unique to cyanobacteria. An appreciable number of eukaryotic, photosynthetic micro-organisms also grow on organic media. The importance of their heterotrophic metabolism under natural conditions has been in dispute for the past 15 years (e.g. Parker, 1961; Sloan & Strickland, 1966; Wright & Hobbie, 1965, 1966; Bennett & Hobbie, 1972; Lylis & Trainor, 1973; White, 1974a,b). Attempts have been made to resolve this issue by laboratory studies and field investigations. Although there is now a clearer understanding of the heterotrophic capabilities of these eukaryotic organisms and the complexity of their natural environment, the basic question of the significance of their heterotrophic metabolism under natural conditions is still unresolved.

Photoautotrophic growth is the predominant metabolic ability of cyanobacteria and under conditions that sustain a high rate of CO_2 fixation, the assimilation of exogenous organic compounds usually makes only a small contribution to the synthesis of cell material (Ingram *et al.*, 1973). In contrast, growth dependent on organic substrates occurs under conditions where CO_2 fixation is greatly reduced (low light intensities) or does not occur at all (complete darkness). Natural populations of cyanobacteria have been found in environments where there is little if any light. These include the sediments of reservoirs (Chernousova *et al.*, 1968) and lakes (Kappers, 1977) and the aphotic zone of the sea (Bernard, 1963). In addition, organisms within macrocolonies of the rivularian type and those harboured by macrophytes such as species of *Gunnera* are also exposed to similar conditions. In such situations the metabolism of cyanobacteria is likely to be heterotrophic and any growth would be dependent on exogenous compounds. An assessment of the significance of cyanobacterial growth on organic compounds in natural environments must take into account a variety of factors. These include the availability of the organic compounds that support the growth of cyanobacteria, the ability of organisms to assimilate these compounds at concentrations which are found in nature, and the interactions between cyanobacteria and other organisms that occupy the same environment.

D-Glucose which supports the heterotrophic growth of more cyanobacteria than any other organic compound is one of the most abundant, free carbohydrates in many environments. The concentration of glucose in the open sea is normally in the range 10^{-7}–10^{-8} M (Degens *et al.*, 1964; Vaccaro *et al.*, 1968). Appreciably higher concentrations are likely in other situations including polluted estuaries and coastal waters where there is sewage outfall, and rivers, lakes and lagoons receiving waste rich in organic material. Relatively high concentrations of organic compounds are also likely in a variety of microenvironments such as the layers of water in immediate contact with aquatic plants, organic-rich sediments and muds, and decaying plant and animal material. In these situations, the concentrations of glucose could be of the order of 5×10^{-6} M or higher (White, 1974a). Saunders (1972) has investigated the ability of a natural population of planktonic cyanobacteria to take up exogenous glucose at concentrations in the range 10^{-5}–10^{-8} M. Using an autoradiographic technique, he demonstrated substrate utilization by three *Oscillatoria* species and one strain of *Arthrospira* at a glucose concentration of 10^{-8} M. Much higher concentrations of substrate (10^{-5} M) were required before assimilation could be demonstrated with *Anabaena flos-aquae* and *Aphanizomenon flos-aquae*. It can be argued, however, that the amount of exogenous substrate taken up under these conditions may be very small and of little, if any, benefit to cyanobacteria. Extrapolating from the data available on the effect of substrate concentration on the uptake of carbohydrates by heterotrophic cyanobacteria (Beauclerk & Smith, 1978; Raboy & Padan, 1978), glucose uptake at 5×10^{-8} M would be of the order of 0.02 μg h^{-1} mg^{-1} dry wt which is insignificant. However, substrate uptake would be much greater and could be of physiological significance to heterotrophic cyanobacteria in those environments where appreciably higher concentrations of glucose are likely. Thus *Anabaena* 29413, isolated originally from an oxidation pond (Tischer, 1965) has been grown in the dark on D-fructose at a concentration of 5×10^{-6} M (Wolk & Shaffer, 1976). In addition, the heterotrophic growth of cyanobacteria may be enhanced by other organic compounds in the natural environment. These are likely to include amino acids which have been shown to stimulate the growth of *Nostoc* strain MAC (Hoare *et al.*, 1971) and of *Plectonema* 73110 (White & Shilo, 1975) on D-glucose. In the highly specialized environment of cyanobacteria in stable symbiotic association with another photosynthetic organism the cyanobacterium probably fixes very little CO_2 and is supplied with organic carbon sufficient for its needs by the other partner in the association (Silvester, 1976; Stewart, 1977). Such an arrangement would mean that the metabolism of the cyanobacterium is essentially heterotrophic.

Observations on field samples (Paerl, 1976) and difficulties encountered in the isolation of pure cultures of cyanobacteria indicate that they coexist in nature with other microorganisms (see Chapter 17). These include chemo-heterotrophic bacteria belonging to the genera *Pseudomonas* and *Caulobacter* as well as actinomycetes (Bershova *et al.*, 1968). The growth of these non-photosynthetic organisms will be wholly dependent on organic compounds which may include those capable of supporting the heterotrophic growth of cyanobacteria. In the 1960s it was presumed that oxygenic photosynthetic microorganisms were unable to compete effectively with chemoheterotrophic bacteria because they did not possess the active transport mechanisms essential for the utilization of organic compounds present in the environment at very low concentrations (Wright & Hobbie, 1965, 1966). This point of view has had to be revised with the discovery of active transport mechanisms, first in eukaryotic algae and more recently in cyanobacteria. The outcome of a competition between two or more microorganisms which possess mechanisms for the active transport of a particular organic substrate is difficult to predict. This can be attempted on the basis of a comparison of the properties of each of the transport mechanisms involved. The limited information available about heterotrophic cyanobacteria suggests that their uptake mechanisms are of lower affinity than those of eukaryotic photosynthetic microorganisms (Table 3.7). As little is known of the properties of the transport mechanisms of the chemoheterotrophic bacteria that occupy the same environments as heterotrophic cyanobacteria, it is necessary to extrapolate from data on organisms which may occupy quite different niches (Table 3.7). Here again the uptake mechanisms of cyanobacteria are of relatively low affinity. These comparisons are, however, made using data obtained from laboratory studies with organisms grown under near-optimum conditions which are likely to be far removed from those that exist in nature.

Besides competition for a common growth substrate, the interrelations between different groups of organisms in particular environments may also include effects which are either beneficial or detrimental to one or more of the organisms present. Thus compounds produced by one organism may prevent, promote, or in the extreme, be essential for the growth of others (Burkholder *et al.*, 1960; Parker, 1961; Vance, 1965). At the present time it is clear that the conditions under which cyanobacterial heterotrophy may be of importance are only likely to exist in specialized environments. It is through studies of these and of the organisms that inhabit them that the natural significance of the heterotrophic metabolism of cyanobacteria will be established.

3.4.4 Why certain cyanobacteria fail to grow on organic compounds

Cyanobacteria can be divided into at least two categories in terms of their response to exogenous organic compounds; some strains will only grow in the

Table 3.7. Affinity of microbial mechanisms for the transport of D-glucose

Organism	Affinity* (M)	Reference
(1) Cyanobacteria		
Anabaena 29413†	1.5×10^{-4}	J.E. More & A.J. Smith (unpubl.)
	4×10^{-5}	Haury & Spiller (1981)
Chlorogloeopsis 6912	1.2×10^{-3}	J.E. More & A.J. Smith (unpubl.)
Nostoc strain MAC	1.2×10^{-3}	J.E. More & A.J. Smith (unpubl.)
Plectonema 73110	2.9×10^{-4}	Raboy & Padan (1978)
Synechocystis 6714	4.3×10^{-4}	Beauclerk & Smith (1978)
(2) Eukaryotic algae and diatoms		
Bracteococcus minor	10^{-5}	Sheath & Hellebust (1974)
Chlamydomonas sp.	2.8×10^{-5}	Bennett & Hobbie (1972)
Chlorella vulgaris	10^{-5}	Komor & Tanner (1971)
Cocconesis sp.	6×10^{-4}	Bunt (1969)
Coscinodiscus sp.	2×10^{-4}	White (1974b)
Cyclotella cryptica	5.8×10^{-5}	Hellebust (1971)
Navicula muralis	1.5×10^{-5}	Jolley *et al.* (1976)
(3) Bacteria		
Achromobacter aquamarinus	4×10^{-8}	Vaccaro & Jannasch (1966)
Flavobacterium sp.‡	4×10^{-4}	Jolley *et al.* (1976)
Pseudomonas aeruginosa	8×10^{-6}	Midgely & Dawes (1973)
Pseudomonas putida	6×10^{-3}	Vicente *et al.* (1975)
Pseudomonas 486	1.3×10^{-5}	Akagi & Taga (1980)
Pseudomonas RP 303	3.2×10^{-6}	Akagi & Taga (1980)
Isolate 13‡	6×10^{-7}	Jolley (1976)
(4) Fungi		
Achlya sp.	4×10^{-5}	Goh & LéJohn (1978)
Aspergillus nidulans	4×10^{-5}	Brown & Romano (1969)
Neurospora crassa	4×10^{-5}	Schneider & Wiley (1971)
Rhodotorula gracilis	5.5×10^{-4}	Kotyk & Höfer (1965)

* Affinity is expressed as the substrate concentration giving half the maximum rate of uptake.
† The figure for *Anabaena* 29413 refers to the uptake of D-fructose.
‡ These bacteria were isolated from the same environment as *Navicula muralis*.

light with CO_2 as carbon source while others will also grow on organic media. It has been suggested that these should be designated 'specialist' and 'versatile' strains respectively in common with analogous categories of chemolithotrophs and methylotrophs (see Smith & Hoare, 1977). Over the past 20 years several theories have been proposed to account for the failure of some cyanobacteria to grow photoheterotrophically or chemoheterotrophically; only the more recent of these will be discussed.

While the specialist condition can no longer be attributed to generalized impermeability to organic compounds, it may be a consequence of functional impermeability to the limited range of organic compounds theoretically capable of providing the cell with sufficient carbon and energy for growth (Pelroy *et al.*, 1972; Raboy *et al.*, 1976; Beauclerk & Smith, 1978). Specialist and versatile strains possess the same complement of enzymes for the metabolism of glucose but they exhibit marked quantitative differences in their ability to convert this substrate to CO_2 (Pelroy *et al.*, 1972). The relatively high rate observed with *Synechocystis* 6714 has been attributed to the presence of a mechanism for the transport of exogenous D-glucose into the cell. Subsequent work has established that several versatile strains capable of growth on this substrate possess mechanisms for the active transport of D-glucose and certain of its analogues into the cell (Beauclerk & Smith, 1978; Raboy & Padan, 1978). A comparison of specialist and versatile strains has shown that the former take up D-glucose at very low rates and that substrate uptake is not affected by any D-glucose analogue including those that block substrate uptake by versatile cyanobacteria (Table 3.8). Although relatively few strains were examined in this survey, the specialist character has been attributed to the failure of potential growth substrates to enter the cell at a rate sufficient to support significant growth. The very low rate of substrate assimilation by these organisms is therefore a result of the intrinsic permeability of the membrane which will be very low for uncharged hydrophilic compounds. Specialist cyanobacteria are therefore naturally occurring counterparts of the cryptic mutants of chemoheterotrophic bacteria

Table 3.8. Influence of D-glucose analogues on the uptake of this carbohydrate by cyanobacteria

	Rate of D-glucose uptake* in the presence of			
Organism	2-deoxy D-glucose	6-deoxy D-glucose	α-methyl D-glucoside	3-O-methyl D-glucose
(1) Versatile strains				
Nostoc strain MAC	93	13	89	27
Synechocystis 6714	76	10	91	16
(2) Specialist strains				
Anabaena 7118	97	87	92	114
Synechocystis 6308	87	88	89	81
Synechocystis 7008	105	90	73	95

* % of control.
Data taken from Beauclerk & Smith (1978) together with unpublished results (J.E. More & A.J. Smith).

which cannot metabolize specific exogenous organic substrates although they possess all of the requisite enzymes within the cell (see Cohen & Monod, 1957). While this accounts for the failure of specialist cyanobacteria to grow on organic compounds in the light and in the dark, it does not explain why some strains that grow photoheterotrophically do not also grow under chemoheterotrophic conditions (Rippka, 1972).

In a review Doolittle (1979) expressed the view that impermeability to potential growth substrates is unlikely to be the only barrier to heterotrophic growth. He has proposed that the metabolic priority of cyanobacteria in the dark is to restrict carbohydrate metabolism in order to ensure the survival of the cell throughout the period of darkness. From this it follows that the cell must be able to distinguish substrate molecules on the basis of their origin if they are to metabolize exogenous glucose at a higher rate than endogenous reserves. Because he considers this unlikely it necessarily follows that versatile cyanobacteria will grow in the dark on organic media at rates much lower than under photoautotrophic conditions. While it is abundantly clear that versatile cyanobacteria grow heterotrophically at relatively low rates, they must be able to distinguish between conditions where an exogenous growth substrate is available and those in which they are wholly dependent on their endogenous reserves. This is probably achieved as it must be in chemoheterotrophic organisms by regulatory mechanisms which respond to exogenous substrate availability. A very limited respiratory capacity could account for the inability of some cyanobacteria to use organic compounds as a source of energy although they can exploit them as a source of carbon during photoheterotrophic growth.

Metabolite pool sizes and the rates of carbon flow through these pools will not be identical in an organism metabolizing endogenous reserves and in one growing under photoautotrophic conditions. Furthermore, it is reasonable to suppose that certain of these differences may have a marked influence on various cell processes such as the synthesis of specific macromolecules. Caution should therefore be exercised in attempting to explain the failure of specialist organisms to grow on organic compounds in terms of effects that are in reality a direct consequence of a metabolism based on endogenous reserves.

3·5 Concluding remarks

While considerable progress has been made during the past decade in the investigation of the intermediary metabolism of cyanobacteria there are many outstanding problems. Among these are considerable gaps in our knowledge of the mechanisms regulating the intermediary metabolism of cyanobacteria in the light and the dark. Likely control points have been proposed in a number of cases but the signalling molecules have yet to be identified with certainty. Little is known about the relationship between photosynthetic and respiratory electron transport in cyanobacteria. Are the two processes quite separate as appears to be the case in *Plectonema* 73110 (Padan & Schuldiner, 1978) or are certain of the carriers shared (Paschinger, 1977; Lockau, 1981) as in photosynthetic bacteria (Jones, 1977a)? Techniques like that described by Murata *et al.* (1981) for the separation of thylakoid membranes and cell envelopes of *Anacystis nidulans* by gradient centrifugation will be of particular value in such studies. In order to interpret the natural distributions of cyanobacteria we need to know more about their metabolism under natural conditions. Of particular interest are the mechanisms that enable an organism such as *Oscillatoria aghardii* to adjust polymer synthesis to a particular diurnal light–dark regime (van Liere, Mur, Gibson & Herdman, 1979). Renewed interest is being shown in the structures resembling cyanobacteria that inhabit various types of eukaryotic cells. It is generally accepted that these cyanelles provide the host cell with a supply of organic carbon produced by the photosynthetic activity of the organelle. What are the benefits of this close association for the cyanelle and how dependent is it on the host cell? Finally the physiology and biochemistry of the varied assortment of organisms classified as apochlorotic counterparts of cyanobacteria have yet to be explored. To what extent does their heterotrophic metabolism resemble that of versatile cyanobacteria? These are likely to be among the problems attracting the attention of physiologists and biochemists in the 1980s.

Acknowledgments

I wish to express sincere thanks to my colleagues Dr L.J. Rogers and Dr G.N. Hutber for their comments on the manuscript during its preparation, to Mr A.K. Jones of the Department of Botany and Microbiology at Aberystwyth for his advice on the section concerned with the significance of cyanobacterial heterotrophy under natural conditions, and to Mrs C. Smith for skilled secretarial services.

CHAPTER 4
NITROGEN FIXATION

H. BOTHE

4.1 Nitrogen-fixing cyanobacteria

The ability to reduce the dinitrogen molecule to ammonia is restricted to prokaryotes where it is relatively scarce but spread widely through the systematic groups. Among these organisms, a prominent role is occupied by the cyanobacteria. They fix N_2 both in the free-living state and in symbiosis with a wide range of partners. Free-living N_2-fixing organisms are often said to contribute rather little to the nitrogen input in the global nitrogen cycle (see Bothe *et al.*, 1982). An exception to this rule are the cyanobacteria which populate many different environments from the Arctic to the Antarctic and might significantly improve the nitrogen balance *in situ* (see Mague, 1977 and Chapter 19). Their success in occupying such a wide habitat range is due to their resistance to extremes of temperature and desiccation and to their modest nutrient requirements; most nitrogen-fixing species thrive photoautotrophically in the absence of combined nitrogen. As regards the symbiotic forms, the contributions of the liverworts and particularly of the lichens to the global nitrogen cycle are virtually unknown but probably of major importance, whereas those of the symbioses with higher plants (*Gunnera*, cycads) are more or less negligible (Stewart *et al.*, 1980). The importance of the water-fern *Azolla* containing the cyanobacterium *Anabaena azollae* has been recognized by farmers of the southeast Asian countries for centuries. *Azolla* is commonly used as a green manure to improve the nitrogen balance in rice fields.

Cyanobacterial N_2-fixation is carried out by unicellular organisms, filamentous, heterocystous forms and by filamentous strains lacking heterocysts (see Stewart, 1973a, 1977; Stanier & Cohen-Bazire, 1977; Haselkorn, 1978; Gallon, 1980; Bothe *et al.*, 1980). The best-known examples are the heterocystous forms (*Anabaena, Nostoc*); they usually differentiate vegetative cells to heterocysts only when grown in the absence of combined nitrogen (Chapter 14). This observation led Fogg (1949) to suggest that heterocysts are the site of nitrogenase, the enzyme complex catalysing the conversion of dinitrogen to ammonia. Much experimental evidence has been accumulated in the past few years that nitrogenase is located in the heterocysts under aerobic growth conditions (see Chapter 14). Fleming & Haselkorn (1973) analysed ^{35}S-labelled proteins from heterocysts and vegetative cells of *Nostoc muscorum* by electrophoresis on polyacrylamide gels and found four unique bands present only in heterocysts. Three of these proteins corresponded to nitrogenase components from *Azotobacter* and *Clostridium*. A rather similar approach using ^{55}Fe and ^{99}Mo labelling and a more intact heterocyst preparation has shown that nitrogenase is located in heterocysts (Peterson & Wolk 1978a). In recent years substantial progress, particularly by Wolk's group, has been made in preparing heterocysts retaining most of the nitrogenase activity. The latest heterocyst preparation from this laboratory (Peterson & Wolk, 1978b) accounted for an average of 60% of the nitrogenase activity of intact filaments. Using the same strain of *Anabaena variabilis*, G. Eisbrenner in our laboratory (unpubl.) obtained heterocyst preparations with similar high activities. In contrast, activity levels of heterocyst preparations of *A. cylindrica* or *Nostoc muscorum* usually did not exceed 10–20% of that of whole filaments. However, virtually no nitrogenase was detected in broken vegetative cells of these organisms. Without exception, all heterocystous cyanobacteria were found to fix N_2 under aerobic conditions.

Several bacteria grow aerobically when supplied with combined nitrogen; after exhaustion of such nitrogen source, they are able to synthesize nitrogenase only under strict oxygen exclusion (e.g. *Klebsiella pneumoniae*, *Bacillus polymyxa*) or under reduced O_2-tensions (microaerobic conditions, e.g. *Mycobacterium flavum* = *Xanthobacter flavus*, *Azospirillum brasilense* and *A. lipoferum*). Similar observations were also made for cyanobacteria, first for *Plectonema boryanum* (Stewart & Lex, 1970). Cultures of this organism synthesized nitrogenase when periodically sparged with an argon–CO_2 mixture to remove air and the O_2 that had been evolved photosynthetically. By introducing more rigorous methods to obtain strict anaerobic conditions, particularly by blocking the photosynthetic O_2-evolution by DCMU, Rippka & Stanier (1978) demonstrated that more than 50% of the filamentous, non-heterocystous forms have the capability to synthesize nitrogenase. A detailed account of the properties of such strains was recently made available by Rippka *et al.* (1979). In all cases, nitrogenase was rapidly and irreversibly inactivated when exposed to air.

Some lines of experimental evidence indicate that the vegetative cells of heterocystous forms express nitrogenase activity under anaerobic conditions. Smith & Evans (1971) observed a drastic increase of nitrogenase levels in extracts of *Anabaena cylindrica* which had been incubated for 4 days under a mixture of 95% argon–5% CO_2. A mutant of *A. variabilis* which lost the capability to form heterocysts was shown to fix N_2 under stringent anaerobic conditions (Rippka & Stanier, 1978). In conclusion, for the heterocystous cyanobacteria, it appears to be the almost generally accepted view that the heterocysts are the sole site of nitrogenase under aerobic growth conditions and that the more rigorously O_2 is excluded from the cultures the more the enzyme is active also in vegetative cells.

N_2-fixation by unicellular cyanobacteria is exceptional. The first report came from Wyatt & Silvey (1969) for *Gloeocapsa*. This observation was confirmed using an axenic culture by Rippka *et al.* (1971) who designated the organism as *Gloeothece* and later added four further axenic strains to the list (Rippka *et al.*, 1979). These authors also confirmed that *Aphanothece* is another example of a unicellular, N_2-fixing cyanobacterium (Singh, 1973). Remarkably, both *Gloeothece* and *Aphanothece* fix N_2 under aerobic conditions similar to the heterocystous forms. *Microcoleus chthonoplastes* is probably an example of a non-heterocystous, filamentous cyanobacterium fixing N_2 in air (Pearson *et al.*, 1979) which, however, has to be verified when the culture is contaminant-free. Examples of unicellular cyanobacteria performing N_2 fixation under anaerobiosis may include three strains of *Synechococcus* (Rippka *et al.*, 1979) which lack detailed characterization at present.

Organisms containing cyanobacterial-like inclusions in lieu of chloroplasts (the cyanelles) deserve special attention, since these are possibly 'missing links' in the evolution of chloroplasts from bacteria. All such organisms (*Cyanophora paradoxa*, *Glaucocystis*, *Glaucosphaera*; Kies, 1980), failed to fix nitrogen under all conditions assayed. A remarkable exception is *Rhopalodia gibba*, a presumptive diatom, which contains both a chloroplast and cyanelle-like inclusions and which was observed to grow in the absence of combined nitrogen (Drum & Pankratz, 1965). A reisolate of this organism was, indeed, found to fix nitrogen (measured by C_2H_2 reduction) in a light-dependent reaction and under microaerobic conditions (Floener & Bothe, 1980). If obtained in pure culture, this alga may hopefully turn out as a model organism for studying the interactions of unicellular, N_2-fixing and photosynthetic organisms embedded in an eukaryotic host. Another unusual, relatively unexplored association exists between the diatom *Hemiaulus membranaceus* and the heterocystous cyanobacterium *Richelia intracellularis* (Kimor *et al.*, 1978, see Chapter 19).

4.2 The nitrogenase complex

Nitrogenase is a complex enzyme with remarkably similar properties in the more than twenty organisms in which it has been investigated. It consists of two iron–sulphur proteins, which, individually, have no detectable enzymic activity. Together they catalyse the reduction of a variety of substrates (N_2, N_3^-, N_2O, HCN, CH_3NC, CH_3CN, C_2H_2, H_3O^+ and, as recently discovered, cyclopropene; see Bothe *et al.*, 1982). The larger protein of nitrogenase has a molecular mass of 200 000–270 000 daltons, contains 18–36 Fe, probably the same amount of acid-labile sulphur, two (or possibly one) Mo atoms and consists of four subunits of two different types ($\alpha_2\beta_2$ structure). This Mo–Fe component binds N_2 as well as the other substrates and catalyses their reduction when supplied with reducing equivalents from the smaller nitrogenase protein. The latter has a molecular mass of ~60 000 daltons, contains a single [4 Fe–4 S^{2-}] cluster and can be separated into two subunits which are always identical, with the exception of those from *Rhodospirillum rubrum* and *Azospirillum.* The Fe protein is that part of the enzyme that accepts reducing equivalents. The transfer of electrons from the Fe protein to the Mo–Fe protein is coupled to ATP hydrolysis. Since the reduction of N_2 to two molecules of ammonia requires the expenditure of 12–15 molecules of ATP *in vitro*, nitrogenase must possess further ATP-hydrolysing sites. ATP is possibly involved in the binding and/or reduction of substrates at the Mo–Fe protein. In this component, the substrates are likely bound to a special subunit (the Mo cofactor) whose existence was recently demonstrated both by genetic and biochemical approaches (see Yates, 1980). The Mo cofactor consists of one molecule of molybdenum, a still unresolved amount of iron and acid-labile sulphur and possibly a protein part, and it may be a subunit common to all Mo proteins. Both nitrogenase components are oxygen sensitive, the Fe protein extremely so. Detailed accounts of the biochemistry of nitrogenase are given by Yates (1980) and by Mortenson & Thorneley (1979).

There is little that is unusual about nitrogenase from cyanobacteria as compared to that from other bacteria. Early investigations demonstrated, in crude cyanobacterial extracts, reductions of ^{15}N (Schneider *et al.*, 1960), of C_2H_2 (Smith & Evans, 1970; Haystead *et al.*, 1970; Bothe, 1970) and of H_3O^+ and CN^- (Haystead *et al.*, 1970) which were dependent upon the supply of reductant ($Na_2S_2O_4$) and ATP. The optimum $Na_2S_2O_4$ concentration was 5–10 times lower than that routinely used in experiments with bacterial nitrogenases (Smith & Evans, 1970; see also Haystead *et al.*, 1970; Hallenbeck *et al.*, 1979). Owing to both the high levels of proteins associated with photosynthesis (specifically phycocyanin) and the occurrence of nitrogenase only in heterocysts which average approximately 5% of the cells of a filament, specific activities of nitrogenase from *Anabaena* or *Nostoc* were rather low in crude extracts compared with those from non-photosynthetic bacteria. Moreover, since nitrogenase turned out to be cold-labile and extremely sensitive to O_2 (Haystead *et al.*, 1970; Bothe, 1970), cyanobacteria have not been suitable organisms for purifying nitrogenase to homogeneity. Recently, two groups independenly of each other (Tsai & Mortenson, 1978; Hallenbeck, *et al.*, 1979), reported the purification of nitrogenase from *Anabaena cylindrica* 629, and the Mo–Fe component was claimed to be homogeneous (Hallenbeck *et al.*, 1979). With such preparations, C_2H_2 reduction was approximately 400 nmol min^{-1} mg $protein^{-1}$ which has to be compared with the 1000–2500 nmol for pure preparations from *Klebsiella*, *Azotobacter* or *Clostridium* (see Yates, 1980 and Chapter 14). Assays on the cross-reaction of the nitrogenase components from cyanobacteria with those from other organisms gave remarkably similar results (Smith, Telfer & Evans, 1971; Tsai & Mortenson, 1978; Nagatani & Haselkorn, 1978; Hallenbeck *et al.*, 1979). The Mo–Fe protein from *Anabaena cylindrica and Plectonema boryanum* was found to cross-react rather effectively with the Fe protein from *Clostridium pasteurianum*, *Azotobacter vinelandii* and *Chlorobium* (measured in *Chloropseudomonas ethylica*). In contrast, the Fe protein from *Anabaena cylindrica* and *Plectonema boryanum* complexed effectively only with the homologous protein, but hardly at all with the Mo–Fe counterpart from *C. pasteurianum*, *A. vinelandii*, *R. rubrum* and *Chlorobium* in *'Chloropseudomonas ethylica'*. Apparently, the Mo–Fe protein has evolved with little specifity, at least with less than the component from *C. pasteurianum* (Tsai & Mortenson, 1978). The Mo cofactor occurs in *Plectonema boryanum*, since acidified extracts from nitrogen-fixing cells of this organism were shown to restore nitrogenase activity to a mutant of *Azotobacter* which lacks it (Nagatani & Haselkorn, 1978). For cyanobacteria, there is no evidence for adenine, ribose and phosphate firmly bound to the Fe protein and nor for an unusual activating factor which was described in *Rhodospirillum rubrum* and *Azospirillum brasilense* by Ludden *et al.* (1978).

4.3 Requirement for reductant and energy

In cyanobacteria it has been suggested that the reducing equivalents for nitrogen fixation may be generated via both photosystems (PS I, PS II) from water, via PS I alone or via dark reactions; ATP may be supplied by cyclic and noncyclic photophosphorylation, oxidative or substrate level phosphorylation. However, considerable experimental evidence has been accumulated to rule out most of these possibilities.

N_2 fixation is light-stimulated in all cyanobacteria investigated. Various authors noted that DCMU affected photosynthetic CO_2 fixation but not

C_2H_2 reduction in the heterocystous *Nostoc* and *Anabaena* (Cox, 1966; Bothe & Loos, 1972). The action spectrum of C_2H_2 reduction was similar to that of PS I (Fay, 1970; Thomas, 1970b). C_2H_2 reduction of *Anabaena cylindrica* proceeded unimpaired in far-red light and in the presence of DCMU where PS II activity was totally blocked (Bothe & Loos, 1972). The acetylene reduction reaction did not show an Emerson-enhancement effect, contrary to photosynthetic CO_2 fixation (Lyne & Stewart, 1973a). All these observations indicated that the electrons for N_2 fixation came from a pool of reductants composed of fixed carbon compounds (Lex & Stewart, 1973). The duration of N_2 fixation depends on the size of this reductant pool which has to be replenished by photosynthetic electron transport. Thus the dependence of nitrogen fixation on photosynthetic water photolysis is indirect. Such a statement is substantiated by the observation that C_2H_2 reduction is severely affected by DCMU in carbon-limited cultures of *Anabaena* (Benemann & Weare, 1974a; Fay, 1976).

These experimental findings are to be expected if heterocysts are the site of nitrogenase under aerobic growth conditions. Heterocysts lack the photosynthetic water-splitting reaction and ribulose-1,5-bisphosphate carboxylase (Tel-Or & Stewart, 1976); consequently they are dependent on a supply of carbohydrates from vegetative cells. The light stimulation of C_2H_2 reduction by intact filaments and by isolated heterocysts is explained by the dependence on ATP which is supplied by cyclic photophosphorylation. In addition, light could also be necessary to transfer the electrons to nitrogenase in a PS I dependent reaction. Alternatively, the electrons could be generated in the dark. These latter two possibilities are difficult to rule out, since an inhibitor specific to PS I is not available. In the non-heterocystous *Plectonema boryanum*, N_2-fixation activity and photosynthetic O_2 evolution were temporarily separated from each other (Weare & Benemann, 1974). Evidence was recently reported (Gallon, 1980) that the generation of reducing equivalents in the unicellular *Gloeocapsa* depends on the operation of both photosystems. Using *Gloeocapsa* grown under a light–dark (16:8 h) regime, Millineaux *et al.* (1981) concluded that acetylene reduction occurred only in the dark phase; in *Anabaena cyclindrica* the major part of the acetylene reduction occurred in the light. The authors suggest that temporal separation of photosynthesis and nitrogen fixation is the way in which a unicellular, non-heterocystous cyanobacterium manages to carry out the two processes. In the dark, oxidative phosphorylation may also provide ATP for nitrogenase operation. N_2-fixation activities are in all cases lower, reaching maximally 70% of the light rate in the heterotrophic *Anabaenopsis circularis* (Bottomley & Stewart, 1976a; Fay, 1976) or 50% in *Anabaena cylindrica* (Weare & Benemann, 1973a,b; Bottomley & Stewart, 1977). Such decreased activity proceeds only for as long as the carbohydrate reserves are not exhausted. N_2-fixation under natural conditions exhibits diurnal variations (see Fay, 1976). Substrate level phosphorylation was demonstrated for *A. cylindrica* but is believed to be of rather limited importance in providing ATP to nitrogenase (Bottomley & Stewart, 1976a). The commonly used cyanobacteria do not ferment organic material to any significant extent.

The immediate electron carrier to nitrogenase is ferredoxin in all organisms with the possible exception of *Azotobacter* (see Bothe *et al.*, 1982). Cyanobacteria contain a soluble, typical plant-type ferredoxin with two Fe and two acid-labile sulphur atoms in the prosthetic group. This was shown to mediate C_2H_2 reduction by *Anabaena* nitrogenase when reduced by illuminated spinach chloroplasts (Bothe, 1970) or by particles from *Anabaena* (Smith, Noy & Evans, 1971). The occurrence of ferredoxin in heterocysts was demonstrated by Tel-Or & Stewart (1976). *Nostoc* strain MAC (Hutson & Rogers, 1975) and *Nostoc verrucosum* (Shin *et al.*, 1977) contained two soluble, separable ferredoxins which have almost identical biological and chemical properties but which differ in their amino-acid composition. This observation has recently be extended to a variety of other species (Ho *et al.*, 1979) indicating that the occurrence of two soluble ferredoxins may be a general phenomenon for cyanobacteria. The physiological role of these proteins and their distribution among heterocysts and vegetative cells are uncertain at present.

When Fe deficiency limits the biosynthesis of ferredoxin, several microorganisms form flavodoxin instead of the iron–sulphur protein (see Bothe, 1977). Flavodoxin contains one molecule of FMN in the prosthetic group per protein molecule and is considered to be the simplest flavoprotein. It was described in *Anacystis nidulans* (Smillie, 1965), *Synechococcus lividus* (Norris *et al.*, 1972) and *Nostoc* MAC (Hutson & Rogers, 1975). It substituted for ferredoxin in all ferredoxin-dependent reactions with similar specific activities. This was true also for C_2H_2 reduction catalysed by *Anabaena* nitrogenase (Bothe, 1969). In catalysis, flavodoxin acts as a one-electron carrier shuttling between the fully reduced and semiquinone form.

Since ferredoxins or flavodoxins are continuously transferring electrons to the nitrogenase complex, they must be reduced by electron donors supplied from cell metabolism. The reaction between donors and carriers is enzyme-mediated. This subject has been thoroughly investigated in *A. cylindrica*, *Nostoc muscorum* (see Bothe *et al.*, 1980; Stewart, 1977) and *Gloeocapsa* (see Gallon, 1980). The most important electron donor is probably NADPH. Using crude extracts from *Anabaena*, Bothe (1970) observed significant C_2H_2 reductions by an electron-donor comprising glucose-6-phosphate, glucose-6-phosphate dehydrogenase, $NADP^+$, NADPH: ferredoxin oxidoreductase, ferredoxin and nitrogenase from *A. cylindrica*. Such a reaction also proceeds in isolated heterocysts (Apte *et al.*, 1978; Lockau *et al.*, 1978). Several authors (see Chapter 14) noted that the activity levels of the key enzymes of the hexose monophosphate shunt, the NADP$^+$-dependent glucose-6-phosphate and 6-phosphogluconate dehydrogenases, are considerably enhanced in heterocysts as compared to vegetative cells (Winkenbach & Wolk, 1973; Lex & Carr, 1974; Apte *et al.*, 1978; Bothe *et al.*, 1980). Heterocysts do not fix CO_2 and are supplied with fixed carbon compounds from vegetative cells (see Chapter 14). The carbon compounds which are transported to the heterocysts have not yet unambiguously been identified (see Chapter 14); they are, however, most likely converted to glucose-6-phosphate which is mainly degraded via the hexosemonophosphate shunt to provide NADPH. The reduction of ferredoxin by NADPH, though unfavourable thermodynamically, should proceed in heterocysts, since otherwise NADPH:ferredoxin oxidoreductase would be redundant in these cells. Both enzymes involved in the glucose-6-phosphate-dependent reduction of ferredoxin, glucose-6-phosphate dehydrogenase (Pelroy & Bassham, 1972; Grossman & McGowan, 1975; Apte *et al.*, 1978; Schaeffer & Stanier, 1978) and NADPH:ferredoxin oxidoreductase (Apte *et al.*, 1978) are subject to complex regulations. Glucose-6-phosphate dehydrogenase in heterocysts is a hysteretic enzyme (Schaeffer & Stanier, 1978) whose activity is negatively affected by NADPH, ATP and dithiothreitol. In addition, an inhibition by light and ribulose-1,5-bisphosphate was demonstrated for crude preparations of the enzyme from vegetative cells but not from heterocysts indicating that the hexosemonophosphate shunt can operate in the dark as well as in the light in the latter cells. NADPH:ferredoxin oxidoreductase is negatively affected by NADP$^+$ (Apte *et al.*, 1978). Thus the NADPH/NADP$^+$ ratio (reduction charge) is probably of major importance in regulating the flow of electrons to nitrogenase. Surprisingly, glucose-6-phosphate dehydrogenase shows greatest activity at reduction charges <0.3, whereas ratios >0.3 are optimal for NADPH:ferredoxin oxidoreductase (Apte *et al.*, 1978). This observation possibly indicates that sources other than glucose-6-phosphate provide the high levels of NADPH required for the transfer of electrons 'uphill' to ferredoxin. Such a source could be isocitrate. Recently, the occurrence of isocitrate dehydrogenase and of several other enzymes of the glycolytic pathway and of the tricarboxylic acid cycle have been demonstrated in heterocyst preparations of *Anabaena* (Bothe *et al.*, 1980), indicating that glycolysis and parts of the tricarboxylic acid cycle may not be as inactive in heterocysts as previously assumed (Stanier & Cohen-Bazire, 1977). C_2H_2 reductions supported by the system isocitrate, isocitrate dehydrogenase, NADP$^+$, NADPH:ferredoxin oxidoreductase, and ferredoxin were reported in crude preparations from *Anabaena* (Smith *et al.*, 1971) and *Gloeocapsa* (Gallon *et al.*, 1972).

An alternative electron donor may be pyruvate. Pyruvate:ferredoxin oxidoreductase in cyanobacteria was first demonstrated for *Anabaena variabilis* (Leach & Carr, 1971) and was characterized in more detail for *A. cylindrica* (Bothe *et al.*, 1974). This iron-sulphur protein catalyses the cleavage of pyruvate and coenzyme A to acetylcoenzyme A and CO_2, and the remaining two electrons reduce ferredoxin. Pyruvate-supported C_2H_2 reduction was demonstrated for intact *A. cylindrica* cells (however, only in the dark under low Po_2; Bennett *et al.*, 1975) and for extracts from the same organism (Codd *et al.*, 1974). However, pyruvate:ferredoxin oxidoreductase still remains to be demonstrated in heterocysts. Several authors failed to observe pyruvate-supported C_2H_2 reductions (Lockau *et al.*, 1978; Eisbrenner & Bothe, 1979) possibly because the pyruvate did not penetrate into the heterocysts. Phosphoenolpyruvate was reported to stimulate C_2H_2 reduction by isolated heterocysts (Bradley & Carr, 1976). Since pyruvate kinase is present in heterocysts (Bothe *et al.*, 1980), phosphoenolpyruvate is likely converted to pyruvate which must be degraded by the phosphoroclastic reaction. Pyruvate:ferredoxin oxidoreductase is the only enzyme catalysing a pyruvate cleavage in *A. cylindrica*. This statement is based on the findings that neither the pyruvate oxidase nor the pyruvate dehydrogenase complex was demonstrable and that the levels of lipoic acid are too low to account for the occurrence of the dehydrogenase complex in *Anabaena* (Bothe & Nolteernsting, 1975). Pyruvate:ferredoxin oxidoreductase has recently also been detected in *Gloeocapsa* (Gallon, 1980).

In addition to glucose-6-phosphate, isocitrate and pyruvate, glycollate, malate and succinate must also be considered since they stimulated light-dependent methyl viologen reductions by lamellar fragments from *Anabaena variabilis* (Murai & Katoh, 1975). As stated above, the transfer of electrons from all these donors to nitrogenase proceeds either in the dark or in a PS I dependent reaction. PS I is definitely involved when H_2 is the electron donor. Detailed investigations showed that H_2 and hydrogenase can only feed in electrons at the donor side of PS I (see 4.6). H_2 gave the highest activity among all electron donors in C_2H_2 reduction experiments with isolated heterocysts (Eisbrenner *et al.*, 1978; Peterson & Burris, 1978).

It has been suggested that the transfer of reducing equivalents to nitrogenase is dependent on the energized state of the membrane (membrane potential) in *Azotobacter* and *Rhizobium* (Laane *et al.*, 1979). Haaker *et al.* (1980) recently speculated that this is also the case in cyanobacteria. This hypothesis, though admittedly attractive, stands and falls on the interpretation of studies with inhibitors, among which the membrane de-energizing agent tetrachloro-2-trifluoromethyl-benzimidazole is particularly important. The crucial experiment, the reduction of the electron carrier (flavodoxin to the fully reduced state or ferredoxin) dependent on the membrane potential, has not been achieved either for *Azotobacter* or *Rhizobium*. Moreover, it should be emphasized that compared with *Azotobacter* or *Rhizobium*, the generation of reductant is different in cyanobacteria. In *Azotobacter* the electron carrier to nitrogenase is possibly a membrane-bound flavodoxin which is believed to be reduced by NADH and a membrane-bound NADH:flavodoxin oxidoreductase. In cyanobacteria, NADH is completely inactive, flavodoxin is only formed under Fe deficiency and the enzymes NADPH:ferredoxin oxidoreductase and pyruvate:ferredoxin oxidoreductase are soluble or loosely attached to membranes. It may be concluded that the modes of electron transfer to the nitrogenase complex appear to differ from organism to organism, and one organism does not necessarily restrict itself to one electron donor.

4.4 Mechanisms to protect nitrogenase from damage by oxygen

All nitrogenases, when isolated, are sensitive to O_2. Aerobic organisms must, therefore, have developed mechanisms to protect the enzyme. In the simplest case the O_2-damaged protein is replaced by synthesis of new nitrogenase, as suggested by Bone (1971b) from experiments with *Anabaena flos-aquae*. Such a mechanism may operate in all nitrogen-fixing cyanobacteria but it is hardly the only reason for the differences which are observed in the responses of the N_2-fixation process towards O_2.

A particular role in protecting nitrogenase from O_2 is often attributed to the heterocysts (see Chapter 14). Indeed, since heterocysts are unable to evolve O_2 photosynthetically, nitrogenase inside these cells must not be protected against this extra gas formation. With regard to the O_2 of air, a special function in binding O_2 was suggested for four unique glycolipids detected in the laminated layer of the heterocyst envelope (Lambein & Wolk, 1973). N_2 fixation has recently been found sensitive to O_2 in mutants lacking these glycolipids (Haury & Wolk, 1978). However, heterocysts do respire and, as mentioned earlier, oxidative phosphorylation can support nitrogen fixation in dark experiments. Therefore O_2 can reach the inside of heterocysts and nitrogenase must be protected therein as in other aerobes. The gas need not necessarily penetrate the heterocyst envelope but may come from the plasma bridges (microplasmodesmata) connecting heterocysts and vegetative cells.

Cyanobacterial nitrogenase has not developed the ability to protect itself by sudden conformational changes when O_2 concentrations inside the cells reach critical levels, as does the enzyme from *Azotobacter* (see Postgate, 1978). The peculiar 2 Fe:2 S^{2-} protein which protects nitrogenase in *Azotobacter* (Scherings *et al.*, 1977) also has not been detected in cyanobacteria. As in *Azotobacter*, respiratory protection might play an important role in heterocysts and presumably in *Gloeocapsa*. Heterocysts have a high capability to reduce tetrazolium-chloride (Drawert & Tischer, 1956; Fay & Kulasooriya, 1972) indicating that they provide a reducing environment. They were reported to have higher respiratory activities than vegetative cells (Fay & Walsby, 1966; Peterson & Burris, 1976a, but see Bradley & Carr, 1976). Complex I of the respiratory chain in cyanobacteria is NADPH-linked, which is in accord with the findings that the main pathway of glucose-6-phosphate degradation is the hexosemonophosphate shunt. Thus respiratory O_2 consumption may remove the gas from the nitrogenase site. The observed O_2 sensitivity of nitrogenase in reductant-limited but not in exponentially growing cultures (Weare & Benemann, 1973b) supports this view. Possibly the cyanide-insensitive branch of respiration which occurs in cyanobacteria (Biggins, 1969) plays a special role in the provision of an anaerobic environment for nitrogenase. In addition, the oxy-hydrogen (Knallgas) reaction has to be considered (see section 4.6). H_2 evolved by nitrogenase is re-utilized in an O_2-dependent reaction in heterocysts. When H_2 was present in assays with intact *A. variabilis* (Bothe *et al.*, 1978) or heterocysts from the same organism (Wolk, 1979b), C_2H_2 reduction was found to be more stable against O_2 addition. Recent investigations (Haury & Wolk, 1978; Gotto *et al.*, 1979b) with O_2-sensitive mutants of *Anabaena* indicate that the protective mechanisms in heterocysts involve different phenomena and that there may be several sites sensitive to O_2. Synthesis of the O_2 protection mechanisms is temporarily separated from that of nitrogenase and heterocysts, as shown with a mutant of *Anabaena* CA (Grillo *et al.*, 1979).

The chemical reactions which lead to damage of both nitrogenase proteins by O_2 have not yet been elucidated. O_2 may produce superoxide radicals or H_2O_2 which destroy the enzyme. Cyanobacteria contain both superoxide dismutase and catalase (Henry *et al.*, 1978). Possibly these two enzymes are particularly associated with nitrogenase within the heterocysts, thereby preventing toxic effects of superoxide and peroxide.

Many bacteria produce slime or mucilage in large amounts, often in response to O_2 (Wilkinson, 1958). Colonies with a massive slime production were found to fix N_2 more actively than those without such material, as shown for *Derxia gummosa* and *Xanthobacter flavus* (Hill, 1971; Yates, 1980). The effect of slime on nitrogenase activity is probably indirect; extracellular polysaccharide formation may diminish the O_2 solution rate in the medium. Wilcockson (1977), however, did not find a correlation between nitrogenase activity and slime production in high or low slime producing strains of *Klebsiella pneumoniae*. Nevertheless, the possibility exists also for cyanobacteria among which many strains produce massive amounts of mucilaginous material.

Carpenter and Price (1976) described a particular O_2 protection mechanism for a marine *Oscillatoria* (*Trichodesmium*). N_2 fixation in this organism is associated with altered vegetative cells located in the centre of the colony (see Chapter 19). In comparison with heterocysts, they do not evolve

O_2 photosynthetically, but they do not possess thick cell walls. Nitrogenase in this *Oscillatoria* is rapidly inactivated by O_2. Apparently, the O_2 tensions in the centre of the colony are low enough to allow nitrogenase to function. For this, a calm sea is necessary as water movement breaks up the filament clusters. In a similar investigation, Bryceson and Fay (1979) observed that N_2-fixation activity of blossoms of *O. erythraea* in the coastal waters of East Africa was directly related to the ability of filaments to aggregate into bundles and was inversely affected by wind speed. In such populations, C_2H_2 reductions were highest in the morning and declined towards midday when light intensities and photosynthetic O_2 production was maximal. Under natural conditions, respiring bacteria associated to cyanobacteria may help to lower O_2 tensions.

4.5 Regulation of nitrogenase biosynthesis and activity

Both biosynthesis and activity of nitrogenase are regulated in N_2-fixing organisms. The modes of regulation appear to differ between organisms and for cyanobacteria have not yet been elucidated in detail.

The N_2 molecule itself does not induce nitrogenase biosynthesis, since the enzyme is formed from protein reserves when the cyanobacteria are kept under argon. If is often stated that the addition of a combined nitrogen source to the cultures suppresses nitrogenase biosynthesis (see Rippka & Stanier, 1978; Brill, 1975). This can be oversimplification, e.g. in *Azotobacter* where nitrate hardly suppresses nitrogenase biosynthesis. The addition of NH_4^+ to *Anabaena cylindrica* abolishes C_2H_2-reduction activity and heterocysts number within 48 h, whereas these are only gradually affected to some extent when the organism is grown on nitrate for weeks (Stewart, Haystead & Dharmawardene, 1975; Bothe & Eisbrenner, 1977). Meeks (1979) has recently shown that *A. cylindrica* lacks an active transport system for nitrate at concentrations higher than 10 μM in the medium. Thus the formation of nitrogenous compounds that repress nitrogenase biosynthesis and heterocysts differentiation is simply limited by the rate of nitrate uptake. In contrast, *Anabaena* 7120 can rapidly utilize concentrations of nitrate above 10 μM; consequently nitrogenase activity and heterocyst number are reduced within hours (Meeks, 1979). In *Plectonema boryanum*, NH_4^+, NO_3^-, glutamine and arginine repress the anaerobic induction of nitrogenase, but aspartate, asparagine and lysine are ineffective (Nagatani & Haselkorn, 1978). *Anabaena* CA grown on NH_4^+ expresses nitrogenase activity and forms heterocysts to an appreciable extent compared to cultures living on N_2 as the nitrogen source (Stacey, Bottomley, van Baalen & Tabita, 1979). Such examples indicate how dangerous are generalizations in this field.

The major pathway of NH_4^+ assimilation comprises glutamine synthetase [L-glutamate: ammonia: ligase (adenosine-5′-diphosphate forming); EC 6.3.1.2] and glutamate synthase [L-glutamine:2-oxoglutarate aminotransferase; (=GOGAT); EC 1.4.7.1.] reactions in N_2-fixing microorganisms including cyanobacteria. Other NH_4^+-assimilating enzymes, e.g. alanine dehydrogenase or glutamate dehydrogenase are of minor physiological significance (see Stewart, 1977). The regulation of nitrogenase biosynthesis has been extensively investigated in *Klebsiella pneumoniae*, where studies with mutants and inhibitors have established that the glutamine synthetase molecule is the key control element. This enzyme is regulated by adenylylation and de-adenylylation which, in turn, are influenced by a complex cascade-regulating mechanism (see Tyler, 1978). The de-adenylylated form of the enzyme is believed to promote nitrogenase biosynthesis at the transcriptional level. There is evidence from studies with methionine sulphoximine, which specifically blocks the action of glutamine synthetase, that this enzyme may also be involved in the control of nitrogenase biosynthesis in cyanobacteria. In the presence of this compound, nitrogenase activity was expressed in *Anabaena* even when growing on NH_4^+ (Stewart & Rowell, 1975). More detailed information came from the biochemical characterization of the purified enzyme from *A. cylindrica* (Sawhney & Nicholas, 1978b) and *Anabaena* CA (Stacey, Van Baalen & Tabita, 1979). The activities of the enzyme from *Anabaena* was regulated by a divalent cation, the energy charge and pH, and was inhibited by a number of amino acids. More importantly, the enzyme from *Anabaena*, in contrast to that from other Gram-negative bacteria, was not controlled by adenylylation. Moreover, Rowell, Enticott and Stewart (1977) observed only poor correlations between the activities of nitrogenase and glutamine synthetase. However, they found an inverse correlation between the intracellular pools of glutamine and aspartate, and therefore concluded that glutamine, or a derived product, rather than glutamine synthetase itself, regulates nitrogenase biosynthesis in *Anabaena*. Glutamine was suggested also to control the pattern of heterocyst formation (Wolk, 1979a).

In addition to glutamine, other factors have to be considered. O_2 not only destroys nitrogenase activity but also represses enzyme biosynthesis in *Klebsiella* (see Brill, 1975; Yates, 1980). A light-regulated nitrogenase biosynthesis in the photosynthetic bacterium *Rhodopseudomonas capsulata* has been described (Meyer *et al.*, 1978). Regulation by O_2, light and other factors such as temperature, pH or nutrient supply may also occur in cyanobacteria. It is difficult in the heterocystous forms to resolve whether such factors primarily impair nitrogenase biosynthesis or heterocyst formation (cf. Bradley & Carr, 1977; Murray & Benemann, 1979).

Molybdenum deficiency severely decreased nitrogenase activity, and addition of the element to starved cultures rapidly restored nitrogenase in *Anabaena cylindrica* (Fay & de Vasconcelos, 1974). According to Nagatani and Haselkorn (1978), *Plectonema boryanum* forms both nitrogenase proteins under Mo starvation, with the Mo–Fe protein in an inactive form. Addition of Mo to cultures resulted in a rapid appearance of the active Mo–Fe protein. Protein synthesis was not involved in this restoration of activity, since chloramphenicol did not affect the process in *P. boryanum* (Nagatani & Haselkorn, 1978) whilst it did in *Klebsiella pneumoniae* (Kennedy & Postgate, 1977). Obviously the effects of Mo starvation on nitrogenase proteins are different from organism to organism. *Clostridium pasteurianum*, for example, in the absence of Mo, did not synthesize the Fe–S protein of nitrogenase in contrast to *Plectonema*. This cyanobacterium appears to lack the molybdenum storage protein which was postulated for *Clostridium* and *Azotobacter* (see Elliott & Mortenson, 1977; Nagatani & Haselkorn, 1978).

Factors influencing nitrogenase activity have been elucidated less than those affecting biosynthesis. NH_4^+ rapidly switches off nitrogenase activity in photosynthetic bacteria (Neilson & Nordlund, 1975) and in *Azotobacter* (Laane *et al.*, 1980) but not in cyanobacteria (Stewart, Haystead & Dharmawardene, 1975; Murray & Benemann, 1979; but see Ohmori & Hattori, 1974). *In vitro*, the reduction of the N_2 molecule requires 12–15 molecules of ATP, and Mg·ADP is a strong inhibitor of nitrogenase. Therefore a fine control of the activity by the ATP/ADP ratio (or alternatively by the energy charge) would be sensible for the intact organisms. However, Bottomley and Stewart (1977) did not find a positive correlation between the ATP pool and nitrogenase activity in *Anabaena cylindrica* and *Anabaenopsis circularis* and concluded that the supply of ATP was not likely to be a critical short-term factor controlling activity. On the other hand, even small changes in the ATP/ADP ratio can effect a large change in activity, as shown for nitrogenase isolated from *Klebsiella* (Thorneley & Cornish-Bowden, 1977). Indeed, the ATP pool may rapidly change also in *Anabaena*. Ohmori and Hattori (1978) showed that the size of the ATP pool was reduced by 40% within 1 min when ammonia was added to N_2-fixing cultures of this cyanobacterium. Nitrogenase activity could be regulated, if not by the ATP/ADP ratio, by the supply of reductant which, in turn, may be dependent on the energized state of the membrane as suggested for *Azotobacter* and *Rhizobium* (Laane *et al.*, 1979). As stated above (see 4.3), this hypothesis so far lacks compelling evidence for any microorganism.

4.6 Hydrogenase–nitrogenase interrelationship

Detailed accounts with rather divergent views have recently been published on this subject (Hallenbeck & Benemann, 1979; Bothe *et al.*, 1978; Lambert & Smith, 1981). Three enzymes catalysing the reaction $2H^+ + 2e = H_2$ must be considered to occur in cyanobacteria:

1 The reversible hydrogenase described for *Clostridium* and other fermentative bacteria catalyses H_2 evolution in a ferredoxin-dependent reaction under physiological conditions. It serves to remove excess reducing equivalents generated during fermentation. It has not been detected in any of the aerobic, N_2-fixing organisms but was purported to occur in cyanobacteria (Hallenbeck & Benemann, 1979).
2 Nitrogenase catalyses the irreversible ATP-dependent reduction of H_3O^+. *In vitro*, this reaction proceeds either in the absence of any other reducible substrate or in the presence of N_2, where the stoichiometry between N_2 reduction and H_2 release is approximately 1 : 1 (see Burns & Hardy, 1975). The extent to which intact organisms evolve H_2 is variable depending on the strain used.
3 Aerobic nitrogen-fixing organisms contain a membrane-bound hydrogenase catalysing only H_2 uptake into the cells. This enzyme has recently been found with surprisingly high activities in *C. pasteurianum* (Chen & Blanchard, 1978) and has been unequivocally demonstrated in cyanobacteria (see Bothe *et al.*, 1978).

H_2 formation by intact cyanobacteria was observed only under nitrogen-fixing conditions, and the addition of $NH_4{}^+$ or $NO_3{}^-$ to the media suppressed N_2 fixation and H_2 formation in a parallel way. In the assays, H_2 evolution was sensitive to carbon monoxide and was reduced by N_2 or C_2H_2 which competed with H_3O^+ for electrons in nitrogenase. The formation of the gas was largely stimulated by light, but can also proceed in the dark, provided low levels of oxygen are present to allow respiration, and is sensitive to uncouplers. All these observations indicate that H_2 must be generated by the ATP-dependent reduction of H_3O^+ catalysed by nitrogenase. It remains to be shown whether or not hydrogenase can catalyse H_2 evolution in the intact organisms.

An exception to this rule is *Oscillatoria limnetica* (see Chapter 9). This organism is capable of performing both H_2S- and H_2-dependent anoxygenic photosynthesis at high rates (Oren *et al.*, 1977). *Oscillatoria limnetica* can also evolve H_2 in a hydrogenase-dependent reaction under anaerobic conditions and in the absence of CO_2 or any other reducible substrate (Belkin & Padan, 1978). Thus *O. limnetica* is similar to *C. pasteurianum* in that it uses hydrogenase as a valve to remove excess reductants.

H_2 evolution by obligate photoautotrophic cyanobacteria has recently attracted special attention in solar energy conversion projects. In particular, Benemann (see Benemann & Weare, 1974b; Weisman & Benemann, 1977) suggested that cyanobacteria might evolve H_2 and O_2 in sufficient quantities to be used as an energy source. Rates which we observed with several cyanobacteria grown and assayed under aerobic conditions (Bothe *et al.*, 1978) were always small. Similarly, Berchthold and Bachofen (1979) found very low H_2 formations with a large series of new isolates of cyanobacteria. H_2

evolution rates can, however, be increased to some extent when cyanobacteria are assayed in the presence of high concentrations of C_2H_2 plus low amounts of CO (Bothe, *et al.*, 1977b; Daday *et al.*, 1977), or under strict anaerobic conditions (Spiller *et al.*, 1978) both of which prevent the re-utilization of H_2 by hydrogenase (see following sections). However, in our hands, maximal rates of H_2 formation never exceeded 20% of those of C_2H_2 reductions (Bothe *et al.*, 1978). Other authors reported higher activities (Weisman & Benemann, 1977; Mitsui & Kumazawa, 1977; Ernst *et al.*, 1979). Since all H_2 formation is catalysed by nitrogenase, C_2H_2 reduction is an indication of the theoretically achievable H_2 formation in intact cyanobacteria. There is a general consensus that C_2H_4 formation in *Anabaena cylindrica*, *A. variabilis* and *Nostoc muscorum* are in the range of 7.5–20 μmol h^{-1} mg chl^{-1}. In comparison, the rates of photosynthetic CO_2 fixation are generally given between 80 and 250 μmol h^{-1} mg chl^{-1} (see Bothe *et al.*, 1980). Such differences between C_2H_4 formation and CO_2 fixation are in accord with the requirements of the cells for carbon and nitrogen. A rate of 20 μmol H_2 formed h^{-1} mg chl^{-1} means that all the electrons flowing to nitrogenase are used to reduce protons and that the organisms do not possess a regulatory mechanism to switch off the energy-expensive H_2 evolutions. It is also speculated that electrons normally utilized for CO_2 fixation could be diverted to reduce H_3O^+. It is obvious that organisms cannot sustain such high H_2 production over a longer period if not supplied with both carbon and nitrogen compounds. There is a slight hope that the new isolates *Anabaena* CA (Stacey *et al.*, 1977) and *Anabaena* sp. TA 1 (Antarikanonda *et al.*, 1980), which were reported to have considerably higher C_2H_2-reduction activities than the strains hitherto used, or mutants selected for higher H_2 production, may have more promise in projects relating to solar-energy conversion.

H_2 evolved by nitrogenase is not necessarily lost to cyanobacteria. Recent investigations have shown that H_2 can be utilized by two different pathways, both of which are catalysed by hydrogenase(s) (Bothe *et al.*, 1977a; 1978; Eisbrenner *et al.*, 1978; Peterson & Burris, 1978; Peterson & Wolk, 1978b; Tel-Or *et al.*, 1978; Tetley & Bishop, 1979; Peschek, 1979a,b,c; Antarikanonda *et al.*, 1980). In the major pathway, exogenous H_2 is consumed in an oxygen-dependent reaction. Such an oxy-hydrogen, or Knallgas reaction involves the respiratory electron chain and supplies the organism with extra ATP. It thereby minimizes the loss of energy caused by H_2 evolution from nitrogenase. In addition, the Knallgas reaction can remove O_2 from the nitrogenase site preventing deleterious effects of this gas (see 4.4). The ATP/2e ratio and the electron acceptor of H_2 and hydrogenase are unknown. Newer data suggest that the electrons from H_2 must enter the respiratory chain close to the plastoquinone (Eisbrenner & Bothe, 1979) or the phylloquinone site (Peschek, 1980).

In the second pathway, H_2 is utilized in a strictly light-requiring but oxygen-independent reaction. This pathway is seen in *Anabaena* limited for carbon reserves (Benemann & Weare, 1974a; Bothe, Tennigkeit & Eisbrenner 1977a) or in isolated heterocysts (Eisbrenner *et al.*, 1978; Peterson & Wolk, 1978b) where H_2 effectively supports C_2H_2 reduction. H_2 also acts as electron donor in the photosynthetic reduction of nitrite (Hattori, 1963; Peschek, 1979b; Bothe *et al.*, 1980), of nitrate by intact cells of *Anabaena* or *Anacystis* (Bothe *et al.*, 1980), and in the NADPH formations by preparations from heterocysts of *Nostoc muscorum* (Eisbrenner *et al.*, 1981). Studies with inhibitors have indicated that the light dependence in all these reactions was not because of a requirement for cyclic photophosphorylation supplying ATP, but because the transfer of electrons from H_2 to the substrates had to proceed via PS I. Such a statement is in line with the findings that H_2 and hydrogenase cannot reduce ferredoxin or other natural electron carriers with redox potential around -400 mV *in vitro* and that hydrogenase does not catalyse H_2 formation *in vivo* (see Bothe *et al.*, 1978). The two pathways of hydrogen utilization are probably not independent of each other. Studies with the inhibitors dibromothymoquinone and tetramethylphenylenediamine (Eisbrenner & Bothe, 1979) indicated that plastoquinone is a common intermediate. Both photosynthetic and respiratory H_2 utilization are presumably located on the thylakoid membranes of cyanobacteria.

Finally a comment is necessary on the number of hydrogenases in cyanobacteria. In addition to the hydrogenase catalysing H_2 uptake, a fully reversible enzyme, favouring H_2-formation under physiological conditions, was purported to occur in cyanobacteria (Tel-Or *et al.*, 1978; Hallenbeck & Benemann, 1979). The main argument for two hydrogenases is the observation that the reversible enzyme, catalysing H_2 formation in the presence of high concentrations of methylviologen and $Na_2S_2O_4$, was present in extracts from vegetative cells and heterocysts with essentially the same specific activities. Contrary to this, the Knallgas reaction has been demonstrated so far only in isolated heterocysts; therefore the uptake hydrogenase is believed to be restricted to nitrogen-fixing cells (Tel-Or *et al.*, 1978; Peterson & Wolk, 1978b). However, it should be pointed out that vegetative cells break easily, whereas heterocysts, owing to their thick wall, are relatively resistant to mechanical and enzymatic treatments. Breakage of vegetative cells easily affects the respiratory electron flow and with it the Knallgas reaction (Eisbrenner *et al.*, 1981). These authors also showed that the more drastic the treatment to break the cells the higher the H_2-evolution activities obtained. Therefore cyanobacteria might contain a single, membrane-bound hydrogenase which becomes more readily accessible to methylviologen and $Na_2S_2O_4$, the more it is solubilized from the membranes. It is not true that the uptake hydrogenase is restricted to nitrogen-fixing cells. It can be demonstrated in the unicellular *Anacystis nidulans* when grown either in air or anaerobically in the presence of H_2 (then with much higher activities; Bothe *et al.*, 1978; Peschek, 1979a,b). The Knallgas reaction is also demonstrable in ammonia grown *Anabaena* when kept anaerobically under H_2 (Eisbrenner *et al.*, 1981). A function for a reversible hydrogenase, catalysing H_2-evolution *in vitro*, is difficult to envisage for cyanobacteria which are able to remove excess of reductants, if at all occurring, by respiration.

There are also recent suggestions for the occurrence of two uptake hydrogenases in cyanobacteria, one catalysing the respiratory-linked H_2 consumption and the other the PS I mediated electron transfer (Peschek, 1979c; Tetley & Bishop, 1979). This possibility cannot be ruled out at present. If respiratory and photosynthetic H_2 consumption share a common segment of electron carriers, as recently suggested (Eisbrenner & Bothe, 1979), a postulate for two uptake hydrogenases is unnecessary. The only hydrogenase unequivocally demonstrated so far in cyanobacteria is apparently membrane-bound, relatively insensitive to exposure to O_2, does not couple with ferredoxin, and catalyses the uptake and the evolution of H_2 when isolated, but is probably a unidirectional enzyme under physiological conditions.

The author is indebted to Deutsche Forschungsgemeinschaft for financial support of his own experiments cited in the text.

CHAPTER 5
PHOSPHATE

F. P. HEALEY

5.1 Uptake

5.1.1 Uptake of orthophosphate

Orthophosphate seems to be available to all microbes and is believed to be the form of phosphorus commonly taken up by autotrophs. Uptake of orthophosphate by cyanobacteria is an apparent hyperbolic function of external phosphate concentration (Fig. 5.1), as it is in other microbes (Medvecsky & Rosenberg, 1971; Rhee, 1973; Chen, 1974; Tilman & Kilham, 1976). This has allowed kinetic parameters for phosphate uptake to be calculated for a few species (Table 5.1). Both the half-saturation values (k) and maximum rates of uptake (V_m) are similar to those of eukaryotic algae, but are generally lower than those of heterotrophic bacteria (Jeanjean, 1969; Fuhs *et al.*, 1972; Rhee, 1973; Chen, 1974; Tilman & Kilham, 1976; Nyholm, 1977).

Table 5.1 illustrates two ways in which the kinetic parameters for phosphate uptake can vary. First, with increasing phosphorus deficiency, the maximum rate of phosphate uptake by *Anabaena flos-aquae* increases, as it

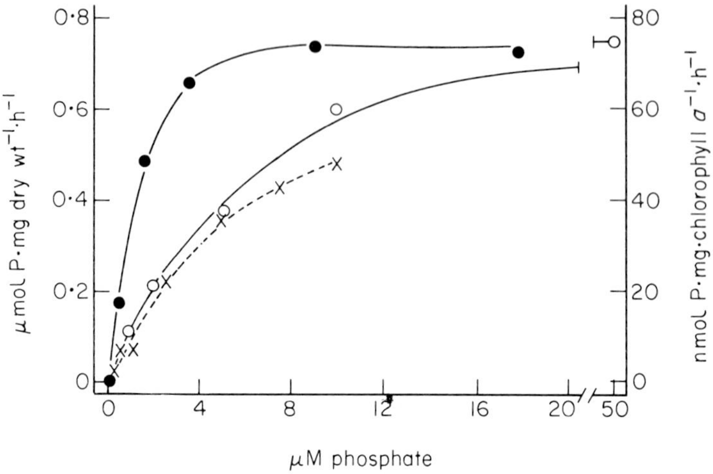

Fig. 5.1. Uptake of orthophosphate as a function of external phosphate concentration. *Anabaena variabilis* (●) as μmol P·mg dry wt^{-1} h^{-1} (Healey, 1973a); *Anacystis nidulans* (○) as nmol P·mg chlorophyll a^{-1} h^{-1} (Falkner *et al.*, 1974); *Oscillatoria thiebautii* (x) as 10^{-2} μmol·mg dry wt^{-1} h^{-1} (McCarthy & Carpenter, 1979).

does in *A. variabilis* (Healey, 1973a; Healey & Hendzel, 1975), *Anacystis nidulans* (Bornefeld *et al.*, 1974), *Nostoc* sp. (Strange *et al.*, 1975), *Plectonema boryanum* (Sicko-Goad *et al.*, 1978), *Synechococcus* sp. (Lawry & Jensen, 1979; Grillo & Gibson, 1979) and in other microorganisms (Healey, 1973b; Chen, 1974). The half-saturation value (k) for phosphate uptake by *Anabaena flos-aquae* also increases with phosphorus deficiency; this is in contrast with *Chlorella pyrenoidosa*, where k is unaffected by phosphorus deficiency (Jeanjean, 1969), and with *Corynebacterium bovis*, where k decreases with phosphorus deficiency (Chen, 1974). Second, phosphorus uptake by *Anabaena variabilis* is affected by the Mg^{2+} concentration present during uptake. With decreasing Mg^{2+} concentration, k rises while the maximum rate of uptake is little affected.

As pointed out by Healey (1980), the ratio V_m/k is a better indicator of competitive position in nutrient uptake than is k alone, a higher ratio implying a steeper slope in the plot of rate of uptake against concentration or higher uptake rates at lowest concentrations. Thus, the competitive ability of *A. flos-aquae* in phosphate uptake increases during growth into phosphorus deficiency, the increase in maximum rate overriding the rise in half-saturation value (Table 5.1). Similarly, the ability of *A. variabilis* to compete for phosphorus is improved by increasing concentrations of Mg^{2+}. The rather high k of *Oscillatoria thiebautii*, combined with its low V_m, suggest it to be a poor competitor for phosphate.

Table 5.1. Kinetic parameters of phosphate uptake

Species	Condition		k*	V_{max}*	V_{max}/k	Reference
Anabaena flos-aquae	Cell P	22	—	—	—	Nalewajko &
	(μg mg	—	0.96	0.18	0.19	Lean (1978)
	dry	14	1.85	0.63	0.34	
	weight^{-1})	7.3	2.05	0.89	0.43	
		—	2.64	1.6	0.61	
		3.6	2.8	1.8	0.64	
Anabaena variabilis†	P-deficient					
	with 300 μmol Mg^{2+}		1.1	0.83	0.75	Healey (1973a)
	with 30 μmol Mg^{2+}		9.5	0.95	0.10	
Anacystis nidulans	Apparently P-sufficient					
	in light		6.0	75		Falkner *et al.*
	in dark		6.0	38		(1974)
Oscillatoria thiebautii†	Natural collection		9.0	0.01‡	0.001	McCarthy & Carpenter (1979)
Synechococcus sp.	P-sufficient		0.2	1.3		Grillo & Gibson
	P-deficient		0.3	50		(1979)

* k (half saturation value) in μM phosphate and V_{max} (saturated rate) in μmol P mg dry wt^{-1} h^{-1}, except for *Anacystis nidulans* where it is nmol P mg chl^{-1} h^{-1}, and *Synechococcus* sp. where it is nmol P mg protein^{-1} min^{-1}.

† Values calculated from fig. 4 of cited paper, using 2 mg dry wt = 1 mg organic C.

‡ Rates given in fig. 4 of cited paper are $10^3 \times$ too high (J.J. McCarthy, pers. comm.).

The values given in Table 5.1 for *Anacystis nidulans* apply only to one of the uptake systems of this microbe. Measurements over a wide range of phosphate concentrations have revealed the presence of three systems nearing saturation about 10, 50, and 100 μmol phosphate, respectively (Simonis *et al.*, 1974). At concentrations above 100 μmol, uptake becomes predominantly passive. Multiple phosphate uptake sites have been found in other microbes as well (Jeanjean *et al.*, 1970; Medvecsky & Rosenberg, 1971; Nyholm, 1977).

In a variety of cyanobacteria, the rate of phosphate uptake is optimal between pH 7.5 and 8.5, declining sharply below neutrality (Fig. 5.2). This is in keeping with the general preference of cyanobacteria for alkaline conditions and suggests the divalent HPO^{2-} as the major species taken up. Washing various cyanobacteria free of culture medium or treatment with a chelator has revealed various cation requirements for maximum rates of phosphate uptake. Stimulation of uptake by Ca^{2+} has been found in *Nostoc verrucosum* (Whitton, 1967) and *Anacystis nidulans* (Falkner, Werdan, Horner & Heldt, 1976), by Ca^{2+} and to a lesser extent by Mg^{2+} in *Plectonema boryanum*

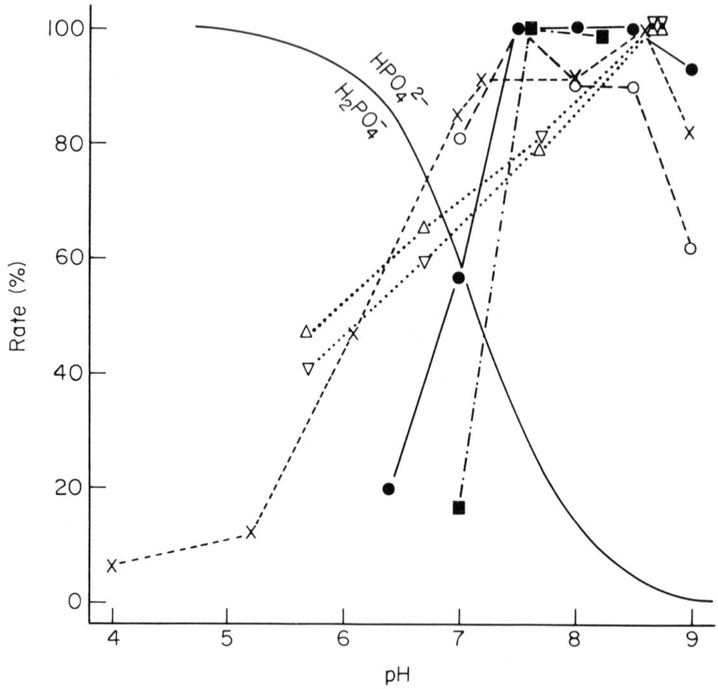

Fig. 5.2. Uptake of orthophosphate as a function of external pH. (●) *Anabaena variabilis* (Healey, 1973a); (×) *Anacystis nidulans* (Ullrich-Eberius & Yingchol, 1974); (□) *Anacystis nidulans* (Falkner *et al.*, 1974); ( ) *Mastigocladus laminosus* (Sperling, Grunewald & Dovala, 1974); (O) *Synechococcus* sp. (Lawry & Jensen, 1979). All rates as a percentage of the maximum shown by each cyanobacterium. The faint line shows the percentage dissociation $H_2PO_4^- \rightleftarrows H^+ + HPO_4^=$ based on ka = 6.2 × 10⁻⁸.

(Sicko-Goad *et al.*, 1978) and by Mg^{2+} in *Anabaena variabilis* (Healey, 1973a). In the last instance, the rate of phosphate uptake was a saturable function of both phosphate and Mg^{2+} concentrations; at low concentrations of both Mg^{2+} and phosphate, a decrease in one could be offset by an increase in the other. In the above studies the cyanobacteria were not grown to be deficient in any cations. Uptake of Mg^{2+} did not accompany that of phosphate by *A. variabilis* and the phosphate uptake rate increased to its maximum within 1 min of addition of Mg^{2+} (Healey, unpubl.). These results suggest that the cations operate outside the cells, presumably by interacting with the phosphate uptake sites to increase their ability to take up phosphate.

The rate of phosphate uptake by cyanobacteria is generally greater in the light than in the dark, with stimulation by light ranging from slight (Batterton & Van Baalen, 1968) or 50% (Healey, 1973a) to 2–3 times (Falkner *et al.*, 1974; Georgiyeva, 1976; Grillo & Gibson, 1979) or as much as 8 times (Simonis *et al.*, 1974). At high concentrations of phosphate (over 1mM) the rate of uptake by *Anacystis nidulans* in the dark exceeds that in the light as passive uptake apparently comes to dominate over active (Simonis *et al.*, 1974). The rate of phosphate uptake by *Anabaena variabilis* increases with temperature to 35°C with a Q_{10} (20–30°C) of 2.1, but rates at 30–35°C can decline with time (Healey, 1973a). Light and temperature dependence of phosphate uptake suggest this to be an active process, at least at lower phosphate concentrations. This is confirmed by the effect of inhibitors. Carbonyl cyanide m-chloro-phenyl-hydrazone (an uncoupler of phosphorylation) and dichlorophenyl-dimethylurea (an inhibitor of photosynthetic electron transport) at 5–10 μmol cause about a 50% inhibition of phosphate uptake by *A. variabilis* (Healey, 1973a) and *Anacystis nidulans* (Simonis *et al.*, 1974), implicating both cyclic and non-cyclic photosynthetic electron flow in the light stimulation of phosphate uptake.

5.1.2 Phosphorus excretion and exchange

In microbial cultures, even when phosphorus-limited, a fraction of the phosphorus provided is found in solution as dissolved organic phosphorus. During growth of *Synechococcus* sp. in batch culture, up to 10% of the phosphorus provided was recovered from the medium as dissolved organic phosphorus (Kuenzler, 1970). In a phosphorus-limited culture of *Anabaena flos-aquae*, 15% of the phosphorus was in dissolved organic form when orthophosphate was depleted and this fell to 1.3% as the culture grew into phosphorus deficiency (Lean & Nalewajko, 1976). In phosphorus-limited chemostat cultures of *A. variabilis*, total dissolved phosphorus always greatly exceeded soluble reactive phosphorus; the difference—a minimal estimate of dissolved organic phosphorus—averaged about 5% of total phosphorus input (Healey & Hendzel, 1975). The extent to which this was due to cell breakage or true excretion is unknown. However, Lean (1973), has found evidence for phosphorus excretion in organic form in natural populations, and more recently, quite high rates of excretion of phosphorus have been reported for cultures, including those of *A. flos-aquae* (Lean & Nalewajko, 1976; Nalewajko & Lean, 1978). Measurements showed gross uptake of phosphate could exceed net uptake by several-fold, implying efflux occurring at rates often greatly exceeding net influx. In contrast to this, Grillo and Gibson (1979) found that *Synechococcus* sp. pre-labelled with ^{32}P released little or no label into media with or without excess phosphate in solution. In agreement with the latter findings others have found evidence for only very limited release of recently taken-up phosphorus by both prokaryotic (Harold & Spitz, 1975) and eukaryotic microbes (Robertson & Button, 1979; Brown & Button, 1979).

5.1.3 Uptake of phosphate from organic compounds

Present evidence suggests that organic phosphorus compounds are not generally taken up intact by autotrophic microbes, but that the phosphate is hydrolysed outside the cell by extracellular phosphatases and the phosphorus taken up as orthophosphate (Kuenzler, 1965). An exception to this situation has recently been found in the uptake of glucose-6-phosphate by *Anabaena flos-aquae* (Rubin *et al.*, 1977). When glucose-6-phosphate labelled with both ^{14}C and ^{32}P is given to cells with negligible extracellular phosphatase activity, uptake of the two labels occurs simultaneously. After uptake, most of the extractable label is in glucose-6-phosphate, as is most of the label remaining in the medium. Uptake of ^{14}C from glucose-6-phosphate occurs at about 10% of the rate of uptake of glucose alone and is not inhibited by high concentrations of glucose. The high concentration of glucose-6-phosphate required to saturate the reaction (over 16 mM) and the low rate of uptake at saturation relative to saturated rates of phosphate uptake reported for other cyanobacteria make doubtful the ecological importance of this uptake system.

Microbes with phosphatase activity located on the external surface of the cell are capable of hydrolysing phosphate from a variety of organic phosphorus compounds. This activity, generally induced by phosphorus deficiency (Table 5.3), is widespread among cyanobacteria. Alkaline phosphatase activity of *Plectonema boryanum* is located in the periplasmic space of the cell wall (Doonan & Jensen, 1977). Phosphatases of some cyanobacteria either are tightly retained in the cell wall or rapidly lose activity once released (Ihlenfeldt & Gibson, 1975b; Doonan & Jensen, 1977) while those of some others are released and remain active in solution (Healey, 1973a; Healey & Hendzel, 1975). In several cyanobacteria, this activity is optimal between pH 8 and 10 (Kuenzler & Perras, 1965; Healey, 1973a; Heath & Cooke, 1975; Ihlenfeldt & Gibson, 1975b; Healey & Hendzel, 1979a). Temperature optima for short-term incubations (about 30 min) are high (40–50°C), with a Q_{10} (30–40°C) of 1.5–2.0 reported for *Anabaena variabilis* (Healey, 1973a; Heath & Cooke, 1975; Healey & Hendzel, 1979a). Maximum activities in *A. variabilis* and *Pseudanabaena catenata* require external Ca^{2+}; in *Anabaena variabilis* the requirement for Ca^{2+} increases as the substrate concentration falls (Healey, 1973a; Healey & Hendzel, 1979a). Alkaline phosphatase activity of *A. variabilis* is competitively inhibited by phosphate and molybdate (Healey, 1973a); acid phosphatase activity (pH 5.8) of *A. cylindrica* and *Anacystis nidulans* is competitively inhibited by molybdate (Talpasayi, 1962).

The alkaline phosphatase of *A. nidulans* hydrolyses various organic phosphorus compounds at different rates (Ihlenfeldt & Gibson, 1975b), suggesting that organic phosphorus compounds differ in their susceptibility to enzymatic attack. Presumably some organic phosphorus compounds are not at all, or only very slowly hydrolysed by microbial phosphatases. These accumulate as organic phosphorus in cultures (as mentioned above) and presumably account for the organic phosphorus found in natural waters. Colloidal phosphorus, produced by the phosphorus metabolism of microflora in lakes, is hydrolysed at a very slow rate (Lean, 1973).

5.2 Phosphorus-limited growth

5.2.1 *Dependence of growth rate on external and internal phosphorus concentrations*

Generally, in nitrogen- or phosphorus-limited continuous cultures of eukaryotic algae, no clear relationship has yet emerged between growth rate and external dissolved nutrient concentration (Rhee, 1973; Caperon & Meyer, 1972; Nyholm, 1976). This may be because growth-rate limiting concentrations of these nutrients are below the limits of present chemical detection (Brown & Button, 1979), but it cannot yet be excluded that there may be no meaningful relationship between growth rate and external nutrient concentration, the nutrient in solution during nutrient-limited growth being that unavailable to the organism (Nyholm, 1976). No relationship was found between growth rate and soluble reactive phosphorus or total dissolved phosphorus in phosphorus-limited chemostats of *Anabaena variabilis* (Healey & Hendzel, 1975), but suitable precautions to limit phosphate uptake between sampling the cultures and filtering the samples were not taken (Robertson & Button, 1979). Phosphorus-limited chemostat cultures of *Oscillatoria agardhii* at 25°C showed near-maximum growth rates down to about 2 μg phosphate-P l^{-1} (Ahlgren, 1977). At lower temperatures (20°C and more so at 15°C), the lowest substrate concentrations (1–2 μg phosphate-P l^{-1}) were associated with growth rates depressed below the maximum (Ahlgren, 1978). Half-saturation values calculated from these data ranged from 0.2 to 1.0 μg phosphate-P l^{-1}. Droop (1968) derived the relationship

$$K = \mu_m Q_0 k / V_m$$

where K = half saturation value for growth, μ_m = maximum growth rate, Q_0 = minimum internal nutrient concentration, k = half saturation value for uptake, and V_m = maximum rate of uptake.

This relationship allows K for growth to be calculated from other more readily measurable kinetic data. Using this formula, the kinetic data for phosphate uptake by *Anabaena variabilis* in the presence of 300 μM Mg^{2+} in Table 5.1, μ_m of 1.15 day^{-1} and Q_0 of 1.3 μg P mg dry wt^{-1} (Healey & Hendzel, 1975), K for growth for this alga is about 0.1 μg phosphate-P l^{-1}. The results for both *Oscillatoria agardhii* and *Anabaena variabilis*, while admittedly tentative, suggest that growth-rate limiting concentrations of phosphate for cyanobacteria, as for eukaryotic algae (Rhee, 1973; Tilman & Kilham, 1976; Droop, 1974a; Brown & Button, 1979), can be expected to be at or below present limits of chemical detection. These values for phosphorus-limited growth, together with the kinetic values for phosphate uptake in Table 5.1, however scanty, offer little support for suggestions that cyanobacteria as a group can be expected to be either more (Bierman, 1976) or less (Lehman *et al.*, 1975) efficient in phosphorus utilization than are eukaryotic algae. The surprisingly high half saturation value for growth (67 $\mu g\ l^{-1}$) reported for *A. oscillarioides* (Lam, 1979) is probably an artifact due to relating the growth rate to the initial phosphate concentration, which was largely depleted before growth started, rather than to the phosphate concentration in solution as growth was occurring.

More clearly established than a relationship between growth rate and external nutrient concentration is the dependence of growth rate on internal concentration of the limiting nutrient. In chemostat studies of both eukaryotic and prokaryotic microorganisms limited by a nutrient other than carbon, either linear or hyperbolic relationships have emerged between steady-state growth rate and the internal concentration of the limiting nutrient (Dicks & Tempest, 1966; Droop, 1968, 1974a; Caperon & Meyer, 1972; Rhee, 1973; Paasche, 1973; Nyholm, 1976, 1977). The steady-state growth rates of both *A. variabilis* and *Oscillatoria agardhii* clearly increase with increasing total internal phosphorus concentration (Fig. 5.3). At higher temperatures (20 and 25°C) no relationship between growth rate and internal phosphorus concentration was reported for *O. agardhii* (Ahlgren, 1978); however, at these temperatures, growth rate was measured over only a narrow range of internal phosphorus concentrations.

5.2.2 *Variation in composition with phosphorus deficiency*

The effect of phosphorus deficiency on the composition of several cyanobacteria is shown in Table 5.2. Most notable is the great variation in internal phosphorus concentrations, which can exceed ten-fold when phosphorus contents supporting maximum and near-minimum growth rates are compared. Minimum internal phosphorus concentrations are similar to those of eukaryotic algae and lower than those of at least some heterotrophic bacteria (Fuhs *et al.*, 1972; Chen, 1974; Healey, 1975). Phosphorus deficiency can also lower the nitrogen content but to a lesser degree, resulting in a rise in the cellular N/P ratio in phosphorus deficiency. During growth into phosphorus deficiency, protein content often falls while that of carbohydrate rises, resulting in a drop in the protein to carbohydrate ratio. *Pseudanabaena catenata* shows this response minimally, if at all. Both RNA and DNA

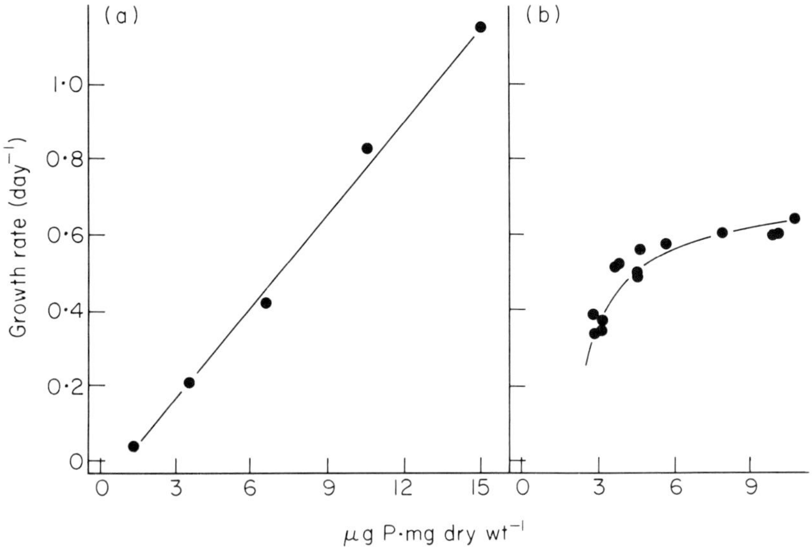

Fig. 5.3. Growth rate of (a) *Anabaena variabilis* (Healey & Hendzel, 1975) and (b) *Oscillatoria agardhii* at 15°C (Ahlgren, 1978) as a function of internal phosphorus concentrations.

contents fall in phosphorus deficiency as does that of chlorophyll *a*. The ATP contents of *Anacystis nidulans*, and *Anabaena cylindrica* fall rapidly with the onset of phosphorus limitation (Batterton & Van Baalen, 1968; Bottomley & Stewart, 1976b). All of the above changes are quite similar to those seen in eukaryotic algae in phosphorus deficiency and, in many instances, in other nutrient deficiencies (Healey, 1973b, 1975).

Growth of *Pseudanabaena catenata* into nitrogen deficiency results in a decline in the phosphorus content per unit dry weight, which is somewhat less than the decline in the nitrogen content; as a result, the N/P ratio of this microbe drops in nitrogen deficiency only slightly (Healey & Hendzel, 1979b).

Increasing phosphorus deficiency decreases the phosphorus content of several pools in *Synechococcus* sp. to similar extents; DNA is exceptional in remaining approximately constant. In nitrogen deficiency, RNA is the only major phosphorus pool to fall in concentration, while the cold acid-soluble and polyphosphate pools rise (Grillo & Gibson, 1979).

5.2.3 *Variation in metabolism with phosphorus deficiency*

The effect of phosphorus deficiency on several physiological rates is shown in Table 5.3. In two species of *Anabaena*, photosynthetic rate declines with cellular phosphorus content, as generally occurs in algae deficient in this or

Table 5.2. Variation in composition with P nutrition*

Species	Condition†	C	N	P	Ext P‡	Protein	CH‡	RNA	DNA	chla‡	PC‡	Lipid	N/P	Protein/CH	Reference
Anabaena	P-sufficient			22	8.0										Nalewajko & Lean (1978)
flos-aquae	P-deficient			3.6	2.2										
Anabaena	P-sufficient	455	105	15.6		544	205	83	13	5.3			6.7	2.7	Healey & Hendzel (1975)
variabilis	37% μ_{max}	480	90	7.0		430	430	34	7.0	3.0			13	1.0	
	3% μ_{max}	380	51	1.3		205	503	18	2.6	0.6			39	0.41	
Anabaena	P-sufficient			8.5§											Senft (1978)
wisconsinense	Zero PS			0.8§											
Anacystis	P-sufficient	510	89	18.3		554¶	338¶					108¶	4.9	1.6	Batterton & Van Baalen (1968)
nidulans	20% μ_{max}	485	66	7.0		414	514					72	9.5	0.81	
	5% μ_{max}	481	51	2.4		320	586					94	21	0.55	
Aphanizomenon	P-sufficient				3.5										Fitzgerald & Nelson (1966)
flos-aquae	P-deficient				0.37										
Gloeotrichia	P-sufficient				4.1										Fitzgerald & Nelson (1966)
echinulata	P-deficient				0.61										
Microcystis	P-sufficient			8.3	4.3										Gerloff & Skoog (1954)
aeruginosa	P-deficient			1.2	0.38										Fitzgerald & Nelson (1966)
Nostoc	P-sufficient				3.6										Fitzgerald & Nelson (1966)
muscorum	P-deficient				0.37										
Oscillatoria	P-sufficient			7.5											Ahlgren (1977)
agardhii	50% μ_{max}			3.2											
Oscillatoria	P-sufficient			10											Feuillade & Feuillade (1975)
rubescens	Zero μ			0.3											
Plectonema	P-sufficient			8.3											Sicko-Goad & Jensen (1976)
boryanum	P-deficient			4.2											
Pseudanabaena	P-sufficient	485	122	16.5	2.8	620	44	59	80	17.8	97	193	7.4	14.1	Healey & Hendzel (1979b)
catenata	77% μ_{max}	574	134	10.7	2.7	695	69	48	81	21.4	128	173	12.5	10.0	
	26% μ_{max}	615	126	5.8	0.62	720	86	46	37	11.4	102	280	21.8	8.4	

* All values are μg mg dry wt^{-1}, except N/P and protein/CH ratios which are by weight.
† P-deficient implies an undefined degree of P deficiency, usually several days' growth of a batch culture in absence of dissolved phosphate; otherwise P deficiency to give the stated growth (μ) or photosynthetic (PS) rates.
‡ Ext P = P extractable with 1 h boiling water; CH = carbohydrate; chla = chlorophyll *a*; PC = phycocyanin.
§ Converted from a cellular volume basis assuming 1 μl = 0.2 mg dry wt (Healey, 1975).

Table 5.3. Variation in metabolism with P nutrition*

Species	Condition	Photosynthesis	Respiration	Phosphate uptake	Phosphatase		Reference
Anabaena flos-aquae	Low N/P input				0.16–0.7	a	Bone (1971a)
	High N/P input				1.5–2.3		
	Early P deficiency	2.3		0.18			Keenan & Auer (1974)
	Advanced P deficiency	1.1		1.8			Nalewajko & Lean (1978)
Anabaena variabilis	P-sufficient			<0.05	0.2–0.5	a	Healey (1973a)
	Advanced P deficiency			0.7–0.9	2.0–6.2		Healey & Hendzel (1975)
	P-sufficient				0.002	b	Healey & Hendzel (1979a)
	Advanced P deficiency				0.1–0.6		
Anabaena wisconsinense	P-sufficient	0.85 (0.65)	(0.17)				Senft (1978)
	Advanced P deficiency	0.05 (0.15)	(0.04)				
Anacystis nidulans	P-sufficient				0.3	a	Ihlenfeldt & Gibson (1975b)
	P-deficient 20 h				2.8		
Aphanizomenon flos-aquae	P-sufficient				1.8	a	Fitzgerald &
	P-deficient				15		Nelson (1966)
Gloeotrichia echinulata	P-sufficient				1.3	a	Fitzgerald &
	P-deficient				14		Nelson (1966)
Microcystis aeruginosa	P-sufficient				0.50	a	Fitzgerald &
	P-deficient				2.8		Nelson (1966)
Nostoc muscorum	P-sufficient				1.1	a	Fitzgerald &
	P-deficient				2.8		Nelson (1966)
Plectonema boryanum	Early P deficiency			0.005			Sicko-Goad,
	Advanced P deficiency			0.6			Jensen & Ayala, (1978)
Pseudanabaena catenata	P-sufficient			<0.05	0.53	a	Healey &
	Advanced P deficiency			1.1	0.50	a	Hendzel (1979b)
	Advanced N deficiency			<0.05	0.05	a	
	P-sufficient				0.002	b	Healey &
	P-deficient				0.003–0.015	b	Hendzel (1979a)

* Rates of light-saturated photosynthesis and dark respiration as μmol O_2 mg dry wt^{-1} h^{-1} (in parentheses as μmol O_2 μg chl a^{-1} h^{-1}); phosphate uptake rates (saturated) as μmol P mg dry wt^{-1} h^{-1}; alkaline phosphatase activities as μmol P mg dry wt^{-1} h^{-1} with various concentrations of p-nitrophenyl phosphate (a) or 10 μM o-methylfluorescein phosphate (b) as substrate.

other nutrients (Healey, 1973b). Like growth rate, photosynthetic rate shows an apparently hyperbolic relationship with phosphorus content (Keenan & Auer, 1974; Senft, 1978). The change in photosynthesis is presumably mediated through the effect of phosphorus deficiency on growth rate—as the growth rate falls due to an insufficient phosphorus supply rate, the rate of photosynthesis must soon follow. As will be described later (5.3.2), neither the dependence of photosynthesis nor that of growth on internal limiting nutrient apply to transient situations where the nutrient concentration increases rapidly. Where it has been measured, in *Anabaena* spp., both the rate of dark respiration (Senft, 1978) and the rate of N_2 fixation (Stewart *et al.*, 1970; Bone, 1971b) fall with phosphorus deficiency.

The potential for phosphate uptake, on the other hand, increases with phosphorus deficiency. This is reflected in elevated rates of phosphate uptake when phosphate is resupplied to the cells. This increase is short-lived, with initial specific rates of phosphate uptake (phosphorus taken up per unit cellular phosphorus) considerably exceeding maximum growth rates, and gradually falling over the first several hours of uptake (Healey, 1973a). The increase in saturated rates of phosphate uptake per unit dry weight can be as great as ten-fold and can begin very early in phosphorus deficiency, increasing rates beginning at about the time soluble reactive phosphorus becomes undetectable, and before exponential growth slows, in batch cultures (Healey, 1973a; Nalewajko & Lean, 1978), and at growth rates only slightly depressed below maximum growth rates in phosphorus-limited chemostat cultures (Healey & Hendzel, 1975). This suggests that an 'all or none' reaction in the de-repression of phosphate uptake by phosphorus deficiency, one in which maximum rates of phosphate uptake are induced by initial declines in internal phosphorus concentration (Fig. 5.4), may operate in microorganisms as well as the more gradual de-repression of phosphate uptake by declining internal phosphorus concentrations described by Rhee (1973). This change seems specific to phosphorus deficiency, not occurring during nitrogen deficiency in *Pseudanabaena catenata* (Healey & Hendzel, 1979b) or *Synechococcus* sp. (Grillo & Gibson, 1979). Although increased rates of phosphate uptake in phosphorus deficiency have usually been measured as increased saturated rates of uptake, rates of phosphate uptake at subsaturating concentrations of phosphate can increase as markedly (Brown & Harris, 1978). Indeed, whenever V_m increases more than k with phosphorus deficiency, rates at all concentrations will increase, since the ratio V_m/k, or the initial slope of rate as a function of concentration, increases (as for *Anabaena flos-aquae* in Table 5.1). The mechanism of this increased potential for phosphate uptake is unknown, but probably involves the synthesis or activation of additional uptake sites. In *Anacystis nidulans* the latter possibility is suggested by the finding that sufficient chloramphenicol to stop growth did not prevent the

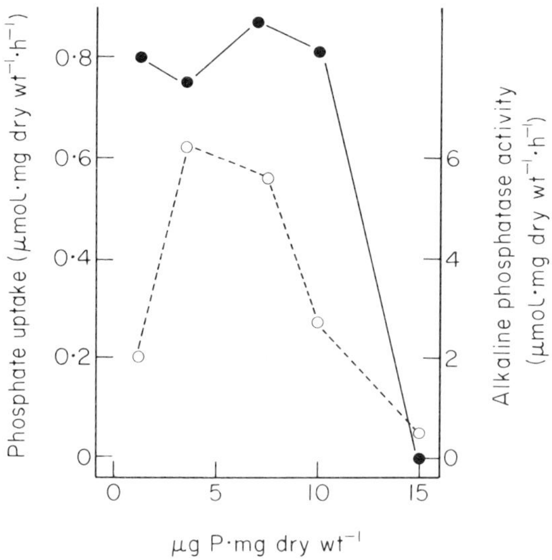

Fig. 5.4. Initial saturated rates of phosphate uptake (●) and alkaline phosphatase activity (○) as a function of internal P concentration of *Anabaena variabilis* (Healey & Hendzel, 1975).

increase in phosphate uptake rates with phosphorus deficiency (Bornefeld *et al.*, 1974).

Like the potential for phosphate uptake, cell-surface alkaline phosphatase activity increases during phosphorus-limited growth in a variety of cyanobacteria (Table 5.3), but not in all. No phosphatase activity was found associated with phosphorus-deficient *Synechococcus* sp. (Kuenzler, 1965), *Coccochloris* sp. (Kuenzler & Perras, 1965) or *Pseudanabaena catenata* (Healey & Hendzel, 1979a,b). Where phosphatase activity is inducible, increases in activity of five- to ten-fold are commonly seen. Onset of increased activity occurs about the time soluble reactive phosphorus is depleted in batch cultures (Healey, 1973a; Ihlenfeldt & Gibson, 1975b) and at high growth rates in phosphorus-limited chemostats (Healey & Hendzel, 1975). Increasing activity with decreasing internal phosphorus concentrations in *Anabaena* (Bone, 1971a; Healey & Hendzel, 1975; Fig. 5.4) suggests control by internal phosphorus concentration. This is confirmed by the observation that antibiotics which inhibit phosphate uptake also induce acid phosphatase activity in *Anacystis nidulans* in the presence of phosphate to the same extent that phosphorus deficiency does (Munoz Calvo & Rodriguez-Lopez, 1978). Growth with various organic phosphorus sources results in alkaline phosphatase activities intermediate between those of phosphorus-sufficient and phosphorus-deficient cultures of *A. nidulans* (Ihlenfeldt & Gibson, 1975b).

Both the increased potential for phosphate uptake at low phosphate concentrations and increased extracellular phosphatase activities can be interpreted as adaptive responses to phosphorus limitation by increasing access to external phosphorus.

5.2.4 *Variation in morphology and ultrastructure with phosphorus deficiency*

Changes in the vegetative morphology of cyanobacteria with phosphorus deficiency include elongated and very short cells in *Plectonema boryanum* (Jensen & Sicko, 1974), and elongated trichomes (failure to release hormogonia) in some Rivulariaceae (Guerts van Kessel *et al.*, 1977). Phosphorus deficiency induces increased hair formation or a narrowing and lengthening of apical cells in several Rivulariaceae (Sinclair & Whitton, 1977a). This is accompanied by a thickening and often a colouring of the sheath. These changes begin before the growth rate is markedly depressed and while polyphosphate bodies are still present, suggesting the structural changes may be an adaptation to increase absorption rather than simply degeneration.

The occurrence of specialized structures—akinetes and heterocysts—are also subject to variation with phosphorus deficiency. Phosphorus deficiency appears to induce akinete formation in *Anabaena cylindrica* (Wolk, 1965b), *Aphanizomenon flos-aquae* (Gentile & Maloney, 1969) and *Gloeotrichia ghosei* (Sinclair & Whitton, 1977a). However, akinete formation can occur up to 4–12% without phosphorus deficiency in natural populations of *Anabaena* spp. (Rother & Fay, 1977, 1979b). Addition of phosphorus to phosphorus-deficient *A. cylindrica* can cause germination of akinetes (Wolk, 1965b). Formation of heterocysts is suppressed by phosphorus deficiency in some but not all heterocystous cyanobacteria (Ogawa & Carr, 1969; Healey, 1973a; Healey & Hendzel, 1975; Guerts van Kessel *et al.*, 1977).

The major ultrastructural change in phosphorus deficiency is the loss of polyphosphate bodies (Jensen & Sicko, 1974; Sinclair & Whitton, 1977a), but expanded thylakoids (Jensen & Sicko, 1974) and loss of gas vacuoles (Reynolds & Walsby, 1975) have also been reported.

5.3 Recovery from phosphorus deficiency

5.3.1 *Rapid phosphate uptake and polyphosphate storage*

Addition of phosphate to phosphorus-deficient cyanobacteria is followed by a series of transitional stages before phosphorus-sufficient growth is resumed. The first is phosphate uptake, which occurs at a rapid rate and can proceed until the phosphorus content of the cells greatly exceeds that of exponentially growing cells. This can be illustrated by the measurement of phosphorus debt

Table 5.4. Increasing ability of *Anabaena variabilis* to store phosphorus as it becomes phosphorus deficient (Healey, 1973a)

P depleted from the medium (h)	SRP* (μM)	Initial cell P (μg mg dry wt^{-1})	P taken up (μg mg dry wt^{-1})	Final cell P (μg mg dry wt^{-1})
−30	3	15	2	17
−7	0.7	15	12	27
20	0.0	10	19	29
40	0.0	5	31	36
60	0.0	5	40	45

* SRP = soluble reactive phosphorus.

during the growth of *Anabaena variabilis* into phosphorus deficiency (Table 5.4). Here, the phosphorus debt is defined as the amount of phosphorus taken up by a unit of biomass during 24 h of darkness. This is sufficient time for the rate of phosphate uptake by *A. variabilis* to fall to zero under conditions where growth is prevented (Healey, 1973a). As *A. variabilis* becomes more phosphorus deficient it acquires the ability to take up and store phosphorus in excess of the phosphorus content required to support its maximum growth rate. Thus, the term phosphorus debt is somewhat misleading, since, as reported in Table 5.4, it represents both the amount of phosphorus required to restore the phosphorus content of exponentially growing cells (actual debt) and storage in excess of that (luxury consumption). Similar results have also been reported for *Plectonema boryanum* (Sicko-Goad & Jensen, 1976) and *Synechococcus* sp. (Grillo & Gibson, 1979). However, this phenomenon may not be general among cyanobacteria, since Batterton and Van Baalen (1968) found the phosphorus content of phosphorus-deficient *Anacystis nidulans* to be restored more slowly.

The phosphorus rapidly taken up in excess of immediate needs is channelled largely into storage as polyphosphate bodies. Labelled phosphate given to phosphorus-deficient *Synechococcus* or *Anacystis* spp. appears first in an acid-soluble pool but moves rapidly into polyphosphate, which accumulates (Niemeyer & Richter, 1969; Grillo & Gibson, 1979). During phosphate uptake by phosphorus-deficient cyanobacteria, polyphosphate bodies form in excess of those found in exponentially growing cells (Jensen, 1968; Stewart *et al.*, 1978). Accumulation of polyphosphate and formation of polyphosphate bodies is presumably mediated by the interaction of polyphosphate synthetase, whose specific activity in *Synechococcus* increases with phosphorus deficiency, and polyphosphatase, whose activity is unaffected by phosphorus deficiency (Grillo & Gibson, 1979). The only other phosphorus compound for which a rapid increase has been recorded is ATP (Batterton & Van Baalen, 1968; Stewart & Alexander, 1971).

The deposition of polyphosphate bodies involves the appearance of electron transparent areas in the nucleoplasm, the formation of porous structures here and the deposition of polyphosphate about and on this structure (Jensen, 1969). Besides phosphate, these bodies contain calcium (Sicko-Goad *et al.*, 1975; Stewart *et al.*, 1978; Lawry & Jensen, 1979), which can be partially replaced by other metals when these are available during phosphate uptake (Crang & Jensen, 1975). One to several polyphosphate bodies can form in each cell, the largest reaching a diameter of 1.5 μm (Jensen & Sicko, 1974). Formation of polyphosphate bodies in cyanophycin and polyhedral bodies has also been reported (Dierksheide & Pfister, 1973; Jensen & Sicko, 1973). Polyphosphate bodies are not formed in heterocysts (Talpasayi, 1963).

The phosphorus content of cyanobacteria is elevated, and polyphosphate bodies are formed, also as a result of sulphur deficiency, but generally not of other (Mg, Ca, Fe, Mo) deficiencies (Lawry & Jensen, 1975; Lawry, 1976; Sinclair & Whitton, 1977a; Vaillancourt *et al.*, 1978; Lawry & Jensen, 1979).

During the rapid uptake of phosphate by phosphorus-deficient cells, both the potential for rapid phosphate uptake and polyphosphate synthetase activity decrease before and as growth resumes (Healey, 1973a; Grillo & Gibson, 1979). This is also seen in the gradual decline in saturated rate of phosphate uptake during measurement of phosphorus debt (Healey, 1973a; Healey & Hendzel, 1976). Alkaline phosphatase activity, on the other hand, appears to be lost by dilution on the resumption of growth (Healey, 1973a; Ihlenfeldt & Gibson, 1975b).

5.3.2 Lag before resumption of growth

Addition of phosphate to phosphorus-deficient cultures often results in no immediate effect on photosynthetic rate (Stewart & Alexander, 1971; Healey, 1979), although this can be variable (Keenan & Auer, 1974). Dark respiration (Stewart & Alexander, 1971; Healey, 1979) and N_2 fixation (Stewart *et al.*, 1970; Stewart & Alexander, 1971) show rapid increases on phosphorus addition. Growth rate does not immediately respond to phosphorus addition; a lag of several hours during which phosphate uptake occurs, and which can increase with the severity of phosphorus deficiency, takes place before the growth rate begins to increase (Healey, 1973a; Ihlenfeldt & Gibson, 1975b). Recently absorbed phosphate moves out of polyphosphate into RNA and other cellular phosphorus components during this lag period (Grillo & Gibson, 1979). The reason for delays in increases in photosynthetic and growth rates probably involves the large changes in composition involved in recovery from phosphorus deficiency, as has been reported for a eukaryotic alga (Healey, 1979).

5.4 Ecological aspects

5.4.1 Phosphorus nutritional status of natural populations

The characteristics of phosphorus-deficient cyanobacteria outlined in 5.2 provide a means of assessing the degree of phosphorus deficiency in natural populations. To date, the phosphorus content and alkaline phosphatase activity of phytoplankton populations have been most widely used. On the basis of the results in Tables 5.2 and 5.3, the values in Table 5.5 provide a reasonable guide for identifying phosphorus deficiency in natural populations of cyanobacteria. In general, these values are quite similar to values suggested for microscopic algae in general (Healey, 1975; Healey & Hendzel, 1979a).

Observations on populations dominated by cyanobacteria show a great diversity of severity of phosphorus deficiency (Table 5.6). Several, such as blooms in Lake Shagawa, several British meres, the Dnieper reservoirs, and some other eutrophic lakes show little or no indication of phosphorus deficiency. Stewart (1971) used lack of stimulation of N_2 fixation by phosphate addition and the presence of polyphosphate bodies to indicate lack of phosphorus deficiency in cyanobacterial blooms in some Scottish lochs. In several lakes in central North America, summer cyanobacterial blooms develop moderate to severe phosphorus deficiency. In some of these, phosphorus deficiency is lacking or slight in spring or early in bloom formation, and becomes more severe later (Table 5.6; Heath & Cooke, 1975). Cyanobacteria with N_2 fixing ability can become more phosphorus deficient than accompanying non-N_2 fixers (Fitzgerald, 1969; Mague *et al.*, 1977); presumably N_2 fixation prevents them from becoming nitrogen deficient, thus allowing phosphorus deficiency to develop further. Variations in phosphorus uptake ability have been used to supplement phosphorus content and phosphatase activity in identifying phosphorus deficiency in natural

Table 5.5. Approximate division points between P-sufficient and P-deficient cyanobacteria on the basis of cellular P and alkaline phosphatase

	P-deficient	P-sufficient
Total cellular P (μg mg dry wt^{-1})	<4	>7
Boiling water extractable P (μg mg dry wt^{-1})	<2	>3
Alkaline phosphatase activity		
p-nitrophenyl phosphate substrate		
μmol mg dry wt^{-1} h^{-1}	>2	<2
o-methyl fluorescein phosphate substrate		
μmol mg dry wt^{-1} h^{-1}	>0.1	<0.02
μmol μg ATP^{-1} h^{-1}	>0.1	<0.02

Table 5.6. Indicators of P nutritional status in natural populations of cyanobacteria

Dominant cyanobacteria	Water body	P-content*		Phosphatase*		Reference
Anabaena circinalis	British mere	4–16	T			Reynolds (1972)
Microcystis aeruginosa	British mere	9–23	T			Reynolds (1973)
Anabaena spp.	British mere	3–17	E	1–2	a	Rother & Fay (1977)
Anabaena spp.	British mere	1–3	E	< 1.0	a	Rother & Fay (1977)
Oscillatoria sp.	Lough Neagh	> 4	T			Gibson & Stevens (1979)
Oscillatoria agardhii	Dutch lake	11–14	T			Zevenboom & Mur (1978)
Oscillatoria agardhii	Lake Norrviken	4.5–6	T†			Ahlgren (1977)
		3	T			
Microcystis aeruginosa	Dnieper reservoirs	6.5–9.0	T			Pomiluyko &
Aphanizomenon flos-aquae		6.9–7.6	T			Stetsenko (1972)
Microcystis aeruginosa	Central Wisconsin	5.2–9.8	T			Gerloff & Skoog
	lakes	1.6–4.0	T			(1954)
						Gerloff & Skoog (1957)
Microcystis sp.	Central Wisconsin	2.4, 1.3	E	0.60, 0.40	a	Fitzgerald &
Lyngbya borgei	lakes	3.5	E	0.85	a	Nelson (1966)
Anabaena sp.		2.5	E	0.60	a	
mixed cyanobacteria		2.1	E	0.70	a	
Anabaena sp.	Lake Mendota	0.6	E	18	a	Fitzgerald (1969)
Aphanizomenon sp.		1.0	E			
Microcystis sp.		2.2, 4.2	E	2.2	a	
Anabaena and *Aphanizomenon* spp.	Lake Shagawa	4.6–12	E			Megard & Smith (1974)
Aphanizomenon flos-aquae	Manitoba lakes	6–9	T‡			Healey & Hendzel
		2–3	T			(1976)
Anabaena and	Manitoba lakes	8–9	T§	0.15	b	Healey & Hendzel
Aphanizomenon spp.		3–9	T	0.1–1.0	b	(1980) (and unpubl.)
mixed cyanobacteria	Pond periphyton	3.4	E	23	a	Hooper-Reid & Robinson (1978)
Oscillatoria thiebautii	Pacific	6.3	T			Mague *et al.*, (1977)
Oscillatoria thiebautii	Atlantic			0.50	b	McCarthy & Carpenter (1979)

* P content, either total (T) or boiling water extractable (E) as μg mg dry wt^{-1}; alkaline phosphatase activity with p-nitrophenyl phosphate as substrate as μmol P mg dry wt^{-1} h^{-1} (a) or with o-methyl fluorescein phosphate as substrate as μmol P μg ATP^{-1} h^{-1} (b).
† Values found before sewage diversion from lake, followed by those found after diversion.
‡ Early bloom values followed by those found late in the bloom.
§ Values found in spring populations followed by those found in summer populations.

populations of cyanobacteria (Healey & Hendzel, 1976, 1980; McCarthy & Carpenter, 1979; Gibson & Stevens, 1979).

5.4.2 *Responses of N_2-fixing cyanobacteria to variations in N/P loading ratio*

Changes in phytoplankton species composition resulting from tube and whole lake enrichments with varying amounts of nitrogen and phosphorus led to the suggestion that low N/P loading ratios (below 5, by weight) can be expected to favour N_2-fixing cyanobacteria (Schindler & Fee, 1974; Schindler, 1975, 1977). This hypothesis is supported by some experiences elsewhere. Enrichment of a large enclosure in a bay of Lake Ontario with phosphorus alone resulted in a bloom of N_2 fixers, while enrichment with both nitrogen and phosphorus ($N/P = 13$) resulted in a bloom of *Microcystis* (Lean *et al.*, 1975). Some lakes with N/P loading ratios greater than 5 support summer blooms of non-N_2-fixing cyanobacteria (Bindloss *et al.*, 1972; Brunskill, 1973). However, other observations show that low N/P loading ratios do not necessarily result in succession to N_2-fixing cyanobacteria. Enrichment of artificial enclosures or ponds with phosphorus alone or with nitrogen and phosphorus at low ratios has often supported increased phytoplankton crops, including cyanobacteria, without a major component of N_2 fixers (Boyd, 1973; Liao, 1977; DeNoyelles & O'Brien, 1978). A Scottish loch with a very high N/P loading ratio, on the other hand, supported blooms dominated by the N_2 fixers *Anabaena* and *Aphanizomenon*, as well as *Microcystis* (Stewart, Tuckwell & May, 1975). Some of these apparent discrepancies may be due to the fact that loading rates to a water body may not accurately reflect the rates at which nutrients are presented to the phytoplankton, since different nutrients can be regenerated from dead and dying particles and from the sediments at different rates (Schindler, 1976). But at least some must be because species succession is only in part controlled by nutrient competition and the ability to fix N_2 can be an advantage only where combined nitrogen is being supplied at limiting or near-limiting rates. This can be illustrated by considering two situations which on the surface seem to support the hypothesis that increased N/P ratios favour the replacement of N_2 fixers by other algae. In both, rises in dissolved N/P ratios were accompanied by non-N_2 fixers replacing N_2 fixers (Sze, 1975; Vanderhoef *et al.*, 1974) but, in both these situations, fixed nitrogen was always present at concentrations expected to saturate growth, so it is difficult to argue that in these situations N_2 fixation was sufficient advantage to maintain the N_2 fixers at low dissolved N/P ratios. It is more likely that some other change accompanying the change in N/P ratio was instrumental in the change in species composition.

5.4.3 Phosphorus sources for cyanobacterial blooms

Eutrophication, with increased loading of phosphorus and other nutrients, is generally accompanied by increased dominance of the phytoplankton by cyanobacteria (Ahl, 1975; Schindler, 1975) and reduction in nutrient loading by the opposite (Cloern, 1976; Gelin & Ripl, 1978). However, it is difficult to argue that cyanobacteria require or are favoured by the higher nutrient concentrations directly because the cyanobacterial blooms often occur when external nutrient concentrations are low or undetectable. On the other hand, rates of nutrient supply may be more important than the actual concentrations. However, in the most eutrophic situations, green algae often replace cyanobacteria (Hammer, 1969; Bush & Welch, 1972; deNoyelles & O'Brien, 1978).

The growth of cyanobacterial populations can be accompanied by major declines in soluble reactive phosphorus (Edmondson, 1969; Wood & Gibson, 1973), suggesting that a part of the phosphorus requirement is derived from previously accumulated dissolved phosphate. But more often changes in dissolved orthophosphate are insufficient to account for the growth of blooms. Here the balance of the phosphorus requirement can be supplied by regeneration (Reynolds, 1976), the action of phosphatases on phosphomonoesters (Heath & Cooke, 1975) and from internally stored phosphorus (Megard & Smith, 1974; Healey & Hendzel, 1976; Stewart *et al.*, 1978).

CHAPTER 6
SUPRAMOLECULAR ORGANIZATION AND COMPOSITION OF MEMBRANES

J. R. GOLECKI & G. DREWS

6.1 Types of membranes and their functions

The photosynthetic apparatus of cyanobacteria is localized on intracytoplasmic membranes (ICM). The organization of these membranes is very similar to that of the chloroplast membranes of higher plants. The structural and functional homology of the photosynthetic membranes, especially between some Rhodophyta and cyanobacteria (Neushul, 1971), has supported the hypothesis of Mereschkowsky (1905), that the ancestors of chloroplasts might have been cyanobacteria becoming endosymbiotic (Chapter 12). The term 'thylakoids' was introduced by Menke for photosynthetic membranes of chloroplasts (1961a) and cyanobacteria (1961b). Thylakoids are flattened membranous sacs carrying photosynthetic activities. In contrast to the thylakoids of chloroplasts of higher plants the thylakoids of cyanobacteria are not enclosed by membranes forming cell organelles (Lang & Rae, 1967). They are arranged in the peripheral space of the cell. The pure morphological term, intracytoplasmic membranes, originally introduced for photosynthetic bacteria, can also be used for the thylakoids of cyanobacteria.

The cytoplasmic membrane (plasmalemma) of the cyanobacteria surrounds the protoplast. In contrast to the photosynthetic bacteria there seems to be no clear morphogenetic relationship between cytoplasmic membrane and the ICM in cyanobacteria. The function and composition of the cytoplasmic membrane is not characterized. A respiratory chain coupled to oxidative phosphorylation has been demonstrated but its location on the cytoplasmic membrane is less clear. Inducible transport systems for the uptake of substrates are not well characterized (see Chapter 3). However, the uptake of organic substrates by some species, the gradient of ions and low molecular weight substances between extra- and intracytoplasmic space and the demonstration of DNA transfer from cell to cell give indirect evidence that the cell envelope of cyanobacteria (cell wall and cytoplasmic membrane) form a diffusion barrier which regulates transport.

Besides the more typical cytoplasmic and thylakoid membranes, mesosome-like structures are observed occasionally (Niklowitz & Drews, 1956; Echlin, 1964a; Allen, 1968b; Edwards *et al.*, 1968; Whitton & Peat, 1969; Ingram & Thurston, 1970; Allen, 1972; Dierkscheide & Pfister, 1973; Van Eykelenburg, 1979). Because such specialized membranes appear infrequently, and their function is unknown, this chapter concentrates on cytoplasmic and thylakoid membranes as such. The cytoplasmic membrane of all cyanobacteria so far examined appears to have the same ultrastructure. The thylakoid membranes, however, show considerable variation in structure and arrangement depending on the species, cell type (vegetative cells, heterocyst, akinete) and culture conditions.

6.2 Arrangement and distribution of thylakoids

The cytoplasmic membrane normally appears appressed to the cell wall as a triple-layered (dark–light–dark; 3.5, 3–4, 3.5 nm) unit membrane (Edwards & Holt, 1968) (Fig. 6.1). In aged cells, and in hypertonic medium, the protoplast is shrunk and the cytoplasmic membrane is wrinkled and partially separated from the cell wall. Cell wall and cytoplasmic membrane are not separated completely, but thin, cordlike structures connect the cell wall and cytoplasmic membrane of the shrunken protoplast (Golecki, unpubl.), a situation similar to that reported for Gram-negative bacteria and called adhesion sites (Bayer, 1968; Mühlradt *et al.*, 1973). Intercellular contact zones of cytoplasmic membranes between neighbouring cells of filamentous cyanobacteria known as microplasmodesmata are discussed in Chapter 13.

Although very sporadic connections between cytoplasmic and intracytoplasmic membranes have been described in unicellular (Allen, 1968b; Edwards *et al.*, 1968) and filamentous cyanobacteria (Pankratz & Bowen, 1963; Schnepf, 1964; Jost, 1965; Smith & Peat, 1967), it has not been demonstrated that intracytoplasmic membranes are formed from the cytoplasmic membrane by invagination as is known to occur in phototrophic

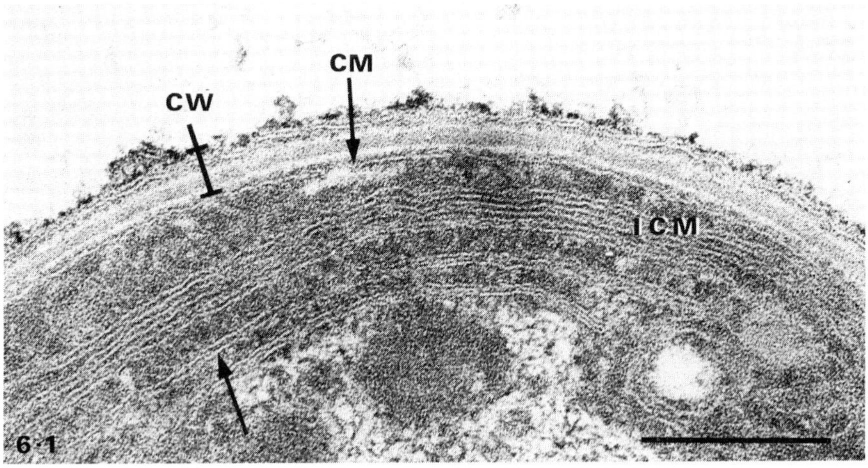

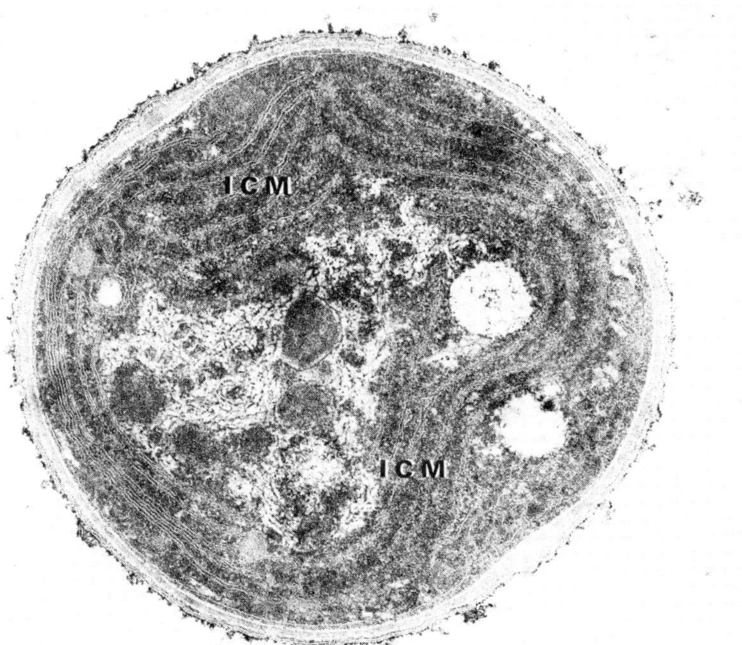

6·2

Fig. 6.1. Section of *Synechocystis* PCC 6714. Cell wall (CW), cytoplasmic membrane (CM) and thylakoids = intracytoplasmic membranes (ICM). The membranes present the appearance of typical 'unit' membranes. The membranes of a single thylakoid are closely apposed to each other and seem to be fused to a single dark line (←). Bar represents (as in all following figures) 200 nm.

Fig. 6.2. *Synechocystis* PCC 6714. Cross-section showing the peripheral localization of parallel-arranged thylakoid membranes.

bacteria (Oelze & Drews, 1972). The rarely observed connections between cytoplasmic and intracytoplasmic membranes may represent only temporary fusions of both membranes. Thus, the growth of the intracytoplasmic membrane system in cyanobacteria seems to be independent from the cytoplasmic membrane (Stanier & Cohen-Bazire, 1977).

In most species the thylakoids are arranged peripherally in three to six layers running parallel to the envelope layer (Figs 6.2; 6.3, 6.4). They form an anastomosing network of concentric shells. This peripheral region has been called chromatoplasm, which is separated from the nucleoplasmic region, the centroplasm (Geitler, 1958). This type of thylakoid arrangement is characteristic of many unicellular and filamentous cyanobacteria, for example *Anacystis nidulans* (Ris & Singh, 1961; Echlin, 1964a; Allen, 1968a). *Synechococcus planctícus* (Drews *et al.*, 1961), *Phormidium* (Drews & Niklowitz, 1956), *Gloeocapsa* (Ris & Singh, 1961), *Anabaena* (Lang & Rae, 1967), *Nostoc muscorum* (Pankratz & Bowen, 1963) and *Chlorogloea fritschii* (Peat & Whitton, 1967). By contrast, in some organisms the thylakoids are orientated perpendicular to the longitudinal cell wall as in *Oscillatoria* (Jost, 1965), *Spirulina platensis*, *S. geitleri* (Busson, 1971) and in *Starria* (Lang, 1977). A radial thylakoid arrangement was demonstrated in *Phormidium retzi* (Niklowitz & Drews, 1957) and *Oscillatoria* (Hagedorn, 1961).

Thylakoids are not always restricted to the cell periphery. In *Chroococcidiopsis* (Komàrek, 1975), *Trichodesmium* (Van Baalen & Brown, 1969) and *Gloeotrichia* (Miller & Lang, 1971) the thylakoid membranes are scattered throughout the cell, in *Trichodesmium* most of the thylakoids are located in the central region of the protoplast. The arrangement of the thylakoids is not strain specific; it can change from cell to cell within the same culture from parallel to a convoluted appearance (Figs 6.5, 6.6).

The amount of thylakoid membranes per cell is variable due to growth conditions (see 6.5) and taxonomic specificity. Two to eight thylakoids in concentric layers have been observed in *Synechocystis* (Fig. 6.2), *Nostoc* (Fig. 6.3) and *Plectonema* (Fig. 6.4), a single peripheral thylakoid membrane is reported from a *Synechococcus* strain by Bailey-Watts *et al.* (1968). The extreme is *Gloeobacter* (Rippka *et al.*, 1974; Guglielmi *et al.*, 1979) a cyanobacterium which lacks thylakoids.

Fig. 6.3. *Nostoc muscorum*. Freeze-fracture preparation of a whole filament with different cross-fractured cells exhibiting up to seven parallel-arranged thylakoid membranes in the cell periphery (←). Arrow in the upper right corner indicates direction of shadowing.

Fig. 6.4. *Plectonema* sp. Parallel-arranged thylakoids and exoplasmic (EF) and plasmic (PF) fracture faces of cytoplasmic and intracytoplasmic membranes.

Fig. 6.5. *Nostoc muscorum*. Cross-fractured cell with convoluted thylakoids. On the exoplasmic fracture faces linear arranged particles are visible (←).

CM
CM
6·3

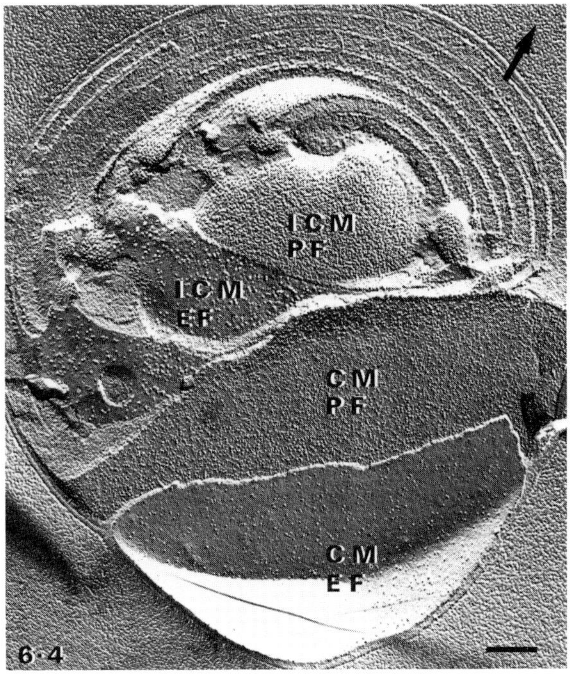
ICM
PF
ICM
EF
CM
PF
CM
EF
6·4

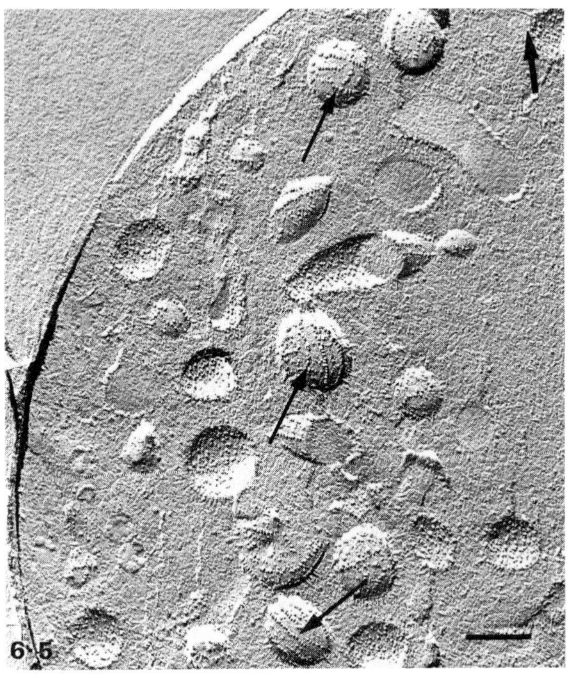
6·5

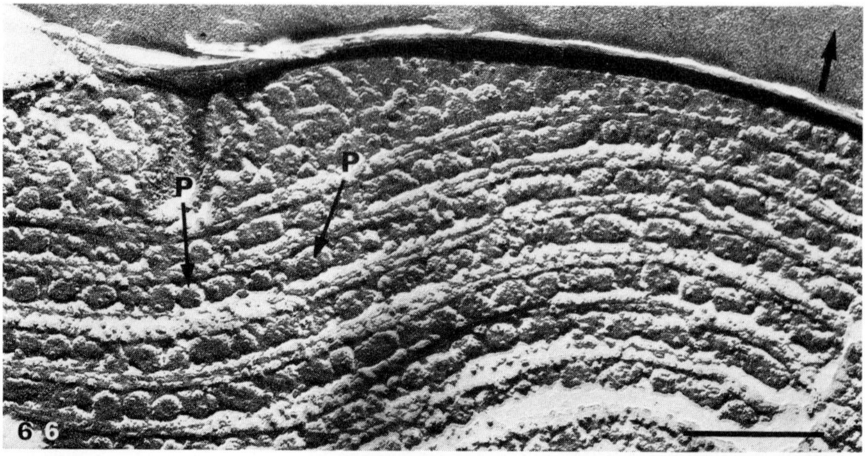

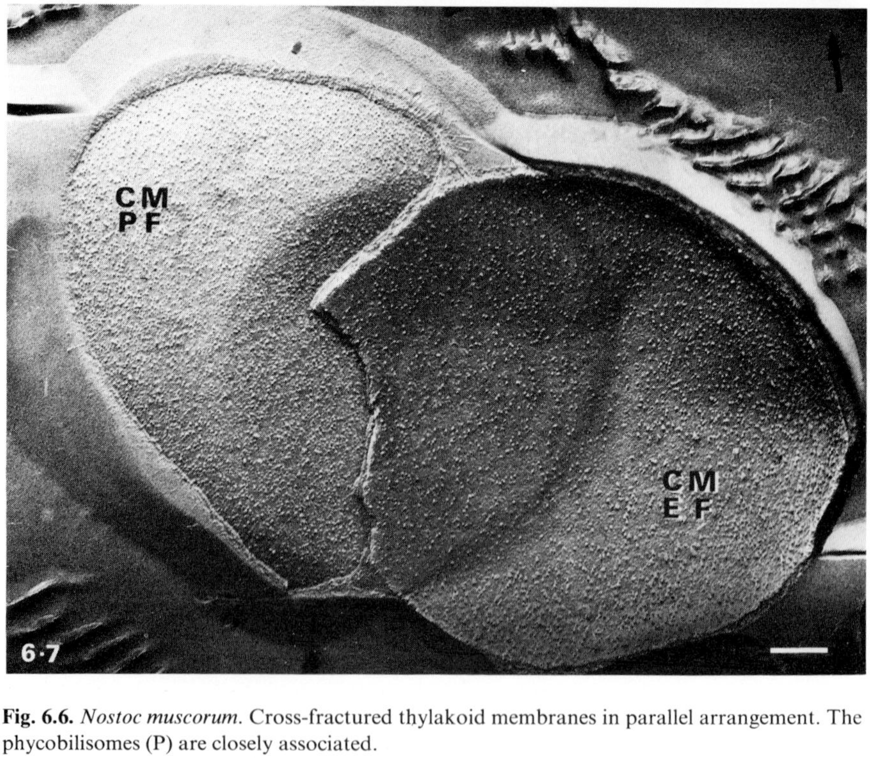

Fig. 6.6. *Nostoc muscorum.* Cross-fractured thylakoid membranes in parallel arrangement. The phycobilisomes (P) are closely associated.

Fig. 6.7. *Anabaena variabilis.* The two corresponding fracture faces of the cytoplasmic membrane with a dense or sparse particle population on the respective plasmic and exoplasmic fracture face (see also Fig. 6.4). No invaginations of the cytoplasmic membranes to form intracytoplasmic membranes are visible. (From Golecki & Drews, 1974)

The ultrastructural appearance of the thylakoid membrane in thin sections depends on fixation (Lefort, 1960; Echlin, 1964b; Pankratz & Bowen, 1963). After osmium tetroxide fixation, a single thylakoid membrane is formed by two parallel electron-dense lines of 3.0 (Lefort, 1960) to 6.5 nm (Pankratz & Bowen, 1963) diameter separated by an electron-transparent space of about 3–5 nm, in extreme 10 nm (Pankratz & Bowen, 1963; Edwards *et al.*, 1968). The membranes of a single thylakoid mostly appear closely apposed to each other (Fig. 6.1). In this case the inner dark lines of the two adjacent membranes appear fused in a single electron-dense line with a greater (5.0–6.0 nm) width (Edwards *et al.*, 1968). But sometimes the two membranes are separated by an electron-transparent zone which may be termed lumen or intrathylakoidal space. The width of this zone can vary, and swollen thylakoids resulting from this are described for different genera of filamentous and, to a lesser extent, unicellular cyanobacteria (Whitton *et al.*, 1971; Ris & Singh, 1961). The variation in the width of the intrathylakoidal space appears to result from the physiological state of the cells and from the osmotic conditions during the fixation procedure (Edwards *et al.*, 1968). Within the intrathylakoidal space different inclusions (Grilli, 1964; Edwards *et al.*, 1968; Jensen & Bowen, 1970) and masses of electron-dense material (Stanier & Cohen-Bazire, 1977) of an unknown nature have been described. On the outer surface of the thylakoids regular rows of electron-dense structures are closely attached. These granules, carrying the light-harvesting phycobiliproteins are known as phycobilisomes (Gantt & Conti, 1966b) and are fully described in Chapter 7.

It is known that thylakoids can fuse with each other and form an anastomosing network (Edwards *et al.*, 1968). In contrast to thylakoids of the chloroplasts in higher plants, the thylakoids of cyanobacteria form stacking regions only to a very limited extent.

During cell division the thylakoids have to be separated and divided into the daughter cells. Although some electron-microscopical studies have indicated that the intracytoplasmic membranes are passively divided by the closing cross wall (Pankratz & Bowen, 1963; Ingram & Thurston, 1970) generally the division of the thylakoids seems to be an active process. Cell division starts by a centripetal growth of cytoplasmic membrane and peptidoglycan layer (Allen, 1968b). The thylakoid cylinder is narrowed at the level of the cross wall by invagination before the septum is formed (Allen, 1968b; Ueda, 1971a), Non-septate mutants of *Agmenellum* contain thylakoids which are grouped into discrete bundles along the longitudinal cell wall without any sign of cell-wall invagination. This suggests that the division of thylakoids is controlled by the division of the chromosome (Ingram & Thurston, 1970).

6.3 Supramolecular architecture of the membrane system

6.3.1 Cytoplasmic membrane

Soon after the introduction of freeze-etching by Steere (1957) and Moor and Mühlethaler (1963) this method was used for the ultrastructural examination of cyanobacteria (Jost, 1965). The great advantage of the method is the avoidance of fixation artefacts and that during fracturing the hydrophobic interior of membranes is exposed (Branton, 1966). The fracture plane of the membrane along the hydrophobic interior exposes two fracture halves, the inner exposing the plasmic fracture face (P face) and the outer exhibiting the exoplasmic fracture face (E face). The two complementary fracture halves of the membrane exhibit intramembrane particles of different sizes. The intramembrane particles are considered to be the structural equivalents of intrinsic proteinaceous complexes. The number of intramembrane particles may be regarded as an indication of the biological activity of the membrane. Membranes having electron transport functions are densely populated with intramembrane particles. It is known from biochemical and physical studies that most membranes are asymmetrically organized, i.e. the functional subunits and the lipids are not equally distributed between the halves of the membranes (Singer & Nicholson, 1972; Singer, 1974; Mühlradt & Golecki, 1975; Verkleij & Ververgaert, 1978). Therefore it is not surprising that the E face and P face of the cytoplasmic membrane show complementary faces of different appearance. The inner leaflet of the cytoplasmic membrane (Figs 6.4 and 6.7), labelled CM PF according to the nomenclature of Branton *et al.*, 1975), is always densely populated with intramembrane particles (2000–3500 particles μm^{-2}), whereas the outer leaflet (Figs 6.4 and 6.7, labelled CM EF) has a lower particle density (500–1000 particles μm^{-2}; Giddings & Staehelin, 1978, 1979; Golecki, 1979a, unpubl.). The particle density (number of particles per membrane area) on fracture faces of the cytoplasmic membrane seems to be influenced by the cell cycle as indicated by results on *Anabaena cylindrica* (Giddings & Staehelin, 1978). Just before cell division the number of intramembrane particles is approximately 42–46% higher than in non-dividing cells. The particles on the P face are in the range of 5–10 nm in diameter with a single maximum at approximately 7 nm. The particles on the E face, however, are on average larger and fall into several size categories (Giddings & Staehelin, 1978; Golecki, 1979a, unpubl.). A correlation of specific membrane functions to distinct particle-size classes is not possible at present (see also Chapter 14). Due to the smaller number of intramembrane particles the E face of the cytoplasmic membrane presents a smooth appearance. Pits and holes on the E face are complementary to particles of the P face. Indentations or protuberances on the fracture faces of the cytoplasmic membrane, which indicate connections to intracytoplasmic membranes, were very seldom demonstrated in any of the cyanobacteria we examined (Golecki, 1977, 1979a) (Fig. 6.7).

6.3.2 *Thylakoid membrane*

The thylakoid structures of cyanobacteria are very similar to the stroma region of higher plants, and especially of red-algal thylakoids (Neushul, 1971). Thylakoid membranes of cyanobacteria present inner fracture faces exhibiting a characteristic arrangement of intramembrane particles both on the plasmic and exoplasmic fracture faces. Similar arrangements were exhibited in strains of unicellular (Lefort-Tran *et al.*, 1973; Lichtlé & Thomas, 1976; Golecki, 1979a), filamentous (Giddings & Staehelin, 1979; Golecki, 1979b) and endosymbiotic cyanobacteria (Lefort-Tran *et al.*, 1973).

The plasmic fracture face of the thylakoid membranes, that is the membrane leaflet adjacent to the plasma or stroma side, exhibits a high particle density 3000–5000 particles μm^{-2} (Lichtle & Thomas, 1976; Armond & Staehelin, 1979; Giddings & Staehelin, 1979; Golecki, 1979a). The area of densely packed particles is regularly interrupted by parallel arranged grooves free of particles. These grooves have a spacing of 35–50 nm (Fig. 6.8). The particles are in the range of 2.5–12 nm in diameter with mean values of 7–9 nm. The corresponding exoplasmic fracture face (ICM EF in Fig. 6.9), that means the membrane leaflet adjacent to the intrathylakoidal space, exposes regular rows of particles in parallel alignment (Jost, 1965; Bourdu & Lefort, 1967; Lefort-Tran *et al.*, 1973; Lichtlé & Thomas, 1976; Giddings & Staehelin, 1979; Golecki, 1979a,b). Particle size histograms show that these aligned particles fall into one size category of 10 nm. The distances between the parallel rows are in the range of 35–50 nm and up to 70 nm. The particle density on the exoplasmic fracture face of the thylakoid membrane is usually lower than on the corresponding plasmic fracture face, but it is higher in thylakoids than on the corresponding E face of the cytoplasmic membrane. This was observed in *Synechococcus lividus* [3832 particles μm^{-2} vs. 1077 particles μm^{-2} (Golecki, 1979a)], *Nostoc* [1961 particles μm^{-2} vs. 1543 particles μm^{-2} (Golecki, unpubl.)], *Plectonema* [1346 particles μm^{-2} vs. 846 particles μm^{-2} (Golecki, unpubl.)] and in *Anabaena* [939 particles μm^{-2} vs. 700 particles μm^{-2} (Giddings & Staehelin, 1979)]. The higher particle density on the thylakoid membranes in comparison to the cytoplasmic membrane agrees with results for intracytoplasmic membranes of photosynthetic bacteria (Golecki *et al.*, 1979; Golecki & Oelze, 1980) and points to the photosynthetic activity of the intracytoplasmic membranes. The characteristic arrangement of parallel particle rows and complementary grooves on the E and P faces respectively of thylakoids corresponds to the arrangement of the phycobilisomes on the outer

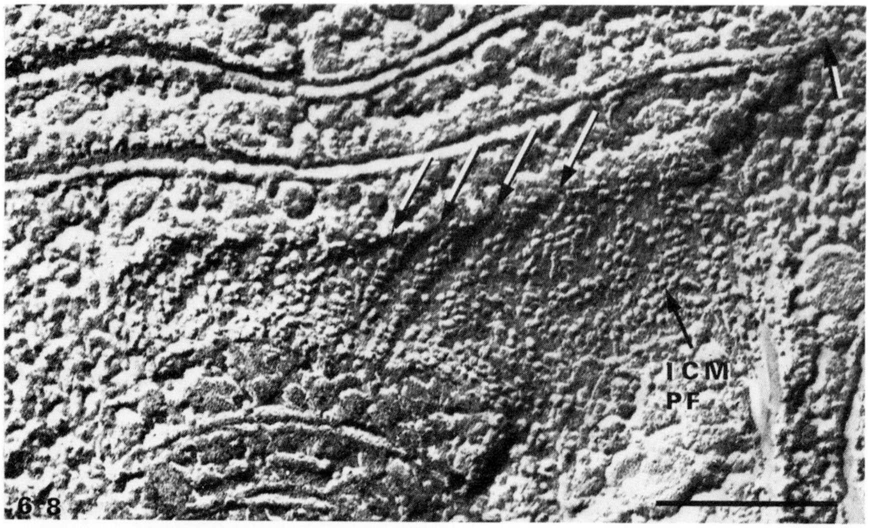

6·8

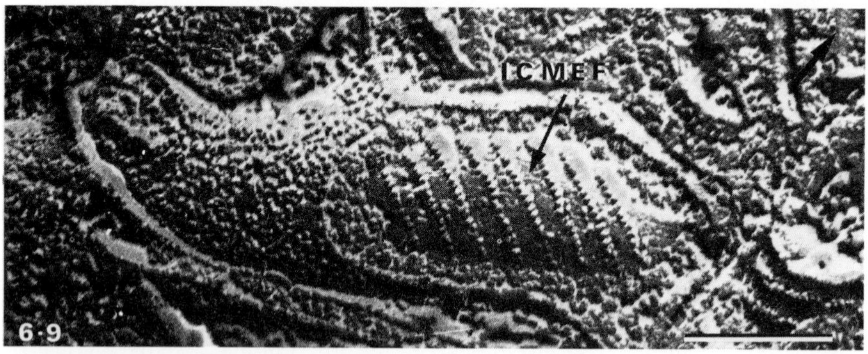

6·9

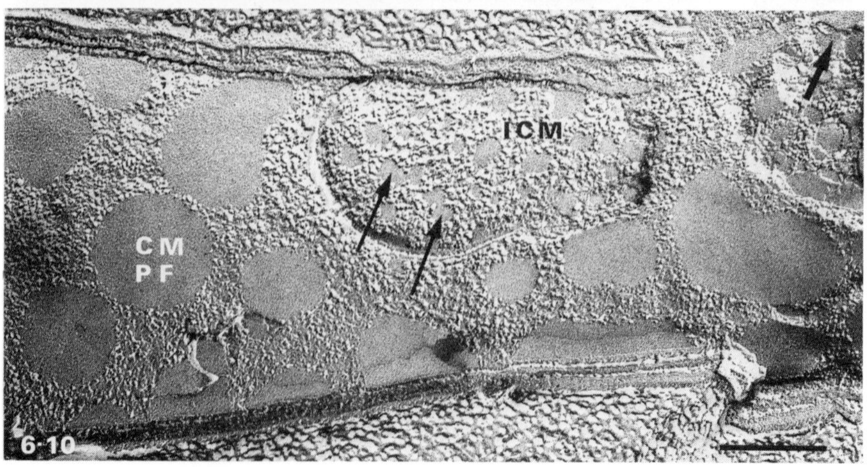

6·10

plasmic surface of the thylakoids (see Figs 6.6, 6.8 and Chapter 7). The rows of phycobilisomes have the same spacing as the particle rows (35–50 nm) and are localized just opposite to the particle rows. The close correspondence of particle rows and phycobilisomes (Bourdu & Lefort, 1967; Lichtle & Thomas, 1976) has suggested that the 10 nm particles on the E face of the thylakoid membrane represent structural equivalents of the photosystem II apparatus (Giddings & Staehelin, 1979; Golecki, 1979a,b).

6.4 Chemical composition of the membrane system

6.4.1 Lipids and fatty acids

The major components of thylakoids of cyanobacteria are proteins, lipids, chlorophyll *a* and carotenoids. The main functional subunits of thylakoids belong to the photosynthetic apparatus which will be described in Chapter 8. Most analyses of proteins and lipids deal with unpurified membrane fractions or whole cells. Analyses on highly purified outer membranes, cytoplasmic and intracytoplasmic membranes have yet to be carried out. However, as demonstrated by Hirayama (1967), the lipid composition of the membrane fraction is very similar to that of the whole cell. Therefore the overall composition of lipids may be representative also of the membrane fraction, because the mass of the isolated thylakoids is much higher as compared to the mass of the isolated cytoplasmic and outer membranes.

The major lipids of cyanobacteria are monogalactosyl diglyceride, digalactosyl diglyceride, sulphoquinovosyl diglyceride and, to a minor degree, the phospholipid phosphatidyl glycerol (Nichols *et al.*, 1965; Hirayama, 1967; Nichols & Wood, 1968a; Fork *et al.*, 1979). The three glycolipids are also present in the chloroplast membranes of higher plants, but they are absent in intracytoplasmic membranes of photosynthetic bacteria (Rhodospirillineae). Membranes of photosynthetic bacteria contain predominantly phospholipids (Kenyon, 1978), but green bacteria (Chlorobiineae) contain monogalactosyl diglyceride especially in the chlorosomes (Constantopoulos & Bloch, 1967;

Fig. 6.8. *Nostoc muscorum*. Plasmic fracture face of a thylakoid exhibiting several particle-free grooves among a dense particle population.

Fig. 6.9. *Anabaena variabilis*. Exoplasmic fracture face of a thylakoid membrane with parallel rows of intramembrane particles possibly representing structural equivalents of PS II. Spacing of the particle rows 35–50 nm.

Fig. 6.10. *Synechococcus lividus*, a thermophilic strain growing at 52°C. Fracture faces of cytoplasmic and intracytoplasmic membranes with a separation of proteins (= particles) and lipids (smooth areas) as a result of freezing the cells from below the phase transition temperature of 30–35°C. (From Golecki, 1979a).

Cruden & Stanier, 1970). Lecithin, phosphatidyl ethanolamine and phosphatidyl inositol are not present in cyanobacteria (Nichols *et al.*, 1965; Nichols & Wood, 1968b).

The fatty-acid patterns of cyanobacteria show great diversity. Some of the cyanobacteria, mostly unicellular, have only saturated (myristic, palmitic and stearic acid) and mono-unsaturated (palmitoleic, trans-3-hexadecanoic and oleic acid) fatty acids, that are characteristic of the other prokaryotes. Occasionally traces of linoleic acid ($C_{18:2}$) have been detected. Other cyanobacteria, mostly filamentous, have fatty-acid patterns similar to chloroplasts of higher plants with a high proportion of polyunsaturated C_{18} fatty acids, especially linolenic acid ($C_{18:3}$) either the α or the γ isomer (Levin *et al.*, 1964; Nichols & Wood, 1968b; Kenyon, 1972; Kenyon & Stanier, 1970). In some strains linoleic ($C_{18:2}$) dominates $C_{18:3}$. There is no strict correlation between the presence of polyunsaturated fatty acids and the cellular organization in filamentous cyanobacteria. Several strains among the unicellular cyanobacteria contain large amounts of polyunsaturated fatty acids (Kenyon, 1972).

6.4.2 *Pigments and elements of electron transport chains*

The only species of chlorophyll detected so far in cyanobacteria is chlorophyll *a*. It is localized in the thylakoid membrane and forms, with its carrier protein, intramembrane complexes. Most of the chlorophyll *a* (90%) is localized in photosystem (PS) I whereas PS II contains only a small fraction (10%) (Mimuro & Fujita, 1977; Tel-Or & Malkin, 1977). The P 700–chlorophyll *a*-protein complex has been isolated from *Phormidium luridum* (Thornber, 1969; Dietrich & Thornber, 1971).

The composition of the P 700–chlorophyll *a*–protein complex in *Phormidium* was 1 mol P 700/40 mol chlorophyll *a*/1.3 mol β-carotene/314 000 g protein (Thornber *et al.*, 1976) which is supported by a chlorophyll *a* to P 700 ratio of 40:1 in PS I of *Synechococcus* (Newman & Sherman, 1978).

The P 700–chlorophyll *a*–protein–SDS complex isolated from *Phormidium luridum* by SDS treatment is photochemically active, has an apparent molecular mass of 11×10^4 daltons and contains two polypeptides of 48 and 46 kD (Hunter & Thornber, 1976).

Among a variety of group-specific carotenoids, especially carotenoid glycosides (Hertzberg *et al.*, 1971), the most common carotenoids of the cyanobacteria are β-carotene accompanied by either or both of the oxycarotenoids zeaxanthin (3,3′-dihydroxy-β-carotene) and echinone (4-keto-β-carotene). Also myxoxanthophyll (1′,2′dihydro-3′,4′-didehydro-3′,1′-dihydroxy-γ-carotene) has been demonstrated as a major xanthophyll carotenoid among numerous investigated strains. Oscillaxanthin (a carotenoid glycoside esterified with rhamnose) provides a useful marker for the detection of past cyanobacterial blooms, owing to its accumulation in lake sediments. See also Stransky and Hager (1970) and Hertzberg *et al.* (1971) and, for a full account of the organization of the photosynthetic unit, Chapters 7 and 8.

6.5 Influence of external factors on development of the membrane system

6.5.1 *Light fluency*

Under low light intensities (0.2–1.0 klx) the number of thylakoids per cell is much higher, compared to high light intensities (7.0–20.0 klx), as demonstrated for *Anacystis nidulans* (Allen, 1968a) and *Chlorogloea fritschii* (Peat & Whitton, 1967; Findley *et al.*, 1970). Under low light three thylakoid layers per cell have been observed in *Anacystis* and four to six in *Chlorogloea.* The area of thylakoids per cell after growth in high light intensities is 62% of that of cells grown at low light intensity. The content of chlorophyll and phycocyanin is directly proportional to the amount of thylakoid membranes (Allen, 1968a) and is reduced in high light to two-thirds. In *Chlorogloea* during high light conditions the thylakoids lose their regular arrangement. They show swelling, separation and vesiculation. The fragmented membranes are no longer parallel to each other and are scattered throughout the cell.

Both the radiation flux and the wavelength of the incident light influence the formation of intra-cytoplasmic membranes. This is indicated by chromatic adaptation (Engelmann, 1902; Tandeau de Marsac, 1977; see Chapters 2 and 7) and also by the variation of the chlorophyll to phycobilin ratio (Boresch, 1922; Jones & Myers, 1965; Evans and Allen, 1973; Tandeau de Marsac, 1977).

In the absence of CO_2 and in the presence of oxygen and high light intensities, cells of *Anacystis nidulans* are reported to lose their intracytoplasmic membranes and become devoid of chlorophyll and phycocyanin (Peschek & Schmetterer, 1978). These bleached cells are still viable and after a 3 day lag-phase they regained their normal pigment content when provided with CO_2.

6.5.2 *Nutritional effects*

Most of the cyanobacteria are obligately photoautotrophic organisms. However, a few species are able to grow in darkness chemoheterotrophically (Fay, 1965; Rippka, 1972; see Chapter 3). Cells of *Chlorogloea fritschii* grown heterotrophically in the dark still contain intracytoplasmic membranes (Peat & Whitton, 1967). The thylakoids of such cells, the culture having been maintained for 3 years in the dark, have changed their originally parallel arrangement to a scattered and irregular pattern (Baulina *et al.*, 1978). No significant ultrastructural changes were observed in an endosymbiotic *Nostoc* strain after heterotrophic growth in the dark (Hoare *et al.*, 1971). This may be due to the relatively short dark-cultivation period of 1–3 weeks. The thylakoids of dark-grown cells are still pigmented. The content of phycocyanin is reduced to 5–10% of photosynthetic cells (Hoare *et al.*, 1971; Evans *et al.*, 1978; Joset-Espardellier *et al.*, 1978) and the activity of PS II is lost (Evans *et al.*, 1978). *Plectonema*, however, retains its phycocyanin after dark growth (Raboy *et al.*, 1976).

Among the nutritional factors the concentration and form of nitrogen in the medium has a strong influence on membrane biogenesis. During nitrogen starvation many cyanobacteria are able to fix molecular nitrogen by a specialized cell type, the heterocyst (see Chapter 14). During cell differentiation the parallel arranged thylakoids become contorted (Lang & Fay, 1971) and arranged in concentric whorls and 'honeycomb'-like structures (Lang, 1965). The contorted membranes are concentrated in the region of the pore channel. Finally these membranes disintegrate and an electron-opaque material rich in arginine (Fogg, 1951) is accumulated, which may represent the reserve material cyanophycin (Simon, 1971).

The supramolecular organization of the membrane system is also modified during heterocyst differentiation, as demonstrated by freeze-fracture electron microscopy on *Anabaena cylindrica* (Giddings & Staehelin, 1978, 1979) and *Nostoc muscorum* (Golecki, unpubl.). The particle density (*i.e.* the number of intramembrane particles per membrane area) in the cytoplasmic membrane is reduced on the plasmic fracture face by one-fourth to one-third. The particle size is slightly reduced. Whereas for the exoplasmic fracture face in *Anabaena* an unchanged particle density was reported, in *Nostoc* the particle number is reduced to approximately 50% of the original value. In both organisms a modification of the particle size is observed due to the loss of distinct particle size classes.

In thylakoid membranes especially, the particle arrangement on the exoplasmic fracture face is reorganized during the heterocyst differentiation. The arrangement of 10 nm particles in parallel rows is replaced by a random distribution (Golecki, unpubl.). The particle density is reduced approximately 24% (*Anabaena*) and 55% (*Nostoc*). This is mainly due to the loss of the aligned 10 nm particles proposed to represent PS II. The loss of the 10 nm particles runs parallel with the loss of phycobilisomes (Fay, 1969a) and PS II activity (Thomas, 1970b; Donze *et al.*, 1972). The particles on the plasmic face of the thylakoid membrane are unchanged in density, but slightly reduced in size and randomly distributed.

Deprivation of CO_2 produces photooxidative death (Abeliovich & Shilo, 1972; Eloff *et al.*, 1976). Besides this dramatic effect, the CO_2 concentration influences the development of the membrane system in cells of *Synechococcus lividus* (Miller & Holt, 1977). After 120 h of CO_2 deprivation the cells were found to be free of chlorophyll *a*, C-phycocyanin and thylakoid membranes. The reintroduction of CO_2 into the bleached culture resulted in a rapid resynthesis of both pigments and thylakoid membranes.

Other nutritional factors such as the deprivation of copper (Lightbody & Krogmann, 1967; Ho *et al.*, 1979), sulphur (Prakash & Kumar, 1971) or iron (Trebst & Bothe, 1966; Öquist, 1971) may cause chlorosis but structural modifications have not been observed. Ageing of cells was accompanied by vesiculation (Echlin, 1964a; Peat & Whitton, 1968; Miller & Lang, 1971; Ermakova *et al.*, 1977) and the formation of lattice-like membrane arrangements (Lang & Rae, 1967) have been reported.

6.5.3 Temperature effects

Change of growth temperature modifies not only the morphology (Kullberg, 1977; Sherman, 1978; van Eykelenburg, 1979; Nikitina *et al.*, 1979) but also the photosynthetic activities (Murata *et al.*, 1975; Ono & Murata, 1979) and the chemical composition of cells (Holton *et al.*, 1964; Kenyon, 1972; Olson & Ingram, 1975; Sherman, 1978; Fork *et al.*, 1979; Furtado *et al.*, 1979). Cells of *Synechococcus* grown at 40°C are longer and more slender than cells grown at 30°C. Many photosynthetic activities of 40°C cells are 50–60% lower when compared with 30°C cells. In *Anacystis* cells, characteristic break points in the Arrhenius plots of several photosynthetic activities were described (Murata *et al.*, 1975; Ono & Murata, 1979). The break points may be related to phase transitions of thylakoid membrane lipids. Protein and lipid patterns are modified when the temperature is changed (Sherman, 1978; Sato *et al.*, 1979; Furtado *et al.*, 1979). In general, with decreasing temperature the chain length and the degree of saturation decrease. The result is a greater fluidity of the membrane and a lower phase transition temperature (Murata *et al.*, 1975). These findings were supported by electron microscopy. Freeze-fractured membranes of *Synechococcus lividus* (Golecki, 1979a) and *Anacystis* (Verwer *et al.*, 1978; Armond & Staehelin, 1979; Furtado *et al.*, 1979) prepared above the phase transition point show a random distribution of intramembrane particles whereas below this point a separation of lipids and proteins (=particles) takes place (Fig. 6.10). On the fracture faces of all membranes, particle-depleted areas, which we interpret to be lipid regions, are found. The sizes of these areas are different in both membranes (Golecki, 1979a). This may be caused by a different chemical composition of the membranes.

6.6 Summary and Outlook

Although photosynthetic bacteria and cyanobacteria have in common a

Table 6.1. Comparison of characteristic membrane components in cyanobacteria, Rhodospirillinae and Chlorobiineae.

	Photochemically active pigment	Light harvesting pigments	Antenna pigment complex	Lipids
Cyanobacteria (Stanier & Cohen-Bazire, 1977)	Chlorophyll *a*	Chlorophyll *a* β-Carotene Zeaxanthin Echinone Myxoxanthophyll	Light-harvesting Chl *a* Phycobiliproteins: Phycocyanin Allophycocyanin Phycoerythrin localized in phycobilisomes	Glycolipids: Monogalactosyldiglyceride Digalactosyldiglyceride Sulphoquinovosyldiglyceride Phospholipids: Phosphatidyldiglyceride
Rhodospirillineae (Oelze & Drews, 1972; Kenyon 1978; Schmidt, 1978)	Bacteriochlorophyll *a* or *b*	Bacteriochlorophyll *a* or *b* Spirilloxanthin Rhodopinal Aliphatic or monocyclic carotenoids	One or two complexes B 870, B 800–850 membrane bound	Phospholipids: Phosphatidylcholine Phosphatidylethanolamine Phosphatidylglycerol Diphosphatidylglycerol
Chlorobiineae (Kenyon, 1978; Schmidt, 1978)	Bacteriochlorophyll *a*	Bacteriochlorophyll *a–e* β-Carotene γ-Carotene Chlorobactene Isorenieratene	Bacteriochlorophyll *c* localized in chlorosomes	Glycolipids: Monogalactosyldiglyceride Glycolipid II Sulpholipid Phospholipids: Phosphatidylglycerol Diphosphatidylglycerol

prokaryotic cell organization and a similar cell wall structure and composition, the intracytoplasmic membranes of cyanobacteria are more related to thylakoids of Rhodophyceae than to the bacterial systems. The major differences between the membrane systems of photosynthetic bacteria and cyanobacteria are listed in Table 6.1.

CHAPTER 7
PHYCOBILISOMES: COMPOSITION AND STRUCTURE

G. COHEN-BAZIRE & D. A. BRYANT

7.1 Introduction

Phycobiliproteins are accessory pigments for the operation of photosystem II (see Chapter 8) in three groups of oxygenic phototrophs: the eukaryotic cryptomonads and rhodophyta and the prokaryotic cyanobacteria (O'Carra

& O'hEocha, 1976; Glazer, 1981). Perhaps contrary to expectations, the two groups of biliprotein-containing eukaryotes show little homology. Each species of cryptomonad contains one type of biliprotein, either phycocyanin or phycoerythrin (Gantt, 1979). The cryptomonad biliproteins are distinctive in their intrathylakoidal location (Gantt *et al.*, 1971), in their aggregation states and in their failure to cross-react immunologically with either rhodophytan or cyanobacterial biliproteins (MacColl & Berns, 1979; Glazer, 1976, 1981).

On the other hand, the rhodophyta and cyanobacteria display a series of very striking homologies with regard to biliproteins. Two classes are universal constituents of red algae and cyanobacteria, phycocyanins and allophycocyanins. A third class, the phycoerythrins, is present in most red algae and in many cyanobacteria. These three pigments, independent of their origin, always cross-react immunologically with pigments of the same class. Among their homologies, the biliproteins of red algae and cyanobacteria share an extrathylakoidal location, being situated in a complex organelle, the phycobilisome. They likewise share a series of characteristic aggregation states and protein sequence homologies.

These chromoproteins have been subject to extreme conservation, presumably as a result of their role in energy transfer. They serve as chemical markers, bridging the realms of eukaryotes and prokaryotes (Stanier, 1974).

7.2 Historical background

Biliproteins absorb light in a broad spectral region, situated between 500 and 650 nm, complementary to the chlorophyll *a*-protein complexes absorption bands (Table 7.1). Their role as agents of radiant energy capture in cyanobacteria was discovered by Engelmann (1883, 1884). The projection of a microspectrum of light on an *Oscillatoria* filament resulted in the accumulation of aerotactic bacteria in spectral regions where oxygen was evolved. These corresponded to regions of absorption of biliproteins and chlorophyll *a*.

Sixty years later, Emerson and Lewis (1942) repeated Engelmann's experiment quantitatively, using the unicellular cyanobacterium *Chroococcus* sp. isolated axenically by Klara Gaffron.* Quantum yield determinations of photosynthetic oxygen production demonstrated the highly efficient contribution of light absorbed by phycocyanin in this process.

On the basis of theoretical considerations, Arnold and Oppenheimer (1950) analysed the possible mechanisms that could explain the very high efficiency (~90%) of energy transfer from phycocyanin to chlorophyll *a* in *Chroococcus* sp. They concluded that energy transfer by internal conversion

* This unique experimental material antedated by almost a decade isolation of the axenic strain of *Synechococcus* sp. (*Anacystis nidulans*) (Allen, 1952). It was subsequently lost.

(analogous to the internal conversion process in radioactivity) was the most probable mechanism. If so, the maximal distance between donor and acceptor molecules (phycocyanin and chlorophyll *a* respectively) could not exceed 4 nm.

Subsequent action spectra for oxygen evolution performed with many red algae and cyanobacteria, not only confirmed that biliproteins were highly efficient light absorbers, but showed that light absorbed by chlorophyll *a* was very inefficient in this process (Haxo & Blinks, 1950; Duysens, 1951, 1952; Blinks, 1954; Haxo, 1960; Lemasson *et al.*, 1973).

Phycobiliproteins represent a considerable fraction (up to 50%) of the total cellular proteins of red algae and cyanobacteria (Myers & Kratz, 1955; Giraud, 1963; Bennett & Bogorad, 1973; Gantt & Lipschultz, 1973, 1974; Tandeau de Marsac, 1977). Before 1965, cytological observations (Brody & Vatter, 1959; Ris & Singh, 1961; Pankratz & Bowen, 1963; Wildon & Mercer, 1963b; Giraud, 1963) of thin sections had revealed the peculiarity distinctive to cyanobacteria and Rhodophyta that the thylakoid membranes were never stacked but separated from one another by a regular space of approximately 80 nm. However, the localization of phycobiliproteins with regard to the thylakoids remained challenging and unresolved (Giraud, 1963; Bogorad *et al.*, 1963).

The solution of the riddle was provided through the utilization of the cross-linking agent glutaraldehyde, as a prefixative for the preparation of specimens for electron-microscopic observations of thin sections (Sabatini *et al.*, 1963). Using this improved procedure for the fixation of the unicellular red alga *Porphyridium cruentum*, Gantt and Conti (1965, 1966b) observed the presence of regular arrays of electron-opaque granules 30–40 nm in diameter, spaced 40–50 nm apart, decorating the outer surfaces of the chloroplast thylakoids. These granules were interpreted as localized accumulations of phycobiliproteins: isolated from broken cells, their shape and dimensions were similar to those observed in thin sections; their spectral properties showed that they contained the total cellular complement of biliprotein pigments. The term 'phycobilisome' was coined for their designation (Gantt & Conti, 1966a).

Phycobilisome-like structures detected in thin sections of the cyanelles* of *Glaucocystis nostochinearum* (Lefort, 1965) were also observed in unicellular and filamentous cyanobacteria (Cohen-Bazire & Lefort-Tran, 1968; Gantt & Conti, 1969) and, without the use of fixatives, in freeze-etched preparations of

* Cyanelles have long been interpreted as endosymbiotic cyanobacteria. Although the cyanelles of the flagellate *Cyanophora paradoxa* are surrounded by a thin peptidoglycan layer (Aitken & Stanier, 1979), the size of their genome is about 10% of the smallest genome of free-living cyanobacteria (Herdman & Stanier, 1977). Genetically speaking, therefore, the cyanelles should be considered as chloroplasts.

Porphyridium cruentum (Branton, 1966), the cyanelles (Bourdu & Lefort, 1967) and of cyanobacteria (Cohen-Bazire & Lefort-Tran, 1968).

7.3 Structure of phycobilisomes in the intact cell

Electron-opaque granules identifiable as phycobilisomes have been described in many red algae (Lichtlé & Giraud, 1970; Wehrmeyer, 1971; Lichtlé, 1973; Waaland *et al.*, 1974; Wehrmeyer & Schneider, 1975; Lichtlé & Thomas, 1976) and practically all cyanobacteria examined (Edwards & Gantt, 1971; Evans & Allen, 1973; Gray *et al.*, 1973; Wildman & Bowen, 1974; Lichtlé & Thomas, 1976; Cosner, 1978; Bryant *et al.*, 1979; Gantt, 1980). In some cyanobacteria (particularly the small unicellular strains) the interthylakoidal space may appear uniformly electron-dense. This can be attributed to the close packing of the phycobilisomes. Cosner (1978) reported that individualized phycobilisomes could be observed in spheroplasts of such cells, after treatment with lysozyme in an osmotically buffered medium.

Phycobilisomes are arranged in short or long rows evenly spaced on the stromal surface of thylakoid membranes. Two main types of phycobilisomes have been observed: the first, found in many red algae (for example in *Porphyridium cruentum*), are ovoid or globular in shape; the second, discoidal or hemidiscoidal, occurs in some red algae (*Rhodella violacea*, *Porphyridium aerugineum*, and *Batrachospermum* sp.) and in most cyanobacteria (see Gantt, 1980). The unique exceptions are the bundle-shaped phycobilisomes of the structurally atypical cyanobacterium *Gloeobacter violaceus* (Rippka *et al.*, 1974; Guglielmi *et al.*, 1979). The cellular biliprotein composition does not determine the shape of phycobilisomes; however, in red algae, phycoerythrin is the main component of all globular types.

Figs 7.1–7.3. Electron micrographs showing various aspects of cyanobacterial phycobilisomes in thin sections of fixed cells.

Fig. 7.1. Thin section of *Pseudanabaena* 7408. The rows of phycobilisomes are seen in cross section, longitudinal section and tangential section. In tangential sections, the rows of closely packed phycobilisomes appear as parallel electron-dense cords, running at an angle to the long axis of the cells (arrows). When the rows of phycobilisomes are cut in cross section, the phycobilisomes are seen in face view. They alternate with one another on the opposing stromal surfaces of thylakoid membranes. Their contour is circular to semicircular with an irregular outline showing electron-dense digitations (×32 000).

Fig. 7.2. Enlargement of the outlined fixed portion of Fig. 7.1 showing details of phycobilisomes seen in face view (×93 000).

Fig. 7.3. Enlargement of part of a thin section of *Synechococcus* sp. showing rows of phycobilisomes cut longitudinally. The phycobilisomes appear as parallel electron-opaque lines 8–10 nm thick perpendicular to the stromal surface of the thylakoids (×93 000).

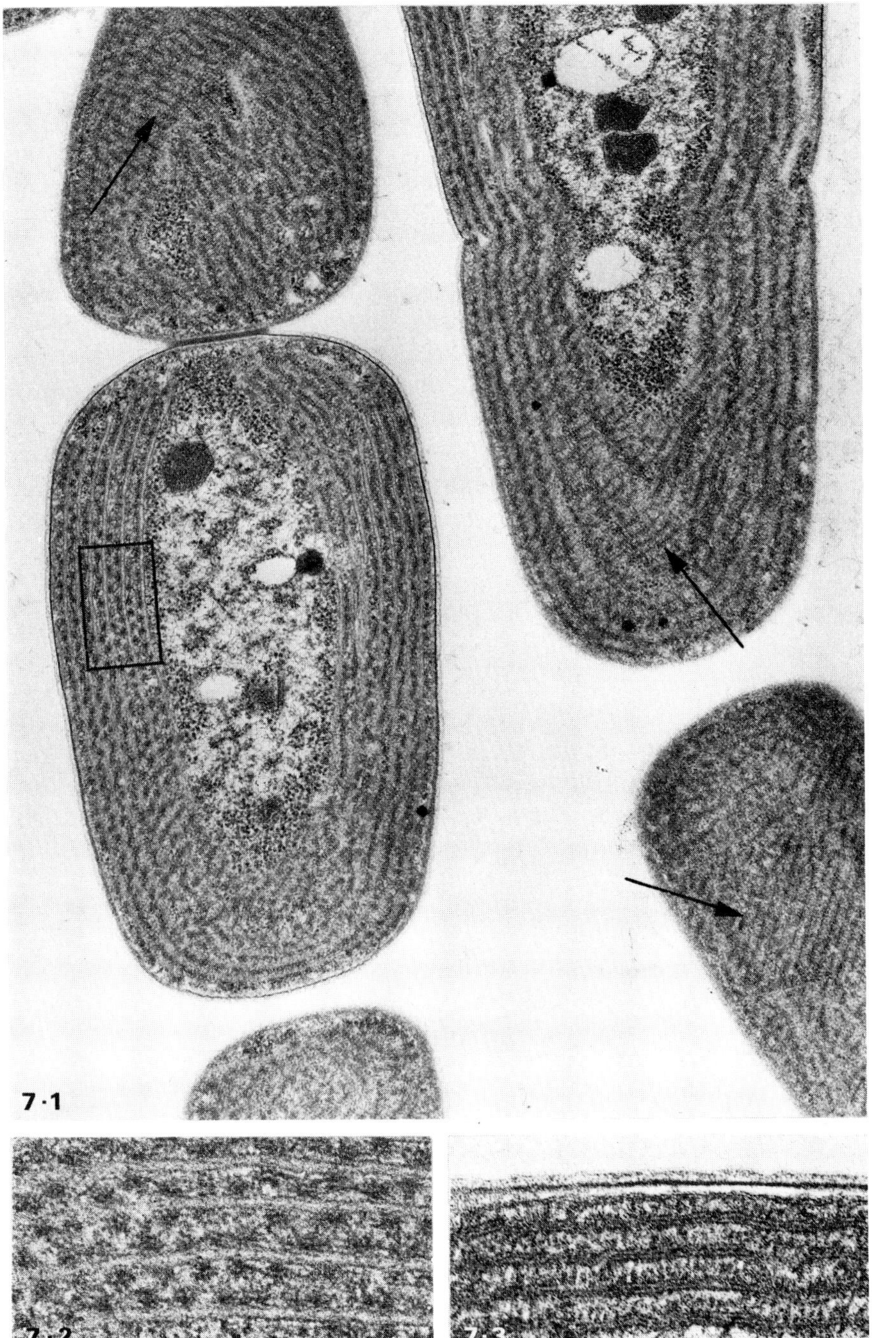
7·1
7·2
7·3

The rows of phycobilisomes in thin sections of *Pseudanabaena* 7408 are particularly clear. They are seen in Fig. 7.1 in various planes of sectioning. When the rows are cut in cross-section, the phycobilisomes appear in face view, and display an irregular, semicircular outline. In this strain, their overall dimensions are 30–35 nm diameter, with a height of 25–30 nm. They are seen to alternate with one another on the opposite stromal surfaces of the thylakoids. Their centre-to-centre spacing is approximately 45–50 nm. In the enlargement shown in Fig. 7.2, finger-like projections radiate from the core of the phycobilisomes. When rows of phycobilisomes are cut in longitudinal sections, they appear as stacks of close-packed, 10–12 nm thick electron-opaque lines, standing perpendicularly on the surface of the thylakoids (see the enlargement shown in Fig. 7.3). When cut tangentially, the rows of phycobilisomes appear as electron-opaque parallel cords, with the same spacing (45–50 nm) as the phycobilisomes seen in face view.

In the freeze-etched preparation shown in Fig. 7.4, the phycobilisomes of *Nostoc muscorum* are visible in face view and in side view; in face view, they appear almost circular with a 40 nm diameter; their thickness in profile view is approximately 8–10 nm.

The distinctive location and structure of *Gloeobacter violaceus* phycobilisomes is apparent in the electron micrographs shown in Figs 7.5 and 7.6. In thin sections the biliproteins form an electron-opaque cortical layer 50–70 nm thick, which immediately underlies the cytoplasmic membrane. This layer is composed of bundles of thin rods, orientated perpendicular to the inner surface of the cytoplasmic membrane. The rod-shaped elements are particularly visible on the freeze-etched replica (Fig. 7.6), and seen as narrow cylinders of irregular length, 12–15 nm in diameter; a 6 nm repeat is detectable along their length and visible on the enlargement shown in Fig. 7.7.

7.4 Isolation of phycobilisomes

A technique for the isolation of phycobilisomes was first devised by Gantt and Lipschultz (1972) for *Porphyridium cruentum*, and later slightly modified for

Fig. 7.4. Electron micrograph of a freeze-etched replica of *Nostoc muscorum* frozen in propane without cryoprotection. Photograph kindly supplied by Dr Golecki (see Chapter 6). The phycobilisomes in face view appear as discs 30–35 nm in diameter disposed in rows in close contact with the stromal surface of the thylakoid membrane. In profile, they appear as rods ~ 10 nm thick and 35–40 nm high (short thin arrows), standing on the stromal surface of the thylakoid. Also visible on this replica are the rows of 10 nm particles characteristic of the (EF) exoplasmic fracture face of thylakoid membranes and the corresponding particle-free grooves among the numerous particles of the (PF) plasmic fracture face. The centre to centre spacing of the phycobilisomes in face view (short thick arrows) is identical to the spacing between the rows of EF particles (long thin arrows) and the grooves of the PF face (long thick arrows). The short arrow at the right hand corner of the photograph indicates the direction of shadowing (× 120 000).

PF
EF

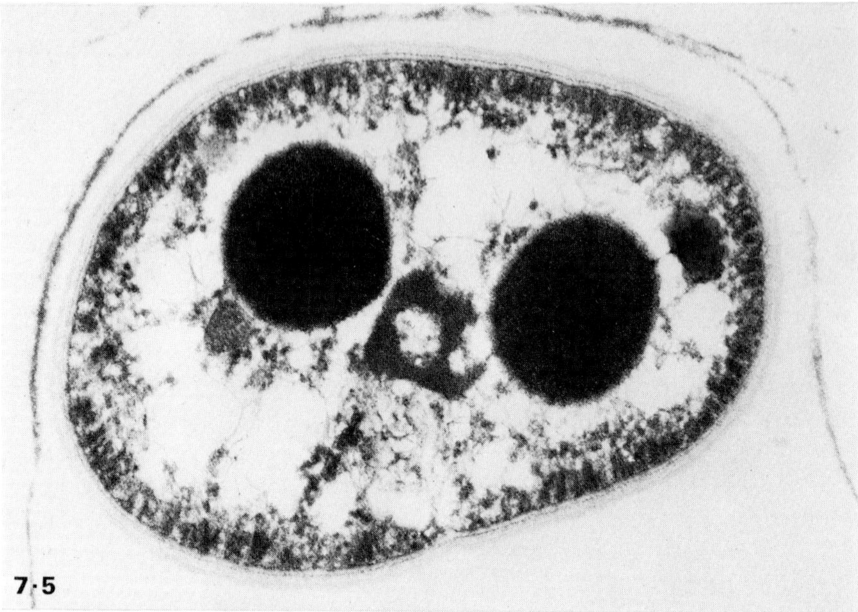

7·5

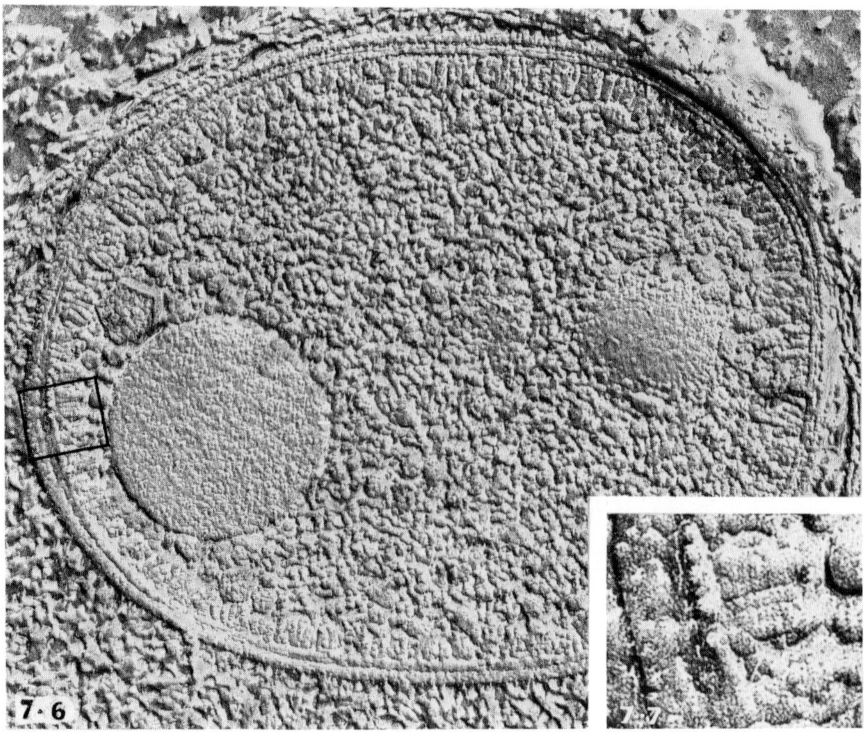

7·6

7·7

Nostoc sp. (Gray & Gantt, 1977; Gantt, 1975). A simplified procedure applicable to most red algae and cyanobacteria for the preparation of high yields of energetically well-coupled phycobilisomes has recently been published (Gantt *et al.*, 1979).

The whole procedure is carried out at room temperature. Cells are suspended in K-phosphate buffer of high molarity (0.75 M) pH 6.8 and broken in a French pressure cell at 10,000 p.s.i. The lysate is incubated with the non-ionic detergent Triton X-100, before being clarified by low-speed centrifugation. The detergent solubilizes the thylakoid membranes and helps to release the phycobilisomes. Another advantage of Triton X-100 is its insolubility in K-phosphate buffers. The phycobilisome-containing low-speed supernatant is carefully removed from under a floating layer containing the Triton-membrane complexes and layered on sucrose step gradients (from 0.25 to 2 M) in 0.75 M K-phosphate buffer. The detergent- and chlorophyll-free phycobilisomes concentrate in the 0.75–1 M sucrose layer after a 3 h centrifugation at 136 000 ***g*** leaving in the upper layers the remaining chlorophyll and any free biliproteins. Any contaminating chlorophyllous material can be eliminated by further incubation of the phycobilisomes with Triton X-100 and a second sucrose gradient centrifugation. Minor modifications of this general procedure have been used by various authors. Stable phycobilisomes from *Anacystis nidulans (Synechococcus* 6301) could not be obtained unless a buffer at pH 8 was used (Yamanaka *et al.*, 1978a). The phycoerythrin-rich phycobilisomes of cyanobacteria were extracted in phosphate buffer of lower molarity (0.65 M) at pH 7, in order to minimize their aggregation and subsequent loss in the low-speed centrifugation step (Bryant *et al.*, 1979). In a recent publication, Glazer *et al.* (1979b) have compared the properties of a number of cyanobacterial phycobilisomes extracted with Triton X-100 and two zwitterionic detergents soluble in K-phosphate buffer (Deriphat 160 and Miranol S2M-SF). The Triton-extracted phycobilisomes of some strains, but not of others, tend to form aggregates leading to abnormally high sedimentation constants and quenching of their fluorescence.

Fig. 7.5. Electron micrograph of a thin section of *Gloeobacter violaceus*. The phycobiliproteins form an irregular electron-opaque layer underlying the inner surface of the cytoplasmic membrane. This layer is composed of cylindrical elements 50–70 nm long and 15–23 nm in thickness projecting at right angles from the cytoplasmic membrane (× 62 000).

Fig. 7.6. Electron micrograph of a freeze-etched replica of *Gloeobacter violaceus* frozen in propane without cryoprotection. The phycobiliprotein layer appears composed of rod-shaped elements 10 nm in thickness and 50–70 nm in length, perpendicular to the inner surface of the cytoplasmic membrane (× 93 000).

Fig. 7.7. Enlarged view of the outlined portion of the cell shown in Fig. 7.6. A regular 6 nm spacing is visible along the length of the phycobiliprotein cylinders (× 193 000).

The effect of temperature on the integrity of phycobilisomes during their preparation and storage seems particularly important. Low temperatures (4–10°C) may lead to uncoupling and loss of allophycocyanin (Gantt *et al.*, 1979). It should be noted that cold shocks have been shown to induce a decrease of energy transfer efficiency (i.e. increase of phycocyanin fluorescence) within the phycobilisomes and between phycobilisomes and chlorophyll *a* in whole cells of *Anacystis nidulans* (Schreiber *et al.*, 1979; Schreiber, 1979). Phycobilisome preparations can remain stable for hours and even days if kept at room temperature as concentrated suspensions in 0.75 M phosphate buffer.

7.5 Protein composition

Methods for the isolation of intact phycobilisomes have allowed studies of their protein composition. The major protein components of phycobilisomes are the phycobiliproteins which account for approximately 85–90% of the total phycobilisome protein (Tandeau de Marsac & Cohen-Bazire, 1977; Yamanaka *et al.*, 1978a). Any attempt to understand phycobilisome structure and function requires a detailed understanding of the properties of these major phycobilisome constituents. The properties of the so-called 'uncoloured' polypeptides, which account for the remaining 10–15% of the total phycobilisome protein, will be described in 7.5.2.

7.5.1 Biliproteins

7.5.1.1 Nomenclature and distribution

For nearly 150 years the aesthetic beauty of the intensely fluorescent blue and red, water-soluble chromoproteins of the cyanobacteria and red algae have captured the attention and scientific imagination of innumerable investigators. Appropriately, the nomenclature for the biliproteins was originally formulated on the basis of the brilliant colour and algal origins of these chromoproteins. Kützing (1843) proposed the names 'phykoerythrin' and 'phykokyan' respectively for the red and blue pigments isolated from cyanobacteria and red algae (from the Greek: *phykos* = seaweed; *erythros* = the colour red; *kyanos* = the colour blue). Kylin (1910) introduced the term 'phycochromoproteids' as a general term for these water-soluble accessory pigments; however, this term was considered to be too general in that it properly includes chlorophyll– and carotenoid–protein complexes as well. The seldom-used term 'bilichromoproteins' (Haxo *et al.*, 1955) was subsequently simplified by O'hEocha (1958) to the currently used biliproteins. Thus, the biliproteins, or phycobiliproteins, are the water-soluble, photosynthetic accessory pigments of which phycocyanins and phycoerythrins are the best-known examples.

Phycocyanins and phycoerythrins were originally designated with a prefix, C or R, to indicate the origins of spectroscopically distinct proteins from Rhodophyta (red algae) or Cyanophyta (cyanobacteria), respectively (Svedberg & Katsurai, 1929). For example, C-phycoerythrins have a single absorption maximum at about 565 nm, while R-phycoerythrins exhibit absorption maxima at about 498, 540, and 565 nm. The prefix B is derived from Bangiales, a primitive order of red algae, and was used by Airth and Blinks (1956) to indicate yet another spectroscopically distinct phycoerythrin which has absorption maxima at 540 and 565 nm. As it has been recognized that biliproteins of a given spectral class can be isolated from both red algae and cyanobacteria, the designations R, C, and B are now applied to biliproteins to indicate visible absorption spectral properties and do not indicate the type or organisms from which they are derived. The marine cyanobacterium *Trichodesmium thiebautii* (Fujita & Shimura, 1974) and the unicellular cyanobacterium *Gloeobacter violaceus* (Rippka *et al.*, 1974) contain phycoerythrins which are spectroscopically similar to the red algal R-phycoerythrins.

As better methods for the isolation and characterization of proteins were developed, and as more cyanobacteria and red algae were examined, other spectroscopic classes and subclasses of the biliproteins were detected. These have included allophycocyanin (Haxo *et al.*, 1955), b-phycoerythrin (Gantt & Lipschultz, 1974), allophycocyanin-B (Glazer & Bryant, 1975), and phycoerythrocyanin (Bryant *et al.*, 1976). The absorption and fluorescence maxima of the various spectroscopic classes of cyanobacterial and red algal biliproteins are presented in Table 7.1.

All cyanobacteria and red algae contain allophycocyanin and C-phycocyanin or R-phycocyanin. Allophycocyanin-B has been demonstrated in numerous cyanobacteria and the red algae *Porphyridium cruentum* and *P. aerugineum* (Glazer & Bryant, 1975; Ley *et al.*, 1977; Zilinskas *et al.*, 1978), but has not been detected in the cyanobacterium *Gloeobacter violaceus* (Guglielmi *et al.*, 1979) nor in the red alga *Rhodella violacea* (Koller & Wehrmeyer, 1974; Koller *et al.*, 1977; Koller *et al.*, 1978). Phycoerythrins are widely distributed among all groups of cyanobacteria and red algae, but are not universally present (Stanier *et al.*, 1971; Rippka *et al.*, 1979; Gantt *et al.*, 1968). Phycoerythrocyanin has a narrow taxonomic distribution, primarily occurring in heterocyst-forming cyanobacteria which do not contain phycoerythrin (Bryant *et al.*, 1976; Rippka *et al.*, 1979).

As pointed out by Glazer (1981), the full complexity of the biliprotein complement of cyanobacteria and red algae is not always revealed by spectroscopic criteria alone. Factors leading to diversity among the phycobili-

Table 7.1. Distribution and properties of biliproteins

Biliprotein	Distribution*	Visible absorption maxima† (nm)	Fluorescence emission maximum† (nm)	Subunit structure of stable assembly form(s)	Approximate molecular mass of stable assembly form(s)‡ (daltons)	Approximate sedimentation coefficient of stable assembly form(s)
Allophycocyanin-B	C, R	671 > 618	675	$(\alpha\beta)_3$	89,000	—
Allophycocyanin	C, R	650	660	$(\alpha\beta)_3$	100,000	5–5.5s
C-phycocyanin	C, R	620	640	$(\alpha\beta)_3$; $(\alpha\beta)_6$	120,000; 232,000	6 S; 11 *s*
R-phycocyanin	R	617 > 555	636	$(\alpha\beta)_3$; $(\alpha\beta)_6$	105,000; 210,000	6 S; 11 *s*
Phycoerythrocyanin	C	570 > 595 (s)§	610	$(\alpha\beta)_3$; $(\alpha\beta)_6$	103,000; 206,000	—
C-phycoerythrin	C	560	577	$(\alpha\beta)_6$	230,000	10 *s*
b-phycoerythrin	R	545 > 563 (s)	570	$(\alpha\beta)_n$	—	—
B-phycoerythrin	R	545 > 563 > 498 (s)	577	$(\alpha\beta)_6\ \gamma$	240,000	—
R-phycoerythrin	C, R	565 > 540 > 498	578	$(\alpha\beta)_6\ \gamma$	240,000	12 *s*

* C, Cyanobacteria; R, red algae.

‡ The positions of the absorption and fluorescence maxima vary depending on the aggregation state and the organism which serves as the source of the protein.

‡ Values vary depending upon the organism which serves as the source of the protein.

§ (s), Shoulder.

proteins of a given spectroscopic class from a single organism could include the following: (a) genetically determined isoproteins; (b) complexing of a phycobiliprotein with additional polypeptides; (c) post-translational modifications. Examples of possibilities (a) and (b) are given below. Although examples of post-translational modification have not been reported, the possibility that such modifications exist should not be disregarded (Glazer, 1981).

Two spectroscopically similar B-phycoerythrins with nearly identical molecular masses but with different amino-acid compositions have been demonstrated in the red alga *R. violacea* (Koller & Wehrmeyer, 1975). Numerous other red algae have been reported to have two spectroscopically similar R-phycoerythrins, one of 250 000–300 000 daltons and a second of about 44 000 daltons (van der Velde, 1973). *Porphyridium cruentum* contains two spectroscopically-distinct phycoerythrins, B-phycoerythrin (240 000 daltons) and polydisperse b-phycoerythrin (Gantt & Lipschultz, 1974; Glazer & Hixson, 1977). The subunit structure of B-phycoerythrin is $(\alpha\beta)_6\gamma$ (Glazer & Hixson, 1977; Abad-Zapatero *et al.*, 1977) and that of b-phycoerythrin is $(\alpha\beta)_n$. Extensive characterization of the subunits of these two phycoerythrins—with respect to molecular mass, amino-terminal sequence, bilin content and immunological properties—suggests but does not prove that the α and β subunits of B- and b-phycoerythrins are the products of the same two genes. The γ subunit carries two phycourobilin and two phycoerythrobilin chromophores (see 7.5.1.2); the α and β subunits carry two and four phycoerythrobilin chromophores, respectively (Glazer & Hixson, 1977). Hence, the presence of the γ subunit in B-phycoerythrin accounts for the distinctive spectroscopic differences near 495 nm for these phycoerythrins.

Certain cyanobacteria capable of chromatic adaptation (see 7.8.3) have been shown to contain two types of phycocyanin (Bryant, 1979, 1981; Bryant & Cohen-Bazire, 1981). One phycocyanin is synthesized constitutively, while the production of the second phycocyanin is specifically induced during growth in red light. The subunits of these two phycocyanins differ with respect to charge, molecular mass, and amino-acid composition (see 7.5.3). *Agmenellum quadruplicatum* (Gray, *et al.*, 1976) and *Phormidium luridum* (Brown *et al.*, 1975) have each been reported to contain two spectroscopically distinct phycocyanins; however, the structural basis for the differences in these two cases is not known.

Several cyanobacteria have been reported to contain chromatographically distinct allophycocyanins (Gysi & Zuber, 1974, 1976; Zilinskas *et al.*, 1978, 1980). The basis for the observed differences in these cases is not clear; the presence of an additional subunit may explain the origin of one of these forms (Zilinskas *et al.*, 1978). Amino-terminal sequencing of three such allophycocyanin forms have thus far revealed no structural differences (Troxler *et al.*, 1979).

7.5.1.2 Chromophore structure and distribution; the nature of the chromophore–protein linkage

The phycobiliproteins owe their intense visible absorption properties to the presence of open-chain tetrapyrrole chromophores which are closely related to the animal bile pigments biliverdin and bilirubin (Lemberg, 1928, 1930; Lemberg & Bader, 1933). These prosthetic groups are commonly referred to as phycobilins, or simply bilins. Unlike the chlorophyll and carotenoid chromophores which are extractable with organic solvents, the phycobilins are bound to their apoproteins by covalent linkages.

A number of techniques have been used to obtain free phycobilins from their respective biliproteins. These include hydrolysis with 12 N HCl (O'hEocha, 1958), refluxing with methanol (O'Carra & O'hEocha, 1966), treatment with HBr in trifluoroacetic acid (Schram & Kroes, 1971), and treatment with Hg^{2+} salts (Williams & Glazer, 1978; Scheer, 1978). Depending upon the treatment employed and upon the phycobiliprotein utilized, different free tetrapyrroles are released as major products. Because of the rather drastic conditions required to liberate the chromophores from their apoproteins, and the ease with which these chromophores undergo a variety of oxidations, isomerizations, and addition reactions, the studies leading to the structure determinations of the phycobilins were particularly challenging. Details of the various interconversions of the phycobilins, as prepared by the above-mentioned methods, and the studies leading to the determinations of their structures are beyond the scope of this chapter. The reader is referred to the reviews of Siegelman *et al.* (1968), Rüdiger (1970, 1975), O'Carra and O'hEocha (1976), Bennett and Siegelman (1979), Glazer (1977), and Scheer (1978).

At present, four different chromophores have been identified in studies with cyanobacterial and red algal biliproteins. The most commonly occurring phycobilins are phycocyanobilin, the blue chromophore of C-phycocyanin and allophycocyanin, and phycoerythrobilin, the red chromophore of phycoerythrins. The structures of the free tetrapyrroles derived from C-phycocyanin (Cole *et al.*, 1967, 1968; Crespi *et al.*, 1962, 1968; Rüdiger *et al.*, 1967; Fu *et al.*, 1979) and phycoerythrins (Chapman *et al.*, 1967; Rüdiger *et al.*, 1967; Fu *et al.*, 1979) by methanolysis have been determined and confirmed by total synthesis (Gossauer & Hirsch, 1974; Gossauer & Klahr, 1979). The α subunit of phycoerythrocyanin carries a third type of chromophore, a phycobiliviolin-type tetrapyrrole of as yet undetermined structure (Bryant *et al.*, 1976). R-phycoerythrins (O'Carra *et al.*, 1964; Glazer &

Hixson, 1977), the γ subunit of B-phycoerythrin (Glazer & Hixson, 1977) and the β subunit of *Gloeobacter violaceus* phycoerythrin (Bryant, unpublished) carry a fourth type of phycobilin, phycourobilin. It should be noted that in this chapter the names of phycocyanobilin, phycoerythrobilin, and phycourobilin are reserved for the protein-bound structures of these tetrapyrroles, as shown in Fig. 7.8.

As initially demonstrated by O'hEocha and coworkers (O'hEocha, 1958, 1960, 1963, 1965; O'hEocha & O'Carra, 1961; O'Carra *et al.*, 1964), under conditions where the normally considerable bilin–protein interactions are abolished, the resultant absorption properties of the polypeptide-bound

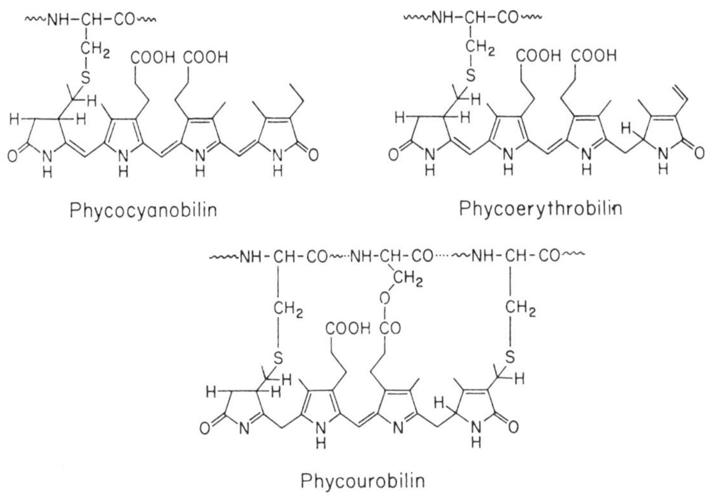

Fig. 7.8. The structure of polypeptide-bound phycocyanobilin, phycoerythrobilin, and phycourobilin. The structure for phycourobilin is that proposed by Killilea and O'Carra (cited in O'Carra & O'hEocha, 1976).

phycobilins closely resemble those of free phycobilins. Absorption spectra for polypeptide-bound phycobilins, determined in 8.0 M urea at pH 2–3, are shown in Fig. 7.9. Under these conditions, the spectral properties of the protein-linked bilins are independent of the nature of the polypeptide chain. For example, allophycocyanin, C-phycocyanin, and allophycocyanin-B have virtually indistinguishable absorption spectra (Glazer & Fang, 1973a; Glazer & Bryant, 1975). Although the phycobilins of denatured phycobiliproteins readily isomerize and degrade to colourless products, (O'hEocha, 1960; Glazer & Hixson, 1975; Bryant *et al.*, 1976), such complications can largely be avoided if the spectroscopic analysis is performed quickly (Glazer & Fang, 1973a; Glazer & Hixson, 1975, 1977; Bryant *et al.*, 1976).

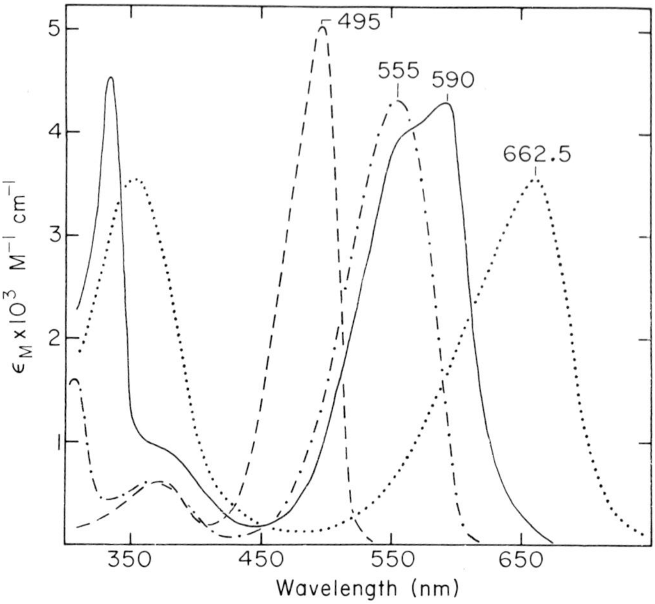

Fig. 7.9. The molar absorption spectra in 8 M urea at pH 3 of polypeptide-bound phycocyanobilin (· · · ·), phycoerythrobilin (– · – ·), phycourobilin(- - -), and the phycobiliviolin-type chromophore of the α subunit of phycoerythrocyanin (——). The spectrum of phycourobilin is plotted as 0.5 ε_M.

Because the proper model compounds were not initially available, attempts to determine molar extinction coefficients for the phycobilins, and thereby the number of bilin groups attached to the polypeptide chains of the phycobiliproteins, required that certain assumptions be made. Glazer and Fang (1973a) calculated the molar extinction coefficient of phycocyanobilin (35 5000 M^{-1} cm^{-1} at 662.5 nm) by assuming that the α subunit of C-phycocyanin carried one such chromophore; in their method, spectra were measured with acidic urea solutions whose protein concentrations were determined by quantitative amino-acid analysis. Likewise, Glazer and Hixson (1975) calculated the molar extinction coefficient for polypeptide-bound phycoerythrobilin (43 300 M^{-1} cm^{-1} at 555 nm) by assuming that the β subunit of R-phycocyanin carries one phycocyanobilin and one phycoerythrobilin chromophore. The correctness of these assumptions has subsequently been established by amino-acid sequence determinations on these proteins (Williams & Glazer, 1978; Freidenreich *et al.*, 1978; Bryant *et al.*, 1978). Recent model compound studies and an independent analysis of phycoerythrobilin-containing chromopeptides are in excellent agreement with these values (Scheer, 1976; Muckle & Rüdiger, 1977; Kufer & Scheer, 1979). The molar extinction coefficient of phycobiliviolin (~43 000 M^{-1} cm^{-1} at 590 nm) was calculated by assuming that the α subunit of phycoerythrocyanin carries one such chromophore (Bryant *et al.*, 1976). The molar extinction coefficient of phycourobilin (104 000 M^{-1} cm^{-1} at 495 nm) was calculated by assuming that the γ subunit of B-phycoerythrin carries two such chromophores (Glazer & Hixson, 1977). The validity of these latter two values remains to be established. A summary of the phycobilin contents of various biliproteins is presented in Table 7.2.

Table 7.2. Chromophore content of biliprotein subunits

	Chromophore content*		
Biliprotein	α subunit	β subunit	γ subunit
Allophycocyanin-B	1 PCB	1 PCB	
Allophycocyanin	1 PCB	1 PCB	
C-phycocyanin	1 PCB	2 PCB	
R-phycocyanin	1 PCB	1 PCB, 1 PEB	
Phycoerythrocyanin	1 PXB	2 PCB	
C-phycoerythrin	2 PEB	3 or 4 PEB	
b-phycoerythrin	2 PEB	4 PEB	
B-phycoerythrin	2 PEB	4 PEB	2 PEB, 2 PUB
R-phycoerythrin	n.d.†	n.d.	n.d.

* The abbreviations for the phycobilins are: PCB, phycocyanobilin; PEB, phycoerythrobilin; PXB, phycobiliviolin-type chromophore of undetermined structure; PUB, phycourobilin.

† n.d., Not determined. The R-phycoerythrin of *Crytonemia obovata* carries 31 PEB and 9 PUB chromophores per $(\alpha\beta)_6\gamma$ (Glazer & Hixson, 1977).

Although a number of bilin–polypeptide linkages have been proposed (see Chapman, 1973; O'Carra & O'hEocha, 1976), thioether linkages are probably universal between phycobilins and apoproteins. The amino-acid sequences around the bilin attachment sites of several phycobiliproteins have been determined recently (see 7.5.1.8). In each case the bilin prosthetic group was covalently linked to the polypeptide chain through a cysteinyl thioether linkage. Glazer *et al.* (1979a) have demonstrated that the bilin-linked cysteines of the phycobiliproteins can be converted with high yield to cysteic acid by hydrolysis at 110°C with 6 N HCl-containing dimethylsulphoxide. This technique, when applied to the phycobiliproteins of a variety of spectroscopic classes, established the equivalence of the number of thioether linkages and the number of phycocyanobilin and phycoerythrobilin groups present.

Much evidence supports a thioether linkage of the polypeptide to ring A of the bilin (see Kost-Reyes *et al.*, 1975; Beuhler *et al.*, 1976; Klein & Rüdiger, 1978, 1979). Proof of such a thioether linkage has been obtained recently for one of the phycocyanobilins attached to the β subunit of *Synechococcus* 6301 C-phycocyanin (Lagarias *et al.*, 1979). As shown in Fig. 7.8, a thioether to the proximal carbon of the ethyl side chain of ring A is the only linkage between the polypeptide chain and the chromophore in this case.

While it has been proposed that there are generally two linkages between the phycobilins and the polypeptide chain (Rüdiger & O'Carra, 1969; Rüdiger, 1970; Kost *et al.*, 1975), presently there is no direct evidence for two linkages to phycocyanobilin (Troxler *et al.*, 1978; Lagarias *et al.*, 1979; Brown *et al.*, 1979). Recent results suggest that most, but perhaps not all, phycoerythrobilin prosthetic groups of C-phycoerythrins are also bound by a single linkage, the cysteinyl thioether bond (Muckle *et al.*, 1978; see Fig. 7.8). There is evidence that phycourobilin is linked to the polypeptide by two thioether linkages (O'Carra & O'hEocha, 1976; Glazer *et al.*, 1979a). The structure for polypeptide-bound phycourobilin proposed by Killilea and O'Carra (cited in O'Carra & O'hEocha, 1976) is shown in Fig. 7.8. More work is required to answer decisively the question of multiple polypeptide-bilin linkages.

The mechanism(s) by which the phycobilins are attached to the polypeptide chains of the phycobiliproteins are unknown. While it is clear that bilin attachment requires the presence of a cysteine residue, this alone is not sufficient. The C-phycocyanins of *Synechococcus* 6301 (Williams & Glazer, 1978; Freidenreich *et al.*, 1978) and *Mastiglocladus laminosus* (*Fischerella* 7603; Frank *et al.*, 1978) contain cysteine residues which are not covalently linked to phycocyanobilins. Although the amino-acid sequence around the chromophore attachment site of the α subunit of C-phycocyanins is homologous to that about one of the chromophore attachment sites of the β subunit, there is no homology between the second attachment site on the β subunit and the other two sequences (see 7.5.1.8). Hence, bilin attachment is probably determined by both linear sequence and local conformation.

7.5.1.3 Subunit structure of biliproteins

The basic structural unit of all phycobiliproteins is a protomer. This is often referred to as 'monomer', αβ, composed of two dissimilar polypeptide chains which generally differ slightly in molecular mass (Glazer & Cohen-Bazire, 1971; Bennett & Bogorad, 1971; O'Carra & Killilea, 1971). Subunit molecular masses, routinely determined by polyacrylamide gel electrophoresis in the presence of sodium dodecylsulphate, range from 14 000 to 23 000 daltons (for extensive lists of subunit molecular mass values, see Bogorad, 1975a; Glazer,

1976). The actual range suggested by the amino-acid sequence data is more narrow, perhaps 17 000–23 000 daltons (Freidenreich *et al.*, 1978; Bryant *et al.*, 1978; Frank *et al.*, 1978; Troxler & Brown 1979). Glazer (1981) has suggested that the mobility of phycobiliproteins, when subjected to gel electrophoresis in the presence of sodium dodecylsulphate, may be anomalous because of the covalently attached phycobilins. A third type of bilin-carrying subunit, the γ subunit with molecular mass about 30 000 daltons, has been detected in certain red algal phycoerythrins (Gantt & Lipschultz, 1974; Glazer & Hixson, 1977).

A number of methods have been described for the preparation of phycobiliprotein subunits. The most commonly used procedure for phycocyanins and phycoerythrins involves chromatography on the weak cation-exchange resin Bio-Rex 70 at pH 3 in the presence of urea and reducing agent (Glazer & Fang, 1973a). Allophycocyanin subunits have been prepared by DEAE-Sephadex chromatography at pH 7–8 in the presence of 8.0 M urea and reducing agent (Gysi & Zuber, 1974; Cohen-Bazire *et al.*, 1977). Other methods applied to biliproteins include preparative gel electrophoresis in the presence of urea (Binder *et al.*, 1972), hydroxyapatite chromatography in the presence of sodium dodecylsulphate (Frank *et al.*, 1975), and gel exclusion chromatography in the presence of sodium dodecylsulphate and reducing agent (Kobayashi *et al.*, 1972).

Many biliproteins and their subunits can be renatured upon removal of the denaturing agent (Murphy & O'Carra, 1970; Glazer & Fang, 1973b; Glazer, Fang & Brown, 1973; Gysi & Zuber, 1974). Glazer and Fang (1973b) exploited this fact in an elegant study in which hybrid C-phycocyanin molecules were constructed from isolated, renatured subunits of *Synechococcus* 6301, *Synechocystis* 6701, and *Anabaena* 6411 C-phycocyanins. The hybrid phycocyanins were virtually indistinguishable from their naturally occurring counterparts with respect to chromatographic, spectroscopic, and isoelectric properties. Similar results were obtained when hybrid allophycocyanins were prepared from renatured subunits of cyanobacterial and red algal allophycocyanins (Cohen-Bazire *et al.*, 1977). These studies indicate that the conformations and inter-subunit contacts of the subunits of allophycocyanins and phycocyanins are highly conserved.

The amino-acid compositions of the phycobiliproteins and their subunits reveal a number of interesting features. Firstly, the amino-acid compositions of all phycobiliproteins, especially those of a single spectroscopic class, are remarkably similar. Moreover, the amino-acid compositions of the α and β subunits of a biliprotein are generally quite similar. A striking feature is the predominance of aliphatic and acidic residues in these proteins, a finding consistent with the observation that all biliproteins are acidic proteins with isoelectric points in the range pH 4.0–5.5 (for tabulations of isoelectric points, see Koller & Wehrmeyer, 1974; Bryant *et al.*, 1976; Bennett & Siegelman, 1979). Most allophycocyanins have been found to contain neither histidine nor tryptophan (Gysi & Zuber, 1974, 1976; Bryant *et al.*, 1976; Hixson, 1976; Cohen-Bazire *et al.*, 1977; Boussiba & Richmond, 1979; Mörschel *et al.*, 1980). The histidine and tryptophan residues of C- and R- phycocyanins generally appear to be confined to the α subunits of these biliproteins (Kobayashi *et al.*, 1972; Glazer & Fang, 1973a,b; Glazer and Hixson, 1975; Glazer, 1976; Freidenreich *et al.*, 1978; Frank *et al.*, 1978).

7.5.1.4 Aggregation behaviour of biliproteins

When cyanobacterial or red algal cells are disrupted in buffers of low to moderate ionic strengths ($\mu < 0.25$), the phycobilisomes dissociate and the constituent phycobiliproteins are released as high molecular weight, water-soluble aggregates. The size of these aggregates depends upon a number of factors including the organism being studied, the type of biliprotein being considered, pH, ionic strength, temperature, and the protein concentration.

The aggregation behaviour of phycocyanins has been studied most extensively and has been shown to be highly dependent upon pH. In the range pH 4.5–6.0 at moderate protein concentration (~ 1 mg ml^{-1}) and at ionic strengths greater than $\mu = 0.01$, the predominant aggregate observed is the hexamer, $(\alpha\beta)_6$, with a sedimentation coefficient of ~ 11 *s* and a molecular mass of $\sim 220\,000$ daltons (Svedberg & Katsurai, 1929; Svedberg & Eriksson, 1932; Eriksson-Quensel, 1938; Hattori & Fujita, 1959; Scott & Berns, 1965; Hattori *et al.*, 1965; Neufeld & Riggs, 1969; Glazer & Cohen-Bazire, 1971). Under these conditions a monomer $(\alpha\beta)$–hexamer $(\alpha\beta)_6$ equilibrium exists which lies far to the side of the hexamer (Berns & Scott, 1966; Berns & Morgenstern, 1966; MacColl *et al.*, 1971; Saito *et al.*, 1971). However, at pH 6.5–8.0, most R- and C-phycocyanins exist in solution predominantly as trimers $(\alpha\beta)_3$ with a sedimentation coefficient of ~ 6 *s* and a molecular mass of $\sim 110\,000$ daltons (Svedberg & Lewis, 1928; Svedberg & Katsurai, 1929; Svedberg & Eriksson, 1932; Eriksson-Quensel, 1938; Hattori & Fujita, 1959; Scott & Berns, 1965; Hattori *et al.*, 1965; Glazer *et al.*, 1973). At these higher pH values, a monomer $(\alpha\beta)$–trimer $(\alpha\beta)_3$ equilibrium favouring the trimer exists at low protein concentrations (Berns & Moregenstern, 1966; MacColl *et al.*, 1971; Saito *et al.*, 1974). However, with increasing protein concentrations, the monomer–hexamer equilibrium appears and becomes increasingly important (MacColl *et al.*, 1971; Mizuno *et al.*, 1975).

Allophycocyanin from a wide range of organisms has been shown to exist in solution as a trimer $(\alpha\beta)_3$ with a sedimentation coefficient of 5.0–5.5 *s* and a molecular mass of $\sim 100\,000$ daltons (Hattori & Fujita, 1959; Glazer &

Cohen-Bazire, 1971; MacColl *et al.*, 1974; Gantt & Lipschultz, 1974; Bryant *et al.*, 1976; Hixson, 1976; Cohen-Bazire *et al.*, 1977; Boussiba & Richmond, 1979; Gysi & Zuber, 1979). The allophycocyanin-B of *Anabaena variabilis* (*Anabaena* 7118) has similarly been shown to exist in solution as a trimer (Glazer & Bryant, 1975). A possible exception is *Cyanidium caldarium* allophycocyanin which has been reported to have a molecular mass of 196 000 daltons (Brown & Troxler, 1977). C-phycoerythrins have been reported to exist in solution as hexamers over a wide range of pH values (Hattori & Fujita, 1959; Bennett & Bogorad, 1971). R- and B-phycoerythrins exist in solution as the unusual structures $(\alpha\beta)_6\gamma$ which have sedimentation coefficients of $\sim 12\ s$ and molecular masses in the range 230 000–300 000 daltons (Svedberg & Lewis, 1928; Svedberg & Katsurai, 1929; Svedberg & Eriksson, 1932; Eriksson-Quensel, 1938; Vaughan, 1964; van der Velde, 1973; Gantt & Lipschultz, 1974; Koller & Wehrmeyer, 1975; Hixson, 1976; Glazer & Hixson, 1977). A summary of the data for native aggregation states, sedimentation coefficients, and molecular mass ranges for the various spectroscopic classes of biliproteins is presented in Table 7.1. For a complete compilation of molecular masses of native cyanobacteria and rhodophytan biliproteins, the reader is referred to the review of Glazer (1981).

Numerous investigators have reported the presence of aggregates larger than the hexamer in crude extracts and purified C-phycocyanin preparations of cyanobacteria and red algae (Scott & Berns, 1965; Berns & Scott, 1966; Lee & Berns, 1968; Kao *et al.*, 1973, 1975; Kessel *et al.*, 1973; Eiserling & Glazer, 1974; Bryant *et al.*, 1979). These aggregates, which have sedimentation coefficients of 19–25 *s*, have been identified by electron microscopy to be short, rod-like stacks of C-phycocyanin hexamers (Kessel *et al.*, 1973; Eiserling & Glazer, 1974; see 7.5.1.6). Such higher aggregates, however, could not be demonstrated in preparations of highly purified *Synechococcus* 6301 C-phycocyanin, even at very high protein concentrations (Eiserling & Glazer, 1974). A number of higher aggregates of phycocyanin or of phycocyanin and phycoerythrin have recently been described (see 7.5.3). One to several uncoloured polypeptides are integral components of each of these complexes, which can occasionally be remarkably resistant to dissociation (Bryant, 1979, 1981; Bryant & Cohen-Bazire, 1981). It has been suggested that these uncoloured polypeptides are necessary for the formation of the rod-shaped, higher aggregates (Bryant *et al.*, 1979). It is possible that the earlier preparations of C-phycocyanin contain small amounts of uncoloured proteins which could have been responsible for the presence of the higher aggregates since many of these studies were performed before the introduction of analytical polyacrylamide gel electrophoresis in the presence of sodium dodecylsulphate as a criterion of protein purity.

7.5.1.5 Spectroscopic properties of biliproteins

The spectroscopic properties of the biliproteins depend upon a number of factors including the properties of the attached phycobilins, chromophore–chromophore interactions, and chromophore–protein interactions. The effects of chromophore–protein interactions can be seen from a comparison of the properties of native and denatured biliproteins. Native biliproteins have narrow, very intense absorption bands in the visible region and weaker absorption bands in the near-u.v.; they exhibit high quantum yields of fluorescence (Scheer & Kufer, 1977; Kufer & Scheer, 1979). The phycobilins of native biliproteins do not form complexes with Zn^{2+} (O'hEocha, 1960; O'hEocha & O'Carra, 1961), are not oxidized during prolonged storage (Kufer & Scheer, 1979), are inaccessible to titration (Vaughan, 1964), and are only partly reduced by dithionite at concentrations greater than 5 mM (Scheer, 1978; Kufer & Scheer, 1979). Molecular orbital calculations (Scheer & Kufer, 1977), as well as studies of model compounds (Kufer & Scheer, 1979), indicate that the bilin chromophores are held in an elongated conformation on the native biliproteins.

The properties of denaturated biliproteins closely resemble those of free bilins (see 7.5.1.2). Free bilins, and probably the phycobilins of denaturated biliproteins, adopt a cyclic conformation in solution (Scheer, 1978). When denaturated, the biliproteins exhibit broad absorption bands which are nearly equal in intensity in the visible and near-u.v. (Glazer & Fang, 1973a; Scheer, 1976; Scheer & Kufer, 1977), and exhibit low quantum yields of fluorescence (Kufer & Scheer, 1979). The phycobilins of denatured biliproteins readily form complexes with Zn^{2+} (O'hEocha, 1960; O'hEocha & O'Carra, 1961), undergo oxidative bleaching (O'hEocha, 1960; Glazer & Hixson, 1975, 1977; Bryant *et al.*, 1976; Kufer & Scheer, 1979), are readily accessible to titration (Vaughan, 1964), and can be completely reduced with 0.5 mM dithionite (Scheer, 1978; Kufer & Scheer, 1979). These results suggest that the chromophores of native biliproteins are buried in the protein matrix.

Further evidence for the influence of the protein on the spectroscopic properties of the chromophores can be seen from an examination of the absorption spectra of C-phycocyanin, allophycocyanin, and allophycocyanin-B. Although each of these proteins contains only phycocyanobilin prosthetic groups (Table 7.2), the absorption maxima span a range of $\sim$60 nm (Table 7.1).

The absorption properties of the biliproteins are highly dependent upon their aggregation state (7.5.1.4). Generally, the long-wavelength absorption maximum is red-shifted and exhibits an increase in absorption coefficient as the biliproteins become more highly aggregated (Hattori *et al.*, 1965; Scott & Berns, 1965; Neufeld & Riggs, 1969; Bennett & Bogorad, 1971; Glazer *et al.*,

1973). Particularly striking examples are provided by allophycocyanin and allophycocyanin-B; the intense, long-wavelength absorption bands of these proteins are characteristic of the trimeric $(\alpha\beta)_3$ assembly forms of these proteins and are greatly diminished when these proteins are dissociated to monomers (Cohen-Bazire *et al.*, 1977; Glazer & Bryant, 1975).

The spectroscopic properties of the subunits and various aggregates of *Synechococcus* 6301 C-phycocyanin have been determined by Glazer *et al.* (1973). The long-wavelength absorption maximum of the α subunit, which carries one phycocyanobilin chromophore, was 620 nm. The β subunit carries two phycocyanobilin groups and has its absorption maximum at 608 nm. The spectrum of monomeric $(\alpha\beta)$ phycocyanin was nearly identical to the sum of the spectra of the isolated α and β subunits. Thus, the combination of the α and β subunits to form the monomer $(\alpha\beta)$ does not lead to significant changes in the protein–chromophore interactions of the individual subunits or to significant chromophore–chromophore (exciton) interactions. The phycocyanin monomer at pH 7 has an absorption maximum at ~615 nm and a molar extinction coefficient of 2.3×10^5 M^{-1} cm^{-1}. The phycocyanin hexamer $(\alpha\beta)_6$, the predominant species at pH 5.5 (see 7.5.1.4), has an absorption maximum at 621 nm and a molar extinction coefficient of 3.33×10^5 M^{-1} cm^{-1}. The values for the trimer $(\alpha\beta)_3$ are intermediate between those of the monomer and the hexamer. Circular dichroism spectra of the trimeric and hexameric species indicate exciton interactions between the phycobilins on different monomers.

Spectroscopic heterogeneity of the chemically equivalent chromophores of phycocyanins or of phycoerythrins can be detected by fluorescence polarization. Generally, the fluorescence emission anisotropy increases at longer wavelengths of the major visible absorption band of the biliproteins (Teale & Dale, 1970; Dale & Teale, 1970; Grabowski & Gantt, 1978). Two chromophore classes exist; these have been designated sensitizing 's' and fluorescing 'f' (Teale & Dale, 1970; Dale & Teale, 1970). The absorption properties of the 's' and 'f' chromophores can be deduced from the fluorescence excitation and fluorescence polarization spectra. For C-phycocyanin the 's' chromophores absorb maximally at 600–615 nm (Teale & Dale, 1970) or ~590 nm (Vernotte, 1971), and the 'f' chromophores absorb maximally at ~630–635 nm (Teale & Dale, 1970; Vernotte, 1971). Furthermore, Dale and Teale (1970) estimated that there are twice as many 's' chromophores as 'f' chromophores for C-phycocyanin. Glazer *et al.* (1973) have suggested that 's' chromophores are those found on the C-phycocyanin β subunit and that the 'f' chromophores are those found on the C-phycocyanin α subunits.

Fluorescence polarization decreases as the size of phycocyanin aggregates increases (Goedheer & Birnie, 1965; Teale & Dale, 1970; Vernotte, 1971; Kessel *et al.*, 1973). Pico-second absorption spectroscopy of C-phycocyanin indicates that the energy transfer from 's' to 'f' chromophores occurs more rapidly in hexamers than in trimers or monomers (Kobayashi *et al.*, 1979). Thus, increased aggregation results in more rapid and extensive depolarization of the absorbed light energy. Fluorescence polarization values for allophycocyanins are significantly smaller than those of C-phycocyanin aggregates of equivalent size (MacColl *et al.*, 1978; Grabowski & Gantt, 1978). It has been suggested that the chromophores of allophycocyanin are positioned for especially favourable intraprotein energy transfer (MacColl *et al.*, 1978).

7.5.1.6 Ultrastructure of biliprotein aggregates

Bryant *et al.* (1976) recently proposed that a disc-shaped trimer, $\sim 3 \times 12$ nm, is the basic unit from which all higher assembly forms of the phycobiliproteins are constructed. A number of experimental observations support this view. When examined by electron microscopy after negative staining, preparations of hexameric C-phycocyanin, $(\alpha\beta)_6$, appear as discs 11–13 nm in diameter (Berns & Edwards, 1965; Eiserling & Glazer, 1974). Edge views of these discs show 6×12 nm rectangles divided into 3 nm halves (Eiserling & Glazer, 1974). Negatively stained preparations of trimeric C-phycocyanin, $(\alpha\beta)_3$, of the red alga *Rhodella violacea*, have recently been reported to show ring-shaped molecules 3.0×10.2 nm which have a large, central hole (Mörschel *et al.*, 1980). Similar studies of B-phycoerythrin, $(\alpha\beta)_6\gamma$, have established that this biliprotein is a disc-shaped molecule 6.4×11.5 nm (Gantt, 1969). *Rhodella violacea* allophycocyanin, $(\alpha\beta)_3$, the only allophycocyanin in solution for which an ultrastructure has been described, is a ring-shaped molecule (3×10 nm) similar in structure to that described above for trimeric C-phycocyanin. The crystalline allophycocyanin of *Anabaena* 6411 is composed of stacks of $\sim 3 \times 12$ nm discs (Bryant *et al.*, 1976).

The results of low resolution X-ray diffraction studies of C-phycocyanin (Dobler *et al.*, 1972; Hackert *et al.*, 1977; Sweet *et al.*, 1977; Fisher *et al.*, 1980) and B-phycoerythrin (Sweet *et al.*, 1977; Abad-Zapatero *et al.*, 1977; Fisher *et al.*, 1980) are generally consistent with the results obtained by electron microscopy. The crystalline C-phycocyanin of *A. variabilis* is a ring-shaped molecule 11 nm in diameter and 3 nm thick, with a central channel approximately 2 nm in diameter (Fisher *et al.*, 1980). The thickness of the crystalline molecule is significantly smaller than that found for C-phycocyanins by electron microscopy; possible explanations for this observation have been discussed in detail (Fisher *et al.*, 1980). The crystalline B-phycoerythrin molecule is 10.7 nm in diameter and 5.5 nm thick, and has a shallow groove about its equator (Fisher *et al.*, 1980). The γ subunit lies on the crystallographic three-fold symmetry axis at the centre of the molecule (Abad-Zapatero *et al.*, 1977; Sweet *et al.*, 1977; Fisher *et al.*, 1980).

Short rods, consisting of stacks of two to ten disc-shaped molecules approximately 12 nm in diameter, have frequently been observed in extracts of cyanobacteria (Kessel *et al.*, 1973; Eiserling & Glazer, 1974; Yamanaka *et al.*, 1978a; Bryant *et al.*, 1979; Guglielmi *et al.*, 1979). These stacks, which are largely composed of C-phycocyanin and occasionally phycoerythrin, usually exhibit strong repeats at ~6 nm intervals along the length of the rods and weaker interperiods at ~3 nm intervals (Eiserling & Glazer, 1974; Bryant *et al.*, 1979). The relationship between these structures and the structure of intact phycobilisomes is discussed in 7.5.3.

7.5.1.7 Immunological relationships of biliproteins

The Ouchterlony double-diffusion technique has been widely used to study the immunological relationships of the various spectroscopic classes of cyanobacterial and red algal biliproteins (see Stanier, 1974; Bogorad, 1975a; MacColl & Berns, 1979). The phycobiliproteins can be divided into three groups on the basis of their immunochemical properties (Glazer, 1981): Group I includes allophycocyanin and allophycocyanin-B; Group II includes C- and R-phycocyanins and phycoerythrocyanin; and Group III includes C, b-, B-, and R-phycoerythrins.

As shown in Table 7.3, which summarizes the data describing these relationships, no cross-reactivity is observed between members of these different groups. The members of Group I are very closely related immunochemically; an antiserum prepared against any allophycocyanin or allophycocyanin-B cross-reacts with any other allophycocyanin or allophycocyanin-B (Glazer *et al.*, 1971; Glazer & Bryant, 1975; Ley *et al.*, 1977; Bryant, 1977). Results obtained for the members of Groups II and III are somewhat more variable; for these groups the source of the immunizing antigen can occasionally affect the results obtained. For example, an antiserum prepared against phycoerythrocyanin cross-reacts with most C-phycocyanins and with R-phycocyanin, but antisera prepared against C- and R-phycocyanins do not cross-react with phycoerythrocyanin (Bryant *et al.*, 1976; Bryant, 1977).

The extreme immunological conservation of the biliproteins might be trivially explained if the bilin chromophores were serving as major antigenic determinants. This, however, does not appear to be the case. Firstly, although both allophycocyanin and C-phycocyanin carry phycocyanobilin chromophores, no cross-reactions between these biliproteins have been detected (Bogorad, 1965; Bennett & Bogorad, 1971; Glazer *et al.*, 1971). Secondly, Vaughan (1964) found that a 20-fold molar excess of a tryptic digest of R-phycoerythrin from *Ceramium rubrum* did not inhibit precipitation of

Table 7.3. Immunological relationships among biliproteins*

	Antiserum to:	Allophycocyanin-B	Allophycocyanin	C-phycocyanin	R-phycocyanin	Phycoerythrocyanin	C-phycoerythrin	b-phycoerythrin	B-phycoerythrin	R-phycoerythrin
		Biliproteins from diverse cyanobacteria and red algae†								
GROUP I	Allophycocyanin-B	+	+	−	−	−	−	−	−	−
	Allophycocyanin	+	+	−	−	−	−	−	−	−
GROUP II	C-phycocyanin	−	−	+	±	−	−	−	−	−
	R-phycocyanin	−	−	±	+	−	−	−	−	
	Phycoerythrocyanin	−	−	±	+	+	−	−	−	
GROUP III	C-phycoerythrin	−	−	−	−	−	+	+	+	+
	B-phycoerythrin	−	−	−	−		±	+	+	+
	R-phycoerythrin		−	−	−		±		+	+

* Data from Vaughan, 1964; Bogorad, 1965; Berns, 1967; Bennett and Bogorad, 1971; Glazer *et al.*, 1971; Glazer and Hixson, 1975; Glazer and Bryant, 1975; Bryant *et al.*, 1976; Bryant, 1977; Gantt and Lipschultz, 1977; Cohen-Bazire *et al.*, 1977; Ley *et al.*, 1977.

† +, Cross-reaction detected; −, no cross-reaction detected; ±, cross-reaction detected with some, but not all, antisera and/or antigens.

either homologous or heterologous phycoerythrins. Thirdly, the fluorescence of the biliproteins is not appreciably quenched by the antigen–antibody reaction (Berns, 1967; Glazer *et al.*, 1971). These results, in agreement with a number of chemical studies (see 7.5.1.5), indicate that the phycobilins do not act as simple haptenic groups because the chromophores are not exposed at the surfaces of the native molecules.

Two studies have examined the immunological properties of phycobiliprotein subunits. Takemoto and Bogorad (1975) reported that an antiserum prepared against the α subunit of *Fremyella diplosiphon (Calothrix* 7601) C-phycoerythrin cross-reacts with the β subunit of this biliprotein, and vice versa. As denatured subunits were used for these studies, the phycoerythrobilin chromophores may have acted as simple haptenic groups in this instance. The immunochemical properties of the renatured subunits of the allophycocyanin of *Synechococcus* 6312 have also been studied (Cohen-Bazire *et al.*, 1977). An antiserum produced against the β subunit of this allophycocyanin cross-reacted with the homologous allophycocyanin and with the renatured β subunits of several other allophycocyanins, but did not cross-react with renatured allophycocyanin α subunits. An antiserum produced against the renatured α subunit cross-reacted with allophycocyanin and with other renatured allophycocyanin α subunits. Moreover, a weak cross-reaction was noted with this antiserum and renatured allophycocyanin β subunits. Thus, the α and β subunits of allophycocyanins appear to share some antigenic determinants, but these determinants were immunogenic for only one of the isolated subunits.

The immunological relationships described above provide a strong indication of the close evolutionary relationship of the cyanobacteria and the red algal chloroplasts. The observation that the major spectroscopic classes of biliproteins do not share common surface determinants suggests that the divergence of the ancestral phycobiliprotein(s), which resulted in the generation of the various spectroscopic classes of these proteins, was an ancient event. The near-identity of the surface determinants within a spectroscopic class indicates that, once the divergence to form these spectroscopic classes had occurred, the surface determinants have changed very slowly. This conservation of surface determinants, which is probably related to the structural restrictions imposed by intermolecular interactions within the phycobilisome, are similarly reflected in studies of the intersubunit contacts of C-phycocyanins and allophycocyanins (see 7.5.1.3) and in studies of the primary structures of the phycobiliproteins (see 7.5.1.8).

7.5.1.8 Amino-acid sequencing of biliproteins

The amino-terminal amino-acid sequences of the α and β subunits of a wide variety of cyanobacterial and red algal phycobiliproteins have been determined. These include those for allophycocyanins (Glazer *et al.*, 1976, Brown & Troxler, 1977; Troxler *et al.*, 1979), C-phycocyanins (Torjesen & Sletten, 1972; Troxler *et al.*, 1975; Harris & Berns, 1975; Williams *et al.*, 1974; Glazer *et al.*, 1976; Frank *et al.*, 1975; Wallin *et al.*, 1978), B-phycoerythrin (Harris & Berns, 1975; Glazer & Hixson, 1977), C-phycoerythrin (Harris & Berns, 1975; Bryant, 1977; Glazer & Hixson, 1977), and R-phycocyanin, phycoerythrocyanin, and b-phycoerythrin (Glazer *et al.*, 1976). The complete amino-acid sequences of the α and β subunits of the C-phycocyanin of the filamentous, heterocyst-forming cyanobacterium *Mastigocladus laminosus* (*Fischerella* 7603) have been determined (Frank *et al.*, 1978). The complete sequence of the α subunit of C-phycocyanin from the red alga *Cyanidium caldarium* has recently been reported (Brown & Troxler, 1977), as has the complete sequence of the β subunit of the C-phycocyanin of the unicellular cyanobacterium *Synechococcus* 6301 (Freidenreich *et al.*, 1978). Additionally, regions containing bilin attachment sites from other C-phycocyanins, *Porphyridium cruentum* R-phycocyanin, *Anabaena variabilis* (*Anabaena* 7118) phycoerythrocyanin, and two C-phycoerythrins have been sequenced (Bryant *et al.*, 1978; Muckle *et al.*, 1978; Brown *et al.*, 1979).

Several important conclusions can be reached from the available data. Firstly, the sequences of all biliprotein subunits fall into two classes—α-type and β-type (Harris & Berns, 1975; Glazer *et al.*, 1976). Secondly, the primary structures of the corresponding subunits of any given phycobiliprotein class are very highly conserved, whether the protein is of eukaryotic or prokaryotic origin (Harris & Berns, 1975; Glazer *et al.*, 1976). For example, the amino-terminal sequences of diverse cyanobacterial and rhodophytan allophycocyanin α- and β-type subunits are virtually identical (Glazer *et al.*, 1976; Brown & Troxler, 1977; Troxler *et al.*, 1979). Similarly, the α subunit of the red algal C-phycocyanin of *Cyanidium caldarium* revealed 85% sequence identity when compared with the α subunit of the C-phycocyanin of *Mastigocladus laminosus* (Brown & Troxler, 1977; Frank *et al.*, 1978). Thirdly, a high degree of homology can be observed, albeit lower than within a given spectral class, when corresponding α and β subunits are compared irrespective of their spectroscopic class (Harris & Berns, 1975; Glazer *et al.*, 1976). Finally, sufficient homology remains between the α- and β-type subunits to establish that the phycobiliproteins constitute a family of proteins descended from a single ancestral gene (Williams *et al.*, 1974; Troxler *et al.*, 1975; Harris & Berns, 1975; Frank *et al.*, 1975).

The sequence homology between the α and β subunits of phycocyanins and phycoerythrins are significantly greater than that between the α and β subunits of these biliproteins and the corresponding allophycocyanin subunits. Zuber (1978) has pointed out that this finding is understandable when one considers the roles that each protein plays in the phycobilisome (7.7). While phycocyanins and phycoerythrins are organized in a similar manner in the peripheral rods (and in certain chromatically adapting cyanobacteria are in fact structurally interchangeable (see 7.8.3.1)), the intermolecular contacts of the allophycocyanin molecules in the core of the phycobilisome are apparently significantly different. None the less, the observed homology suggests that certain aspects of the three-dimensional structures will be conserved among the members of all biliprotein classes irrespective of the type and number of chromophores carried. This is reflected in the similarities observed in the aggregation, assembly, and ultrastructural properties of the phycobiliproteins (see 7.5.1). Sufficient differences exist among the various phycobiliprotein classes to lead to distinct immunological groups (see 7.5.1.7).

Glazer (1976) and Zuber (1978) have discussed hypothetical schemes for the evolution of the biliproteins. The central element of these schemes is duplication of an ancestral gene followed by mutational events leading to divergence, this process thereby leading to the formation of the ancestral genes for the α- and β-type polypeptides. Subsequent gene duplication events, again followed by divergence, could thus have generated the present-day diversity of spectroscopically distinct biliproteins through the attachment of different numbers of chromophores or structurally different chromophores onto basically similar polypeptides. Mechanisms for the evolution of a second binding site, as found on the β subunit of C-phycocyanin, have been discussed (Frank *et al.*, 1978; Zuber, 1978). A comparison of the bilin-attachment sites on the β subunits of C- and R-phycocyanins suggest how a spectroscopically distinct biliprotein can be generated by the substitution of one chromophore type for another (Bryant *et al.*, 1978).

7.5.2 *Uncoloured polypeptides of the phycobilisomes*

Initial investigations indicated that the phycobiliproteins accounted for the total protein composition of isolated phycobilisomes (Gantt & Lipschultz, 1974; Gray *et al.*, 1973; Gray & Gantt, 1975). Tandeau de Marsac and Cohen-Bazire (1977) analysed the polypeptide composition of phycobilisomes from eight species of cyanobacteria by polyacrylamide gel electrophoresis in the presence of sodium dodecylsulphate. All of the phycobilisomes examined contained a small number (four to nine) of uncoloured polypeptides, which accounted for 10–15% of the total phycobilisome protein. The uncoloured polypeptides can be arbitrarily divided into two classes on the basis of apparent molecular masses: Group I, 70 000–120 000 daltons and Group II, 25 000–70 000 daltons.

Convincing evidence that these colourless polypeptides are truly components of the phycobilisomes was obtained by exploiting the phenomenon of complementary chromatic adaptation (see 7.8.3.1). Tandeau de Marsac and Cohen-Bazire (1977) found that like the major phycobilisomal constituents, phycocyanin and phycoerythrin, in cyanobacterial strains capable of chromatic adaptation, the synthesis of certain of the Group II polypeptides is similarly regulated by the wavelength of light to which the cells are exposed during growth. This observation suggests that the photocontrolled polypeptides interact specifically with the phycobiliprotein (either phycocyanin or phycoerythrin) to which their synthesis is coupled.

An immunological technique was used to demonstrate the presence of uncoloured polypeptides in the phycobilisomes of *Synechococcus* 6301 (Yamanaka *et al.*, 1978a). Phycobilisomes, whose sedimentation properties had been altered by the binding of anti-allophycocyanin antibodies, were isolated on sucrose gradients. The phycobilisome–antibody complex contained all of the colourless polypeptides present in native phycobilisomes. Substitution of the zwitterionic detergents Miranol S2M-SF and Deriphat 160 for the nonionic detergent Triton X-100 had no effect on the uncoloured polypeptides found associated with cyanobacterial phycobilisomes from *Synechococcus* strains 6301 and 6312 (Glazer *et al.*, 1979a). Uncoloured polypeptides have also been reported to be components of the phycobilisomes of the red algae *Rhodella violacea*, *Porphyridium cruentum*, and *P. aerugineum* (Koller *et al.*, 1978; Yamanaka *et al.*, 1978a).

The uncoloured polypeptides have been postulated to play a number of roles in the structure and assembly of the phycobilisome (Tandeau de Marsac & Cohen-Bazire, 1977; Bryant *et al.*, 1979). The polypeptides of Group I, which may also be components of the thylakoids, could serve to attach the phycobilisomes to the thylakoids. These components are probably a part of the allophycocyanin-containing cores of the phycobilisomes and may also play some role in the stabilization and/or assembly of this core unit.

The polypeptides of Group II are found in the peripheral rods of the phycobilisomes (Bryant *et al.*, 1979; Bryant, 1979, 1981; Bryant & Cohen-Bazire, 1981). These polypeptides may facilitate the binding of the rods to the core, or control and stabilize the interactions of phycocyanin with phycoerythrin or other phycocyanin molecules. The uncoloured polypeptides of Group II may be necessary for the formation of the peripheral rod structures (Bryant *et al.*, 1979; see 7.5.4.1 and 7.5.3). Although all phycobiliproteins are acidic proteins with isoelectric points in the range pH 4–5.5, all of the uncoloured polypeptides of *Synechococcus* 6301 phycobilisomes are basic. It has been suggested that electrostatic interactions between these basic polypeptides and the acidic biliproteins could significantly stabilize the phycobilisome assembly (Glazer *et al.*, 1979c).

Glazer and coworkers have reported that phycobilisomes of *Synechocystis* 6701 mutants, which are totally or partially lacking in phycoerythrin, are also missing two uncoloured polypeptides of Group II (Glazer *et al.*, 1979c; Glazer, 1981). The assembly and function of the residual structures are unaffected by the removal of the phycoerythrin and these polypeptides. Similar results were obtained with wild-type *Synechocystis* 6701 through growth in chromatic illumination; growth in red light suppresses the formation of phycoerythrin and the same two uncoloured polypeptides, but the absence of these components likewise had no effect on the assembly of the residual components (Bryant *et al.*, 1979; Glazer, 1981). These results support the hypothesis that certain of the uncoloured polypeptides of Group II interact specifically with individual biliproteins.

Recent developments from Glazer's laboratory reinforce this hypothesis (Lundell *et al.*, 1981). Four of the five uncoloured polypeptides—linker polypeptides—associated with the phycobilisomes of *Synechococcus* 6301 (Yamanaka *et al.*, 1978a) have been isolated, purified and their amino-acid compositions determined. The three linker polypeptides of molecular masses 33K, 30K and 27K have similar amino-acid compositions, but their peptide maps differ. The amino-acid composition of the 75K polypeptide bears no relation to the three others. A function for the 33K polypeptide is proposed: the stabilization of the phycocyanin hexamers at pH 8.

7.5.3 Multiprotein complexes from phycobilisomes

Bryant (1979, 1981); Bryant & Cohen-Bazire (1981) have recently described two multiprotein complexes isolated from the phycobilisomes of LPP* (*Lyngbya–Plectonema–Phormidium* Group; Rippka *et al.*, 1979) 7409, a chromatically adapting cyanobacterium which can control the rates of synthesis of both phycocyanin and phycoerythrin (see 7.8.3). By utilizing DEAE–cellulose chromatography at pH 5.5, a complex of phycocyanin and phycoerythrin (molar ratio 1 : 1) was purified from dissociated phycobilisomes prepared from cells grown in green light. This complex, which has a sedimentation coefficient of 17 *s*, contained the five uncoloured polypeptides of Group II which occur in the phycobilisomes of this organism when grown in green light. After fixation and negative staining, the complex appeared in the electron microscope as short stacks consisting of two $\sim 6 \times 12$ nm discs.

Phycobilisomes, prepared from cells of *Pseudanabaena* 7409 grown in red light, contained a very stable complex consisting of two types of phycocyanin, the three uncoloured polypeptides of Group II which occur in the red-light phycobilisomes of this organism, and a previously undetected polypeptide with a molecular mass of $\sim$ 10 000 daltons. This complex had a sedimentation coefficient of 19–21 *s*, although approximately 10% of the complex had a sedimentation coefficient of 24 *s* or greater. When examined in the electron microscope after fixation and negative staining, the complex appeared as short stacks of two to four disc-shaped subunits 12 nm in diameter and $\sim$6 nm thick.

The subunits of the two phycocyanins of the red-light complex were purified by ion-exchange chromatography. The four subunits (two α and two β subunits) differed with respect to charge as well as molecular weight. The amino-acid compositions of the two α and two β subunits are similar, but none the less distinctive; each of the four subunits appears to be a distinct gene product. One pair of the subunits is identical in charges and molecular weights to those found in green light. These results indicate that one phycocyanin is synthesized constitutively (i.e. irrespective of the wavelength of light to which the cells are exposed during growth) while the second phycocyanin is specifically induced during growth in red light.

* This strain has now been reclassified as *Pseudanabaena* 7409 (Guglielmi, pers. comm. and hereafter in the text will be so designated.

A complex similar to those isolated from *Pseudanabaena* 7409 has been isolated from the phycobilisomes of the alga *Rhodella violacea* (Koller *et al.*, 1978). The ~20 *s* particles, called 'tripartite units', contain two types of B-phycoerythrin and phycocyanin (molar ratio 2:1) but no allophycocyanin, and consist largely of stacks of three disc-shaped subunits. Gantt (1980) has also recently described complexes of phycoerythrin, phycocyanin, and uncoloured proteins which were prepared from the phycobilisomes of a *Nostoc* species and the red alga *Porphyridium sordidum*.

Ultrastructurally, these complexes are virtually identical to those of the peripheral rods of isolated intact phycobilisomes (7.7). This similarity strengthens the arguments that the phycocyanin, phycoerythrin (when present), and Group II linker polypeptides are contained in the peripheral rods of the phycobilisomes while the allophycocyanin and the linker polypeptides of Group I are contained in the core of the phycobilisomes (Bryant *et al.*, 1979). The rod-shaped, ~19–25 *s* particles commonly detected in crude extracts of cyanobacteria and red algae and in dissociated phycobilisome preparations probably arise from the peripheral rods of the phycobilisomes as well (see 7.5.1.4).

7.6 Physical properties of isolated phycobilisomes

The main criteria for the characterization of isolated phycobilisomes are their absorption and fluorescence emission spectra.

The absorption spectra of phycobilisomes in the visible region reflect the biliprotein composition of the cells from which they have been prepared. It should be noted, however, that the absorption spectra of intact and dissociated phycobilisomes (after dialysis in buffers of low concentration) are slightly different. These differences can be attributed to changes in the absorption properties of phycoerythrin and allophycocyanin, whereas little or no change occurs in the region of phycocyanin absorbance (Yamanaka *et al.*, 1978a; Bryant *et al.*, 1979). In the region of phycoerythrin absorbance, the spectral differences are similar to those observed between concentrated and dilute solutions of the purified pigment (Bennett & Bogorad, 1971; Gantt & Lipschultz, 1974; Gray & Gantt, 1977). In intact phycobilisomes, the absorption maximum of phycoerythrin can be shifted by as much as 10 nm towards longer wavelengths. An enhanced absorption of allophycocyanin in intact phycobilisomes has also been observed for the phycobilisomes of the red alga *Rhodella violacea* (Koller *et al.*, 1977) and of cyanobacteria (Bryant *et al.*, 1979). These differences in absorption properties probably reflect changes in chromophore environment or chromophore–chromophore interactions which vary with the aggregation state of the biliproteins (see 7.5.1.5).

The most meaningful criterion used to ascertain the integrity of phycobilisomes is provided by their fluorescence emission spectra (Gantt & Lipschultz, 1973; Grabowski & Gantt, 1978; Bryant *et al.*, 1979; Gantt *et al.*, 1979). Fast kinetic studies of energy transfer performed with intact cells of *Porphyridium cruentum* have confirmed the following energy transfer sequence: phycoerythrin → phycocyanin → allophycocyanin → allophycocyanin-B → chlorophyll *a* (Porter *et al.*, 1978). When phycobilisomes are detached from the thylakoid membrane, this energy transfer chain is interrupted and excitation energy is dissipated as fluorescence of the last components in the transfer chain: allophycocyanin and allophycocyanin-B (Searle *et al.*, 1978).

Whether phycoerythrin or phycocyanin-rich, the major fluorescence emission maximum of intact phycobilisomes at room temperature, independent of the wavelength used for excitation, is situated between 667 and 675 nm and between 678 and 685 nm at liquid nitrogen temperature. This long wavelength fluorescence emission can be satisfactorily accounted for by the fluorescence of allophycocyanin and that of the small amount of allophycocyanin-B generally present in these organelles (Glazer & Bryant, 1975; Ley *et al.*, 1977; Searle *et al.*, 1978). In phycoerythrin-rich phycobilisomes, a minor fluorescence emission band of phycoerythrin around 580 nm is generally observed. The high efficiency of energy migration within the phycobilisomes is reflected by the very low degree of fluorescence polarization, values between +0.02 and −0.02 being observed, compared to that of individual component biliproteins, for which the degree of polarization is two to twelve times greater (Grabowski & Gantt, 1978). Similarly, the fluorescence quantum yields of phycobilisomes (observed values between 0.60 and 0.68) are similar to that of allophycocyanin in solution (0.68) (Grabowski & Gantt, 1978).

Dilution of phycobilisomes in low ionic strength buffers leads to the dissociation and uncoupling of their component phycobiliproteins (Gantt *et al.*, 1976a,b; Gantt *et al.*, 1979). In dissociating cyanobacterial phycobilisomes, a shift of the long wavelength fluorescence (670–675 nm) to shorter wavelength (660 nm) characteristic of allophycocyanin occurs rapidly (30 s in 50 mmol K phosphate buffer at pH 6.8), indicating the rapid uncoupling of allophycocyanin-B. This shift is followed by a gradual increase in the fluorescence of the major component phycobiliproteins.

The molecular masses and sedimentation coefficients of phycobilisomes extracted from several cyanobacteria and two red algae, calculated from their phycobiliprotein content and/or by ultracentrifugation analysis are given in Table 7.4. The disc-shaped phycobilisomes of cyanobacteria or of the red alga *Rhodella violacea* have molecular masses in a range extending from 4.5 to 8×10^6 daltons. The estimated molecular masses of the phycobilisomes of *Porphyridium cruentum* (globular) and *Gloeobacter* (bundle-shape) fall in the range of 15 to 23×10^6 daltons.

Table 7.4. Sedimentation coefficients and estimated molecular masses of phycobilisomes

Source			s_{20}, W	Molecular mass: $\times 10^6$ daltons
Cyanobacteria				
Synechococcus	6301*	(Glazer *et al.*, 1979b)	61–63	ND
Synechococcus	6312*	(Glazer *et al.*, 1979b)	55	ND
Synechococcus	6312†	(Glazer *et al.*, 1979b)	48.2	ND
Synechococcus	6312‡	(Bryant *et al.*, 1979)	35–40	4·5–5
Pseudanabaena	7409‡	(Bryant *et al.*, 1979)	47	7–8
Synechocystis (green light)	6701‡	(Bryant *et al.*, 1979)	47	7–8
Synechocystis (red light)	6701‡	(Bryant *et al.*, 1979)	ND	4·5
Anabaena	6411	(Bryant, 1977)	42	4·5
Gloeobacter violaceus		(Guglielmi, Cohen-Bazire & Bryant, unpublished)	ND	15–20
Rhodophyta				
Rhodella violacea		(Koller *et al.*, 1978)	ND	5.6–6
Porphyridium cruentum		(Gantt, 1975)	ND	19–23

ND. Not determined.
*† Corrected sedimentation coefficients. Phycobilisomes extracted with *, Triton X-100; Deriphat.
‡ Estimated values.

7.7 Structure of phycobilisomes

A model for the structural organization of *Porphyridium cruentum* and *Nostoc* sp. phycobilisomes has been proposed from the study of the sequential release of their component biliproteins during dissociation experiments (Gantt *et al.*, 1976a,b; Gantt & Lipschultz, 1977). Using monospecific antisera and ferritin-conjugates, Gantt and Lipschultz (1977) demonstrated by immuno-electron microscopy the location of the major biliproteins in *Porphyridium cruentum* phycobilisomes still attached to thylakoid membrane fragments. The proposed model is consistent with energy transfer requirements: allophycocyanin forms a core closest to the thylakoid membrane, surrounded by a layer of phycocyanin, with phycoerythrin situated at the periphery.

The size and complexity of *P. cruentum* phycobilisomes has hitherto precluded a detailed analysis of their fine structure. Electron micrographs of negatively stained preparations reveal 'finger-like projections' at their periphery, consisting of stacks of disc-shaped subunits (Gantt, 1975, 1980).

A detailed fine structure has been obtained only for the much simpler disc-shaped phycobilisomes of the red alga *Rhodella violacea* (Mörschel *et al.*, 1977; Koller *et al.*, 1978) and of several cyanobacteria (Bryant *et al.*, 1979; Glazer *et al.*, 1979b).

Selected views of glutaraldehyde-fixed, negatively stained cyanobacterial phycobilisomes are shown in Figs 7.10–7.16.

The phycobilisomes shown in Figs 7.10 to 7.13 have been extracted from two phycoerythrin-synthesizing chromatically adapting strains (see 7.8.3.1): *Pseudanabaena* 7409 and *Synechocystis* 6701. In both strains, phycoerythrin synthesis is promoted by green light and inhibited by red light. Red light provokes an increased synthesis of phycocyanin in *Pseudanabaena* 7409 whereas in *Synechocystis* 6701, phycocyanin synthesis remains the same as in green light (Tandeau de Marsac, 1977). As a result, the biliprotein content of *Pseudanabaena* 7409 is approximately the same after growth in green or red light whereas that of *Synechocystis* 6701 is reduced in red light compared to green light (see Table 7.5). The phycobilisomes from red- and green-light grown *Pseudanabaena* 7409 are shown in Figs 7.10 and 7.11 and those from red- and green-light grown *Synechocystis* 6701 in Figs 7.12 and 7.13, respectively.

Figs 7.14 and 7.15 present phycobilisomes from two different *Synechococcus* strains which do not synthesize phycoerythrin, *Synechococcus* 6312 (Fig. 7.14) and *Synechococcus* 6301 Fig. 7.15). Fig. 7.16 shows the more complex phycobilisomes of *Gloeobacter violaceus*.

These phycobilisomes all share a common feature: rod-shaped elements 12 nm wide showing strong repeats at 6 nm intervals and weaker ones at 3 nm intervals perpendicular to their long axis. The length of the rods varies in

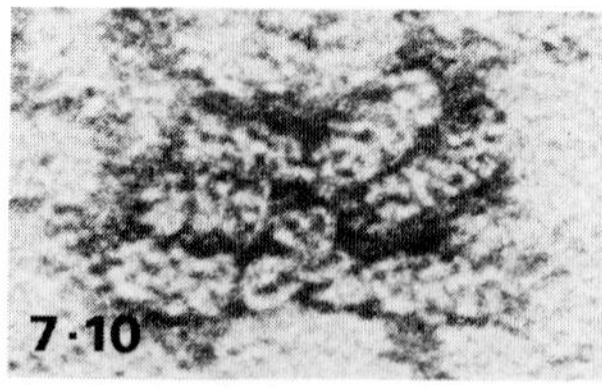

7·10

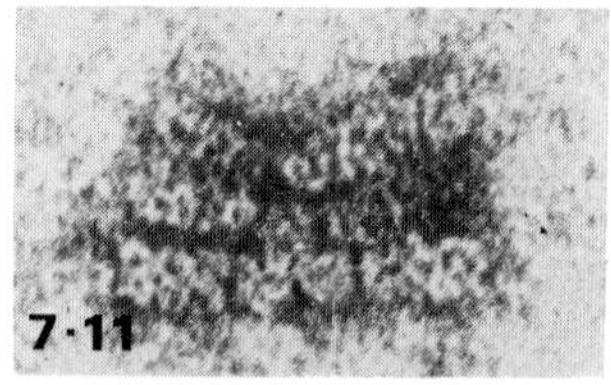

7·11

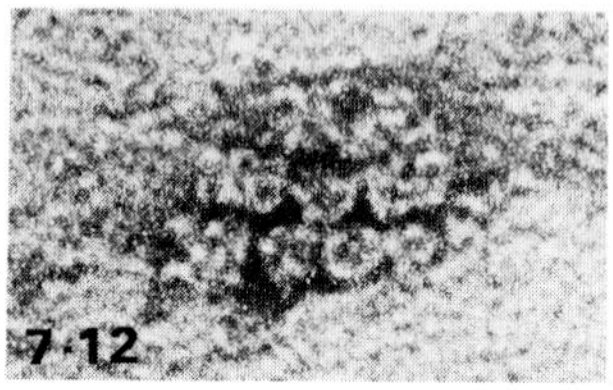

7·12

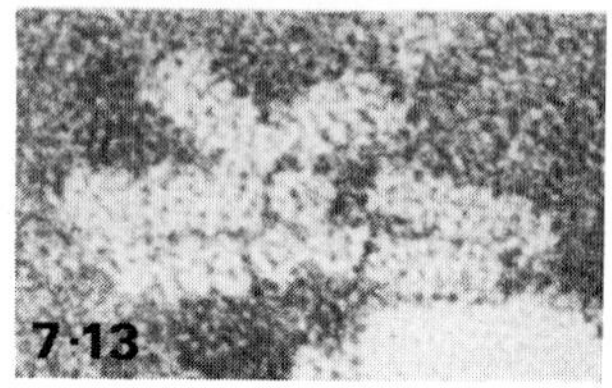

7·13

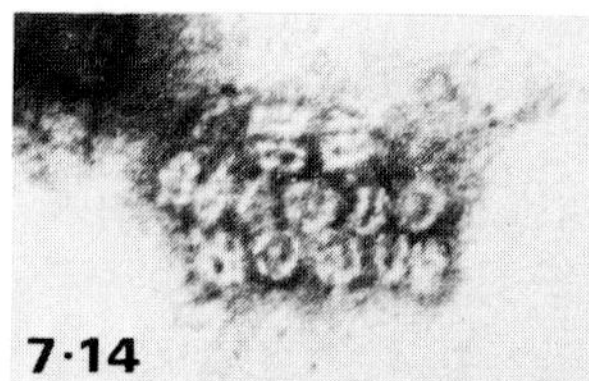

7·14

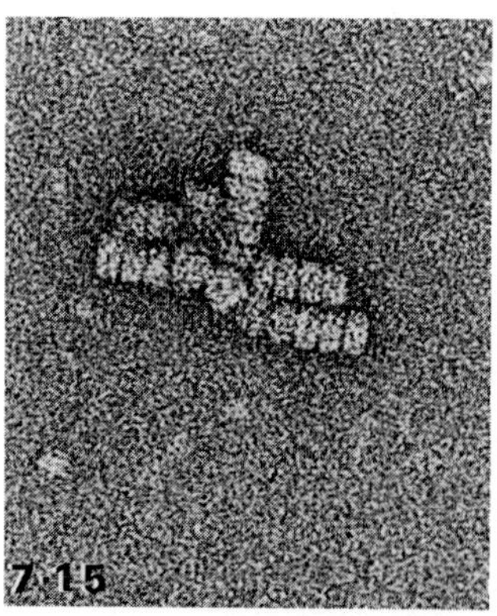

7·15

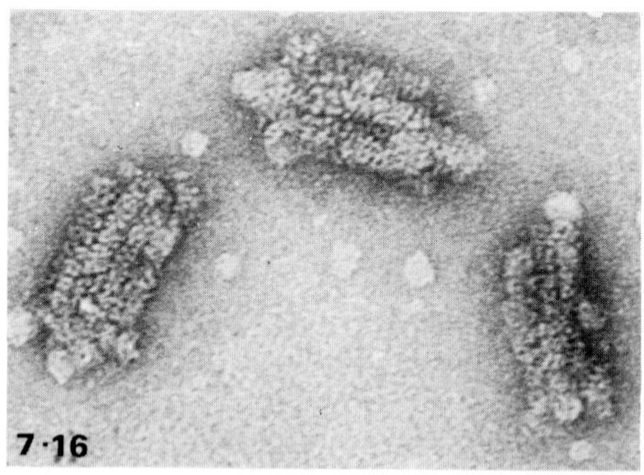

7·16

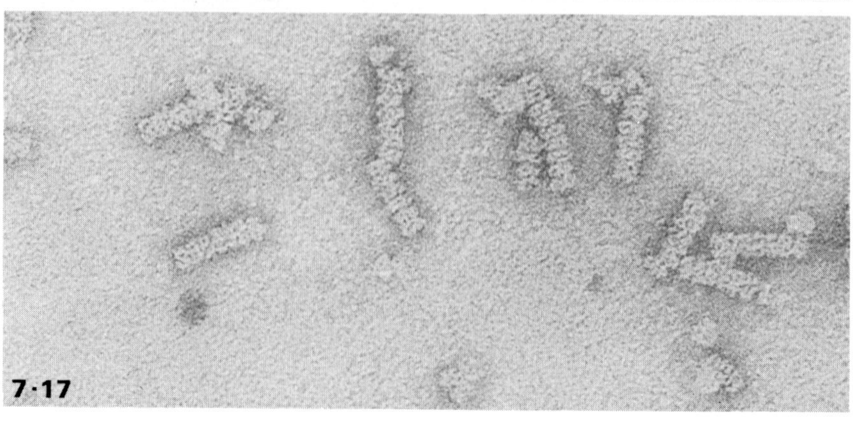

7·17

different strains and with the conditions of cultivations. In dimensions, the discs which compose the rods correspond to those of hexamers of phycocyanin or phycoerythrin (see 7.5.1.6). The rod-shaped assemblies are similar in structure to the stacks of phycocyanin hexamers described by Eiserling and Glazer (1974) in extracts of *Synechococcus* 6301 and to those described by Kessel *et al.* (1973) in crude extracts of *Calothrix parietina.*

The phycobilisomes shown in Figs 7.10–7.15 have the same fundamental structure. Their contour is roughly semicircular and reminiscent of the shape of discoidal phycobilisomes in face view observed in thin sections. Their centre is occupied by three ring-shaped subunits 10–12 nm in diameter forming a triangular core in Figs 7.10–7.14, or by two ball-shaped subunits ~ 10 nm in diameter in the phycobilisomes of *Synechococcus* 6301 (Fig. 7.15). From two sides of the core, forming a semicircle, radiate the rod-shaped assemblies. A structure composed of six rods radiating from the core has been interpreted as representing an intact phycobilisome (Bryant *et al.*, 1979). A detailed model of cyanobacterial phycobilisome structure has been proposed by these authors based on the comparison between the observed structure and the biliprotein molar composition of phycobilisomes with a triangular core such as those shown in Figs 7.10–7.14. Examination of the data presented in Table 7.5 shows the close correspondence between the molar ratios of phycocyanin/allophycocyanin or of (phycocyanin + phycoerythrin)/allophycocyanin and

Figs 7.10–7.15. Selected views of negatively stained hemidiscoidal phycobilisomes (× 300 000).

Figs 7.10 and 7.11. Phycobilisomes from red (Fig. 7.10) and green (Fig. 7.11) light grown cells of *Pseudanabaena* 7409.

Figs 7.12 and 7.13. Phycobilisomes from red (Fig. 7.12) and green (Fig. 7.13) light grown cells of *Synechocystis* 6701.

Fig. 7.14. Phycobilisome from *Synechococcus* 6312. The core of these phycobilisomes is 'triangular', composed in face view by three ring-shaped subunits. The rod-shaped structures radiating from two sides of the core are formed by the coaxial stacking of double discs, 6 nm thick. The number of 6 nm double discs in the rods varies with the molar ratios of phycocyanin/allophycocyanin or of (phycocyanin + phycoerythrin)/allophycocyanin, found in these phycobilisomes (see Table 7.5 and Sections 7.7 and 7.8.3.1).

Fig. 7.15. Phycobilisome from *Synechococcus* 6301. The core is composed of two ball-shaped subunits. The rod-shaped assemblies radiating from the core are similar in structure to those observed in the phycobilisomes with a triangular core (× 300 000). Electron micrograph kindly provided by Dr A. N. Glazer.

Fig. 7.16. Bundle-shaped phycobilisomes isolated from *Gloeobacter violaceus* (× 300 000).

Fig. 7.17. Rod-shaped assemblies isolated from disintegrated *Gloeobacter* phycobilisomes (× 300 000).

the average number of disc-shaped subunits in the peripheral rods of intact phycobilisomes of various biliprotein compositions. These data are consistent with the location of phycocyanin and phycoerythrin in the rods while the core is composed of allophycocyanin. If this is the case, the core must contain 12 trimeric units of allophycocyanin. From the occasional observation of such phycobilisomes standing with their base 'up', the basal subunits of the core were seen to consist of stacks of two double discs, 6–7 nm thick, 10–12 nm in diameter, separated by a distance of 1–2 nm.

Photographs of the scale model proposed by Bryant *et al.* (1979) for the structure of hemidiscoidal phycobilisomes is seen in Fig. 7.18 in face view and Fig. 7.19 with its base 'up'. A model, less detailed but similar in general features, has been proposed for the phycobilisomes of *Rhodella violacea* by Koller *et al.* (1978).

The model shown here reflects the pigment composition and distribution of a phycobilisome from *Pseudanabaena* 7409 grown in green light (see Table

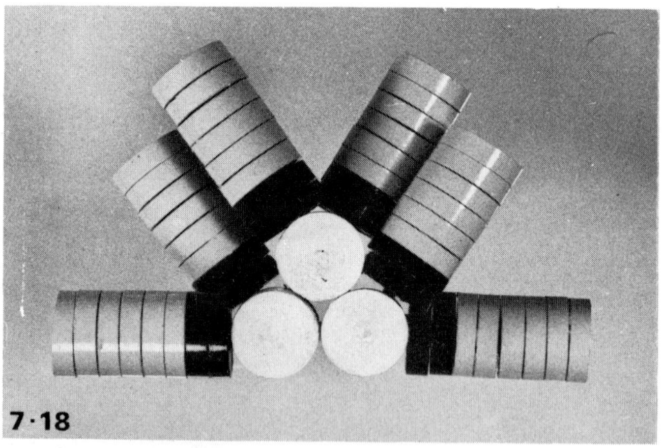

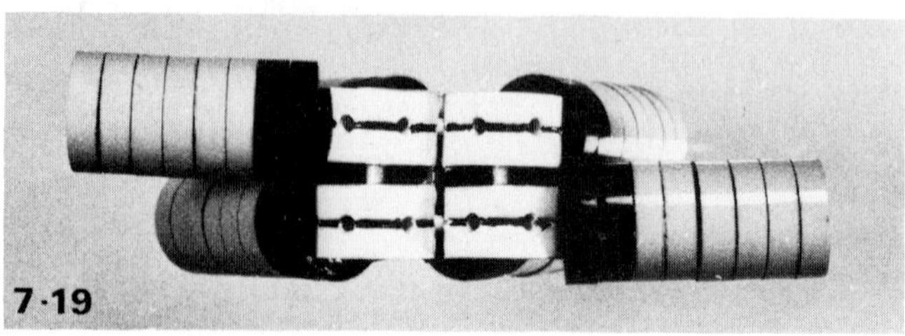

Figs 7.18 and 7.19. Photographs of two aspects of the model of a hemidiscoidal phycobilisome presented by Bryant *et al.* (1979). The model reflects the pigment composition and distribution of a phycobilisome from *Pseudanabaena* 7409 grown in green light. The six white double discs represent allophycocyanin molecules. The peripheral rods are stacks of disc-shaped hexamers $2(\alpha\beta_3)$ of phycocyanin (black discs) and phycoerythrin (grey discs).

Table 7.5. Relative molar phycobiliprotein composition and peripheral rod length of phycobilisomes

	AP*	PC*	PE*	PC/AP or PC+PE/AP	Average number of discs (6 nm thick) in peripheral rods†
Synechococcus 6312	1	1.5–2	0	1.5–2	1–2
Synechocystis 6701 (red light)	1	1.63	0.27	1.90	2.2
Synechocystis 6701 (green light)	1	1.70	1.98	3.68	3.60
Pseudanabaena 7409 (red light)	1	3.44	0.04	3.48	3.45
Pseudanabaena 7409 (green light)	1	0.94	2.97	3.91	3.74
Rhodella violacea‡	1	1	2	3	3

* Molar values estimated spectrophotometrically (Bryant *et al.*, 1979).
† Average values obtained from 25–50 phycobilisomes.
‡ Values estimated in Koller *et al.* (1978).

7.5). The six white double discs (6 × 12 nm) represent 12 trimeric molecules of allophycocyanin. The six peripheral rods represent stacks of disc-shaped hexamers $(\alpha\beta)_6$: the black discs correspond to hexamers of phycocyanin and the grey discs to hexamers of phycoerythrin. Each peripheral rod is in contact with one double disc of allophycocyanin at an equivalent site, if one assumes a six-fold symmetry for each double disc of the core. Steric hindrance allows the positioning of only six rods around two sides of the triangular core. Equivalent rod-binding positions at the base of the core would be blocked when the phycobilisome is in contact with the thylakoid membrane. Numerous observations support the proposed specific locations of the major phycobiliproteins within the phycobilisomes. Rod-shaped aggregates structurally identical to the peripheral rods of *Rhodella violacea* were isolated and characterized by Koller *et al.* (1978) (see 7.5.3). These 'tripartite units' contained phycocyanin and phycoerythrin in the same proportions as intact phycobilisomes, i.e. 1:2, but were devoid of allophycocyanin. Rod-shaped elements made up of two to four stacked disc-shaped subunits (~6 × 12 nm) have been isolated from dissociated phycobilisomes of *Pseudanabaena* 7409 grown in red or green light (see 7.5.3). These complexes composed of isophycocyanins or of phycocyanin and phycoerythrin, accompanied by their specific linker polypeptides, did not contain any allophycocyanin or allophycocyanin-B.

In the model described above, the position of allophycocyanin-B is not shown. This biliprotein, detected in many cyanobacterial and red algal phycobilisomes, represents between 0.5 and 1% of their biliprotein complement, which would correspond to one trimeric unit per phycobilisome (Glazer & Bryant, 1975; Ley *et al.*, 1977). One trimeric molecule of allophycocyanin-B could be accommodated between the double discs of allophycocyanin, in the centre of the core, in a position bridging allophycocyanin and the chlorophyll molecules of the thylakoid, a location consistent with its postulated role in energy transfer.

Although this model may represent the structure of many cyanobacterial phycobilisomes, it cannot be applied as such to the phycobilisomes of *Synechococcus* 6301 (Glazer *et al.*, 1979b) shown in Fig. 7.15 in which the core is composed of two discs or ball-shaped subunits.

Negatively stained phycobilisomes of *Gloeobacter violaceus* are shown in Fig. 7.16. They consist of bundles of six long rod-shaped elements, 50–70 nm long, made up of eight to ten double discs 6 nm thick connected to a base of ill-defined structure. The rods resemble the cylindrical elements observed in freeze-etched preparations (Fig. 7.7). Their fine structure with a strong repeat at 6 nm intervals and a weaker one at 3 nm is identical to that of the peripheral rods of hemidiscoidal phycobilisomes. Isolated rods identical in fine structure to those of whole phycobilisomes (Fig. 7.17) contain equimolar amounts of phycocyanin and phycoerythrin, but no allophycocyanin. The structure at the base of the phycobilisomes is therefore assumed to contain allophycocyanin; this biliprotein has been calculated to represent 15–18% by weight of the biliprotein content of *Gloeobacter* phycobilisomes (Guglielmi *et al.*, 1979). It should be noted that the fluorescence emission spectrum of *Gloeobacter* phycobilisomes shows a main maximum at 663 nm, which corresponds to the fluorescence emission of allophycocyanin. Allophycocyanin-B seems to be absent from *Gloeobacter* phycobilisomes (Guglielmi *et al.*, 1979).

In summary, all phycobilisomes, whether from red algae or cyanobacteria, seem to be built from modular elements: rod-shaped assemblies of coaxially stacked hexameric molecules of phycocyanin or of phycocyanin and phycoerythrin, converging on core structures composed of allophycocyanin. Linker polypeptides appear to be essential elements for the assembly of these organelles but their specific locations remain to be determined.

7.8 Effects of environmental factors on phycobilisomes

The phycobiliprotein content and composition of cyanobacteria can be affected by numerous environmental factors (Fritsch, 1945; Halldal, 1970; Stanier & Cohen-Bazire, 1977). These factors include temperature, carbon dioxide concentration (Halldal, 1970; Eley, 1971; Goedheer, 1976; Miller &

Holt, 1977), phosphate concentration (Ihlenfelt & Gibson, 1975b), sulphate starvation (Schmidt; pers. comm.), nitrogen depletion (Fujita & Hattori, 1960a,b; Allen & Smith, 1969; Lau *et al.*, 1977; Yamanaka & Glazer, 1980), light intensity (Halldal, 1958; Myers & Kratz, 1955; Allen, 1968a) and light wavelength (see Bogorad, 1975a for a review; Tandeau de Marsac, 1977).

Among the above-mentioned factors, the influence of light intensity and light wavelength and the effects of nitrogen depletion have been studied in most detail.

7.8.1 Nitrogen starvation

Allen and Smith (1969) determined quantitatively the amount of phycocyanin in the non-nitrogen-fixing cyanobacterium *Synechococcus* 6301, growing photosynthetically in a medium containing limiting amounts of nitrate. The depletion of nitrate coincided with the arrest of protein synthesis and the rapid disappearance of measurable phycocyanin, while the chlorophyll content of the cells was little affected. The colour of the culture turned from emerald-green to yellow-green. Addition of nitrate to the chlorotic culture restored protein synthesis immediately, and phycocyanin synthesis after a lag period of a few hours.

Under conditions of nitrogen starvation, the disappearance of measurable phycocyanin coincides with the coordinated degradation of both α and β subunits of phycocyanin, but protein synthesis continues, since labelled amino acids are incorporated into cellular proteins (Lau *et al.*, 1977).

Yamanaka and Glazer (1980) have recently analysed in some detail the effects of nitrogen deprivation on the phycobilisomes of *Synechococcus* 6301. The first stage of nitrogen starvation (0–48 h) is manifested by a preferential loss of phycocyanin and of one linker polypeptide (30K) (see 7.5.2). Phycobilisomes become smaller, their sedimentation velocity decreases accompanied by an increase in the ratio of allophycocyanin to phycocyanin, but the remaining particle shows unimpaired energy transfer. Prolonged nitrogen starvation leads to the degradation of two other linker polypeptides (33K and 75K) and presumably to the detachment of the phycobilisomes from the thylakoid membranes, followed by their complete breakdown. In the cyanobacterium *Pseudanabaena* 7409, nitrogen starvation provokes a decrease in the number of phycobilisomes which parallels the simultaneous degradation of phycoerythrin and phycocyanin. The phycobilisomes extracted after 72 h of starvation have the same apparent sedimentation coefficient as those of unstarved cells; two of their linker polypeptides are partly degraded (Tandeau de Marsac, 1978, and unpubl.).

The demonstration of continued protein synthesis at the expense of biliprotein breakdown in cells deprived of nitrogen suggests the operation of specific proteolytic enzyme(s). Such proteolytic activity has been demonstrated associated with the membrane fraction of *Anabaena* strains (Foulds & Carr, 1977; Wood & Haselkorn, 1977). The detergent-solubilized enzyme of *Anabaena* degrades specifically the subunits of phycocyanin and allophycocyanin, as well as three other polypeptides among the soluble proteins of the same organism (Wood & Haselkorn, 1979). Biliprotein breakdown under nitrogen starvation is a direct function of the cellular metabolic activity, which depends in turn both on light intensity and the availability of carbon dioxide. The biliprotein proteolytic activity assayed *in vitro* correlates with the rate at which biliproteins are degraded *in vivo* (Wood & Haselkorn, 1979). Inhibition of protein synthesis by chloramphenicol or methionine deprivation in methionine auxotrophic mutants prevents phycobiliprotein and phycobilisome degradation (Wood & Haselkorn, 1979; Yamanaka & Glazer, 1980). The addition of chloramphenicol after the onset of phycobilisome breakdown arrests the process, indicating a high turnover rate for the specific proteolytic enzyme(s) (Yamanaka & Glazer, 1980).

In summary, the available data relevant to the effects of nitrogen depletion on phycobilisomes suggest that: (1) biliproteins represent a major and specific nitrogenous reserve; (2) biliprotein degradation is mediated by an inducible, membrane-bound protease (or proteases) with a high turnover rate; (3) the activity of such enzyme(s) is directly related to the metabolic activity of the cells; (4) chlorophyll *a*, the only photopigment remaining after biliprotein destruction in nitrogen-starved cells, assumes the role of primary light receptor for the operation of PS II (Lemasson *et al.*, 1973).

7.8.2 *Light intensity*

Within certain limits, an inverse correlation between light intensity and photopigment content is general among photosynthetic organisms. Cyanobacteria are no exception to this rule (Myers & Kratz, 1955; Allen, 1968a). As might be expected, the ratio of phycocyanin to chlorophyll can vary widely in response to simultaneous changes in light intensity, temperature and carbon dioxide partial pressure (Goedheer, 1976). The 'inverse chromatic control' observed by Jones and Myers (1965) in *Anacystis nidulans* grown in red light has also been interpreted by the authors as an intensity control. Such studies have not included electron-microscopic observations or quantitative extractions of phycobilisomes; is it the number of phycobilisomes or their size that determines the observed variations? Unpublished observations of Yamanaka and Glazer (in Glazer, 1981) suggest that an increase in phycocyanin to allophycocyanin ratio in *Synechococcus* 6301 growing under low light intensity is reflected by a lengthening of the peripheral rods in the phycobilisomes of this strain (see Fig. 7.15).

In the red algae, *Griffithsia pacifica* (Waaland *et al.*, 1974) and *Porphyridium cruentum* (Brody & Emerson, 1959; Gantt, 1975) the ratio of phycoerythrin to chlorophyll *a* is inversely related to light intensity. *Griffithsia pacifica* grown at 50 ftc contains twice as much phycoerythrin per unit chlorophyll *a* than alga grown at 300 ftc; correlatively, the number of phycobilisomes per unit surface of thylakoids is nearly double but the biliprotein composition of the alga remains the same (Waaland *et al.*, 1974). Similarly, the increased ratio of phycoerythrin to chlorophyll in *Porphyridium cruentum* growing at low light intensity reflects an increase in the density of phycobilisomes on the surface of the thylakoids; their size and shape remains the same and the relative proportions of their component biliproteins is unchanged (Gantt, 1975).

7.8.3 Light wavelength: chromatic adaptation

7.8.3.1 Effect on phycobilisome structure

Whereas light intensity adaptation is general among phototrophs, chromatic adaptation is specific to cyanobacteria and, among them, to certain species capable of C-phycoerythrin synthesis* (Bogorad, 1975a; Tandeau de Marsac, 1977).

In chromatically adapting strains, growth in green light stimulates the synthesis of the red-coloured phycoerythrin (absorbing in the green) whereas when growth occurs in red light it is the blue protein, phycocyanin (absorbing in the red), which is the dominant biliprotein.

The earlier observation of Boresch (1922), that phycoerythrin synthesis is a necessary but not sufficient attribute for the phenomenon to take place, was confirmed by Tandeau de Marsac (1977) in an extensive survey of 44 phycoerythrin-synthesizing strains belonging to five taxonomic sections (Rippka *et al.*, 1979). Three groups were thus defined. The first group does not adapt chromatically; the ratio of phycoerythrin to phycocyanin remains constant whether cells are grown in green, red or white light. In strains of group II, phycoerythrin synthesis alone is subjected to chromatic control; its synthesis is promoted in green light, repressed in red light, whereas the rate of phycocyanin synthesis is constant under both light regimes. As a result, in strains of this group, the cellular amount of phycobiliprotein after growth in red light is reduced compared to that in green light.

In strains of group III the synthesis of both phycoerythrin and phycocyanin is chromatically modulated; in green light, the differential rate of phycoerythrin synthesis is high and this biliprotein becomes, after several

* Phycoerythrocyanin-containing cyanobacteria do not adapt chromatically. Growth at low intensities induces a relative increase in the phycoerythrocyanin cellular content (Bryant, unpubl.).

generations, the dominant pigment. In red light, phycoerythrin synthesis is repressed and that of phycocyanin increases; this increased rate of phycocyanin synthesis reflects the expression of structural genes coding for this protein, which are not expressed during growth in green light (Bryant, 1979, 1981; Bryant & Cohen-Bazire, 1981; see also 7.5.3). As a result, in strains of group III, the total amount of cellular biliproteins remains relatively constant under both light regimes.

Chromatically adapting strains growing in white light have a biliprotein composition intermediate between that in red and green light. When grown in the dark, chemoheterotrophic strains have a biliprotein composition similar to that after growth in red light (Fujita & Hattori, 1960b; Tandeau de Marsac, 1977).

The arrest of phycoerythrin synthesis when a culture is shifted from green to red light does not lead to its destruction; the existing phycoerythrin is diluted out by subsequent growth. Conversely, a shift from red to green light elicits the *de novo* synthesis of this biliprotein (Bennett & Bogorad, 1973). The early kinetics of phycoerythrin synthesis after a shift from red to green light has been followed by fluorescence spectroscopy and ^{35}S labelling in *Fremyella diplosiphon* (Gendel *et al.*, 1979). Synthesis of phycoerythrin measured spectrofluorometrically begins after a lag of approximately 90 min, at which time ^{35}S-labelled polypeptides characteristic of phycoerythrin are detected. Addition of rifamycin at the time of the shift stops phycoerythrin synthesis immediately. However, once the synthesis has started, it continues with a half life of 45 min after addition of rifamycin or transfer of the cells to red light. These experiments demonstrate that chromatic regulation operates at the transcriptional level.

The effects of chromatic adaptation on phycobilisome structure was used by Bryant *et al.* (1979) to determine the location of the various phycobiliproteins in negatively stained phycobilisomes. The phycobilisomes of *Synechocystis* 6701 extracted from cells grown in red or green light are shown in Figs 7.12 and 7.13 respectively. In this strain which belongs to chromatic adaption group II, the reduced synthesis of phycoerythrin in red light is not compensated by an increased synthesis of phycocyanin. The red-light phycobilisomes are correspondingly smaller. This difference in size is reflected in a lower number of subunits (6 nm thick) in the peripheral phycobilisome rods. Whereas this number is 3.6 on the average in green-light phycobilisomes, it falls to 2.2 in red-light phycobilisomes (see also Table 7.5). By contrast, in the phycobilisomes of *Pseudanabaena* 7409, belonging to chromatic adaptation group III, the number of disc-shaped subunits of the peripheral rods is almost identical whether they come from green- or red-light grown cells; the hexameric subunits of phycoerythrin in 'green-light phycobilisomes' are replaced by an almost equivalent number of phycocyanin hexamers in the 'red-light phycobilisomes' (Table 7.5; Figs 7.10, 7.11).

7.8.3.2 Action spectra

Action spectra for chromatic adaptation have been determined for three cyanobacterial strains: one unicellular, *Synechocystis* 6701 (Tandeau de Marsac *et al.*, 1979) and two filamentous strains closely related to each other, *Tolypothrix tenuis* (Fujita & Hattori, 1962a; Diakoff & Scheibe, 1973) and *Fremyella diplosiphon* (Haury & Bogorad, 1977; Vogelmann & Scheibe, 1978), both recently classified by Rippka (pers. comm.) as *Calothrix* 7101 and 7601, respectively.

The techniques used by most authors are closely similar and derived from those employed by Fujita and Hattori (1960a,b) in their pioneering study of chromatic adaptation in *Tolypothrix tenuis*. Cultures are starved of combined nitrogen under photosynthetic conditions until a major fraction of the measurable biliproteins have been degraded; during this period, the cells accumulate large quantities of glycogen (Fujita & Hattori, 1962b; Allen & Smith, 1969; Tandeau de Marsac *et al.*, 1979). Provided with a nitrogen source, such suspensions are able to synthesize phycobiliproteins *de novo* in complete darkness under aerobic conditions, using the stored glycogen as a source of reducing power, carbon and energy. Such synthesis takes place over periods ranging from 16 to 48 h.

Fujita and Hattori (1960a,b) discovered that the pattern of biliproteins synthesized in the dark depends on the wavelength of light received by the suspension immediately prior to dark incubation. After a brief 'red' irradiation (of the order of minutes), phycocyanin alone is synthesized whereas after a brief 'green' irradiation, phycoerythrin and phycocyanin are both synthesized. These effects are repeatedly reversible; the last irradiation determines the pattern of biliprotein synthesis in the dark. The chromatic effects are clearly photochemical, being independent of the temperature used during the irradiations (Fujita & Hattori, 1962b).

Using brief monochromatic irradiations in the ranges of 320–720 nm, action spectra for chromatic adaptation in all three strains show unmistakable similarities. The main maxima for the promotion of phycoerythrin synthesis are at 540–550 nm and 640–650 nm for its inhibition; in the two chromatic adaptors of Group III (*T. tenuis* and *Fremyella diplosiphon*) the inhibition of phycoerythrin synthesis is correlated with an increased phycocyanin synthesis.

In conclusion, these results suggest that chromatic adaptation in cyanobacteria is under the control of a single photoreversible master-pigment analogous to, but spectrally distinct from, phytochrome, the photomorphogenic biliprotein of plants (Scheibe, 1972; Vogelmann & Scheibe, 1978). Once activated, this master-pigment has a relatively long half-life (Ohki & Fujita, 1978) and appears to act at the level of transcription (Gendel *et al.*, 1979;

Scheibe, 1979), not only of the genes coding for their specific linker polypeptides (Tandeau de Marsac & Cohen-Bazire, 1977).

7.9 Phycobilisomes and thylakoid membrane organization

Efficient energy transfer from the phycobilisomes to the reaction centres of PS II requires that the phycobilisomes, in particular the allophycocyanin–allophycocyanin-B of the core, be precisely orientated with and in close proximity to the thylakoid-bound chlorophyll *a* of this photosystem. In cyanobacteria and red algae, a small fraction of the total chlorophyll *a* is associated with PS II—16% in *Anacystis nidulans* (Wang *et al.*, 1977) and only 5% in *Porphyridium cruentum* (Ley & Butler, 1976)—whereas the bulk of the chlorophyll is associated with PS I.

Freeze-etching studies of cyanobacterial and red algal thylakoids indicate a direct relationship between the distribution of the 10 nm intramembrane particles observed on the exoplasmic fracture face (EF) of the thylakoids and the distribution of phycobilisomes on their surface (Lichtlé & Thomas, 1976; see Staehelin *et al.*, 1978a for a review). This correspondence is striking in the electron micrograph of freeze-fractured *Nostoc muscorum* shown in Fig. 7.4, where the phycobilisomes and the complementary fracture-faces of the thylakoids are exposed. The centre-to-centre spacing between the phycobilisomes is identical to the spacing between the rows of 10 nm EF particles, and the grooves among the particles of the complementary PF face.

Several lines of evidence support the conclusion that the 10 nm EF particles of cyanobacterial and red algal thylakoids represent the protein–chlorophyll complexes associated with PS II. Firstly, PS II is inactive in heterocysts (see Chapter 14); correlatively, the 10 nm particles are absent from the exoplasmic fracture faces of the thylakoids in the heterocysts of *Anabaena cylindrica* (Giddings & Staehelin, 1979). Secondly, the mutant III-C of *Cyanidium caldarium*, which is incapable of phycobiliprotein synthesis, contains approximately twice the number of 10 nm particles on the EF face of its thylakoids as the wild type. This is correlated in the mutant with a near doubling of PS II centres relative to PS I centres (Wollman, 1979).

If the 10 nm particles on the thylakoid EF faces really represent the protein–chlorophyll complexes of PS II, how does their density correlate with the concentration of phycobilisomes?

Computations by Lichtlé and Thomas (1976) and Staehelin *et al.* (1978) show that the ratio of phycobilisomes to EF particles can vary from unity, to one phycobilisome per four EF particles.

In the freeze-etched replica of *Nostoc muscorum* shown in Fig. 7.4, the spacing between adjacent EF particles is identical to that of individual phycobilisomes seen in profile view. In this particular case there seems to be a one-to-one relationship between phycobilisomes and the 10 nm EF particles.

For the large ovoid phycobilisomes of the red algae *Porphyridium cruentum* and *Griffithsia pacifica*, a ratio of one phycobilisome for four EF particles would be sterically conceivable. However, comparisons of phycobilisome densities estimated in thin sections of fixed cells and the density of EF particles in freeze-etched replicas may be misleading. A reliable relationship can be established only when, as in the case of *Nostoc muscorum*, the phycobilisomes as well as the EF particles can be observed on the same specimen.

7.10 Summary and conclusions

It is instructive to compare David Chapman's chapter on biliproteins, published in 1973 in *The Biology of Blue-Green Algae*, with the current status of this field. A flood of original and detailed observations has been devoted to the biliproteins, in particular to their chemistry. Such recent findings include: subunit structure; aggregation states; ultrastructure of these aggregates; X-ray diffraction analysis; immunological relationships; the great majority of chromophore assignments and their attachment to the apoproteins. New phycobiliprotein types have been discovered; allophycocyanin-B and phycoerythrocyanin. A rapid start has been made on the problem of amino-acid sequence, which reveals the fundamental homology of the entire family of biliproteins. This mass of solidly established information concerning the biliproteins was either unknown, vague or controversial when Chapman wrote his chapter eight years ago.

The macromolecular composition of the phycobilisome has proved to be unexpectedly complex: from four to eight minor constituents have been revealed by SDS–PAGE, accounting for some 10–15% of their total protein content. These colourless linker polypeptides may have a cardinal role—as yet largely unexplored—in the assembly and organization of this organelle. Based on a conjunction of techniques—the known aggregation patterns of the biliproteins, precise spectrophotometric techniques and high resolution electron microscopy—a plausible, detailed model of phycobilisome structure has been proposed. Nevertheless, major questions remain unanswered. What are the sites for apoprotein and chromophore synthesis, so manifestly coordinately regulated in all cyanobacteria? What is the precise role of the linker polypeptides? Is the phycobilisome self-assembled, with the aid of the linker polypeptides? Is the assembly initiated at specific sites determined by the chlorophyll-proteins of PS II in the thylakoids? Many of these questions might find an answer through an effective means of genetic analysis of cyanobacteria.

Acknowledgments

The authors wish to express their thanks to the many colleagues who kindly provided manuscripts prior to publication. The preparation of this chapter was aided in part by grants from the Centre National de la Recherche Scientifique (E.R.A. 398) and the Délégation Générale à la Recherche Scientifique et Technique (Décision d'aide No. 77.7.1751).

CHAPTER 8
PHOTOSYNTHESIS

K. K. HO & D. W. KROGMANN

8.1 Introduction

This review is centred on the biochemical and biophysical aspects of photosynthetic energy conversion in the cyanobacteria. The details of the photosynthetic process provide overwhelming evidence for the similarity of cyanobacteria and eukaryotic plants and algae with respect to photosynthesis. The term photosynthetic bacteria has been used to include many organisms that use light and a chemical reductant other than water. These bacteria have a single photosystem and some unique types of molecules as parts of their photosynthetic machinery. While the cyanobacteria may, in some instances, show the ability to use hydrogen sulphide instead of water in an anoxygenic type of photosynthesis, all of the cyanobacteria seem to use a molecular machinery of photosynthesis which is remarkably similar to that found in eukaryotic oxygen-producing organisms. Oxygenic photosynthesis is shared by eukaryotes and cyanobacteria and, in all cases examined in detail, seems to involve the use of very similar molecules. Anoxygenic photosynthesis is shared by some cyanobacteria (Chapter 9) and all of the other prokaryotes that have been called 'photosynthetic bacteria'. However, the anoxygenic photosynthetic bacteria are quite distinct from cyanobacteria in their molecular architecture. It seems appropriate to find a new distinguishing name for the 'photosynthetic bacteria' to set them apart from the cyanobacteria.

There are two features of the photosynthetic apparatus in cyanobacteria which distinguish it from the chloroplast structures in most eukaryotic plants. Cyanobacteria contain phycobiliproteins organized in phycobilisome structures on the surface of the photosynthetic membrane (Chapter 7). This feature is shared with the eukaryotic red algae but has been replaced by the chlorophyll-*b* containing, light-harvesting protein complex (CP II) in other eukaryotes. The CP II complex is integrated into the chloroplast membrane and does not appear as a distinct structure above the membrane surface. The second distinguishing feature is the organization of photosynthetic membranes. In cyanobacteria, these membranes pervade the cytoplasm in varying arrangements and densities (Chapter 6). In eukaryotes, the photosynthetic membranes are packaged within a distinct chloroplast envelope membrane and are often stacked in grana structures. While there have been serious efforts to identify a 'stacking factor' in chloroplast membranes, it is not yet clear that stacking is the result of a special gene product. Stacking could result from the modification of a membrane-surface protein. These two differences between the eukaryotes and the cyanobacteria seem small in comparison to the great similarities in the details of photosynthetic electron transfer and phosphorylation mechanics shared among them. Two earlier publications (Krogmann, 1973, 1977) cover the work published before 1975 on photosynthesis in the cyanobacteria.

8.2 Electron transport assemblies in membranes and membrane fragments

Figure 8.1 shows the photosynthetic electron transfer sequence as it is presently understood. There are additional participants or intermediates in this process but, because their existence is inferred from physical measurements and is not yet identified with specific molecules, they are not included in the diagram. It is convenient to first review work on cell-free systems of high complexity that contain the whole electron transfer sequence or large sub-assemblies of it and then go on to studies of individual participants. Sigalat and de Kouchkovsky (1975) used the very gentle procedure of digesting *Anacystis nidulans* with lysozyme to produce protoplasts which were then ruptured by osmotic shock. The broken cell material was fractionated by centrifugation and yielded fractions corresponding to Photosystem (PS) I and PS II particles (de Kouchkovsky & Sigalat, 1975). This is remarkable since the resolution of the two photosystems usually requires extensive physical disruption or detergent treatment of the membrane. Further, PS I characters reside in a 'light fraction' of what are likely to be small particles and PS II characters are found in a heavy fraction. This result is like the general experience of fractionating the photosystems from higher plant chloroplasts

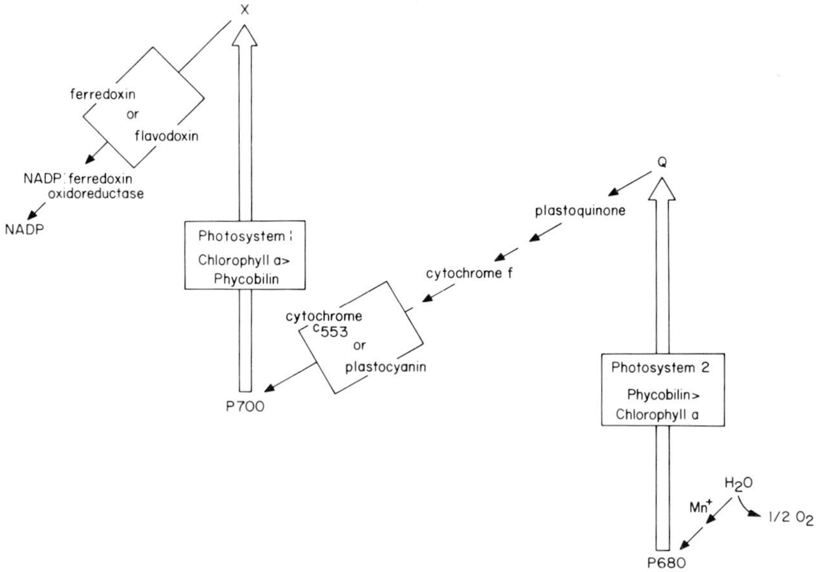

Fig. 8.1. Photosynthetic electron transfer sequence.

and is in contrast to the results of Ogawa *et al.* (1969) who reported PS II in a small particle fraction and PS I in a heavy fraction of large membrane fragments from *Anabaena variabilis*. The PS I fraction from *Anacystis* appears to be enriched in carotenoids but this might be due to the inclusion of osmiophilic globules which are likely to be concentrated in this fraction by the procedures used. The PS I fraction has a low ratio of chlorophyll to P700—36:1. This ratio indicates that the P700 has been separated from much of the bulk chlorophyll. Usually one obtains this kind of P700 enrichment only after detergent treatment and arduous fractionations. Binder *et al.* (1976) have described a preparation with more conventional properties from *Phormidium luridum*. Lysozyme digestion produced spheroplasts in excellent yield and these authors made the useful discovery that maintenance of spheroplast morphology is not equivalent to maintaining good electron transport activity. Mannitol as an osmoticum maintained the appearance of the spheroplasts but the membranes isolated from these spheroplasts had low photosynthetic activity. In contrast, KCl osmoticum both preserved the spheroplasts and allowed isolation of active membranes. The concentration of KCl for optimum electron transport activity is 0.25 M and $MgCl_2$ or $CaCl_2$ do not work as well. After rupture of the spheroplast, $MgCl_2$ or $CaCl_2$ are required to maintain PS II activity and KCl does not serve in their stead. McSwain *et al.* (1976) made similar observations on the effect of divalent cations on membranes of *Nostoc muscorum*. Both these studies confirm the earlier observations of Susor and Krogmann (1966) on the stabilizing effects of divalent cations on PS II. The damage caused by the absence of divalent cations seems to be localized near the reaction centre of PS II. Washing the membrane fraction causes some loss of electron transport activity and a more drastic loss of cyclic photophosphorylation. PS I activity was diminished by the release of a protein from the membranes and activity could be restored by readdition of this factor. The circumstances of the experiment strongly suggest that this factor is either plastocyanin or cytochrome c_{553}. (See subsequent section on these proteins.) While the authors report experiments in which plastocyanin gave incomplete restoration of activity, it is possible that the plastocyanin used by Binder *et al.* (1976) was from a different source and so less effective than the protein from *Phormidium luridum*. It is also possible that *P. luridum* protein is a new and different constituent. Ono and Murata (1978) used lysozyme and brief sonication on *Anacystis nidulans* to obtain a membrane fraction quite similar to that of Binder *et al.* (1976). Yamaoka *et al.* (1978) isolated photosynthetic membranes from a thermophilic *Synechococcus* which grows best at 60°C. These membranes showed a high rate of photosynthetic electron transfer activity and the reaction had an optimum rate at 50°C—a temperature which would destroy the activity of preparations others have made from cells grown at lower temperatures. Thus the photosynthetic membranes of this organism are thermophilic—stable and optimally active at elevated temperatures. Katoh and Gantt (1979) made an elegant improvement in the isolation of intact membranes from *Anabaena variabilis*. Lysozyme treatment followed by disruption in a French press in the presence of sucrose, phosphate, citrate and albumin led to the isolation of vesicles with good rates of electron transport and ratios of phycobilin to chlorophyll like those found in the intact cell. There was efficient transfer of excitation energy from phycocyanin to PS II in these preparations.

Evidence is appearing which suggests special effects of calcium ions on the photosynthetic membranes of cyanobacteria. Piccioni and Mauzerall (1978a) have amplified their earlier observations on the stimulatory effects of calcium ions on membranes isolated from *Phormidium luridum*. These effects are seen only when the cells are ruptured in a French press. Gentle lysis does not reveal the calcium effect. While both calcium and magnesium ions stimulate turnover of reaction centres, calcium ion seems to specifically stimulate the activation of photosynthetic units in PS II. Brand (1979) has found a calcium stimulus to electron flow in preparations of *Anacystis nidulans*. Cells broken in a French press show maximum activity when calcium ion is present during cell breakage and during the measurement of activity. Calcium was not required for stabilization of the membranes in the period between cell breakage and assay of activity. Yu and Brand (1980) have advanced the interesting idea that calcium ion is exposed to only one surface of the photosynthetic membranes *in vivo* (the inner side of the vesicle) where it stimulates PS II. Magnesium ion is thought to act on the other membrane surface (the outside of the vesicle) where it facilitates photophosphorylation and where its presence may stabilize PS II.

The fractionation of photosynthetic membranes into subparticles which are enriched for one or the other photosystem usually requires detergents. Newman and Sherman (1978) used digitonin to disperse the photosynthetic membrane of *Synechococcus cedrorum* and then fractionated on sucrose gradients. One upper band enriched in PS II and a lower band enriched in PS I were obtained. The particles were purified by DEAE-cellulose column chromatography and the activities in electron transfer and the concentrations of key components were measured. Valuable data are presented on the types of proteins found in each photosystem by SDS gel electrophoresis. Stewart and Bendall (1979) used the non-ionic detergent Nonidet P40 on membranes from the thermophilic *Phormidium laminosum* and obtained small vesicles that have a decreased P700 to chlorophyll ratio and an increased oxygen evolution capacity per chlorophyll. These vesicles are thus enriched for PS II; PS I is recovered in a pelleted fraction. The work of both Newman and Sherman (1978) and Stewart and Bendall (1979) finds PS I in a difficult-to-sediment fraction of either low density or small particle size or both. PS II is in an easily sedimented fraction of high density or easily aggregated or large particle size material. These results are like the early work of Ogawa *et al.* (1969). All three used detergents and are in contrast to the work of Sigalat and de Kouchkovsky (1975) who achieved fractionation without detergents. The photosynthetic membranes of heterocysts of *Anabaena cylindrica* have been characterized in an excellent paper by Tel-Or and Stewart (1977). This work demonstrated the presence of all of the recognizable constituents in the pathway from plastoquinone to NADP. Only PS II was absent (Chapter 14).

8.3 Reducing reactions from Photosystem I

A major goal of photosynthetic energy transduction is to provide reduced NADP for the continual reduction of carbon dioxide. The enzyme which delivers electrons to NADP is a flavoprotein called ferredoxin–NADP oxidoreductase. Recently, the three-dimensional structure of the spinach enzyme was solved (Sheriff & Herriott, 1981). This accomplishment, together with the three-dimensional structures of several other photosynthetic catalysts to be mentioned later, gives promise of the day when we may know the complete chemical structures of every molecule in the photosynthetic membrane. This knowledge will help us to understand not only the interactions of individual catalysts but also how the membrane is assembled from its constituent parts. An amino-acid sequence or primary structure for the oxidoreductase has not been done yet due to some difficulties with the pure protein. Gozzer *et al.* (1977) described five ionic forms, resolvable by isoelectric focusing, of the enzyme from spinach. Ellefson and Krogmann (1979) confirmed this and added a sixth form. The latter authors found that these ionic variants seemed to be in equilibrium with one another. That another sort of complication might occur in cyanobacteria was indicated by the work of Masaki *et al.* (1979) who reported on two forms of ferredoxin-NADP oxidoreductase from *Spirulina platensis*. The two forms of the enzyme, resolved by chromatography on DEAE-cellulose, have different N-terminal residues and slightly different peptide maps so they could not be in equilibrium with one another. One form might be a slightly modified variant resulting from proteolysis of the parent form or there may be two distinct structural genes for these enzymes. Hutber *et al.* (1978) described the purification of ferredoxin-NADP oxidoreductase from *Nostoc* strain MAC and failed to detect two forms of the enzyme despite very careful DEAE-cellulose column chromatography. However, the five equilibrating forms of the enzyme have recently been found with this preparation from *Nostoc* (Rogers, 1979, pers. comm.). That this very unusual series of equilibrating isozymes is found in both a cyanobacterium and a higher plant suggests that other structural similarities will be found. It is thus likely that the three-dimensional structure of the spinach oxidoreductase will be very helpful in understanding the enzyme from cyanobacteria.

The reduction of carbon dioxide is not the only destination of reducing power generated in photosynthesis. A task of substantial magnitude is the reduction of nitrate when this compound is used as the source of nitrogen. Special enzymes, which like ferredoxin-NADP oxidoreductase are reduced by ferredoxin, must be produced for the reduction of nitrate and nitrite. As might be expected for a branch point in a metabolic pathway, there is regulation of both the production of the enzymes and the catalytic activity of the enzymes that compete for electrons from ferredoxin. Losada and his colleagues have characterized this regulation of divergent pathways to NADP or to nitrate and nitrite in several cyanobacteria (Manzano *et al.*, 1976). In *Anacystis nidulans* both nitrate and nitrite reductases are adaptive enzymes whose synthesis is repressed by ammonia. Both enzymes are bound to the photosynthetic membranes but can be rather easily released by mechanical disruption. Both enzymes are reduced by ferredoxin. The enzymes can be coupled through ferredoxin to the photosystem in the membranes and so catalyze the photoreduction of nitrate and nitrite. Ortega *et al.* (1976) found a similar system for nitrate reduction in *Nostoc muscorum* and, working with Losada, they found that ammonia inhibits the photoreduction of nitrate (Ortega *et al.*, 1977a). This nitrate reductase is reversibly inactivated by its reduction in a ferredoxin-dependent reaction. Nitrate protects against this

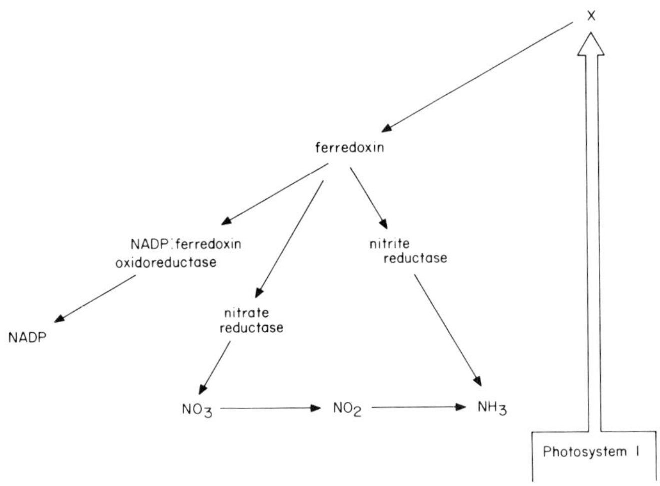

Fig. 8.2. Photosynthetic reduction of nitrate.

inactivation and oxidation reverses the inactivation. This group, in another publication (Ortega *et al.*, 1977b) showed that reduced ferredoxin inhibits the reduction of both nitrate and NADP by reductive inactivation of the respective reductases. They suggest that ammonia causes this inhibition indirectly by uncoupling photophosphorylation which increases the production of reductants from the photosystems. While they demonstrate that ammonia does increase the flow of reductant from water through the photosystems, caution is needed before attributing this effect to uncoupling since phosphorylation by some cyanobacteria seems extremely resistant to amine uncoupling (Duane *et al.*, 1965). Manzano *et al.* (1978) have begun the purification of nitrate reductase. A detailed comparison of the structures of nitrate and nitrite reductases from cyanobacteria with their complex counterparts from higher plants should prove quite interesting.

Ferredoxin is a small, non-haem iron-sulphur protein which receives electrons from PS I and transfers them to various reductases. Because this enzyme is held very loosely, if at all, to the photosynthetic membranes, it was the first of the photosynthetic catalysts to be purified and characterized. One wonders if this loose association of ferredoxin with the membrane has some physiological meaning. Ferredoxin is reduced by a constituent of PS I which is firmly embedded in the membrane. Ferredoxin reduces in turn the NADP, nitrate and nitrite reductases mentioned above and these enzymes, while not intrinsic membrane proteins, are surely held in place on the membrane surface more tightly than ferredoxin. Ferredoxin probably serves in other reactions not yet fully documented in cyanobacteria by enzyme isolation studies but suggested by their occurrence in other photosynthetic organisms—the reduction of proteins, the reduction of sulphate, the reductive carboxylation of organic molecules, etc. Perhaps as a common reductant for many processes, ferredoxin needs to be quite mobile in seeking out the various oxidants which require its electrons.

The function of ferredoxin in the cell is obviously crucial and provision is made for maintaining this function when ferredoxin cannot be synthesized due to lack of iron. In algae, cyanobacteria and clostridia, ferredoxin can be replaced under conditions of iron deficiency by a flavoprotein—flavodoxin. Fitzgerald *et al.* (1977) have documented this replacement of ferredoxin by flavodoxin in iron-deficient cultures of *Nostoc* strain MAC and have characterized the flavodoxin induced in these cells as having a molecular weight of 20 000 and a single FMN prosthetic group. It will be interesting to compare amino-acid sequences, when they become available, of flavodoxins from algae and cyanobacteria with those of other bacteria. That this substitution of a flavoprotein for ferredoxin occurs in nature is suggested by the failure to find ferredoxin in some natural blooms of cyanobacteria (Ho *et al.*, 1979) and provides a cautionary note to those who might conclude that a specific protein did not occur in a given organism since they could not find it. The gene may not always be expressed.

It is evident from many kinds of experiments that ferredoxin has been stable over a long period of evolution. The functional stability of this molecule was confirmed in studies of Andrew *et al.* (1975). These authors isolated ferredoxin from *Anabaena flos-aquae* and found it to be catalytically active in a barley chloroplast system. This substantiates earlier experiments which indicate that ferredoxins from cyanobacteria and higher plants are interchangeable in their reactions with both the reductant in PS I and the oxidant-ferredoxin-NADP oxidoreductase from either higher plants or cyanobacteria. This part of the photosynthetic electron transfer chain has changed little since the divergence of eukaryotes and prokaryotes.

The discovery of two distinct ferredoxin molecules in some cyanobacteria and some higher plants raises interesting questions. Hutson *et al.* (1978) have characterized the two ferredoxins of *Nostoc* strain MAC and found them present in the same ratio of one form to the other (ferredoxin I/II = 3) whether the cells were grown autotrophically in the light or heterotrophically in the dark. The molecules had the same light-absorption spectra and isoelectric points but differed by approximately 100 mV in their midpoint potentials. Ferredoxin I had a midpoint potential of −350 mV and was more effective in an NADP photoreduction assay while ferredoxin II had a midpoint potential of −455 mV and was more active in a pyruvate decarboxylation assay from the phosphoroclastic system of *Clostridium pasteurianum*. The low midpoint potential of ferredoxin II from *Nostoc* strain MAC is the lowest recorded in the extensive series of measurements made on ferredoxins from higher plants, algae and cyanobacteria (Cammack *et al.*, 1977). Shin *et al.* (1977) found these two ferredoxins in *N. verrucosum* but here the ratio of forms I/II = 1/0.84. The two forms of ferredoxin from *Aphanothece sacrum* have been sequenced (Hase *et al.*, 1978c) and the difference in primary structure leave no doubt that there are two distinct structural genes for these molecules. Not all organisms are found to possess two types of ferredoxin but this may reflect some still unrecognized variable in growth conditions. Matsubara *et al.* (1979) have found that ferredoxin I of *Aphanothece sacrum* is more closely related to the ferredoxins of green algae while ferredoxin II is more closely related to the ferredoxins of other cyanobacteria and red algae. The sequences of ferredoxins from higher plants suggest that gene duplication has occurred more than once during the evolution of this protein. Sequences are known for ferredoxins from *Nostoc muscorum* (Hase *et al.*, 1976b) from *Spirulina platensis* (Wada *et al.*, 1975) and *S. maxima* (Tanaka *et al.*, 1975). A recent sequence of ferredoxin from a red alga (Takruri *et al.*, 1978) has confirmed the suggestion raised by earlier work that the ferredoxin of *Spirulina* is more similar to the eukaryotic ferredoxins than to ferredoxins from some of the cyanobacteria. This suggests that ferredoxin sequences may prove to be useful in interpreting the evolution of the cyanobacteria (Chapter 12). A similar indication of the distinct difference in primary structure of cytochrome c_{553} from *S. maxima* as compared to cytochromes c_{553} from other cyanobacteria will be discussed in a later section.

While sequence comparisons allow the most detailed insight into the relations of molecules from different species, valuable information can be obtained from much simpler immunological comparisons. Tel-Or *et al.* (1977a) used an antibody prepared against ferredoxin from *S. maxima* to examine the similarity of ferredoxins from many other sources. This study indicated that cyanobacteria have a ferredoxin more like the algal ferredoxins than the ferredoxins from other bacteria which is not surprising in view of other chemical data. A very interesting pattern of interrelatedness appeared in comparisons among the cyanobacterial ferredoxins. A span going from ferredoxins most similar to *S. maxima* ferredoxin to least similar shows the source organisms in the following order: *S. platensis*; *Chlorogloea fritschii*; *Mastigocladus laminosus*; *Anabaena variabilis*; *A. flos-aquae*; *Microcystis flos-aquae*; *Anacystis nidulans*; *Anabaena cylindrica*; *Aphanocapsa* 6714; *Nostoc* strain MAC type II; *Nostoc* strain MAC type I.

Ferredoxins from thermophilic cyanobacteria are heat stable and hold the promise of helping us to understand how proteins can be made to withstand thermal denaturation. Koike and Katoh (1979) have documented the stability at high temperatures of ferredoxin from a thermophilic *Synechococcus*. Hase *et al.* (1978b) have worked out the amino-acid sequence of a ferredoxin of unusual thermal stability from the thermophile, *Mastigocladus laminosus*. The primary structure of this ferredoxin does not reveal any clues which would account for its thermal stability. The tertiary structures of these molecules should reveal the especially strong interactions that make them resistant to heat denaturation.

Happily, the X-ray diffraction analysis of the three-dimensional structure of ferredoxin is well under way and soon the crystal structures of the molecules from several species will be solved (Ogawa *et al.*, 1977; Tsukihara *et al.*, 1978 a,b). The three-dimensional structure of ferredoxin can then be studied in relation to the three-dimensional structure of its *in vivo* oxidant ferredoxin-NADP oxidoreductase and the precise interactions between this pair of reaction partners will be worked out. The three-dimensional structure of flavodoxin from *Anacystis nidulans* is also available (Smith *et al.*, 1978) which will allow a precise understanding of how this molecule can substitute for ferredoxin in transferring electrons to the oxidoreductase.

The compound which reduces ferredoxin has not been solubilized, purified or characterized in the usual chemical sense. It is clear that this compound, or some near-neighbour to it, is especially vulnerable to oxidation by molecular oxygen in the cyanobacteria (Honeycutt & Krogmann, 1972). While ferredoxin and ferredoxin-NADP oxidoreductase normally pass electrons efficiently to NADP, both are subject to oxidation by oxygen. All three autooxidizable components are potential sources for the destructive superoxide anion radical. Cells which produce oxygen have a special obligation to protect themselves from the superoxide anion radical by synthesizing the enzyme superoxide dismutase to remove this radical very rapidly. While cyanobacteria have superoxide dismutases in both the cytosol and the photosynthetic membranes, it is interesting to note that the heterocysts of *Anabaena cylindrica* which lack the ability to produce oxygen, have only one-tenth the superoxide dismutase of vegetative cells which do produce oxygen (Henry *et al.*, 1978). Okada *et al.* (1979) have shown that the cytosol enzyme contains iron while the superoxide dismutase bound to the photosynthetic membranes is a manganese-containing enzyme. Lumsden and Hall (1975) had suggested that the original manganese protein may have had a dual role in the oxygen evolving step of PS II, where manganese is surely involved, and in the manganese-containing superoxide dismutase needed to detoxify oxygen.

Shilo and his colleagues (Abeliovich *et al.*, 1974; Steinitz & Shilo, 1976) have provided evidence for an elegant relation between the enzyme superoxide dismutase and the sudden photooxidative destruction of blooms of cyanobacteria. These workers have demonstrated that susceptibility to photooxidative destruction at high concentrations of oxygen in light is related to the loss of superoxide dismutase activity.

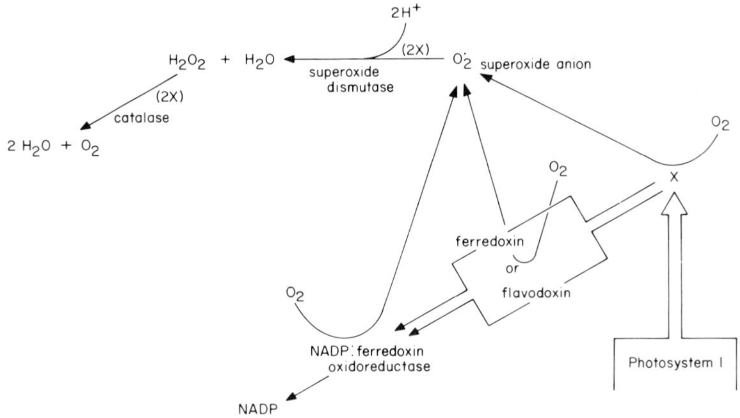

Fig. 8.3. Relationship of superoxide dismutase to oxygen evolved in photosynthesis.

8.4 Photosystem I

PS I contains two functionally distinct kinds of elements. One is an array of pigment molecules dispersed over a large surface area to collect light energy which is funnelled by physical processes into a reaction centre. The second kind of element is the reaction centre where a chemical change in molecules occurs. Here the light energy is used to drive an electron away from a molecule—P700—leaving it oxidized. The electron is driven into an acceptor molecule—called X in the absence of any firm chemical identity—so that the acceptor molecule becomes reduced. It is convenient to discuss the light-absorbing system first, then X, the molecule reduced by light, which in turn reduces ferredoxin. Finally, P700, which is oxidized by light to give its electron to X, will be discussed.

Chlorophyll seems to be the predominant light-absorbing pigment of PS I while the phycobilins are the predominant energy collectors of PS II. There have been several efforts to estimate the amount of chlorophyll in each of the photosystems and the results are gratifyingly similar despite the different methods used and the plasticity of cyanobacteria. Mimuro and Fujita (1977) used delayed light emission from intact cells and the measurement of some partial reactions of photosynthetic electron transport in isolated membrane fragments of *Anabaena variabilis*. This led to an estimate of 91–95% of the chlorophyll residing in PS I. Tel-Or and Malkin (1977) used several optical techniques on whole cells, spheroplasts and membrane fragments isolated from spheroplasts of *Phormidium luridum*. The cells and spheroplasts were found to have 90% of the chlorophyll, 90% of the carotenoid and 15% of the phycocyanin in PS I. Wang *et al.* (1977) did careful action spectra for the two photosystems in normal *Anacystis nidulans* and three variants with altered pigment ratios. The normal cells showed 84% of the chlorophyll and 40% of the phycocyanin in PS I. It was clear that energy absorbed by phycocyanin in PS II is delivered to the reaction centre by a small number of chlorophyll molecules. Studies on the variants in pigment composition showed that the amount of chlorophyll in PS I is variable but the small amount of chlorophyll in PS II is not altered. Myers *et al.* (1978) have carefully documented the ability of cells to adjust the ratio of chlorophyll to phycocyanin in response to the wavelength of light used to grow the cells. Growth of *A. nidulans* in light preferentially absorbed by chlorophyll gives rise to a lowered concentration of chlorophyll. Just as the concentration of bulk chlorophyll in PS I of cyanobacteria is variable, the number of PS I reaction centres may be variable as well. Kawamura *et al.* (1979) have measured the number of reaction centres in each photosystem of several cyanobacteria. When cells are grown at low light intensity, the ratio of PS I to PS II reaction centres is much higher than in cells grown at high light intensity. There are indications that PS I is the variable while PS II remains constant.

The chlorophyll in PS I is mainly engaged in an antenna-like function of collecting energy. This chlorophyll is associated with a specific protein. The study of chlorophyll–protein complexes has become a very active area of research and has been capably reviewed by Thornber *et al.* (1979). In addition to the bulk chlorophyll-protein complex of the antenna type, there is a very small amount of chlorophyll–protein at the reaction centre, P700, which undergoes chemical change. There may be other complexes whose function is to channel energy from the antennae to the P700. Rusckowski & Zilinskas (1979) have found four major chlorophyll–protein complexes from *Nostoc* sp. using SDS polyacrylamide electrophoresis. One appears to be a PS I reaction centre complex and another may be from the PS II reaction centre. Here the individual types of proteins are sorted out on the basis of molecular weight and at high concentrations of strong detergent, one is dealing with individual species of polypeptides in the molecular weight range of 40 000–70 000. Another approach is to use a milder detergent treatment to isolate an aggregate of minimum complexity which retains some photochemical activity. A nice example of this is the work of Malkin *et al.* (1976) who fractionated the photosynthetic membranes of *Phormidium luridum*. When Triton X-100 was used, a particle was obtained which showed P700 photooxidation activity and the light-induced electron spin resonance signals of iron-sulphur centres which are associated with neighbours of PS I. When particles were prepared using the stronger detergent, sodium dodecylsulphate, the iron-sulphur signals had been stripped away. Klein and Vernon (1977) used Triton X-100 to isolate a P700–enriched particle from *Anabaena variabilis*. This particle, when subjected to SDS-gel electrophoresis, yielded four polypeptides. The largest polypeptide contained chlorophyll and could be resolved into smaller polypeptides on extended detergent treatment. Alberte and Thornber (1978) simplified the procedure for isolating the PS I reaction centre by using brief exposure to a combination of detergents and then fractionating the dissolved material on hydroxylapatite gel. This gave a P700-containing fraction which had two polypeptide species with molecular weights near 50 000 and two very small polypeptides. There is a very interesting report of the inclusion of allophycocyanin in a PS I complex from *Chlorogloea fritschii* (Pullin *et al.*, 1979). In this preparation, the detergents used are of the milder sort but the ratio of chlorophyll *a* to P700 is clear evidence that the reaction centre has been substantially purified. The allophycocyanin does not reveal itself in the absorption spectrum but does appear in the fluorescence spectrum. Lipid extraction releases the allophycocyanin suggesting hydrophobic binding.

In the reaction centre, one expects to find the participants in the light-driven electron transfer reaction—P700 and its oxidant. P700 is known from its light-induced change in absorbance at 700 nm, which is surely a photooxidation, but it has not been purified as a homogenous chemical entity. Perhaps harsh detergent treatments distort P700 or simply remove its oxidant X so that photochemical activity is lost. Now there are reports of multiple forms of P700 in the photosynthetic membranes of cyanobacteria (Hiyama *et al.*, 1977). Hoarau *et al.* (1977) have direct spectroscopic evidence for three species of P700 in detergent-derived membrane fragments. There is no suggestion of what function these multiple forms might serve.

The reaction partner of P700 is X—a substance which receives an electron when P700 is oxidized. The biophysical recognition of X is made possible by electron spin resonance spectroscopy which detects the electron that arrives from P700. The arrival of this electron is very rapid and temperature-independent, as is expected of a photochemical reaction. This measurement is made a bit more complex by the appearance of two light-induced spin resonance signals attributed to two iron-sulphur moieties in equilibrium with X. These iron-sulphur groups are not the primary electron acceptor but are very quickly reduced by it. These signals have been found in cyanobacteria and seem quite similar to the ones found in higher plants (Evans, Cammack & Evans, 1976; Cammack *et al.*, 1979a,b; Hiyama *et al.*, 1979). The iron-sulphur centres have been further characterized by Mössbauer spectroscopy providing further evidence of their similarity to ferredoxin (Evans *et al.*, 1979). The electron spin resonance signals at the reaction centre were elegantly made to reveal new details by the use of deuterated *Synechococcus lividus* (Thurnauer *et al.*, 1979).

8.5 Reactions between the photosystems

PS I receives electrons from a series of dark reactions which are, in their turn, supplied with electrons from PS II. Plastocyanin and cytochrome c_{553} supply electrons to P700 either directly or through a compound which has not yet been chemically characterized. There is kinetic evidence from the very sophisticated absorption spectroscopy measurements of Bouges-Bocquet and Delosme (1978) that, in *Chlorella*, an intermediate carrier may transfer electrons from plastocyanin to P700. It will be of considerable interest to learn if such an intermediate exists in cyanobacteria, since, as discussed below, there seems to have been some important evolutionary change in the photosynthetic apparatus at this site.

It is important to understand that plastocyanin and cytochrome c_{553} are alternative routes of electron transport to P700. Wood (1978) was the first to realize that in algae and cyanobacteria growing under conditions of copper deficiency, plastocyanin, a copper protein, was replaced by the iron–haem protein cytochrome c_{553}. Böhme and Böger (1978) promptly confirmed this important observation. Generally, plastocyanin is not present in dense natural blooms of cyanobacteria and can only be found in laboratory cultures where special care is taken to make copper available (Ho *et al.*, 1979). The complete amino-acid sequence for plastocyanin from *Anabaena variabilis* has been published (Aitken, 1975) and there is a partial sequence for the plastocyanin from *Plectonema boryanum* (Aitken, 1976). Markley *et al.* (1975) used nuclear magnetic resonance spectroscopy to show the great structural similarity between plastocyanins from higher plants and *Anabaena variabilis.* This provides an experimental link to the three-dimensional structure of plastocyanin which has been solved by X-ray crystallographic methods for a molecule from a higher plant (Colman *et al.*, 1978).

Cytochrome c_{553} can substitute for plastocyanin *in vivo* so these two rather different types of catalysts must have some important structural similarities. Gomez-Lojero and Perez-Gomez (1978) have characterized in fine detail this cytochrome from *Spirulina maxima.* A three-dimensional structure of cytochrome c_{553} will be very interesting to compare to the extant structure of plastocyanin. At present, the only documented similarities are of size—each has about one hundred amino-acid residues—and of oxidation-reduction potential—about 350 mV (Yamanaka *et al.*, 1978b; Makovkina *et al.*, 1978). When isolated from thermophilic organisms, the cytochrome is quite heat stable (Enami, 1978; Koike & Katoh, 1979). There are conflicting observations on the light-induced absorbance changes at the alpha band region of cytochrome c_{553} in isolated photosynthetic membranes. Tsujimoto *et al.* (1976) observed light-induced oxidation of a *c*-type cytochrome in membrane fragments of *Nostoc muscorum* while Kawamura and Fujita (1979) failed to see *c*-type cytochrome changes in a preparation from *Anabaena variabilis*. The confusion is compounded by the fact that both cytochromes c_{553} and cytochrome *f* have virtually identical alpha absorption bands. The failure of Kawamura and Fujita to see cytochrome c_{553} changes is easily explained since their preparation had an ample amount of plastocyanin and so would not be expected to contain cytochrome c_{553}. Their failure to observe any absorption change in this region is a failure to observe cytochrome *f* and that is very puzzling.

Cytochrome c_{553} is the natural analogue of and functional replacement for plastocyanin. The cytochrome has a modest size, good stability and is easily released from the membranes and easily purified. The cytochromes from several species of cyanobacteria have been sequenced (Ambler & Bartsch, 1975; Aitken, 1977; Cades *et al.*, 1979). A comparison of amino-acid sequences of this cytochrome from more genera of cyanobacteria would provide an interesting insight into the taxonomy and evolution of these organisms. The size, absorption spectra and oxidation-reduction potential of the cytochrome is quite uniform among cyanobacteria and several eukaryotic algae. However, the isoelectric point of this cytochrome varies from pH 9 in some cyanobacteria to pH 4 in others. The cytochrome in eukaryotes is uniformly an acidic protein. Susor & Krogmann (1966) found the cytochrome c_{553} of *Anabaena variabilis* was a basic protein. Lightbody and Krogmann (1967) then noted that plastocyanin of *A. variabilis* is a basic protein in contrast to the acidic plastocyanin of higher plants. Thus the functionally interchangeable components from the same organism have the same isoelectric point. That this might have a functional significance was suggested by the work of Mutuskin *et al.* (1976). These investigators found polyene antibiotics inhibited the PS I activity of chloroplasts from bean leaves and this inhibition could be reversed by plastocyanin or cytochromes c_{553} from various sources. They noted that acidic cytochromes were more effective than basic cytochromes in reversing this inhibition. Davis *et al.* (1980) used a purified, PS I-enriched particle from spinach and measured the ability of various protein electron donors to reduce P700. Acidic proteins like spinach plastocyanin and algal cytochromes c_{553} worked well if mangesium ion was present. Basic proteins such as plastocyanin or cytochromes from some cyanobacteria had a greater affinity for reduction of spinach P700 and magnesium ions interfered with this. Clearly ionic charge is important in the reduction of P700 and perhaps the purification of the P700-enriched particle removes factors which magnesium ion can replace and allows the negatively charged eukaryotic proteins to react with negatively charged areas near the eukaryotic P700. Basic proteins from prokaryotes would react very readily with these negatively charged areas. Plastocyanin and cytochrome c_{553} have apparently evolved in parallel fashion from basic to acidic proteins. Perhaps the photosynthetic membrane has had to evolve as well to accommodate this change. It is interesting to note that among the cyanobacteria, *Spirulina maxima* and *Microcystis aeruginosa* have acidic cytochromes c_{553}. The amino-acid sequence data mentioned earlier had indicated that the ferredoxin of *Spirulina maxima* was more like the ferredoxin of eukaryotic algae than the ferredoxins of some cyanobacteria. The cytochrome isoelectric point data contain the same suggestion—that *S. maxima* is more like a eukaryote than a prokaryote in the structures of its photosynthetic catalysts.

The next participant in the chain of reactants which brings electrons from PS II to PS I is cytochrome *f*. For many years there has been confusion over the identities of cytochrome c_{553} and cytochrome *f*. The two cytochromes have very similar absorption spectra and oxidation-reduction potentials. It was assumed that cyanobacteria and algae had a type of cytochrome *f* that was smaller and more easily released from the membrane than cytochrome *f* of higher plants. Wood (1977) clarified this by demonstrating that cyanobacteria and algae had, in addition to the small, easily solubilized cytochrome c_{553}, another cytochrome of higher molecular weight that was tightly bound to the membrane and was the counterpart to higher plant cytochrome *f*. This cytochrome *f* was isolated in large amounts from cyanobacteria (Ho *et al.*, 1979). Sequence work has begun and shown that the ten positions at the N-terminus of the cytochromes from *Spirulina maxima* and *Aphanizomenon flos-aquae* are identical to one another and are very similar (four of the first ten residues identical) to the N-terminal sequence of cytochrome *f* from spinach (Ho & Krogmann, 1980). However, there are important differences between cytochromes *f* from spinach and cyanobacteria. Spinach cytochrome *f* is more readily oxidized by spinach plastocyanin while *Spirulina* cytochrome *f* clearly prefers to react with the protein oxidant of *Spirulina*.

Beyond cytochrome *f*, the biochemical description of photosynthetic electron transfer becomes far less clear. There is evidence from studies with higher plant materials that cytochrome *f* is reduced by a non-haem iron-sulphur protein. This evidence comes from measurements of electron spin resonance signals and from characterization of some mutations altering the photosynthetic apparatus. Experiments have been reported using cyanobacteria and they resemble eukaryotes in this respect.

Plastoquinone is the next electron carrier in the sequence. Cyanobacteria and higher plants seem to be identical with respect to plastoquinone function (Lightbody & Krogmann, 1966). No new evidence about plastoquinone function in cyanobacteria or eukaryotes has been reported.

There are spectroscopic data indicating that other components respond to irradiation of the photosystems. For a time it was thought that cytochrome b_{559} was an electron carrier in the pathway near PS II. Now there is kinetic evidence that this carrier is not a part of the pathway but is somehow responsive to light in what may be a diversionary side reaction. Kawamura and Fujita (1979) saw cytochrome b_{559} in the light-dark difference spectra of photosynthetic membrane fragments from *Anabaena variabilis*. Tsujimoto *et al.* (1976) found that cytochrome b_{559} was photooxidized by PS II at room temperature in membrane fragments of *Nostoc muscorum*. This fits with earlier data of Fujita (1975) who suggested that part of the cytochrome b_{559} photooxidation was due to PS II and part to PS I. There are many indications that cytochrome b_{559} in higher plants exists in two forms—one of high, the other of low oxidation-reduction potential—and that these two forms are on either side of PS II. The function of cytochrome b_{559} remains a mystery.

Cytochrome b_6 has been observed in the membrane fragments of *N. muscorum* (Knaff, 1977) and seemed to be both photooxidized and photoreduced by PS I. Influences of ADP on the rates of oxidation and reduction of cytochrome b_6 suggest it may be a catalyst in a special cyclic path around PS I to generate ATP.

Must all electrons reach the cytochromes from PS II? There are some interesting experiments which suggest that organic acids might serve as electron donors instead of PS II. Murai and Katoh (1975) found that glycolate, malate, succinate and isocitrate could donate electrons to PS I via cytochrome *f* in membranes from *Anabaena variabilis*. The rates of these reactions were low compared to the rates of photosynthetic electron transfer but the experiments provided a plausible connection between respiration and photosynthesis. Sorensen and Halldal (1977) found another indication that such a reaction might occur *in vivo*. They measured action spectra of CO_2 fixation and glycolate excretion by *Anacystis nidulans* and found that blue light depressed glycolate excretion relative to CO_2 fixation. They reasoned that PS II was poorly excited and so electrons were supplied to PS I by the oxidation of glycolate. Halldal and Holmen (1979) have repeated this hypothesis in explaining a correlation between enhancement of CO_2 fixation and decreased glycolate excretion but here the interpretation is less direct.

8.6 Photosystem II

Like PS I, PS II contains a large antenna pigment array for the collection of incident photons. The principal pigments are phycobiliproteins and their organization into the phycobilisome structure is the subject of the previous chapter in this volume. The organization of the antenna pigments should facilitate the transfer of energy by physical resonance processes from phycocyanin through allophycocyanin, then through chlorophyll to the reaction centre where a pigment called P680 undergoes net loss of an electron as a result of receiving this energy. Wang and Myers (1977) have observed a reverse energy transfer from chlorophyll to phycobilin in 3-(3,4-dichlorophenyl)-N,N′-dimethyl urea (DCMU)-poisoned *Anacystis nidulans*. Paths of energy migration relate to the mechanism and efficiency of energy utilization. In the cyanobacteria, this takes on a special significance since the phycobilisomes are discrete morphological units. This is in contrast to the reaction centre of PS I in cyanobacteria and both photosystems in eukaryotes which are served by antenna pigment arrays in a topological continuum on the membrane surface. Does each antenna unit serve a single reaction centre in the membrane beneath the phycobilisome? Diner and Wollman (1979) studied the eukaryote *Cyanidium caldarium* which also has phycobilisomes. They estimated that there are 40 chlorophyll molecules in each PS II reaction centre and that energy transfers could occur among four centres in close proximity to one another. Diner (1979) then estimated that only half of the reaction centres were connected to phycobilisomes. Mimuro and Fujita (1978) measured the efficiency of energy transfer in PS II of cyanobacteria and estimated that there are 20 chlorophyll *a* molecules per reaction centre. Since there is poor efficiency in energy transfer from reaction centres which have been closed by chemical reaction to reaction centres which are open, they concluded that the phycobilisomes are functionally discrete as well as morphologically discrete units.

There are tantalizing indications that carotenoids may be associated with PS II. Newman and Sherman (1978) found carotene in their PS II enriched particle. Szalonti and Csatorday (1979) found a decay in the resonance Raman spectra of membrane-bound carotenoids accompanying changes in phycobilins induced by nitrate starvation of *Anacystis nidulans*. They suggested that carotenoids may be involved in coupling the phycobilisome to the photosynthetic membrane. Suzuki and Fujita (1977) have found a photobleaching of carotenoids associated with PS II in membranes from *Anabaena variabilis*. The reaction is stimulated by ferricyanide and some inhibitors of PS II. Fujita (1978) found the reaction was stimulated by high concentrations of oxygen and suggested that oxygen removed electrons from a slow, reversible step near the PS II reaction centre. A different type of evidence for carotenoid participation in PS II comes from the work of Schmid *et al.* (1977). These workers prepared antisera against lutein and neoxanthin and found these antisera caused agglutination of membrane preparations from *Nostoc muscorum* and *Oscillatoria chalybea*. The agglutination indicates that the carotenoids are on the surface of the membrane. Furthermore, these antisera inhibited electron transfer on the oxygen evolving side of PS II. The response of *O. chalybea* in the presence of various electron donors suggested that the antiserum to lutein affected a different site to that inhibited by the antiserum to neoxanthin. These results do not demand that carotenoids be identified as electron carriers. It may be that deformation by antibody interaction with carotenoids on the membrane surface distorts the orientation of electron carriers nearby. The isolation of a carotenoid protein complex from a number of cyanobacteria (Holt & Krogmann, 1981) may help to clarify the role of carotenoids in the cyanobacterial membrane.

There have been several biophysical probes of the reaction centre constituents in PS II. Piccioni and Mauzerall (1978b) have identified special characteristics of PS II in membrane fragments prepared when cyanobacteria are broken open with the French press. These preparations show DCMU-insensitive oxygen evolution and a special component in the oxygen-evolving system with an oxidation-reduction potential of 450 mV. There is a change in the absorbance spectrum of photosynthetic membranes at 550 nm that is associated with PS II and this change has been characterized in cyanobacteria by Fujita (1976). This absorbance change may be due to a change in state of membrane carotenoids near Q in PS II. The extent of change in response to light suggested that the number of PS II reaction centres might be half that of PS I reaction centres in autotrophically grown *Anabaena variabilis*. Kawamura *et al.* (1979) have measured the ratio of PS I to PS II reaction centres in several genera of cyanobacteria grown in high and low light intensities. The number of PS II reaction centres remains constant while the number of PS I centres increases at lower light intensities. Ley *et al.* (1975) measured the flash kinetics of oxygen evolution in *Anacystis nidulans* and concluded that the same mechanism of charge separation and water photolysis occurred here as in spinach chloroplasts.

There are a few hopeful signs that the biochemistry of PS II may become accessible. Bendall (1978) used the thermophilic *Phormidium laminosum* as a source of photosynthetic membranes which were then fractionated with detergents to yield a particle which produces oxygen via PS II. Stewart and Bendall (1979) have shown that this particle is devoid of P700 and is enriched for PS II activity. England and Evans (1981) have modified the above procedure by including calcium ions so as to adapt it to an easy isolation of PS II particles from *A. nidulans* which are active in oxygen evolution. The thermal stability of PS II activity in membranes isolated from thermophilic cyanobacteria varies with the organism studied (Stewart & Bendall, 1980, Bohl & Binder, 1980, Yamaoka *et al.*, 1978). Hurt & Hauska (1981) have isolated an enzyme complex which connects PS II to PS I in that it is a plastoquinol-plastocyanin oxidoreductase containing the appropriate cytochromes and non-haem iron protein.

The effects of specific inhibitors emphasize the similarity of PS II in cyanobacteria and higher plants. Codd and Cossar (1978) demonstrated that DCMU has the same site of inhibition of PS II in *Anabaena cylindrica* as in spinach, and they, like Schmid and Lehmann-Kirk (1977), showed that diphenyl carbazide would reverse the inhibition by supplying electrons to the photosystem just as it does in spinach chloroplasts. Sofrova *et al.* (1976a) found that trypsin inhibits PS II activity in cell-free preparations from *Plectonema boryanum* just as it inhibits this activity in spinach chloroplasts. Furthermore the trypsin inhibition can be reversed by several artificial electron donors as is the case in spinach (Sofrova *et al.*, 1976b). Oren *et al.* (1979) have shown that sulphide inhibits electron transport between water and PS II in several cyanobacteria and in tobacco chloroplasts.

At the physiological level PS II in cyanobacteria seems extremely malleable. The next chapter of this volume discusses the replacement of PS II by sulphide. Organotrophic growth causes the virtual disappearance of PS II. Ohki and Katoh (1975) adapted *Anabaena variabilis* to grow on glucose and casamino acids to the point where CO_2 was not required and CMU did not inhibit. Growth was still proportional to light intensity. These cells showed normal chlorophyll levels, an 80% decrease in phycobilin content and they had lost the ability to evolve oxygen. Next Katoh and Ohki (1975, 1976) found that removal of nitrate from the photoorganotrophic medium hastened the disappearance of phycobilisomes and oxygen-evolving activity. Photosynthetic membranes from these cells could not use diphenylcarbazide to bypass the inhibited site suggesting that an integral part of PS II was lacking. These authors found that restoration of nitrate to the cells would allow resynthesis of biliprotein in the dark but light was needed to make the newly synthesized biliprotein effective in energy transfer of absorbed light into PS II. This final step of integrating the antenna pigment into the photosystem was not dependent on protein synthesis. Harnischfeger and Codd (1978) have discovered a similar, light-driven reinsertion of phycobilins into PS II of *Anacystis nidulans*. In this series of experiments, the phycobilisomes seem to become detached from the membrane energy transfer system in a few minutes of darkness and were reinserted in a few minutes of illumination. The light-driven reinsertion of phycobilins was not inhibited by DCMU and so was not dependent on PS II electron transport. Light might be needed to supply ATP via PS I or to alter the conformation of the photosynthetic membrane. Csatorday (1978) has described the fluorescence emission of the biliproteins as they are resynthesized on nitrate addition to nitrate starved *Anacystis nidulans*. Evans *et al.* (1978) demonstrated the two-stage recovery of PS II in *Chlorogloea fritschii* after prolonged organotrophic growth in darkness. They obtained evidence for the stability of PS I under these conditions by electron paramagnetic resonance spectroscopy. The most dramatic effects of environmental alteration on cyanobacteria are those reported by Miller and Holt (1977). They found that *Synechococcus lividus*, when deprived of carbon dioxide for 120 h, lost chlorophyll, phycocyanin and photosynthetic membranes. Readdition of carbon dioxide allowed a rapid resynthesis of the photosynthetic apparatus.

8.7 Photophosphorylation

Cyanobacteria synthesize ATP in light by the same mechanism used by higher plants. A proton gradient is formed as a result of light-driven electron transport and the proton gradient then drives the synthesis of ATP. Proton gradients are formed at two or possibly three sites (Velthuys, 1978) in the electron-transfer sequence but the mechanism of proton translocation is still a matter of speculation. Falkner, Horner, Werdan and Heldt (1976) demonstrated a light-driven movement of protons from the cytoplasm into the thylakoid space in *Anacystis nidulans*. This gradient was of substantial magnitude and could be abolished by the uncoupler CCCP. Padan and Schuldinger (1978) described the proton gradient formation in *Plectonema boryanum* and Raboy and Padan (1978) showed the relation of the proton gradient to active transport of hexoses in this organism. Matsamoto and Nishimura (1978) have provided a cautionary note with their demonstration that measurements of the proton gradient in lysozyme-treated cells of *Plectonema boryanum* are influenced by the tonicity of the medium in which the cells are suspended. They suggest there may be two distinct compartments for proton accumulation in these cells. Coupling factor is an aggregate of five different types of polypeptides and is required for photophosphorylation in higher plant chloroplasts. The coupling factor aggregate can be reversibly removed from the chloroplast membrane with attendant loss and restoration of ATP synthesis and proton gradient-forming ability. Under proper circumstances, the coupling factor can be converted to an ATP-hydrolysing enzyme system. Most of these properties are emerging in the description of ATP synthesis in cyanobacteria suggesting that there is remarkable conservation of detail in the evolution of the photosynthetic apparatus. The coupling factor from *Spirulina platensis* resembles that of spinach in the way it is released from and reinserted in the membranes; in possessing at least four of the five types of subunits; and in its ATPase activation (Owers-Narhi *et al.*, 1979). The coupling factor of *S. platensis* differs from the spinach coupling factor in its temperature stability. The spinach enzyme complex is cold labile and must be purified at room temperature. The *S. platensis* enzyme complex loses activity at room temperature amd must be purified at 4°C. Binder and Bachofen (1979) isolated the coupling factor from the thermophile *Mastigocladus laminosus* and found four of the five subunit types seen in higher plant preparations. The ATPase activity of this preparation had a temperature optimum at 5°C and the coupling factor complex resembles the spinach preparation in that it is stable at room temperature and cold labile. This interesting difference in temperature stability in the coupling factors from *Spirulina platensis* and *Mastigocladus laminosus* might prove an advantageous situation in understanding cold lability of protein complexes.

Tel-Or and Stewart (1976) showed the capacity for photophosphorylation in heterocysts of *Anabaena cylindrica*. Bottomley and Stewart (1976b) obtained very good evidence for the role of photophosphorylation *in vivo* by measuring ATP pools and shifts during dark-light steady-state transitions. Imafuku and Katoh (1976) made similar measurements to study light induced inhibition of respiration in *A. variabilis*.

8.8 Relationship of growth temperature to photosynthesis

In the above sections there have been many examples of interest in the relation of growth temperature to specific photosynthetic catalysts. The cyanobacteria are especially well suited for such studies since they may exhibit psychrophilic, mesophilic, or thermophilic growth habits and since the mesophiles respond to differences in growth temperature by altering the composition of their photosynthetic membranes. What follows is a collection of fascinating observations on the effects of temperature on photosynthetic structures in cyanobacteria. Schreiber *et al.* (1979) found that exposure of *Anacystis nidulans* to temperatures below 5°C led to a loss of energy transfer from the phycobilins to chlorophyll and suggested that this was due to a detachment of the phycobilisomes from the photosynthetic membrane. Perhaps this is related to the observations of Gantt *et al.* (1979) that phycobilisomes seem to be cold labile and must be isolated from broken cells by centrifugation at room temperature. Schreiber (1979) examined the low-temperature induced loss of energy coupling of phycobilisomes in lysozyme-treated cells and found that divalent cations reduced the phycobilin uncoupling and monovalent cations enhanced it. This lead to the reasonable suggestion that negative surface charges favour phycobilisome detachment or disorganization in the cold. A clearer idea of how low temperatures cause the reorganization of complex protein structures is needed. While low temperature may disorient the phycobilins of a mesophile, it is unlikely that this would occur in a psychrophile and so a comparison of the three-dimensional structures of the phycobilins might be instructive. Experience, noted earlier, suggests that knowledge of the tertiary and quarternary structure of the proteins will be required to understand thermophily and psychrophily. A preliminary attempt at such a comparison is seen in the work of Adams *et al.* (1979). These authors examined the aggregation behaviour of a phycocyanin isolated from the psychrophile *Microcoleus vaginatus*. While aggregation behaviour might reasonably reflect alteration of protein structure in adaptation to temperature stress, this was not evident in this case. The aggregation behaviour of the protein from this psychrophile is like that of mesophiles. However, thermophiles produce a phycocyanin with altered aggregation properties. Murata, Fork and their colleagues have worked for several years on the relation of photosynthetic membrane lipid composition to the physical properties of these membranes. The photosynthetic membranes of *Anacystis nidulans* respond to growth temperatures by changing the extent of unsaturation and the chain length of the fatty acyl groups in the membrane lipid. Fork *et al.* (1979) recently documented this change in a detailed lipid analysis of thermophilic *Synechococcus lividus* grown at 55°C and at 38°C. The results parallel those found in *Anacystis nidulans* grown at 38°C and 25°C. When the cells are grown at a lower temperature, there is an increase in unsaturation and decrease in chain length of fatty acyl residues giving an increase in fluidity to these lipids. Murata *et al.* (1975) first used a lipophilic electron spin label to report on the fluidity of membranes of *A. nidulans* grown at 38°C or at 28°C. The membranes from 38°C grown cells showed a lipid transition at 24°C while membranes from cells grown at 28°C showed the transition at 13°C. The transition was thought to be one of conversion of a liquid crystal state to a mixed solid–liquid crystalline state. Fluorescence probes in the membrane gave similar results as did the temperature dependence of fluorescence yield of chlorophyll *a*. The Arrhenius plots of P700 reduction in these cells showed similar growth-temperature induced differences. Similar results have been obtained with the temperature dependence of delayed fluorescence of chlorophyll *a* (Ono & Murata, 1977), with a partial reaction of electron transport through PS II (but not with a reaction through PS I), with photophosphorylation (Ono & Murata 1979) and with cytochrome *f* reduction (Fork *et al.* 1979). *Anacystis nidulans* is uniquely suitable for this work since *Anabaena variabilis* does not respond to change in growth temperature by such dramatic alterations of membrane lipid (Sato *et al.*, 1979). All of the above phenomena—energy transfer through chlorophylls, electron transfer from water through PS II to cytochrome *f* and some part of the ATP synthesizing mechanism—are sensitive to the solidification of membrane lipid in their immediate vicinity. There is an interesting effect of growth temperature on the sensitivity of *Anacystis nidulans* to low temperature. Rao *et al.* (1977) found that growth at 40°C made the cells susceptible to killing by a 5 min exposure to 0°C. This was not true for cells grown below 34°C. They found the acquisition of sensitivity to low temperature is slow and comparable to the rate of growth on shifting the culture from 25°C to 39°C. By contrast, the loss of cold sensitivity is rapid on shifting the culture from 39°C to 25°C implying a rapid alteration of the critical components. Brand (1977) found that chilling the sensitive cells causes an alteration in the carotenoid region of the absorption spectrum. Glutaraldehyde treatment, which should lock the membrane into a rigid configuration by cross-linking various components, prevented this cold-induced decrease in light absorption. Brand suggested that low temperature had induced an aggregation of the carotenoids in the membrane. Thus cold-induced spectral change was found in membrane fragments prepared from high-temperature grown cells and in liposomes prepared from a total lipid extract of these cells. Thus the saturated long-chain fatty-acyl lipids which solidify in the cold provide the driving force for aggregation of carotenoids in the membrane. Brand *et al.* (1979) found that chilling of cold-sensitive cells causes a marked alteration of the structure of the photosynthetic membrane surface as seen with the electron microscope. Verwer *et al.* (1978) had noticed this cold-induced aggregation of particles on the photosynthetic membranes of *Anacystis nidulans* grown at 38°C, and they suggested that the gradual solidification of lipids drives the aggregation of intramembrane particles. Furtado *et al.* (1979) have confirmed the freeze-fracture electron-microscopy work on particle aggregation and have used differential scanning calorimetry to establish the phase transitions of the membranes by this direct measurement technique. Clearly growth-temperature induced alterations in membrane lipid composition is a powerful probe of the organization of photosynthetic constituents.

CHAPTER 9
ANOXYGENIC PHOTOSYNTHESIS

E. PADAN & Y. COHEN

9.1 Introduction

Many strains of cyanobacteria have recently been shown to be capable of shifting between anoxygenic bacterial-type photosynthesis and oxygenic plant-type photosynthesis (Cohen, Padan & Shilo, 1975; Garlick *et al.*, 1977; Castenholz, 1976, 1977; Padan, 1979a,b). This capacity appears to bridge the gap between the strict aerobic phototrophic physiology of eukaryotic algae and plants and that of the strict anaerobic phototrophic bacteria. This apparent continuum at the physiological level suggests a similar continuum in the evolution of the phototrophic organisms (Padan, 1979a). Moreover, it explains the distribution of photosynthetic organisms in the very common, but hitherto little investigated, aquatic systems in which alternations between aerobic and anaerobic conditions prevail (Padan, 1979b).

9.2 Facultative sulphide-dependent anoxygenic photosynthesis of *Oscillatoria limnetica*

9.2.1 Sulphide and photosystem I-dependent anoxygenic photosynthesis

Oscillatoria limnetica, isolated (Cohen, Padan & Shilo, 1975) from the sulphide-rich layers of the Solar Lake (near Elat, Israel) (Cohen, Krumbein & Shilo, 1977), performs DCMU-sensitive, oxygenic photoassimilation of CO_2 (Chapter 8). When even a very low concentration (0.1–0.2 mM) of sulphide is added to the oxygenic system, immediate inhibition of CO_2 photoassimilation is obtained. However, if exposure to light is continued for about 2 h in the presence of a high sulphide concentration (3 mM), photoassimilation reappears and this regenerated photoassimilation is insensitive to DCMU (Cohen, Padan & Shilo, 1975). It is therefore anoxygenic, independent of photosystem (PS) II and driven by PS I with sulphide as an electron donor.

The possible participation of PS II in anoxygenic CO_2 photoassimilation has been excluded definitively (Cohen, Padan & Shilo, 1975) by the results of experiments in which PS II was simply not activated (rather than being inhibited) in the presence of far red light. Oxygenic photosynthesis, requiring operation of both photosystems, was drastically inhibited (red drop), whereas anoxygenic photosynthesis with sulphide was fully operative. Furthermore, if both photosystems could contribute to the reaction, the enhancement in quantum yield of the photoassimilation would be predicted with respect to that obtained with only PS I in operation. However, the 'enhancement phenomenon' was observed only with the oxygenic photosynthesis (Oren *et al.*, 1977).

The 2 h period of incubation in the presence of sulphide and light that is required for the anoxygenic photosynthesis may indicate that an induction process is involved. Indeed, in the presence of chloramphenicol, anoxygenic photosynthesis is not induced (Oren & Padan, 1978).

Sulphide is oxidized to elemental sulphur by *O. limnetica* according to the following stoichiometric relationship (Cohen, Jørgensen, Padan & Shilo, 1975):

$$2H_2S + CO_2 \rightarrow HCHO + 2S + H_2O$$

Elemental sulphur expelled from the cells was observed as refractile granules either free in the medium or adhering to the cyanobacterial filaments. Apart from this, the appearance of the cells under the electron microscope is the same under both anoxygenic and oxygenic conditions (A. Oren & S. Holt, pers. comm.).

9.2.2 *Sulphide and PS I-dependent hydrogen evolution and nitrogen fixation*

When CO_2 is eliminated from the reaction system, sulphide donates electrons to hydrogen evolution in *O. limnetica* (Belkin & Padan, 1979). This reaction occurs only in cells adapted to sulphide in the presence of CO_2, and it is dependent on light and sulphide while being insensitive to DCMU. Like the CO_2 photoassimilation reaction, it requires an induction period of 2 h and it is driven by PS I. However, whereas CO_2 photoassimilation is fully induced after 2 h incubation in the presence of sulphide, the capacity for hydrogen evolution, although initiated, is very low at this time, and for full induction an additional 46 h incubation in the presence of sulphide is required. It is concluded that in the sulphide-dependent hydrogen evolution an adaptation step(s) is required in addition to that needed for the sulphide-dependent CO_2 photoassimilation.

The hydrogen evolution reaction is accelerated by FCCP which uncouples phosphorylation from electron transport in cyanobacteria as in other organisms. The operation of the reaction under this condition, in which the cells are deprived of ATP, indicated that hydrogenase rather than nitrogenase is involved. Furthermore, the acceleration observed in the presence of the uncoupler implies that electrons from sulphide are donated directly to the electron transport chain and their rate of flow is controlled by a coupling site(s).

When sulphide induction is carried out without combined nitrogen in the medium and in an atmosphere of nitrogen gas, *O. limnetica* acquires the capacity to reduce acetylene to ethylene (Belkin *et al.*, 1979). This reaction is detected after a 15 h induction period and reaches its maximal activity at hour 48. Like hydrogen evolution and anoxygenic photoassimilation of CO_2, acetylene reduction is driven by PS I and sulphide serves as an electron donor. The acetylene reduction reaction is sensitive to uncouplers which suggests that nitrogenase is involved in the reaction. This suggestion is strengthened by the inhibition exerted by ammonia both during the induction process as well as in the assay.

It is thus concluded that electrons driven by PS I from sulphide can be channelled either to CO_2 fixation, hydrogen evolution or nitrogen fixation.

9.2.3 *Reversible inhibition of PS II by sulphide*

In the presence of low sulphide concentration (0.1–0.2 mM) far below the one (3.5 mM) allowing optimal anoxygenic CO_2 photoassimilation in *O. limnetica*, PS II is inhibited (Oren *et al.*, 1979). In many photosynthetic systems, e.g. *Aphanothece halophytica*, *Anacystis nidulans*, *Chlamydomonas* sp., and isolated chloroplasts, sulphide (0.1–0.5 mM) acts as an electron-transport inhibitor at a site preceding PS II (Oren *et al.*, 1979). Thus after exposure to sulphide in the presence of PS II light, the fluorescence yield of PS II decreased to the non-variable value, but the full yield was restored upon addition of DCMU. This kind of inhibition is observed in the presence of hydroxylamine which acts on a site before PS II (Mohanty *et al.*, 1971). In the dark, sulphide affected the variable fluorescence of the phototrophic systems in a similar way to that obtained by N_2 flushing, yielding anaerobiosis of the system, i.e. the variable fluorescence remained at a high constant value (for review of fluorescence changes in photosynthetic systems, see Papageorgiou, 1975).

If sulphide is excluded, *Oscillatoria limnetica* immediately shifts to oxygenic photosynthesis (Oren & Padan, 1978), which must therefore be constitutively present in the cells. Hence the photosynthetic system of *O. limnetica* operates facultatively both oxygenically and anoxygenically.

9.2.4 Utilization of absorbed quanta in anoxygenic and oxygenic photosynthesis

The concentrations in the cell of light-absorbing pigments, chlorophyll *a*, phycocyanin and carotenoids, are identical in both photosynthetic conditions (Oren *et al.*, 1977). Quantum-yield spectra of oxygenic and anoxygenic photosynthesis show (Oren *et al.*, 1977) that whereas only a narrow sector is utilized in the oxygenic type, the entire absorbed spectrum is utilized in anoxygenic photosynthesis. The drop in efficiency of oxygenic photosynthesis at both blue and red ends of the spectrum is marked. This limited range of utilization of quantum energies in oxygenic photosynthesis of *O. limnetica* is similar to that of other cyanobacteria and eukaryotic algae with similar pigment composition (Jones & Myers, 1964; Lemasson *et al.*, 1973); it is markedly different, however, from that of eukaryotic algae and plants containing chlorophyll *b* as the light-harvesting system. In these, almost the entire absorbed spectrum is utilized in the oxygenic photosynthesis, with the exception of the far 'red drop' (Govindjee *et al.*, 1968). Hence, quantum utilization in the photosynthetic system of *O. limnetica* appears to be oriented less towards oxygenic photosynthesis and permits efficient utilization of the total absorbed spectrum. This photosynthetic pattern of facultative alternation between anoxygenic and oxygenic photosynthesis by *O. limnetica* may be considered to represent an evolutionary link between the anoxygenic bacterial type and the oxygenic plant type of photosynthesis. There is also an alternative possibility that the photosynthetic pattern of *O. limnetica* has been secondarily evolved. An investigation of the frequency of occurrence of this photosynthetic pattern among cyanobacteria may help to distinguish between these alternative interpretations.

9.3 Sulphide-dependent anoxygenic photosynthesis among cyanobacteria

9.3.1 Distribution of anoxygenic photosynthesis

Various cyanobacteria have been tested for their capacity for anoxygenic photosynthesis, including strains from culture collections, new isolates (Garlick *et al.*, 1977; Utkilen & Castenholz, 1979; Pescheck, 1978) and natural populations (Castenholz, 1976, 1977). All the strains—30 in number—can carry out oxygenic photosynthesis inhibited by DCMU and not operative at wavelengths above 700 nm. Sixteen of them (including *O. limnetica*) are also capable of anoxygenic photosynthesis with sulphide, driven by PS I. In these latter cases photoassimilation of CO_2 has been demonstrated in the presence of sulphide, DCMU and/or of light wavelength 700 nm. The property of sulphide-dependent anoxygenic photosynthesis appears in different groups of cyanobacteria in filamentous as well as unicellular forms (Castenholz, 1977; Garlick *et al.*, 1977; Padan, 1979b). Moreover, it is not confined to particular ecosystems, geographical locations or culture conditions (Padan, 1979b).

9.3.2 Comparison of anoxygenic photosynthesis among cyanobacteria

Anoxygenic photosynthesis of *Aphanothece halophytica* was compared with that of *Oscillatoria limnetica*; in both, an induction process is involved and the stoichiometric relations of the reaction are the same (Garlick *et al.*, 1977).

A comparison of rates of anoxygenic photosynthesis with respect to sulphide concentration was carried out among *Lyngbya* 7104, *Aphanothece halophytica*, and *Oscillatoria limnetica* (Garlick *et al.*, 1977). In all three, the pattern of dependence on sulphide concentration was similar, generating an optimum curve rather than a saturation curve. The drop in photosynthetic rates at high sulphide concentrations is presumably caused by sulphide toxicity, an effect recognized both in phototrophs and heterotrophs (Castenholz, 1976); the observed rates may thus not be the true ones. Nevertheless, the highest rate of oxygenic photosynthesis is similar to that of anoxygenic photosynthesis in both *O. limnetica* (1–2 μmol C mg^{-1} protein h^{-1}) and *Aphanothece halophytica* (0.5–1 μmol mg protein h^{-1}). While the trend of dependency on sulphide is similar, both the affinity toward sulphide and the tolerance for sulphide are different. Thus, each strain shows a different range of sulphide concentration at which anoxygenic photosynthesis can be performed. The range, measured at pH 6.8, is for *Lyngbya* 7104, 0.1–0.3 mM, for *Aphanothece halophytica*, 0.1–1.5 mM, and for *Oscillatoria limnetica* 0.7–9.5 mM the highest. Furthermore, in each case this range is markedly affected by the pH of the medium (Padan, 1979b) which governs the dissociation of hydrogen sulphide.

9.3.3 *Comparison of cyanobacterial anoxygenic photosynthesis with that of bacteria*

Differences in pH-dependent sulphide ranges are known to constitute a determining factor in the ecology of photosynthetic sulphur bacteria (Baas Becking & Wood, 1955; Pfennig, 1975). At pH 7, the Chlorobiaceae thrive at the highest sulphide concentrations, 4–8 mM, as does *O. limnetica*. The Chromatiaceae do best at 0.8–4 mM, and Rhodospirillaceae at 0.4–2 mmol. The extracellular deposition of sulphur by *O. limnetica* and *Aphanothece halophytica* is carried out in a similar way by sulphide-oxidizing photosynthetic bacteria (Chlorobiaceae and some of the Chromatiaceae: Pfennig, 1977). However, the pattern of sulphide oxidation of these cyanobacterial strains seems thermodynamically inefficient; photosynthetic bacteria oxidize sulphide to sulphate (Pfennig 1975, 1977) gaining eight electrons for each H_2S molecule, while *Oscillatoria limnetica* and *Aphanothece halophytica* remove only two electrons per molecule. The oxidation of sulphide to sulphur may be a by-product in the sulphide oxidation process of sulphur phototrophic bacteria (Trüper, 1973). Finally, other sulphur-containing electron donors which are efficiently used by photosynthetic bacteria (Gromet-Elhanan, 1977; Pfennig, 1975, 1977) do not seem to serve the photosynthetic system of the sulphide-utilizing cyanobacteria. *Anacystis nidulans*, however, can oxidize thiosulphate to sulphate (Utkilen, 1976).

As in certain bacteria and heterocystous cyanobacteria, in both *Oscillatoria limnetica* and *Aphanothece halophytica* molecular hydrogen has been shown to be an efficient electron donor for CO_2 in an anoxygenic reaction driven by PS I. Hydrogen evolution dependent on sulphide oxidation has been demonstrated in the former strain (Belkin & Padan, 1978, 1979). Anoxygenic hydrogen metabolism is thus present both in these nonheterocystous sulphide-utilizing cyanobacteria and in heterocystous species (Bothe *et al.*, 1977b; Tel-Or *et al.*, 1977b; Weisman & Benemann, 1977).

9.3.4 *Sulphide toxicity to oxygenic and anoxygenic photosynthesis*

In cyanobacteria, sulphide appears more toxic to oxygenic photosynthesis than to anoxygenic photosynthesis; in *Oscillatoria limnetica* the former is inhibited by 0.1 mM sulphide (Oren *et al.*, 1979) while the latter is inhibited only by 4 mM (at pH 6.8) (Garlick *et al.*, 1977; Padan, 1979b). These differences have also been observed in other facultative anoxygenic strains (Padan, 1979b). This difference in toxicity implies that in the higher range of sulphide concentrations characteristic of a given strain, oxygenic photosynthesis will make only a minor contribution, if any, to photoassimilation of CO_2. However, at the lower range, oxygenic photosynthesis may occur simultaneously to anoxygenic photosynthesis. An example of such a case has been demonstrated in *O. amphigranulata* (Utkilen & Castenholz, 1979) and occasionally also in the chemocline of the Solar Lake (Jørgensen *et al.*, 1979a). Low sulphide concentrations (0.5 mM) are toxic to eukaryotic phototrophs and cyanobacterial strains which do not carry out anoxygenic photosynthesis (Castenholz, 1976; Knobloch, 1969; Padan, 1979b; Howsley & Pearson, 1979).

The greater sulphide tolerance of anoxygenic photosynthesis operating with PS I confers a selective advantage on species of cyanobacteria which can perform anoxygenic photosynthesis over those which cannot, in sulphide-rich habitats. This advantage is obtained even if facultative anoxygenically photosynthesizing cyanobacteria strains cannot grow under purely anaerobic conditions because they can at least maintain themselves in the presence of sulphide. However, the most successful strains are capable of growth under these conditions.

9.4 Facultative anaerobic phototrophs vs. obligate anaerobic or aerobic phototrophs

9.4.1 Gap between eukaryotic phototrophs and phototrophic bacteria in phototrophic physiology

Phototrophic bacteria differ markedly from eukaryotic algae in the mechanism of phototrophic growth. Phototrophic bacteria have an obligate anaerobic phototrophic growth physiology, characteristic for obligate as well as for facultative anaerobic photosynthetic bacteria which can grow aerobically in the dark (Cohen-Bazire *et al.*, 1957; Lascelles, 1959; Pfennig, 1975, 1977). In contrast, the majority of, if not all, eukaryotic phototrophs cannot grow anaerobically either in light or dark (Kessler, 1974); hydrogen-dependent photoreduction of CO_2 could not be shown to support anaerobic photoautotrophic growth of eukaryotic algae (Gaffron, 1944). Although aerobic photoheterotrophic growth does occur in eukaryotic algae (Kessler, 1974; Simonis & Urbach, 1973), neither anaerobic photoheterotrophic growth nor anaerobic heterotrophy have been unequivocally demonstrated in these organisms (Gibbs, 1962; Kessler, 1974; Simonis & Urbach, 1973); hence, eukaryotic phototrophs appear to be obligate aerobes.

The inability of eukaryotic algae to photoassimilate CO_2 and to grow in the presence of sulphide may be explained by the toxicity of sulphide to either the photosynthetic systems or some other system. However, it is not evident why eukaryotic algae do not carry out anaerobic growth, either heterotrophically or phototrophically, using H_2 as an electron donor. There are at least two fundamental prerequisites for anaerobic growth; the first is that all of the essential enzymatic reactions should be capable of proceeding in the absence of oxygen. The second is a capacity to function under reducing conditions. Eukaryotic phototrophs possess a number of enzymes which require O_2, e.g. those involved in fatty acid or sterol synthesis, or in phenol oxidation (Goad & Goodwin, 1972; Hamberg *et al.*, 1974; Wood, 1974). Some of these reactions may not be replaceable by alternative anaerobic pathways. Oxygen-dependent pathways exist in *Saccharomyces*, which will grow anaerobically only if supplied with essential products such as sterols (Andreasen & Stier, 1953). No anaerobic growth experiments of this kind have been carried out with eukaryotic phototrophs.

Reducing conditions, from −200 mV to −300 mV with sulphide or H_2 present, are often encountered in anaerobic habitats (Baas Becking & Wood, 1955; Cohen, Krumbein & Shilo, 1977; Fenchel & Riedl, 1970). The capacity for growth in a particular redox potential depends on the ability of the cell to maintain the appropriate internal redox potential. The evolution of hydrogen in anaerobically growing phototrophic bacteria is an example of an adaptation to reducing conditions (Gest, 1972). In this context, it is possible that the stringent dependence of eukaryotic algae on photoaerobic conditions for growth is due to their inability to maintain their redox potential under anaerobic conditions (Gibbs, 1962).

Both of the requirements for anaerobic growth are met by the phototrophic bacteria; they do not need oxygen, and they tolerate low external redox potentials. In fact, oxygen is toxic to several metabolic steps in obligate anaerobic phototrophic bacteria (Frenkel, 1970), as in other obligate anaerobes (McCord *et al.*, 1971). Negative control exerted by oxygen on the photosynthetic system of facultative anaerobic bacteria is also a well-known phenomenon (Cohen-Bazire *et al.*, 1957; Lascelles, 1959). Finally, the photosynthetic system of both facultative and obligate anaerobic bacteria appears to operate most efficiently when poised at negative redox potential (Frenkel, 1970) as expected for a cyclic system operating without change in net oxidoreduction potential.

9.4.2 *Facultative anoxygenic phototrophic physiology*

In *Oscillatoria limnetica*, once induction to utilization of sulphide in anoxygenic CO_2 photoassimilation has taken place, it can readily carry out anaerobic photoautotrophic growth (Oren & Padan, 1978) and the anaerobic photoautotrophic growth rate (2 days, doubling time) is similar to the aerobic photoautotrophic rate. Once sulphide is excluded, *O. limnetica* immediately shifts to photoaerobic growth with oxygenic photosynthesis, and the anoxygenic photosynthetic capacity is diluted out by growth. Hence *O.*

limnetica is a facultative anaerobic photoautotroph, which readily shifts between aerobic photoautotrophic and anaerobic photoautotrophic growth patterns. It therefore possesses an intermediate photoautotrophic physiology which bridges the gap between that of eukaryotic algae and phototrophic bacteria.

In carrying out anaerobic growth in conditions of high sulphide concentrations (3.5 mM; pH 6.8), *O. limnetica* must both cope with the low redox potential and dispense with oxygen in its metabolism. Studies of fatty-acid composition in unicellular cyanobacteria have found species with poly-unsaturated as well as with saturated and mono-unsaturated fatty acids (Kenyon, 1972), along with an oxygen-dependent desaturation mechanism (Nichols *et al.*, 1965). In almost all filamentous cyanobacteria analysed by Kenyon *et al.* (1972), there was an abundance of unsaturated fatty acids; an oxygen-dependent mechanism for desaturation has been demonstrated in *Anabaena variabilis* (Bloch *et al.*, 1966). Only some species of *Spirulina*, possibly related to the anoxygenic photosynthesizing strain (Castenholz, 1977), were found to contain saturated and mono-unsaturated fatty acids (Kenyon *et al.*, 1972). Recently Oren, Dvir and Padan (unpubl.) have shown that *Oscillatoria limnetica*, grown either aerobically or anaerobically, lacks poly-unsaturated fatty acids as do anaerobic growing bacteria; like these latter it does not require an oxygen-dependent desaturation mechanism. In contrast, the anoxygenic photosynthesizing cyanobacteria *Aphanothece halophytica*, *Lyngbya* 7104 and *Lyngbya* 7004 contain polyunsaturated fatty acids (Kenyon *et al.*, 1972) and *Aphanothece halophytica* does possess an oxygen-dependent desaturating mechanism (Oren, Dvir, Padan, unpubl.). Thus anoxygenic photosynthesis need not necessarily be linked with anaerobic growth and anaerobic fatty acid metabolism. This physiological spectrum, representing different degrees of anaerobiosis, may be a useful criterion for the taxonomy of cyanobacteria (Kenyon *et al.*, 1972; Stanier & Cohen-Bazire, 1977).

Although *Oscillatoria limnetica* apparently does not grow anaerobically in the dark, it can maintain itself in darkness by reversing the anoxygenic reaction, forming H_2S from elemental sulphur. There are also indications of glucose fermentation in *O. limnetica* (Oren & Shilo, 1979).

Eukaryotic algae or phototrophic bacteria are restricted by their physiologies to habitats with (exclusively) photoaerobic or photoanaerobic conditions respectively. In habitats with fluctuating conditions, a cyanobacterium with an *O. limnetica* type of facultative anoxygenic photosynthesis should clearly have a selective advantage. In fact, the more frequent and marked the fluctuations, the more prominent the cyanobacteria are with respect to the other phototrophic organisms (Padan, 1979b).

9.5 Ecological impact of facultative anoxygenic photosynthesis of cyanobacteria

9.5.1 *Instability of sulphide in ecological systems*

Since sulphide is readily oxidized by oxygen, the maintenance of a particular sulphide concentration in a habitat is influenced by the oxygenic photosynthetic activity and the diffusion of oxygen from the overlying atmosphere. Sulphide will therefore often accumulate in the deep aphotic zone where oxidation processes will be limited to oxidation by colourless sulphur bacteria and restricted by the availability of suitable electron acceptors (Sorokin, 1964; Kuenen, 1975). This ecological niche or biotope is relatively stable and is characterized by a slow rate of sulphide turnover, as in the Black Sea (Sorokin, 1964).

On the other hand, when light penetrates into the sulphide-rich layer, the oxidation processes are markedly accelerated and a sharp chemocline of H_2S and O_2 is observed. Furthermore, the sulphide turnover rate is much faster under such a situation and a wide fluctuation of sulphide concentration and even temporal build up of oxygen in this zone may be encountered (Jørgensen, Kuenen & Cohen, 1979). These oscillations may occur seasonally (Kusnezov, 1970) or diurnally (Ganf & Viner, 1973) or even during shorter time periods (Jørgensen, Kuenen & Cohen, 1979). Cyanobacteria having the capacity for facultative anoxygenic photosynthesis have an obvious ecological advantage in environments alternating between aerobic and anaerobic conditions. An attempt is made here to evaluate the contribution of cyanobacterial activity in these biotopes, namely hot sulphur springs, cyanobacterial mats or stromatolites and stratified waterbodies.

9.5.2 *Cyanobacteria in hot sulphur springs*

Downstream of hot sulphur springs a sulphide gradient is established and may serve as a model for the study of the effect of sulphide, as well as other environmental factors such as temperature, pH and chemical composition, on the distribution of microorganisms (Brock, 1978). Such studies have been carried out in different geographical locations by Schwabe (1936, 1960), Czurda (1940–41), Anagnostidis and Golubić (1966), Castenholz (1969, 1976, 1977) and Brock (1978).

The filamentous photosynthetic bacterium *Chloroflexus* (Pierson & Castenholz, 1974) was found to predominate in most upstream water of hot springs in New Zealand at sulphide concentrations between 6 μM and 2 mM and at temperatures of up to 60–70°C (Castenholz, 1976), forming a unibacterial thick orange-green mat. Downstream, at temperatures lower than 56°C, but still with high sulphide concentrations (around 2 mM), *Oscillatoria amphigranulata* replaced *Chloroflexus* as the predominant mat-forming organism. *Oscillatoria amphigranulata* formed the top layer of the mat with *Chloroflexus* below. *Synechococcus* and *Synechocystis* also occurred at high sulphide concentrations (up to 1.5 mM), but at lower temperatures. *Mastigocladus laminosus* predominated only at lower sulphide concentrations of up to 0.2 mM.

In situ CO_2 photoassimilation experiments done under various sulphide concentrations, using DCMU and different light wavelengths may explain the pattern of distribution of cyanobacteria observed (Castenholz, 1977). The rate of CO_2 photoassimilation of *Oscillatoria amphigranulata* collected from high sulphide waters—upstream—was three times higher in the presence of sulphide concentrations of up to 2 mM compared to that of the sulphide-free control. This activity was only slightly inhibited by the addition of DCMU or in the presence of light of wavelength 700 nm. Furthermore *O. amphigranulata* grown *in situ* at high sulphide concentrations was found to have less phycocyanin and phycoerythrin, indicating a lower activity of PS II and a PS I-dependent anoxygenic photosynthesis. On the other hand, cyanobacteria found downstream at lower sulphide concentrations had higher phycobilin contents and CO_2 photoassimilation activity was inhibited by H_2S and by DCMU. Anoxygenic photosynthetic activity could then be demonstrated after long pre-incubation in the presence of sulphide. Axenic cultures of *O. amphigranulata* were found by Utkilen and Castenholz (1979) to have a doubling time of 6 h under 0.6–1.0 mM sulphide, with or without DCMU. Similar growth rates were observed for oxygenic conditions under 1% CO_2+99% N_2 sparging. *Synechococcus* found at lower sulphide concentrations appears to tolerate sulphide, but cannot use it. HTF *Mastigocladus*, which occurs only at very low sulphide concentrations, was found to be very sensitive to sulphide (Schwabe, 1960).

The sulphur springs in Iceland exhibit similar patterns of distribution of cyanobacteria, but since the sulphide content of these springs is much lower, the niche occupied by *Oscillatoria amphigranulata* in New Zealand is absent. *Synechococcus* and *Phormidium* are found at sulphide concentrations of up to 5 μM. HTF *Mastigocladus* is the predominant organism by far; since it is sulphide sensitive, its upper distribution is determined by the *in situ* sulphide level (Schwabe, 1960; Castenholz, 1976).

In Yellowstone's hot springs *Chloroflexus* forms a mat at 50–65°C and 30–60 μM sulphide, often under an upper layer of the green phototrophic bacterium *Chlorobium*. *Spirulina labyrinthiformis* appears together with *Chloroflexus* at relatively high sulphide concentrations of up to 90 μM. At lower sulphide concentrations of up to 22 μM and at 60°C, *Synechococcus lividus* predominates, while at very low sulphide concentrations and 50°C more cyanobacterial species develop, including *Mastigocladus laminosus*, *Phormidium*, *Oscillatoria*, *Pseudanabaena* and *Synechococcus minervae*. Of the cyanobacteria found in Yellowstone Park only *Spirulina labyrinthiformis* seems to carry out anoxygenic photosynthesis. Sulphide concentrations of up to 1.6 mM were found to stimulate CO_2 photoassimilation slightly, and this activity continued in the presence of 7 μmol DCMU or under 700 nm light. Oxygenic photosynthesis was completely inhibited by 0.4 mM hydroxylamine, but 80% of this inhibition was relieved upon addition of 2 mM sulphide (Castenholz, 1977). *Spirulina labyrinthiformis* appears to shift readily to oxygenic photosynthesis. No other reduced sulphur compounds can replace sulphide for the stimulation of CO_2 photoassimilation and most compounds inhibit photosynthesis (Castenholz, 1977). The behaviour of *Phormidium* found at sulphide concentrations of less than 60 μM was investigated by Weller *et al.* (1975). Photosynthesis was inhibited by 0.2 mM sulphide in both the presence and absence of DCMU, but stimulated by lower sulphide concentrations, as well as by cysteine, thioglycollate or by merely sparging with N_2. This cyanobacterium was found to be inhibited markedly by oxygen, explaining the effect of the reducing conditions.

Hunter's Hot Spring in Oregon, USA, provides another example of cyanobacterial adaptation to sulphide. Here *Synechococcus lividus* and *Oscillatoria terebriformis* occupy the stream at very low sulphide concentrations of up to 3 μM. *Oscillatoria terebriformis* was insensitive to moderate sulphide concentrations of up to 0.8 mM, but DCMU completely blocked CO_2 photoassimilation in the presence of sulphide. However the same species collected from nearby springs with a higher sulphide concentration (65 μM) showed both tolerance and some sulphide utilization in anoxygenic photosynthesis. Under these conditions, there was also partial operation of PS II (Castenholz, 1977).

Sulphur springs provide a relatively stable biotope with respect to sulphide since the geothermal source continuously supplies sulphide. The sulphide concentration decreases in a steady gradient downstream with little temporal fluctuation. This biotope allows the development of various cyanobacteria with different specialized adaptations to sulphide.

9.5.3 Cyanobacteria in sediments

Unlike hot sulphur springs, the sulphide source in most other sulphide-rich environments is biogenic and undergoes considerable temporal fluctuation. Marked temporal sulphide fluctuations are observed in cyanobacterial mats and stromatolites, which often develop in close association with reduced sediments and have a wide global distribution covering: tidal marine sediments (Fenchel & Riedl, 1970), hypersaline marshes (Krumbein & Cohen,

1974; Krumbein *et al.*, 1977; Howarth & Teal, 1978), marine upwelling regions (Gallerdo, 1976), mangrove swamps and rice paddies (Stewart, 1973b), freshwater bodies (Kusnezov, 1970), and hot springs (Brock, 1978).

Light penetration in cyanobacterial mats is restricted to the upper 1–2 mm in hot springs (Brock, 1978), 2–3 mm in Solar Lake mats (Jørgensen, Revsbech, Blackburn & Cohen, 1979) and up to 8 mm in coastal sediments (Fenchel & Straarup, 1971). This very narrow photic zone is accompanied by a high rate of microbial sulphur cycling (Jørgensen & Cohen, 1977) and creates steep gradients along micrometre ranges, giving rise to methodological problems in the study of the dynamics of oxygen sulphide fluctuations. The development of redox electrodes for *in situ* Eh measurements (Whitfield, 1969) and sulphide electrodes (Berner, 1963) allowed the first information on the changes in reducing conditions in the mat. A method for the determination of *in situ* rate of sulphide production through sulphate reduction was developed by Jørgensen and Fenchel (1974). Later Blackburn *et al.* (1975), using sulphide electrodes inserted to various depths, were able to demonstrate gradients of sulphide oxidation rates. The recent application of oxygen microelectrodes in sediments (Revsbech *et al.*, 1980) and the combination of the latter with Eh and sulphide electrodes (Jørgensen, Revsbech, Blackburn & Cohen, 1979) permits a more complete understanding of the dynamics of sulphur cycles in sediments. For an overall understanding of this system further methods have still to be developed: microsulphide and micro-pH electrodes, *in situ* methods for CO_2 assimilation rate and methods for studying the *in situ* H_2S oxidation pathway.

Annual fluctuations in sulphide and oxygen in marine sediments in the Limfjorden, Denmark, have been described by Jørgensen (1977a). At the end of summer, the oxygen content of the water column drops and the sediment is reduced up to the surface. Under these conditions *Beggiatoa* appear at the top of the sediment above purple phototrophic bacteria and cyanobacteria. Anoxic conditions occur only for a short period and during the rest of the year the sediment is oxygenated well below the photic zone. In the oxic sediment, however, reduced micro-niches may occur where a high rate of sulphate reduction creates temporal fluxes of sulphide within the oxidized sediment (Jørgensen, 1977b).

Sulphate reduction rate and sulphide oxidation processes in marine sediment model systems were studied by Jørgensen and Fenchel (1974) and Blackburn *et al.*, (1975) respectively. Both investigations indicate a close association of sulphur and carbon cycles. The maximum activity of sulphide oxidation as well as sulphate reduction occurs at the upper few millimetres of the sediment, where sulphide production is directly related to photosynthates released by cyanobacteria and diatoms. *In situ* measurements in the Solar Lake mats (Jørgensen & Cohen, 1977) show that dissolved organic matter produced by apparent anoxygenic photosynthesis of *Phormidium* and *Microcoleus* allows sulphate reduction to occur even far below the photic zone, where this reaction is limited by the availability of suitable organic carbon sources.

The diurnal cycle of microgradients of oxygen and sulphide and its effect on the microbial photosynthesis in the cyanobacterial mat of the Solar Lake was studied by Jørgensen, Revsbech, Blackburn and Cohen (1979). Their experiments were done in mats below the peripheral shallow waters which are oxygenated throughout the year. The diurnal development of oxygen and sulphide microgradients on a summer day is shown in Fig. 9.1, which illustrates the two extreme situations. The upper millimetre of the sediment is exposed to a very high light intensity (1000 μE m^{-2} s^{-1}), which creates photoinhibition conditions at the very top of the sediment (Fig. 9.1a,d). At 1–2.5 mm, high oxygen concentrations occur during most hours of the day. However, during the last hour when light intensity dropped to 250 μE m^{-2} s^{-1} sulphide reappeared in the photic zone. Similar conditions occur early in the morning, but oxygen is quickly built up and persists throughout the day. Yet H_2S accumulates just below the photic zone. Hence, the cyanobacteria *Phormidium*, *Microcoleus* and *Aphanothece halophytica* found in the photic zone at 2.5 mm encounter within 24 h 450% saturation of oxygen and 2 mM sulphide. The light intensity in winter dropped to only 20% of its summer value; the same general diurnal cycle took place, but the photic zone was shifted 1.5 mm upwards.

The maximal rates of alternation between oxygenated and fully reduced conditions in the sediment can be even faster, as shown by an experiment in which isolated fresh sediment cores were transferred from dark to high light intensity (Fig. 9.2a) and vice versa (Fig. 9.2b). The oxygen concentration decreased rapidly after turning off the light, and oxygen was depleted completely after 10 min. Following the shift, an increase of sulphide was observed after 40 min and reached the maximum dark steady state concentration after 180 min. On re-illumination the sulphide concentration gradually decreased, but only after a full hour was a significant oxygen increase detected and maximum oxygen concentrations were reached only after 4 h of exposure to light. Eventually, a steady state was reached, with the sulphide gradient from below overlapping the overlying oxygen.

The drastic diurnal changes between sulphide and oxygen in the photic zone must be a very selective environmental factor against both strict anaerobic and strict aerobic phototrophic microorganisms. Indeed cyanobacteria which have the physiological flexibility to cope with changing aerobic–anaerobic conditions are by far the most abundant organisms in this environment. Of the cyanobacteria forming the mat, *Microcoleus* and *Aphanothece halophytica* carry out anoxygenic photosynthesis (Garlick, Oren

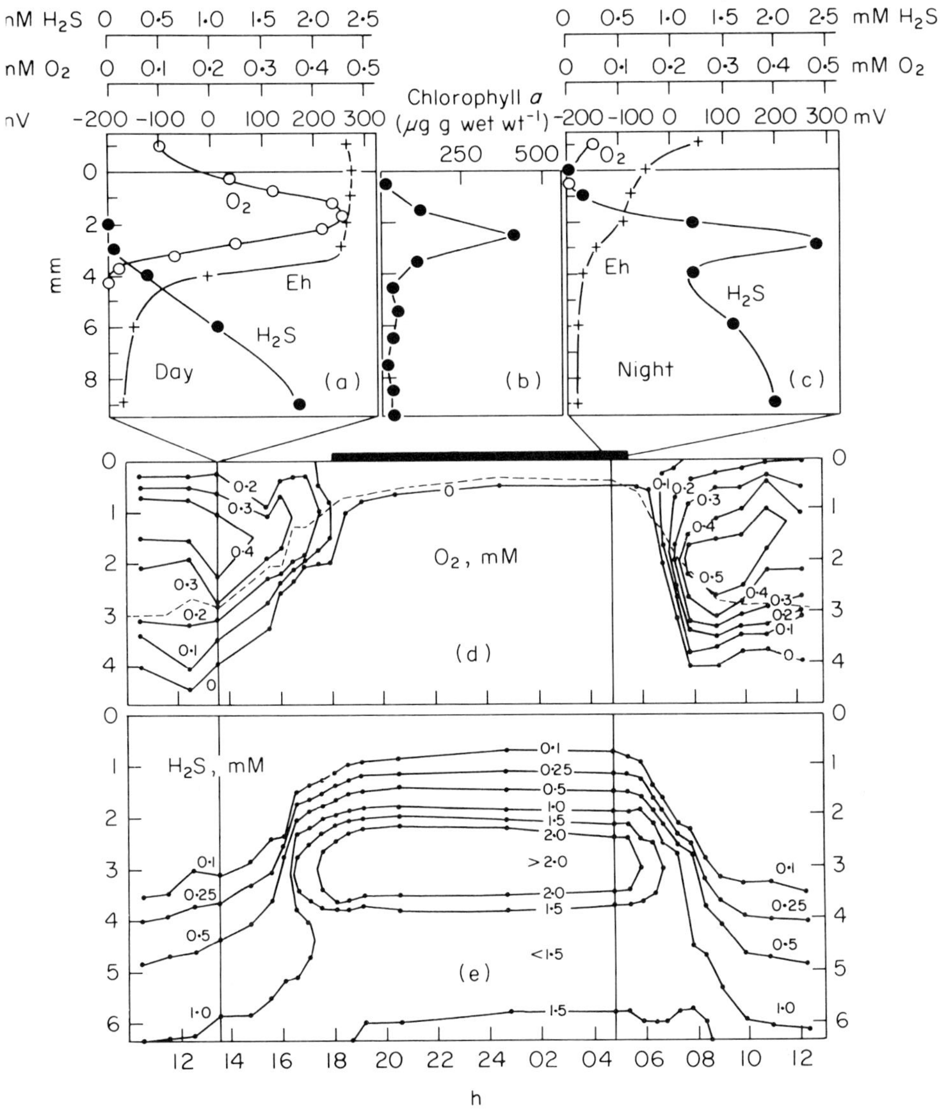

Fig. 9.1. Diurnal cycles in cyanobacterial mats showing the daily extremes. (a) Redox potential (Eh), O_2 and H_2S in the pore water at noon (13.35). (b) Chlorophyll *a* distribution along sediment core. (c) Redox potential O_2 and H_2S in the pore water at night (04.30). (d) Diurnal cycle of O_2 (isopleths). (e) Diurnal cycle of H_2S (isopleths). All measurements done September 1979. (Redrawn from Jørgensen, Revsbech, Blackburn & Cohen, 1979)

& Padan, 1977). In the cyanobacterial mat oxygen appeared when H_2S levels were reduced below 0.1–0.2 mM. The same sulphide level was found by Oren and Padan (1978) to be the minimum concentration needed to block PS II in *Oscillatoria limnetica*. Since alternations between high oxygen and high sulphide occur twice daily, similar shifts are expected between oxygenic and anoxygenic photosynthesis. However, it is impossible at this stage to differentiate accurately between sulphide oxidation activities of cyanobacteria and those of the accompanying *Chloroflexus* and purple phototrophic bacteria (Jørgensen, Revsbech, Blackburn & Cohen, 1979).

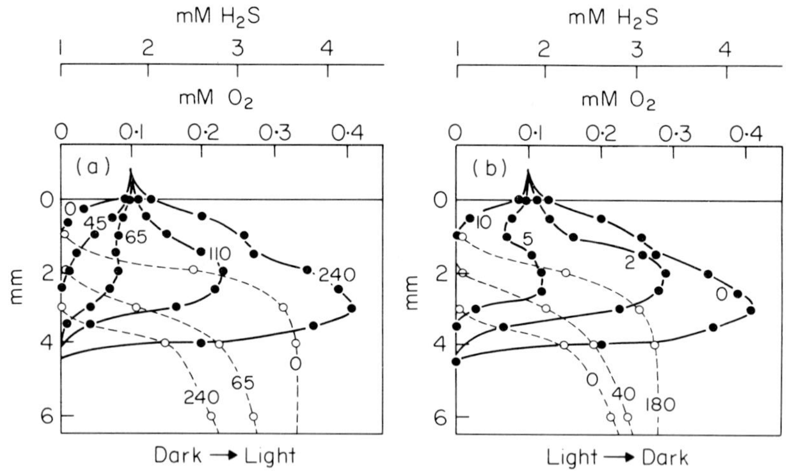

Fig. 9.2. Vertical distribution of (——) O_2 and (– – –) H_2S in a mat core at different times (minutes) after change (a) from dark to light or (b) vice versa. (Redrawn from Jørgensen, Revsbech, Blackburn & Cohen, 1979)

Because of the methodological problems in sediments there are few data available on the sulphide oxygen gradients in the vast areas of the world covered by cyanobacterial mats. The activity by burrowing fauna may introduce oxygenated water well into the reduced sediments and this further complicates the sulphide oxygen microgradient (Fenchel & Riedl, 1970). Nevertheless, the similar environmental conditions and the universal distribution of predominant cyanobacterial species indicate that similar processes occur in other stromatolitic environments. Vast recent cyanobacterial mats are described for Shark Bay, Western Australia (Playford & Cockbain, 1976), the Persian Gulf (Kinsman & Park, 1976), freshwater environments like Green Lake, New York (Eggleston & Dean, 1976) and open marine tidal stromatolites in the Caribbean (Gebelein, 1976), the Red Sea (Krumbein, 1979) and the general marine 'thiobios' (Fenchel & Riedl, 1970).

9.5.4 Cyanobacterial anoxygenic photosynthesis in stratified waterbodies

In stratified lakes an oxygen–sulphide gradient is established in the chemocline,* the stability of which is controlled mainly by the mixing pattern of the lake. The most labile chemocline occurs in polymictic lakes, like the shallow eutrophic Lake George in equatorial Uganda, which has been the subject of detailed study (Baxter *et al.*, 1965; Talling *et al.*, 1973; Viner & Smith, 1973; Ganf & Viner, 1973; Ganf & Horne, 1975; Greenwood, 1976). The lake is stratified from late morning to evening. During the evening and throughout the night the lake is completely mixed. Drastic daily fluctuations in oxygen concentration were observed with an occasional drop below saturation level. Yet only when the thermocline persisted for several days did the hypolimnion become fully reduced. The almost exclusively cyanobacterial community consisting of *Microcystis*, *Anabaenopsis*, *Lyngbya*, *Aphanocapsa* and *Chroococcus* will thus be exposed to photoanaerobic conditions for short periods of time during the early morning. Since these oscillations are short and their amplitude rarely allows sulphide build-up, anoxygenic photosynthesis cannot play an important role in the primary production of these lakes. However, the occasional exposure to sulphide may perhaps select for sulphide-resistant and possibly even sulphide-utilizing strains.

In the range from oligomictic to meromictic lake, sulphide is often built up in the hypolimnion and persists for long periods (Kusnezov, 1970). The contribution of sulphide-dependent CO_2 photoassimilation to the overall primary production in these lakes ranges from 3% in the oligomictic Fayetteville Greenlake, New York, USA (Culver & Brunskill, 1969) to 91% in the monomictic Solar Lake, Sinai (Cohen, Krumbein & Shilo, 1977).

In the Solar Lake, a brine-filled basin of 140 × 50 m and 5 m depth on the coast of the Red Sea, a pycnocline builds up in September due to seepage of seawater to the surface. Solar heating produces a mesothermic temperature curve with a maximum of up to 65.5°C at 2–3 m depth which decreases to about 40–45°C at the bottom (Cohen, Krumbein, Goldberg & Shilo, 1977). Under these conditions sulphide accumulates at the hypolimnion to concentrations over 3.5 mM, and the Eh ranges from +390 mV at the surface to −185 mV at the bottom. Maximum stratification is achieved in April–May and then, due to the increasing evaporation rate, the pycnocline disappears and the lake becomes mixed. During the short summer holomixis the mesothermal temperature profile disappears and the whole water column is oxygenated.

The annual development of primary production activity in the lake is

* For definition of limnological terms, see Hutchinson (1967).

closely related to sulphide accumulation in the hypolimnion and can explain the seasonal succession of cyanobacteria leading to the distinct vertical distribution at the height of the stratification (Cohen, Krumbein & Shilo, 1977). In years when the maximum H_2S concentration at the bottom reached 1.2 mM *Chromatium violescens* occupied the chemocline above a lower bacterial plate of the green sulphur bacterium *Prosthecochloris*, while at the bottom, where H_2S concentrations were highest, *Oscillatoria limnetica* predominated. The oxic epilimnion is occupied by *Aphanothece halophytica*, *Aphanocapsa littoralis* and several diatoms. During holomixis the sulphur phototrophic bacteria disappear and the previously stratified cyanobacteria become distributed homogeneously. In years when the maximum H_2S concentrations are highest (reaching 3.5 mM), the sulphur-phototrophic bacteria disappear and the stratified lake is governed by cyanobacteria. *Aphanothece halophytica* appears in the epilimnion above a distinct, narrow plate of *Dactylococcopsis* (Walsby & Cohen, in prep.). The chemocline and upper hypolimnion is populated by *Phormidium*, while *Oscillatoria limnetica*, which appears in low numbers at the chemocline, builds up a dense bloom towards the bottom together with *Pseudanabaena*.

Of the cyanobacteria occurring in Solar Lake during stratification *Aphanothece halophytica*, *Phormidium* and *Oscillatoria limnetica* are capable of anoxygenic photosynthesis, but require different optimal sulphide concentrations of 0.08–0.4 mM, 0.45–0.9 mM and 3.5 mM, respectively (Garlick *et al*., 1977). The different sulphide requirements presumably explain the vertical distribution of these cyanobacteria in the lake during stratification. Sharp and diurnally fluctuating gradients of oxygen and sulphide occur at the chemocline of Solar Lake during stratification, where *Phormidium* and *Oscillatoria limnetica* proliferate (Jørgensen, Kuenen & Cohen, 1979; Fig. 9.3). Sulphide can be traced 10–20 cm up into the oxygenated region at night, but descends away from the oxygen layer at noon. This pattern is complicated by the secondary peak of oxygen which builds up within the sulphide zone in the late morning and disappears in the evening (Fig. 9.3b,d). The shift occurs at sulphide concentrations of 5–10 μM. At this sulphide concentration, CO_2 photoassimilation was stimulated by the addition of DCMU. It is suggested that at this low sulphide concentration both oxygenic and anoxygenic photosynthesis occur concomitantly. The addition of DCMU halts O_2 production and this increases the availability of sulphide for PS I-dependent anoxygenic photosynthesis. As a result the efficiency of CO_2 photoassimilation increases. Sulphide oxidation by cyanobacteria results in the build-up of elemental sulphur during daytime. Elemental sulphur disappears during the night, probably due to its reduction through the anaerobic respiration of the cyanobacteria (Oren & Shilo, 1979). Below the O_2–H_2S borderline the photosynthetically produced elemental sulphur reacts chemically with

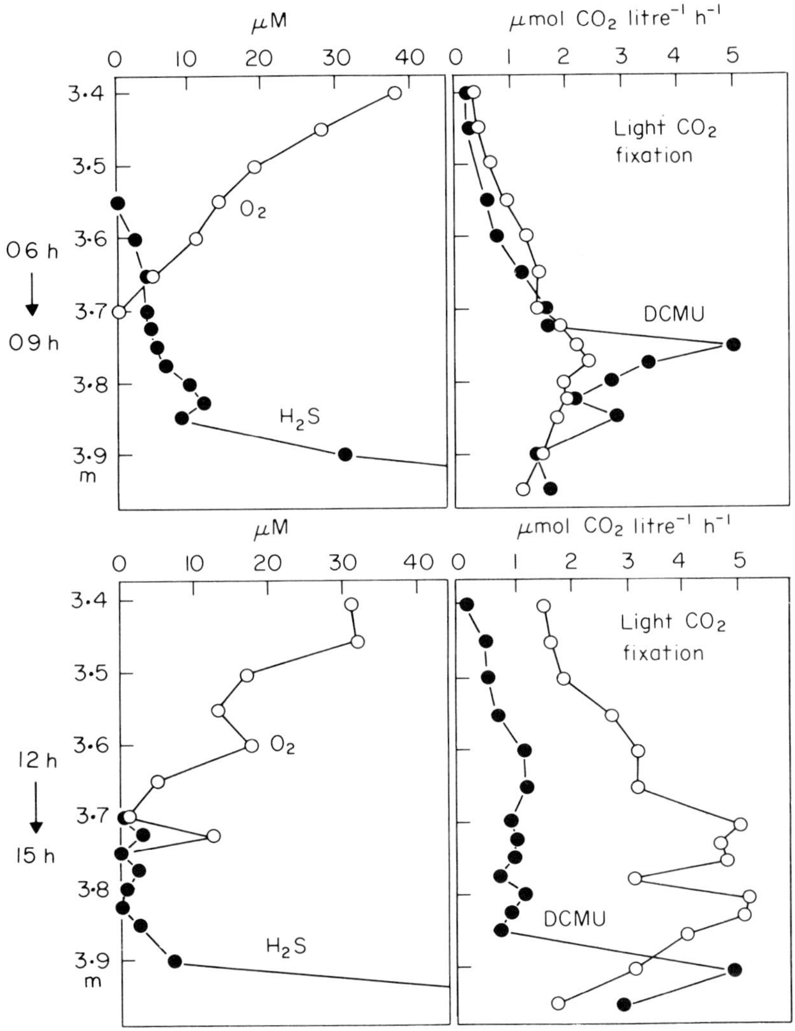

Fig. 9.3. Distribution of oxygen and sulphide and light-dependent CO_2 fixation with or without DCMU in chemocline of Solar Lake (Jørgensen, Kuenen & Cohen, 1979).

sulphide to form high polysulphide concentrations of up to 150 μM (Cohen *et al.*, 1980).

Cyanobacteria have been found to occur in many other sulphide-rich layers in stratified lakes, including Lake Kononjer, Mariji, USSR (Gorlenko *et al.*, 1978) with blooms of *O. prolifica* and volcanic lakes in India (Singh, 1961), where *O. agardhii* and *Spirulina platensis* form thick blooms under 0.3 mM sulphide.

Cyanobacteria have for a long time been cited as indicators of eutrophic, organic-carbon loaded waterbodies (Pearsall, 1932). The oxygen regime in such systems is highly unstable and exhibits wide O_2 fluctuations and periodic sulphide build-up (Abeliovich & Shilo, 1972). Even sporadic short sulphide and oxygen alternations may serve as an important selective force against the strict aerobic or anaerobic photoautotrophs and will thus promote cyanobacteria. Recently, Belkin *et al.* (1979) have demonstrated both sulphide and light-dependent nitrogen fixation and hydrogen evolution (Belkin & Padan, 1979) in *Oscillatoria limnetica*, as well as hydrogen utilization for CO_2 photoassimilation (Belkin & Padan, 1978). The ecological contribution of these activities is still to be determined.

9.6 Concluding remarks

The discovery of facultative anoxygenic photoautotrophic physiology in cyanobacteria clarifies certain previously obscure aspects of their ecology, namely their occurrence and predominance in habitats alternating between photoaerobic and photoanaerobic conditions. The dual capacities of oxygenic photosynthesis and facultative anoxygenic photosynthesis permit cyanobacteria to occupy an intermediate ecological and physiological position unsuitable for organisms with a strictly anaerobic phototrophic physiology or with a strictly aerobic physiology. In fact, preference for low oxygen tension and for low redox potential appears to be characteristic of many cyanobacteria (Abeliovich & Shilo, 1972; Brock, 1973; Fogg *et al.*, 1973; Stewart & Pearson, 1970; Weller *et al.*, 1975).

The interlinking position of cyanobacteria in the phototrophic world is compatible with the facts that they are prokaryotes and among the oldest organisms, dating from the early Precambrian period of 3.6×10^9 years ago (Chapter 21). Significantly, two of the sulphide-rich ecosystems with abundant cyanobacteria, marine littoral sediments (Fenchel & Riedl, 1970) and the hot sulphur springs (Castenholz, 1976, 1977) may represent old ecosystems which may even predate the oxidized biosphere. Cyanobacteria retained the anoxygenic option with chlorophyll *a* and PS I, while developing plant-type oxygenic photosynthesis with PS II. Indeed, it is widely accepted that cyanobacteria represent the predecessors of certain contemporary chloroplasts of red algae (Margulis, 1968; Stanier & Cohen-Bazire, 1977) as evidenced by the great similarity between cyanobacteria and chloroplast structure, composition and function of the oxygenic photosynthetic system and nucleic acid composition. Recent work on the mechanism of energy transductions in the cyanobacterial membrane adds further support to this hypothesis (Padan & Schuldiner, 1978; Raboy & Padan, 1978).

Acknowledgments

The authors thank Professors M. Shilo and A.E. Walsby. The writing of this review was supported by the German Federal Ministry for Research and Technology.

CHAPTER 10
CELL–WATER AND CELL–SOLUTE RELATIONS

A. E. WALSBY

10.1 Introduction

Water is the major constituent of all living cells. It is the solvent in which most of the cell constituents are transported and the medium in which many of the biochemical reactions take place. Cell–water relations occupy a central position in physiological studies on eukaryotic plants but they have been largely neglected in the cyanobacteria. The principal reason for this has been lack of tools to do the work; the cells of cyanobacteria are too small to manipulate with the gadgets used by plant physiologists, and experiments with plasmolysed cells, which established the subject of osmotic relationships in higher plants, either do not work or may yield misleading results with cyanobacteria, as with other bacteria. However, some novel methods have been developed in recent years which have permitted all the main facets of cell–water relations to be investigated in cyanobacteria and other prokaryotes. These are the subject of the first part of this review.

Solute relations go hand-in-glove with water relations because the thermodynamic activity of water is affected by the presence of substances dissolved it it. Many of the important advances in the understanding of solute uptake have been made with bacteria. The techniques developed are in this case directly applicable to cyanobacteria and yet, again, there have been only a few investigations with these organisms. These are reviewed in the second part.

Perhaps it is the large number of unfamiliar terms that are used in the quantitative description of water and solute relations that has discouraged cyanobacteriologists from becoming immersed in this subject. I have therefore prefaced each section with a brief explanation of the general principles involved in the hope that it will make this review more comprehensible to the general reader.

10.2 Cell–water relations

10.2.1 General principles

The few investigations carried out into the cell-water relations of cyanobacteria indicate that they are similar in principle to those of plants and animals i.e., the cell is bounded by a semipermeable membrane which allows water to diffuse freely between the inside and outside but restricts the passage of many dissolved solutes. Excellent reviews of this topic are given by Dainty (1969a, 1976). The following summarizes the essential points.

The pressures which drive water from one system to another have been described in many terms but the one generally accepted nowadays is water potential, Ψ, defined by the expression

$$\Psi = (\mu_w - \mu_w^*)/\bar{V}_w \qquad (1)$$

where μ_w is the chemical potential of water in the system, which has units of (free) energy per mole (kg m^2 s^{-2} mol^{-1}), μ_w^* is the chemical potential of pure liquid water at atmospheric pressure and the same temperature, and $\bar{V}_w$ is the partial molar volume of water (18×10^{-6} m^3 mol^{-1}). Hence the dimensions of Ψ are kg m^{-1} s^{-2}, pressure. According to thermodynamic arguments, the water potential depends on the hydrostatic pressure p, the presence of solutes and the presence of solid–liquid and gas–liquid interfaces, related in the equation

$$\Psi = p - \pi - \sigma, \qquad (2)$$

where π is the osmotic pressure generated by the solute and σ the matric pressure due to attraction at interfaces. The matric pressure may have overriding importance in systems like drying soils, but in cells suspended in water it is probably small and can be neglected.

When two aqueous systems are separated by an air space or a semipermeable membrane water molecules will diffuse between them and this diffusion will result in net transport of water if the systems differ in water potential:

$$\Delta\Psi = \Delta p - \Delta\pi - \Delta\sigma. \tag{3}$$

This equation can be used to compare the potential of water inside and outside a cell. If we neglect σ and represent Δp *by* P (the hydrostatic pressure above the ambient pressure when the solution outside the cell is at the ambient pressure) then, when $\Delta\Psi$ is zero, there will be no net water movement and it follows from equation (3) that

$$P = \pi_{\mathrm{i}} - \pi_{\mathrm{e}} \tag{4}$$

where π_{i} is the osmotic pressure of the solution inside and π_{e} outside the cell. The excess hydrostatic pressure P, which balances the difference in osmotic pressure, is known as the cell turgor pressure. In plant cells, and in cyanobacteria, the turgor pressure arises because the osmotic swelling of the cell is limited by the mechanical properties of the enclosing cell wall. The degree of swelling for a given pressure increase is defined by ε, the volumetric elastic modulus of the cell; the speed of swelling is determined by the rate of entry of water under a given driving pressure. This is usually described by L_{p}, the hydraulic conductivity of the cell, a term which is dependent on ε.

10.2.2 Cell membrane and wall layers

Electron microscopy has shown that cyanobacteria are bounded by a plasma membrane also known as the plasmalemma, enclosed by a multilayered wall (Chapter 13).

Each of these layers potentially offers a barrier to the movement of water and solutes into or out of the cell but the principal barrier to diffusion is thought to be the innermost, the plasmalemma. This is concluded by analogy with higher plant cells, which plasmolyse when placed in hypertonic solutions of non-penetrating solutes. The cells of certain cyanobacteria can be made to plasmolyse (see chapter 4 of Fogg *et al.*, 1973), but it is more usual for the whole cell to shrink when placed in a hypertonic solution of a non-penetrating solute. This seems to indicate that the plasmalemma adheres rather firmly to the adjacent cell wall layers. Thus, even if plasmolysis occurs, the plasmolytic method will give an overestimate of the osmotic pressure of the turgid cyanobacterial cell. One cell type that does appear to plasmolyse readily is the heterocyst but in this case it is the cell wall which separates from the enclosing heterocyst envelope.

One of the functions of the cell wall is to prevent excessive osmotic swelling (Section 10.2.6). The mechanical strength of the wall mainly resides in the peptidoglycan layer (Chapter 13).

10.2.3 Determination of the volume of cell water

It is necessary to know the volume of cell water in a sample in order to estimate not only its osmotic pressure, but also its pH (10.3.2.1), its ionic strength (10.3.3.2–3), the accumulation in it of other solutes (10.3.3.4), and the membrane potential of the cells (10.3.3.1).

All the methods used for determining the cell water volume are based on corrections to the packed cell volume of centrifuged suspensions of cyanobacteria, made ostensibly by subtracting from the total volume the volumes of the cell solids and the interstitial water (ideally all of the water which lies outside the plasmalemma of the cells, including that permeating the sheath and cell wall layers). The crudest corrections are simply 'guestimates' made by assuming that the cell water will occupy the same proportion of the pellet (e.g. 50%) as found in the case of other organisms (Paschinger, 1977). This is risky. E.M.Allison and I have found wide variation, from 15% to 42% in the case of a single organism, *Anabaena flos-aquae*, grown under different conditions, though admittedly this variation was partly due to variable amounts of sheath material, which may be less of a problem in other species.

The basic method of estimating the amount of interstitial water present in the pellet is to add a water-soluble substance to the cell suspension before centrifugation and subsequently to estimate the amount of the substance in known volumes of the pellet and supernatant (e.g. Rottenberg *et al.*, 1972). It is assumed that the substance will maintain the same concentration in both. Radioactively labelled substances are often used and their concentrations determined by counting, usually after a secondary separation of the cells from the interstitial water in the pellet sample. Full details are given by Walsby and Xypolyta (1977). It is important that the labelled substance should not be absorbed by the cells. This should be carefully checked by showing that no label remains after rinsing the cells with unlabelled medium. Such details are seldom recorded. Substances with high molecular weight, such as ^{14}C-labelled inulin (Dewar & Barber, 1973), dextran (Grant & Walsby, 1977) and polyethylene glycol (Raboy & Padan, 1978) have been chosen to avoid uptake problems, though there is a danger these may not penetrate the sheath or cell wall up to the plasmalemma. In this respect ^{14}C-sucrose may be better (Falkner, Horner, Werdan & Heldt, 1976; Guarino & Cohen 1979a; Allison & Walsby, 1981), as this will not be excluded from wall structure by molecular sieving and appears not to penetrate the cells (see 10.3.1), or be physically absorbed. The volume of the extracellular space can also be estimated by using a dye, which is measured by colorimetry; Craig and Budd (1979) used nigrosine, and Miller *et al.* (1976) used blue dextran. The principal problems here will be physical absorption of the dye on the pelleted material and the release of pigments by the cyanobacteria.

Cell solids usually occupy a relatively small proportion of the whole and errors in estimating it will be less important. Again 'guestimates' have been made, at 15% of cell volume (Craig & Budd, 1979). Another approach is to estimate the volume of solids from a dry-weight determination and from the assumption that the solids have an average density of 1.3 g cm^{-3} (Dinsdale & Walsby, 1972). However, the best way is to determine the total water content of the sample by ^3H-labelled water (Falkner, Horner, Werdan & Heldt, 1976; Raboy & Padan 1978) or the difference between the wet and dry weights of the pellet. By the latter method I have estimated the solids to occupy about 12% of cell volume in *Anabaena flos-aquae*. The cyanobacteria used in the other investigations cited below are described in later sections (10.3–10.4).

It is obvious from the foregoing that errors in estimating the volume of cell water will tend to be large if the sample contains a large proportion of interstitial (extracellular) water, though this is unavoidable when the cells are heavily sheathed. A good way of removing much of the extraneous water is to centrifuge the cells or filaments through a layer of silicone oil. This can be done very rapidly with small samples in a 'microfuge', a high-speed microcentrifuge. This technique in combination with the use of ^3H-labelled water, to determine the total water volume, and ^{14}C-labelled sucrose to determine the extracellular water volume, probably provides the best method of estimating the volume of cell water (see Rottenberg, 1979).

10.2.4 Osmotic pressure

The osmotic pressure of an aqueous solution is given by the expression

$$\pi = -RT \ln a_w / \bar{V}_w \qquad (5)$$

where R is the universal gas constant (8.31 J mol^{-1} K^{-1}), T the absolute temperature (K) and $\bar{V}_w$ as defined above (10.2.1). The term a_w is the activity of water in solution, given by the ratio p'/p'_o where p' is the vapour pressure over the solution and p'_o is that over pure liquid water at the same temperature. Since a_w is always less than unity, its logarithm is negative; hence the minus sign in eqn (5). The dimensions of π are those of pressure (kg m^{-1} s^{-2}) equivalent to Pascals (Pa) in the SI system. (10^5 Pa is equivalent to 1 bar, a convenient but non-SI unit which is roughly equal to 1 atmosphere.)

The osmotic pressure generated is dependent on the number of free solute molecules in solution and for an ideal substance is about 24.4 bars for a 1 molal solution at 20°C. (At low concentrations there is little difference between molality and the more convenient molarity.) Solutions of salts which dissociate to give *n* ions will develop an osmotic pressure *n* times the molar equivalent, depending on the activity of the ion. This is often less than unity; e.g. in many potassium salts the ion has an activity of about 0.9.

The osmotic pressure of the solution in which the cell is suspended (π_e) can, in theory, be calculated from the known composition of the solution or measured by osmometer. The commercial osmometers actually measure related caligative properties, either the freezing-point depression or the decrease in vapour pressure. Many synthetic freshwater media contain salts of low total concentration; for instance BG 11 (Stanier *et al.*, 1971) is less than 1.0 mM and would develop a negligible osmotic pressure (<0.05 bar) even if each salt was totally dissociated. Media for marine species, on the other hand, have very high osmotic pressures. For seawater itself π_e is about 28 bars.

Brock (1975a) used NaCl solutions of different concentrations to investigate the effect of external osmotic pressures on the growth and photosynthesis of a *Microcoleus* sp. from a desert crust. The maximum photosynthetic rate was achieved in a solution giving a water potential of -18 bars. Unwetted samples of the crust equilibrated in an atmosphere of similar water potential showed virtually no activity, indicating that contact with fresh liquid water is important. In investigating the water-potential effects of solutions it is advisable to use a number of different solutes in case of specific effects, e.g. toxicity to or uptake by the organism; these effects may vary with different species. For example, NaCl is toxic to certain freshwater planktonic organisms at concentrations isotonic with sucrose solutions that are tolerated (unpublished results).

10.2.4.1 Measurement of internal osmotic pressure

In order to determine the internal osmotic pressure, π_i, of the cells it is necessary first to concentrate them into a dense pellet by centrifugation and to estimate the proportion, y, of the total water present which is outside the cell membrane (see 10.3.1.1). The cells in the pellet sample are then disrupted by sonication or passage through a French pressure cell and, after clarifying the suspension by high-speed centrifugation, the osmotic pressure, π_t, of the total suspension is measured. The value of π_i is then calculated from the expression

$$\pi_i = (\pi_t - y\pi_e)/(1-y). \qquad (6)$$

Using an essentially similar method, Dewar & Barber (1973) estimated the internal osmolarity of *Anacystis nidulans* to be about 0.46 osmolar, equivalent to an osmotic pressure of 11.4 bars. Similar determinations by R.G.W. Jones and myself show a much lower value, just under 4 bars, in *Anabaena flos-aquae*.

10.2.5 Turgor pressure

Turgor pressure is the difference in hydrostatic pressure between the inside and outside of the cell and to measure it a pressure probe is required. Plant physiologists have developed miniature probes connected to pressure transducers but they cannot be used with cells smaller than 20 μm in diameter, which excludes virtually all prokaryotes. Bacteriologists sometimes measure internal osmotic pressures and then call them turgor pressures. The only direct method of measuring turgor pressure in prokaryotes is by using their gas vesicles, if they have them, as pressure probes (Walsby, 1971, 1980). The method requires an understanding of gas vesicle structure and its pressure relationships.

10.2.5.1 Gas vesicles as pressure probes

The gas vacuoles of cyanobacteria are made up of stacks of hollow cylindrical structures called gas vesicles (Bowen & Jensen, 1965). Detailed reviews of their structure and function are given by Walsby (1972, 1978b).

There may be as many as 10 000 gas vesicles in a single cell (Lehmann & Jost, 1971). The gas vesicle has the shape of a hollow cylinder of about 70 nm diameter with a conical cap at each end. The structure is made of protein (Walsby & Buckland, 1969; Jones & Jost, 1970), probably a single species of protein which is arranged along ribs that are orientated at right angles to the long axis of the cylinder (Jones & Jost, 1970). The protein has a crystalline arrangement, in the wall of the structure, which is stabilized by hydrogen bonding between a β-pleated sheet with chains running at an angle of 35° to the long axis of the gas vesicle. There are two layers of the β-sheet within the wall, which may be intermeshed in a way which gives stability to the overall structure (Blaurock & Walsby, 1976).

The wall of the gas vesicle is permeable to gas and, as a consequence, gases dissolved in the surrounding aqueous solution diffuse freely into and out of the structure. The space inside the gas vesicle, which must be created by the way in which the constituent protein molecules self-assemble, therefore usually has air in it at atmospheric pressure. The wall may allow water vapour to enter the gas space but it is impermeable to liquid water, which is kept out by surface tension (Walsby, 1969). Partition experiments have shown that the outer surface of the gas vesicle wall is hydrophilic (Walsby, 1971). Comparisons by neutron diffraction of collapsed gas vesicles in a dry state and those in a hydrated state show that one of the surfaces is hydrophobic (Worcester,

Blaurock & Walsby, in prep. see also Walsby 1978b) and this must, therefore be the inner surface.

The gas vesicle wall is a rigid structure (Walsby, 1971). When the intact gas vesicle is subjected to a pressure of 1 bar (10^5 Pa) its volume decreases by only 1 part in about 650 (Walsby, in prep.). Hence, pressures applied to the structure are not transmitted to the gas inside, whose pressure is determined by the chemical potential of gases dissolved in the suspending solution (Walsby, 1969). However, when a certain critical pressure (C) is exceeded the gas vesicle collapses irreversibly to a flattened envelope. The gas that was inside escapes by diffusion into the surrounding solution without first forming a bubble (Walsby, 1971). Each individual gas vesicle has its own characteristic critical collapse pressure, C, which is defined as the minimum pressure difference across the gas vesicle wall which is required to collapse the structure. This pressure difference is equal to the total pressure acting on the outer surface less the pressure of the gas inside and in an equilibrated suspension, can therefore be determined as the pressure in excess of the ambient air pressure which just causes the gas vesicle to collapse (Walsby, 1971, 1980).

The critical pressure of gas vesicles can be determined by turbidity measurements, as the amount of light they scatter decreases markedly when they collapse. An instrument for measuring the turbidity change, as pressure is applied, is described by Walsby (1973). Measurements show that the critical pressure of gas vesicles in a population of cells is normally distributed in a precise fashion about a well-defined value and microscope observations indicate that the same is true of the critical pressure variation of gas vesicles in individual cells. The reproducibility of the critical pressure distribution enables it to be used to measure the hydrostatic pressure inside cells, as described below (10.2.5.2–10.2.7).

10.2.5.2 Turgor pressure measurement

The critical pressure of gas vesicles isolated from a cyanobacterium like *Anabaena flos-aquae* varies from about 4.5 to 7.5 bars, symmetrically about a mean value, $\bar{C}$, of 6 bars (see Fig. 10.1). Inside the turgid cells gas vesicles are already subjected to the turgor pressure, P, and the mean pressure required to cause their collapse, known as the apparent mean critical pressure, $\bar{C}_a$, is therefore less. The cell turgor pressure can be calculated from the relationship

$$P = \bar{C} - \bar{C}_a. \qquad (7)$$

A plot of the collapse-pressure curves, from which these mean values are determined, is shown in Fig. 10.1. In practice, the value of $\bar{C}$ is obtained using cells suspended in a hypertonic solution of sucrose (or some other solute that

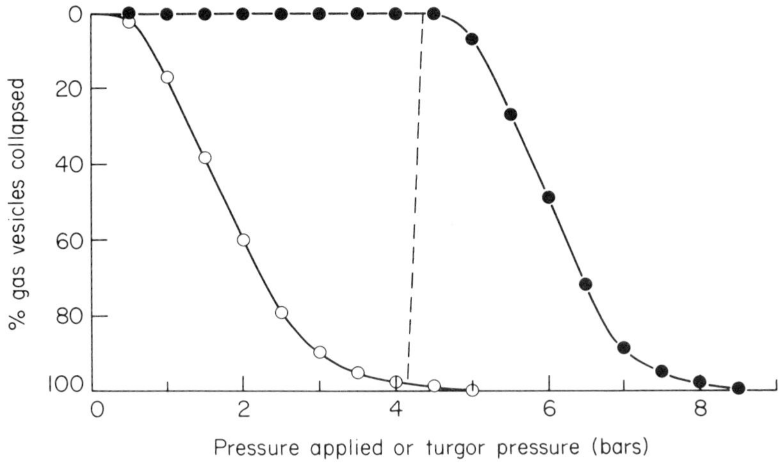

Fig. 10.1. Collapse of gas vesicles with pressure in *Anabaena flos-aquae* suspended in ○, culture medium, ●, 0.5 M sucrose solution. The broken line, which represents the distance between equivalent points on the two collapse-pressure curves, shows the cell turgor pressure at different states of gas vesicle collapse. (Redrawn from Walsby, 1980)

has been shown not to penetrate the cells), which abolishes cell turgor pressure (Walsby, 1971). Full details of the method, including possible pitfalls, are given by Walsby (1980).

One of the complications is that as the gas vacuoles are collapsed the turgor pressure of the cells will fall slightly. (The reasons for this are discussed below in Sections 10.2.6–7.) This is seen if the difference between the two curves is plotted as shown by the broken line in Fig. 10.1. The best estimate of the original turgor pressure in the cells is therefore given by extrapolating this line to the value of zero collapse (the top of Fig. 10.1).

Measurements of turgor pressure by this technique have shown that it varies according to the conditions in which the cells are grown. In *Anabaena flos-aquae* it ranges from 2 to 4.5 bars (Walsby, 1971, 1978b; Dinsdale & Walsby, 1972), depending on the light intensity in which the organism is grown. The turgor pressure may rise to exceed the *C* of the weaker gas vesicles in the cells and cause their collapse (Dinsdale & Walsby 1972; Grant & Walsby 1977). The significance of this in buoyancy regulation is discussed in Chapter 2 by van Liere & Walsby. Turgor pressures within a similar range have been encountered in *Oscillatoria agardhii* (Walsby & Klemer, 1974), *Microcystis aeruginosa* (Reynolds, 1973), *Aphanizomenon flos-aquae* (Konopka *et al.*, 1978), *Gloeotrichia echinulata* and several other planktonic species of *Anabaena* (E.M. Allison & A.E. Walsby, unpubl.). The changes in turgor pressure can be attributed partly to organic products of photosynthesis

(Grant & Walsby, 1977) and partly to variations in uptake of inorganic ions, particularly potassium (Allison & Walsby, in prep.).

A particular advantage of the gas vacuole system of turgor pressure determination is that measurements can be made rapidly (at about 2 min intervals) and with dilute suspensions. They can, therefore, be used to investigate short-term changes under normal physiological conditions (Grant & Walsby, 1977) and to demonstrate transient changes which would not be detectable by other means (Walsby, 1980 and unpubl.). One thing that remains to be done, however, is to carry out simultaneous measurements of turgor pressure, by the technique described here, and cell osmotic pressure, by the method described in 10.2.4.1. Perhaps a comparison of the two measurements might provide the first estimates of σ, the matric pressure (eqn 2).

10.2.6 Volumetric elastic modulus of the cell

All cell walls are stretchable to a degree, and for this reason the cell volume will rise if the turgor pressure increases. The relationship between the pressure change, ΔP, and the relative volume change, $\Delta V/V_o$, is given by ε, the volumetric elastic modulus

$$\varepsilon = (\Delta P/\Delta V)V_o \tag{8}$$

where V_o is the cell volume, usually taken at zero turgor pressure. This relationship has been investigated in *Anabaena flos-aquae* (Walsby, 1980) using the following indirect method for determining the volume change.

The cyanobacterium was suspended in sucrose solutions giving a series of increasing osmotic pressure. In each case $\bar{C}_a$ was determined and from this the turgor pressure was calculated (eqn 7). It was found that the decrease in turgor pressure was less than the increase in external osmotic pressure and from this it was shown that there had been an increase in the internal osmotic pressure (eqn 4). This occurred as a result of shrinkage by water loss. Since the osmotic pressure is proportional to the concentration of solutes and this is inversely proportional to the volume, the change in cell water volume could be calculated. These changes could be related to the total volume of the cell by making suitable corrections for the volumes of cell solids and gas vacuoles. Full details are given by Walsby (1980). A graph of the volume change related to the turgor pressure change, calculated by this method, is shown in Fig. 10.2. The elastic modulus, calculated from the gradient of the curve, is seen to have a constant value, 10.6 bars, over the physiological range of turgor pressure. This value is lower than that found in the cells of many plants and indicates that the cell wall is relatively more easily stretched in this cyanobacterium. The stretchiness of the cell as a whole depends on the elastic properties of the wall

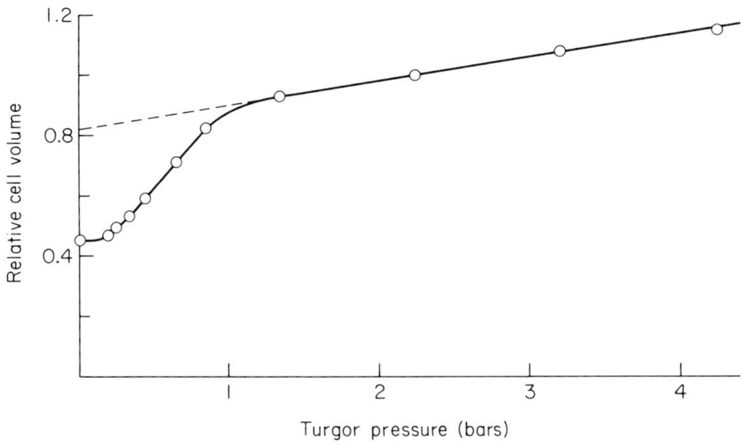

Fig. 10.2. The change in relative cell volume with turgor pressure in *Anabaena flos-aquae*. The changes are brought about by suspending the cells in a series of sucrose solution of increasing concentration, and the relative cell volume is the ratio of the total cell volume to the initial volume of cell water. The volumetric elastic modulus (ε) of the cell is calculated from the slope of the straight line and intercept with the ordinate. (Redrawn from Walsby, 1980)

material and on the thickness of the wall relative to the diameter of the cell. It should now be possible to determine these features.

Knowledge of the cell elastic modulus itself is needed for defining the exact relationship between osmotic pressure and turgor pressure. The magnitude of ε also determines the amount of water loss which occurs on dehydration (see Zimmermann, 1978) and, if cyanobacteria are adapted in the same way as higher plants, ε may be different in xerotolerant (or halotolerant) and sensitive forms.

10.2.7 Hydraulic conductivity of the cell surface

Any change in turgor pressure or osmotic pressure which unbalances the osmotic relations of the cell (i.e. unbalances eqn 4) will cause water to be driven into or out of the cell so that equilibrium is restored. The rate at which water flows through the cell surface under this pressure differential is described by L_p, the hydraulic conductivity of the membrane, or more precisely, of the membrane and wall layers in series. This is defined by the equation

$$L_p = \frac{\ln 2}{\tau} \frac{V}{A} \frac{1}{(\varepsilon + \pi)} \tag{9}$$

where V is the cell volume, A is the surface area of the cell through which the water is driven and τ is the half-time for equilibrium to be restored: ε and π are as defined above. (Thus the units of L_p are m s^{-1} Pa^{-1}: see Dainty, 1969a.)

In theory L_p could be measured from the rate of volume change that ensues on changing the osmotic pressure of the solution bathing the cell. In practice this always leads to large errors because it is impossible to mix the new solution up to the cell surface without delays occurring, and these 'unstirred layers' may offer greater resistance to the passage of water than does the cell membrane itself (see Dainty, 1969a, 1976).

The same problem has been encountered when attempting to measure the value of L_p in gas-vacuolate cyanobacteria by following the rate at which the cell turgor pressure (and hence the cell volume) decreases when the cell suspension is mixed rapidly into hypertonic solutions of various solutes (Walsby, 1980). The delay in equilibration is less in solutions of low viscosity, e.g. glycine compared with sucrose, but even so the time taken for mixing up to the cell surface provides the main component of this delay. This source of error is discussed here because it will also affect the kinetic investigation of all other rapid processes when the experiment depends on changes following the addition of substances to a cell suspension.

Fortunately, there is a method for determining L_p that is free of the 'unstirred layer' error. When gas vesicles are collapsed by pressure the ensuing decreases in cell volume must result in an immediate decrease in turgor pressure, because, as explained above (10.2.6) the ratio of $\Delta P/\Delta V$ is constant over the physiological range. As this will then throw the osmotic equilibrium out of balance, water will enter to replace much of the lost volume and the turgor pressure will rise again. Not quite all of the lost volume (and turgor) is regained because the entering water will dilute the cytoplasmic solution, thereby decreasing the value of π_i. Quantitative evidence for these changes is provided by the degree of turgor pressure decrease when gas vesicles are progressively collapsed by pressure (Walsby, 1980) already referred to (10.2.5.2, Fig. 10.1). The rate at which the water enters can itself be determined from chart recordings of the time taken for a proportion of a population of gas vesicles to collapse when a pressure is suddenly applied. Isolated gas vesicles collapse immediately when subjected to their critical pressure. In turgid cells, on the other hand, only a proportion of the gas vesicles present collapse when a given pressure is applied, as the total pressure on the gas vesicles immediately decreases as cell turgor pressure falls. More vesicles then collapse as the turgor pressure rises, owing to the entry of water into the cells. Since the ensuing collapse, the turgor pressure rise and the change in cell volume are all interdependent variables, the rate of water entry can be calculated (Fig. 10.3), and from this an estimate of L_p can be obtained. Full details are given by Walsby (1980).

The estimate of L_p in *Anabaena flos-aquae*, 0.14 μm s^{-1} bar^{-1}

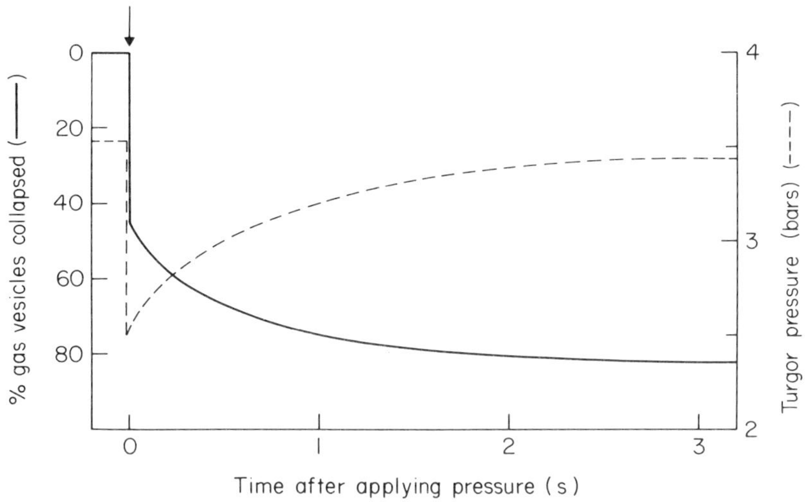

Fig. 10.3. The time course of gas vesicle collapse in turgid cells of *Anabaena flos-aquae* subjected (arrow) to a pressure of 3.2 bars (——) and the changes in cell turgor pressure (– – –) calculated to result from the change in cell volume which the gas vesicle collapse causes. (Redrawn from Walsby, 1980)

(1.4×10^{-12} m s^{-1} Pa^{-1}), is of the same order as that obtained using recent pressure-probe techniques in giant algal cells (see Zimmerman, 1978). If independent measurements could be made on the rate of *diffusive* transfer of water, by using tritiated H_2O, comparison would reveal the nature of the membrane permeability, by pores or by solution (see Dainty, 1969a). As discussed by Walsby (1980) all methods of estimating L_p from turgor pressure changes assume instantaneous contraction of the cell wall as the cell volume changes, something which could be checked by comparing results using cells of different sizes.

10.3 Cell–solute relations

The osmotic pressure of the cytoplasmic solution is generated mainly by substances of low molecular weight. These substances may be synthesized from other low molecular weight compounds or formed by the breakdown of macromolecules. This section considers the mechanisms by which substances enter or are excluded from cyanobacteria, paying particular attention to those substances which may contribute significantly to the total osmolarity of the cytoplasm, but excluding inorganic carbon (Chapter 3) phosphate (Chapter 5) and combined nitrogen. Each of these substances is taken up by a specific process.

10.3.1 Passive diffusion of uncharged molecules

If the cell membrane is permeable to a substance, the concentration of the substance on the inside and outside of the cell will reach the same value at equilibrium, assuming the activity coefficient is the same inside and out. A change in either concentration will result in net diffusion of the solute across the membrane at a rate determined by its permeability constant, k, to the solute. Experimentally, this can be determined by raising the external concentration of the solute by a concentration S_o, and monitoring in some way the rise in concentration within the cell, S_i, with time, t. The permeability constant is then given by

$$k = \frac{1}{t}\frac{V}{A} ln\left(\frac{S_o}{S_o - S_i}\right) \quad (10)$$

where V and A have the same meanings as before, eqn (9). In theory, it might be possible to measure the uptake rate by using radioactively labelled tracers as has been done with substances that are actively taken up by cyanobacteria (Beauclerk & Smith, 1978; Raboy & Padan, 1978). In practice, there may be insurmountable difficulties. First, the amount of radioactivity outside the cells will vastly exceed that inside since the volume of solution outside will be much greater, and the concentration of the solute inside the cells will never exceed that outside. This makes it essential to remove the external solution very thoroughly by rinsing, giving rise to a second problem, that the labelled solute will be leached from the cells during the rinsing procedure.

An alternative approach is to monitor the rise in turgor pressure as the solute enters from a solution of known osmotic pressure. This was first done, indirectly, by Elo (1937) who observed that the filaments of *Oscillatoria princeps* became bent when placed in hypertonic solutions, presumably owing to uneven shrinkage by osmotic water loss, but subsequently straightened again in solutions of penetrating solutes. The time for recovery was used to calculate the permeability constants. Some of the straightening times are so short that there must be considerable errors from the delay in mixing the solution up to the filament surface (see 10.2.7), but for many years Elo's work remained the sole source of information on solute uptake by cyanobacteria.

The changes in turgor pressure as the solute penetrates the cell can be monitored directly and more accurately by the gas vacuole method (Walsby, 1980). With *Anabaena flos-aquae* the filaments were suspended in a 0.15 osmolar solution of the substance, which causes a substantial decrease in turgor pressure but maintains sufficient turgor to prevent excessive shrinkage. As the solute penetrates the turgor rises, asymptotically approaching its original value in water; the rise is indicated by a decrease in $\bar{C}_a$ as shown in Fig. 10.4. The time course of the decrease in critical pressure $\bar{C}_a$, follows a negative exponential and the time taken for $\bar{C}_a$ to fall to half-way between its initial value and its final value is equal to τ', the half time for penetration at which $S_i = \frac{1}{2}S_o$. Substituting in eqn (10) it follows that

$$k = \frac{\ln 2}{\tau'} \frac{V}{A}. \qquad (11)$$

Using this method the permeability of *A. flos-aquae* cells to over a dozen compounds was investigated. The cells were virtually impermeable to sucrose, and glucose and glycine permeated very slowly, with a half-time for penetration of about 40 h. The kinetics of penetration by the sugar alcohol series (Fig. 10.4, Table 10.1) indicated passive uptake with the smaller molecules permeating faster, as expected. The rapid penetration of the

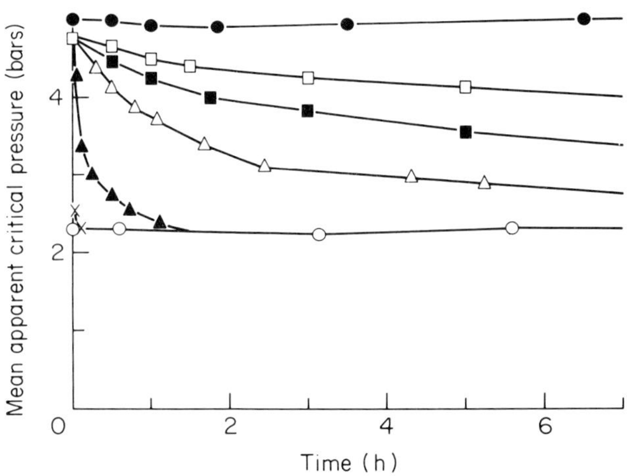

Fig. 10.4. The penetration of various sugars and polyols indicated by the time course of critical pressure changes. Changes observed with cells suspended in ○, water, and in 0.15 M solutions of ●, sucrose, □, glucose, ■, mannitol, △, xylitol, ▲, erythritol and ×, glycerol. (Redrawn from Walsby, 1980)

smallest of this series, glycerol, was measured by a modified procedure in which the cells were held under a pressure which by itself was too small to collapse the gas vesicles, but which brought about progressive collapse as the solute permeated the cells and raised the turgor pressure (Walsby, 1980). Evidently the cells are far too permeable to glycerol for this substance to be accumulated to balance high external salt concentrations, as occurs in the green alga *Dunaliella* (Ben-Amotz & Avron, 1973), since it would diffuse outward as well as inward. The outward diffusion can also be demonstrated by the gas vacuole method; on resuspending, in fresh medium, cells that have imbibed a solute, the turgor pressure is seen to fall again as the solute leaks out

Table 10.1. The permeability of *Anabaena flos-aquae* cells to various substances determined by related turgor pressure changes (Walsby, 1980)

Substance	Half-time for penetration (τ')/h	Permeability constant (k)/μm h^{-1}
Sucrose	100	0
Glucose	40	0.2
Mannitol	6	0.12
Xylitol	2	0.37
Erythritol	0.15	4.9
Glycerol	0.003	265
Hexamine	0.006	132
Glycine	50†	0
Proline	5.5	0.13
Glycine–betaine	15*	0.05
K^+Cl^-	0.85*	0.86
Na^+Cl^-	100	0

* Non-exponential, active uptake.
† Initially no uptake, followed by slow penetration after 7 h.

(Walsby, 1980). This technique could be used to check the theory that certain extracellular products, such as glycollic acid, are lost from the cell owing to the permeability of its plasma membrane.

The gas vacuole method described here can also be used for following the penetration of substances which are accumulated by other mechanisms (see 10.3.2 and 10.3.3.2). It is restricted in use to those gas-vacuolate cyanobacteria which have a good initial turgor pressure but it has the advantage that it permits measurements with dilute cell suspensions and does not require lengthy determination of cell water volume in the sample.

10.3.1.1 Permeability to gases

It is usually assumed that cell membranes are freely permeable to all of the atmospheric gases. Certainly oxygen, carbon dioxide and nitrogen must be taken up by cells because they are known to be metabolized. It does not necessarily follow that the uptake is by passive diffusion. However, other experiments with gas vesicles show this to be so. It has been demonstrated that the gas vesicles are freely permeable to gases and that the gas inside them comes rapidly to passive equilibrium with the gas in the suspending medium, both in the case of isolated gas vesicles and gas vesicles inside cells (Walsby, 1969, 1971). It therefore follows that the gas must also equilibrate passively across the cell plasma membrane. Experiments with gas-vacuolate heterocysts have shown that they are less permeable to oxygen and nitrogen than are vegetative cells (author, in preparation).

10.3.2 Uptake by ion trapping of weak electrolytes

If an ionizable compound is added to a suspension of a cell whose plasma membrane is permeable only to the undissociated molecule, the concentration of the undissociated compound will become the same on both sides of the membrane whereas the concentrations of the corresponding ions will depend on the concentration of protons inside (i) and out (o):

$$\frac{[H^+]_i}{[H^+]_o} = \frac{[BH^+]_i}{[BH^+]_o} = \frac{[A^-]_o}{[A^-]_i} \tag{12}$$

(see Heldt *et al.*, 1973). In the case of weak bases, such as methylamine, this effect can result in considerable accumulation of the total amount (i.e. $CH_3NH_3{}^+ + CH_3NH_2$) of the substance inside the cell when the internal pH is lower than the external. The reverse is true when the internal pH is higher. Weak acids such as acetate are accumulated in the opposite sense (see eqn (12)), i.e. inside the cell when the internal pH is higher.

The accumulation of weak acids and bases by this mechanism, which Guarino and Cohen (1979a) have called ion trapping, can be manipulated to generate considerable increases in the internal osmotic pressure of cells. Shilo and Shilo (1962) demonstrated that the chrysophyte *Prymnesium parvum* could be made to swell osmotically to such an extent that it lysed when placed in solutions of ammonium salts at high pH or acetate at low pH. This observation has now been applied to provide a method of controlling this toxic organism in fish ponds (Shilo, 1971). Guarino and Cohen (1979a) showed that ion trapping resulted in the toxic accumulation of another cationic substance, the polyamine putrescine, by the cyanobacterium *Anacystis nidulans* but in this case the toxicity is due to the substance causing dissociation of ribosomes, apparently by displacing the Mg^{2+} essential to their stability (Guarino & Cohen, 1979b).

10.3.2.1 Determination of intracellular pH

Ion trapping provides a means of measuring the internal pH of a cell. If a radioactively labelled base, such as ^{14}C-methylamine, is let to equilibrate in a suspension in which the volume of cell water has been determined, the internal pH can be calculated from the following expression, derived from equation (12)

$$pH_i = pH_o + \log_{10}\frac{[CH_3NH_3{}^+]_o}{[CH_3NH_3{}^+]_i}. \tag{13}$$

The total amounts of methylamine inside and outside the cell are determined from the radioactivity of cells and medium after separating them by filtration; as with other passively distributed substances there is a danger of loss by leaching during the procedure (see 10.3.1). The proportion of the methylamine outside that is un-ionized can be calculated from the external pH and from knowledge of the pK of methylamine. If the concentration of the un-ionized methylamine is assumed to be the same inside and outside the cells, the ratio of the ionized form can be calculated and, by substitution into eqn (13), the internal pH can be calculated. For details see Rottenberg (1979).

Falkner, Horner, Werdan & Heldt (1976) used the ion-trapping method to measure the pH in *Anacystis nidulans*. They supplied two ^{14}C-labelled compounds, the cationic methylamine (MA^+) and the anionic 5,5-dimethyl-oxazolidine–2,4-dione (DMO^-) and found that, paradoxically, both were accumulated in the cells. They concluded that MA^+ was accumulated preferentially in the acidic thylakoid space and the DMO^- preferentially in the more alkaline cytoplasm, using the analytical procedure developed by Heldt *et al.* (1973) with chloroplasts. The volume of the total cell water was determined by subtracting the sucrose impermeable space from the packed cell volume, and the volume of thylakoid space, about 7% of the cell volume, by planimetric measurements on electron-micrographs. Their determinations showed that the pH of the cytoplasm rose from 6.9 in the dark to 7.5 in the light, while that of the thylakoid space fell from 5.2 to 4.6. This increase in ΔpH between the two compartments, from 1.7 to 2.9, indicates the existence of a light-dependent proton pump which drives protons from the cytoplasm to the thylakoid space. Falkner and his coworkers found that the pH of the cytoplasm was not additionally affected by changing the pH of the culture medium over the range 6.5 to 8.0, suggesting that the cell membrane is impermeable to protons. An alternative explanation is that the internal pH is maintained by an active process.

The increase in cytoplasmic pH on illumination of the cells has been confirmed in *A. nidulans* and *Plectonema boryanum* by Matsamoto and Nishimura (1977) using quite a different method which depended on finding an external pH at which no proton efflux occurred when the cell membrane was rendered permeable to protons by addition of an ionophore, nigericin. The change was from pH 7.4 in the dark to 8.5 in the light. Padan and Schuldiner (1978) have independently confirmed the changes in the thylakoid space, using isolated photosynthetic lamellae from *P. boryanum*. The ΔpH between inside and outside rose from 0.75 to 1.76 on exposure to light, according to measurements with ^{14}C-methylamine, while the accumulation of another basic compound, 9-aminoacridine, estimated by the fluorescence quenching method of Schuldiner *et al.* (1972), indicated rather larger values of

ΔpH. They found that the ΔpH across the thylakoid membrane is sensitive to external pH, unlike that across the plasma membrane (Fig. 10.5). Kallas and Dahlquist (1981) used another method, involving ^{31}P nuclear magnetic resonance analysis, to show that the internal pH in *Synechococcus* is buffered when the external pH is changed.

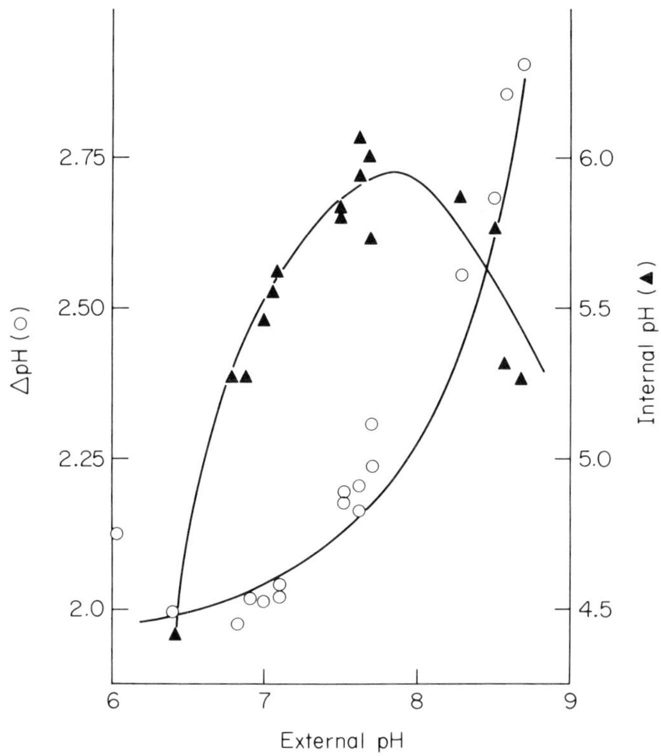

Fig. 10.5. Variation of internal pH (○) and ΔpH (▲) of photosynthetic membranes isolated from *Plectonema boryanum* as a function of the external pH, according to the accumulation of 9-aminoacridine. (Redrawn from Padan and Schuldiner, 1978)

The significance of these measurements for cell–water relations is that the cyanobacterial cell water cannot be considered as a single pool. The thylakoid space can occupy a significant fraction of the cell and the thylakoid membranes may form a barrier to the diffusion of ions, though there has been no direct investigation of the concentration of ions other than H^+ inside them. The osmotic pressure of the thylakoidal sap is presumably the same as that of the cytoplasmic sap since it is unlikely that the thylakoids are able to withstand hydrostatic pressure differences.

10.3.3 Ion uptake

Inorganic ions may account for a substantial proportion of the cytoplasmic osmotic pressure. They may be specifically accumulated in the cell or excluded from it by a number of mechanisms. An explanation of the processes first requires an understanding of the factors that influence the distribution of charged ions across a membrane. A lucid account is given by Dainty (1969b). The essential features are summarized here.

The passive distribution of a species of ion on two sides of a membrane is influenced not only by the concentration of the ion but also by the electrical potential. The dual effects are combined in the electrochemical potential, $\bar{\mu}$. This is defined approximately by the expression

$$\bar{\mu} = \bar{\mu}^* + RT \ln M + z\mathrm{F}\psi \tag{14}$$

where $\bar{\mu}^*$ is the chemical potential of the ion in the standard state (at atmospheric pressure, in the absence of any matric potential or electrical potential, and at unit concentration), M is the molar concentration, z the valency of the ion, F the Faraday constant (9.65×10^4 A s mol^{-1}) and ψ is the electrical potential in volts (J A^{-1} s^{-1}). Hence each term in eqn (14) has the dimensions of J mol^{-1}. (The exact equation also contains terms describing the effects of hydrostatic pressure and matric potential.) When a specific ion, j, is passively distributed on two sides of a membrane then at equilibrium

$$\bar{\mu}_j^i = \bar{\mu}_j^o \tag{15}$$

from which it follows that

$$\bar{\mu}_j^* + RT\ln M_j^i + z\mathrm{F}\psi^i = \bar{\mu}_j^* + RT\ln M_j^o + z\mathrm{F}\psi^o \tag{16}$$

where i denotes the inside and o the outside values. Cancelling $\bar{\mu}_j^*$ and rearranging gives

$$z\mathrm{F}(\psi^i - \psi^o) = RT(\ln M_j^o - \ln M_j^i). \tag{17}$$

Therefore,

$$\psi^i - \psi^o = \frac{\mathrm{RT}}{z_j\mathrm{F}} \ln \frac{M_j^o}{M_j^i} \tag{18}$$

The difference in potential across the membrane, ψ^i–ψ^o, is usually described as the Nernst potential, E_j^N, and solving numerically for RT/F and taking $\log_{10}$ gives the Nernst equation

$$E_j^N = \frac{58}{z_j} \log \frac{M_j^o}{M_j^i} \text{ mV} \tag{19}$$

where T is 20°C. The potential is negative on the inside of the membrane when there is excess of a cation inside the cell membrane. For example, if the concentration of a passively distributed univalent cation inside the cell is ten times higher than the outside concentration the Nernst or membrane potential calculated from eqn (19) will be −58 mV. This will be true of protons when they are passively distributed and since the log $(M^{o}_{H^+}/M^{i}_{H^+})$ is equal to the $-\Delta$pH it follows that a ΔpH of one unit will be equivalent to a potential of −58 mV (i.e. negative inside the cell when the pH is lower inside the cell).

While the membrane potential will result in the unequal passive distribution of certain ion species it must, in the first place, be generated by the active unequal distribution of one or more ion species. By simply analysing the concentrations of specific ions inside the cell it can be shown that a potential exists but this does not, by itself, establish what is generating it. In bacteria and chloroplasts it is the transport of protons, driven by redox chains, by proton-translocating ATPase or by photosynthetically generated reactions (see Harold, 1977, for a good review). The evidence available suggests that protons also provide the primary transport system in cyanobacteria.

10.3.3.1 Proton translocation and membrane potentials

The outward translocation of protons from cyanobacterial cells was first investigated by Scholes *et al.* (1969). They suspended *Anabaena variabilis* in a solution containing 150 mM KCl, 3mM glycylglycine and 5mM $MgCl_2$, pH 6.0, and allowed the cell proton gradient to collapse by leaving the preparation under anaerobic conditions in the dark, i.e. under conditions allowing neither respiration nor photosynthesis. When the suspension was exposed to light there was a rapid drop in pH. A more gradual pH drop was observed when oxygen was supplied to the suspension in the dark. These results indicated that protons were pumped out of the cell under conditions supporting photosynthesis and respiration. The conditions used for these experiments, rather low pH and high potassium concentration, may have stressed the cyanobacterium.

Surprisingly, there has been no other direct investigation of this outward proton flow in cyanobacteria though it has been tacitly accepted in a number of studies. Paschinger (1977) proposes that in *Anacystis nidulans* the export of protons is driven by a proton-translocating ATPase in the plasmalemma and that its activity is stimulated by supply of ATP from photosynthesis or respiration. His evidence is that N,N′-dicyclohexylcarbodiimide (DCCD), which is known to inhibit such ATPase activity, collapses the proton gradient across the membrane. It also decreases the membrane potential which Paschinger therefore concludes is directly dependent on the ATPase-driven ion pump. (Although a residual potential remains, this can be explained by diffusion out of accumulated potassium; see 10.3.3.2.) He determined the membrane potential, -117 mV, by measuring the accumulation of triphenylmethylphosphonium cations ($TPMP^+$) in the cytoplasm. Unlike most other cations the charged $TPMP^+$ freely permeates membranes as its charge is surrounded by hydrophobic groups. It is therefore not accumulated by ion trapping (which depends on the impermeability of the membrane to the ion but not the uncharged molecule) but is accumulated solely as a consequence of the membrane potential (see Rottenberg, 1979, for details of this method).

In contrast to the above, Raboy and Padan (1978) have proposed that in *Plectonema boryanum* the electrochemical gradient of protons across the cell membrane is driven directly by photosynthetic or respiratory electron transport. They found that glucose transport, which is coupled to the $\Delta\bar{\mu}_{H^+}$ (see 10.3.4) is greatly reduced in the dark by KCN, which inhibits respiration, and in both light and dark by FCCP, which uncouples both respiratory and photosynthetic electron transport. The chemiosmotic glucose transport was insensitive to DCCD, however. Further investigations are required to see if the mechanism of maintaining the $\Delta\bar{\mu}_{H^+}$ across the cell membrane is different in the two organisms.

Häder (1978b) has performed the remarkable feat of inserting a microelectrode into the cell of *Phormidium uncinatum* and measuring its membrane potential directly. The values he obtained, -32 mV in the dark changing to -34 mV in the light, are considerably lower than those determined by other indirect techniques in other organisms, but they appear to be reproducible.

The light-stimulated increase in ΔpH across the thylakoid membrane, already described in 10.3.2.1, provides evidence for the existence of a proton pump here which is driven directly by the photosynthetic reaction. According to Mitchell's (1968) chemiosmotic theory, the return of protons across the membrane to the cytoplasm is responsible for the synthesis of ATP. Padan and Schuldiner (1978) found that the electrochemical potential of protons, $\Delta\bar{\mu}_{H^+}$, across the thylakoid must be generated almost totally by the ΔpH as the membrane potential, $\Delta\psi$, was virtually zero. This is similar to the situation in chloroplast lamellae.

It seems, then, that both the photosynthetic membranes and plasmalemma transport protons out from the cytoplasm in the light. They may do so by the same or different mechanisms. It will be interesting to see what the mechanism is in *Gloeobacter violaceus* where the plasmalemma is the photosynthetic membrane (Rippka *et al.*, 1974). This is being investigated by E. Padan (pers. comm.).

10.3.3.2 Potassium uptake and sodium extrusion

The first investigations of other ion concentrations in cyanobacteria were made by Dewar and Barber (1973). They determined the total Na^+ and K^+ ion concentrations in centrifugally concentrated pellets of *Anacystis nidulans* and then calculated the concentrations of these ions in the cell water, whose volume was estimated by the method described in 10.2.3. Under standard culture conditions with 20 mM K^+ and 13 mM Na^+ in the culture medium they found that the concentration of the K^+ ion in the cells was over seven times greater than the external concentration, while that for Na^+ was about three times less. The Nernst potentials calculated in each case were −50 and +25 mV respectively. When the external concentrations of K^+ and Na^+ were varied reciprocally from 0 to 40 mmol (so that the total of the two was always 40 mmol) the internal levels of the ions varied in the same direction as their

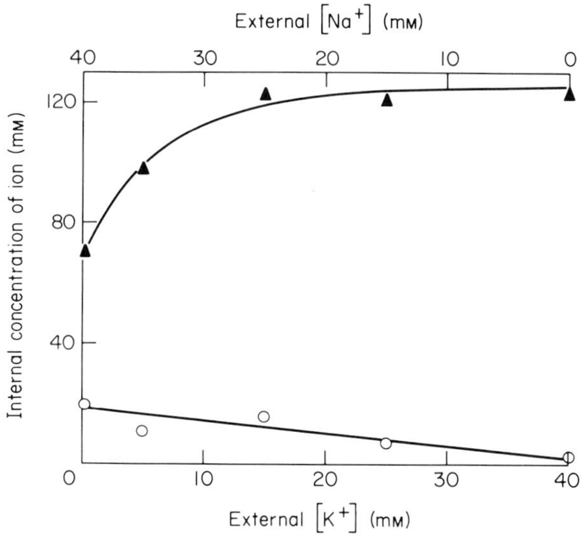

Fig. 10.6. Effect of varying the external concentrations of Na^+ and K^+ on the internal concentration of Na^+ (○) and K^+ (▲) in *Anacystis nidulans*. The total external concentration of the two ions was 40 mmol in each case. (Redrawn from Dewar and Barber, 1973)

external concentration (Fig. 10.6) but K^+ was always higher inside the cell (varying from 70 to 120 mmol) whereas Na^+ was always lower (varying from 0 to 20 mmol). Dewar and Barber (1973) found that radioactively labelled K^+ exchanged freely between the inside and outside of the cell and they therefore proposed that K^+ is passively distributed, indicating a membrane potential equal to $E^N_{K^+}$ i.e. −50 mV. This being so, Na^+ must have been actively extruded.

These findings were essentially confirmed by Paschinger (1977) using the same organism, though he observed a rather larger membrane potential driving a greater passive accumulation of K^+. When the proton translocation that maintains the potential is collapsed by DCCD (as described in 10.3.3.1) an efflux of K^+ was demonstrated and this itself generated a small potential. Paschinger (1977) confirmed the active extrusion of Na^+ from the cells and attributed to it a Na^+/H^+ antiport system, since it, too, is no longer maintained in the presence of DCCD.

The levels of K^+ found in *A. nidulans* by Dewar and Barber (1973) would generate an osmotic pressure of over 2 bars, and must, therefore, contribute similarly to the cell turgor pressure. It was found by the gas vesicle method described above (10.6.1) that a substantial rise in turgor pressure of *Anabaena flos-aquae* occurs when it is placed in high concentrations (75 mM) of KCl and this eventually leads to cell lysis. No rise occurs, however, in similar concentrations of NaCl (Walsby, 1980). The uptake of K^+ from more 'physiological' concentrations, 0.4–4 mM, also contributes to the light-driven turgor pressure changes in this organism (Allison & Walsby, 1981). Detailed investigations of the role of Δ_ψ in K^+ transport in *A. variabilis* have also been made by Reed *et al.* (1981a,b).

10.3.3.3 Chloride uptake

Dewar and Barber (1974) investigated chloride uptake in *Anacystis nidulans* by monitoring the accumulation of ^{36}Cl-labelled chloride from a 0.2 mM solution. By relating the radioactivity absorbed to the packed cell volume of the sample they estimated the internal chloride concentration to be 2.8 mM, but this may be an underestimate as it is not clear if their volume measurements were corrected for interstitial water. The ratio of internal to external chloride concentration was calculated to be equivalent to a Nernst potential of +67 mV and they concluded that chloride uptake must therefore be active, since the ion was accumulated against a negative membrane potential. Chloride uptake was stimulated by light. These findings were confirmed by Craig and Budd (1979) using the same organism and methods, but with better standardized conditions. They found a greater degree of accumulation, 6 mM inside from a 0.1 mM external Cl^- solution, which may partly be explained by a better estimate of cell water volume, and they obtained more clear-cut evidence of light stimulation. They showed that Cl^- uptake was inhibited by carbonyl cyanide m-chlorophenyl hydrazone (CCCP), an ionophore which inhibits phosphorylation by short-circuiting proton return through the membrane. Craig and Budd suggest that the active uptake of Cl^- is driven by ATP; uptake by H^+/Cl^- symport is also considered but thought to be less likely because estimates of the ΔpH across the plasmalemma (Falkner, Horner, Werdan & Heldt, 1976; Masamoto & Nishimura, 1977) appear to be too small to account for it. Zarou and Tromballa (1981) also support the involvement of an ATP-dependent pump in chloride uptake.

10.3.4 Active uptake of uncharged molecules

Both the active uptake of organic nutrients and the concentrations of intermediary metabolites in cyanobacteria impinge on their cell–water relations but detailed discussion of them is beyond the scope of this review. A study by Raboy and Padan (1978) on uptake of glucose and its non-metabolizable analogue α-methylglucoside (αMeGlc) has shown that these sugars are actively accumulated by a mechanism that also involves the proton electrochemical gradient ($\Delta\bar{\mu}_{H^+}$) described above. Either component of the $\Delta\bar{\mu}_{H^+}$ could be linked to the transport of αMeGlc since it continued in the presence of 100 mM KCl and valinomycin, which collapses the $\Delta\psi$, or KCl and nigericin, which collapses the ΔpH, but it was inhibited in the presence of both ionophores. This indicates that the sugars are transported by symport with H^+ ions returning through the plasmalemma. The mechanism appears to be specific for glucose and its analogue. The techniques used by Raboy and Padan (1978) would be applicable to the investigation of active uptake of inorganic ions and of other organic substances.

Beauclerk and Smith (1978) also demonstrated the active uptake of glucose and of an analogue, in the cyanobacteria *Aphanocapsa* 6714 and *Nostoc* strain MAC, but they did not investigate the mechanism of uptake.

In both of the above investigations the rate of uptake was related to the amount of cell material (expressed as weight of protein or of dry material) without reference to the area of membrane through which it passed. This is satisfactory for the purposes of relating uptake rate to the overall assimilation rate in growth but not for comparing the kinetics of uptake in different organisms; because cell volume is proportional to r^3 and the area is proportional to r^2, large cells will fill up more slowly than small ones that have membranes actively transporting substances at the same rate. The appropriate units to use for active uptake are mol min^{-1} per unit area of cell surface.

10.4 Halophiles

Up until this point all of the organisms considered have been cyanobacteria which grow, perhaps exclusively, in culture media of low salinity. However, there are many species that grow in seawater, which has an osmotic pressure of about 28 bars, and a number of halophiles which grow in brines with osmotic pressures exceeding 100 bars. Some of these halophilic cyanobacteria have now been brought into culture. One of them, variously referred to as *Coccochloris elabens* (Kao *et al.*, 1973), *Aphanocapsa halophytica* (Brock, 1976) or *Aphanothece halophytica* (Yopp *et al.*, 1978) is able to grow in culture media containing NaCl at concentrations ranging from 1 to 4 M. The intensively investigated hypersaline pool, Solar Lake in Sinai, is dominated by cyanobacteria (Chapter 9), of which a number have been brought into culture and found similarly to require NaCl but to tolerate a wide range of concentrations (Y. Cohen, A. Oren & A.E. Walsby, unpubl.). Two of these organisms, identified as *Aphanothece halophytica* and a new species of

Dactylococcopsis, possess gas vacuoles and it is therefore possible to determine their turgor pressure. My measurements indicate that their cells possess a low turgor pressure of about 0.5 bars, even when growing in concentrated salt solutions. It follows that the cytoplasm must contain substances whose total osmolarity slightly exceeds that of the suspending medium. The eukaryotic green alga *Dunaliella* does this by accumulating glycerol while the prokaryotic halobacteria accumulate potassium, sodium and roughly equivalent amounts of chloride. In halophilic cyanobacteria inorganic ions also seem to provide the bulk of the internal osmoticum. Miller *et al.* (1976) found that in *Aphanothece halophytica* cells grown in 2 M NaCl, the internal K^+ concentration was over 1 M, while the internal Na^+ concentration was only 0.027 M. The divalent cations Ca^{2+} and Mg^{2+} together accounted for less than 0.1 M. If these analyses are correct then the total concentration of ions inside the cell is sufficient to balance only about half the external concentration. The substances which provide the rest of the osmotic balance have yet to be investigated. Yopp *et al.* (1979) showed that the same organism rapidly plasmolysed when transferred from culture media containing 1 M NaCl to media containing 4 M NaCl. Gradual deplasmolysis then occurred as the cells accumulated K^+. There are various indications that the K^+ uptake is by an active process: the deplasmolysis time was decreased in the light and increased by previous incubation in the dark, both anaerobiosis and inhibitors preventing the synthesis of ATP.

If halophilic cyanobacteria have high concentrations of ions inside them their proteins and other cell constituents must be able to function at high ionic strength. There have been many demonstrations of this in halobacteria but only one so far in cyanobacteria. Kao *et al.* (1973) showed that the aggregation state of c-phycocyanin isolated from *Coccochloris elabens* was similar in 2 N NaCl to that of C-phycocyanin from mesophiles (non-halophiles) in the absence of salt. It is more likely that the phycocyanin in the cells is in an environment of high KCl rather than NaCl; the effects of the two salts are probably equivalent but this should be checked.

Halophiles seem to be able to tolerate drastic changes in water potential. They are bound to encounter such changes in their natural habitats where brines are subjected to concentration by evaporation or dilution by rain and flooding. Much remains to be resolved on the mechanism of halotolerance.

Acknowledgments

I would like to thank Prof. E. Padan for detailed discussions on the ionic relations of cyanobacteria, and Professor M. Shilo and Dr. A. Oren for discussions on halophiles. This chapter was written while the author was supported by the Royal Society–Israel Academy Visiting Professorship in Prof. Shilo's department at the Hebrew University, Jerusalem.

CHAPTER 11

EVOLUTION AND GENETIC PROPERTIES OF CYANOBACTERIAL GENOMES

M. HERDMAN

11.1 Introduction

The collective properties of the cyanobacteria (oxygenic photosynthesis, chromatic adaptation, nitrogen fixation and capacity for cellular differentiation by the formation of heterocysts, akinetes, baeocytes and hormogonia), many of which are unique to this group of prokaryotes, make them attractive organisms for genetic research. Unfortunately, no single organism has been found which exhibits all of these properties and the additional desirable characteristics: rapid growth, the formation of homogeneous suspension in liquid culture, reproducible viability on plates, ease of replica plating and high frequency of genetic exchange.

Difficulties initially encountered in obtaining good viability on plates have long been solved for most organisms by the work of Van Baalen (1965a) and Allen (1968c). The early paucity of axenic cultures no longer exists (for a recent summary of axenic strains, together with their major properties, see Rippka *et al.*, 1979). The persistent myth that isolation of auxotrophic mutants from cyanobacteria is impossible (Asato & Folsome, 1969; Houghton, 1971; Kaney & Dolack, 1972) was without foundation and contributed little to the development of cyanobacterial genetics.

This chapter is a critical attempt to review current knowledge of cyanobacterial genetics and to remove much of the mythology. Cyanobacterial genetic studies have been previously reviewed by Houghton (1972), Van Baalen (1973), Delaney *et al.* (1976) and Ladha and Kumar (1978a); nucleic acid structure and function was reviewed by Leach and Herdman (1973). The reader is referred to Doolittle (1979) for an excellent review of the structure, genetics and expression of the cyanobacterial genome.

11.2 Terminology

11.2.1 Generic terminology

The literature concerning cyanobacterial genetic research is confused by the different taxonomic names appended to the same strain by different authors, resulting from the present state of cyanobacterial taxonomy (see Rippka *et al.*, 1979). For example, strains named as *Anacystis nidulans* (Tx20, UTEX 625 and UTEX 1550), *Synechococcus leopoliensis* CCAP 1405/1 and *Synechococcus* PCC 6301 (ATCC 27144) all originated from the single isolate of Kratz and Myers (1955a) and are therefore identical. The name *Agmenellum quadruplicatum* implies membership of a different genus for a strain which has been described as similar in most respects to *Synechococcus* PCC 6301 (Rippka *et al.*, 1979; but see 12.3.4.3). To permit the comparison of results from different laboratories, and repetition of the work, the alternative names of the organisms reviewed in this chapter are listed in Table 11.1, together with their reference numbers in the major culture collections. For ease of tabulation, the provisional classification of Rippka *et al.* (1979) is employed in Tables 11.3 to 11.5, although the generic names originally employed in the publications reviewed are used in the text.

11.2.2 Genetic terminology

The nomenclature employed in describing cyanobacterial mutants is often confusing. Rules of genetic terminology are well defined for other bacteria, and cyanobacterial genetic studies would be more easily understood universally if the same rules were applied. It would be convenient if the proposals of Demerec *et al.* (1966) were adopted. It would be useful if phenotype and genotype symbols followed, wherever possible, those in current usage in bacterial systems; the majority can be obtained from the current linkage maps of *Escherichia coli* (Bachmann & Low, 1980), *Bacillus subtilis* (Henner & Hoch, 1980) and *Salmonella typhimurium* (Sanderson & Hartman, 1978) and relevant genotypes are listed in Table 11.2. Clearly, some gene functions unique to cyanobacteria require a new terminology. Het^- appears to be commonly employed as the phenotype symbol for mutants defective in heterocyst differentiation (but see 11.5.5.2 (b) for the important distinction between Het^- and Nif^-). The word 'akinete' is preferable to 'spore' (Chapter 15), in order to avoid confusion with bacterial endospores, and thus Aki^- would conveniently replace Spo^- or Spr^-. In choosing symbols for new mutants defective in photosynthetic pigments, care should be taken to avoid duplication; the genotype *chl* denotes chlorate resistance and would not be appropriate for chlorophyll-deficient mutants. In common with accepted practice, plasmid nomenclature would conveniently follow the framework proposed by Novick *et al.* (1976).

11.3 Genome structure and evolution

11.3.1 Cyanobacterial genomes

As first demonstrated by cytochemical studies (see Fuhs, 1973) and confirmed by electron microscopy (Hopwood & Glauert, 1960; Ris & Singh, 1961; Pankratz & Bowen, 1963; Leak, 1967), the nuclear regions of cyanobacteria are comparable with those of other prokaryotes in lacking both a nuclear membrane and compact chromosomes of the eukaryotic type. The apparent absence of histones (Makino & Tsuzaki, 1971) was consistent with the prokaryotic nature of cyanobacteria. The assumption that the fibrils observed in electron micrographs of the nuclear regions were composed of DNA was confirmed by their sensitivity to DNase (Fuhs, 1963) and incorporation of

Table 11.1. Alternative generic designations of the strains discussed

Organism	Culture collection number		
	ATCC†	UTEX‡	CCAP§
Section I*			
Synechococcus PCC¶ 6301	27144		
Anacystis nidulans			
Anacystis nidulans Tx20		625, 1550	
Synechococcus leopoliensis			1405/1
Synechococcus PCC 6311	27145		
Anacystis nidulans 6311		1549	
Synechococcus PCC 7942**			
Anacystis nidulans R2			
Synechococcus PCC 7943**			
Anacystis nidulans 602			
Synechococcus PCC 73109	29404		
Agmenellum quadruplicatum BG-1			
Synechococcus PCC 7002	27264		
Agmenellum quadruplicatum PR-6			
Synechococcus PCC 7003	27265		
Coccochloris elabens 17a			
Synechococcus PCC 6307	27147		
Coccochloris peniocystis		1548	
Synechococcus PCC 6908	27146		
Synechococcus cedrorum		1191	
Synechocystis PCC 6714	27178		
Aphanocapsa 6714			
Synechocystis PCC 6308	27150		
Gloeocapsa alpicola		1598	1430/1
Synechocystis PCC 7005	27153		
Microcystis aeruginosa			1450/1
Gloeothece PCC 6909	27152		
Gloeocapsa sp.		795	1430/3
Section II*			
Chroococcidiopsis PCC 6712	27176		
Chlorogloea sp.			1411/2
Section III*			
LPP group PCC 7602**			
Phormidium luridum		426	1462/2
LPP group PCC 6306	27894		
Plectonema boryanum		581, 1542	1463/1
LPP group PCC 73110	29407		
Plectonema boryanum		594	

Table 11.1 (continued)

Organism	Culture collection number		
	ATCC	UTEX	CCAP
Section IV*			
{ *Anabaena* PCC 7938**			
Anabaena cylindrica		381	
Anabaena inequalis			1446/1c
{ *Anabaena* PCC 7936**			
Anabaena doliolum (R.N. Singh)			
{ *Anabaena* PCC 7937**			
Anabaena variabilis 1444	29143		
Anabaena flos-aquae		1444	1403/13
{ *Anabaena* PCC 7118	27892		
Anabaena variabilis (Myers)			
{ *Anabaena* PCC 7120	27893		
Nostoc muscorum G			
Nostoc muscorum 7120			
{ *Anabaena* PCC 7119	29151		
Nostoc muscorum UW			
{ *Calothrix* PCC 7601**			
Fremyella diplosiphon		481	1429/1
{ *Calothrix* PCC 7103	27905		
Nodularia 7103			
Nodularia sphaerocarpa		583	
{ *Calothrix* PCC 7101	27914		
Tolypothrix tenuis			1482/3a
Section V*			
{ *Chlorogloeopsis* PCC 6912	27193		
Chlorogloea fritschii			1411/1
{ *Fischerella* PCC 73103	29114		
Fischerella muscicola		1301	1427/1

* Sections I–V, broadly equivalent to traditional taxonomic orders, are as defined by Rippka *et al.* (1979).
† American Type Culture Collection.
‡ Culture Collection of Algae at the University of Texas (formerly IU).
§ Cambridge Culture Collection of Algae and Protozoa.
¶ Pasteur Culture Collection of Cyanobacteria.
** Recent additions to the PCC (R. Rippka, pers. comm.).

Table 11.2. Chromosomal genetic markers of cyanobacteria

Gene symbol	Mnemonic	Phenotype or enzyme deficiency	Reference
ace	Acetate	Requirement	
aki	Akinete	Defective in akinete formation or structure	
amp	Ampicillin	Resistant to ampicillin or penicillin	
arg	Arginine	Requirement*	
aro	Aromatic	Resistant to aromatic amino acids or their analogues	
azi	Azide	Resistance	
bio	Biotin	Requirement	
can	Canavanine	Resistance	
chl	Chlorate	Resistance	
cml	Chloramphenicol	Resistance	
cys	Cysteine	Deficient in sulphate reduction	
ery	Erythromycin	Resistance	
*fad*D	Fatty acid degradation	Resistance to propionate (acetate thiokinase)	Smith & Lucas (1971)
fil	Filamentous	Formation of filaments by uni–cellular spp.†	
fru	Fructose	Resistance	
fur	5-Fluorouracil	Resistance	
*gln*A	Glutamine	Glutamine synthetase	Fisher *et al.* (1981)
glt	Glutamate	Requirement	
gnd		6-Phosphogluconate dehydrogenase	Doolittle & Singer (1974)
het	Heterocyst	Defective in heterocyst formation, function or structure	
kan	Kanamycin	Resistance	
*leu*A	Leucine	isopropylmalate synthetase	Singer & Doolittle (1975)
mdh		Malate dehydrogenase	Herdman *et al.* (1980)
met	Methionine	Requirement or resistance to analogues	
mit	Mitomycin	Resistance to mitomycin C	
nar	Nitrate	Nitrate reductase‡	
neo	Cryneomycin	Resistance	
nfr	Nitrofurazone	Resistance	
nic	Nicotinic acid	Requirement	
*nif*D		Nitrogenase subunit, structural gene	Mazur *et al.* (1980)
*nif*K		Nitrogenase subunit, structural gene	Mazur *et al.* (1980)
*nif*H		Nitrogenase reductase, structural gene	Mazur *et al.* (1980)
nir	Nitrite	Nitrite reductase	
pab	p-Aminobenzoate	Requirement	

Table 11.2 (continued)

Gene symbol	Mnemonic	Phenotype or enzyme deficiency	Reference
pmr	Polymixin	Resistance to polymixin B	
*phe*A	Phenylalanine	Prephenate hydratase	Kaney (1973)
pho	Phosphate	Polyphosphate kinase	Vaillancourt *et al.* (1978)
pur	Purine	Adenine requirement	
pyr	Pyrimidine	Uracil requirement	
*pyr*B	Pyrimidine	Aspartate transcarbamylase	Currier & Wolk (1978)
rif	Rifampicin	Resistance§	
ser	Serine	Requirement for serine or glycine	
str	Streptomycin	Resistance	
thi	Thiamine	Requirement	
*trp*A	Tryptophan	Tryptophan synthetase A protein	Ingram *et al.* (1972)
zwf	Zwischenferment	Glucose-6-phosphate dehydrogenase	Doolittle & Singer (1974)

Genotype symbols, wherever possible, are taken from those in current use in *Escherichia coli* (Bachmann & Low, 1980), *Bacillus subtilis* (Henner & Hoch, 1980) or *Salmonella typhimurium* (Sanderson & Hartman, 1978); they do not necessarily correspond to those originally published.
* The single mutant of this class isolated to date responded to ornithine (Orkwiszewski & Kaney, 1974)
† Several classes of *fil* mutations have been described (see text); terminology should change as new data become available.
‡ The genotype *nar* has been employed in place of *chl*, because few mutants defective in nitrate reductase have been tested for chlorate-resistance.
§ The genotype *rif* is retained in place of the conventional *rpo*B, since RNA polymerase structure has never been studied in *rif* strains.

^{3}H-thymidine (Leak, 1965). Studies performed during the past 25 years have failed to demonstrate any unusual structure of the cyanobacterial genetic apparatus (see Leach & Herdman, 1973). The presence of the four normal nucleotide bases (Biswas, 1956, 1960), and the characteristic hyperchromicity of cyanobacterial DNA during thermal denaturation (Kaye *et al.*, 1967; Craig *et al.*, 1969) supported the double-stranded nature of the genome.

Early studies of the base composition of cyanobacterial DNA (Edelman *et al.*, 1967; Stanier *et al.*, 1971) have been extended to an analysis of 176 strains, representing at least 22 genera (Herdman *et al.*, 1979b). The overall range of DNA base composition, 35–71 mol % GC (Table 11.3), corresponds to the diversity found in other bacteria. This diversity, however, is confined to two major groups: unicellular cyanobacteria dividing by binary fission (35–71 mol % GC), and filamentous, aheterocystous cyanobacteria (40–67 mol % GC).

All other strains, including the heterogeneous pleurocapsalean and heterocystous forms, contain DNA with a narrow (38–47 mol % GC) range of base composition. Within the unicellular cyanobacteria, DNA base composition is a useful taxonomic criterion: the genus *Synechococcus* may be subdivided into three subgroups and *Synechocystis* into two (Table 11.3). Such subdivision is consistent with other phenotypic properties of the strains (Rippka *et al.*, 1979) and the marked discontinuities are sufficient to warrant generic separation.

The cyanobacteria in their entirety differ from all other bacteria in exhibiting an extremely wide range of genome size (genetic complexity). All other bacteria examined (Fig. 11.1 (b)) contain genomes of relatively uniform size, $1{\cdot}0 \times 10^9$ to $3{\cdot}6 \times 10^9$ daltons mol. wt, (Gillis *et al.*, 1970; Wallace & Morowitz, 1973) with the exception of a group of *Mycoplasma* strains which contain smaller genomes (Bak *et al.*, 1969). The few other apparent exceptions (Bak *et al.*, 1970) are generally thought (Gillis *et al.*, 1970; Pemberton, 1974) to be incorrect.

The first quantitative studies of cyanobacterial genetic complexity revealed genome sizes similar to those of other bacteria: $2{\cdot}27 \times 10^9$ daltons and $2{\cdot}47 \times 10^9$ daltons mol. wt for *Anacystis nidulans* UTEX 625 and *Anabaena cylindrica* CCAP 1403/1, respectively (Herdman & Carr, 1974) and $2{\cdot}2 \times 10^9$ to $2{\cdot}8 \times 10^9$ daltons for *Agmenellum quadruplicatum* PR–6 (Roberts *et al.*, 1977). As in other bacteria, and in contrast to eukaryotic organisms, reiterated DNA sequences were absent. The study of 128 cyanobacterial strains (Herdman *et al.*, 1979a) revealed a much wider diversity in genetic information content ($1{\cdot}6 \times 10^9$ to $8{\cdot}6 \times 10^9$ daltons genome mol. wt (Table 11.3)) which, in general, was correlated to the morphological and/or physiological complexity of the organisms. The majority of unicellular strains contained small genomes ($1{\cdot}6 \times 10^9$ to $2{\cdot}7 \times 10^9$ daltons). Some unicellular strains, however, contained larger genomes (up to $5{\cdot}2 \times 10^9$ daltons), the majority being capable of fixing atmospheric nitrogen under anaerobic conditions and those (*Gloeothece* strains) with the largest genomes being able to fix nitrogen under aerobic conditions. In general, filamentous organisms possessed large genomes, and the heterocyst-forming strains showed a wider variation in genome size ($3{\cdot}2 \times 10^9$ to $8{\cdot}6 \times 10^9$ daltons) than any other group, accompanied by extensive variation in morphological complexity and physiological properties (Rippka *et al.*, 1979). Indeed, some *Calothrix* strains, together with the single strain of *Scytonema* examined, contained larger genomes than any other prokaryote and possessed the same genetic information content as some microbial eukaryotes.

A number of strains contained DNA which renatured rapidly (Herdman *et al.*, 1979a). Such DNA can be interpreted either as reiterated chromosomal DNA or small extrachromosomal elements (phages or plasmids) present in high copy number. Phage-like particles were observed in two of these strains;

Table 11.3. Mean DNA base composition and genome sizes of cyanobacteria*

Genus	Range of mean DNA base compositions (mol% G+C)	Range of genome sizes (daltons × 10^{-9})
Section I		
Synechococcus I†	39–43(5)‡	2.6–4.2(5)‡
II	47–56(14)	1.6–3.1(10)
III	66–71(9)	2.0–2.6(5)
Synechocystis I	35–37(5)	2.1–2.5(4)
II	42–48(12)	1.8–3.5(9)
Gloeobacter	64(1)	2.7(1)
Gloeocapsa	40–46(4)	2.9–3.5(4)
Gloeothece	40–43(5)	5.0–5.2(2)
Chamaesiphon	47(2)	3.6–3.8(2)
Section II		
Dermocarpa	38–44(6)	3.1–4.4(2)
Xenococcus	44(2)	4.0(1)
Dermocarpella	45(1)	3.3(1)
Myxosarcina	43–44(2)	3.5(1)
Chroococcidiopsis	40–46(8)	3.3–4.8(4)
Pleurocapsa group	39–47(11)	3.1–3.4(2)
Section III		
Spirulina	44–53(2)	2.5(1)
Oscillatoria	40–50(10)	2.5–4.4(6)
Pseudanabaena	44–52(8)	2.1–3.8(7)
LPP group	42–67(25)	2.6–5.2(24)
Section IV		
Anabaena	38–44(6)	3.2–3.9(6)
Nodularia	41(1)	3.3(1)
Cylindrospermum	42–47(3)	5.7–6.2(2)
Nostoc	39–46(13)	4.0–6.4(11)
Scytonema	44(1)	7.4(1)
Calothrix	40–44(11)	5.1–8.6(10)
Section V		
Chlorogloeopsis	42–43(2)	4.2–5.2(2)
Fischerella	42–46(7)	3.6–4.8(4)

* Summarized from Herdman *et al.* (1979a,b); generic classifications from Rippka *et al.* (1979).

† Sub-groups of genus.

‡ Numbers in parentheses indicate the number of strains examined.

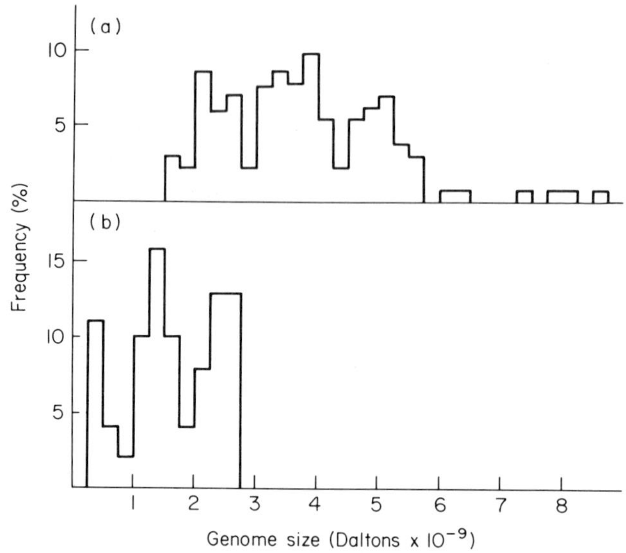

Fig. 11.1. Frequency distributions of the genome sizes of (a) cyanobacteria (after Herdman *et al.*, 1979a and author, unpubl.) and (b) all other bacteria listed in the literature.

three (PCC 6701, 7005 and 7120; Tables 11.4, 11.5) contain plasmids. The remainder have not been further examined.

It has been suggested that bacterial genomes evolved by a series of duplications of a small ancestral genome (Wallace & Morowitz, 1973; Zipkas & Riley, 1975; Riley & Anilionis, 1978). The cyanobacterial genomes occupied a discontinuous frequency distribution (Fig. 11.1(a)), falling into four groups of size ($\times 10^9$ daltons): (a) 1·6–2·9 (b) 2·9–4·4 (c) 4·4–5·8 and (d) 6·0–8·6, respectively (Herdman *et al.*, 1979a). The mean genome sizes within each group were ($\times 10^9$ daltons, $\pm$standard deviation): $2{\cdot}21 \pm 0{\cdot}31$, $3{\cdot}56 \pm 0{\cdot}37$, $5{\cdot}00 \pm 0{\cdot}39$ and $7{\cdot}40 \pm 0{\cdot}95$, respectively. These may correspond to genomes which evolved by a series of fusions of a primitive genome (of $1{\cdot}2 \times 10^9$ daltons mol. wt) to create new chromosomes containing two ($2{\cdot}4 \times 10^9$), three ($3{\cdot}6 \times 10^9$), four ($4{\cdot}8 \times 10^9$) and six ($7{\cdot}2 \times 10^9$) copies of the ancestral genome. The redundant DNA created by fusion would be available for mutational change and so create novel genetic information. The groups of different genome size, together with the accompanying diversity of morphological and physiological features, are compatible with genome evolution by duplication in cyanobacteria. However, it is difficult to understand the function of the large excess of DNA in the 'complex' strains, whose novel characteristics probably demand no more than 100 new gene functions. In comparing this phenomenon with the genetic information content of eukaryotes, Doolittle

(1979) posed the question, still unanswered, 'is it possible that complex ("advanced") prokaryotes also carry large amounts of non-informational DNA?'

All other aspects of cyanobacterial genetic structure and function reveal the similarity of these organisms to other bacteria. DNA replication is controlled and related to the cell-division cycle, and total cellular DNA content varies with growth rate (Mann & Carr, 1974; Parrot & Slater, 1980; Herdman *et al.*, 1982). RNA polymerase is inhibited by antibiotics which inhibit the bacterial enzyme (see Leach & Herdman, 1973) and the polymerases from *Fremyella diplosiphon* and *Anacystis nidulans* UTEX 625, although differing slightly from that of *Escherichia coli* (Miller & Bogorad, 1978), recognize bacteriophage lambda promoters (Miller *et al.*, 1979). A histone-like DNA-binding protein in *Synechocystis* PCC 6701 appears in all respects to be similar to the *Escherichia coli* HU protein (Haselkorn & Rouvière-Yaniv, 1976; Aitken & Rouvière-Yaniv, 1979). The long-awaited discovery that the genome of *Anabaena variabilis* (Myers strain), isolated as a folded chromosome, is circular (Dmitrieva & Kozlov, 1979), and the probability that genome replication is bidirectional in *Anacystis nidulans* (Delaney & Carr, 1975), confirm the typical bacterial nature of the cyanobacterial genetic apparatus.

11.3.2 Cyanobacterial plasmids

Roberts and Koths (1976) performed the first thorough study of cyanobacterial plasmids, in *Agmenellum quadruplicatum* PR-6. Six different sizes (see Table 11.4) of covalently closed circular (CCC) DNA were observed (and confirmed by van den Hondel *et al.*, 1979; Lau *et al.*, 1980a), the molecular weights from electron-microscopic analysis and from gel electrophoresis showing good agreement. The frequency of the different plasmids varied with culture age; the two smallest plasmids were most frequent in post-exponential phase cells but decreased in number in exponentially growing cells (Roberts & Koths, 1976). Such dependency of plasmid copy number on culture age (and therefore cell metabolic state) has unfortunately not been examined in other cyanobacterial strains.

Subsequently, plasmids were characterized in a large number of cyanobacteria in various laboratories, as summarized in Tables 11.4 and 11.5. At least half of the cyanobacteria examined contain plasmids, and the majority of these contain several different types. The overall size range of the plasmids (1.9–75 Md) is substantial. Where examined, the base composition of plasmid DNA was significantly different from that of chromosomal DNA (Roberts & Koths, 1976; Lau & Doolittle, 1979).

Although Kumar *et al.* (1967) suggested that loss of streptomycin

Table 11.4. Cyanobacterial plasmids in unicellular strains

Genus and strain*	DNA base ratio mol% GC†	No. of plasmids	Size of plasmids (Md)	Reference	Published strain designation (if different)
Synechococcus					
6301	55.1	2	37, 4.7‡	Simon (1978)	*Anacystis nidulans* UTEX 625
		2	27.5, 5.3‡	Lau & Doolittle (1979)	*Anacystis nidulans* UTEX 625
			30.5, 5.1§	Lau & Doolittle (1979)	*Anacystis nidulans* UTEX 625
		2	32.7, 5.3‡§	van den Hondel *et al.* (1979)	
6311	54.8	1	5.0‡§	Friedberg & Seijffers (1979)	*Anacystis* 6311
		2	33, 5.3	van den Hondel & van Arkel (1980)	
6908	55.6	2	32.4, 5.3‡§	van den Hondel *et al.* (1979)	
		2	30.5, 5.3‡§	Lau & Doolittle (1979)	*Synechococcus cedrorum* UTEX 1191
7942¶	—	2	33, 5.3‡	van den Hondel *et al.* (1980)	*Anacystis nidulans* R-2
7943¶	—	2	33, 5.3	van den Hondel & van Arkel (1980)	*Anacystis nidulans* 602
6707	68.4	2	32.7, 5.3‡	van den Hondel *et al.* (1979)	
7002	49.1	6	75, 24.7, 20.1 10.3, 6.5, 3.0§	Roberts & Koths (1976)	
		6	as above	van den Hondel *et al.* (1979)	*Agmenellum quadruplicatum* PR-6
		5–7	≥ 50, 26, 10.6, 6.2, 3.5‡	Lau *et al.* (1980a)	*Agmenellum quadruplicatum* PR-6
73109	49.0	4–5	≥ 50, 10.6, 5.9‡	Lau *et al.* (1980a)	*Agmenellum quadruplicatum* BG-1
7003	49.3	5	≥ 50, 33, 3.5, 2.9, 1.9‡	Lau *et al.* (1980a)	*Coccochloris elabens* 17AR
7425	48.6	2	24, 5.4‡§	van den Hondel *et al.* (1979)	
6312	50.2	1	15.9‡§	van den Hondel *et al.* (1979)	
Synechocystis					
6714	47.9	1	30‡	Lau *et al.* (1980a)	
7005	45.4	1	2‡§	van den Hondel *et al.* (1979)	
6701	35.8	8	40–2‡	Lau *et al.* (1980a)	

* PCC strain numbers from Rippka *et al.* (1979).
† From Herdman *et al.* (1979b).
‡ Estimated by gel electrophoresis.
§ Estimated from contour length in the electron microscope.
¶ Recent additions to the PCC (R. Rippka, pers. comm.).
Plasmids were not detected in *Synechococcus* PCC 6715 (van den Hondel *et al.*, 1979), *Synechococcus* PCC 6307 and 7502, *Synechocystis* PCC 6711, *Gloeobacter* PCC 7421, *Gloeothece* PCC 6909, *Chroococcidiopsis* PCC 6712, *Microcystis aeruginosa* NRC-1, *Gloeocapsa alpicola* W and approximately one half of newly isolated strains (Lau *et al.*, 1980a).

Table 11.5. Cyanobacterial plasmids in filamentous strains

Genus and strain*	No. of plasmids	Size of plasmids (Md)	Reference	Previous strain designation (if different)
Section III				
Oscillatoria				
limnetica	1	5.1§	Friedberg & Seijffers (1979)	
LPP Group				
7602†	2	9.0, <2.1‡	Simon (1978)	*Phormidium luridum* UTEX 426
6306	1	9.4‡	Simon (1978)	*Plectonema boryanum* UTEX 581
Section IV				
Anabaena				
7938†	5	74, 53, 40, 24, 14‡	Simon (1978)	*Anabaena cylindrica* UTEX 381
7936†	3	36, 26, 5.7‡	Simon (1978)	*Anabaena doliolum* (R.N. Singh)
7937†	2	32, 25‡	Simon (1978)	*Anabaena variabilis* UTEX 1444
7120	4	74, 42, 33, 3.3‡	Simon (1978)	*Nostoc muscorum* 7120
Nostoc				
7524	3	28, 8, 4‡§	Reaston *et al.* (1980)	
Calothrix				
7601†	4	48, 40, 32, 24‡	Simon (1978)	*Fremyella diplosiphon* UTEX 481

* PCC strain numbers from Rippka *et al.* (1979)
† Recent additions to the PCC (R. Rippka, pers. comm.).
‡ Estimated from gel electrophoresis.
§ Estimated from contour length in the electron microscope.
Plasmids were not detected in *Pseudanabaena* PCC 6901, *Nostoc* MAC, *Chlorogloeopsis* PCC 6912, *Anabaena flos-aquae* (3 strains) (Lau *et al.*, 1980a) and *Cylindrospermum* (2 strains), *Nodularia* PCC 7103, *Tolypothrix* PCC 7101 and *Fischerella* PCC 73103 and 7115 (Simon, 1978).

resistance, after subculture of *Anacystis nidulans* Tx20 in the presence of proflavine, resulted from the curing of a plasmid, drug resistance has never been shown to be plasmid coded. Indeed, the low resistance of this strain to streptomycin does not support this interpretation. No function has as yet been ascribed to any cyanobacterial plasmid. Spontaneous loss of the 5.3 Md plasmid from *Synechococcus* PCC 6301 (Lau & Doolittle, 1979) and of the 5.9 Md plasmid from *Agmenellum quadruplicatum* BG–1 (Lau *et al.*, 1980a) did not cause any phenotypic change, and organisms which lacked plasmids showed not detectable metabolic defects (Simon, 1978). It was clearly shown that the plasmids of *Anabaena* 7120 do not carry any identifiable *nif* genes (Mazur *et al.*, 1980). However, the 5 Md plasmid of *Synechococcus* PCC 6301 was transcribed *in vivo* (Lau *et al.*, 1980b) and some plasmid restriction fragments were transcribed differentially in light and dark (W.F. Doolittle, pers. comm.). Information of this importance is unfortunately lacking for other plasmids.

Identical restriction enzyme digestion patterns were produced in each of the plasmids of similar size in *Synechococcus* strains PCC 6301 and 6908 (van den Hondel *et al.*, 1979; Lau & Doolittle, 1979) and 6311 (van den Hondel & van Arkel, 1980), indicating the close relatedness of the plasmids. This is not surprising, since the strains themselves are closely related (Rippka *et al.*, 1979). However, the same restriction patterns were observed in the plasmids of *Synechococcus* PCC 6707 (van den Hondel *et al.*, 1979), a strain which differs from the others by 13 mol % GC (Herdman *et al.*, 1979b) and may in fact belong to a separate genus. The occurrence of identical plasmids in such divergent strains may indicate their transferable nature, and intergeneric transfer may be possible. Furthermore, the diverse origin of strains 6301 and 6707 (Texas & California, respectively, see Rippka *et al.*, 1979) indicates the widespread geographical occurrence of these plasmids. *Anacystis nidulans* R-2 and 602 (*Synechococcus* PCC 7942 and 7943), of unknown DNA base composition and geographical origin, contained identical plasmids to the strains listed above (van den Hondel & van Arkel, 1980).

By contrast, hybridization experiments with the plasmids from the similar strains *Agmenellum quadruplicatum* BG-1 and PR-6, together with the related (Rippka *et al.*, 1979) strain *Coccochloris elabens*, showed no complete plasmid homology even between those of similar size (Lau *et al.*, 1980a). Surprisingly, however, significant degrees of *interplasmic homology* were observed, even between plasmids of different size classes. Such regions of homology were not confined to single restriction enzyme digest fragments, suggesting that they may be analogous to the transposable elements carried by many bacterial plasmids (Lau *et al.*, 1980a). Since the strains were isolated in Puerto Rico, New York and Connecticut (Rippka *et al.*, 1979), it would appear that these elements are extremely widespread in nature.

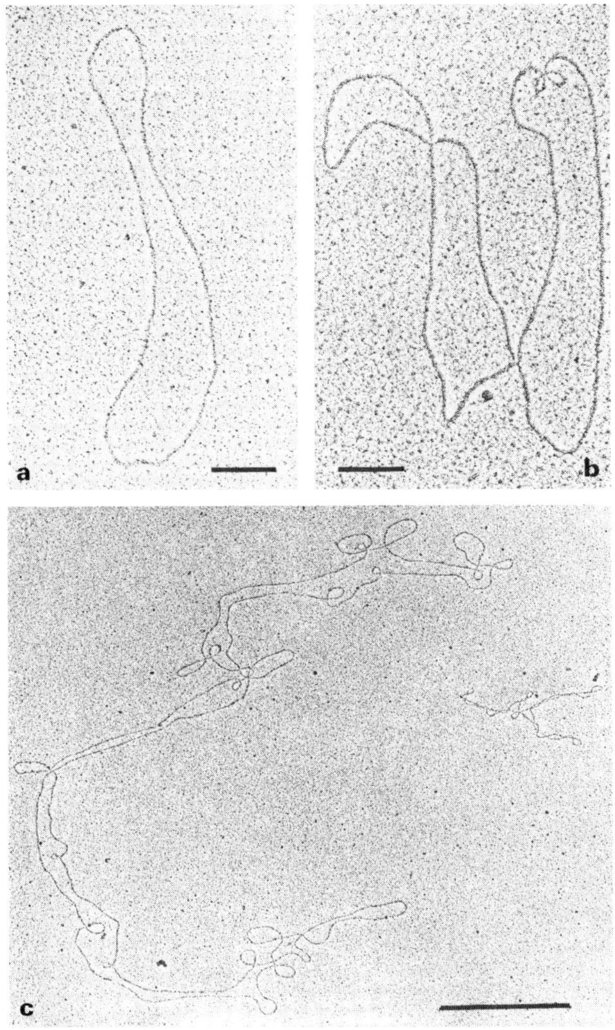

Fig. 11.2. Electron micrographs of plasmids isolated from *Nostoc* PCC 7524: (a) pDU1, (b) pDU2, (c) pDU3 (left) with pDU1 (right). Bar markers represent 0.25 μm. (after Reaston *et al.*, 1980).

The three plasmids (pDU1, pDU2 and pDU3, see Fig. 11.2) of *Nostoc* PCC 7524 had molecular weights of 4, 8 and 28 Md, respectively. Digestion of pDU1 and pDU2 with endoR. *Hind*II revealed identical restriction patterns. Thus, pDU2 appears to be a dimer of pDU1 and is the first example of a dimeric plasmid in cyanobacteria (Reaston *et al.*, 1980).

Table 11.6. Sequence specificities of cyanobacterial restriction endonucleases

Organism*	Culture collection no.				Enzyme†	Sequence‡	Reference
	ATCC	UTEX	CCAP	PCC			
Section I							
Agmenellum quadruplicatum	27264			7002	*Aqu*I (*Ava*I)	C↓PyCGPuG	Lau & Doolittle (1980)
Section III							
Microcoleus sp.					*Mst*I	TGCGCA	Gingeras *et al.* (1978)
Section IV							
Anabaena cylindrica	27899	629	1403/2a	7122	*Acy*I	GPu↓CGPyC	de Waard *et al.* (1978)
Anabaena oscillarioides			1403/11		*Aos*I (*Mst*I)	TGC↓GCA	de Waard *et al.* (1979)
					*Aos*II (*Acy*I)	GPu↓CGPyC	
Anabaena subcylindrica	29211	377	1403/4b	6309	*Asu*I	G↓GNCC	Hughes *et al.* (1980)
					*Asu*II (*Mla*I)	TT↓CGAA	de Waard & Duyvesteyn (1980)
					*Asu*III (*Acy*I)	GPu↓CGPyC	
Anabaena variabilis	27892			7118	*Ava*I	C↓PyCGPuG	Murray *et al.* (1976) Hughes & Murray (1980)
					*Ava*II	G↓G($\genfrac{}{}{0pt}{}{A}{T}$)CC	Sutcliffe & Church (1978)
					*Ava*III	ATGCAT	Roizes *et al.* (1979)
Anabaena variabilis UW¶					*Avr*I (*Ava*I)	C↓PyCGPuG	de Waard & Duyvesteyn (1980)
					*Avr*II	CCTAGG	
Anabaena sp.	29208			7108	*Ast*'W'I (*Acy*I)	GPu↓CGPyC	de Waard & Duyvesteyn (1980)
Section V							
Mastigocladus laminosus	29161		1447/1	7414	*Mla*I	TT↓CGAA	Duyvesteyn & de Waard (1980)

* Name as published, not necessarily valid (see 11.2.1); *Agmenellum quadruplicatum* is *Synechococcus* PCC 7002, *Mastigocladus laminosus* is *Fischerella* PCC 7414 according to Rippka *et al.* (1979).
† If two enzymes are isoschizomers, the prototype (the first enzyme isolated) is shown in parentheses.
‡ The recognition sites are shown as the sequence of only one strand of DNA, written from 5′→3′. The point of cleavage is represented by an arrow (↓). Thus, the *Ava*I recognition sequence is an abbreviation of
5′-C↓PyCGPu G-3′
3′-G PuGCPy↑C-5′
and the *Aos*I sequence represents 5′-TGC↓GCA-3′
3′-ACG↑CGT-5′.
Bases in parentheses ($\binom{A}{T}$) indicate that either base may occur in that position; thus *Ava*II cleaves 5′-G↓GACC-3′ or 5′-G↓GTCC-3′. The abbreviations Py and Pu indicate pyrimidine (C or T) and purine (A or G), respectively. A position indicated by N may be filled by any base. The absence of an arrow shows that the precise cleavage site has not been determined.
§ From Rippka *et al.* (1979).
¶ UW, University of Wisconsin.

11.3.3 Restriction endonucleases

Sequence specific endonucleases (restriction enzymes) have been characterized in only a small number of cyanobacteria. A survey of their sequence specificities is given in Table 11.6. As in other bacterial groups (see Roberts, 1978, 1980), isoschizomers (enzymes recognizing the same nucleotide sequence) are of common occurrence even in unrelated strains. For example, *AquI* and *AvaI* recognize and cleave the sequence C↓PyCGPuG and are produced by a unicellular and a filamentous cyanobacterium, respectively. Since two members of the same bacterial species often contain different restriction endonucleases (see Roberts, 1980) these enzymes are not useful taxonomic markers, despite claims to the contrary (de Waard & Duyvesteyn, 1980).

An example of the probable *in vivo* action of a restriction endonuclease, and of the associated modification process, was described by Currier and Wolk (1979): following propagation in *Anabaena* 7120, cyanophage N-1 initially produced plaques at only low efficiency on *A. variabilis* UTEX 1444 but these rare progeny subsequently showed an increased efficiency of lysis. These results are consistent with a small fraction of the phage population escaping restriction in the new host and then becoming resistant as a result of methylation in the new modification pattern. Doolittle (1979) has commented on the presence of methylated bases in cyanobacterial DNA and the probability that they reflect the presence of restriction–modification systems.

11.4 Mutation and repair

11.4.1 Mutagens

Apparent difficulties in isolating mutants (especially auxotrophs) of cyanobacteria have led to repeated suggestions that these organisms are particularly resistant to mutagens. In organisms in which a thorough analysis of the effects of different mutagens has been performed, it is becoming evident that different organisms have varying responses to the mutagens employed and that at least one efficient mutagen can be found for each strain.

The nitrosoamines nitrosomethyl urea (NMU) and N-methyl-N′-nitro-N-nitrosoguanidine (NTG) were efficient mutagens for a strain of *A. nidulans* whereas the alkylating agents diethyl sulphate (DES), methyl methanesulphonate (MMS) and ethyl methanesulphonate (EMS) (Shestakov & Jevner, 1968) and u.v., X-rays, mitomycin C, acridine orange and hydroxylamine (Shestakov, 1972) were less effective in this organism. Induction of cyanophage-resistant and *str* mutations of *Plectonema boryanum* UTEX 594 showed differential responses to the mutagens employed (NTG, u.v., acriflavine, 2-aminopurine and caffeine), NTG being the most effective in inducing the former and acriflavine the latter (Singh & Kashyap, 1977). *Aphanocapsa* 6714, as measured by the formation of *aro* mutants resistant to p-fluorophenylalanine, was most susceptible to u.v.; NMU, NTG, EMS, euflavine and ^{60}Co γ irradiation were relatively inefficient although they caused substantial killing (Astier *et al.*, 1979). *Gloeocapsa alpicola* CCAP 1430/1 was similarly most susceptible to u.v. (Lambert *et al.*, 1980) as judged by the induced frequencies of *str* mutants, EMS, mitomycin C, NTG and MMS being inefficient mutagens. By contrast, *Synechococcus* PCC 6301 was more readily mutagenized by NTG and EMS than by u.v. (Herdman *et al.*, 1980) as shown by the induced frequencies of a broad range of auxotrophic mutants. It is therefore of fundamental importance that a comparison of a range of potential mutagens be performed for each new strain studied if reasonable success is to be achieved.

Ionizing radiation appears to have been largely neglected as a mutagen. The only quantitative survival curves were described by Bazin (1969) and Asato (1971) in *A. nidulans* Tx20 and by Astier *et al.* (1979) in *Aphanocapsa* 6714, after ^{60}Co irradiation. It is still uncertain whether ionizing radiations are useful mutagenic agents in cyanobacteria, although Asato (1971) isolated a variety of pigment mutants at frequencies as high as 1×10^{-3} following γ irradiation of *A. nidulans* Tx20.

NTG appears to have been the most widely used mutagen in cyanobacteria. Care must be taken in interpreting the nature of mutants isolated following exposure to this compound, in view of its known ability to induce multiple mutations at closely linked sites in other organisms. Less potent mutagens may be preferable.

Where examined, mutagens affect cyanobacterial DNA in exactly the same way as in other organisms. Ultraviolet irradiation (260 nm) induces the formation of thymine dimers (Tang & Asato, 1978); EMS (Delaney & Carr, 1975) and NTG (Asato & Folsome, 1970) act specifically at the DNA replication fork, producing only a transient peak of mutation frequency as the replication fork passes through the gene during synchronized replication.

It is important to distinguish between the direct lethal effects of the mutagen on the cells and the production of toxic compounds in the suspension medium during exposure. Srivastava and Kumar (1969) demonstrated that u.v. irradiation of culture medium produced compounds which inhibited growth when the medium was subsequently inoculated with cells of *Anabaena doliolum*. Clearly, therefore, cells should be irradiated in a suitable buffer, as is standard practice in bacterial genetic studies.

11.4.2 DNA repair and protection phenomena

Bacteria possess a number of mechanisms for the repair of premutational damage in DNA (Hayes, 1968; Witkin, 1969a,b; Bridges, 1972), of which several have been demonstrated in cyanobacteria.

11.4.2.1 Photoreactivation

Photoreactivation of u.v.-irradiated cyanophage LPP-1 occurred only under white or blue (360–500 nm) light, whereas chromosomal DNA of the host (*Plectonema boryanum* UTEX 594) was additionally repaired under red (> 540) or black (300–400 nm) light (Wu *et al.*, 1967). This difference remains unexplained, although it is noteworthy that the phage DNA was repaired *in vitro* by cell extracts in black light. By contrast, Singh (1975a) observed strong photoreactivation of both LPP-1 and host chromosomal DNA in the same strain in black light (355 nm), blue (380–480 nm) and white, but not in red (620–700 nm), green (480–560 nm) or yellow light. *Fischerella muscicola* behaved similarly, whilst *Cylindrospermum* exhibited photoreactivation under all of the above wavelengths. The discrepancy in the effects of black light may be explained by the findings of Singh (1975a) with *Plectonema boryanum* UTEX 594 and phage LPP-1, Buckley and Houghton (1976) with *Gloeocapsa alpicola* CCAP 1430/1 and Amla (1979) with *Anacystis nidulans* UTEX 625 and phage AS-1, who demonstrated that light of this wavelength (more correctly termed near-u.v.) is itself lethal at high doses.

Photoreactivation was also demonstrated in *Agmenellum quadruplicatum* PR-6 (Van Baalen, 1968). The action spectrum in this organism showed a peak at 425–450 nm, no activity in the 500–700 region, and very little in the black region (300–400 nm) (Van Baalen & O'Donnell, 1972). Other organisms studied include *Anacystis nidulans* UTEX 625 (Asato & Folsome, 1969; Asato, 1972; Amla, 1979) and the cyanophage AS-1 (Asato, 1976; Amla, 1979); *Synechocystis aquatilis* 428 (Zhevner & Shestakov, 1972); *Aphanocapsa* 6714 (Astier *et al.*, 1979); *Gloeocapsa alpicola* CCAP 1430/1 (Buckley & Houghton, 1975; Williams *et al.*, 1979); *Anabaena doliolum* (Srivastava *et al.*, 1971); *Anabaena variabilis* (Singh, 1978a); and *Nostoc* PCC 7524 (author unpubl.).

The action of a photoreactivating enzyme in extracts of *Plectonema boryanum* UTEX 594 was demonstrated in the *Haemophilis* transformation assay (Werbin & Rupert, 1968) and the enzyme has been purified from *A. nidulans* UTEX 625 (Saito & Werbin, 1970; Minato & Werbin, 1972). The purified enzyme had an action spectrum resembling that of *A. quadruplicatum* measured *in vivo*, with a sharp peak at 436 nm and no activity above 500 nm, and was additionally shown to cleave pyrimidine dimers both *in vitro* (Saito & Werbin, 1970) and *in vivo* (Tang & Asato, 1978). A photoreactivation system, operating in white but not red (> 560 nm) light also appears to be capable of repairing γ irradiation damage in *A. nidulans* UTEX 625 (Asato, 1971).

Measurement of the photoreactivation of DNA *in vivo* may be complicated by the fact that a separate system may repair lethal damage to the photosynthetic apparatus (Van Baalen, 1968; see also Van Baalen, 1973) and thus data obtained from survival curves may not accurately reflect DNA repair alone.

11.4.2.2 Dark repair

The accumulated evidence suggests that dark repair processes also operate on cyanobacterial DNA *in vivo*. Sensitivity to u.v. irradiation increased (under conditions where photoreactivation was inoperative) following caffeine treatment of *Anabaena doliolum* (Srivastava, 1970; Srivastava *et al.*, 1971), *Anacystis nidulans* UTEX 625 (Asato, 1972), *Plectonema boryanum* UTEX 594 (Singh & Kashyap, 1977) but not in *Aphanocapsa* 6714 (Astier *et al.*, 1979), and following acriflavine treatment of *Anacystis nidulans* (Singh strain) (Singh, 1968), *A. nidulans* UTEX 625 (Singh *et al.*, 1969), *Fischerella muscicola* (Singh & Singh, 1972a) and *Anabaena variabilis* (Singh, 1978a,b). Caffeine and acriflavine prevent excision repair of u.v.-induced damage in other bacteria (Setlow, 1964; Clarke, 1967) and it has been generally assumed (but never demonstrated) that they act similarly in cyanobacteria.

Cell-free extracts of *Anacystis nidulans* 602 contained enzymes, specific for double-stranded DNA, which repaired u.v. and X-ray inactivated transforming DNA of bacteriophage T4 *in vitro*. Lesions induced by X-rays were repaired by an NAD-dependent polynucleotide ligase; u.v.-induced damage was repaired by a DNA polymerase (indicated by dependence on Mg^{2+}, NAD and the four deoxynucleoside triphosphates) and the ligase (Shestakov *et al.*, 1975).

γ Radiation produced single-strand breaks in the DNA of *A. nidulans* UTEX 625 (Dmitriev & Grodzinsky, 1975) which were subsequently repaired. The radiosensitizing agents iodoacetic acid and p-chloromercuribenzoate both increased the number of breaks during irradiation and prevented their repair.

NTG-induced damage of *Agmenellum quadruplicatum* PR-6 was repaired during incubation of the cells in low light (Stevens & Van Baalen, 1969), suggesting the presence of a dark repair system for this presumably different damage, although this repair was not observed in *Plectonema boryanum* UTEX 594 (Singh & Kashyap, 1977). Increased sensitivity to the mutagenic effect of EMS was observed following incubation of mutagenized cells of *Anacystis nidulans* UTEX 625 in caffeine, although the lethal effect was unchanged (Delaney & Carr, 1975). Since photoreactivation did not repair

EMS-induced mutational damage, the repair system was suggested to be a dark repair process which, unlike that operating on u.v.-induced damage, repaired lethal lesions but was error prone.

11.4.2.3 Mutants altered in repair capacity

Mutants which showed increased or decreased sensitivity to u.v. light have been isolated from several cyanobacteria and are listed by Ladha and Kumar (1978a). The u.v.-sensitive mutant of *A. nidulans* (Singh strain) which was insensitive to the post-irradiation effect of acriflavine (Singh, 1968), was presumably defective in the dark (excision) repair system. Ultraviolet-sensitive mutants of *Synechocystis aquatilis* 428 were simultaneously sensitive to alkylating agents and X-rays (Zhevner & Shestakov, 1972) and thus behaved like *hcr* mutants of *Escherichia coli* which are defective in excision repair. A second mutant, showing increased sensitivity to the lethal effect of u.v. irradiation but reduced mutability, possibly resembles *E. coli exr* mutants which lack post-replication repair. Mutants defective in photoreactivation have also been isolated in *Synechococcus aquatilis* 428 (Zhevner & Shestakov, 1972) and *Anacystis nidulans* UTEX 625 (Asato, 1972, 1976).

A mutant of *Gloeocapsa alpicola* CCAP 1430/1 resistant to u.v. showed an increased carotenoid content in comparison to the wild type (Buckley & Houghton, 1976). During irradiation with near-u.v. (310–380 nm) the level of carotenoids in the wild type fell rapidly, while those of the mutant remained unchanged. The resistant strain became sensitive to UV after growth under conditions which resulted in decreased carotenoid content; although chlorophyll levels were similarly reduced, it was concluded that carotenoids serve a protective function against the effects of u.v. irradiation.

11.5 Isolation of mutants

11.5.1 Mutants resistant to antibiotics and other compounds

A summary of cyanobacterial mutants resistant to antibiotics, isolated prior to 1978, is given by Ladha and Kumar (1978a); relevant phenotypes are included in Table 11.2. The majority of these mutants were isolated in *Anacystis nidulans* UTEX 625 and related strains, and most have served as markers in studies of gene transfer (11.6). Clonally pure resistant mutants are easily isolated with spontaneous mutation frequencies in the range 10^{-7}–10^{-9} and mutationally induced frequencies of 10^{-4}–10^{-6} (see Ladha & Kumar, 1978a). The nature of drug resistance has not apparently been examined in any mutant.

Mutants resistant to other toxic compounds have been isolated equally easily in *A. nidulans* UTEX 625 including *fur* (Bazin, 1971) and *fad* (Smith & Lucas, 1971). The latter were altered in acetate thiokinase, resulting in resistance to propionate, and are therefore equivalent to *fad*D mutants of *Escherichia coli*. Mutants (*aro*) resistant to the aromatic amino-acid analogues p-fluorophenylalanine or 2-amino-3-phenylbutanoic acid were of three phenotypes (Phares & Chapman, 1975). The most common exhibited normal levels of DAHP synthetase activity and may be equivalent to the *aro*P or *aro*T permease mutants of *E. coli*. A second class showed low levels, and in the third the enzyme was less sensitive to feedback inhibition by L-tyrosine. By contrast, *aro* mutants of *Agmenellum quadruplicatum* BG-1 resistant to the analogues 4-fluorophenylalanine or β-2-thienylalanine, or to phenylalanine itself, showed increased sensitivity of DAHP synthetase to feedback inhibition by aromatic amino acids (Jensen *et al.*, 1974), indicating that different controls may operate in the aromatic amino-acid biosynthetic pathways in the two organisms; a small proportion of the phenylalanine-resistant mutants appeared to be permease mutants with a normal DAHP synthetase. Tyrosine-resistant mutants (Herdman & Carr, 1972) were presumably of the Aro phenotype.

Regulatory mutants, resistant to amino-acid analogues and excreting tryptophan, phenylalanine, tyrosine, methionine or arginine were isolated from four cyanobacteria (*Anabaena* 29151, *Synechococcus* 602 and Tx20, *Synechocystis* 29108) (Hall *et al.*, 1980). Such mutants did not support a halo of cross-fed wild-type growth on analogue-containing agar but did cross-feed relevant *Bacillus subtilis* auxotrophs, indicating that the excreted amino acid entered the wild type cells too inefficiently to overcome the inhibitory effects of the analogue. However, *Synechocystis* 29108 phenylalanine-excreting mutants, in which DAHP synthetase was no longer sensitive to feedback inhibition by phenylalanine, cross-fed a lawn of *Synechococcus* 602 *phe*A1 auxotrophs on minimal medium (Hall & Jensen, 1980).

Aphanocapsa 6714 mutants *aro*, *fru* and *fad*, and mutants resistant to 3-(3,-4-chlorophenyl)-1,1-dimethylurea (DCMU) (11.5.4) have been described (Astier *et al.*, 1979; Astier *et al.*, 1980).

Hentschel *et al.* (1978) isolated an ethionine-resistant (*met*) mutant of *Plectonema boryanum* UTEX 594 which was simultaneously resistant to DL-norleucine. This mutant appeared to be pleiotropic since the uptake of both methionine and ethionine was depressed by approximately 60% while methionine, but not ethionine, continued to be incorporated into protein. *Nostoc* PCC 7524 *can* mutants were defective in arginine transport but not utilization (author, unpubl.).

Nostoc muscorum G mutants *azi* (Singh & Singh, 1978), *chl* (Singh & Sonie, 1977; Singh, Sonie & Singh, 1977) and mutants resistant to tungsten (Singh, Vaishampayan & Sonie, 1978), have been described and a *chl* mutant of *N.*

calcicola was reported (Singh, Singh & Singh, 1972). Rai (1978) isolated *cml* and *fur* mutants of *Anabaena ambigua*.

Unlike mutants resistant to antibiotics, the above strains were mostly isolated in attempts to explore specific metabolic pathways, and several will be described in the relevant sections below. In using mutants resistant to analogues in studies of this type, care should be taken to exclude the possibility that any observed phenotypic change is caused by alteration in the amino acid transport systems.

Mutants resistant to cyanophage occurred spontaneously or following mutagenic treatment in *Plectonema boryanum* UTEX 594 (Singh *et al.*, 1972; Singh & Kashyap, 1977), *Anacystis nidulans* UTEX 625 (Delaney & Carr, 1975) and *Nostoc muscorum* G (Padhy & Singh, 1978). Another class of resistance mutants, altered in their response to mutagens, has been described in 11.4.2.3.

11.5.2 Temperature-sensitive mutants

Most temperature-sensitive mutants isolated in cyanobacteria have been listed previously (Delaney *et al.*, 1976; Ladha & Kumar, 1978a) and are not considered here. Some recently isolated mutants are described in 11.5.4.

11.5.3 Auxotrophic mutants

The isolation of auxotrophic mutants has been the most difficult single step in the development of cyanobacterial genetics, and therefore current induction and selection techniques are summarized below. Abbreviations are listed in Table 11.2.

NTG has, until recently, been the only mutagen employed in auxotroph induction in cyanobacteria, but it is now clear that different mutagens should be employed in order to recover a broad spectrum of mutants (Herdman *et al.*, 1980).

Nar^- mutants were easily recoverable in *Anacystis nidulans* UTEX 625 (Van Baalen, 1965b; Herdman & Carr, 1972), *A. nidulans* R2 (Shestakov & Jevner, 1968) and *Agmenellum quadruplicatum* PR-6 (Stevens & Van Baalen, 1970) even in the absence of enrichment. Following penicillin enrichment, Nar^- mutants were recovered at frequencies as high as 44% among the survivors (Herdman *et al.*, 1980). The reason for the high frequency of such mutants, in contrast to other phenotypes, is far from clear, but their recovery may be facilitated by their residual growth and yellow colony morphology displayed in the absence of a suitable nitrogen source. By contrast, Nir^- mutants have rarely been isolated in *Agmenellum quadruplicatum* PR-6 (Stevens & Van Baalen, 1970) and *Anacystis nidulans* UTEX 625 (Herdman *et al.*, 1980).

Other phenotypes have been recovered only following the application of penicillin selection techniques and include: Trp$^-$ *Agmenellum quadruplicatum* BG-1 (Ingram *et al.*, 1972); Phe$^-$ *Synechococcus cedrorum* UTEX 1191 (Kaney, 1973); and, from *Anacystis nidulans* UTEX 625, Ace$^-$, Bio$^-$, Cys$^-$, Met$^-$ and Phe$^-$ (Herdman & Carr, 1972); Met$^-$ (Delaney *et al.*, 1973); Phe$^-$ and Arg$^-$ (Orkwiszewski & Kaney, 1974); Leu$^-$ (Singer & Doolittle, 1975); and Pyr$^-$ (uracil-requiring) (Stevens *et al.*, 1975). All of the above mutants were selected on media containing only one, or several, potential growth factors. A chemically defined complete medium suitable for growth of *A. nidulans* UTEX 625 permitted the isolation of three additional phenotypes (Pab$^-$, Pur$^-$, requiring adenine, and Thi$^-$) following NTG mutagenesis and two (Ser$^-$ and Mdh$^-$) following EMS treatment (Herdman *et al.*, 1973). A variation of this medium led to the isolation of a wide range of auxotrophs of *Anabaena variabilis* UTEX 1444 (Currier *et al.*, 1977). Among 80 mutants recovered, 13 displayed the phenotypes Met$^-$, Pyr$^-$ (uracil), Pur$^-$ (adenine), Bio$^-$ or Nic$^-$. Most (65) required a source of inorganic nitrogen, and many of these produced heterocysts with aberrant morphology. A useful technique employed in the study of this filamentous organism was the sonication of filaments to yield approximately unicellular fragments for mutagenesis, penicillin selection and plating (see Wolk & Wojciuch, 1973).

A generalized procedure suitable for the isolation of auxotrophs of *Anacystis nidulans* UTEX 625 has been described (Herdman *et al.*, 1980). Even in the absence of penicillin selection, auxotrophs were recovered relatively easily on complete medium (0.11% of survivors): Nar$^-$, Bio$^-$, Thi$^-$ and Cys$^-$ induced by NTG, Nir$^-$ and Bio$^-$ by EMS and Nir$^-$ by u.v. Penicillin enrichment, using the 2 h pre-enrichment starvation period in minimal medium previously suggested (Herdman & Carr, 1972) increased the frequency of recovery of some phenotypes and permitted the isolation of Ser$^-$ and Mdh$^-$ mutants following EMS mutagenesis. Other phenotypes, not enriched by this procedure, were recovered following longer (16 h) starvation periods. The use of repeated cycles of penicillin enrichment produced a range of auxotrophs including Nir$^-$, Bio$^-$, Thi$^-$, Pab$^-$, Pyr$^-$ (uracil), Pur$^-$ (adenine) and Glt$^-$, at recovery frequencies from 0.1% (Pur$^-$) to 77.7% (Thi$^-$) in the fourth cycle of enrichment. It was demonstrated that different mutants require different pre-enrichment starvation periods in order to cease growth completely; variation in the length of such periods is therefore important if a wide range of mutants is to be obtained. The use of different mutagens, and repeated cycles of penicillin selection, are also to be recommended (Herdman *et al.*, 1980).

A mutant of *Nostoc muscorum* UW deficient in sulphate utilization was isolated by the useful technique of selection for selenate resistance (Singh, 1977, 1978) and resembled the Cys$^-$ mutants of *Anacystis nidulans* UTEX 625

(Herdman & Carr, 1972). These mutants, together with the Pho$^-$ mutants of *A. nidulans* (Vaillancourt *et al.*, 1978), degrade phycocyanin during starvation, indicating that nitrogen deprivation (Allen & Smith, 1969) is not the only trigger for the use of this storage material.

A mutant of *Anabaena* strain CA was capable of growth under N_2-fixing conditions but was inhibited by exogenous nitrate or ammonia in the presence of N_2 (Gotto *et al.*, 1980). Although glutamine synthetase activity appeared normal, the absence of growth on exogenous NH_4^+ suggested an alteration in the enzyme which resulted in progressive inactivation or repression of synthesis by NH_4^+. Presumably, levels of NH_4^+ produced by N_2 fixation were insufficient to effect the enzyme.

With the exception of the strains described above, auxotrophs have been isolated in few cyanobacteria. Ultraviolet-induced mutants requiring glucose for growth were described in *Anabaena cycadeae* and *A. doliolum* (Singh & Singh, 1964a,b). In the latter organism, the only survivor of DES treatment showed, surprisingly, five phenotypic changes: false branching, bulbous cells, heterotrophy, loss of nitrogen fixation and a black–brown colour (Sharma & Tiagi, 1975). A mutant of *Phormidium mucicola* (Srivastava, 1969) was reported to have a glucose requirement which could be satisfied by either casein hydrolysate, acetate or a mixture of vitamins, in view of which it is not surprising that the mutant remains to be characterized. Indeed, few of the isolated auxotrophs have been completely characterized; where possible, their known phenotypes (enzymes affected) are given in Table 11.2. Mutants defective in nitrogen fixation are considered in section 11.5.5.2 (b).

11.5.4 Mutants defective in photosynthetic pigments or activity

Colonies displaying a yellow phenotype are commonly recovered in *Agmenellum quadruplicatum* PR-6 (Stevens & Van Baalen, 1969, 1970), *Anacystis nidulans* UTEX 625 (Van Baalen, 1965b; Herdman & Carr, 1972), *Nostoc* PCC 7524 (author, unpubl.) and other strains (see Van Baalen, 1973). Care must be taken in interpreting these mutants, since they are commonly defective in nitrogen assimilation (e.g. Nar$^-$) and consequently degrade phycocyanin (PC) unless a suitable source of nitrogen (e.g. NH_4^+ or glutamine) is provided to overcome the mutational block.

Mutants altered in their contents of photosynthetic pigments have been isolated in *Anacystis nidulans* UTEX 625 and Tx20 (Van Baalen, 1965b; Asato & Folsome, 1969; Asato, 1971; Stevens & Myers, 1976; Delaney & Carr, 1975; Wang *et al.*, 1977), in an unspecified strain of *A. nidulans* (Shestakov & Jevner, 1968) and in *Agmenellum quadruplicatum* PR-6 (Stevens & Van Baalen, 1970). Yellow mutants usually contained reduced levels of PC (Asato & Folsome, 1969; Stevens & Myers, 1976), while blue mutants exhibited increased PC and decreased chlorophyll (Chl) contents (Stevens & Myers, 1976). Loci giving rise to a yellow phenotype following EMS mutagenesis mapped at three distinct sites, and those giving blue phenotypes mapped at two sites, on the genome (Delaney & Carr, 1975). All of these sites were distinct from those involved in the early stages of nitrogen assimilation (Nar^-, Nir^-).

Studies of two NTG-induced pigment mutants of *Synechocystis* PCC 6701 (Williams *et al.*, 1980) and one of *Synechococcus* 6301 (Yamanaka *et al.*, 1980) correlated their reduced phycobiliprotein content (phycoerythrin (PE) and PC in the former, PC in the latter) with altered phycobilisome structure. The demonstration of the simultaneous absence of non-pigmented phycobilisome polypeptides (33.5 Kd in the mutants of PCC 6701, 33 Kd and 30 Kd in the mutant of PCC 6301) illustrates the value of a thorough approach to the examination of cyanobacterial pigment mutants.

Temperature-sensitive mutants of *Synechococcus cedrorum* UTEX 1191 selected by their high fluorescence at 40°C (Sherman & Cunningham, 1977) showed, at the restrictive temperature, abnormal fluorescence kinetics, slow growth rates and decreased oxygen evolution, and were suggested to be defective in photosynthetic membrane function. Subsequently, three classes of temperature-sensitive high fluorescence mutants were described (Sherman & Cunningham, 1979) on the basis of their fluorescence-induction kinetics. All of the mutants showed altered photosynthetic membrane protein patterns when grown at the higher temperature.

Another interesting approach is the finding by Myers *et al.* (1980) that continued culture of *A. nidulans* Tx20 over many generations in far-red (> 650 nm) illumination led to the spontaneous selection of mutants with altered pigment composition. Under these conditions the wild type grew slowly as a result of inefficient light harvesting. Stepwise and concomitant increases in growth rate and Chl:PC ratio were observed; after such transitions, mutant colonies of different colours were identified on agar. Of six such mutants examined, all grew faster than the wild type in far-red illumination. Two showed reduced PC and increased allophycocyanin (APC) (which absorbs at 650 nm, see Chapter 7); others showed increased Chl and reduced PC contents. There would thus appear to be two selective pressures applied by this procedure: one causing loss of PC (which, due to its inefficiency of light harvesting in the far-red, becomes redundant), the second resulting in an increase in pigments which do absorb efficiently in this region. *Agmenellum quadruplicatum* PR-6, which was able to grow well in far-red illumination, produced no mutants under these conditions.

An alternative approach to the isolation of mutants altered in thylakoid function was adopted by Astier *et al.* (1980) in *Aphanocapsa* 6714. Of two mutants resistant to DCMU, an inhibitor of electron transport on the acceptor side of photosystem (PS) II, one probably showed decreased cellular permeability to the compound, the isolated thylakoids remaining DCMU sensitive; in the second, the isolated thylakoids retained DCMU resistance, suggesting an alteration in their structure. Further analysis (Astier & Joset-Espardellier, 1981) demonstrated that the thylakoids of the second mutant contained the same proteins as the wild type when grown in the absence of DCMU, but differed in that a 33 Kd protein had disappeared from the isolated membranes when the mutant cells were grown in the presence of the inhibitor. The absence of the protein conferred DCMU resistance and a decrease in the rate of electron transfer from the primary (Q) to the secondary (B) acceptors of PS II; however, the growth rate of the cells was not impaired. The DCMU-binding protein could be degraded in the isolated thylakoids of the wild type by treatment with trypsin, again conferring DCMU resistance *in vitro*.

Heterotrophic growth of *Anabaena variabilis* UTEX 1444 was exploited to produce six mutants defective in photoautotrophy (Shaffer *et al.*, 1978). Four were impaired in CO_2 fixation, three showing low PS II activities as a consequence of PC deficiency. The others exhibited enhanced sensitivity to light, for unknown reasons.

A mutant of *Plectonema boryanum* UTEX 594 resistant to photooxidation showed low levels of Chl and PC (Steinitz *et al.*, 1979). Water-soluble pigments, absorbing in the region 459–490 nm, were present in the mutant but not the wild type, and a dramatic shift in the ratio of the two superoxide dismutase enzymes occurred.

A unique feature of some cyanobacteria is their ability to undergo chromatic adaptation (see Chapter 7). Mutants defective in this process have been isolated from *Calothrix* PCC 7601 (R. Rippka, S. Beguin & G. Cohen-Bazire, pers. comm.) and included strains (1) defective in regulation (constitutive for either PC_2 or PE), (2) producing extremely low levels of all phycobiliproteins under all light conditions, but nevertheless being capable of chromatic adaptation, and (3) producing low levels of PE and PC but overproducing APC. Furthermore, mutants isolated from a *Pseudanabaena* sp. (PCC 7409) were either (1) constitutive for PE but unable to induce PC_2 synthesis, or (2) constitutive for PC_2 but unable to induce PE. Pseudorevertants from class (1) mutants resembled class (2), and vice versa. These results strongly suggest a common controlling gene in chromatic adaptation and are in marked contrast to the hypothesis of Rohatgi and Singh (1979) (based on the isolation of a single class of mutant overproducing PC with reduced levels of PE) who concluded that no common control for phycobiliprotein synthesis exists in *Aphanothece stagnina*.

11.5.5 Mutants altered in cellular differentiation

11.5.5.1 Unicellular cyanobacteria

Two common classes of filamentous (*fil*) mutants are easily recovered from unicellular strains. First characterized by Kunisawa and Cohen-Bazire (1970) in *Anacystis nidulans* UTEX 625, class I mutants underwent unequal binary fission to produce cells varying from one to ten times normal length; class II mutants rarely divided, giving coenocytic filaments of cells up to 100-fold normal length. Similar mutants were isolated in the same organism (Asato & Folsome, 1969; Gupta & Kumar, 1970; Herdman & Carr 1972) and in *A. nidulans* 602 (Shestakov & Khyen, 1970; Zhevner *et al.*, 1973). In UTEX 625 the two mutations mapped at different sites on the genome (see Delaney & Carr, 1975).

Septate and serpentine mutants of *Agmenellum quadruplicatum* BG-1 have been extensively characterized (Ingram & Thurston, 1970; Ingram & Van Baalen, 1970); the former were impaired in the final stage of septum formation, the latter divided at low frequency and resembled the *Anacystis nidulans* class II mutants.

In a temperature-sensitive mutant of *Agmenellum quadruplicatum* BG-1 the filamentous phenotype was expressed only at the permissive temperature (Ingram & Fisher, 1973a); in both this and a mutant which divided normally only in dense cultures (Ingram & Fisher, 1973b) wild-type morphology was restored by the addition of cell-free medium from the wild type, suggesting the production of a diffusible effector of cell division (Ingram & Fisher, 1973a,b). A second temperature-sensitive mutant produced long filaments under normal growth conditions at 39°C (resulting from infrequent cell division during rapid growth) whereas nutrient limitation caused rapid cell division and the formation of abnormally small cells (Ingram *et al.*, 1975).

11.5.5.2 Filamentous cyanobacteria

11.5.5.2 (a) General

Padan and Shilo (1969) first described a short trichome mutant of the non-heterocyst-forming *Plectonema boryanum* UTEX 594. For unknown reasons, a short trichome morphology often accompanies the mutational loss of heterocysts in *Anabaena* and *Nostoc* strains (author unpubl.; see also Stewart & Singh, 1975; Padhy & Singh, 1978). Conversely, a long trichome Het^- mutant of *Cylindrospermum* sp. has been described (Singh, 1976a).

Induced mutants of *Nostoc linckia* (Singh & Tiwari, 1969) and *Anabaena doliolum* (Dhar, 1979) were reported to have developed the ability to form true branches. Altered filament morphology was described in mutants of *A. ambigua* (Kale & Talpasayi, 1969), *A. doliolum* (Sinha & Kumar, 1973) and *Nostoc linckia* (Tiwari, 1978a).

Mutants lacking gas vacuoles arose spontaneously in *Anabaena flos-aquae* (Walsby, 1977) and *Anabaenopsis raciborskii* (Das & Singh, 1976). Such mutations arise at high frequency both in the unicellular *Microcystis aeruginosa* and in *Aphanizomenon flos-aquae* (PCC 7905) (R. Rippka, pers. comm.). Surprisingly, a mutant strain containing gas vacuoles was reported by Singh (1978c) in the normally non-gas vacuolate *Anabaena variabilis*.

11.5.5.2 (b) Heterocysts and nitrogen fixation

Many mutants of these types are described in Chapter 14. Het$^-$ mutants, usually showing not even the early stages of differentiation, arise spontaneously in cultures maintained in the presence of a source of combined nitrogen (Rippka *et al.*, 1979). The mutant outgrows the wild type after several transfers, becoming the dominant member of the population. This process occurs rapidly in nitrate-grown cultures of *Nostoc* PCC 7524 (author, unpubl.), an organism which produces heterocysts in such media (Sutherland *et al.*, 1979) for an obvious reason: an increased growth rate results when the filaments are composed only of vegetative cells all of which (unlike heterocysts) are capable of cell division. Het$^-$ mutants appear to arise equally easily, however, under the same growth conditions in organisms in which heterocyst differentiation is repressed in the presence of nitrate; the growth advantage of such mutants is unknown. The widely employed *Anabaena variabilis* (Myers strain) appears to be a spontaneous mutant of this type, since Het$^+$ revertants have been isolated (Rippka & Stanier, 1978).

The competitive advantage of a Het$^-$ mutant of *Aphanizomenon flos-aquae* in the presence of nitrate was demonstrated in light-limited continuous culture, where a sudden replacement of the heterocystous wild type by the spontaneously-arising mutant was observed (Zevenboom *et al.*, 1981). The isolated mutant exhibited higher growth rates than the wild type at all light irradiance values tested. To achieve the same growth rate, the wild type required specific light uptake rates varying from 1.3 to 1.9-fold greater (depending on light irradiance) than the mutant. It is probable that part of this extra energy demand is required for the processes of heterocyst differentiation and nitrogen fixation; it may be suggested that the wild type should also require additional energy to maintain a fixed growth rate since, unlike the mutant, only 91% of the cellular population (the vegetative cells) are capable of division. In comparing the energy demand of the wild type and Het$^-$ mutant, it would be appropriate to exclude the possibility that the mutant carries a second advantageous mutation which increases the efficiency of utilization of light energy (which could easily arise in strains recently isolated from nature and subsequently exposed continuously to laboratory culture conditions). This could be examined in continuous cultures of the strains grown in the presence of NH_4^+, where both should show equal efficiencies (unless other changes had indeed occurred) under conditions where heterocyst differentiation and nitrogenase activity are repressed.

Most cyanobacterial Het⁻ mutants have been claimed to be simultaneously Nif⁻, supporting suggestions that the heterocyst is the only (or major) site of nitrogen fixation. Unfortunately, the Nif⁻ phenotype has frequently been identified from the inability of the mutants to reduce acetylene, or to grow at the expense of atmospheric nitrogen, under aerobic conditions (Singh, Singh & Singh, 1972; Sinha & Kumar, 1973; Singh, 1976a,b; Singh & Dikshit, 1976; Singh, 1978c; Zevenboom *et al.*, 1981).

Since many mutants defective in heterocyst differentiation are incapable of aerobic nitrogen fixation but can fix nitrogen anaerobically or under carefully defined microaerobic conditions (Rippka & Stanier, 1978; Haury & Wolk, 1978; Gotto *et al.*, 1979b; Grillo *et al.*, 1979; Rippka *et al.*, 1979) they must be defective in the heterocyst oxygen protection mechanism for nitrogenase or in other functions required for heterocyst activity under aerobic conditions. It is therefore inappropriate to term them Nif⁻. It is hoped that a standard terminology, Het⁻, is adopted for all mutants incapable of nitrogen fixation under aerobic conditions but which do fix under anaerobic conditions, the phenotype Nif⁻ being reserved for mutants which are additionally incapable of nitrogen fixation even under anaerobic conditions. Het⁻ therefore applies not only to mutants incapable of producing heterocysts, but also to those defective in their function.

The establishment of strict anaerobic conditions for the nitrogenase assay is therefore fundamentally important and must replace aerobic growth experiments. Carefully controlled microaerobic conditions achieved by flushing the cultures at a specified rate with N_2:CO_2 or Ar:CO_2 may be satisfactory, and indeed have demonstrated the Nif⁺ nature of mutants of *Anabaena* CA which were incapable of nitrogen fixation under aerobic conditions (Gotto *et al.*, 1979b; Grillo *et al.*, 1979). Undefined microaerobic conditions (Stewart & Singh, 1975; Ladha & Kumar, 1977a,b; Singh, Ladha & Kumar, 1977; Padhy & Singh, 1978) may have been inadequate. True anaerobic nitrogen fixation can be assayed easily in heterotrophic strains by the methods of Ladha and Kumar (1976) and Rippka and Waterbury (1977); for obligate autotrophs, the second method of Rippka and Waterbury (1977) would seem appropriate. These techniques have identified Nif⁻ mutants (in both Het⁺ and Het⁻ strains) of *Nostoc linckia* (Ladha & Kumar, 1976), and the Het⁻ Nif⁺ nature of both *Anabaena variabilis* (Myers strain) (Rippka & Stanier, 1978) and of spontaneous mutants of *Nostoc* PCC 7524 (Rippka & Herdman, unpubl.).

Incomplete characterization of supposed Nif⁻ mutants has led to suggestions that *het* and *nif* genes share a common genetic determinant (a regulatory gene). An elaborate model (the heterocystulon) was proposed (Singh, Singh & Singh, 1972) on the basis of the occurrence of Het^- Nif^- mutants (determined aerobically) and it was further suggested that the heterocysts were *not* the sites of nitrogen fixation. The latter conclusion was also reached by Sinha and Kumar (1973) from an aerobic study of supposed Het^+ Nif^- mutants which, in reality, may have been Het^- Nif^+.

Singh, Ladha and Kumar (1977) proposed a common genetic determinant of which mutation produced Het^- Nif^- mutants of *Nostoc muscorum* G and *N. linckia*. The reversion frequency to wild type was consistent with a single mutation. It was concluded that 'our Het^- Nif^- class of mutants, therefore, cannot be regarded as mutants in one of the *het* genes as they fail to show microaerobic nitrogenase activity'. Although this phenotype may indeed have arisen by mutation in a common regulatory gene, the undefined conditions employed for nitrogenase assay do not exclude the possibility that the mutants were truly Het^- Nif^+. Singh and Singh (1978) described similar results which were interpreted as showing close genetic linkage of *het* and *nif* genes, again based on undefined conditions (see also 11.6.1).

Making the assumption that *nif* and *his* genes are closely linked as in other nitrogen-fixing bacteria, and that 'segregation of *nif* might concomitantly affect the nearby loci including *his*', Yadava and Singh (1979) examined the sensitivity to histidine of colonies of *N. muscorum* UW which grew slowly on atmospheric nitrogen. A histidine-sensitive strain grew on sources of combined nitrogen in the presence or absence of histidine, but on N_2 only in the absence of histidine. An extensive model proposed 'exchange between parts of *nif* and *his* genes resulting in configurations that would show regulatory response of the two sets of genes to each other's signal . . . . or of both sets of genes to the signal of histidine biosynthesis only depending upon the fusion of the operator regions with the structural parts in different configurations'. It was further suggested that a *nif* plasmid was involved. In view of the positions of the *nif* genes in *Anabaena* PCC 7120, which are dispersed over a large part of the genome (Mazur *et al.*, 1980; see 11.8), and the close relatedness of PCC 7120 with *Nostoc muscorum* UW (*Anabaena* PCC 7119) (Rippka *et al.*, 1979), it seems unlikely that models based on the *Klebsiella* linked *nif–his* gene arrangement will prove to be valid. In addition, using a powerful *nif* DNA probe in hydridization experiments, Mazur *et al.* (1980) found no evidence to support the existence of *nif* plasmids.

A useful method for obtaining a specific class of Nif^- mutants is the selection for chlorate resistance (Singh & Sonie, 1977). Some Chl mutants were Nar^-, as expected from repeated observation in other bacteria; a second class was Nif^- and a third simultaneously Nar^- Nif^-. This suggested a component common to both pathways (Singh & Sonie, 1977) which, in view of the *chl*D (*nar*D) mutations in other bacteria, should be the molybdenum-processing system for production of the active Mo cofactor. This interpretation is supported by the requirement of *Nostoc muscorum* G for Mo when grown with atmospheric nitrogen or nitrate, but not in other nitrogen sources (Singh, Vaishampayan & Singh, 1978; Singh, Vaishampayan & Sonie, 1978). The latter authors also described a mutant resistant to tungstate (a Mo antagonist) which showed absolute requirement for tungstate for growth on N_2 or nitrate.

Singh and Dikshit (1976) isolated Nar^- and Nir^- mutants of *Anabaena doliolum*, identified by their inability to grow on nitrate or nitrite, respectively. Since most (four out of seven) were simultaneously Nif^- (so classified by their failure to grow on N_2 under aerobic conditions) it was suggested that a gene common to both nitrogenase and nitrate assimilation was involved. While this may be true (see above) the selection procedure employed (failure to grow on NO_3^- or NO_2^- under a normal atmosphere) should only produce double mutants: Het^- Nar^- or Het^- Nir^-. $Het^+(Nif^+)$ strains would grow at the expense of N_2 and would not be identified as Nar^- or Nir^-. Growth on NO_3^- or NO_2^- under $Ar:CO_2$, followed by screening under N_2 alone, would be desirable to isolate mutants of these phenotypes.

A potentially useful approach to the genetic study of heterocyst differentiation and function arises from the fact that azide is an alternative substrate for nitrogenase (Schollhorn & Burris, 1967). Mutants which constitutively produce nitrogenase should detoxify this compound and thus, among *azi* mutants some should be *nif* constitutive and (if selected under aerobic conditions) constitutive heterocyst producers. A mutant of this class was described in *Nostoc muscorum* G (Singh & Singh, 1978) although a similar concentration of azide stimulated the growth of *Cylindrospermum majus* (Sharma & Kumar, 1973). Azide is a highly efficient mutagen in bacteria (Kleinhofs & Smith, 1976), and thus results based on a single isolate should be interpreted with care.

Mutants of *Anabaena* CA, selected as resistant to methionine sulphoximine or 7-azatryptophan, were defective in the oxygen-protection mechanism of nitrogenase in the heterocysts (Gotto *et al.*, 1979b). Although the relationship between oxygen sensitivity and resistance to the inhibitors is not yet understood, this approach is clearly useful and may indicate that, in addition to the role of heterocyst wall glycolipids (determined in oxygen-sensitive mutants of *A. variabilis* UTEX 1444 by Haury and Wolk, 1978), other protection mechanisms may exist. Such mechanisms presumably operate in the unicellular *Gloeothece* strains which have the unique ability of aerobic nitrogen fixation (see Rippka *et al.*, 1979). The isolation of a sheathless mutant of *Gloeothece* PCC 6909, which was still capable of aerobic nitrogen fixation (R. Rippka, pers. comm.), shows that the sheath plays no role in oxygen protection. Further study of this organism, together with oxygen-sensitive mutants of heterocystous cyanobacteria, should elucidate the alternative protection mechanism.

11.5.5.2 (c) Akinetes

Mutants defective in akinete differentiation arose spontaneously in *Cylindrospermum majus* (Singh & Sinha, 1965) and in *Anabaena doliolum* following u.v. (Singh, 1967) or NTG (Dhar, 1979) treatment (Chapter 15). The simultaneous mutational loss of the ability to produce both heterocysts and akinetes in *Cylindrospermum* sp. (Singh, Singh & Singh, 1972), *Wollea bharadwajae* (Singh, 1976b) and *Anabaena variabilis* (Singh, 1978c) suggests a common component in these differentiated cells. This probably lies in the polysaccharide envelopes, which show extensive similarities in the organisms studied (Chapters 14 and 15). The simultaneous reappearance of akinetes and heterocysts in revertants of a Het$^-$ mutant of *Wollea bharadwajae* (Singh, 1976b), and the altered akinete envelope structure in a Het$^-$ mutant of *Nostoc* PCC 7524 (Sutherland *et al.*, 1980) support such a common component.

As expected from the extensive differences between heterocysts and akinetes, however, their differentiation processes can mutate independently. Het$^+$ Aki$^-$ mutants have been described in *Anabaena doliolum* (Singh & Kumar, 1971; Singh & Dikshit, 1976; Dhar, 1979), *A. variabilis* (Singh, 1978c), *Nostoc linckia* (Singh, Tiwari & Singh, 1972) and *Gloeotrichia ghosei* (Singh, Tiwari & Singh, 1972). Conversely, Het$^-$ mutants of *Anabaena doliolum* (Singh & Kumar, 1971; Sinha & Kumar, 1973), *A. variabilis* (Singh, 1978c), *Cylindrospermum* sp. (Singh, Singh & Singh, 1972; Singh, 1976a) and *Nostoc* PCC 7524 (Sutherland *et al.*, 1980) are often capable of producing akinetes.

Aki$^-$ mutants of *Nostoc* PCC 7524 fell into several phenotypic classes, varying from complete loss of cellular differentiation (inability to accumulate cyanophycin, absence of the thick cell wall and multilayered envelope) to mutants defective in formation of one or more envelope layers (author, unpubl.).

11.6 Gene transfer

11.6.1 Studies in mixed culture

Early studies (reviewed by Van Baalen, 1973; Delaney *et al.*, 1976; Ladha & Kumar, 1978a) of gene transfer in cyanobacteria were performed exclusively in liquid medium and, although suggestive of genetic recombination, no statistically validated results were supplied.

The occurrence of genetic transfer in *Anacystis nidulans* was suggested by the appearance of double *str amp* colonies following mixed culture of the two singly resistant parental strains (Singh *et al.*, 1966). All (222) of the supposed recombinants, however, arose on only one of the twelve selective plates employed, and this report requires further confirmation.

The first convincing demonstration of genetic transfer was that of Bazin (1968); mixed cultures of *A. nidulans* UTEX 625 *pmr* and *str* mutants gave rise to the double resistant phenotype, enumerated on agar plates, at frequencies significantly greater than the spontaneous mutation frequency.

The transfer of a functional *nif*$^+$ gene from the wild type *Nostoc muscorum* UW to a *nif str* recipient was suggested following mixed incubation of the two strains under conditions where the wild type was rapidly killed (Stewart & Singh, 1975). The selected recombinant genotype (*nif*$^+$ *str*) apparently arose at frequencies 400-fold greater than the spontaneous mutation rates. It seems probable, however, that the recombination frequency was overestimated; of the two recombinants studied, one grew slowly in the absence of combined nitrogen (but reduced acetylene after a lag of 2 days) while the second showed virtually no growth and only minimal levels of acetylene reduction. Other recombinants were described as being intermediate between these two extremes. Furthermore, the *nif* mutant employed was also *het* and, since all *nif*$^+$ recombinants were simultaneously *het*$^+$, it was suggested that there is close genetic linkage between these genes. Since the conditions of nitrogenase assay were not defined, this suggestion, although plausible, was not substantiated; the *het* mutant may have been *nif*$^+$ (see 11.5.5.2 (b)). Additionally, the transfer of *str* from mutant to wild type survivors could have occurred, producing Nif$^+$ Het$^+$ colonies resistant to streptomycin in the absence of *nif het* linkage.

Padhy and Singh (1978) described genetic transfer in the same strain. Although the Het$^-$ Nif$^-$ mutants, characterized by their inability to grow in the absence of combined nitrogen under 'microaerobic' conditions (achieved by overlaying the colonies with agar) were again possibly Het$^-$ Nif$^+$, recombination frequencies approximately 100-fold greater than spontaneous mutation frequencies were observed. Resistance to streptomycin and to cyanophage N-1 were other genetic markers transferred in addition to the *het*$^+$ gene. The suggestion that the Het$^+$ strains were donors, and Het$^-$ strains were recipients, may be invalid; transfer could have occurred in the reverse, or even both, directions. It was again suggested that there is close linkage between *het* and *nif* genes, or that there is a common regulatory gene, because both characters were apparently always transferred simultaneously. However, since the degree of microaerobic conditions in the two layers of agar was stated to be unknown, this suggestion remained unsubstantiated because the true phenotype of the mutants was not elucidated.

Therefore, despite suggestions of *het–nif* linkage, further work must clearly be performed under conditions where Nif$^-$ mutants are unambiguously identified (see 11.5.5.2 (b)). Genetic linkage of these genes may be considered unlikely, since the *nif* genes themselves are dispersed on the genome of *Anabaena* PCC 7120 (Mazur *et al.*, 1980; see 11.8).

Although genetic exchange in mixed cultures may occur by a variety of methods, only one attempt has been made to elucidate the nature of the transfer process. Recombination in mixed cultures of *A. nidulans* UTEX 625 continued when the parental strains were grown in the classical U-tube of Davis (1950) in which the strains were separated by a filter which was impermeable to the cells but allowed free passage of the culture medium (Herdman, 1973a,b). Gene transfer was the result of transformation (see 11.6.3) by nucleic acids released into the medium.

11.6.2 Transduction

Evidence of a phage-mediated transfer of streptomycin resistance from a mutant of *Plectonema boryanum* to the wild type was suggested by Singh and Singh (1972b). The low frequencies of transfer from (2×10^{-7} to 4×10^{-10}) remain to be confirmed. Transduction has not been reported in any other cyanobacterium, despite the wide range of potentially useful virulent and temperate cyanophages available; this probably reflects a lack of suitable genetic markers in the host strains.

11.6.3 Transformation

Transformation, mediated by DNA extracted from the donor cells, was first conclusively demonstrated in *Anacystis nidulans* 602 (Shestakov & Khyen, 1970). Frequencies of transfer of *str*, *ery* and *fil* markers (0.004–0.07%) were significantly greater than the spontaneous mutation frequency of the recipients and the process was completely inhibited by DNase. Uptake appeared to be light-dependent and transformants started to appear after incubation with DNA for 30 min, reaching a maximum after 5 h. The system was exploited to show that the phenotype of a filamentous radiosensitive mutant arose from two independent mutations, separable by recombination (Mitronova *et al.*, 1973). Herdman (1973a,b) demonstrated the transfer of *bio*$^+$, *nar*$^+$, *met*$^+$, *cys*$^+$ and *str* genes in *A. nidulans* UTEX 625 by transformation at frequencies of up to 0.01% and confirmed the sensitivity of the process to inactivation by DNase. In contrast, transformation was insensitive to RNase. In the same organism, Orkwiszewski and Kaney (1974) obtained transfer of *str*, *phe*$^+$ and *arg*$^+$ markers at frequencies as high as 0.1%, and showed a peak of transformation during the transition of the recipient culture from the exponential to the post-exponential phase of growth. The optimum exposure to DNA at 20–40 μg ml^{-1} was 4 h at pH 8.0; other workers have used different exposure times and DNA concentrations.

Transformation in *Aphanocapsa* 6714, involving transfer of parafluorophenylalanine resistance at frequencies of 1×10^{-5} to 1×10^{-6}, also showed a peak during the growth phase transition (Astier & Espardellier, 1976); importantly, these authors described a modified lysis method for this organism. Devilly & Houghton (1977) transferred *str* in *Gloeocapsa alpicola* CCAP 1430/1 at a frequency of 0.3%, using a modified procedure involving treatment with $CaCl_2$. Intergeneric transfer of the marker to *Anacystis nidulans* CCAP 1405/1 occurred at equally high frequencies whilst *rif* resistance was transferred at only low (7.5×10^{-7}) frequency. Stevens and Porter (1980) observed transfer of *str* and *nir*$^{+}$ markers in *Agmenellum quadruplicatum* PR-6 at frequencies of 5×10^{-4} to 4×10^{-3}, the frequency further increasing ten-fold following $CaCl_2$ treatment. Unlike the experiments with *Anacystis nidulans* and *Aphanocapsa* 6714 described above, a peak of competence was not observed at a specific point in the growth of the recipient culture. In short-term experiments, light was not essential for DNA uptake.

The genetic exchange system in *Anacystis nidulans* UTEX 625, which operated in the absence of cell–cell contact, was the result of accumulation of nucleic acid molecules in the culture medium. When cell-free culture supernatants of any mutant strain were mixed with cells of a second strain, recombinants arose at frequencies up to 0.1% (Herdman 1973a,b). Reciprocal crosses gave gene exchange, indicating the absence of a unidirectional donor–recipient relationship. This process was not transduction, since it occurred following deproteinization of the supernatants. Gene transfer was incompletely sensitive to DNase (maximum 60% inhibition) and its similar sensitivity to RNase suggested the involvement of a DNA–RNA complex. The nucleic acids were present in culture medium in quantities in the order of 2 ng DNA and 15 ng RNA per ml (Herdman & Carr, 1971a) and retained transforming activity following concentration and purification. The accumulation of nucleic acids in bacterial cultures is quite common (Mauzy *et al.*, 1955; Catlin, 1956, 1960; Catlin & Cunningham, 1958; Ottolenghi & Hotchkiss, 1960; Takahashi, 1962) and they are often active in transformation. The presence of RNA, necessary for efficient transformation, is unusual; RNA appears to protect the DNA, giving higher linkage values than transformation with chemically extracted DNA (Herdman, 1973a,b). A similar phenomenon was observed by Devilly and Houghton (1977) using nucleic acid preparations of *Gloeocapsa alpicola* CCAP 1430/1 in which the RNase step had been omitted during DNA extraction. Transformation by the mixed nucleic acid preparation was inhibited 84% by DNase treatment and 58% by RNase, again suggesting a function of RNA in efficient transformation.

Transformation of *Anacystis nidulans* R-2 with the *Escherichia coli* plasmid pR146 (containing the transposon Tn901, carrying a β-lactamase gene for ampicillin resistance) gave rise to resistant mutants at low frequency (10^{-8}–10^{-9} per μg of plasmid DNA); the same mutants were obtained at high frequency (10^{-4}–10^{-5} per μg of plasmid DNA) following transformation with a recombinant cyanobacterial plasmid carrying Tn901 (van den Hondel *et al.*, 1980; see 11.8). Such demonstrations of transformation with plasmid DNA have resulted from the introduction of genetic markers into cyanobacterial plasmids, and other examples are given in 11.8. Most plasmids, though probably transformable, do not carry suitable markers (11.3.2).

11.7 Genetic mapping

The two transformation systems of *Anacystis nidulans* UTEX 625 have been exploited in the construction of a limited genetic linkage map (Herdman, 1973b). Transformation with chemically extracted DNA established close linkage of only two markers (*bio*-1 and *str*-2). Extracellular nucleic acids permitted more extensive mapping, owing to the increased linkage values; markers which were unlinked in the former system showed high (minimum 87%) cotransfer indices in the second system. A similar situation is found in transformation of *Bacillus subtilis* with extracellular DNA (Ephrati-Elizur, 1968). Accurate map distances were not determined because recombination caused, at high frequency, mutation in genes which were wild type in both parental strains; the consequent loss of many undetectable recombinants prevented calculation of linkage values. The analysis of the reassortment of outside markers showed that mutation occurred only at the site of recombination, and the inability to separate mutation from recombination by a variety of methods suggested that some step of the recombination process was error-prone (Herdman, 1973a,b). The mutations induced behaved as transversions on the basis of their reversion patterns with different mutagens (Delaney *et al.*, 1976).

Following selection for induced streptomycin resistance, mutation associated with recombination was not observed in the same organism by Orkwiszewski and Kaney (1974) or in *Agmenellum quadruplicatum* PR-6 (Stevens & Porter, 1980), although the latter authors described additional colony colour phenotypes amongst the transformants. Mutagenesis may therefore depend on incubation conditions although, since the mutagenic event occurs specifically at the site of recombination (Herdman, 1973a) and must therefore be closely linked to the gene transferred, selection for a single mutant phenotype (Str) may be insufficient to detect the process.

A second system of genetic mapping, which measures the position of a gene relative to the chromosomal replication origin from its mutation frequency during synchronized genome replication (temporal mapping) has also been exploited in *Anacystis nidulans* UTEX 625 (Asato & Folsome, 1970; Delaney & Carr, 1975). The pulsed application of NTG (Asato & Folsome, 1970) or EMS (Delaney & Carr, 1975) gave sharp peaks of mutation frequency (corresponding to the time of replication of the gene) for 6 and 19 markers respectively, measured as reversion of auxotrophs or forward mutation to drug resistance or pigment mutation. EMS, used for the first time in this bacterial mapping method (Delaney & Carr, 1975), is particularly useful because it is rapidly inactivated by sodium thiosulphate and therefore the exposure time is easily controllable.

The two maps occupied markedly different time scales (420 min and 60 min, respectively) due to different chromosome replication rates. When drawn on the same scale, however, good agreement was obtained for the positions of three markers common to both: *pmr*A, *blu*A and *yel*A (see Delaney & Carr, 1975). Map positions of *sna* and *fil* did not correspond, although both are involved in cell division (corresponding, respectively, to the class I and class II mutants of Kunisawa and Cohen-Bazire, 1970; see 11.5.5.1), indicating that these genes are not linked. The markers *met*-2 and *met*-4 were similarly unlinked, as were *pmr*A and *pmr*B (Delaney & Carr, 1975). Yellow and blue mutants affected in phycocyanin synthesis mapped at at least four different and well-separated positions (Delaney & Carr, 1975), perhaps indicating the absence of the clusters of related genes which occur in other bacteria. Only *bio*-1 and *bio*-2 appeared to be clustered, although the possibility that these mutations both lay in the same gene cannot be excluded. A suppressor of *bio* (*sup*A) mapped at a separate site (Delaney & Carr, 1975).

Experiments in which ^{32}P-labelled rRNA from the same organism was hybridized with restriction enzyme digestion products of total DNA demonstrated that the rRNA genes were arranged in two clusters, 16S, 23S and 5S genes being linked within each cluster as in *Escherichia coli* (W.F. Doolittle, pers. comm.). By contrast, the *nif* structural genes of *Anabaena* PCC 7120 were dispersed over a relatively large part of the genome (Mazur *et al.*, 1980; see 11.8), thus differing from the *Klebsiella nif* gene arrangement. With the above exceptions, little is known about the linkage relationships of related genes.

11.8 New approaches

The necessary prerequisite for gene cloning, the recognition and isolation of specific DNA fragments, has been achieved in several cases. Following the discovery that at least some of the *nif* genes of *Anabaena* PCC 7120, *A. variabilis* UTEX 1444 and *Plectonema boryanum* UTEX 594 are homologous with those of *Klebsiella pneumoniae* (Mazur *et al.*, 1980; Ruvkun & Ausubel, 1980), cloned fragments of *Klebsiella pneumoniae nif* DNA were used as probes in DNA–DNA hybridizations with restriction endonuclease fragments of *Anabaena* PCC 7120 DNA, permitting the identification of the *Anabaena nif* structural genes (Mazur *et al.*, 1980). Despite the extensive sequence homology between the structural genes *nif* K, D and H of *Anabaena* 7120 and *Klebsiella*, the gene order was substantially different and the *Anabaena nif* structural genes were dispersed over at least 30 kb in contrast to the *Klebsiella* genes carried on a 6 kb probe. None of the resident plasmids of *Anabaena* 7120 hybridized with the *nif* gene probes of either *Anabaena* or *Klebsiella*, and thus the *nif* genes must be chromosomally located.

A 10 kb *Anabaena Eco*RI fragment containing *nif*H and approximately half of *nif*D was cloned in the recombinant plasmid λgt-An154 (Mazur *et al.*, 1980). The cyanobacterial *nif*H gene was transcribed *in vivo* and the product was detected in *Escherichia coli* as a protein of mol. wt 33 000 which was selectively precipitated by antibody raised against the *Anabaena* nitrogenase reductase (unpublished results quoted by Mevarech *et al.*, 1980).

A 1.8 kb *Hind* III fragment, derived from the 10 kb fragment, contained the entire coding portion of the *Anabaena nif*H gene plus approximately 0.75 kb of DNA upstream from this gene (Mevarech *et al.*, 1980). The nucleotide sequence of this fragment was determined and the gene was identified within the sequence by its ATG initiation codon and TAG termination codon 900 b.p. away. The deduced amino acid sequence of the product was consistent with a mol. wt of 33 000 and showed striking similarity with the nitrogenase-reductase amino-acid sequence of *Clostridium pasteurianum* and *Azotobacter vinelandii*, especially within the sequences around five cysteine residues. Regions resembling bacterial promoters were not found, consistent with the observation that the gene was only expressed in *E. coli* if it was in the correct orientation in λgt-An154 to be transcribed from the λ leftward promoter (Mevarech *et al.*, 1980). The sequences of cyanobacterial promoters remain unknown.

The two clusters (11.7) of *Anacystis nidulans* rRNA genes were cloned in *Escherichia coli*. Two different DNA fragments each contained (in order) 16S, 23S and 5S rRNA genes, with a tRNA gene in the spacer region between the 16S and 23S genes (Tomioka *et al.*, 1981).

The absence of known cyanobacterial plasmid functions (see 11.3.2) has prevented their use as cloning vehicles in cyanobacterial strains themselves. This problem has been overcome by the introduction of an ampicillin resistance (Ap) marker into a plasmid of *Anacystis nidulans* R-2 by van den Hondel *et al.* (1980), creating a potential cloning vehicle which is traceable. The *Escherichia coli* plasmid pR146, carrying the Ap resistance transposon Tn901, was transformed into the *Anacystis nidulans* R-2 cells. Ap-resistant cells did not contain pR146; the original 33 Md plasmid (Table 11.4) pUH25 was present but the second resident plasmid (pUH24, 5.3 Md) had been

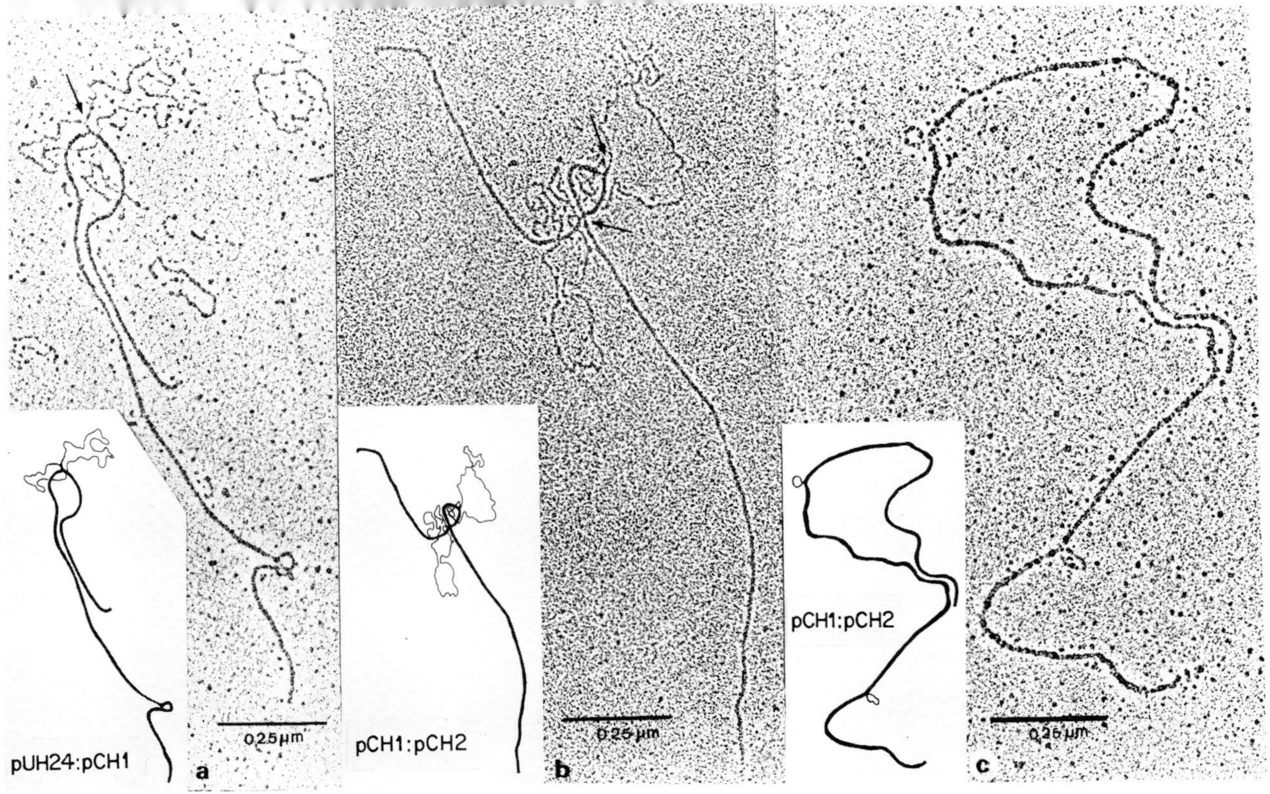

Fig. 11.3. Heteroduplex analysis of the insertion of Tn901 into pUH24. (a) Heteroduplex of *Xho* I-cleaved pUH24 and pCH1. The transposon is seen as a single loop of single-stranded DNA; the arrow indicates the short double-stranded stem generated by the two inverted repeat sequences of Tn901. (b) Heteroduplex of pCH1 and pCH2 DNA showing non-annealed insertion loops of Tn901 at a distance of approximately 0.2 μm. (c) Heteroduplex of pCH1 and pCH2 DNA in which the DNA strands of the transposon are annealed (the region between the single-strand loops) indicating that the orientation of Tn901 is the same in both plasmids. The loops result from the different insertion sites of the transposon. (From van den Hondel *et al.*, 1980)

replaced by one of 8.3 Md. Transformation of the latter (pCH1) into wild-type cells produced Ap-resistant cells which again contained pUH25 and an 8.3 Md plasmid but had lost pUH24. Restriction enzyme digests of the 8.3 Md plasmids in the recipients were identical with those of the plasmid used for transformation, confirming that Ap resistance was located on this plasmid.

Restriction enzyme analysis of five independent 8.3 Md recombinant plasmids showed that all contained Tn901, inserted at four different sites, but with the same orientation, in different plasmids. This was confirmed by the visualization of Tn901 as an insertion loop following hybridization of pUH24 with pUH24::Tn901 after their cleavage to linear molecules with *Xho* I (Fig. 11.3). The observed contour length and molecular weight of the recombinant plasmid were consistent with the sum of Tn901 (3.1 Md) and pUH24; the insertion loop was equivalent to 3.1 Md. The entire Tn901 was therefore present in the hybrid plasmid which itself had lost no genetic information (van den Hondel *et al.*, 1980). Since the transposition of Tn901 could complicate the use of the 8.3 Md plasmids as cloning vehicles, part of the transposon was removed by treating pCH1 with *Bam*HI, which produced one cut inside and one outside Tn901. The resulting fragments (2.8 Md and 5.5 Md) were separately circularized by ligation and the new 5.5 Md plasmid pUC1 transformed Ap resistance. This plasmid contained only one of the inverted repeats of Tn901 so that the transposon was immobilized, retained the single *Xho* I cleavage site of pCH1 and contained a single *Bam*HI site. These characters make pUC1 an excellent cloning vector in a unicellular cyanobacterium.

The *Escherichia coli* plasmid pACYC184 (carrying chloramphenicol (Cm) and tetracycline (Tc) resistance genes) was cut with *Bam*HI within the Tc gene to produce a linear molecule; following ligation to *Bam*HI-treated pUC1, a recombinant plasmid was obtained by selection for doubly (Cm, Ap) resistant *E. coli.* The resistance genes were subsequently expressed following transformation of the plasmid into *Anacystis nidulans* R-2. The plasmid (pUC0104, 8.2 Md) contains a single *Eco*RI restriction site for cloning purposes and is the first known plasmid capable of replication in both *Escherichia coli* and a cyanobacterium (Kuhlemeier *et al.*, 1981).

The expression of Ap (β-lactamase) and Cm resistance in *A. nidulans* R-2 is the first report of the activity within a cyanobacterium of genes from another organism. Furthermore, the transposition of Tn901 from pR146 to pUH24 indicates that transposition functions are expressed in cyanobacterial cells. Conversely, the detection of the *Anabaena* 7120 *nif*H gene product in *Escherichia coli* demonstrates cyanobacterial gene expression in another organism. The *Anabaena* 7120 gene for glutamine synthetase (*gln*A) was not only transcribed, but also produced a *functional enzyme*, following cloning in *Escherichia coli* (Fisher *et al.*, 1981). A 7.5 kb *Hin*dIII fragment of *Anabaena*

DNA, which hybridized with ^{32}P-labelled λ811 DNA (carrying the *E. coli* *gln*A gene) was cloned on phage λ in *E. coli* ET8051 (which had undergone a large deletion including *gln*A, the closely linked regulatory gene *gln*G, and flanking sequences). The protein synthesized (mol. wt 51,500) was immunologically identical to the purified *Anabaena* glutamine synthetase, and distinguishable from that of *E. coli.* The complementation of the *E. coli gln*A deletion mutant by the *Anabaena gln*A gene is the first demonstration of the activity of a cyanobacterial gene product in a foreign cell. The *Anabaena gln*A gene was transcribed even when the λ promoters were repressed, suggesting that the *Anabaena* DNA fragment contained a sequence that functioned as a promoter in *E. coli.* Unlike the *E. coli* enzyme, the cyanobacterial glutamine synthetase in ET8051 was neither adenylylated nor subject to repression. The observations that the gene was not subject to control even when transferred to an *E. coli gln*G^+ background, and that λAn503 carrying the 7.5 kb fragment did not restore the regulatory $GlnG^+$ phenotype to ET8051, may explain the lack of control of glutamine synthetase in the cyanobacterial cell itself.

11.9 Concluding remarks

The construction of plasmid vectors suitable for cloning cyanobacterial DNA both in a unicellular cyanobacterium and in *Escherichia coli*, together with techniques available for the identification of specific cistrons, creates the possibility of a thorough analysis of these genes. Clearly, these techniques will be extended to other organisms, to permit the study of cellular differentiation, photosynthesis and chromatic adaptation. By their nature, many of these processes can only be studied by cloning the genes into cyanobacterial mutants which lack the appropriate function. Thus, detailed studies of mutagenesis and transformation in a variety of cyanobacteria remain an urgent necessity.

Acknowledgments

I thank Chantal Astier, W.F. Doolittle, Françoise Joset-Espardellier, R. Haselkorn, C.A.M.J.J. van den Hondel, J.A. Houghton and J. Reaston for supplying manuscripts prior to publication, and Mme S. Beguin, Dr G. Cohen-Bazire, Rosmarie Rippka, C.A.M.J.J. van den Hondel and C.J. Kuhlemeier for access to unpublished results. I am indebted to A. van der Ende for providing the original photographs for Fig. 11.2, F. Joset-Espardellier and C.A.M.J.J. van den Hondel for critical comments on the manuscript, and to Rosmarie Rippka for patient help with the strain histories and taxonomy.

CHAPTER 12
MOLECULAR EVOLUTION

W. F. DOOLITTLE

12.1 Introduction

12.1.1 Coming to terms with the cyanobacteria

Arguments about the inclusion of phycobiliprotein-containing oxygenic–photosynthetic prokaryotes among the bacteria are arguments about taxonomy, not phylogeny (Lewin, 1979; Stanier *et al.*, 1978). No one now doubts that these organisms, whatever we call them, are fundamentally like those prokaryotes we know as bacteria (Stanier & Cohen-Bazire, 1977; Doolittle, 1979). In this chapter, all oxygenic–photosynthetic prokaryotes with phycobiliproteins will be called 'cyanobacteria', the term 'algae' will be used only for eukaryotes, and the term 'bacteria' will be used, purely for literary convenience, to designate all prokaryotes (including *Prochloron*) which are not cyanobacteria. Its use in this way should not be taken to mean that the cyanobacterial:bacterial division is the most profound one among the prokaryotes. There are other divisions within the prokaryotes which are deeper, and the most profound is probably that between the 'archaebacteria' and the rest of the prokaryotes (Woese *et al.*, 1978; Woese & Fox, 1977; Doolittle, 1980)

12.1.2 Their evolutionary importance

There are three good reasons to be interested in cyanobacterial evolution. (i) The fossil record for cyanobacteria is unquestionably more complete than is the fossil record for any other group of prokaryotes (see Chapter 21 and Knoll [1977]; Schopf [1974a,b, 1980]). (ii) Cyanobacteria are commonly held responsible for the conversion of the earth's atmosphere from anaerobic to aerobic, an event of signal importance to subsequent cellular evolution (Margulis *et al.*, 1976; but see also Towe, 1978). (iii) There are strong homologies between cyanobacteria and the plastids of eukaryotic photosynthesizers. Whether we attribute this to the gradual 'autogenous' development of entire eukaryotic algal cells from cyanobacteria or to the acquisition by non-photosynthetic 'protoeukaryotes' of cyanobacteria-like endosymbionts destined to become plastids, we cannot deny the pivotal role played by cyanobacteria in the origin of eukaryotic cells.

There are three questions to ask. (i) What is the proper phylogenetic position of the cyanobacteria within the prokaryotes? (ii) What phylogenetic relationships exist within the cyanobacteria? (iii) What evolutionary relationship do cyanobacteria bear to eukaryotic photosynthesizers?

12.1.3 Approaches and limitations

These questions can in principle be answered by appeals to the fossil record, to data from comparative physiology and biochemistry, or to measures of homology between nucleic acids and proteins. Each approach has limitations.

The fossil record, although extensive (Knoll, 1977; Chapter 21) cannot always easily be interpreted. Controversy exists even about which late Precambrian microfossils are eukaryotic and which prokaryotic (Knoll & Barghoorn, 1975; Schopf & Oehler, 1976). Golubic and Barghoorn (1977) warn that 'identifications and comparisons with extant counterparts are almost invariably restricted to morphology as only one of many biological identification criteria, and conclusions about biological functions and diversity of ancient microorganisms need to be treated with caution. . .'.

Comparative biochemical and physiological data are more extensive, and a number of prokaryotic phylogenies have been constructed from them (for example, Margulis, 1970, 1981; Broda, 1975; Olson, 1970, 1978; Ragan & Chapman, 1978; Ochiai, 1978). Such approaches only provide plausible evolutionary schemes. Their validity must be established through the fossil record or molecular sequence data. They are subject to bias in the selection of certain metabolic processes as of key importance, and require *ad hoc* assumptions about the course of evolution of these processes and about true homology (at the genomic level) between apparently similar processes in different groups.

Measurements of evolutionary relatedness derived either directly from complete sequences of nucleic acids and protein or indirectly from molecular hybridization or immunological data are in principle not subject to such bias, do not require *ad hoc* assumptions of the same sort, and do provide more certain information on the relatedness of genomes. *Ad hoc* assumptions must be made, however, about how data should be handled. Recently, the validity of prokaryotic phylogenies constructed from cytochrome sequences has come under vigorous attack from workers responsible for generating much of the primary data (Ambler, Daniel, Hermoso, Meyer, Bartsch & Kamen, 1979; Ambler, Meyer & Kamen, 1979). Phylogenetic trees derived from sequence studies will be invalid if (i) paralogous, rather than orthologous, molecular homologues are compared (Wilson *et al.*, 1977; Demoulin, 1979; Uzzell & Spolsky, 1981), (ii) there are significant differences in rates of fixation of mutations within different lineages, (iii) sequences have diverged so far that measures of similarity are at equilibrium, and additional changes are as likely to produce convergent as divergent substitutions (Ambler, Meyer & Kamen, 1979; Romero-Herrera *et al.*, 1979), or (iv) 'horizontal' (inter-lineage) gene transfer has occurred. It is currently fashionable to attribute to the latter process (as mediated by transmissible plasmids and phages) a major role in prokaryotic evolution (Reanney, 1978; Richmond, 1979).

In spite of all these difficulties, phylogenies derived from sequence comparisons (e.g. Schwartz & Dayhoff, 1978, 1981; Hori & Osawa, 1979; Almassy & Dickerson, 1978; Dickerson, 1980a; Woese & Fox, 1977) usually do appear sensible. In this chapter, the support which molecular sequence data lend to hypotheses about cyanobacterial evolution (most of which have been motivated by biochemical, physiological, morphological or palaeontological considerations) will be examined.

12.2 The position of the cyanobacteria within the prokaryotes

12.2.1 Possible phylogenies

Prokaryotic taxonomists (Pfennig, 1977; Trüper & Pfennig, 1978; Stanier, 1977; Stanier & Cohen-Bazire, 1977) recognize three principal groups of photosynthetic prokaryotes: the green bacteria (Chlorobiineae), the purple bacteria (Rhodospirillacae), and the oxygenic–photosynthetic prokaryotes (cyanobacteria and 'prochlorophytes'). Prokaryotic taxonomies are now of necessity phenetic, not phylogenetic (Sneath & Sokal, 1973). Nevertheless, it seems reasonable to convert Pfennig's (1978) phenetic assertion that the green and purple bacteria 'constitute two remarkably different groups of photoautotrophic prokaryotes on the basis of their cytological properties' into the phylogenetic assumption that each of these groups is monophyletic. That is, it seems reasonable to assume that the two kinds of green bacteria—the obligately photoautotrophic green sulphur bacteria (Chlorobiaceae) and the facultatively heterotrophic green filamentous bacteria (Chloroflexaceae)—are evolutionarily more closely related to each other than to either of the two kinds of purple bacteria—the obligately photoautotrophic purple sulphur bacteria (Chromatiaceae) and the facultatively heterotrophic purple non-sulphur bacteria (Rhodospirillaceae). This assumption is equivalent to the assumption that the cladogram shown in Fig. 12.1 is topologically correct, in a phylogenetic sense.

It also seems reasonable to assume that the cyanobacteria and prochlorophytes are more closely related to each other than either of them is to any of the green or purple bacteria (Withers *et al.*, 1978b). If photosynthesis (photosystem (PS) I) arose only once, then the oxygenic–photosynthetic prokaryotic lineage now represented by the cyanobacteria and prochlorophytes must have diverged from the rest of the photosynthetic prokaryotes at one of the seven points indicated in Fig. 12.1.

The fossil record will probably never allow us to divine the point of cyanobacterial (and prochlorophytan) divergence. Comparative biochemical

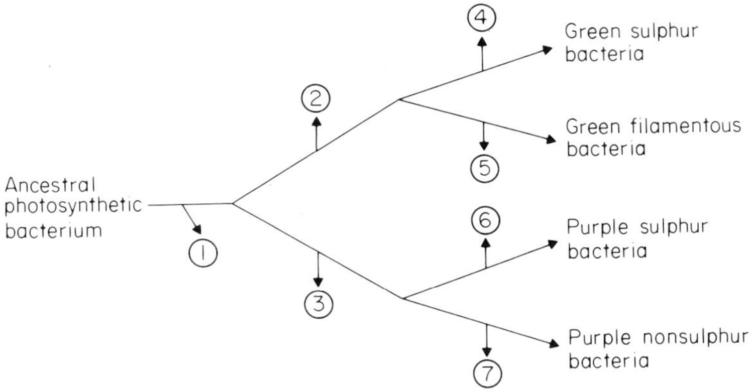

Fig. 12.1. Possible phylogenetic relationships of photosynthetic prokaryotes. Green (sulphur and filamentous) and purple (sulphur and non-sulphur) bacteria are assumed to represent two separate monophyletic lineages. Possible positions of divergence of the oxygenic–photosynthetic prokaryotic lineage now represented by the cyanobacteria and *Prochloron* are indicated by numbers.

and physiological data might, but quite different phylogenetic trees have been proposed by different authors arguing on the basis of such data. Three are discussed below.

(i) Olson (1970, 1978) and Mauzerall (1978) suggest a divergence at position 1 (Fig. 12.1), arguing that the ability to synthesize chlorophyll *a* should have preceded the ability to synthesize bacteriochlorophylls. Olson favours an early divergence of cyanobacteria and all other photosynthetic prokaryotes (other than *Prochloron*) from a photoheterotrophic ancestor using chlorophyll *a*, but lacking PS II. Mauzerall (1978) 'would go a step further, and suggest that the modern photosynthetic bacteria arose from a simplification of a primitive unicellular cyanophyte'. Support for this suggestion might be derived from an examination of the ranges of chromosomal G + C contents shown by the various groups of photosynthetic prokaryotes (Herdman *et al.*, 1979b; Trüper & Pfennig, 1978). The range shown by the chlorophyll *a*-containing prokaryotes (35–71 mol%) is greater than, and encompasses, the ranges shown by green bacteria (45–58 mol%), purple sulphur bacteria (45–70 mol %) and purple non-sulphur bacteria (61–72 mol%). This is at least consistent with the notion that the chlorophyll *a*-containing prokaryotes (cyanobacteria and *Prochloron*) are the most ancient photosynthetic group. But all chlorophyll *a*-containing prokaryotes, regardless of G + C content, use PS II-dependent non-cyclic photophosphorylation, and it seems inherently reasonable to assume that PS II is a more recent evolutionary development than PS I (Margulis, 1970; Oslon, 1970, 1978; Schopf, 1974a; Broda, 1975).

(ii) Kenyon (1978) suggests that oxygenic–photosynthetic prokaryotes diverged at position 2, arguing on the basis of complex lipid composition that '*Chloroflexus* and the Chlorobiaceae are more closely related to each other and to the cyanobacteria than are the other groups of photosynthetic bacteria'. Similarly, Staehelin, Golecki, Fuller & Drews (1978) suggest, on grounds of morphology and motility, that '*Chloroflexus* represents a transitional organism, intermediate in various ways, between the green sulphur bacteria and the oxygenic cyanobacteria' (thus favouring divergence at positions 2 or 5). However, it appears likely that the green bacteria differ from purple bacteria and cyanobacteria in fixing CO_2 by a major pathway other than the Calvin cycle, a distinction which may mark them as 'primitive' (Pfennig, 1977, 1978; Sirevåg, 1979).

(iii) Partly on the basis of the latter evidence, Dickerson (1980a) has favoured a divergence of the oxygenic–photosynthetic prokaryotes at position 3 (Fig. 12.1). This scheme (or schemes in which cyanobacteria diverge at position 7) is popular (Margulis, 1970; Broda, 1975; Schwartz & Dayhoff, 1978, 1981; Ochiai, 1978), but must contend with the fact that both the Rhodospirillaceae and Chloroflexaceae possess the energetically advantageous oxidative tricarboxylic acid cycle (Fuller, 1978; Sirevåg & Castenholz, 1979), whereas cyanobacteria characteristically lack it, due to a deficiency of α-ketoglutarate dehydrogenase (see Smith & Hoare, 1977). It remains unclear whether a 'broken' tricarboxylic acid cycle is a 'primitive' or a derived trait (Uzzell & Spolsky, 1974).

There is much more relevant biochemical and physiological data than can be summarized here (see Ragan & Chapman, 1978; Chapman & Ragan, 1980; Clayton & Sistrom, 1978; Broda, 1975). A detailed consideration of it would not, however, lead to any uniquely satisfying decision on the proper position of the divergence of oxygenic–photosynthetic and green and purple bacterial evolutionary lineages.

12.2.2 Molecular evidence

12.2.2.1 *c*-Type cytochromes

Respiratory and photosynthetic *c*-type cytochromes from most of the recognized species of the Rhodospirillaceae, two of the Chlorobiaceae and a variety of non-photosynthetic bacteria have been sequenced (Ambler, Daniel, Hermoso, Meyer, Bartsch & Kamen, 1979; Ambler, Meyer & Kamen, 1979; see Dickerson 1980a for other references). Complete sequences are available for the cytochromes c_6 of the cyanobacteria *Synechococcus* 6312, *Spirulina maxima, Plectonema boryanum* and *Anacystis nidulans* (Ambler & Bartsch, 1975; Aitken, 1976, 1977, 1979 and E. Margoliash, cited in

Dickerson, 1980a), and for several eukaryotic algal plastids (Schwartz & Dayhoff, 1981).

Analyses of tertiary structure and primary sequence have allowed Dickerson (1980a,b; see also Almassy & Dickerson, 1978) to assign most of the *c*-type cytochromes to one of four structural subgroups: L(large), M(medium), S(small) and F(cyanobacterial and algal). Members of each subgroup show more primary sequence homology with each other than they do with members of other subgroups, although subgroups L and M are clearly closely related. The Rhodospirillaceae contain cytochromes of subgroups L, M and S, distributed among recognized species in a way which is inconsistent with their classification in Bergey's Manual. Ambler and collaborators (Ambler, Daniel, Hermoso, Meyer, Bartsch & Kamen, 1979) conclude from this that cytochromes *c* cannot be used to reconstruct bacterial phylogenies. However, Dickerson (1980b) and Woese *et al.* (1980) point out that the phylogenetic relationships indicated are consistent with relationships among the Rhodospirillaceae suggested both by 16S ribosomal RNA sequence comparisons (see below) and by analysis of photosynthetic membrane structure (Trüper & Pfennig, 1978). It is necessary to remember that Bergey's is a manual of determinative bacteriology, not a specific proposal on bacterial phylogeny.

Cyanobacterial and algal cytochromes c_6 are clearly homologous to the other *c*-type cytochromes, but show no preferential homology to L, M or S subgroups of Rhodospirillacean cytochromes c_2, which indicates that the oxygenic–photosynthetic prokaryotes diverged from the non-sulphur purple bacteria long before this group became internally diversified. The sequenced cytochromes c_{555} from *Chlorobium thiosulfatophilum (C. limicola* forma sp. *thiosulfatophilum)* and *Prosthecochloris aestuarii,* although again clearly homologous to other cytochromes *c*, show no especially strong homology to subgroups L, M, S or F, suggesting an early divergence of green sulphur bacteria from both non-sulphur purple and cyanobacterial lineages. Dickerson (1980a,b) favours a cyanobacterial divergence at position 3 in Fig. 12.1, while admitting that sequence data do not prove such a divergence. Schwartz and Dayhoff (1978, 1981), using a matrix method and an alignment procedure which does not (as does Dickerson's) make use of tertiary structural information, have produced a phylogeny for *c*-type cytochromes which does indicate a divergence at this point. Demoulin (1979) and Uzzell and Spolsky (1981) question whether the cytochromes used are indeed orthologous. If they are paralogous, the phylogenetic relationships between organisms indicated by Schwartz and Dayhoff might be incorrect. This question needs to be resolved, and sequences for cytochromes of *Chloroflexus*, purple sulphur bacteria and *Prochloron* must be obtained.

12.2.2.2 Ferredoxins

Ferredoxins are iron–sulphur proteins unique among sequenced proteins in being present in all well-characterized prokaryotes and eukaryotes. They can be divided into two major classes, those with 4Fe+4S centres and those with 2Fe+2S centres. The former can be further divided into those which contain eight and those which contain four iron atoms (8Fe and 4Fe). 4Fe+4S centre ferredoxins of the 8Fe type are found in the allegedly primitive clostridia, and among photosynthetic prokaryotes, in *Chlorobium*, *Chromatium* and certain non-sulphur purple bacteria. *Chromatium* and non-sulphur purple bacteria contain, in addition, 4Fe ferredoxins. All cyanobacterial ferredoxins, as well as most ferredoxins from algae and plants, are of the 2Fe+2S type, although 2Fe+2S ferredoxins are not confined to these groups (Hall *et al.*, 1977). The cyanobacteria *Aphanothece sacrum* (Hase *et al.*, 1976a, 1978c) *Nostoc muscorum* (Hase *et al.*, 1976b) and *Nostoc verrucosum* (Shin *et al.*, 1977) contain two species of 2Fe+2S ferredoxins, as do some plants (Schwartz & Dayhoff, 1978, 1981). Immunological studies show significant cross-reactivity between cyanobacterial, algal and plant ferredoxins but no homology between any of these and the 8Fe (4Fe+4S) or 4Fe(4Fe+4S) ferredoxins of bacteria, including *Chromatium* and *Chlorobium* (Tel-Or *et al.*, 1977a).

Hall *et al.* (1977) suggest that the 8Fe(4Fe+4S) ferredoxins of clostridia and green sulphur bacteria are primitive, and that the 2Fe+2S ferredoxins of cyanobacteria and eukaryotic photosynthesizers arose from 4Fe(4Fe+4S) ferredoxins of the type now found in the sulphate-reducing anaerobe *Desulfovibrio*, this in turn having derived from ferredoxins like those in *Chromatium*. The implied point of cyanobacterial divergence is position 6 in Fig. 12.1. On the other hand, Schwartz and Dayhoff (1978, 1981) have produced a phylogenetic tree in which *Chlorobium* and *Chromatium* ferredoxins diverge together and rather early from the common ancestral anaerobic heterotrophic stock, while ferredoxins from *Desulfovibrio* and the cyanobacteria diverge more recently. This result is reconcilable with the assumptions implicit in Fig. 12.1 only if cyanobacteria are presumed to have diverged at position 1. Rates of ferredoxin sequence change are clearly not uniform; there is as much divergence among the ferredoxins of higher plants as among those of the cyanobacteria, although the latter are unquestionably more ancient. Schwartz and Dayhoff (1978) suggest that 'unlike the situation in eukaryotes, the rate of acceptance of point mutations in prokaryotic sequences is very uneven, and cannot provide a useful evolutionary clock'. An additional reason for caution in interpreting ferredoxin sequences is provided by *Halobacterium halobium*. Although halobacteria do show light-driven ATP synthesis, most authors would agree that this novel 'photosynthetic' mechanism was independently developed in the halobacterial lineage (Broda, 1975; Bayley, 1979), which is one of the archaebacterial lineages and certainly very remote from true photosynthetic prokaryotic lineages (Magrum *et al.*, 1978). Nevertheless, *H. halobium* contains a 2Fe+2S ferredoxin which resembles in sequence those of cyanobacteria, algae and plants much more closely than does any other (photosynthetic or non-photosynthetic) bacterial ferredoxin (Hase *et al.*, 1977, 1978a; Schwartz and Dayhoff, 1978, 1981). Unless this is due to convergence of unprecedented proportions, it may constitute the strongest available evidence for horizontal (interspecific or indeed 'inter-kingdom') gene-transfer involving a gene for an indispensable biochemical function.

12.2.2.3 RuBP carboxylase

Although few primary sequence data are available for the RuBP carboxylases (ribulose-1,5-bisphosphate carboxylases) of photosynthetic prokaryotes, considerable structural and functional information on these crucial CO_2-fixing enzymes has been assembled, and recently reviewed by McFadden (1978), McFadden and Purohit (1978) and Fuller (1978). McFadden distinguishes two types of RuBP carboxylase: O-enzymes, which are considered primitive and which contain only large (L) *c.* 55 000 dalton catalytic subunits in L_2, L_6 or L_8 aggregates; and T-enzymes, which contain an additional small (S) subunit of unknown function and *c.* 12 000–18 000 daltons. These are characteristically L_8S_8 aggregates, and are considered a more recent evolutionary development. McFadden has proposed (McFadden, 1978; McFadden & Purohit, 1978) that RuBP carboxylases have evolved in the direction $L_2 \rightarrow L_6 \rightarrow L_8 \rightarrow L_8S_8$. The scheme is appealing, but, as pointed out by Tabita *et al.* (1979), the distribution of RuBP carboxylase types among photosynthetic (or other CO_2-fixing) prokaryotes is consistent with no coherent evolutionary tree, unless each recognized group is considered polyphyletic. An L_2 enzyme is found in *Rhodospirillum rubrum*, which is not likely on the other grounds to have diverged at an early stage from all other photosynthetic lineages. L_6 enzymes are found in some cyanobacteria and chemolithotrophs, whereas L_8S_8 enzymes are found in other cyanobacteria, *Chromatium*, *Ectothiorhodospira* (purple sulphur bacteria) and *Rhodopseudomonas sphaeroides,* which also contains an L_6 species (McFadden & Purohit, 1978; Gibson & Tabita, 1977; Tabita *et al.*, 1979). Most or all plastid RuBP carboxylases are of the L_8S_8 variety, although only the large subunit is plastid genome-encoded (Gillham, 1978). Either T (two subunit) enzymes have evolved separately in parallel lineages, or the possession of a second, small subunit is in fact a primitive trait which has been repeatedly lost.

12.2.2.4 Ribosomal RNAs

Woese and his collaborators pioneered the use of partial sequence characterization (T1-oligonucleotide cataloguing) of 16S ribosomal RNAs (rRNAs) as a tool for reconstruction of prokaryotic phylogenies (e.g. Pechman & Woese, 1972; Fox *et al.*, 1977; Woese & Fox, 1977). This technique involves the complete enzymatic sequencing of all guanylate (Gp)-terminated residues released from ^{32}P-labelled 16S rRNA by digestion with ribonuclease T1. It yields data (oligonucleotide catalogues) which can be analysed by standard numerical taxonomic techniques to give similarity coefficients (S_{AB}s) which presumably reflect the evolutionary relatedness of the organisms (or organelles) from which the rRNAs were isolated (Sneath & Sokal, 1973; Bonen & Doolittle, 1976, Doolittle & Bonen, 1981). Woese and coworkers have applied this method to a large number of bacterial species, including members of all the photosynthetic bacterial groups other than *Prochloron*, while Bonen and Doolittle (1975, 1976, 1978), Doolittle and Bonen (1981), and Bonen *et al.* (1979) have used the same approach to analyse cyanobacterial 16S rRNAs.

Woese *et al.* (1980) report from such studies that (i) the Rhodospirillaceae are deeply divided into two major groups which, although inconsistent with generic assignments made in Bergey's Manual, agree with inferences drawn by Dickerson from sequences of cytochromes *c*, (ii) the purple sulphur bacteria *Chromatium vinosum* and *Thiocapsa pfennigii* form a distinct group about equally related to each of the two Rhodospirillacean groups, and the order of branching for the three lineages cannot be precisely determined, (iii) these three lineages are, however, more closely related to each other than to the cyanobacteria and plastids, which form a coherent cluster, and (iv) *Chlorobium limicola* and *Chloroflexus aurantiacus* are quite remote from all other groups and, surprisingly, from each other. The divergence for cyanobacteria which these data most favour is that indicated as position 3 in Fig. 12.1, that which is also favoured by Dickerson on the basis of cytochrome *c* comparisons.

12.2.3 Conclusions

We have yet too few data, and these are in insufficient agreement, to allow a final decision about phylogenetic relationships within the photosynthetic prokaryotes. Divergence of the oxygenic–photosynthetic prokaryotes at position 3 of Fig. 12.1 is defensible on physiological and biochemical grounds, and is consistent with available rRNA and cytochrome data, but not with ferredoxin data.

It seems reasonable to suggest that the surviving lineages of oxygenic photosynthetic prokaryotes, the cyanobacteria and *Prochloron*, are more closely related to each other than they are to other photosynthetic lineages (Withers *et al.*, 1978b) and diverged from a common, chlorophyll *a*-containing oxygenic photosynthetic ancestor which lacked phycobiliproteins. These latter would then have developed as accessory pigments for PS II in the cyanobacterial lineage, while a (possibly more ancient?) chlorophyll *a*/*b*-protein complex serves this function in prochlorophytes. (D.J. Chapman & M.A. Ragan (pers. comm.), on the other hand, suggest that prochlorophytes have secondarily lost the ability to produce biliproteins.) Sequence data on cytochromes and rRNAs from *Prochloron* are clearly needed, not only to clarify the relationship between these two groups, but to provide some minimal estimate of the antiquity of oxygenic photosynthesis.

12.3 Phylogenetic relationships within the cyanobacteria

12.3.1 Taxonomies and possible phylogenies

Taxonomies and phylogenies are not the same. As Sneath and Sokal (1973) point out, 'one branch of systematics, classification, has the relatively unexciting but supremely important function of ordering nature into a practical and generally useful system. This can best be done on the basis of a phenetic classification. . . . Systematics in the wider sense also has the more challenging task of understanding the mechanisms which brought about this order.' Stanier and coworkers (Rippka *et al.*, 1979) have proposed a useful new classification system for the cyanobacteria (summarized in Table 12.1). They recognize five principal subgroups ('Sections'). Section I comprises all unicellular cyanobacteria reproducing by binary fission (*Gloeobacter, Gloeothece, Synechococcus, Gloeocapsa* and *Synechocystis*) or budding (*Chamaesiphon*). Section II comprises unicells or cellular aggregates which reproduce by multiple fission, producing 'baeocytes' (*Dermocarpa, Xenococcus, Dermocarpella, Myxosarcina, Chroococcidiopsis* and the 'Pleurocapsa group'). Section II contains filamentous species without heterocysts or akinetes (*Spirulina, Oscillatoria, Pseudanabaena* and the LPP (*Lyngbya, Phormidium, Plectonema*) group). Section IV includes all heterocystous filamentous cyanobacteria in which intercalary cell division is always perpendicular to the trichome axis (*Anabaena, Nodularia, Cylindrospermum, Nostoc, Scytonema* and *Calothrix* (including *Tolypothrix* and *Rivularia*)). Section V comprises the heterocystous forms in which cell division can occur in more than one plane (*Fischerella* (*Mastigocladus*) and *Chlorogloeopsis*). Although this classification scheme is phenetic, we can legitimately ask two sorts of phylogenetic questions about it: (i) are the groups recognized by Stanier and coworkers monophyletic or polyphyletic, and (ii) what are the evolutionary relationships between these groups?

Table 12.1. Taxonomic scheme of Rippka *et al.* (1979), including chromosomal G+C contents estimated by Herdman *et al.* (1979b)

Cell type and reproductive mode		Section (G+C range)	Recognized genera	Chromosomal G+C content of genera
Unicells,	Reproduction by	I	*Synechococcus*	39–43, 47–56, 66–71
single or	binary fission	(35–71)	*Synechocystis*	35–37, 42–48
aggregated	or budding		*Gloeobacter*	64
			Gloeothece	40–43
			Gloeocapsa	40–46
			Chamaesiphon	47
	Reproduction by	II	*Dermocarpa*	38–44
	multiple fission	(38–47)	*Xenococcus*	44
			Dermocarpella	45
			Myxosarcina	43–44
			Chroococcidiopsis	40–46
			Pleurocapsa group	39–47
Filaments	Trichomes contain	III	*Spirulina*	44–54
	only vegetative	(40–67)	*Oscillatoria*	40–50
	cells		*Pseudanabaena*	44–52
			LPP group (*Lyngbya, Phormidium, Plectonema*)	42–67
	Capable of heterocyst	IV	*Anabaena*	38–43
	formation.	(38–47)	*Nodularia*	41
	No true branching		*Cylindrospermum*	42–47
			Nostoc	39–45
			Scytonema	44
			Calothrix (*Tolypothrix, Rivularia*)	40–44
	Capable of heterocyst	V	*Chlorogloeopsis*	42
	formation.	(42–46)	*Fischerella* (*Mastigocladus*)	42–46
	True branching			

Earlier speculations on phylogenetic relationships within the cyanobacteria based on morphological comparisons have been reviewed by Desikachary (1973) and Fogg *et al.* (1973). There is general agreement that unicellular forms gave rise to filamentous ones. Schemes in which heterocystous species of Sections IV and V evolved either directly from non-heterocystous members of Section III (Fritsch, 1942), or separately from unicellular species (Tilden,

1933), or in which members of Section III derived 'retrogressively' from the more 'advanced' forms (Section IV and V, see Desikachary & Padmaja, 1970; Chadefaud, 1951) have all been proposed. Species of Section II are variously seen as advanced descendants of Section I (*Dermocarpa*, *Dermocarpella*) or retrogressive descendents of Section V (*Pleurocapsa*).

There are few limitations on morphology-based evolutionary schemes, especially for a group in which morphology so easily varies with growth conditions. The incongruity of morphology-based taxonomic groupings with molecular sequence-based phylogenies discussed above for the Rhodospirillaceae should be sufficient reason for caution in constructing evolutionary scenarios from morphological data alone. Support from the fossil record, comparative biochemical and physiological data, and molecular sequence homology measurements are required and, in all but a few cases, lacking.

12.3.2 The fossil record

Knoll (1977) and Schopf (1974a,b, 1980; see also Chapter 21) have reviewed the appearance, in the fossil record, of apparent representatives of modern cyanobacterial groups. Although there is uncertainty (Schopf, 1976) about the biogenicity of the oldest (> 3100 million years old) microfossils identified as cyanobacteria-like (Schopf & Barghoorn, 1967), and the earliest (2500–2800 million years old, Bulawayan) stromatolites may possibly be of bacterial rather than cyanobacterial origin (Doemel & Brock, 1974; Knoll, 1979; but see Awramik, 1977), there seems little doubt about the cyanobacterial nature of microfossils present in stromatolities of the Transvaal sequence (2250 million years old) and more recent deposits (Nagy, 1974; MacGregor *et al.*, 1974). Both Knoll (1977) and Schopf (1974a,b) consider that chroococcalean (Section I) microfossils, although hard to identify, appear in the record before those of Section III (Oscillatoriaceae) or IV (Nostocaceae, Scytonemataceae and Rivulariaceae), while *Pleurocapsa*-like (Section II) or *Stigonema*-like (Section V) forms appear only fairly recently (< 1500 million years ago).

12.3.3 Biochemical and physiological data

For a group of such morphological and genetic (as evidenced by the range of G+C contents) diversity and apparent antiquity, the cyanobacteria show a remarkable degree of metabolic and nutritional monotony. There are, admittedly, species to species variations in (i) facultative ability to grow on organic substances in the light (photoheterotrophy; Rippka *et al.*, 1979), (ii) facultative ability to grow exclusively on organic substances in the dark (chemoheterotrophy; Stanier, 1973; Raboy *et al.*, 1976) and (iii) facultative ability to survive in darkness by fermentation or anaerobic respiration (Oren

& Shilo, 1979). However, photoautotrophy remains the dominant nutritional mode. There are also variations in (iv) the ability to use sulphide or H_2 as electron donor in PS II-independent anoxygenic photosynthesis (Belkin & Padan, 1978; Peschek, 1978), (v) the ability to produce phycoerythrin, (vi) the ability to synthesize nitrogenase and/or fix N_2 anaerobically (Rippka *et al.*, 1979; Rippka & Waterbury, 1977), (vii) the ability to fix N_2 aerobically without heterocysts (Wyatt & Silvey, 1969) and (viii) the ability to form heterocysts. All but the last two traits are distributed through recognized cyanobacterial taxa in a manner which makes no phylogenetic sense. For instance, both facultatively photoheterotrophic and obligately autotrophic strains are found in all sections, facultative chemoheterotrophy is known in strains of (at least) Sections I, III and IV, anaerobic dark survival is found in representatives of Sections III and IV (Oren & Shilo, 1979; Hoare *et al.*, 1971), and strains capable or incapable of anaerobic N_2 fixation are found in groups I–III, as are strains capable or incapable of producing phycoerythrin. Facultative sulphide-dependent anoxygenic photosynthetic growth has been demonstrated for members of both groups I and III (Garlick *et al.*, 1977, Padan, 1979a).

This whimsical distribution of traits suggests most strongly that all but aerobic N_2 fixation are primitive ones (present in the last common ancestor of all cyanobacteria), although the possibilities that phylogenetic structure within the cyanobacteria is not at all congruent with taxonomic structure, or that some of these traits are plasmid-determined, cannot be eliminated. In any event, these traits, which are the only major ones in which cyanobacteria are known to vary, provide no certain basis for a phylogeny.

One biochemical character which does correlate reasonably well with the taxonomic groupings of Stanier and coworkers is fatty-acid composition. Photosynthetic bacteria (including *Prochloron*; Lewin, 1981: Perry *et al.*, 1978) produce few (if any) polyunsaturated fatty acids, although they are, like other bacteria, capable of synthesizing a variety of monounsaturated fatty acids (Kenyon, 1978; Ragan & Chapman, 1978; Chapman & Ragan 1980). Some, but not all, cyanobacteria can synthesize polyunsaturated fatty acids, which suggests that this ability is a derived trait arising since the first appearance and initial divergence of cyanobacterial lineages, and that fatty-acid composition might be useful in cyanobacterial phylogenetic analysis (Holton *et al.*, 1968; Kenyon & Stanier, 1970): available data suggest that it is (Kenyon & Stanier, 1970; Kenyon, 1972; Kenyon *et al.*, 1972; Wood, 1974).

Of the unicellular genera recognized by Stanier and coworkers in Section I, *Synechococcus* appears the most diverse (39–71 mol % G+C (Herdman *et al.*, 1979b)). It was divided by Stanier *et al.* (1971) into nine clusters on the basis of cell size, maximal growth temperature, phycocyanin/chlorophyll *a* absorbance ratios, and chromosomal G+C content. Members of clusters 1, 3, 4 and

5 all showed little (≤2%) if any polyunsaturated fatty acids, whereas those of clusters 2, 6, 7, 8 and 9 showed polyunsaturated fatty acids (16.2, 18.2, 18.3α and 18.3γ) representing from 23 to 61% of the total fatty-acid content. All but the sole representative of cluster 2 are of similar (48–50 mol %) G+C content. *Synechocystis* (Section I) strains divided neatly into two groups: those with G+C contents from 35 to 37 mol % and up to 4% polyunsaturated fatty acids, and those with 42 to 48 mol % G+C and 42–55% polyunsaturated fatty acids (predominantly 18.2 and 18.3γ). Filamentous cyanobacteria of Sections III, IV and V (*Chlorogloeopsis*) all show high (27–60%) polyunsaturated fatty acid levels (with the exception of two isolates of *Spirulina* and one of *Hapalosiphon*; Kenyon & Stanier, 1970; Holton *et al.*, 1968; Kenyon *et al.*, 1972; Wood, 1974).

It seems likely, as Kenyon and Stanier (1970) and Holton *et al.* (1968) suggest, that the ability to synthesize polyunsaturated fatty acids first arose within a subgroup of Section I, and that filamentous cyanobacteria (and some plastids) descend from members of this group. The nature of the 18.3 trienoic fatty acid (α- or γ-linolenic acid) in strains showing this degree of desaturation may also be of phylogenetic significance, although RNA sequence data discussed below indicate that Nostocalean forms which show primarily α-linolenic acid may be derived from *Synechocystis*-type unicells in which the predominant 18.3 fatty acid is γ-linolenate.

12.3.4 Molecular evidence

12.3.4.1 Proteins

The number of sequenced cyanobacterial ferredoxins and cytochromes c_6 is too limited to allow construction of a detailed phylogenetic scheme (see 12.2.2.1–12.2.2.3). However, these data together do suggest (i) that cyanobacteria of Sections IV (*Nostoc muscorum*) and V (*Mastigocladus laminosus*) diverged recently from each other and (ii) that the Section III genera *Plectonema* and *Spirulina* are more closely related to each other than to cyanobacteria of Sections I (*Anacystis nidulans*, *Synechococcus lividus*), IV or V, and may have arisen from a unicellular stock different from that giving rise to Sections IV and V (Schwartz & Dayhoff, 1981). The cross-reactivity of antibodies against *Spirulina maxima* ferredoxin is strong with ferredoxins from *S. platensis* and strains of Section V, and weak with strains of Section I, but the pattern of cross-reactivity with ferredoxins from strains of Sections III and IV are not easily interpreted phylogenetically (Tel-Or *et al.*, 1977a). The distribution of T and O type RuBP carboxylases is similarly difficult to explain in evolutionary terms.

12.3.4.2 DNA base composition and DNA:DNA hybridization

Although DNA G+C contents, when similar, do not necessarily indicate phylogenetic affinity, and the reasons for evolutionary change in G+C content remain unclear (Singer & Ames, 1970; Woese & Bleyman, 1972; Bak *et al.*, 1972; Herdman *et al.*, 1979b), divergence in G+C content is evidence for phylogenetic divergence. The range of G+C contents shown by an entire group can be taken as a rough and minimal measure of its evolutionary diversity and antiquity (Stanier *et al.*, 1976; Ragan & Chapman, 1978).

Bacterial chromosomal DNAs span the range of G+C contents from *c.* 25 to 75 mol %, near the limits allowable by the genetic code (Woese & Bleyman, 1972). Cyanobacterial chromosomal DNA G+C contents (see Herdman *et al.*, 1979b, and earlier literature reviewed therein) span nearly as great a range (35–71 mol%; see Table 12.1). However, this diversity is confined almost entirely to the single Section I 'genus' defined (by morphology and reproductive mode; Rippka *et al.*, 1979) as *Synechococcus*, and to the 'LPP-group' of Section III. The *Synechococcus* range (39–71 mol %) is much greater than that shown by most recognized bacterial 'genera' (Normore & Brown, 1970). Three clusters within *Synechococcus* can be defined by G+C content: low (39–43 mol %), intermediate (45–56 mol %) and high (66–71 mol %) (Herdman *et al.*, 1979b; see Table 12.1). The Section I 'genus' *Synechocystis* is similarly divisible into two clusters (35–37 mol % and 42–48 mol %). Other genera of Section I show more restricted ranges of base composition, although they are admittedly represented by fewer strains.

Three of the four remaining major cyanobacterial groups recognized by Stanier and coworkers are of relatively restricted base composition (38–47 mol % for Section II, 38–47 mol % for Section IV, and 42–46 mol % for Section V), in spite of the much greater morphological and developmental diversity shown by their members. Cyanobacteria of Section III show a broader range (40–67 mol %) of G+C contents, most of the variability lying within the LPP (*Lyngbya*, *Phormidium*, *Plectonema*) group (42–67 mol %).

It seems hazardous to draw from these data conclusions other than three obvious ones. (i) The (heterocystous) cyanobacteria of Sections IV and V may have diverged from the unicellular or non-heterocystous filamentous forms which gave rise to them more recently than the various unicellular lineages diverged from each other. (ii) Cyanobacteria of Section III are either polyphyletic or began to diversify at some intermediate time. (iii) Retrogressive evolution of all non-heterocystous filamentous cyanobacteria (Section III) from Stigonematalean (Section V) or Nostocalean (Section IV) species (Desikachary, 1973) appears unlikely.

Lachance (1979, and Lachance & Stanier, in prep.) has used DNA:DNA hybridization to assess phylogenetic relationships within Sections IV and V. Four major groups could be defined in this way. The genera *Chlorogloeopsis* (two strains), *Nodularia* (two strains), *Cylindrospermum* (three strains) and *Fischerella* (eight strains) each formed reasonably tight clusters (as defined by percentage cross-hybridization) within one or another of these groups. However, strains identified by the criteria of Rippka *et al.* (1979) as belonging to *Nostoc*, *Calothrix* or *Anabaena* were each found in at least two of the four groups. These genera thus appear polyphyletic. None of the heterocystous strains of Sections IV and V showed any homology detectable by DNA : DNA hybridization with the unicellular cyanobacteria of Section I or the non-heterocystous filamentous cyanobacteria of Section III (M.-A. Lachance, pers. comm.). Together these results seem to imply that the ability to produce heterocysts arose only once in cyanobacterial evolution, and that all heterocystous species are part of a phylogenetically restricted assemblage, whose phenetic division on the basis of morphology is misleading.

Stam and Venema (1977) and Stam (1981) have used DNA:DNA hybridization to measure relatedness of 27 strains variously identified as belonging to the (Section III) 'LPP' genera *Lyngbya*, *Plectonema*, *Phormidium* and *Schizothrix*, as well as other strains identified as belonging to *Nostoc*, *Anabaena*, *Tolypothrix*, *Mastigocladus* and *Anacystis* (*Synechococcus*). Eighteen LPP-type strains which showed a limited range of chromosomal G+C contents (45–49 mol %) gave cross-hybridization values not significantly different from 100%. These strains showed limited (17–22%) cross-hybridization with other non-heterocystous filamentous species, although these were also formally identified as *Phormidium* or *Schizothrix*, and had similar G+C contents. DNAs of *Nostoc*, *Anabaena*, *Tolypothrix*, *Mastigocladus* and *Anacystis* gave no detectable hybrids with DNAs of the LPP type.

12.3.4.3 T1 oligonucleotide catalogues of 16S ribosomal RNAs

Bonen and Doolittle have applied the technique of partial sequence characterization (T1 oligonucleotide cataloguing) described above (12.2.2.4) to the 16S rRNAs of eight cyanobacterial strains of groups I, IV and V (Bonen & Doolittle, 1975, 1976, 1978; Bonen *et al.*, 1979). The results are summarized in the dendrogram shown in Fig. 12.2. They suggest that (i) *Synechococcus* strains 6301 (*Anacystis nidulans*, 55.1 mol % G+C) and 7502 (39 mol % G+C) are, as expected, only distantly related, (ii) *Agmenellum quadruplicatum* (49 mol % G+C) although considered a *Synechococcus* by Rippka and coworkers, is closely related to a *Synechocystis* strain (6714) of similar G+C content (47.4 mol %), (iii) *Nostoc* (Secton IV) and *Fischerella* (Section V) are more closely related to each other than are members of the high (42–48 mol %) and low (35–37 mol %) G+C clusters of *Synechocystis* (strains 6714 and

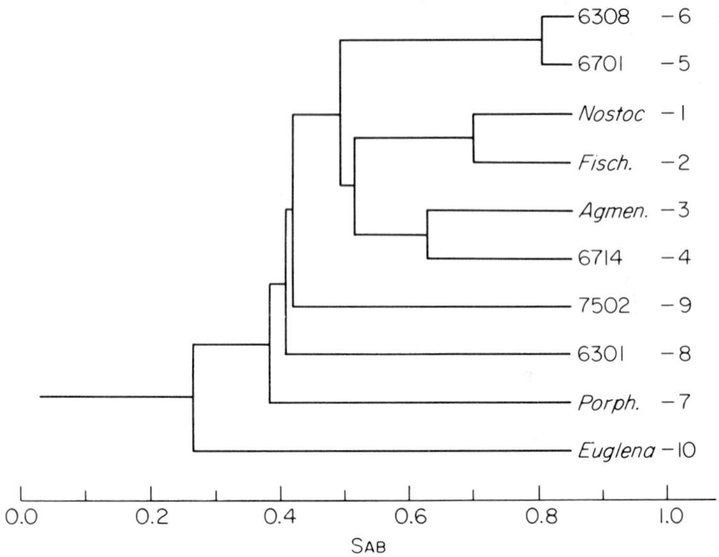

Fig. 12.2. A dendrogram representing relationships between the 16S rRNAs of eight cyanobacteria and the plastids of *Porphyridium cruentum* and *Euglena gracilis*. Strains 6308, 6701 and 6714 are described as *Synechocystis* species by Rippka *et al.* (1979). Strains 6301, 7502 and *Agmenellum quadruplicatum* are considered by them to belong to the genus *Synechococcus*. The *Nostoc* strain represented is that usually described as strain MAC. The *Fischerella* strain was described by L.O. Ingram as *Fischerella ambigua* (pers comm.). Figure reprinted from Bonen *et al.* (1979), with permission from the National Research Council of Canada.

6308/6701, respectively) and (iv) Section IV and V filamentous cyanobacteria seem to have arisen from within the high G+C *Synechocystis* cluster.

It is of interest that only this high G+C group, among *Synechocystis* strains, shows the high fatty-acid levels characteristic of filamentous species. Even though the 18.3 fatty acids of the unicellular species are predominantly 18.3γ, while filamentous species more often show 18.3α, some high G+C *Synechocystis* strains do produce both isomers (Kenyon *et al.*, 1972). Furthermore, high fatty-acid levels were found, among *Synechococcus* strains, predominantly in those of the intermediate (47–56 mol%) G+C cluster, and among these, only in strains which show G+C contents of less than 50%. A scheme in which Section IV and V cyanobacteria, high G+C *Synechocystis* strains, and those *Synechococcus* strains of intermediate G+C and high fatty-acid content together share a common ancestor more recent than that which any of them shares with most other unicellular species seems tenable (Fig. 12.3). It should be tested by DNA:DNA hybridization or 16S rRNA cataloguing. The latter method (which is the more sensitive) should also be applied to representatives of Sections II and III, as there are currently few molecular sequence data bearing on the phylogenetic positions of those groups.

12.4 Relationships between cyanobacteria and eukaryotes

12.4.1 Fundamental premises and alternative explanations

Homologies between oxygenic photosynthesis as it is practised by cyanobacteria and oxygenic photosynthesis as it is practised by eukaryotic photosynthesizers are many and profound (Stanier, 1970, 1974; Margulis, 1970, 1981; Allsopp, 1969; Ragan & Chapman, 1978; Broda, 1975; Olson, 1978; Mauzerall, 1978; Raven, 1970; Uzzell & Spolsky, 1974; Dodson, 1979; Whatley *et al.*, 1979). No one has recently (if ever) suggested that oxygenic photosynthesis arose independently in prokaryotes and eukaryotes or (except whimsically, Stanier, 1970) that oxygenic–photosynthetic prokaryotes descend from eukaryotic algae. Thus eukaryotes 'inherited' the ability to perform oxygenic photosynthesis from prokaryotes, and specifically from cyanobacteria, prochlorophytes, or other oxygenic–photosynthetic lineages derived from the common ancestor of these (Raven, 1970; Margulis, 1970; Stanier, 1974; Taylor, 1976).

There are two ways this could have happened. (i) Eukaryotic algae could have evolved 'directly' ('gradually' or 'autogenously') from oxygenic–photosynthetic prokaryotes by some complex series of intracellular rearrangements which resulted in the sequestration of some of the genetic information for photosynthetic functions within a membrane–bounded plastid. The rest of this information, and that required for cytoplasmic functions, would have remained within a separate membrane-bounded structure, the nucleus. Hypotheses of this sort, here called *autogenous origin hypotheses* (after Taylor, 1976), take several forms (Allsopp, 1969; Uzzell & Spolsky, 1974; Cavalier-Smith, 1975; Bogorad, 1975b; Taylor, 1976). However, all must suppose that the genomic lineage to which the plastid genomic lineage is most closely related, phylogenetically, is that of the nucleus within the same cell. (ii) Alternatively, eukaryotic algae could have arisen from non-photosynthetic 'protoeukaryotic' cells which first engulfed and later enslaved free-living oxygenic–photosynthetic prokaryotes, of which modern plastids are the degenerate descendants (Mereschowsky, 1905; Margulis, 1970; Raven, 1970; Stanier, 1970, 1974; Whatley *et al.*, 1979; Taylor, 1979). Endosymbiotic-origin hypotheses of this sort are here called *xenogenous origin hypotheses* (again after Taylor, 1976). They all must assume that the genomes of plastids are less closely related to those of the nuclei of the cells in which they reside than they are to those of modern representatives of the oxygenic–photosynthetic prokaryotes which gave rise to them, if any such survive.

12.4.2 How the xenogenous origin hypothesis might be proven

The xenogenous (endosymbiotic) hypothesis for the origin of plastids enjoys current favour (Taylor, 1979; Whatley *et al.*, 1979; Dodson, 1979; Doolittle, 1980), largely through the efforts of Margulis (1970, 1975a,b, 1981). Much of the available cyanobacterial molecular sequence data is directed at proving it. However, most such data (including most of that showing close affinity between cyanobacterial and algal cytochromes, ferredoxins and RuBP carboxylases discussed in 12.2.2.1–12.2.2.3) only confirms the close relationship of cyanobacteria and plastids, which is predicted by both sorts of hypothesis. There are only two ways to 'prove' the xenogenous origin hypothesis for plastids, and neither is perfect.

The first is to show that plastid genomes are less closely related to nuclear genomes resident within the same cell than they are to prokaryotic (or specifically cyanobacterial) genomes. This can be done only with sequence information on proteins or nucleic acids which have identifiable homologues separately encoded by plastid, nuclear and prokaryotic genomes. No sequenced proteins have identifiable plastid gene-coded, nuclear gene-coded and prokaryotic gene-coded homologues. The ribosomal and transfer RNAs (rRNAs and tRNAs) do (Gillham, 1978), and it is to them that we should turn. However, even when it has been shown (see below) that plastid rRNAs or tRNAs are more similar in sequence to prokaryotic than to nuclear gene-coded (cytoplasmic) rRNAs or tRNAs, autogenous origin hypotheses can always be salvaged with an auxiliary assumption. This assumption is that plastid RNAs have changed little since the intracellular divergence of nuclear and plastid genomes, while their (nuclear gene-coded) cytoplasmic homologues have experienced rapid evolutionary divergence. It is not clear why this should be so, but the assumption is not inherently illogical. It can be defeated only by demonstrating that the nuclear gene-coded RNA is more closely homologous to the RNA of some prokaryote which is not an oxygenic photosynthesizer; that is, by identifying the prokaryotic genomic lineage ancestral to the nuclear genomic lineage, and showing that it is not an oxygenic–photosynthetic one. Such an identification has not been, and probably never will be, made (Woese & Fox, 1977; Darnell, 1978; Doolittle, 1980).

The second way to 'prove' the xenogenous origin hypothesis requires no information on nuclear gene-coded products, but does require the inherently reasonable assumption (Taylor, 1976) that all of the features (histones, chromosomal organization, endomembrane and microtubular systems, 80S ribosomes) which define various algal groups (with the possible exception of dinoflagellates) as eukaryotic (Dodson, 1979; Roberts *et al.*, 1976) reflect a single common origin for their nuclear-cytoplasmic components. If this is assumed, and if it can then be shown by protein or nucleic acid sequencing data that the plastids of one group of algae arose from one class of oxygenic–photosynthetic prokaryotes, while those of a second group arose from a different class, then of necessity one or the other (or both) plastids are of endosymbiotic origin. A simpler way of saying this is that autogenous origin hypotheses cannot accommodate evidence for a polyphyletic origin of plastids (Taylor, 1976).

This discussion may seem unnecessarily belaboured, but adherents of autogenous origin hypotheses (Cavalier-Smith, 1975; Uzzell & Spolsky, 1974, 1981) remain articulate and vocal, and arguments become increasingly involuted. Evidence bearing on the two different sorts of proof will be presented below.

12.4.3 Proofs of the first sort: genomes of plastids are more closely related to those of cyanobacteria than to those of nuclei

12.4.3.1 RNA:DNA hybridization data

RNA:DNA hybridization experiments have provided several successful demonstrations that plastid ribosomal RNAs (rRNAs) are homologous in sequence to those of cyanobacteria (or other prokaryotes) and not homologous to nuclear gene-coded cytoplasmic rRNAs (see reviews by Phillips & Carr, 1977, 1981). Pigott and Carr (1972) showed that total ^{32}P-labelled rRNAs from cyanobacteria hybridized strongly to *Euglena* chloroplast DNA, highest values being obtained with *Gloeocapsa alpicola* (Section I). Bacterial (including purple non-sulphur bacterial) rRNAs hybridized only weakly to this DNA, and *Euglena* cytoplasmic rRNA hybridized not at all. Gruol *et al.* (1975) produced similar results, showing further that *Euglena* chloroplast and cytoplasmic rRNAs formed hybrids only with chloroplast DNA or nuclear DNA (respectively), and that rRNA from the Section I cyanobacterium *Anacystis nidulans* formed specific hybrids only with the former DNA. Phillips and Carr (1975) used a variety of purified 5S rRNAs labelled *in vitro* with ^{125}I, in hybridization experiments with *Euglena* chloroplast DNA. All prokaryotic 5S rRNAs employed showed more extensive hybridization than did *Euglena* cytoplasmic 5S, with highest values again being obtained using the Section I cyanobacterium *Gloeocapsa alpicola*.

12.4.3.2 16S ribosomal RNA catalogue data

The T1-oligonucleotide cataloguing techniques described in 12.2.2.4 and 12.3.4.3 have been applied by Bonen and Doolittle (1975, 1976, 1978; see also Bonen *et al.*, 1979; Doolittle & Bonen, 1981) to the plastid 16S and cytoplasmic 18S rRNAs of the red alga *Porphyridium cruentum*, by Zablen *et al.* (1975) to the plastid 16S of *Euglena gracilis*, and by C.R. Woese, G.M. Bowman and T.A. Dyer (pers. comm.) to the plastid 16S and cytoplasmic 18S rRNAs of *Lemna* (duckweed). Similar information is also available on the cytoplasmic 18S rRNAs of wheat (Bonen *et al.*, 1977), yeast and mammalian cells (Woese & Fox, 1977). Catalogue comparisons together allow the following conclusions (see Fig. 12.2). (i) The plastid 16S rRNAs of *Porphyridium*, *Euglena* (and *Lemna*) are all more closely homologous to cyanobacterial 16S rRNAs than they are to any bacterial 16S rRNA, although they are demonstrably related to these latter. (ii) *Porphyridium* plastid 16S rRNA is especially closely related to cyanobacterial 16S rRNA, and only minor modifications in clustering techniques would result in the inclusion of this RNA within the cyanobacterial assemblage (G.E. Fox & C.R. Woese, pers. comm.). (iii) None of the plastid RNAs shows homology detectable by this method to the cytoplasmic 18S rRNAs contained within the same or different eukaryotic cells, nor in fact is significant homology demonstrable between any eukaryotic 18S rRNA and any prokaryotic 16S rRNA (Woese & Fox, 1977).

12.4.3.3 Complete sequences of rRNAs

Dyer and Bowman (1979) have reported the complete sequence of the 5S rRNA of *Lemna* chloroplast, and compared it to the previously determined sequences of the cytoplasmic 5S rRNA of *Lemna* and to various prokaryotic 5S rRNAs. The *Lemna* chloroplast 5S rRNA shows only 50% primary sequence homology to the cytoplasmic 5S of the same species, but 63% homology with *Escherichia coli* 5S and 75% homology with cyanobacterial 5S (*Anacystis nidulans*; Corry *et al.*, 1974).

Schwarz and Kössel (1980) have completed the entire sequence of the cloned 16S rRNA gene of *Zea mays* chloroplast, and find it to be 74% homologous to that of a 16S rRNA gene of *Escherichia coli* (Brosius *et al.*, 1978). This is a remarkably high degree of sequence conservatism. Although no eukaryotic cytoplasmic 18S rRNA has yet been fully sequenced, catalogue data indicate that only a very much lower level of homology can be expected (Bonen & Doolittle, 1976; Woese & Fox, 1977) The 3′ terminal sequence of the plastid species (5′-UGGAUCACCUCC$_{OH}$-3′) is identical to that of *E. coli*, and presumably therefore similarly interacts with complementary sequences in plastid messenger RNAs.

12.4.3.4 Conclusion

In all cases in which homologous gene products coded for by nuclear genomes, plastid genomes and prokaryotic (cyanobacterial) genomes can be compared, plastid molecules are considerably more similar in sequence to prokaryotic (and specifically cyanobacterial) molecules than they are to cytoplasmic molecules. (These cases include not only the rRNAs discussed above but also the $tRNA^{Phe}$ of *Euglena* [Larue *et al.*, 1977]). This result can only be reconciled with autogenous origin hypotheses by invoking the auxiliary *ad hoc* assumption discussed in 12.4.2, but is directly predicted by xenogenous origin hypotheses.

12.4.4 Proofs of the second sort: plastids are of polyphyletic origin

If it can be shown that the plastids of different major groups of eukaryotic photosynthesizers arose from different oxygenic–photosynthetic prokaryotic lineages, then autogenous explanations can only be maintained by assuming that nuclei arose separately in each of these lineages, and that their many similarities reflect convergence on an unprecedented scale (Taylor, 1976). Both RNA and protein sequence data support the notion that plastids are of polyphyletic origin.

12.4.4.1 Ribosomal RNAs

Porphyridium plastid 16S rRNA is much more closely related to cyanobacterial 16S rRNAs than either it or they are to *Euglena* (or *Lemna*) plastid 16S rRNA (Fig. 12.2; C.R. Woese, pers. comm.). The last common ancestor of *Porphyridium* plastids and cyanobacteria must have been a free-living prokaryote (unless cyanobacteria are feral plastids). Thus the last common ancestor of *Porphyridium* and *Euglena* plastids must also have been a free-living, oxygenic–photosynthetic prokaryote; the two plastids are of independent (diphyletic) origin, each having separately evolved from once free-living prokaryotes.

12.4.4.2 Biliproteins

The work of Glazer, Stanier, Cohen-Bazire and collaborators has shown that the phycobiliproteins (see Chapter 7) found, in prokaryotes, only among cyanobacteria and, in eukaryotes, only among red and cryptomonad algae, comprise a family of evolutionary homologues (Bryant *et al.*, 1978; Glazer & Apell, 1977; Glazer, 1976, 1977; Stanier, 1974). Various lines of evidence (immunological cross-reactivity, heterologous subunit reassociation, amino-acid sequences, intracellular organization) show that red algal and cyanobacterial phycobiliproteins are much more like each other than either is like cryptomonad phycobiliproteins. Thus cryptomonad and rhodophytan plastids are likely to have arisen independently, although from a similar (phycobiliprotein-containing) stock which, if it had modern free-living descendants, would be considered cyanobacterial.

12.4.4.3 Cytochromes c_6 and ferredoxins

Schwartz and Dayhoff (1978, 1981) have analysed in some detail the phylogenetic relationships between plastids and cyanobacteria inferred from the comparison of sequences of ferredoxins and cytochromes c_6. Ferredoxin sequences provide good evidence for polyphyly. This evidence is gratifyingly consistent with the popular notion that rhodophytan plastids are of cyanobacterial origin (Stanier, 1974), while plastids of chlorophytes and

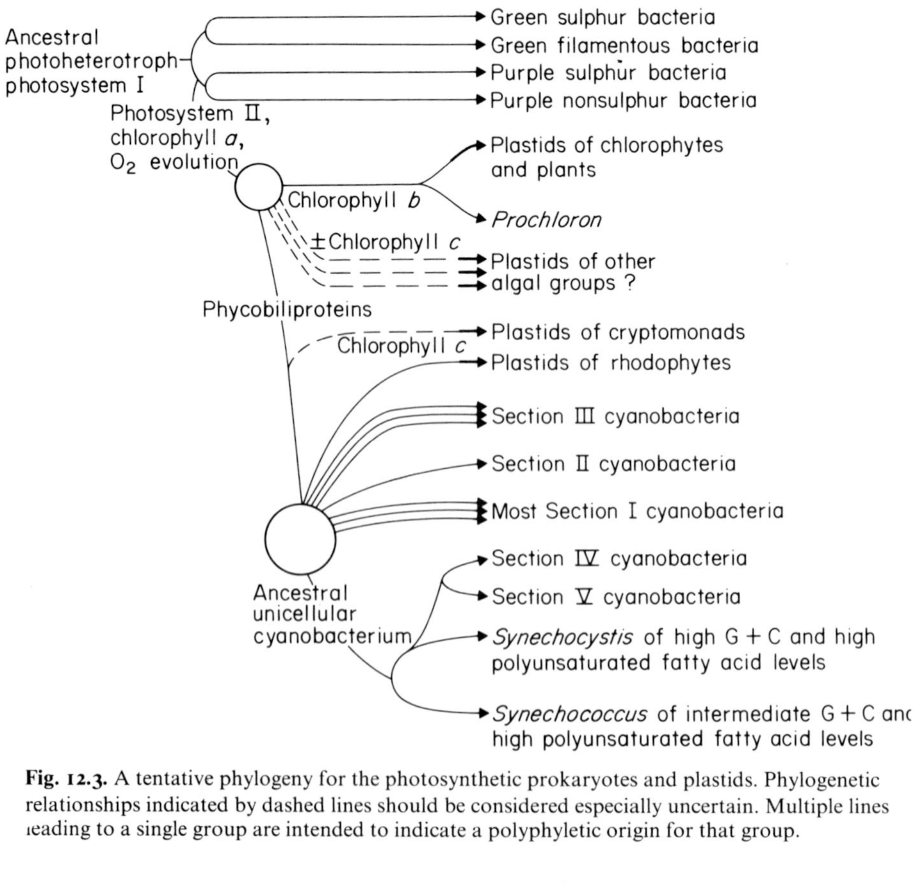

Fig. 12.3. A tentative phylogeny for the photosynthetic prokaryotes and plastids. Phylogenetic relationships indicated by dashed lines should be considered especially uncertain. Multiple lines leading to a single group are intended to indicate a polyphyletic origin for that group.

higher plants derive from a chlorophyll *a*- and *b*-containing phycobiliprotein-deficient oxygenic–photosynthetic prokaryote (Raven, 1970) like *Prochloron* (Withers *et al.*, 1978b). Although a ferredoxin gene duplication apparently occurred early in the cyanobacterial lineage, and 'local' duplications have since occurred in various plastid lineages, sequence data allow clear distinction of orthologues and paralogues (Schwartz & Dayhoff, 1978, 1981). The data indicate one endosymbiotic event occurring within the cyanobacteria which gave rise to *Porphyra* and *Cyanidium* (red algal) plastids and another, which can be viewed as outside the cyanobacteria, which gave rise to the plastids of *Scenedesmus* and higher plants.

The cytochrome c_6 data are more perplexing. They imply that the chlorophyll *c*-containing plastids of the phaeophyte *Alaria esculenta* share a common plastid ancestor with rhodophytan plastids, and that *Euglena* plastids and those of the chlorophyll *c*-containing chrysophyte *Monochrysis lutheri* share a common plastid ancestor which would have to have been cyanobacterial in origin. These implications are at variance with other data suggesting a common, and non-cyanobacterial, ancestor for phaeophytan and chrysophytan plastids (Ragan & Chapman, 1978). It is possible that the growing evidence (Whatley *et al.*, 1979; Whatley, 1981; Gibbs, 1981) that the plastids of these two groups (and perhaps of all algal groups other than rhydophytes and chlorophytes) are in fact the descendants of endosymbiotic *eukaryotic* algae (and not of free-living prokaryotes) will lead to a resolution of this dilemma. It remains unclear what form this resolution might take.

12.5 Summary

More data are clearly needed before any of the questions asked in 12.1.2 can be answered. However, the temptation to propose a specific phylogeny for photosynthetic prokaryotes and plastids is too strong to resist. Such a phylogeny is presented in Fig. 12.3. It rests only in part on the molecular sequence data presented here, is obviously derivative in many ways from phylogenies proposed by others, and involves too many *ad hoc* assumptions to enumerate. It should be regarded as tentative.

Acknowledgments

I am very much indebted to R.M. Schwartz, C.R. Woese, R.E. Dickerson, M.-A. Lachance and W.T. Stam for providing me with data and ideas in advance of publication, and to M.A. Ragan and D.J. Chapman for critical comments on the text. Work from my laboratory described in this chapter has been supported by the Medical Research Council of Canada and the Natural Science and Engineering Research Council of Canada.

CHAPTER 13
FUNCTION, STRUCTURE AND COMPOSITION OF CELL WALLS AND EXTERNAL LAYERS

G. DREWS & J. WECKESSER

13.1 Types of cell envelope layers and their functions

The terminology for biological structures has been strongly influenced by the history of the discovery of these structures. Since the fine structure and macromolecular composition of the cyanobacterial cell wall was found to be similar to that of Gram-negative bacteria, the terminology introduced for the cytology of the Gram-negative bacterial cell envelope will be used (see Weckesser *et al.*, 1979) instead of the provisional terminology used earlier. The term 'cell envelope' comprises all layers surrounding the cytoplasm. That means—from inside to outside—'the cytoplasmic membrane', the 'cell wall' ('peptidoglycan layer' plus 'outer membrane') and the 'external layers' surrounding the cell wall (Fig. 13.1). There is no uniform terminology for the external layers (Stanier & Cohen-Bazire, 1977). We will call a structured external layer a 'sheath', and undefined, unstructured zones 'slimes'. The terms are independent of whether one or more cells are surrounded by the external layer.

The cell wall is involved in determination and maintenance of cell shape, size and arrangement of cells. Its synthesis is correlated with the process of growth, division and differentiation of cells. The export and import of substances can be controlled by the cell wall (Braun & Hantke, 1977).

Receptors for cyanophages are localized in the outer membrane (Samimi & Drews, 1978). The cell envelope may also be involved in the process of gliding motility in cyanobacteria (see Halfen & Castenholz, 1971a, and Chapter 16). The external layers can increase considerably the thickness of the cell envelope and are thought to protect the cell from unfavourable environmental conditions. The external layers contribute to the formation of multicellular organisms, e.g. of *Calothrix* or *Chlorogloeopsis* (Stanier & Cohen-Bazire, 1977).

The first section of this article describes the fine structure of the cell wall and external layers, while in the second section the contribution of the cell wall formation to the process of cell division and other morphogenetic events will be characterized. In the third section, the present knowledge on the chemical composition of cell wall and external layers is summarized.

13.2 Cytology and supramolecular organization

13.2.1 Cell wall

The fine structure of the cell wall of cyanobacteria is of the Gram-negative type (Edwards *et al.*, 1968; Allen, 1968b; Jost, 1965; Butler & Allsopp, 1972). The innermost layer of the cell wall, the electron-opaque or 'peptidoglycan layer' ('L II layer' or 'murein sacculus' in the older terminology) overlays the cytoplasmic membrane. In ultrathin sections, the peptidoglycan layer is separated from the cytoplasmic membrane by an electron-transparent space (='L I layer', Figs 13.1–13.3). Fibrillar material has been described for the L I layer of *Spirulina platensis* (Van Eykelenburg, 1977). Generally, this layer is believed to be an artifact, formed by differential shrinkage of cell envelope layers and of the protoplast during fixation and embedding.

The peptidoglycan layer of most cyanobacteria is thicker than in most Gram-negative bacteria (Figs 13.2 and 13.8), and the width varies between 1 and 10 nm (mean value 5 nm) and can go to 200 nm in *Oscillatoria princeps* (Halfen & Castenholz, 1971a). The peptidoglycan layer is resistant to treatment with hot sodium dodecylsulphate (4%, 100°C) whereas both the outer membrane and the cytoplasmic membrane are solubilized by this detergent (Fig. 13.4; Golecki, 1977). By contrast, the peptidoglycan layer is completely or partially removed by lysozyme treatment (Jensen & Sicko, 1971b; Golecki & Drews, 1974). Isolated peptidoglycan layers show a smooth surface (Golecki, 1977) and conserve the shape of the cell (Höcht *et al.*, 1965; Fig. 13.4b).

Regularly arranged discontinuities have been observed in peptidoglycan layers of many cyanobacteria. Pores which average 13 nm in diameter are located in single circumferential rows on either side of every cross wall and are formed from the inside outward (Frank *et al.*, 1962a; Lamont, 1969a; Lang,

1977). Pit-like pores which average 70 nm diameter are uniformly distributed over the cell surface with an average centre-to-centre spacing of 200 nm (Shukovsky & Halfen, 1976; Lang, 1977). In *Starria zimbabweënsis* the outer membrane invaginates into and lines the depressions (Lang, 1977), whereas in *Oscillatoria* the cytoplasmic membrane evaginates into large pores (Lamont, 1969a; Halfen & Castenholz, 1971a; Shukovsky & Halfen, 1976). It is suggested that the large pores allow diffusion of sheath precursors and permit close contact with, and steady ATP supply to, the postulated contractile fibrils of the outer membrane (Lamont, 1969b; Halfen & Castenholz, 1971b). The junctional pores are involved in trichome breakage in oscillatorian cyanobacteria (Lamont, 1969a). Pores have also been observed in cross walls (Fig. 13.4b; Metzner, 1955).

Freeze fracture electron microscopy shows the peptidoglycan layer as a small, smooth edge between the fracture planes in the cytoplasmic and outer membrane (CW I; Golecki, 1977; Fig. 13.5; Sm̆arda *et al.*, 1979).

The outer membrane of the cell wall ('L IV layer' in the older terminology) appears as a double-track structure in micrographs of ultrathin sections (diameter 7–10 nm), separated from the peptidoglycan layer by an electron-transparent space (Butler & Allsopp, 1972) ('L III layer') (Figs 13.1–13.3). This space (L III layer) was observed in ultrathin sections by transmission electron microscopy, and seems to be an artifact of fixation, because outer membrane and peptidoglycan layer are tightly connected in freeze fracture preparations and in ultrathin sections of isolated cell wall preparations (J. Golecki, unpubl.). Total preparations of cell walls from *Spirulina platensis* reveal fibrillar material after shadowing. It was proposed to be associated with the L III layer (Van Eykelenburg, 1977). Similar material was observed in *Oscillatoria princeps* and *Oscillatoria animalis* (Halfen & Castenholz, 1971a). The outer membrane exhibits in ultrathin section a number of evaginations thought to represent sites of extrusion of material from the cytoplasm through the wall into the slime (Butler & Allsopp, 1972).

Fig. 13.1. Schematic representation of the cyanobacterial cell envelope. E, cell envelope; CM, cytoplasmic membrane; CW, cell wall (PL, peptidoglycan layer + OM, outer membrane); S, external layer (slime or sheath).

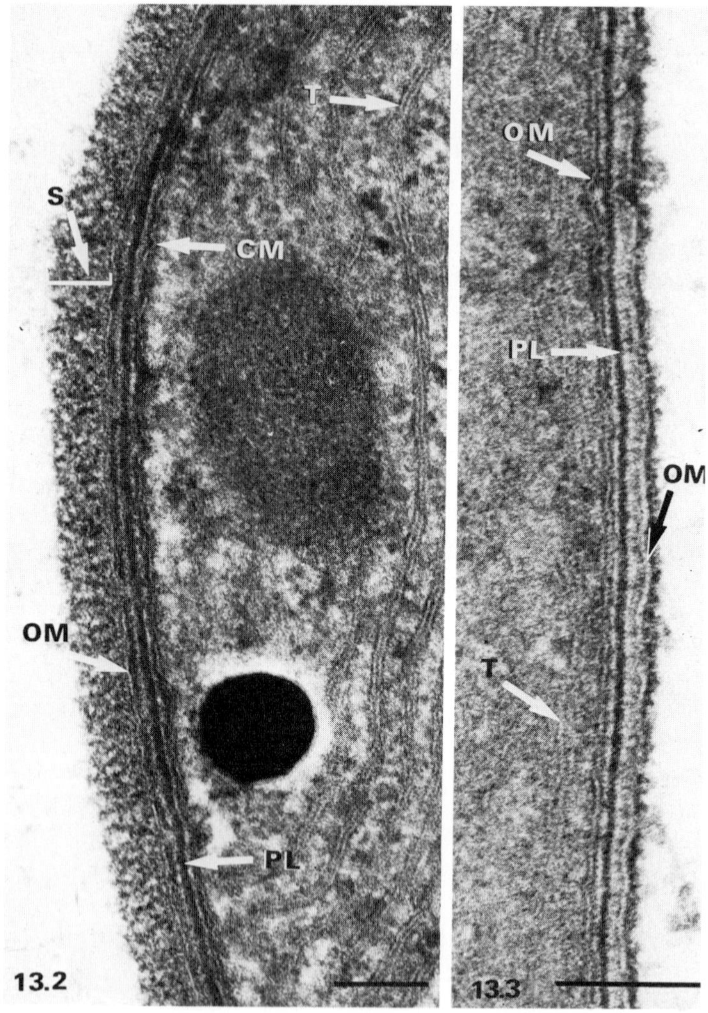

Fig. 13.2. *Gloeocapsa alpicola*. CM, cytoplasmic membrane; PL, peptidoglycan layer; OM, outer membrane; S, sheath; T, thylakoids. Micrograph by R. Jank-Ladwig. Bar = 0.1 nm.

Fig. 13.3. *Anacystis nidulans* (*Synechococcus*). For symbols see Fig. 13.2. Micrograph by R. Jank-Ladwig. Bar = 0.1 nm.

Within the outer membrane of *Synechococcus*, particles were detected on the convex fracture plane (plasmic fracture face, PF; CW-3 fracture plane, Golecki, 1977) by freeze-fracture electron microscopy (Fig. 13.6). They are frequently attached to each other and form triangular structures with a central hole of 3 nm diameter (Golecki, 1977). The exoplasmic fracture face (Ed; concave) of the CW-3 plane appears to be smooth. Sometimes the particles are

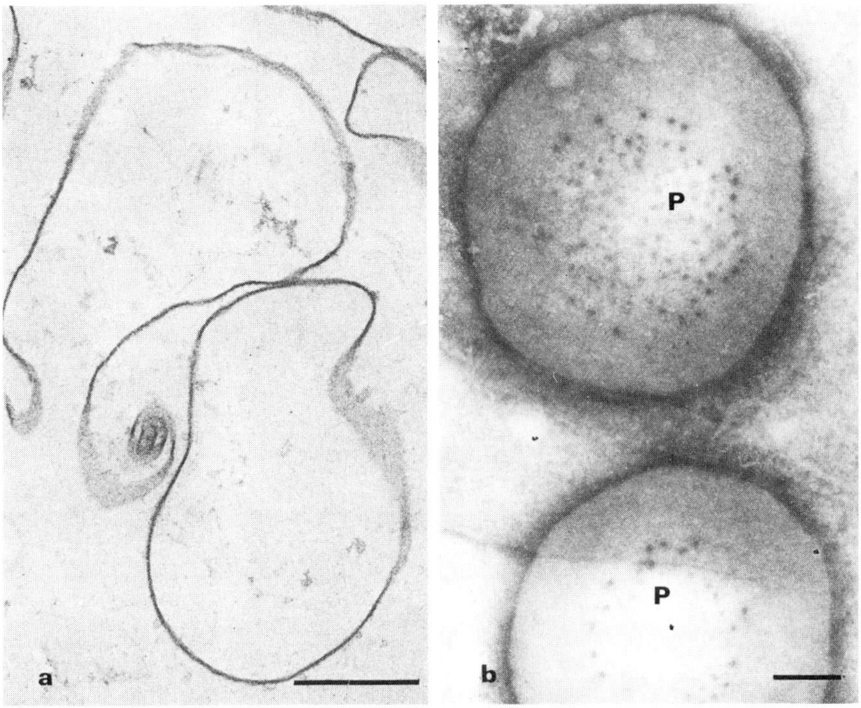

Fig. 13.4. (a) Peptidoglycan layer (murein sacculus) isolated from cell envelopes by treatment with hot sodium dodecylsulphate. Bar = 200 nm. From Golecki (1977). (b) Peptidoglycan layer of the cross wall after treatment with hot sodium dodecylsulphate. P, Pores in the peptidoglycan layer. Preparation and micrograph, J.R. Golecki, Bar = 200 nm.

continued into fibres (15 nm long, 6 nm in diameter). It is assumed that these fibres are produced during the fracturing process and, accordingly, are thought to be artifacts (Fig. 13.6; Golecki, 1977).

Below the CW-3 fracture plane, still within the outer membrane, a CW-2 plane is exposed by freeze-fracture technique. This plane is densely covered with flat particles of 6 nm diameter in *Anacystis nidulans* (*Synechococcus*) cell walls (Golecki, 1977) (Fig. 13.6a); a comparable particle layer has been detected in cell walls of *Anabaena variabilis* (Fig. 13.5), *Synechococcus lividus*, *Microcystis aeruginosa* and *Escherichia coli* (Golecki & Drews, 1974). The mean distance of particles is approximately the same as for *Anacystis nidulans*. The appearance of two fracture faces within the outer membrane is exceptional and needs further studies.

Particles of 10 nm diameter form a hexagonal lattice on the outer surface of the membrane of *Synechocystis aquatilis* (W 5). Particles of the same size, but not regularly arranged, are localized in a fracture face of the outer

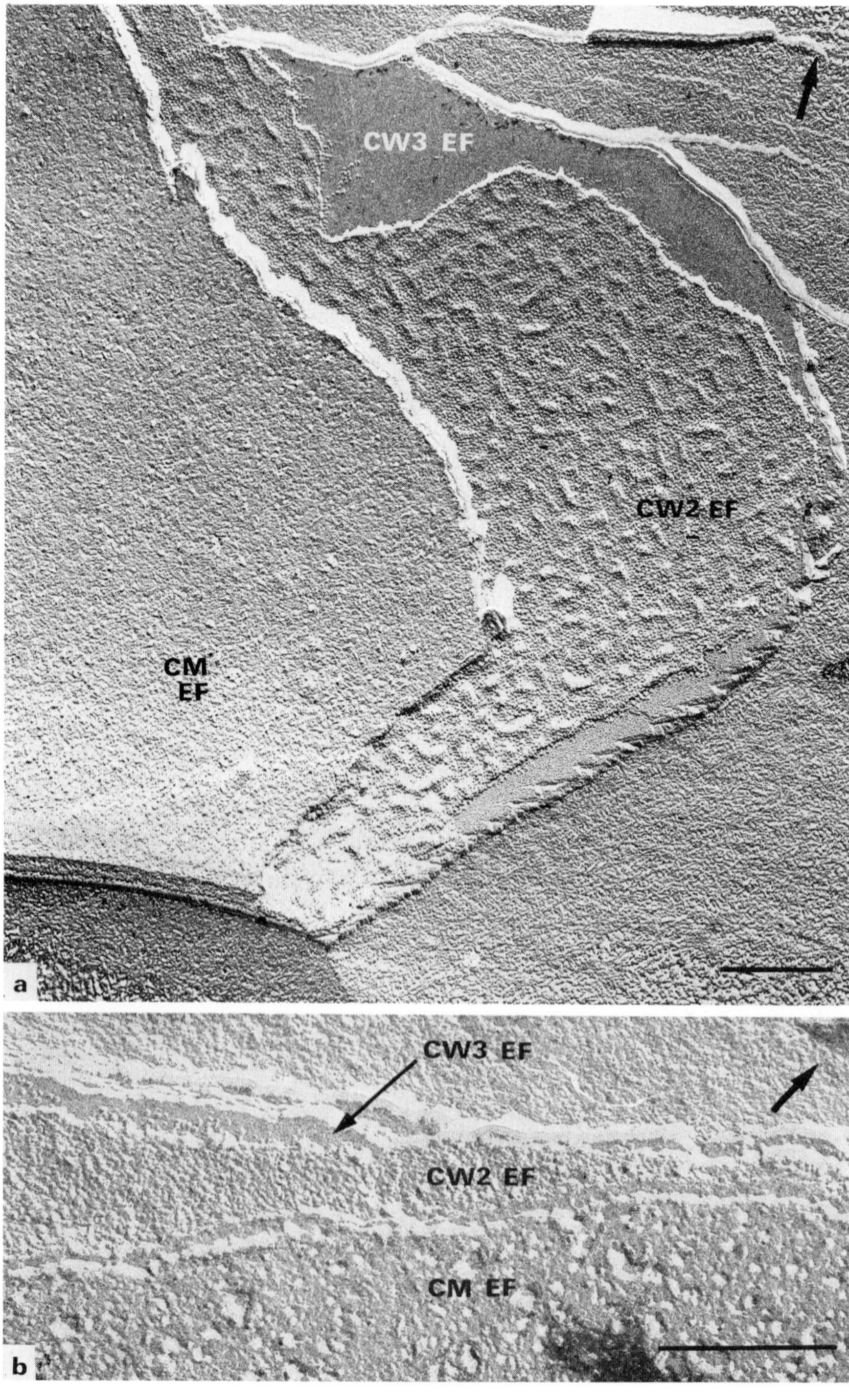
CW3 EF
CW2 EF
CM
EF
a
CW3 EF
CW2 EF
CM EF
b

membrane (W_4; Sm̆arda *et al.*, 1979), which corresponds to the CW-3 layer shown in cell walls of *Anacystis nidulans* (= *Synechococcus*) by Golecki (1977).

Membrane particles, which are exposed by freeze fracture electon-microscopy, are thought to consist of proteins. However, particles of the outer membrane, showing complementary pits on the opposite fracture face, are possibly of lipid origin (Verkleij *et al.*, 1979).

13.2.2 External layers

The Gram-negative cell wall may be surrounded by one or more additional layers of considerable thickness (up to 1 μm) (Figs 13.2, 13.7, 13.8). These layers have been named cell wall layer, sheath, capsule, or slime. The preparations obtained by various techniques may influence the delicate structure of the cell envelope in different ways and difficulties in terminology have been discussed recently (Drews, 1973; Martin & Wyatt, 1974; Stanier & Cohen-Bazire, 1977).

Pleurocapsalean cyanobacteria synthesize a fibrous material deposited outside but closely adherent to the outer membrane (Fig. 13.8). The material is produced and laid down after cell division upon the whole surface of the outer membrane. Consequently, after some generations the cells are surrounded by one or several fibrous external layers (Waterbury & Stanier, 1978; Fig. 13.9). Some unicellular cyanobacteria such as *Gloeobacter*, *Gloeothece*, and *Gloeocapsa* also form external layers (Fig. 13.2). The layers are deposited initially adjacent to the outer membrane of a single cell. After division and growth of the enclosed cells, the external layer becomes expanded and detached from the outer membrane until it is eventually ruptured (Kessel & Eloff, 1975). In *Gloeothece*, the enclosed cells synthesize new layers. Thus, aggregates of cells are formed in which singles, doubles or multiples of cells are wrapped by envelope layers. This type of formation of the multi-sheathed aggregates may indicate that the external cell envelope layers are expandable and synthesized during the growth period of the cell cycle but not during cell division (Waterbury & Stanier, 1978).

Cells of *Chlorogloeopsis fritschii* (previously known as *Chlorogloea fritschii*) undertake cell division in more than one plane during their development. Some of the resulting cell aggregates are enveloped by thick multi-layered sheaths which are ruptured after growth and division of cells.

Fig. 13.5. (a) *Anabaena variabilis*. Exoplasmic fracture faces exposing cytoplasmic membrane (CM), peptidoglycan layer (PL) and inner-membrane fracture faces of the outer membrane. Preparation by V. Speth, micrograph J.R. Golecki. Bar = 200 nm. (b) Larger magnification of exoplasmic fracture faces (EF) of the cell envelope of *Anabaena variabilis*. CW-2 and CW-3 are EF within the outer membrane. For other symbols see (a). Bar = 200 nm. From Golecki and Drews (1974)

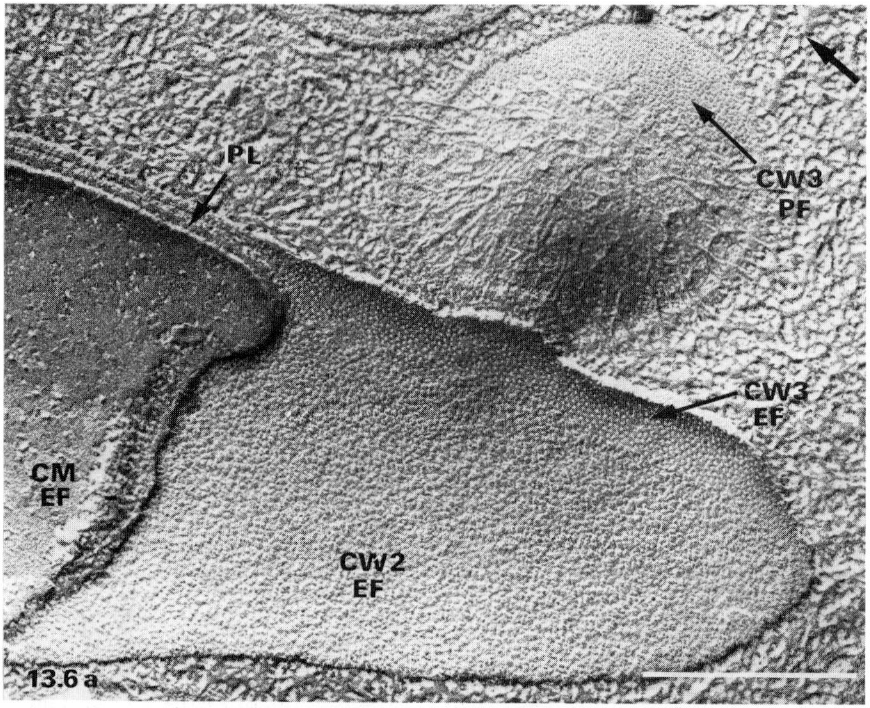
PL
CW3
PF
CW3
EF
CM
EF
CW2
EF
13.6 a

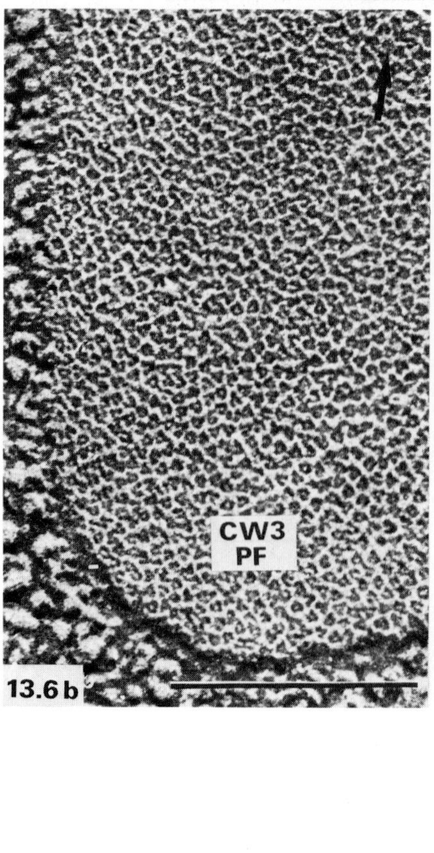
CW3
PF
13.6 b

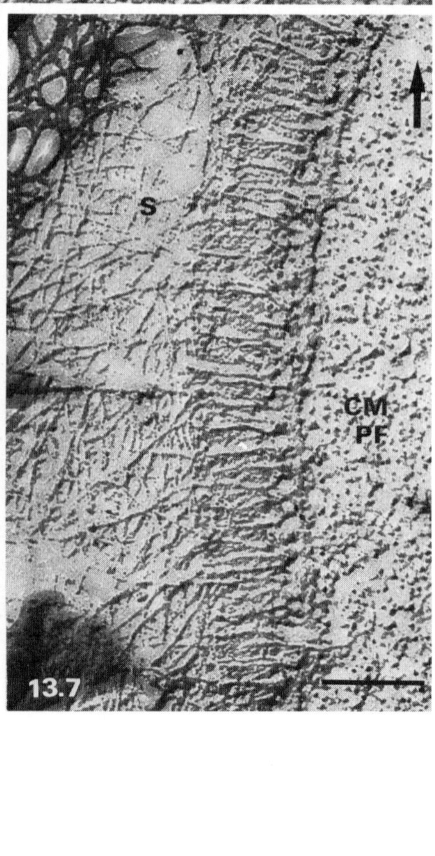
S
CM
PF
13.7

The material is fibrous (Evans, Foulds & Carr, 1976; M. Schrader *et al.*, 1982).

Some cyanobacteria excrete slime or mucilage which becomes dispersed around the organism and, to an extent, partially dissolved in the culture medium. The cells or trichomes are not regularly arranged in the slime mass (Martin & Wyatt, 1974). Slimy investments are frequently produced by planktonic cyanobacteria; they can be visualized on agar plates by Indian ink either as broad, amorphous shrouds insoluble in water or as regular zones around the cells with a definite delimitation (Martin & Wyatt, 1974). Slimy investments do not reflect the shape of the cell and have no well-defined structure. In ultrathin sections of *Synechocystis aquatilis* CB3 (Lounatmaa *et al.*, 1980) radial mucilaginous filaments were observed on the cell surface (Sm̆arda *et al.*, 1979). The projections observed in *S. aquatilis* penetrate the cell wall. An additional external layer seems to cover the outer membrane (Lounatmaa *et al.*, 1980). Many trichome-forming cyanobacteria are surrounded by a tubular sheath of fibrous fine structure (Lamont, 1969b; Frey-Wyssling & Stecher, 1954; Tuffery, 1969; Jost, 1965; Thurston & Ingram, 1971; Halfen & Castenholz, 1971a; Butler & Allsopp, 1972). The microfibrils in the mucilaginous investments of motile oscillatorean cyanobacteria lie roughly parallel and comprise a right-handed (*Oscillatoria chalybea*) or a left-handed (*Lyngbya*) helical array. It is proposed that the microfibrils are oriented by shear in a zone external to the outer membrane during helical motion (Schulz, 1955; Lamont, 1969b). Although the mechanism of gliding movement still remains an unsolved problem (Chapter 16), it seems to be clear that the slime production accompanies movement but does not cause it (Doetsch & Hageage, 1968; Halfen & Castenholz, 1971a,b). It is not clear whether the fibrils localized between the peptidoglycan layer and the outer membrane (Halfen & Castenholz, 1971) or on the cell wall (Halfen, 1973) are responsible for gliding movement (Jarosch, 1964; Halfen & Castenholz, 1971a; Halfen, 1973).

Sheaths formed by immotile trichomes (Rippka *et al.*, 1979), reveal also a fibrillar substructure (Tuffery, 1969; Thurston & Ingram, 1971). The fibrils seem not to be oriented or they radiate from the cell wall surface (Fig. 13.7; Echlin, 1964b; Leak, 1968; Frey-Wyssling & Stecher, 1954; Metzner, 1955).

Fig. 13.6. (a) *Anacystis nidulans*. Envelope layers exposed by freeze-fracture electron microscopy. CW-2 and CW-3 fracture faces proposed to be within the outer membrane (OM). The diameter of particles is about 7 nm. PL, peptidoglycan layer; exoplasmic fracture face (EF) of cytoplasmic membrane (CM) covered with numerous particles. The particles in the CW-2 fracture plane are tightly arranged (21 500 particles/μm^2). Bar = 200 nm. From Golecki (1977). (b) Plasmic fracture face (PF) of the CW-3 layer in the outer membrane at higher magnification. Bar = 100 nm. From Golecki (1977)

Fig. 13.7. *Anabaena variabilis*. Sheath (S) of fibrillar structure. Freeze-fracture electron microscopy. Bar = 200 nm. From Golecki and Drews (1974)

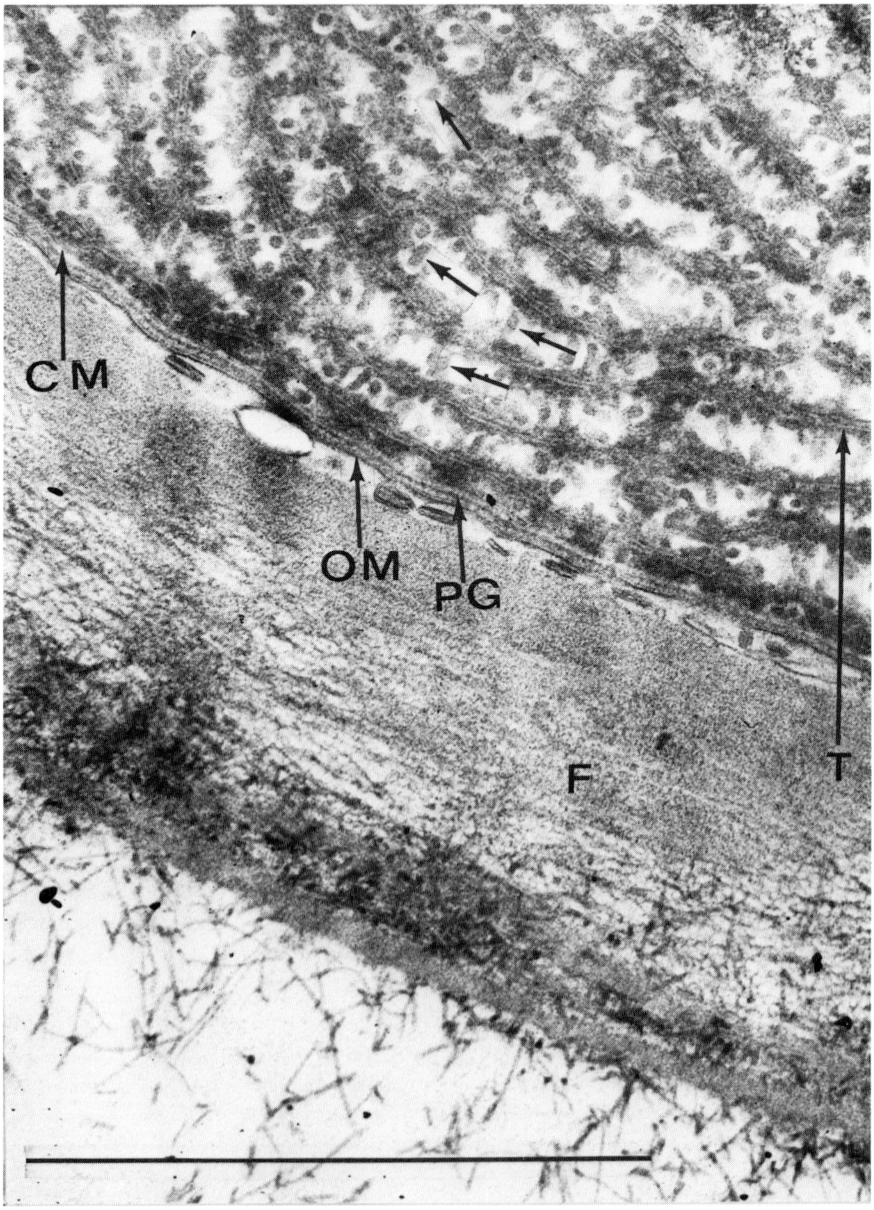
CM
OM
PG
F
T

During fixation, staining and embedding, the natural water content and the charge of the slime is changed. Thus structures seen in ultrathin sections may be more artificially modified than structures obtained by freeze etch technique. In addition, culture conditions may influence the formation of external envelope layers. Data on physical properties and chemical composition of slime and sheath material are needed before their function can be understood. A thin crystalline layer, consisting of regularly arranged hexagonal subunits, is formed external to the outer membrane of *Microcystis marginata*. The lattice constant was estimated to be about 13 nm (Kessel, 1978).

13.3 Cell division and separation

Three types of cell division have been observed among cyanobacteria: in the first, simultaneous invagination of all cell wall layers occurs (constrictive type, similar to that of Gram-negative bacteria; Figs 13.10, 13.11); in the second, a septum is formed by invagination of the cytoplasmic membrane and the peptidoglycan layer prior to the invagination of the outer membrane (septum-type) (Fig. 13.11). The difference between both types is more gradual than clear-cut. Cell division of chroococcacean and pleurocapsalean cyanobacteria, which are reproduced by binary fission, is initiated by invagination of the cytoplasmic membrane and peptidoglycan layer (Allen, 1968b). Outer membrane and external layers follow immediately, or delay this constrictive invagination (Allen, 1968b; Ingram & Thurston, 1970; Thurston & Ingram, 1971; Butler & Allsopp, 1972). In the constrictive type, cell division and cell separation are combined.

In many organisms of the heterocystous subgroups (*Anabaena*, *Nostoc*, *Cylindrospermum*, *Scytonema*, *Stigonema*, *Calothrix*, *Nodularia*) in which cell division is of the septum type, constriction follows septum formation after a delay. In the final step of cell division, i.e. cell separation, peptidoglycan–lytic enzymes seem to be active. The middle portion of the peptidoglycan cross septum of *Agmenellum quadruplicatum* was found to be more sensitive to lysozyme than was the layer immediately adjacent to the cytoplasmic membrane (Ingram & Aldrich, 1974).

In some oscillatorian cyanobacteria such as *Microcoleus vaginatus*, *Oscillatoria princeps* and *Phormidium uncinatum*, hormogonia formation, i.e. separation of short trichomes, is realized by transcellular breakage of trichomes. The peptidoglycan layer tears along a zipper-like circumferential set of junctional pores which are localized closely to the septa in the

Fig. 13.8. Electron micrograph of a thin section of part of a large vegetative cell of *Dermocarpa*. F, fibrous outer wall layer, 750 mm thick and of variable density; OM, outer membrane and PG, peptidoglycan layer forming the cell wall; CM, cytoplasmic membrane; T, thylakoids. Bar = 1.0 μm. From Waterbury and Stanier, (1978)

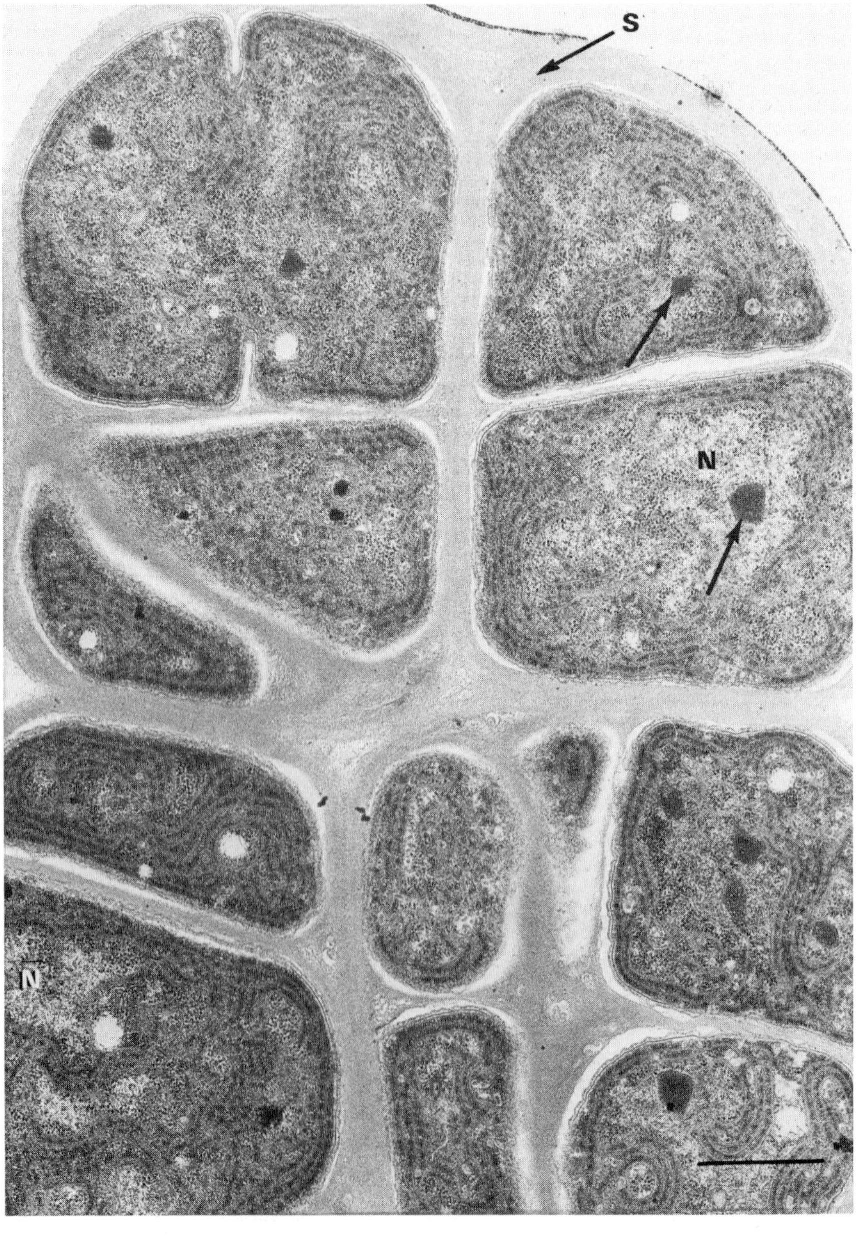
S
N
N

longitudinal walls (Frank *et al.*, 1962a; Lamont, 1969a; Halfen & Castenholz, 1971a). Tearing of the murein sacculus appears to be necessary and sufficient for transcellular breakage. This process involves the death of the cell whose wall is torn (Lamont, 1969a). In trichomes of other species, such as *Trichodesmium erythraeum*, the peptidoglycan layer of the longitudinal wall is lysed at the site adjacent to transverse wall attachment (Van Baalen & Brown, 1969). These weak points may facilitate cell breakage. The third type of cell division, the budding process, is established in *Chamaesiphon* (Rippka *et al.*, 1979).

A stimulation of peptidoglycan synthesis was observed during cross-wall formation (Dickens & Ingram, 1976). There was, however, no correlation between septum formation or cell separation and the rate of turnover of the peptide portion.

In cell populations of *Synechococcus*, synchronized by light–dark change, the new cell wall formation (cell division) starts 30 min after the completion of the previous cell division. Septum formation and cell division can be arrested by addition of chloramphenicol early in the cell cycle and addition of penicillin G (Gleason & Oaka, 1978).

Metabolites are transported from cell to cell in filamentous multicellular cyanobacteria (Chapter 14). Therefore, it was not surprising that morphological evidence for intercellular connections has been discovered. Pore-like structures were found by different electron-microscopical techniques (Fig. 13.4b; Drawert & Metzner, 1956; Hagedorn, 1961; Pankratz & Bowen, 1963; Wildon & Mercer, 1963a; Butler & Allsopp, 1972; Bisalputra *et al.*, 1975). Microplasmodesmata which connect adjacent cells, were demonstrated by freeze fracture technique in cells of *Anabaena cylindrica* (Giddings & Staehelin, 1978). The outside diameter of microplasmodesmata in *Anabaena cylindrica* is 20 nm. Thus, they are considerably smaller than those of the plasmodesmata of eukaryotic plants. It was speculated that in cyanobacteria the continuity of the cytoplasmic membrane is not preserved within the microplasmodesmata. Instead, the microplasmodesmata may represent here separate channel-forming elements that bridge the space between adjacent plasma membranes (Giddings & Staehelin, 1978).

Determination and maintenance of cell shape in cyanobacteria is an unsolved problem. However, the mechanisms that operate in the expression of

Fig. 13.9. *Chroococcidiopsis* strain 7432. Electron micrograph of a thin section. During binary fission the cross wall is formed by invagination of cytoplasmic membrane and peptidoglycan layer (PG) and subsequent ingrowth of the outer membrane (OM). After delineation of the cross wall, each daughter cell, now completely enclosed by PG and OM, begins to synthesize sheath material (S), which is laid down evenly over the entire surface of the OM. Consequently daughter cells are pushed apart from one another by intercalation of new S-material, but remain enclosed by the common sheath synthesized by the parental cell. Bar = 1.0 μm. From Waterbury and Stanier (1978)

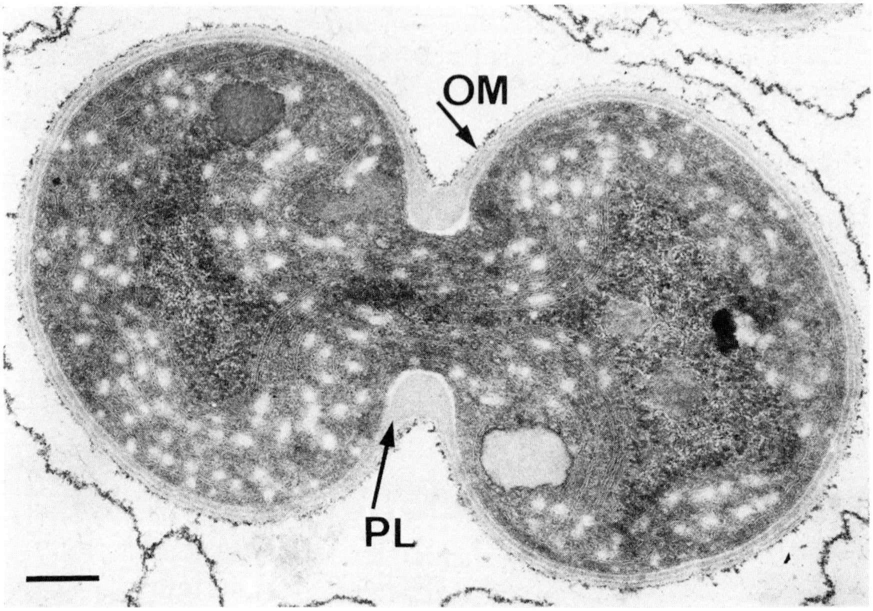

Fig. 13.10 *Synechocystis* PCC 6714. The cell is divided by the constrictive type of cell division. All cell wall layers ingrow simultaneously. The peptidoglycan layer is thickened in the division plane. PL, peptidoglycan layer; OM, outer membrane. Preparation and micrograph, J.R. Golecki. Bar = 200 nm.

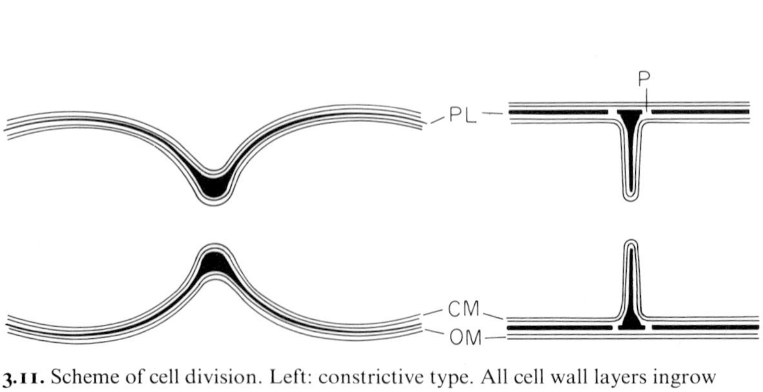

Fig. 13.11. Scheme of cell division. Left: constrictive type. All cell wall layers ingrow simultaneously. Right: Oscillatorian type; cytoplasmic membrane (CM) and peptidoglycan layer (PL) form a cross septum. The outer membrane (OM) is not involved in cell division. Cell separation is realized by transcellular breakage due to tearing of murein sacculus along a circumferential set of junctional pores (P) adjacent to the cross septum.

cellular shape must do so in the assembly of the cell envelope, since cell wall-free cells form protoplasts. Different macromolecules of the bacterial Gram-negative cell wall have been proposed to be responsible for cell shape (Henning, 1975).

Weidel and Pelzer (1964) suggested that the murein sacculus is analogous to the crustacean's outer-skeleton. Specificity for shape generation, however, cannot solely reside in the peptidoglycan layer, because this covalently linked structure does not represent a self-assembly system (Henning, 1975).

The cross-walls of *Spirulina*, a helical cyanobacterium, have a sectorial fold. Its size is related to the magnitude of their trichome pitch in different species. It is proposed that the fold in the cross-wall as observed *in vitro* must be ascribed to anisotropic shrinkage during isolation. The anisotropy might be due to the polysaccharide chains of the peptidoglycan layer running in the circumferential direction and the peptide side-chains running in the radial direction. These properties of peptidoglycan may determine the *in vivo* helical shape of the trichome and allow for flexibility and variability in shape (van Eykelenburg, 1979; van Eykelenburg *et al.*, 1980). Other components of the cell wall and especially the thick and structured sheaths might be of importance for maintenance of shape.

13.4 Chemical composition

In contrast to earlier ideas that pectins and cellulose might be constituents of the cyanobacterial cell wall (Metzner, 1955), components typical of the bacterial cell wall such as peptidogylcan and lipopolysaccharide are found in cyanobacteria (for reviews see Echlin & Morris, 1965; Lang, 1968; Wolk, 1973; Drews, 1973; Stanier & Cohen-Bazire, 1977; Weckesser *et al.*, 1979). The similarity to the Gram-negative type of cell wall, as indicated by the fine structural studies, was further established by recent analytical–chemical work as will be discussed in the following paragraphs.

13.4.1 Peptidoglycan

The presence of peptidoglycan in cyanobacteria was indicated by the sensitivity of cyanobacterial cells to specific enzymes and antibiotics. Treatment with penicillin or lysozyme inhibits their growth (Foter *et al.*, 1953; Frank *et al.*, 1962b) and induces lysis or formation of spheroplasts (Fuhs, 1958; Crespi *et al.*, 1962; Drews & Meyer, 1964; Biggins, 1967; Fulco *et al.*, 1967; Gusev *et al.*, 1970; Jensen & Sicko, 1971b; Lindsey *et al.*, 1971; Jones & Yopp, 1979). Cell envelopes (Drews and Meyer, 1964) or murein sacculi (Höcht *et al.*, 1965; Golecki & Drews, 1974) isolated from cyanobacteria were degraded by lysozyme.

Analytical–chemical data on peptidoglycan from cyanobacteria were first obtained when Work and Dewey (1953) found diaminopimelic acid in cyanobacterial cells and later by quantitative analysis of isolated peptidoglycan from *Phormidium unicinatum* cells (Frank *et al.*, 1962a; Table 13.1). Höcht *et al.* (1965) purified peptidoglycan by phenol extraction, and treatment with proteases and polysaccharide hydrolases. Molar ratios of glucosamine:muramicacid:alanine:glutamic acid:diaminopimelicacid of 1:1:2:1:1 were found with *Phormidium foveolarum* and *Tolypothrix tenuis*. Thus, a close structural similarity to the A-type of peptidoglycan from Gram-negative and some Gram-positive bacteria was indicated (Schleifer & Kandler, 1972).

The peptidoglycan from some cyanobacteria, including that of a halophilic strain (Jones & Yopp, 1979), has a higher glucosamine content than the A-type of Gram-negative bacteria. Ratios up to 2:1 = glucosamine:muramic acid were observed (Table 13.1). Although larger fragments of lysozyme-treated peptidoglycan of *Anacystis nidulans* still showed this high glucosamine:muramic acid ratio, smaller fragments had respectively 1.5:1 or 1:1 ratios (Höcht *et al.*, 1965). Structural studies are needed to clarify whether the additional glucosamine is a constituent of the peptidoglycan.

Additional amino sugars were found in peptidoglycan preparations of

Table 13.1. Composition of peptidoglycan fractions from cyanobacteria (mol ratios)

Peptidoglycan from	Mur	$GlcNH_2$	Ala	Glu	DAP
Anacystis nidulans (Drews & Meyer, 1964)	1.1	1.3	2.4	1.5	1.0
Anacystis nidulans (Höcht *et al.*, 1965)	0.9	1.9	2.0	1.1	1.0
Anacystis nidulans (Golecki, 1977)	0.9	1.5	1.9	1.1	1.0
Synechococcus lividus (M. Schrader, unpubl.)	0.9	2.1	2.1	1.2	1.0
Synechocystis PCC 6714 (U. Jürgens, pers. comm.)	1.1	1.9	2.4	1.5	1.0
Aphanothece halophytica (Jones & Yopp, 1979)	0.6	1.1	1.8	1.0	1.0
Chlorogloea fritschii (Drews & Meyer, 1964)	+	+	+	+	+
Phormidium uncinatum (Frank *et al.*, 1962a,b)	0.6	1.0	2.0	0.9	1.0
Phormidium foveolarum (Höcht *et al.*, 1965)	0.8	1.1	1.8	1.1	1.0
Tolypothrix tenuis (Höcht *et al.*, 1965)	0.8	1.1	1.9	1.1	1.0

+, Present but not quantified.

cyanobacteria; for *Anacystis nidulans* they were determined to be galactosamine (Drews & Meyer, 1964; Höcht *et al.*, 1965) and mannosamine (Katz, 1975; Golecki, 1977). Mannosamine is present in protein- and lipopolysaccharide-free peptidoglycan fractions of *Synechocystis* PCC 6714 (U. Jürgens, pers. comm.). It is not clear whether the amino sugar is a constituent of peptidoglycan (Höcht *et al.*, 1965). This is also true for mannose, found in peptidoglycan from *Anacystis nidulans* (Höcht *et al.*, 1965; Golecki, 1977) or from *Synechocystis* PCC 6714 (U. Jürgens, pers. comm.).

13.4.2 Proteins and lipoprotein

Amino acids, indicating protein, were found in varying amounts in cell envelope fractions of cyanobacteria (Frank *et al.*, 1962a,b; Drews & Meyer, 1964; Höcht *et al.*, 1965; Katz, 1975; Golecki, 1977). It is difficult, however, to confidently attribute them to the cell wall or more specifically to the outer membrane. Disruption of cells under mild conditions and fractionated centrifugation may not completely remove the cytoplasmic membrane (Drews & Gollwitzer, 1965; Golecki, 1977). Breaking of the cells with high mechanical power and treatment with detergents may result, on the other hand, in a partial loss of outer-membrane material.

Evidence for outer-membrane proteins of cyanobacteria was obtained from SDS-polyacrylamide gel electrophoresis of SDS or phenol extracts from cell envelopes of *Anacystis nidulans* (Golecki, 1977). Four major protein bands were detected. One of them (44 kdaltons) ran in the same position as a major protein of the *Escherichia coli* outer membrane, another one had an apparent molecular mass of 10 kdaltons, similar to the lipoprotein of *E. coli* cell walls.

Cell envelopes of *Anacystis nidulans* treated with SDS or trypsin contained more than 20% amino acids in addition to peptidoglycan (Drews & Meyer, 1964; Golecki, 1977); those of *Synechocystis* PCC 6714 at least 3% (U. Jürgens, unpubl.). In phenol-treated cell envelopes of *Phormidium foveolarum*, *Tolypothrix tenuis* and *Anacystis nidulans*, 25–35% protein was found (Höcht *et al.*, 1965). Lysozyme treatment of *Anacystis nidulans* peptidoglycan yielded a (C_4) fraction containing more than 8% of non-peptidoglycan amino acids (Höcht *et al.*, 1965). A protein, resembling envelope proteins of extremely halophilic bacteria in its chemical composition, was found in *Aphanothece halophytica* cell walls (Jones & Yopp, 1979). It remains to be studied whether cyanobacteria contain polypeptides covalently bound to peptidoglycan.

There is no evidence as yet for the presence of lipoprotein in cyanobacterial cell walls. No fatty acids were present in the SDS-insoluble cell envelope fraction of *Synechocystis* PCC 6714 and *Synechococcus lividus* OH-53S (U. Jürgens & M. Schrader, pers. comm.).

13.4.3 Lipids

A high lipid content was found in cell envelopes of *Anacystis nidulans* (Drews & Meyer, 1964; Drews & Gollwitzer, 1965). Isolated cell envelopes of *Synechococcus lividus* OH-53S and *Synechocystis* PCC 6714 both contain approximately 4% fatty acids. No lipid analyses, however, are available as yet. The fatty acid and lipid analyses on whole-cell extracts (Kenyon, 1972; Zepke *et al.*, 1978) do not allow specific assignments to lipids of the cell wall. Interestingly, the phosphate content of cell envelopes of cyanobacteria is low (0.1% of cell envelope dry weight in *Anacystis nidulans* (Drews & Meyer, 1964; Drews & Gollwitzer, 1965); 0.1 and 0.2% in *Synechococcus lividus* OH-53S and *Synechocystis* PCC 6714, respectively (M. Schrader, U. Jürgens, unpubl.)). This might indicate a low phospholipid content of the outer membrane assuming that no lipids were lost during the isolation of cell envelopes.

13.4.4 Lipopolysaccharides

In 1965, Höcht *et al.* described an 'amorphous' fraction obtained from crude cell walls of *Anacystis nidulans* by phenol extraction. It resembles the lipopolysaccharide of Gram-negative bacteria in physico-chemical properties, although no lipid and glucosamine were detected in this material which is rich in the neutral sugars mannose, galactose and glucose. The difficulty was overcome when Weise *et al.* (1970) extracted whole cells of *Anacystis nidulans* KM with hot phenol-water. They succeeded in the isolation of a highly aggregated polymer which could be reversibly disaggregated on treatment with 1% SDS and which was clearly shown to consist of polysaccharide with covalently bound lipid A. Mild acid hydrolysis cleaved off lipid A with the concomitant release of 2-keto-3-deoxyoctonate known to be the molecule connecting lipid A with the polysaccharide moiety of O-antigens of Gram-negative bacteria.

Major similarities between lipopolysaccharides from cyanobacteria and Gram-negative bacteria have been revealed since that time for a number of unicellular and filamentous strains of cyanobacteria. There is, however, considerable structural diversity in the lipid A moiety, even between closely related species, such as *Phormidium* and *Schizothrix* (Tables 13.2, 13.3). Whereas *Anabaena variabilis* contains a typical lipopolysaccharide (Weckesser *et al.*, 1974), such a polymer was not detected in *Anabaena flos-aquae* (Wang & Hill, 1977).

Sugar constituents known so far for cyanobacterial lipopolysaccharides comprise typical O-specific sugars (Table 13.2). The various O-methyl sugars found support the idea that sugar O-methyl ethers are characteristic constituents of lipopolysaccharides of both phototrophic bacteria and

Lipopolysaccharide from	2-O-Methyl-xylose	2,3,4-Tri-O-methyl-arabinose	2-O-Methyl-mannose	3-O-Methyl-D-mannose	4-O-Methyl-D-mannose	2-O-Methyl-fucose	3-O-Methyl-L-rhamnose	2,3-Di-O-methyl-fucose	6-O-Methyl-heptose	Xylose	Arabinose	Glucose	Galactose	Mannose	D-Glycero-L-glucoheptose	Rhamnose	Fucose	3,6-Dideoxy-D-arabinohexose	2-Keto-3-deoxyoctonate	Glucuronic acid	Galacturonic acid	Glucosamine	Galactosamine	Mannosamine	Unknown sugars
Synechococcus spp., chemotype I (Schmidt *et al.*, 1980a; Katz *et al.*, 1977)												●	●	●		●	●	●				●		●	●
Synechococcus spp., chemotype II (Schmidt *et al.*, 1980a; Katz *et al.*, 1977)		●										●	●	●			●					●			
Synechococcus spp., chemotype III (Schmidt *et al.*, 1980a; Katz *et al.*, 1977)				●	●							●	●	●			●		●			●		●	
Agmenellum quadruplicatum (Buttke & Ingram, 1975)										●		●	●	●		●			●			●			
Synechocystis PCC 6714 (Schmidt *et al.*, 1980b)												●	●	●			●					●			
Synechocystis PCC 6803 (Schmidt *et al.*, 1980b)	●					●		●				●	●	●			●					●			
Synechocystis PCC 6807 (Schmidt *et al.*, 1980b)			●						●			●	●	●	●	●	●			●	●	●			
Synechocystis PCC 6308 (Schmidt *et al.*, 1980b)			●									●	●	●	●	●	●			●	●	●			
Aphanothece halophytica (Jones & Yopp, 1979)												●	●	●			●								
Phormidium spp. (Mikheyskaya *et al.*, 1977)										●	●	●	●	●		●			●			●	●		●
Schizothrix calcicola (Keleti *et al.*, 1979)										●		●	●	●		●						●			
Anabaena variabilis (Weckesser *et al.*, 1974)							●					●	●	●		●						●			

Table 13.3. Fatty acids in lipopolysaccharides from various cyanobacteria

Lipopolysaccharide from	β-C_{12}OH	β-C_{14}OH	β-C_{16}OH	β-C_{17}OH	β-C_{18}OH	Branched β-OH fatty acids	C_{10}	C_{12}	C_{14}	C_{15}	C_{16}	$C_{16:1}$	C_{18}	$C_{18:1}$	$C_{18:2}$	C_{22}	Unknown fatty acids
Synechococcus spp. (Katz *et al.*, 1977; Schmidt *et al.*, 1980a)		(●)	●	(●)		●			(●)		●	●	●				●
Agmenellum quadruplicatum (Buttke & Ingram, 1975)		●	●		●						●					●	
Synechocystis spp. (Schmidt *et al.*, 1980b)		●	●			●			(●)		●	●	●				●
Aphanothece halophytica (Jones & Yopp, 1979)							●	●	●		●	●	●				●
Phormidium spp. (Mikheyskaya *et al.*, 1977)											●	●		●			
Schizothrix calcicola (Keleti *et al.*, 1979)	●		●						●	●	●	●		●	●		
Anabaena variabilis (Weckesser *et al.*, 1974)		●	●		●						●						●

(●) Present in a few strains only.

cyanobacteria (Weckesser *et al.*, 1979). As shown in Table 13.2, three different sugar chemotypes were revealed on investigation of lipopolysaccharides from nine *Synechococcus* strains (Weckesser *et al.*, 1979; Schmidt *et al.*, 1980a). Strains with a high GC (mol%) quotient of 66–71 belong to chemotype I, those with a lower GC (mol%) quotient of 48–55 to chemotype III. Chemotype II comprises two strains, one with a high (65.7) and one with a low (50.0) GC (mol%) quotient. The data thus support the idea that lipopolysaccharide composition can be used for taxonomic studies on cyanobacteria.
cated. Its validity, however, needs the investigation of additional strains.

The polysaccharide moiety of the *Anacystis nidulans* KM O-antigen has been studied in detail. It represents a D-mannose polymer with small amounts of additional sugars (Weise *et al.*, 1970; Katz *et al.*, 1977). From a methylation analysis it was concluded that it contains 1→3 bound mannose trisaccharides interlinked by 1→4 bonds (Katz *et al.*, 1977). 3-O-Methyl-D-mannose occupies terminal positions at the non-reducing end of these O-chains, 4-O-methyl-D-mannose those at their reducing ends (Tharanathan *et al.*, 1978). In the *Anabaena variabilis* O-antigen, containing L-acofriose, L-rhamnose, D-mannose, D-glucose and D-galactose (Weckesser *et al.*, 1974), equal amounts of terminal and chain-linked positions were revealed for both L-acofriose and L-rhamnose (Tharanathan *et al.*, 1978).

A typical core region, characteristic of enterobacterial lipopolysaccharides, was not obtained on gel-filtration of the polysaccharide moiety of the *Anacystis nidulans* O-antigen (Katz *et al.*, 1977). Moreover, the lack of L-glycero-D-mannoheptose and often also of 2-keto-3-deoxyoctonate in cyanobacterial lipopolysaccharides (Table 13.2) indicates major structural differences in this region of the polymer.

Attempts to isolate lipid A from cyanobacterial lipopolysaccharides known so far are hampered by the low content or lack of 2-keto-3-deoxyoctonate. Applying even 10% acetic acid (100°C, 6 h), lipid A is only partially (*Anacystis nidulans* KM; Katz *et al.*, 1977) or not at all (*Anabaena variabilis*; Weckesser *et al.*, 1974) split off from the polysaccharide moiety. Nevertheless, the *Anacystis nidulans* KM lipid A has been shown to have structural similarities to lipid A of Gram-negative bacteria in that glucosamine with amide-bound long-chain β-hydroxy fatty acids forms its backbone structure (Katz *et al.*, 1977). On the other hand, there are characteristic differences. First, the negligible phosphorus content in most of the known cyanobacterial lipopolysaccharides excludes phosphorylated glucosamine disaccharides (Galanos *et al.*, 1977). Second, β-hydroxypalmitic acid, rarely detected in lipid A from Gram-negative bacteria is by far the predominant fatty acid in most cyanobacterial lipopolysaccharides studied (Table 13.3), and even longer-chained β-hydroxy fatty acids such as β-hydroxy stearic acid were found. Lipid A from *Synechococcus* and *Synechocystis* strains contain rare branched (iso and ante-iso) β-hydroxy fatty acids (Schmidt *et al.*, 1980a,b). Some of the unsubstituted fatty acids may be bound to contaminating lipids. This is especially likely for lipopolysaccharides extracted into the phenol phase of phenol-water extracts such as those of chemotype I of *Synechococcus* (Schmidt *et al.*, 1980a).

The lipid moiety of lipopolysaccharide fractions of *Phormidium* strains amounts to only 1.5% of lipopolysaccharide dry weight (Mikheyskaya *et al.*, 1977) and the lipopolysaccharide of *Aphanothece halophytica* lacks β-hydroxy fatty acids (Jones & Yopp, 1979). The data indicate that these strains do not contain a typical lipid A moiety.

Lipopolysaccharides from cyanobacteria are part of the cell wall. This is indicated by their presence in cell envelope fractions (Weise *et al.*, 1970) and their serological activity in O-antisera. Presumably they are localized in the outer leaflet of the outer membrane. O-Titres published for lipopolysaccharides from most cyanobacteria are low (Schmidt *et al.*, 1980a,b). The reason for this is not known. It is also not known which polymers of the outer membrane will provide the necessary negative charge since many lipopolysaccharides from cyanobacteria such as that of *Anabaena variabilis* (Weckesser *et al.*, 1974) do not contain acidic constituents.

The participation of lipopolysaccharides in the process of adsorption of cyanophages onto the cell surface confirms their localization in the outer membrane (Samimi & Drews, 1978). Specificity in infection of different strains corresponds with lipopolysaccharide chemotyping (Schmidt *et al.*, 1980a; Table 13.2). Only cells of chemotype III propagate the phage; those of chemotype I still adsorb the phage and become lysed, the phage however is not propagated. Isolated lipopolysaccharide of chemotype III but not of chemotypes I or II is able to inactive the phage. The specificity resides in the polysaccharide and not in lipid A. The receptor complex is known to contain proteins (Samimi & Drews, 1978).

Lipopolysaccharides from cyanobacteria are not pyrogenic to rabbits and not lethally toxic to mice (Weise *et al.*, 1970; Katz *et al.*, 1977; Mikheyskaya *et al.*, 1977; Keleti *et al.*, 1979; Schmidt *et al.*, 1980a). A possible correlation between low phosphorus content and low endotoxic activity of lipid A has been suggested (Weckesser *et al.*, 1979). In contrast to a weak local Schwartzman reaction of the *Agmenellum quadruplicatum* lipopolysaccharide (Buttke & Ingram, 1975), lipopolysaccharide from *Schizothrix calcicola* induced *Limulus* lysate gelation and positive local Schwartzman reaction (Keleti *et al.*, 1979). Low toxicity to mice may not necessarily mean little toxicity to man. Whether lipopolysaccharide contributes to the toxicity of some cyanobacteria (Sykora *et al.*, 1980) will have to be studied.

13.4.5 Polysaccharides

Heteropolysaccharides are regular constituents of cell walls of cyanobacteria. In a number of strains they have been shown to be part of lipopolysaccharides (Table 13.2). Although with other strains no fatty-acid analyses have been performed (*Anacystis nidulans* (Drews & Gollwitzer, 1965; Höcht *et al.*, 1965), *Phormidium foveolarum* and *Tolypothrix tenuis* (Höcht *et al.*, 1965), *Anabaena cylindrica* (Dunn & Wolk, 1970)), it is reasonable to assume that some of the sugars found are constituents of lipopolysaccharides. In *Spirulina platensis*, a homoglucan was found in cell wall fractions (Van Eykelenburg, 1978). Curie-point pyrolysis mass spectrometry revealed a rare 1,2-glucan in β-configuration.

Recently, at least two different polysaccharides were detected in the cell envelope of *Chlorogloeopsis* PCC 6912 (Schrader *et al.*, 1982). One polysaccharide, characterized by galactose, among other sugars, is exclusively part of the fibrous sheath, the other one is presumably a component of the cell wall and contains 6-O-methyl-D-mannose but no galactose. The fatty-acid content of the two polymers remains to be established. The cells investigated cannot as yet be ascribed to a defined cell type in the complex developmental cycle of

Table 13.4. Sugar composition of extracellular polysaccharides from various cyanobacteria

Extracellular polysaccharide from	Xylose	Arabinose	Ribose	Glucose	Galactose	Mannose	Fucose	Rhamnose	Glucuronic acid
Anacystis nidulans (Sangar & Dugan, 1972)				●	●	●			
Palmella mucosa (Moore & Tischer, 1964a, b)		●		●			●		●
Nostoc sp. (Moore & Tischer, 1964a,b)		●		●			●		●
Nostoc sp. (Hough *et al.*, 1952)	●			●	●			●	●?
Nostoc sp. (Mehta & Vaidya, 1978)	●			●					●
Anabaena cylindrica (Bishop *et al.*, 1954)	●	●		●	●			●	●
Anabaena cylindrica (Dunn & Wolk, 1970)	●			●	●	●			
Anabaena flos-aquae (Moore & Tischer, 1965)	●		●	●					●

Chlorogloeopsis (Evans, Foulds & Carr, 1976), although different stages of akinetes formation predominate.

Lipid-free, high-molecular-weight polysaccharides were extracted from cells of *Anabaena variabilis* (glycogen; Weckesser *et al.*, 1974) and from *A. flos-aquae* (a glucan and a heteropolysaccharide; Wang and Hill, 1977) using the phenol-water procedure. Nothing is known, however, about the localization of these polymers in the cell. Studies on extracellular polysaccharides from cyanobacteria are restricted essentially to comparative studies of chemical composition (Table 13.4). The sequence of $^{14}CO_2$ incorporation into intra- and extracellular polysaccharides from *A. flos-aquae* led to the suggestion that the polymer is synthesized intracellularly and is then secreted into the medium (Moore & Tischer, 1965). Active excretion of polysaccharides from *Nostoc* sp. is discussed by Mehta and Vaidya (1978). Moreover, these authors suggest selective secretion due to the finding of a different chemical composition of cell wall and extracellular polysaccharides in this organism.

13.5 Summary and outlook

The data reported in this chapter allow a first insight into the general architecture and composition of the cell wall of cyanobacteria. It seems to be clear that cyanobacteria have a cell wall type which is closely related to that of Gram-negative bacteria. This is verifiable by the demonstration of an outer membrane containing lipopolysaccharide and a peptidoglycan layer. As far as is known this is true for both the unicellular and the filamentous forms.

In spite of the similarities between the cell wall of cyanobacteria and Gram-negative bacteria, there are substantial differences between the cell walls of both of these groups of prokaryotes. First, the peptidoglycan layer in cyanobacteria is thicker due to formation of a three-dimensional net of peptidoglycan. Second, the sugar composition of peptidoglycan fractions indicates the additional presence of a polysaccharide possibly bound to peptidoglycan. Third, the low phosphorus content of cell envelopes of cyanobacteria might indicate a low phospholipid content of this outer membrane. Fourth, the data available as yet give no evidence for the presence of lipoprotein covalently bound to peptidoglycan in cyanobacteria. Fifth, there are major differences in the most conservative region of lipopolysaccharide, lipid A, between cyanobacteria and Gram-negative bacteria.

Clear conclusions require extensive further analytical work on cell wall composition: representative strains of each group of the wide spectrum of various cyanobacteria must be studied. The biology and chemical composition of slime and sheath materials are other broad and open fields. Studies on the macromolecular organization of the cyanobacterial cell envelope are important for our understanding of the taxonomy, phylogeny and ecology of the cyanobacteria

CHAPTER 14
HETEROCYSTS

C. P. WOLK

14.1 Introduction

Heterocysts of filamentous cyanobacteria (Fig. 14.1) are differentiated cells which are specialized for aerobic fixation of dinitrogen. In some species they may also function in a supplementary way in perennation (see 14.5, 14.6), and mechanically, e.g. as sites of filament breakage (Borzi, 1878; Turchini, 1918; Serpette, 1948; de Puymaly, 1957; Allsopp, 1968). Apparently already present in Precambrian cyanobacteria, over 2×10^9 years ago (Nagy, 1974; see Fay, 1973), they presumably evolved during the period when the atmosphere of the earth was becoming markedly aerobic. Because of their role in dinitrogen fixation, heterocysts are of agronomic importance (Singh, 1961; Peters, 1977; Singh, 1977; Watanabe *et al.*, 1977); and because their differentiation is easily followed in the microscope, and involves intercellular controls, they have assumed increasing importance as developmental models (Wolk, 1975, 1979a,b; Haselkorn, 1978; Carr, 1979).

14.2 Localization of nitrogen fixation

The correlations that, with rare exceptions, cyanobacteria capable of aerobic fixation of dinitrogen have heterocysts, and that ammonium, which apparently represses nitrogenase, inhibits the formation of heterocysts (Fogg, 1949), led Fay and co-workers (1966, 1968) and Pringsheim (1968a,b) to suggest that heterocysts are important in dinitrogen fixation. The results of many indirect, often ingenious experiments (Fay *et al.*, 1968; Stewart *et al.*, 1969; Stewart & Lex, 1970; van Gorkom & Donze, 1971; Neilson *et al.*, 1971; Weare & Benemann, 1973a; Fleming & Haselkorn, 1973) were consistent with the idea that under aerobic conditions, the nitrogenase activity of heterocyst-forming cyanobacteria is restricted to the heterocysts, although other results seemed not to corroborate that idea (Ohmori & Hattori, 1971; Wolk & Wojciuch, 1971b; Kurz & LaRue, 1971). Perhaps most graphically persuasive that heterocysts are foci of nitrogen fixation in intact filaments was the work of van Gorkom and Donze (1971). These authors showed that when nitrogen-starved filaments of *Anabaena cylindrica* deficient in biliproteins were re-exposed to N_2 under aerobic conditions, biliprotein reformation—indicative of nitrogen sufficiency—was observed first adjacent to heterocysts. Concordantly, Donze (1973) observed gradients of phycocyanin fluorescence, decreasing away from heterocysts, in N_2-grown *A. cylindrica*, and Singh (1976a) observed such gradients of phycocyanin and cell division in a long-filament mutant of *Cylindrospermum* with terminal heterocysts. However, Fay and Kulasooriya (1972) showed that upon exposure of intact filaments to nitroblue tetrazolium, the dye is reduced in vegetative cells adjacent to heterocysts. None of the indirect experiments ruled out one hypothesis suggested by this observation (see also Hirosawa and Wolk, 1979a), namely, that a substantial fraction of the total N_2 reduction might take place in vegetative cells adjacent to heterocysts, perhaps as a result of some interaction with heterocysts (see Wolk, 1970). The issue was to be resolved by direct measurements of nitrogenase activity in isolated heterocysts.

The nitrogenase activity of an organism growing on N_2 may be estimated from its doubling time. If *Anabaena cylindrica* doubles in 18 h, its maximum rate of growth on N_2, $N = N_0 2^{t/(18\cdot60)}$, where N is cell nitrogen and t is min. Nitrogen is fixed at a rate $dN/dt = N_0 \ln 2\, 2^{t/(18\cdot60)}/18\cdot60$, i.e. 642×10^{-6} mmol N fixed/(mmol cell N·min) = 321 nmol N_2 fixed/(mmol cell N·min) = 23 nmol N_2 fixed/(mg cell N·min). If C_2H_2 reduction is *c.* four times as rapid as N_2 fixation (Peterson & Burris, 1976b), if heterocysts have about as much nitrogen per cell as does a vegetative cell, if heterocysts have all of the nitrogenase activity of the filaments, and if 1 cell in 12 is a heterocyst, the heterocysts should reduce acetylene at a rate of about 1100 ($\simeq 23\times4\times12$)

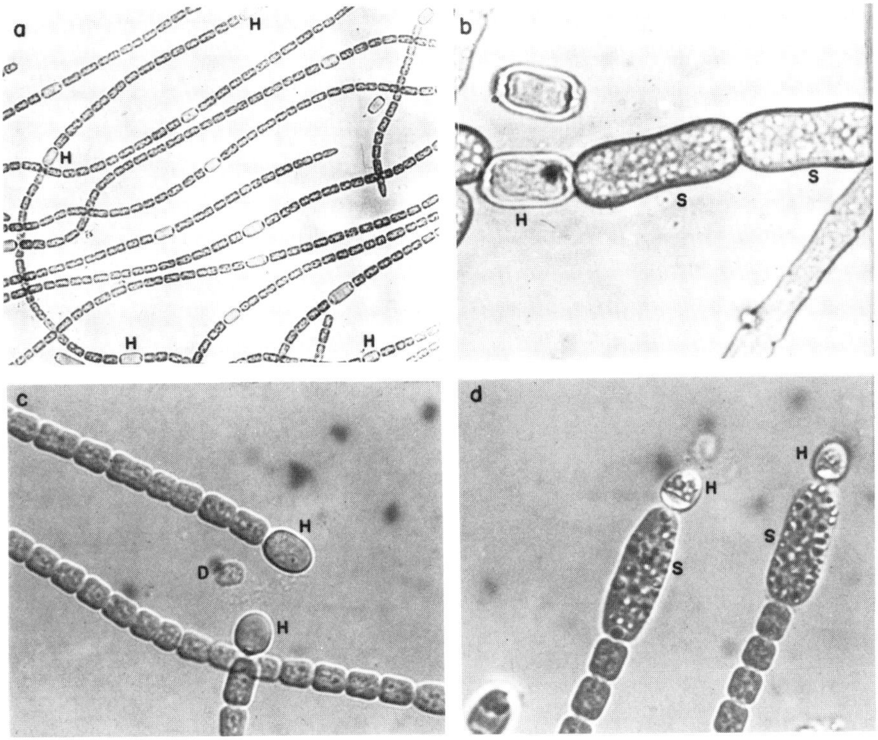

Fig. 14.1. Heterocysts (H) and akinetes (S). In *Anabaena cylindrica* (a,b), heterocysts are present at semi-regular intervals along the filaments. In *Cylindrospermum licheniforme* (c,d), heterocysts normally form at a terminal position to either side of a vegetative cell which has died (D) approximately midway between two pre-existing heterocysts. In both species, akinetes (shown mature in (b), immature in (d)) normally form adjacent to the heterocysts. (From Wolk (1979b), with permission.)

nmol/(mg cell N·min). Heterocysts from N_2-fixing cyanobacteria with shorter doubling times (see Peterson & Wolk, 1978b, and especially Gotto, Tabita & Van Baalen, 1979a, and Bottomley *et al.*, 1979) would be predicted to reduce acetylene more rapidly or to be more closely spaced.

Starting with very low rates [0·23 nmol C_2H_2/(mg N·min): Stewart *et al.*, 1969], heterocysts of increasing nitrogenase activity were isolated (Tel-Or & Stewart, 1976; Peterson & Burris, 1976a; Thomas *et al.*, 1977; Peterson & Wolk, 1978b). The nitrogenase activity, over 1000 nmol C_2H_2/(mg N·min) (see Simon, 1979), of heterocysts isolated by only one method from one organism routinely accounts for a majority of the nitrogenase activity of the intact filaments from which the heterocysts are derived (Peterson & Wolk, 1978b): an average of 60% or 70–75% depending on whether comparison is made with uncentrifuged or less-active, centrifugally harvested filaments. The isolation procedure makes use of two relatively gentle but synergistic techniques for differential destruction of vegetative cells: short (30 min) treatment with lysozyme, followed by cavitation in a sonic cleaning bath. The organism, *Anabaena variabilis*, a strain variously called A-37, UTEX 1444 and ATCC 29413, was chosen for study because it had been shown to withstand fragmentation to single cells (Wolk & Wojciuch, 1973).

A protein band corresponding to nitrogenase was identified after electrophoresis of *A. variabilis* proteins on anaerobic, non-denaturing gels, on the basis of a number of criteria, including the content of molybdenum and non-heme iron in that protein band, and its lability to oxygen, suppression by ammonium, and subunit-polypeptide molecular weights. Quantitation of the protein by incorporation of ^{55}Fe showed that, averaged over three experiments, 91% of the nitrogenase of intact filaments could be recovered in the purified, isolated heterocysts (Peterson & Wolk, 1978b; see Fig. 14.2). The above results permit the firm assertion that all, or nearly all, of the nitrogenase and nitrogenase activity of whole, aerobically grown filaments of *A. variabilis* is associated with the heterocysts. Although one is tempted to generalize, it remains important to ascertain whether a high recovery of nitrogenase can be demonstrated directly in heterocysts of other cyanobacteria as well. Whether nitrogenase is preferentially localized in heterocysts in microaerobically grown filaments of heterocyst-forming cyanobacteria remains unanswered.

A challenge for the future is to devise conditions permitting very prolonged nitrogenase activity by isolated heterocysts (Wolk, 1979b). Heterocysts isolated from *A. variabilis* are capable of synthesizing proteins, including nitrogenase, from radioactive amino acids, but the rate at which they do so decreases rapidly (Janaki & Wolk, 1982). Raising the osmotic strength of the medium or adding amino acids to the suspension medium prolongs the nitrogenase activity, possibly by prolonging protein synthesis (Giddings & Wolk, 1981). Because heterocysts are non-growing cells, turnover may provide all or nearly all of the amino acids and ribonucleotides which they may require while present within intact filaments. However, once they have been isolated, certain such metabolites may leak from their poles, and therefore have to be replaced if the heterocysts are to continue to be metabolically active. For example, heterocysts of *A. variabilis* appear unable to synthesize methionine (Giddings *et al.*, 1981).

14.3 Cytological aspects of differentiation

14.3.1 The envelope

Structural and metabolic differences between heterocysts and vegetative cells may enhance the effectiveness of heterocysts as sites for fixation of dinitrogen. The most conspicuous structural difference is the presence, in heterocysts, of a

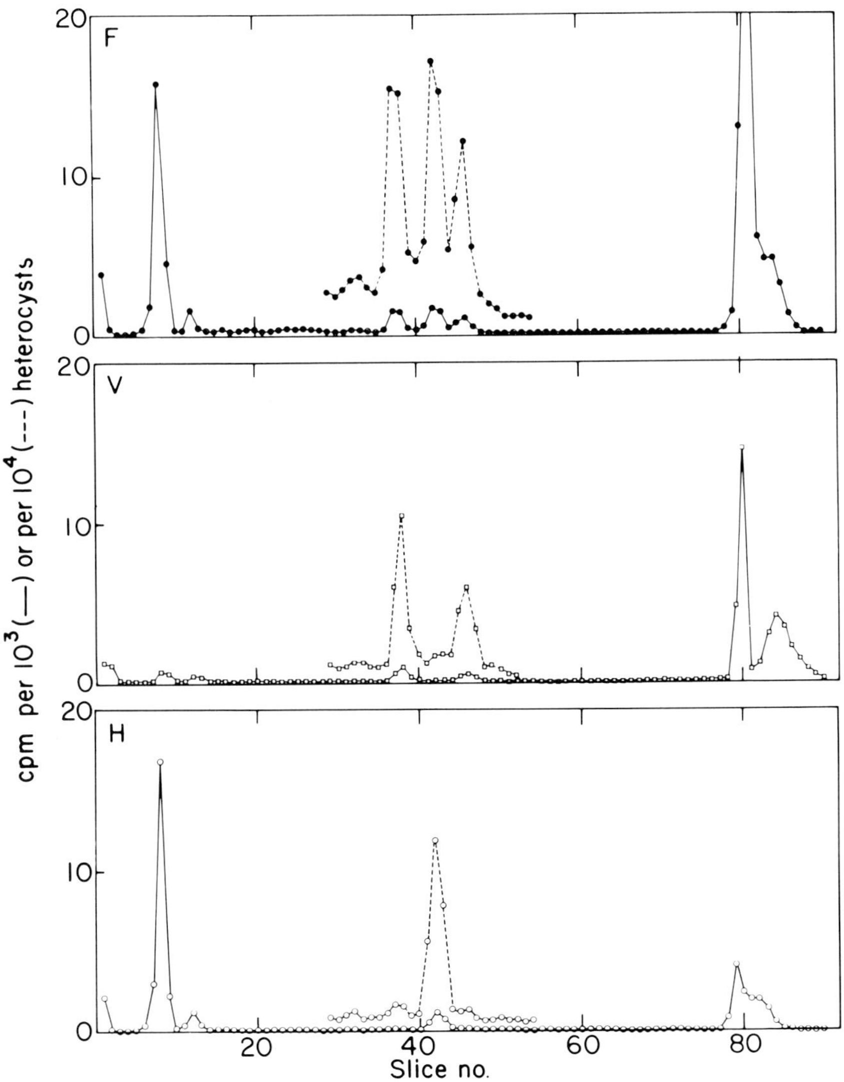

Fig. 14.2. Distribution of radioactivity in nondenaturing electrophoretograms of soluble extracts of filaments (F), heterocysts (H), and a vegetative cell fraction (V), from a ^{55}Fe-labelled culture of *Anabaena variabilis*. Counts per minute of ^{55}Fe are expressed on the basis of equal numbers of heterocysts; i.e. they are calculated for equal amounts of filaments. In this experiment, the extract of heterocysts accounted for 103%, and the extract of vegetative cells for 6%, of the nitrogenase (the band centred at slice no. 8), calculated on the basis of the extract from filaments that had not been incubated with lysozyme. The band at slice no. 42 is nitrogenase reductase, the extremely oxygen-labile Fe-protein of the nitrogenase complex, which was recovered with about 22% lower efficiency in the isolated heterocysts. (From Peterson and Wolk (1978b), with permission.)

thick envelope (Fig. 14.1). This envelope comprises an inner laminated layer (L, Fig. 14.3), a central homogeneous layer (H, Fig. 14.3), and an outer fibrous layer (Lang & Fay, 1971). The inner layer consists of glycolipids (Winkenbach *et al.*, 1972), the structures of certain of which have been identified (Fig. 14.4; Bryce *et al.*, 1972; Lambein & Wolk, 1973; see also Abreu-Grobois *et al.*, 1977). The most abundant glycolipid in heterocysts of *A. cylindrica*, 1-(O-α-I)-glycopyranosyl-3,25-hexacosanediol, is present, but is not always the most plentiful glycolipid, in the heterocysts of other species and genera (Lorch & Wolk, 1974; see also Nichols & Wood, 1968a, and Wahal *et al.*, 1973b). Semi-quantitative data suggest that these glycolipids are absent from vegetative cells (Walsby & Nichols, 1969; Wolk & Simon, 1969). The laminated layer has been reported to have a periodicity of 3.5–4.0 nm, approximately the length of the main glycolipid component, after isolation (Winkenbach *et al.*, 1972), but 7–8 nm *in situ* (Golecki & Drews, 1974; Granhall, 1976; Giddings & Staehelin, 1978).

Certain mutants of *A. variabilis* which have nitrogenase activity under microaerobic, but not under aerobic, conditions are deficient in the envelope

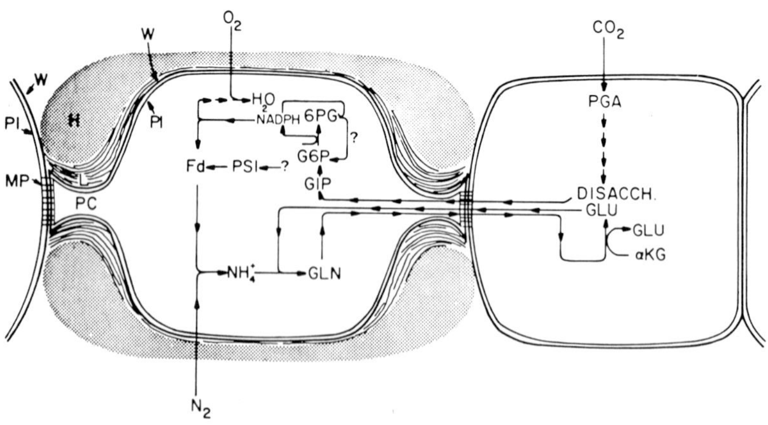

Fig. 14.3. Diagram illustrating the principal structural differences and known interactions between a heterocyst (at left) and a vegetative cell (at right). Outside the wall (W) of the heterocyst is an envelope consisting principally of a laminated, glycolipid layer (L) and a homogeneous, polysaccharide layer (H). Microplasmodesmata (MP) join the plasma membranes (Pl) of the two types of cells at the end of the pore channel (PC) of the heterocyst. A disaccharide formed by photosynthesis in the vegetative cells moves into heterocysts, and may then be metabolized to glucose-6-phosphate and oxidized by the oxidative pentose phosphate pathway. Pyridine nucleotide (NADPH) reduced by this pathway can donate electrons to O_2 to maintain reducing conditions within the heterocysts, and can reduce ferredoxin (Fd). Ferredoxin can also be reduced by PS I. Reduced ferredoxin can donate electrons to nitrogenase, which reduces N_2 to NH_4^+. Glutamate produced principally by vegetative cells reacts with the NH_4^+ to form glutamine. The glutamine moves into the vegetative cells, where it reacts with α-ketoglutarate to form two molecules of glutamate. (From Wolk (1979b), with permission.)

glycolipids. Revertants capable of aerobic fixation of dinitrogen form the lipids (Haury & Wolk, 1978). Thus, it may be that the glycolipid layer normally reduces the rate of diffusion of oxygen into heterocysts to such an extent that most of that which does enter is reduced by oxidases present in the heterocysts. Alternatively, the effects of the mutations on nitrogenase activity may be independent of their effects on glycolipid synthesis. That some oxygen does enter heterocysts is suggested by the observation that in the presence of ammonium, which presumably represses synthesis of nitrogenase, the nitrogenase activity of filaments of *A. cylindrica* decreases rapidly in air but remains relatively stable under anaerobic conditions (Ohmori & Hattori, 1974). Heterocysts formed by *Anabaena* strain 7120 grown under strictly anaerobic conditions appear either to lack, or to have an irregularly formed, glycolipid layer—anaerobiosis may prevent hydroxylation of the aglycones

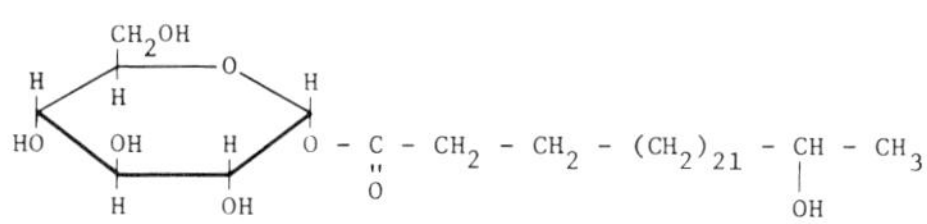

25-hydroxyhexacosanoic acid (1-α-D-glucopyranose) ester

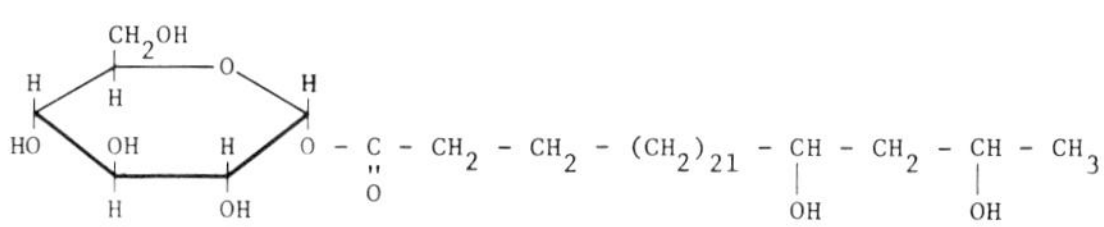

25,27-dihydroxyoctacosanoic acid (1-α-D-glucopyranose) ester

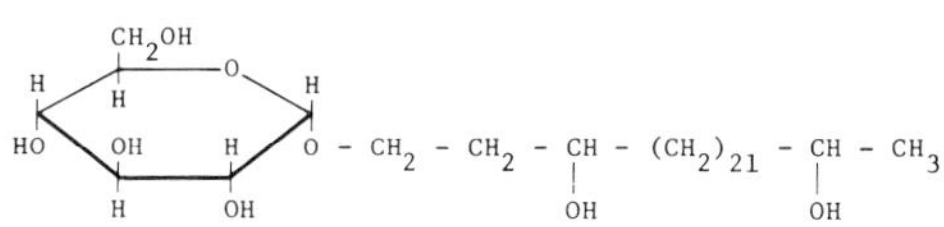

1 - (O-α-D-glucopyranosyl) - 3,25-hexacosanediol

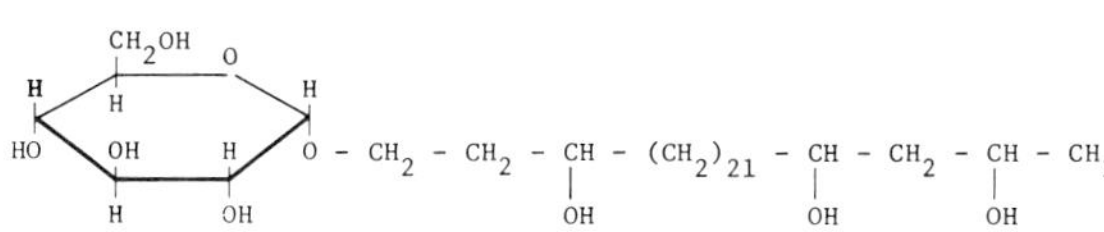

1 - (O-α-D-glucopyranosyl) - 3,25,27-octacosanetriol

Fig. 14.4. Envelope glycolipids of *Anabaena cylindrica*. The carbohydrate moiety is glucose in about 90% of the molecules and galactose in the remainder. (From Wolk (1975), with permission.)

of the glycolipids—and the nitrogenase activity of the organism is oxygen-sensitive (Rippka & Stanier, 1978).

The glycolipid layer surrounds the protoplast of the heterocyst except at the junctions to adjacent vegetative cells. The presence of this layer may therefore prevent lipophobic substances from passing directly between the inside of the heterocyst and the external milieu, and may thereby constrain heterocysts to interact with adjacent vegetative cells (Winkenbach *et al.*, 1972). However, this conjecture has yet to be directly tested.

The homogeneous envelope layer consists largely of polysaccharide (Dunn & Wolk, 1970; Winkenbach *et al.*, 1972); the outermost, fibrous layer, which is not always evident (Winkenbach *et al.*, 1972), may consist of uncompacted strands of the same polysaccharide (Cardemil & Wolk, 1976). The polysaccharide from the envelopes of heterocysts of *A. cylindrica* consists of concatenations of *c.* 38 units such as shown in Fig. 14.5 (Cardemil & Wolk, 1976, 1979). Polysaccharides very similar to the material just described are present in the envelopes of the heterocysts of *A. variabilis* (ATCC 29413) and *Cylindrospermum licheniforme* (Cardemil & Wolk, 1981b). The envelopes of akinetes (see 15.2) in both species of *Anabaena*, but not in *Cylindrospermum licheniforme*, contain polysaccharides which appear to be identical to the

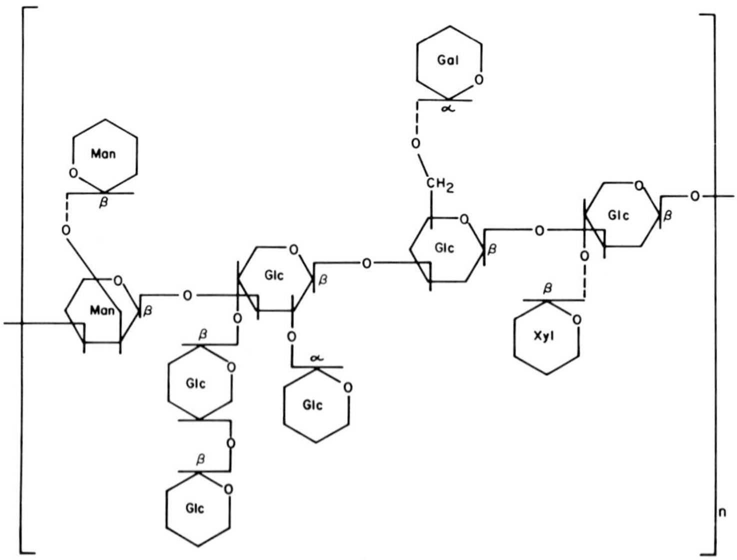

Fig. 14.5. Approximate subunit structures of the polysaccharides from the envelopes of heterocysts and spores of *A. cylindrica*. Linkages indicated by dashed lines are present attached to only about half of the (residual) units shown. An alternative position is possible for the terminal sugar of the disaccharide branch and for the galactosyl residue. Unlinked carbon atoms 6 have been omitted so as to simplify the drawing. (From Cardemil and Wolk (1979), with permission.

polysaccharides in the envelopes of their heterocysts (Cardemil & Wolk, 1976, 1979, 1981b). Upon incubation of isolated heterocysts of *Anabaena variabilis* with [^{14}C]mannose, the envelope polysaccharide was labelled, probably via phosphoglycolipid intermediates (Cardemil & Wolk, 1981a).

The polysaccharide layer of the envelope may protect against degradation or dissolution of the glycolipid layer. However, a combination of carbohydrases appear able to digest the latter despite the presence of the former (Granhall, 1976). In certain revertible mutants incapable of aerobic fixation of dinitrogen, the carbohydrate layer of the envelope of each heterocyst has very thin regions (Currier *et al.*, 1977): perhaps the polysaccharide layer also normally impedes the ingress of oxygen.

14.3.2 The protoplast

Heterocysts, compared with vegetative cells, are normally pallid, because of a deficiency of biliproteins (Fay *et al.*, 1968; Fay, 1969a; Wolk & Simon, 1969; Thomas, 1970b). However, chlorophyll and carotenoids are present (Fay, 1969a; Wolk & Simon, 1969; Thomas, 1970; Winkenbach *et al.*, 1972), and biliproteins sometimes are (Fritsch, 1951; Thomas, 1972; Ladha & Kumar, 1977a; Peterson *et al.*, 1981). Contrary to the earlier reports (Walsby & Nichols, 1969; Wolk & Simon, 1969; Fay, 1969a), the acyl lipids and glycosylated carotenoids of the membranes of vegetative cells *are* present in the membranes of heterocysts. Their apparent absence was a consequence of their rapid degradation during isolation of damaged heterocysts at room temperature (Winkenbach *et al.*, 1972). α-Linolenic acid may be absent from heterocysts (Walsby & Nichols, 1969) or present in reduced amount (Yamamoto, 1972).

Upon electrophoresis, detergent-disrupted membranes from heterocysts show two bands of chlorophyll-bearing protein found also in vegetative cells, but lack the third band which is associated with photosystem (PS) II in the latter cells (Reinman & Thornber, 1979). Membranes from the two types of cells appear to have certain other proteins in common (Fleming & Haselkorn, 1973; Sallal & Codd, 1977). However, the most abundant particles (not phycobilisomes) within the exoplasmic face of freeze-fractured thylakoids from vegetative cells have sizes of 7.5 and 10 nm, whereas the sizes of those from heterocysts are 5.5 and (less abundant) 7.5 nm (Giddings & Staehelin, 1979).

A difference was also observed between the plasmalemmae of vegetative cells and heterocysts: a 5.5-nm particle present in the former was much less frequent in, or absent from, the latter (Giddings & Staehelin, 1978). The end membrane of a heterocyst of *A. cylindrica* is joined to that of an adjacent vegetative cell by about 50 plasmodesmata-like structures, each less than 20 nm in outside diameter (Giddings & Staehelin, 1978; see also Wildon &

Mercer, 1963a; Lang & Fay, 1971). Haselkorn (1978) has postulated that the interactions between vegetative cells and heterocysts are controlled by 'pumps' situated between the two types of cells, but this proposal awaits experimental test. Material often found in the polar regions of heterocysts has the ultrastructural appearance and solubility properties of cyanophycin (Lang & Fay, 1971; Lang *et al.*, 1972; Wolk, 1973), and like cyanophycin (Simon & Weathers, 1976), is rich in arginine (Fogg, 1951). Phosphoenolpyruvate carboxylase and carbamyl phosphate synthetase, present in isolated heterocysts, may be involved in the synthesis of this material (Carr, 1979). Enzymes responsible for both the synthesis and breakdown of cyanophycin have been reported to be enriched in heterocysts (Gupta & Carr, 1981).

Other cytochemical characteristics of heterocysts may be noted (see also 14.5). DNA is distributed as in a vegetative cell (Leak, 1965) or more diffusely (Fogg, 1951; Fay & Lang, 1971). In view of the mechanical constraint imposed by the heterocyst envelope, it seems highly unlikely that mature heterocysts can divide, although heterocysts may differentiate from vegetative cells during the process of cell division (see Kumar, 1963; Ladha & Kumar, 1975; Gupta & Dashora, 1975; Singh & Dikshit, 1976). Incorporation of [^{3}H]thymidine into the nuclear region of heterocysts (Leak, 1965) may therefore be attributable to repair synthesis or conceivably to amplification of some segment of the genome. Structures resembling ribosomes are observed in the electron microscope (Wildon & Mercer, 1963a; Leak & Wilson, 1965). A heterocyst of *A. variabilis* contains more RNA and nearly as much DNA as a vegetative cell (Simon, 1979, 1980). It seems probable that the maintenance of nitrogenase activity by mature heterocysts in the presence of oxygen requires continued synthesis of nitrogenase (Weare & Benemann, 1973a; Ohmori & Hattori, 1974). A number of other, unidentified proteins continue to be synthesized by mature heterocysts (Fleming & Haselkorn, 1974; Janaki & Wolk, 1982).

14.4 Metabolic aspects of differentiation, and chemical interactions between cells

14.4.1 Nitrogen metabolism and interactions

Present knowledge of heterocyst biochemistry (Table 14.1) should increase rapidly, now that metabolically active, isolated heterocysts are available. The specific activities of many enzymes of nitrogen metabolism differ little between heterocysts and vegetative cells (Table 14.1). This is true, in particular, of glutamine synthetase, the initial enzyme of ammonium assimilation during growth on N_2 (Lawrie *et al.*, 1976; Wolk *et al.*, 1976; Meeks *et al.*, 1978).

However, the glutamine synthetase of *Anabaena cylindrica* and other cyanobacteria is subject to allosteric regulation by many effectors (e.g. alanine, serine, glycine and aspartate; glyoxylate; ADP and AMP; NH_4^+, Mg^{++} and pyrophosphate: Stewart, Haystead & Dharmawardene, 1975; Sawhney & Nicholas, 1978a; Stacey, Van Baalen & Tabita, 1979; Tuli & Thomas, 1980), so that it is currently not possible to deduce *in vivo* rates of catalysis by this enzyme from *in vitro* measurements.

Most of the N_2-derived nitrogen incorporated into glutamine is further metabolized by glutamate synthase (Wolk *et al.*, 1976; Meeks *et al.*, 1978). However, heterocysts metabolize $[^{13}N]N_2$ and $^{13}NH_4^+$ essentially only to glutamine; and probe sonication renders glutamate synthase activity non-sedimentable much more rapidly than it does glutamine synthetase activity, and about as rapidly as vegetative cells are destroyed (Thomas *et al.*, 1977). The conclusion that glutamate synthase appears to be restricted to vegetative cells is important. It implies that glutamine formed from N_2-derived ammonium—principally in heterocysts (Thomas *et al.*, 1977; Meeks *et al.*, 1977)—must move into vegetative cells in order to undergo amide transfer, and the glutamate from which the glutamine is synthesized is derived ultimately, in large part, from vegetative cells (Fig. 14.3; Thomas *et al.*, 1977). Because glutamate synthase is easily assayed, it may be hoped that its reported localization will soon be subject to confirmation. Bothe and coworkers (Bothe *et al.*, 1980) have reported the remarkable observation that although isocitrate dehydrogenase is active in nitrate-grown vegetative cells of *A. cylindrica*, in N_2-grown filaments the enzyme activity is virtually restricted to heterocysts. If true *in situ*, it would appear to follow that the α-ketoglutarate required as a substrate for glutamate synthase in vegetative cells is largely synthesized in heterocysts! Glutamate–oxaloacetate transaminase has been shown histochemically to be present in a gradient in *Nodularia*, with maximal activity in the vegetative cells to either side of a heterocyst (Barlow & Scott, 1975, 1976). Based on the assumption that heterocysts account for all of the N_2 fixation by aerobically grown *Anabaena cylindrica*, mathematical analysis shows that nitrogen moves out of heterocysts with a half-time of ≤ 1.5 min, and moves from cell to cell along the filaments, while being simultaneously metabolized, with a characteristic time of <5 s (Wolk *et al.*, 1974, 1976).

14.4.2 Carbon metabolism and interactions

Heterocysts within intact filaments fix little or no CO_2 (Wolk, 1968; Stewart *et al.*, 1969) because they lack RuBP carboxylase (RuBPC: Winkenbach & Wolk, 1973; Codd & Stewart, 1977a; see Table 14.1). Energy and reducing equivalents are thereby conserved for fixation of dinitrogen. Heterocysts also lack (i) phosphoribulokinase, which generates the substrate for RuBPC

Table 14.1. Relative activities of enzymes of heterocysts and vegetative cells of aerobically grown *Anabaena cylindrica*

	Specific activity in heterocysts much greater than in vegetative cells	Relative specific activity close to 1:1	Specific activity in heterocysts much lower than in vegetative cells
Enzymes of nitrogen metabolism			
Nitrogenase	Peterson & Wolk (1978b) (and see 14.2)		
Glutamine synthetase		Stewart, Haystead & Dharmawardene (1975) Thomas *et al.* (1977) Dharmawardene *et al.* (1973)	
Glutamate synthase			Thomas *et al.* (1977)
Alanine dehydrogenase (NADH)		Stewart, Haystead & Dharmawardene (1975) Scott & Fay (1972)	
Glutamic acid dehydrogenase (NADPH)		Stewart, Haystead & Dharmawardene (1975)	
Glutamate-oxalo-acetate transaminase		Stewart, Haystead & Dharmawardene (1975) Scott & Fay (1972) (see also 14.4.1)	
Glutamate-pyruvate transaminase		Stewart, Haystead & Dharmawardene (1975)	
Enzymes of carbohydrate metabolism			
Reductive pentose phosphate cycle			
RuBP carboxylase			Winkenbach & Wolk (1973) Stewart & Codd (1975) Codd & Stewart (1977c), Codd *et al.* (1980)
RuBP oxygenase			Codd *et al.* (1980)
Ru5P kinase			Codd *et al.* (1980)
F-1,6-diphosphate aldolase			Lex & Carr (1974)
Phosphoglycollate phosphatase			Codd *et al.* (1980)

Glycollate dehydrogenase			Codd *et al.* (1980)
GAP dehydrogenase		Bothe *et al.* (1980)	Winkenbach & Wolk (1973), Lex & Carr (1974)
3-phosphoglycerate kinase			Bothe *et al.* (1980)
Oxidative pentose phosphate cycle			
G6P dehydrogenase	Winkenbach & Wolk (1973), Lex & Carr (1974) Bothe *et al.* (1980), Apte *et al.* (1978)		
6PG dehydrogenase	Winkenbach & Wolk (1973), Lex & Carr (1974)		
Phosphoglucoisomerase		Bothe *et al.* (1980)	
Other			
Hexokinase		Winkenbach & Wolk (1973), Lex & Carr (1974)	
Phosphofructokinase			Lex & Carr (1974)
Enolase		Bothe *et al.* (1980)	
Pyruvate kinase			Bothe *et al.* (1980)
Electron transfer pathway			
'Uptake' hydrogenase	Peterson & Wolk (1978a)		
Classical, reversible hydrogenase		Eisbrenner *et al.* (1978), Tel-Or *et al.* (1978)	
PS I		Wolk & Simon (1969), Bradley & Carr (1971) Donze *et al.* (1972), Tel-Or & Stewart (1977)	
PS II			Bradley & Carr (1971, 1976), Donze *et al.* (1972) Tel-Or & Stewart (1977)
Fd-NADP$^+$ oxidoreductase		Apte *et al.* (1978)	
NADPH oxidase		Bradley & Carr (1976)	
Miscellaneous			
Isocitrate dehydrogenase	Bothe *et al.* (1980)		
Superoxide dismutase		Daday *et al.* (1977), Henry *et al.* (1978)	
Catalase		Henry *et al.* (1978)	
Arg-poly(asp) synthetase	Gupta & Carr (1981)		
Cyanophycinase	Gupta & Carr (1981)		

(Codd *et al.*, 1980); (ii) the structures ('carboxysomes') in which, it appears, much of the RuBPC of vegetative cells is concentrated (Stewart & Codd, 1975); (iii) the oxygenase activity of RuBPC; and (iv) phosphoglycollate phosphatase and glycollate dehydrogenase, which further metabolize the phosphoglycollate produced by that oxygenase activity (Codd *et al.*, 1980).

Despite the absence of CO_2 fixation, carbohydrate deposition is extensive: *c.* 43% of the dry weight of a heterocyst is accounted for by sugar residues in the envelope (Dunn & Wolk, 1970; Lambein & Wolk, 1973; Cardemil & Wolk, 1976). This deposition depends upon an intercellular transfer, to heterocysts, of products of CO_2 assimilation by vegetative cells—a transfer which has been directly demonstrated (Wolk, 1968). If deposition exceeds gluconeogenesis, carbohydrate *per se* must be imported into heterocysts from vegetative cells. Although a constituent of the gluconeogenic pathway (and also the glycolytic and reductive pentose phosphate pathways), glyceraldehyde-3-phosphate dehydrogenase, has been reported by two laboratories to be essentially undetectable in heterocysts of *A. cylindrica* (Winkenbach & Wolk, 1973; Lex & Carr, 1974), another group has reported little difference in its activity between heterocysts and vegetative cells (Bothe *et al.*, 1980). The extent to which heterocysts have the capacity to synthesize the sugars which they incorporate is therefore uncertain.

In contrast to the enzymes considered so far, heterocysts have much higher activities of two enzymes of the oxidative pentose phosphate pathway, glucose-6-phosphate dehydrogenase and 6-phosphogluconate dehydrogenase, than have vegetative cells (Winkenbach & Wolk, 1973; Lex & Carr, 1974; Apte *et al.*, 1978). Just how much higher is a matter of instructive disagreement. When a comparison is made of the specific activity of an enzyme in heterocysts with its specific activity in vegetative cells, its specific activity is normally measured with isolated heterocysts, and either with the proteins released when vegetative cells are broken or with whole filaments. Correction must be made for the possible loss of enzyme from heterocysts and concomitant appearance of that same amount of enzyme in the 'released' fraction during heterocyst isolation, and for the contribution by heterocysts to the assays starting with whole filaments. Apte *et al.* (1978) determined the specific activities of glucose-6-phosphate dehydrogenase in isolated heterocysts, in 'vegetative cells' and in whole filaments of *A. cylindrica*: 120.8, 16.9, and 42.0 nmol/(min·mg protein), respectively. These results can be explained only if the heterocysts contained about 24% of the protein in filaments [$42.0 \cong 0.24\ (120.8) + 0{\cdot}76\ (16.9)$], or if some of the enzyme leaked out of heterocysts and appeared in the vegetative cell lysate. Because heterocysts normally comprise only 6–8% of the cells in that organism, leakage presumably occurred, as might have been expected on the basis of electron micrographs of heterocysts damaged by isolation (Fay & Lang, 1971). Ratios of specific activities of 6–7-fold for glucose-6-phosphate dehydrogenase, uncorrected for possible leakage (Lex & Carr, 1974; Apte *et al.*, 1978), may therefore have been underestimates. By following the kinetics of solubilization of activity during cavitation, and by further correcting for the observed frequency of heterocysts in intact filaments, Winkenbach and Wolk (1973) estimated that there was 60–70-fold higher activity of glucose-6-phosphate dehydrogenase and of 6-phosphogluconate dehydrogenase per heterocyst than per vegetative cell. However, the regulation of cyanobacterial glucose-6-phosphate dehydrogenase involves slow relaxation between states of very different activity (Schaeffer & Stanier, 1978), and the kinetics of that enzyme differ in heterocysts and vegetative cells (Lex & Carr, 1974). Until these properties are taken into account, no determination of the relative specific activity of this enzyme in heterocysts and vegetative cells can be considered definitive. To complicate the matter further, the activity of the enzyme is also controlled by ATP, NADPH and RuBP, so that the *in vitro* and *in vivo* activities need not be equal. No such considerations are known to apply to 6-phosphogluconate dehydrogenase. It is possible that high activities of both enzymes serve to maintain a very high ratio of NADPH to $NADP^+$, so as to permit the NADPH to reduce ferredoxin, which could, in turn, serve as electron donor to nitrogenase. This is, the *potential* fluxes measured *in vitro* (Winkenbach & Wolk, 1973) may be much greater than the actual metabolic fluxes catalysed by these enzymes *in vivo*. Because one enzyme of the oxidative pentose phosphate cycle, fructose-1,6-diphosphate aldolase, appears to be absent from heterocysts (Lex & Carr, 1974), ribulose-5-phosphate (or some product of its metabolism) may be returned to vegetative cells rather then being recycled.

The activities of hexokinase detected in heterocysts (Winkenbach & Wolk, 1973) were too low to account for the electron flow to N_2 calculated from the measured doubling time of the cultures. Thus, if electrons flow to nitrogenase via the oxidative pentose phosphate cycle, the electron source entering heterocysts from vegetative cells (see below) would appear not to be glucose or fructose. Carr and Bradley (1973) stated, without corroboratory data, that enzymes of the glyoxylate cycle are present in heterocysts of *A. cylindrica*, some at higher specific activity than in vegetative cells. To date, enzymes of carbohydrate metabolism in heterocysts have been studied only in *A. cylindrica*: it would be very desirable to learn whether the results are valid for heterocysts in general.

As discussed above, and in 14.4.3.2, electrons and carbon move from vegetative cells to heterocysts. What substance or substances mediate this transport? Jüttner and Carr (1976) exposed filaments of *A. cylindrica*, predigested with lysozyme, to brief pulses of $^{14}CO_2$, and then isolated the heterocysts rapidly in the cold by passage through a French press. After the shortest pulse reported, 10 s, the principal compounds labelled in vegetative cells were intermediates of the reductive pentose phosphate cycle, whereas the principal compound labelled in heterocysts was a glucose-containing disaccharide tentatively identified as maltose. Their method unfortunately favours identification of those of the substances moving from vegetative cells to heterocysts which increase in concentration rapidly within vegetative cells. Corroboratory data, including a demonstration that the substance identified can support a high level of nitrogenase activity by isolated heterocysts (see Wolk, 1979a), are therefore necessary before it can be concluded that in the *steady state*, a disaccharide is the principal substance transferring reducing equivalents into heterocysts.

Sulphate reduction appears to be restricted to vegetative cells of *A. variabilis*, but sulphide, cysteine and glutathione (among other sulphur-containing metabolites) are present in heterocysts (Giddings *et al.*, 1981). Certain cyanobacteria are known to be able to use S^z as electron donor to PS I (Garlick *et al.*, 1977). The potential occurrence of a cycle in which a sulphur compound is oxidized in heterocysts and re-formed by reduction in vegetative cells may merit consideration as a possible source of electrons for heterocysts *in vivo*.

14.4.3 Pathways of electron transfer and generation of ATP

14.4.3.1 In the dark

Activity of nitrogenase is dependent upon a 'low potential reductant' and a supply of energy, presumably from ATP (see Chapter 4). *Anabaena variabilis* can grow heterotrophically in the dark, fixing nitrogen (and making normal heterocysts), at about 25% of the rate at which it grows in the light (Wolk & Shaffer, 1976), and even obligately photoautotrophic *A. cylindrica* can fix nitrogen in the dark (Donze, 1973). By what pathways are electrons and ATP supplied to nitrogenase in heterocysts in the dark? With cyanobacterial nitrogenase *in vitro* (Smith *et al.*, 1971) and with heterocysts isolated by probe cavitation (Lockau *et al.*, 1978), glucose-6-phosphate, 6-phosphogluconate and isocitrate can serve as electron donors provided that $NADP^+$ (and in the case of the *in vitro* reaction, also ferredoxin) is present. In heterocysts in the dark, ferredoxin-$NADP^+$ reductase may therefore function to reduce not $NADP^+$, but ferredoxin. Because fluoroacetate—a metabolic product of which inhibits isocitrate formation—decreases acetylene reduction, Lex and Stewart (1973) suggested that isocitrate dehydrogenase is important *in vivo*. However, fluoroacetate from alternative suppliers was not equally effective (Lyne & Stewart, 1973b) and fluoroacetate may act by interrupting the formation of α-ketoglutarate, so that ammonium accumulates (Gallon *et al.*,

1978). Which pathway of electron transfer to nitrogenase predominates in the dark is therefore unknown.

In the dark or at low intensities of light, oxygen stimulates the nitrogenase activity of intact filaments (Wolk, 1970; Weare & Benemann, 1973a; Benemann & Weare, 1974a; Donze *et al.*, 1974; Bothe, Tennigkeit & Eisbrenner, 1977a) and (in the presence of H_2) heterocysts (Peterson & Wolk, 1978b), suggesting that oxidative phosphorylation supplies the requisite ATP. Coupling of phosphorylation to the oxyhydrogen (Knallgas) reaction has been directly demonstrated (Peterson & Burris, 1978; see also Bothe, Tennigkeit & Eisbrenner, 1977a). However, it is not known to which pathway, in the absence of H_2, ATP formation is coupled *in vivo*. Oxidative phosphorylation has been detected with endogenous substrates (Scott & Fay, 1972; Tel-Or & Stewart, 1976); NADPH and NADH are oxidized (Scott & Fay, 1972; Carr & Bradley, 1973; Peterson & Burris, 1976a); and—in the presence of $NADP^+$—glucose, fructose-6-phosphate, glucose-6-phosphate, 6-phosphogluconate, pyruvate, and isocitrate increased the respiratory rate of isolated heterocysts of *Anabaena* strain 7120 (Peterson & Burris, 1976a).

14.4.3.2 Effects of light

The deficiency of biliproteins in heterocysts (14.3.2) was interpreted as implying a lack of PS II (Fay *et al.*, 1968; see Amesz & Duysens, 1962). Light absorbed by phycocyanin enhanced reduction of CO_2, but not acetylene, during irradiation of *A. cylindrica* by light absorbed by chlorophyll (Lyne & Stewart, 1973a). However, this result does not permit one to distinguish whether PS II, or only absorption by biliproteins, is deficient in the cells housing nitrogenase activity. Manganese, thought to be required for oxygen production, was also found to be deficient, and PS II activity to be lacking, in isolated heterocysts (Bradley & Carr, 1971, 1976; Donze *et al.*, 1972; Tel-Or & Stewart, 1975, 1977; Almon & Böhme, 1980; Peterson *et al.*, 1981). The presumptive absence of PS II from heterocysts implies that these cells do not generate oxygen which might inactivate nitrogenase (Fay *et al.*, 1968), and that reducing equivalents used by nitrogenase in heterocysts must be derived ultimately from the vegetative cells.

There are similar amounts of chlorophyll and numbers of reaction centres in PS I per heterocyst and per vegetative cell (Wolk & Simon, 1969; Donze *et al.*, 1972; Tel-Or & Stewart, 1977), but consistent with the apparent absence of PS II from heterocysts, the ratio of cell chlorophyll : P_{700} is lower in them than in vegetative cells (Donze *et al.*, 1972; Cammack *et al.*, 1976; Alberte *et al.*, 1980; Almon & Böhme, 1980; Peterson *et al.*, 1981). Stoichiometric relationships among cytochromes c_{553}, f_{557}, b_{559} and b_{563}, P_{700} and plastoquinone have been reported for filaments (3.2 : 1 : 1 : 2.1 : 2.3 : 17.2) and heterocysts

(2.5:1:0:2.3:2.5:7.2) of *Nostoc muscorum* by Almon and Böhme (1980). By contrast, it had earlier been reported that heterocysts of *Anabaena cylindrica* contain cytochrome b_{559}, as well as other of the constituents found by Almon and Böhme, and also ferredoxin and iron-sulphur centres (Cammack *et al.*, 1976; Tel-Or & Stewart, 1977). Vitamin K_1 has been reported also to be present, but tocopherol and tocopherol quinone to be absent (Carr & Bradley, 1973).

Action spectra of light-stimulated reduction of acetylene by whole filaments (Fay, 1970) and isolated heterocysts (Peterson *et al.*, 1981) show activity attributable to PS I. Even residual biliproteins in isolated heterocysts appear to transfer energy to the long-wavelength fluorescing (PS I) form of chlorophyll-*a* (Peterson *et al.*, 1981). Does light generate reductant, ATP, or both? Heterocysts isolated by probe cavitation express nitrogenase activity in the dark if supplied with both ATP and either dithionite (Stewart *et al.*, 1969; see also Wolk & Wojciuch, 1971a) or glucose-6-phosphate plus $NADP^+$ (Lockau *et al.*, 1978), but not if supplied with only ATP or only a source of reductant. However, illuminated heterocysts express nitrogenase activity when supplied with neither ATP nor reductant (Lockau *et al.*, 1978; Peterson & Wolk, 1978b; and see also Wolk & Wojciuch, 1971a). Thus, it appears that intact heterocysts can generate ATP and a low-potential reductant photosynthetically. However, it is possible that the populations of probe-isolated heterocysts used in those experiments contained a porous subpopulation which responded to exogenous supplements, and an intact subpopulation which, impermeable to ATP and reductant, generated either one *or* the other in response to illumination. It therefore cannot be concluded that heterocysts both photoreduce and photophosphorylate *in vivo*.

Heterocysts isolated by probe cavitation can photoreduce ferredoxin (Apte *et al.*, 1978) and nitrogenase (Lockau *et al.*, 1978) in the presence of ascorbate and either dichlorophenolindophenol or diaminodurene as electron donor. Metronidazole, thought to inhibit ferredoxin-linked reactions, selectively inhibits photostimulated reduction of acetylene by intact filaments of *Anabaena* (Tetley & Bishop, 1979) and by isolated heterocysts (Eisbrenner & Bothe, 1979). Thus, photoreduced ferredoxin may be the electron donor to nitrogenase, but alternatively, ferredoxin may be dark-reduced. Because metronidazole did not inhibit glucose-6-phosphate/$NADP^+$-dependent acetylene reduction by isolated heterocysts in the dark (Eisbrenner & Bothe, 1979), the target of the metronidazole is uncertain. Dibromothymoquinone (DBMIB), an inhibitor of plastoquinone-mediated reactions, selectively inhibits acetylene reduction by intact filaments of *A. cylindrica* (Bothe & Loos, 1972) and the H_2-dependent photoreduction of acetylene and dark reduction of O_2 by isolated heterocysts (Eisbrenner & Bothe, 1979). In the presence of the artificial electron carrier phenazine methosulphate, heterocysts can photophosphorylate (Tel-Or & Stewart, 1975, 1976, 1977). However, it remains undetermined whether there is normally non-cyclic electron flow via PS I to nitrogenase in heterocysts; if so, what the electron donor and electron transport pathways are to PS I; and whether cyclic or non-cyclic photophosphorylation, or both, takes place.

Superoxide may be generated in heterocysts, as in vegetative cells, by reduction of O_2 by PS I. Superoxide dismutase, which prevents build-up of this toxic anion, is present at approximately the same specific activity in heterocysts as in vegetative cells (Daday *et al.*, 1977; Henry *et al.*, 1978). The physiological significance of preferential localization within heterocysts of other reductions, viz., of nuclear emulsion, silver ions and triphenyltetrazolium chloride (Tischer, 1957; Talpasayi, 1967; Stewart *et al.*, 1969; Thomas & David, 1972; Fay & Kulasooriya, 1972), and of oxidations of phenylenediamine and benzidine (Tischer, 1957; Wahal *et al.*, 1973a), is unclear (see caveats by Carr and Bradley, 1973 and Peters, 1975).

14.4.3.3 Metabolism of hydrogen

Hydrogen gas is an excellent electron donor to nitrogenase in isolated heterocysts in the light (Peterson & Wolk, 1978b; Eisbrenner & Bothe, 1979; see also Wolk & Wojciuch, 1971a, Benemann & Weare, 1974a, and Bothe, Tennigkeit & Eisbrenner, 1977a). No organic electron donor is known to be as effective. Heterocysts which express nitrogenase activity in response to ATP and dithionite in the dark show no such dark response to ATP and H_2 (Wolk & Wojciuch, 1971a; Eisbrenner *et al.*, 1978), suggesting that H_2 can serve as electron donor to nitrogenase only in the light, i.e. that H_2 donates its electrons to PS I in heterocysts. However, it remains possible that the penetration of the heterocysts by ATP was dithionite-dependent.

Three types of hydrogenase activity are known in cyanobacteria: nitrogenase, which produces hydrogen; a classical, reversible hydrogenase (see Chapter 4); and an 'uptake' hydrogenase which can reduce O_2 and certain artificial electron acceptors, but for which the immediate natural electron acceptor is unknown. The specific activity of the classical, reversible hydrogenase, with dithionite and methyl viologen as electron donors, was similar in broken heterocysts and vegetative cells of *Anabaena cylindrica* and *Nostoc muscorum* (Tel-Or *et al.*, 1978; Eisbrenner *et al.*, 1978).

Uptake of hydrogen in the dark by three aerobically grown strains of *Anabaena* [0.60 to 1.74 μmol/(mg chl·hr)] is greatly reduced or eliminated when cultures are grown with ammonium, and so lack heterocysts (Peterson & Wolk, 1978a; Eisbrenner *et al.*, 1978; Daday *et al.*, 1979; see also Tel-Or *et al.*, 1978). Uptake hydrogenase was 'solubilized' (i.e. rendered incapable of sedimenting at $1000 \times g$ in 4 min) in close parallel with sonic destruction of heterocysts, very little or no activity being detected in extracts of vegetative cells. Eighty-six and 84% of the total *in vitro* hydrogenase activity of filaments of *Anabaena* 7120 (Peterson & Wolk, 1978a) and *Cylindrospermum licheniforme* (Hirosawa & Wolk, 1979a), respectively, was recovered in isolated heterocysts. These results seemed to indicate that all or virtually all of the uptake hydrogenase of aerobically grown filaments was localized in the heterocysts. Rates measured *in vitro* were far higher than rates measured *in vivo*, and were at least six times greater than the expected rates of nitrogenase-catalysed production of H_2. However, the observation (Eisbrenner *et al.*, 1978) that the non-heterocystous (Myers) strain of *Anabaena variabilis*, grown and assayed aerobically, consumed H_2 at a rate of 3 μmol/(mg chl·hr) calls into question the idea that the uptake hydrogenase is, under aerobic conditions, restricted to heterocysts. That uptake hydrogenase activity which is in heterocysts permits recycling of reducing equivalents 'wasted' by nitrogenase; salvages some of the energy lost, by coupling to oxidative phosphorylation (Peterson & Burris, 1978); and helps to maintain nitrogenase active (Bothe, Tennigkeit & Eisbrenner, 1977a; Peterson & Wolk, 1978b), presumably by reducing oxygen.

14.5 The process of differentiation

Heterocysts normally arise in filaments when exogenous fixed nitrogen is depleted or, during growth on N_2, midway between two pre-existing heterocysts when these become sufficiently widely spaced. If the heterocysts are taken to be the sole sites of N_2 fixation, with intervening cells metabolizing a fraction of the N_2-derived nitrogen which reaches them, the cell midway between two heterocysts may also be undergoing nitrogen deprivation at the time that it is stimulated to differentiate. Seeming—perhaps real—exceptions to the rule that heterocysts form in response to a stepdown in nitrogen nutrition will be discussed below. However, it is useful for the moment to consider heterocyst differentiation as the culmination of a process which begins as a non-developmental response to nitrogen deprivation.

Akinyanju and Smith (1979) observed a marked but transient increase in ppGpp in the filaments 30 min after *Anabaena cylindrica* was deprived of fixed nitrogen, whereas Adams *et al.* (1977) and Rogerson *et al.* (1978) detected no such increase. Cellular cyclic AMP increased *c.* four-fold over 4 h when *A. variabilis* was deprived of fixed nitrogen (Hood *et al.*, 1979). Within hours, two proteases are activated: one degrades phycobiliproteins specifically; the other, a Ca^{++}-dependent protease, degrades numerous other proteins which are lost during heterocyst development (Wood & Haselkorn, 1976, 1979, 1980; see also Foulds & Carr, 1977, and Ownby *et al.*, 1979). In consequence, biliproteins are lost throughout the filaments, as are the activities of Ru5P kinase and RuBP carboxylase (Bradley & Carr, 1977). The synthesis of a number of polypeptides characteristic of heterocysts appears to be initiated in vegetative cells as well as in presumptive heterocysts, but then to continue only in developing heterocysts (Fleming & Haselkorn, 1974). It is as if all cells were undergoing metabolic changes making them more similar to heterocysts. After 2.3 h (a small fraction of the doubling time), there is still no morphological distinction between presumptive heterocysts and all other cells in *A. cylindrica*, but if the filaments are re-supplied with fixed nitrogen at this time, certain cells continue to differentiate morphologically to the stage of being 'proheterocysts', heterocyst progenitors. (In certain mutants, heterocyst differentiation proceeds only to the stage of proheterocysts (Wilcox *et al.*, 1975b; Singh, 1976a).) The time of commitment to *completion* of differentiation, 5.0 ± 1.0 h, agrees closely with the time at which proheterocysts are first observed, 5.5 ± 0.5 (Bradley & Carr, 1976). Proheterocysts which form upon re-addition of ammonium do not revert to being vegetative cells (Kulasooriya & Fay, 1974). However, surgically isolated proheterocysts, even those at a very advanced stage of differentiation, can undergo such a reversion (Wilcox *et al.*, 1973b). (This observation raises the question as to whether heterocysts which have been reported to germinate (e.g. Brand, 1901; Geitler, 1921, 1932; Hollerbach, 1928; Fogg, 1944; Desikachary, 1946; Wolk, 1965b; Singh, Tiwari & Singh, 1972) were in fact fully mature.) 'Early proheterocysts' could be isolated from *Anabaena* 7120 9 h following nitrogen stepdown by incubating filaments with lysozyme for 2–3 h in the presence of 20 mM EDTA. A number of proteins abundant in vegetative cells were already absent from the isolated proheterocysts, presumably as a result of the action, *in vivo*, of the proteases already mentioned (Fleming & Haselkorn, 1974).

In *A. catenula*, proheterocysts pass through seven stages of morphological differentiation (Wilcox *et al.*, 1975b). (i) Traces of a fibrous envelope layer surround a cell not otherwise morphologically differentiated. (ii) The fibrous layer is completed, and the junctions to adjacent cells decrease in diameter. (iii) The homogeneous envelope layer starts to form, and the cell elongates at its poles. (iv) Deposition of the homogeneous layer is extensive; the protoplast assumes its final shape in the region of the pore channel (see Fig. 14.3); and the thylakoids become contorted, often appearing vesiculate. (v) The laminated layer of the envelope is deposited near the poles of the cell, and contorted membranes appear to be concentrated in the polar regions. (vi) Deposition of the laminated layer continues, and a 'plug' of material may form in the pore channel. (vii) Extensive proliferation of lamellae takes place, resulting in (viii) mature heterocysts. Each of the stages (i) to (vii) lasts about 3 h. This sequence has been corroborated for *A. azollae* (Lang, 1965) and *A. cylindrica* (Kulasooriya *et al.*, 1972; see also Fogg, 1951, and Wahal *et al.*, 1974). Glycogen granules and polyphosphate granules decrease in number, and polyhedral bodies (carboxysomes; see 14.4.2) are lost during development (Kulasooriya *et al.*, 1972; Talpasayi & Bahal, 1967), whereas the wall of the vegetative cell from which the heterocyst differentiated is apparently retained (Fay & Lang, 1971; Lang & Fay, 1971; Wilcox *et al.*, 1973b; Dunn *et al.*, 1971). The minimum duration of morphological differentiation during exponential growth of *A. cylindrica* was calculated to be about 4.3 h (Dunn *et al.*, 1971), whereas morphological differentiation requires about 1.7 h, or *c.* 45% of the doubling time, in *Anabaena* strain CA (data of Bottomley *et al.*, 1979, calculated by the equation in Dunn *et al.*, 1971).

The appearance of nitrogenase activity lags about 1.5–2 h behind the appearance of mature heterocysts in *Anabaena* CA (Stacey, Bottomley, Van Baalen & Tabita, 1979), and lags not at all in *Anabaena* 6411 (Neilson *et al.*, 1971) and *Anabaena* 7120 (Fleming & Haselkorn, 1973; see Haselkorn, 1978); for *A. cylindrica*, Bradley and Carr (1976) found no lag, whereas Kulasooriya *et al.* (1972) reported a lag of 12 h under aerobic conditions. Thus, under aerobic conditions, nitrogenase activity appears not to precede morphological maturity of heterocysts. (Under anaerobic conditions, nitrogenase can be active in the absence of heterocysts (Rippka & Stanier, 1978; see also Chapter 4).) The suggestion that upon nitrogen stepdown, synthesis of nitrogenase polypeptides takes place initially in all cells (Fleming & Haselkorn, 1974) was based on use of one-dimensional gels, in which those polypeptides could not be distinguished from others of closely similar molecular weight. If synthesis of nitrogenase is repressed, in part, by oxygen, it may take place only in cells which are sufficiently anaerobic. However, in one mutant of *Anabaena* CA, synthesis of nitrogenase precedes the development of full protection against atmospheric oxygen (Grillo *et al.*, 1979). How the increases in activity—in at least certain cases, syntheses (Wood & Haselkorn, 1976, 1979, 1980)—of the other enzymes required for heterocyst differentiation are coordinated remains to be elucidated.

14.6 Controls over differentiation

14.6.1 Control of the formation of heterocysts

14.6.1.1 Fixed nitrogen and nitrogen stepdown

Fogg's (1949) finding that fixed nitrogen inhibits the appearance of heterocysts has been confirmed in many species of cyanobacteria (see, for instance, references in Wolk, 1973). Ammonium suppresses the formation of intercalary heterocysts in *Nostochopsis lobatus*, whereas terminal heterocysts continue to initiate differentiation but subsequently revert to a vegetative condition (Tiwari, 1978b). Heterocyst formation by *Anabaena* CA, in contrast to other organisms, is much more strongly inhibited by nitrate than by ammonium, but the effect of nitrate is reversed by methionine sulphoximine (see below; Bottomley *et al.*, 1979), suggesting that the inhibition is mediated by intracellular ammonium. In *Anabaena* L-31, the inhibition by nitrate of heterocyst formation in continuous cultures varies with the dilution rate, no inhibition being observed at high rates of dilution and growth (Thomas & David, 1971). Heterocyst formation is stimulated by modifications of the environment which accentuate nitrogen deficiency, e.g., elimination of molybdenum (Fogg, 1949; Donze, 1973; Fay & de Vasconcelos, 1974), addition of vanadium (Fay & de Vasconcelos, 1974), incubation under Ar-CO_2 (Carr & Bradley, 1973; de Vasconcelos & Fay, 1974), or incubation in the presence of certain amino acids which reduce the specific activity of nitrogenase (David & Thomas, 1980). Certain mutants of *Nostoc muscorum* and *Anabaena variabilis* form heterocysts in the presence of concentrations of ammonium which suppress heterocyst formation in wild-type cultures; nitrogenase activity is regulated separately (Singh & Singh, 1978; Wolk, 1979b). (Suppression, by ammonium, of gradients of cell size in the Rivulariaceae (Fay *et al.*, 1968; Tyagi, 1978b; Rai *et al.*, 1978; Pattnaik & Singh, 1978) has been taken as evidence for nitrogen fixation by heterocysts. However, nitrate can suppress heterocyst formation but leave such gradients (Sinclair & Whitton, 1977a,b; Rai *et al.*, 1978).)

There remains the question, however, whether fixed nitrogen can wholly suppress heterocyst formation, or whether certain cells are differentiated in a latent, i.e., not morphologically manifest, manner and complete morphological differentiation when transferred from fixed nitrogen. Presumptive differentiation *is* possible: as discussed above, filaments of *A. cylindrica*, transferred out of and then again into the presence of fixed nitrogen, show a commitment to differentiation which precedes morphological differentiation (Bradley & Carr, 1976). Remarkably, that commitment can be expressed in the light in the presence of 80 μmol DCMU, or in the dark in the presence of 2 mmol NH_4^+, 1 μg/ml proflavin, 2 μg/ml rifampicin, or 10 μg/ml 5-fluorouracil (Bradley & Carr, 1977). (In mutants showing no morphological evidence of differentiation following nitrogen stepdown (Singh, Singh & Singh, 1972; Wilcox *et al.*, 1975b; Ladha & Kumar, 1976; Singh, 1976a; Singh, Ladha & Kumar, 1977; Padhy & Singh, 1978), have early biochemical steps of the process taken place?) Wolk and Quine (1975) argued that some undifferentiated cells may be more disposed than others to differentiate upon nitrogen stepdown. Concordantly, it is normally the smaller of two daughter cells of *A. catenula* which becomes a heterocyst (Mitchison & Wilcox, 1972).

When *Cylindrospermum licheniforme* grown with fixed nitrogen is subjected to nitrogen stepdown, the resulting pattern of heterocyst formation is qualitatively different (multiple intercalary heterocysts are formed, and single intercalary heterocysts are formed at relatively high frequency) from that during growth on N_2 (Wolk & Quine, 1975). It therefore cannot be argued that a normal pattern has merely been obscured by fixed nitrogen. Wilcox's (1970) observation that a pattern persisted, at least ephemerally, during growth on fixed nitrogen may mean only that presumptive heterocysts were completing their differentiation. Because a revertible mutant of an *Anabaena* (heterocysts normally intercalary) with the form of a *Cylindrospermum* (nearly all heterocysts terminal) could be isolated (Wilcox *et al.*, 1975b), the basic mechanisms governing heterocyst formation presumably do not differ in these two genera. That is, the pattern of loci of heterocyst formation is probably determined 'regulatively' upon nitrogen stepdown in both of these genera, rather than being rigidly predetermined. The semi-regularity of the resulting pattern then implies that very immature heterocysts inhibit nearby cells from differentiating (Wolk & Quine, 1975).

Spaced groups (or bunches) of proheterocysts appear in *Anabaena catenula* upon nitrogen stepdown. Only one cell per group completes differentiation; the others revert to being vegetative cells (Wilcox *et al.*, 1973a). Perhaps related are the observations that 7-azatryptophan (Mitchison & Wilcox, 1973; Bothe & Eisbrenner, 1977; Agrawal & Kumar, 1978; Stacey, Bottomley, Van Baalen & Tabita, 1979), β-2-thienyl-alanine (Rogerson, 1979), rifampicin (Wolk & Quine, 1975; Wilcox *et al.*, 1975a; Tyagi, 1978a; Dikshit *et al.*, 1979; Grover & Puri, 1979), methionine sulphoximine (Singh, Ladha & Kumar, 1977), methionine sulphone and DL-β-hydroxynorvaline (Kumar & Kumar, 1980) induce the formation of groups of contiguous heterocysts in a variety of cyanobacteria. Mutants with that same morphological trait have been isolated (Singh & Tiwari, 1969; Singh, Tiwari & Singh, 1972; Wilcox *et al.*, 1975b; Ladha & Kumar, 1977b; Singh, Ladha & Kumar, 1977). It has also been reported that exposure to a critical concentration of mitomycin C can elicit production of pairs of heterocysts (Tyagi, 1974b).

14.6.1.2 Inhibitory fields

Compared with vegetative cells from unfragmented filaments, vegetative cells detached from heterocysts by fragmentation of filaments are stimulated—both absolutely and relative to the alternative fate of cell division—to differentiate into heterocysts. No such stimulation occurs in fragments containing presumptive or already-differentiated heterocysts. The stimulation in vegetative fragments can therefore only be attributed to detachment from heterocysts, and not, for example, to injury (Wolk, 1967; cf. also Singh, Tiwari & Singh, 1972; Wolk & Quine, 1975; and van de Water & Simon, 1978). These results showed conclusively that heterocysts, both mature and immature, inhibit nearby vegetative cells in the same filament from differentiating into heterocysts. The pattern of heterocyst formation during growth on N_2 must, therefore, be regulative rather than, e.g., only the result of differential cell divisions. Wilcox *et al.* (1973a,b) and Mitchison *et al.* (1976) interpreted the results of surgical experiments (for other interpretations of these, see Wolk, 1979b) and observations on the fate of groups of proheterocysts (see above), as supporting these conclusions. It follows that the mechanisms governing the pattern of spaced heterocysts can serve as a paradigm for studies of inhibitory fields in plant and animal development.

Wolk and Quine (1975) recognized that these mechanisms might be reflected in the distribution of distances between heterocysts arising upon nitrogen stepdown. They determined the distributions expected, based on the assumption that heterocysts arise randomly in space and time, subject (or not) to the restriction that none can form within an inhibitory zone propagating outwards from each newly determined proheterocyst. By considering different kinetics of propagation of inhibition, and comparing the observed with the predicted distributions, they concluded that a propagating zone of inhibition was essential (see also Wolk, 1979b), that inhibition could not (as does an action potential) propagate linearly with time, and that prediction and observation matched most closely if inhibition involved diffusion of a substance being metabolized. Similar pattern formation is seen when certain mutants unable to fix nitrogen are subjected to nitrogen stepdown (Wilcox *et al.*, 1975b; Currier *et al.*, 1977; Singh & Sonie, 1977; Ladha & Kumar, 1977b; Singh, Ladha & Kumar, 1977).

Two such diffusion-based mechanisms could account for what Mitchison *et al.* (1976) called a 'zone of inhibition' around heterocysts. Either heterocysts produce a differentiation inhibitor which is metabolized by vegetative cells, or vegetative cells produce a differentiation stimulator which is metabolized by heterocysts. A vegetative cell of *Cylindrospermum licheniforme*, unless very close to a heterocyst, will differentiate into a heterocyst if, by cutting of the filament, it becomes the terminal cell. Cells much farther from the nearest heterocyst do not differentiate in an intact filament. Differentiation cannot, therefore, be controlled by the accretion within vegetative cells of a differentiation-stimulatory substance produced by those cells and metabolized by heterocysts. Were that idea correct, intercalary heterocysts should normally have been formed, but were not. Rather, it appears that an inhibitor (or inhibitors) which diffuses from heterocysts to vegetative cells and can be lost through the cut ends of filaments (see Wilcox *et al.*, 1973a,b) is central to the control of the spacing of heterocysts (Reddy & Talpasayi, 1974; Wolk & Quine, 1975). In addition, because very short filaments are slower than longer ones to produce a heterocyst, a substance (or substances) facilitating differentiation—one which can also be lost through cut ends, and which need not be specific for the differentiation process—may be produced by vegetative cells.

14.6.1.3 The inhibitory interactions

The idea that N_2-derived ammonium is the inhibitor (Fay *et al.*, 1968) has been superseded by the working hypothesis that glutamine may be the inhibitor (Thomas *et al.*, 1977), for several reasons. First, some organisms maintain a pattern of spaced heterocysts in the presence of ammonium ion (Bottomley *et al.*, 1979); even *Anabaena cylindrica*, in which the repression by ammonium was first noted (Fogg, 1949), produces a low frequency of heterocysts continuously in the presence of ammonium under certain experimental conditions (Wolk, 1970). Second (see 14.4.1), glutamine, rather than ammonium, is the principal known product of heterocysts. Third, in the presence of methionine sulphoximine (MSX; Stewart & Rowell, 1975) and hydroxylysine (Ladha *et al.*, 1978), inhibitors of glutamine synthetase, heterocyst formation takes place even in the presence of exogenous ammonium. (The effects of MSX can be reversed by glutamine and other amino acids (Ownby, 1977; Ladha & Kumar, 1978b), even if glutamine is added hours after addition of MSX, so that in that case the effect of glutamine cannot be attributed to prevention of uptake of MSX (Haselkorn *et al.*, 1980).) Fourth, glutamine can decrease the frequency of formation of heterocysts in *Anabaena* and *Cylindrospermum*, although the decrease may be attributable to ammonium formed by its hydrolysis (Wolk, 1979b).

Even if glutamine is the critical inhibitory substance produced by mature heterocysts, it may be that the substance which passes from vegetative cell to vegetative cell, resulting in a gradient of inhibition, or the substance which controls differentiation intracellularly, is not glutamine itself, but one or more derivatives of glutamine. For example, the induction of heterocysts by certain analogues of tryptophan may be attributable in part to their inhibitory effect on glutamine-dependent synthesis of tryptophan (Bottomley *et al.*, 1980) rather than on synthesis of glutamine itself (Stacey, Bottomley, Van Baalen & Tabita, 1979). The ability of a developing heterocyst to inhibit further heterocyst formation is developed very early during differentiation, hours before glutamine is produced from N_2, and very roughly at a period of extensive proteolysis in the proheterocyst. This proteolysis may, by yielding glutamine or a spectrum of amino acids, mimic inhibitor production by mature heterocysts or by vegetative cells from heterocyst-derived glutamine (Fleming & Haselkorn, 1974).

Anabaena cylindrica and *Anabaena* 29411 maintain a normal pattern of spaced heterocysts during growth in the absence of atmospheric N_2 but in the presence of nitrate (Wolk, 1970; Sutherland *et al.*, 1979; and see also Sinha &

Kumar, 1973). In these instances, inhibition by mature heterocysts cannot be attributed to N_2-derived glutamine. Perhaps nitrate reductase activity is high in heterocysts, or as a consequence of allosteric effects, glutamine synthetase activity is much higher in heterocysts than in vegetative cells, so that a gradient of glutamine is still in existence. Alternatively, the hypothetical inhibitor could be a minor and perhaps very specialized product of heterocysts.

Inhibition by mature heterocysts is related to heterocyst spacing, and inhibition by immature heterocysts is presumably also related to the absence of bunching of heterocysts. The mechanisms of the two inhibitions need not be the same; that they are at least related is suggested by the fact that 7-azatryptophan, β-2-thienyl-dl-alanine, and rifampicin, all of which are known to elicit bunching, lead to decreased spacing of the bunches. The decrease in spacing is apparently not a consequence of changed growth rate (Mitchison & Wilcox, 1973). Significantly, these substances also induce the formation of heterocysts in the presence of low but differentiation-preventing concentrations of ammonium (Rogerson, 1979; Dikshit *et al.*, 1979; Bottomley *et al.*, 1980). On the basis of this observation, Rogerson (1979) has suggested that the inhibition controlling spacing is related to the inhibitory effect of ammonium (contrast Singh, Ladha & Kumar, 1977), and therefore that vegetative cells also synthesize the same inhibitor at a level regulated by the supply of fixed nitrogen. It is also possible that azatryptophan, β-2-thienyl-alanine and rifampicin affect not the production of the inhibitor but the synthesis of one or more proteins, in vegetative cells, which mediate the response of those cells to the inhibitor (compare Ownby *et al.*, 1979). Concordant with this idea is the observation that of a group of tryptophan analogues thought to reduce the synthesis of tryptophan by inhibiting anthranilate synthetase, only those reputed to be readily charged to tRNA and incorporated into protein induced heterocyst formation (Bottomley *et al.*, 1980). The question of why proheterocysts do not 'turn off' their own differentiation as well as the differentiation of contiguous cells remains, despite speculation (Wilcox *et al.*, 1973a; Wolk, 1979b), an enigma.

14.6.2 Role in akinete formation

There is no evidence that heterocysts play a stimulatory role in the nearly simultaneous akinete formation (Chapter 15) of all vegetative cells between two heterocysts, as occurs in *Anabaena variabilis* (see Singh, Tiwari & Singh, 1972), nor—evidently—do they play a role when akinetes form in the absence of heterocysts (Singh, 1975b, 1976a). Akinetes may form approximately midway between pre-existing heterocysts, as in *A. doliolum* (Tyagi, 1974a), perhaps by redirected differentiation of a presumptive heterocyst (Geitler, 1925; compare Kulasooriya *et al.*, 1972, and Tyagi, 1974b; see also Sutherland *et al.*, 1979). In certain organisms, heterocysts may inhibit the differentiation of adjacent cells into akinetes (Singh, Tiwari & Singh, 1972; and see Wolk, 1965b). Akinete formation adjacent to heterocysts, as in *Anabaena cylindrica* and *Cylindrospermum licheniforme* (Fig. 14.1), appears to involve an inductive effect by heterocysts: vegetative cells of *A. cylindrica* which are detached from heterocysts by shaking of a culture, although apparently uninjured, fail to differentiate into akinetes (Wolk, 1965a, 1966). In *C. licheniforme*, hydrogen gas stimulates akinete formation adjacent to heterocysts in the presence of oxygen and the presence or absence of nitrogen gas. Because uptake hydrogenase activity appears to be localized in the heterocysts in this organism (see 14.4.3.3), heterocysts may mediate the effect of hydrogen on the cells which form akinetes (Hirosawa & Wolk, 1979a). (Whether the akinete-eliciting substance produced by *C. licheniforme* (Fisher & Wolk, 1976; Hirosawa & Wolk, 1979a,b; see Chapter 15) plays a role in the inductive interaction is unknown.) The formation of akinetes by vegetative cells adjacent to heterocysts can therefore serve as a model for embryogenic, intercellular inductions.

CHAPTER 15
AKINETES

J. M. NICHOLS & D. G. ADAMS

15.1 Introduction

While this chapter is concerned with the akinetes of cyanobacteria, there are a number of other structures associated with reproduction which are worthy of brief mention.

Hormogonia consist of short filaments of undifferentiated cells characteristic of filamentous cyanobacteria, in which they are a major means of reproduction. In *Oscillatoria* and *Cylindrospermum*, hormogonia are formed by the disintegration of whole filaments, while in branching forms (e.g. *Scytonema*, *Stigonema*) they form at the tips of branches. In some cases breakage of trichomes is aided by specialized cells (necridia) which become biconcave in shape due to lysis and dehydration and permit fragmentation of the trichome and subsequent liberation of hormogonia (see Fogg *et al.*, 1973). Hormogonia usually move away before developing into new filaments, although they occasionally germinate *in situ* within the sheath of the parent filament. Hormocysts develop either terminally or at intercalary positions in some species such as *Westiella laminosa*. They consist of short filaments of very granular cells, surrounded by a common condensed sheath, and as such might be regarded as an intermediate between hormogonia and akinetes (Fogg *et al.*, 1973). They become detached from the main filament by breakage and may then germinate at one or both ends (Allsopp, 1968).

Exospores and endospores are products of cell fission and are characteristic of the two orders Chamaesiphonales and Pleurocapsales (see Waterbury & Stanier, 1978). Exospores develop by successive cleavages from the apical pole of an elongated vegetative cell (e.g. *Chamaesiphon*). Endospores are formed in, for example, *Dermocarpa* (another member of the Chamaesiphonales) by multiple fission of a large spherical vegetative cell and are subsequently released by rupture of the parental wall and eventually form new vegetative cells. All members of the Pleurocapsales reproduce by this mechanism of multiple fission and endospore release, and Waterbury and Stanier (1978) have applied the term baeocyte to the endospore of this order of cyanobacteria.

The earlier literature contains descriptions of akinetes from a variety of cyanobacteria, including some unicellular forms (Geitler, 1932). *Chlorogloeopsis fritschii* changes from an aseriate or filamentous cell type to that of a large unicell when held in non-growing conditions (Evans, Foulds & Carr, 1976). It is not certain how comparable such descriptions are. Akinetes of some members of the Nostocaceae, Rivulariaceae and of Stigonemataceae are specialized single cells which may serve as means of perennation in these organisms (for reviews see Wolk, 1965a, 1973; Nichols & Carr, 1978; Adams & Carr, 1981). They are usually much larger than vegetative cells, although their size can vary considerably, even within a single culture (see, for example, Fig. 15.1h) and have a granular appearance under the light microscope (Fig. 15.1c–f), due to the accumulation of large numbers of cyanophycin granules (see 15.2). The shape of akinetes also varies considerably, being, for example, elongated in *Anabaena cylindrica* (Fig. 15.1c and d), but almost spherical in *Anabaena* CA (Fig. 15.1e–h).

Akinetes were reported as early as 1856 by Carter who described a large granular cell (which he called a sporangium) which always developed next to a heterocyst (Fig. 15.2). A number of other descriptions of akinetes from different species of cyanobacteria followed (Bary, 1863; Rabenhorst, 1865; Janczewski, 1874; Beck v. Mannagetta, 1898; Brand, 1903) and illustrations from some of these texts are shown in Fig. 15.2, making an interesting comparison with contemporary photomicrographs (e.g. Fig. 15.1).

15.2 Structure and chemical composition

The major differences between akinetes and vegetative cells are readily observed under the light microscope. The akinete is usually larger, its cell wall thicker and its cytoplasm more granular than the vegetative cell (see Fig. 15.1). Numerous studies have employed the electron microscope to discern the structural and cytoplasmic changes associated with the differentiation of an akinete from a vegetative cell (Wildon & Mercer, 1963a; Leak & Wilson, 1965; Miller & Lang, 1968; Clark & Jensen, 1969; Jensen & Clark, 1969; Wildman *et*

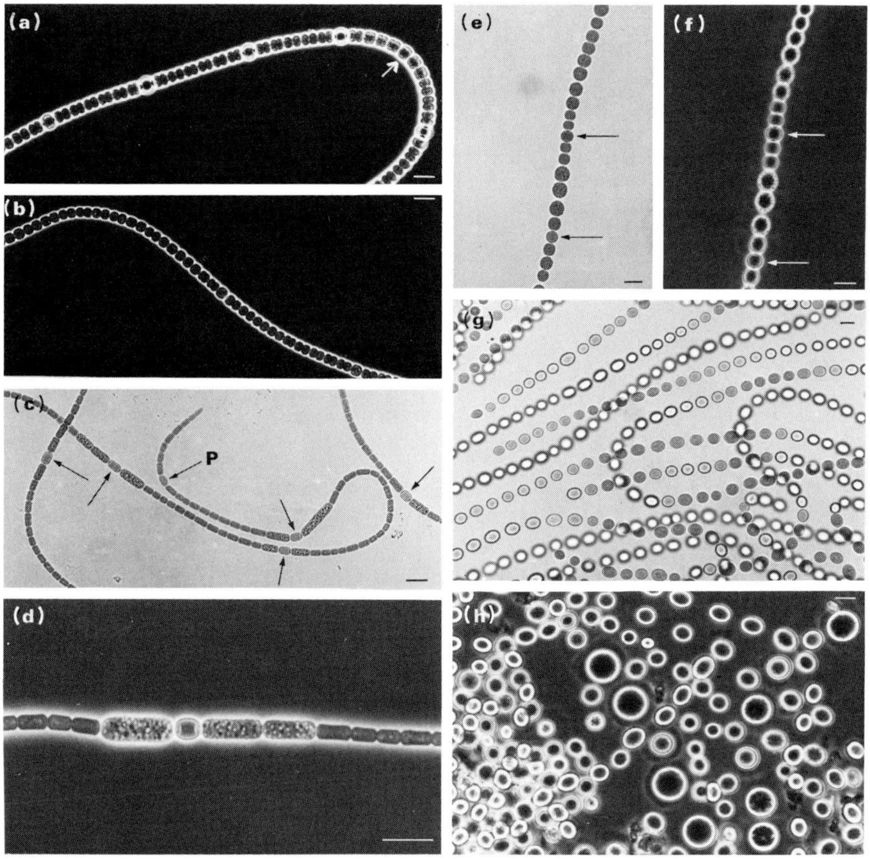

Fig. 15.1. The akinetes of *Anabaena* sp. (a) *Anabaena* CA grown in the absence of fixed nitrogen, showing the regular spacing of heterocysts and an early proheterocyst (arrow). (b) *Anabaena* CA grown in the presence of nitrate (17 mM) showing the suppression of heterocyst development. (c) *A. cylindrica* grown in the absence of fixed nitrogen. A single proheterocyst (p) and several heterocysts (arrows) are indicated and large, granular akinetes can be seen on one or both sides of many of the heterocysts. (d) Part of a filament of *A. cylindrica* showing a heterocyst with akinetes on either side. (e) and (f) *Anabaena* CA grown in the absence of fixed nitrogen, showing heterocysts (arrows) and granular akinetes developing mid-way between them. (g) *Anabaena* CA grown in the presence of nitrate, showing filaments completely transformed into akinetes. Although most have become separated from one another, the shape of the original filaments can still be seen. (h) As (g) but a more vigorously shaken culture in which all the akinetes have become separated. The variation in the size of the akinetes is clearly visible. Bar markers represent 10 μm. All phase contrast except c, e and g which are bright field.

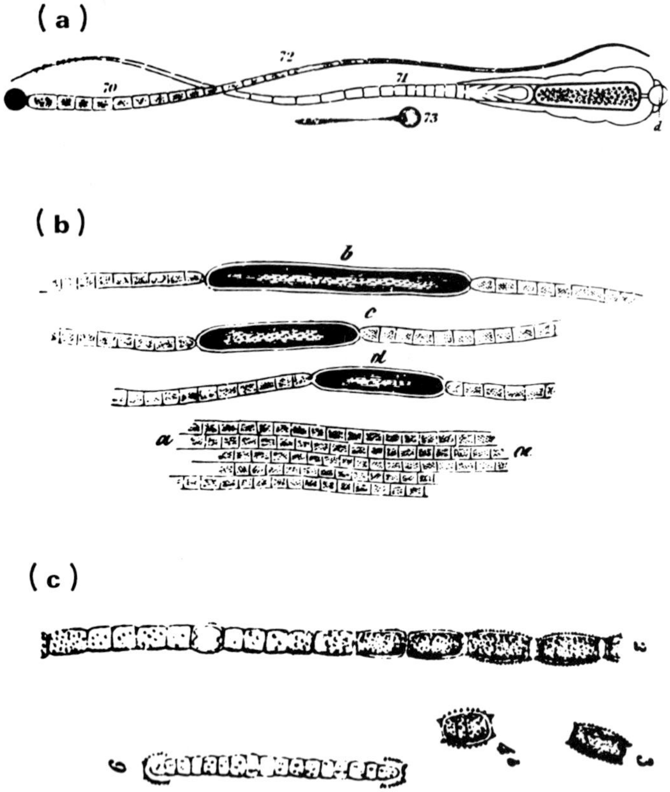

Fig. 15.2. Early drawings of akinetes in cyanobacteria, taken from (a) Carter (1856), (b) Rabenhorst (1865) and (c) Janczewski (1874).

al., 1975; Sutherland *et al.*, 1979; Grilli Caiola & de Vecchi, 1980). In *Cylindrospermum* sp. enlargement of the vegetative cell was accompanied by condensation of the mucilaginous sheath and deposition of a dense fibrillar substance in localized areas within it (Miller & Lang, 1968). As enlargement neared completion this fibrillar layer became continuous and thicker. Additional layers were then deposited between the fibrillar layer and the inner investment, the sheath remaining outside the fibrillar layer. The sheath contained a mucilagenous or pectinaceous substance, whereas the fibrillar layer was considered to be a polyglucoside, containing no lipids, cellulose or chitin. Concomitant with the development of the envelope was the deposition of large numbers of cyanophycin granules in the cell periphery. The glycogen granules, polyhedral bodies and thylakoids of the vegetative cell were all retained.

An increase in abundance of cyanophycin granules and enlargement of the vegetative cell at the onset of akinete differentiation in *Cylindrospermum* sp. was also observed by Clark and Jensen (1969) and Jensen and Clark (1969). More advanced akinetes began to accumulate a fibrous coat between the outer layer of the cell wall and the sheath, and there followed the deposition of an electron-dense substance within the fibrous coat. However, in contrast to Miller and Lang (1968), these authors suggested that the fibrous coat was not a simple condensation of the mucilagenous sheath of the vegetative cell, since the akinete coat was fibrous, thick and dense compared with the vegetative cell sheath and therefore required new synthesis. Clark and Jensen (1969) observed a separation of the fibrous coat from the cell wall by an electron transparent layer at an advanced stage in akinete development, whereas Miller and Lang (1968) indicated that separation of the cell wall from the spore coat occurred immediately before germination of the akinete.

The electron-transparent layer observed by Clark and Jensen (1969) and Jensen and Clark (1969) during akinete development, was also described by Sutherland *et al.* (1979) in *Nostoc* 7524. Soon after the deposition of a loose fibrillar material, it became separated from the cell wall by an electron-translucent layer of variable thickness. Other similarities with *Cylindrospermum* were evident in that a third layer formed between the outer fibrillar and the electron-transparent layer. This was described as being moderately electron-dense material and uniform in width (Fig. 15.3). Both groups of workers also reported an increase in thickness of the cell wall during akinete development. This was shown to be due to the thickening of the intermediate layer, which in *Nostoc* 7524 increased from 10 to 70 nm, the total thickness of the wall increasing from 50 to 110 nm (Fig. 15.3). The mature akinete of *Anabaena variabilis* also has a thickened coat, formed by condensation and thickening of the fibrillar sheath of the filaments (Braune, 1980; Fig. 15.6c, d).

Recently, Grilli Caiola and de Vecchi (1980) have examined akinete development in *Nostoc* strains isolated from different cycad coralloid roots. These species do not form akinetes when they are in symbiosis, but do so when free in nature and in culture. Akinetes showed structural and ultrastructural features similar to those of *Cylindrospermum* sp. (Miller & Lang, 1968; Clark

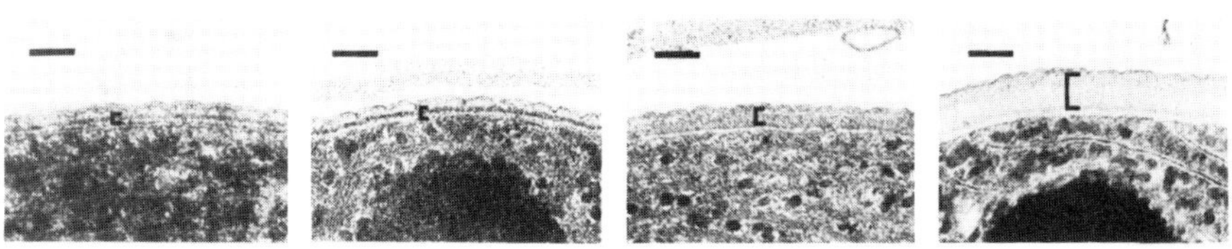

Fig. 15.3. Changes in the cell wall of *Nostoc* 7524 during akinete development. Increase in thickness of the cell wall is due to the thickening of the intermediate wall layer (indicated by brackets) which also becomes progressively less electron-dense. Bar markers represent 100 nm. (Reproduced with permission from Sutherland *et al.*, 1979)

& Jensen, 1969), the cell wall being similar to that of vegetative cells, and external to that, two multi-layered envelopes. The external envelope was mucilaginous and laminar; the internal envelope was approximately 0·15 μm wide, electron-dense and fibrous with a peripheral, more compact and electron-dense laminar zone. The authors observed two morphologically distinct akinetes. The cytoplasm of the first was similar to that of vegetative cells, in that it was electron-dense with thylakoids distributed throughout the whole cell, very numerous ribosomes, numerous polyglucoside granules, small cyanophycin granules and lipid bodies. This type of akinete differentiated, either free in the medium or in chains. In the second type, the cytoplasm contained large and numerous cyanophycin granules, ribosomes, few polyhedral granules and a large nucleoplasmic area. The thylakoids were concentric or branched and peripheral. These akinetes differentiated within the trichomes of vegetative cells. Grilli Caiola and de Vecchi (1980) suggested that the first type of akinete could represent the stage of vegetative cells which have temporary growth difficulties and were able to germinate immediately, the second type representing the mature, quiescent akinete, since it was able to survive for a long time in unfavourable growth conditions.

The elaborate, multilayered akinete envelope is therefore common to *Cylindrospermum* sp., *Nostoc* 7524, *Anabaena cylindrica* (Wildon & Mercer, 1963a; Leak & Wilson, 1965), *A. variabilis* (Braune, 1980) and *Nostoc* sp. (Grilli Caiola & de Vecchi, 1980) but was not found in *Aphanizomenon flos-aquae* (Wildman *et al.*, 1975), in which only a narrow, structured sheath was detected. The change from vegetative cell to mature akinete involved an increase in cell size, gradual disappearance of gas vesicles, narrowing of intrathylakoidal spaces and an increase in compactness of the protoplasts. The length of the cell increased from 10 μm to 50–80 μm. Polyhedral bodies were still present and cyanophycin granules, absent in vegetative cells, increased in number and size. There was a proliferation of ribosomes and glycogen granules, while lipid bodies were reduced in size and number. By contrast, Miller and Lang (1968) were unable to detect ribosomes in mature akinetes of *Cylindrospermum* sp.

In *Nostoc* 7524 the polyhedral and lipid bodies and polyglucoside content of the cells did not change visibly during akinete differentiation (Sutherland *et al.*, 1979). The most pronounced changes were an eight-fold increase in cyanophycin granule content per cell and a 16-fold increase in glycogen per cell. The per cell values of total dry weight, carbon and chlorophyll-*a* doubled, nitrogen content increased by 17%, while phycocyanin was unchanged. Glycogen accumulation in cells undergoing akinete formation has also been demonstrated in *Anabaena* sp. by Sarma and Kantu (1979). The lipid and fatty acid contents of *A. cylindrica* akinetes have been shown to be respectively 15 and 30% less than vegetative cells, the fatty acids being 20–30% more saturated in akinetes (Yamamoto, 1972).

There is a dearth of information on the pigment content of akinetes. Whereas Fay (1969a) found that akinetes of *A. cylindrica* lacked the accessory photopigment phycocyanin, *in vivo* absorption spectra of akinetes and vegetative cells of the same organism have been shown to be very similar (Wolk & Simon, 1969). Sutherland *et al.* (1979) showed that, on a per cell basis, the phycocyanin content of akinetes of *Nostoc* 7524 was the same as vegetative cells, although as a percentage of dry weight it had decreased by 50%. Fay (1969a) found that chlorophyll was largely replaced by pheophytin in akinetes and that they contained a larger proportion of xanthophylls than the intact filaments. In *Nostoc* 7524 the chlorophyll-*a* content as a percentage of dry weight remained constant but doubled per cell (Sutherland *et al.*, 1979).

Wolk and coworkers have determined the nature of the polysaccharides in the cellular envelopes of akinetes of *Anabaena cylindrica* (Dunn & Wolk, 1970; Cardemil & Wolk, 1976, 1979). The wall of the vegetative cell is composed predominantly of amino compounds (65%) with a mannose-rich carbohydrate comprising only 18% of the dry weight. By contrast 62% of the heterocyst and 41% of the akinete wall is carbohydrate, the amino content being 4% and 24% respectively. The carbohydrate is similar in heterocysts and akinetes in that the ratio of glucose:mannose:galactose:xylose is approximately 75:20:3:4 (cf. vegetative cells 35:50:5:8). During akinete differentiation therefore, a glucose-rich polysaccharide is produced, that is different from the polysaccharide component of not only the vegetative cell, but also the sheath in which the ratio of sugars is 47:25:6:21. The sheath is also water-soluble whereas the akinete envelope is highly insoluble. It seems unlikely therefore, that the new envelope of the akinete is derived from the sheath (cf. Miller & Lang, 1968). Cardemil and Wolk (1979) have shown that the repeating sequences of sugars in the polysaccharide of akinetes and heterocysts appear to be identical. Repeating units contain one mannosyl and three glycosyl residues all linked by $\beta(1\rightarrow3)$ glycosidic bonds, with glucose, xylose, galactose and mannose present in side branches. One of every three glucosyl residues in the backbone is branched at both positions 2 and 4, and approximately half of the remaining glucosyl residues have a side chain. Because of the similarity between the polysaccharides of heterocysts and akinetes the authors suggested there must be identical processes leading to their biosynthesis.

Comparison of the nucleic acid content of akinetes and vegetative cells has produced widely differing values. Whether this is a reflection of true interspecies variation, or differences in the methods employed to estimate the nucleic acids remains to be seen. Ueda (1971b) and Ueda and Sawada (1972), using an *in vivo* fluorescent dye procedure, determined that akinetes from

Anabaena baltica contained more DNA than vegetative cells and that akinetes of *Cylindrospermum* contained 30 times more. Simon (1977a) utilizing a chemical assay procedure, showed that the DNA content of *Anabaena cylindrica* akinetes was approximately equal to that of vegetative cells taken from an exponentially growing culture, and was equivalent to 57 genome copies. The akinete, however, contained four times as much RNA as the vegetative cell, these values giving an RNA:DNA ratio of 54 for the akinete, compared with 13 for the vegetative cell. Protein content was also three times higher in the akinete. By contrast, Sutherland *et al.* (1979), found that in *Nostoc* 7524 the RNA and DNA content per cell did not change on differentiation of a vegetative cell into an akinete. The amount of DNA corresponded to 10 genome equivalents and the RNA:DNA ratio was approximately 20. It should be remembered, however, that the relative concentration of DNA and RNA of akinetes and vegetative cells would be expected to depend upon the growth rate of the organism, as this is known to alter the nucleic acid content of vegetative cells (Leach *et al.*, 1971; Mann & Carr, 1974; Adams & Carr, unpubl.). It should also be noted that in *Aphanizomenon flos-aquae* the development of akinetes was accompanied by an increase in ribosome and polyribosome content (Wildman *et al.*, 1975).

15.3 Metabolism

Our poor knowledge of akinete metabolism is a reflection of the difficulty of obtaining preparations of akinetes which are homogeneous in age. Studies with *Nostoc* 7524 (Sutherland *et al.*, 1979) and *Anabaena variabilis* (Braune, 1980) have shown that well synchronized akinete differentiation can be obtained and this should greatly facilitate metabolic studies of both 'resting' and germinating akinetes. One might predict that the metabolism of the former, if active at all, would be extremely limited.

Akinetes of *A. cylindrica* isolated by extrusion through a French pressure cell and differential centrifugation in sucrose, have been shown to fix CO_2 at approximately 15% of the rate obtained for intact filaments (Fay 1969b) which is consistent with the absence of the accessory photopigment phycocyanin from akinetes of the same organism (Fay, 1969a; cf. reduced levels of phycocyanin in heterocysts). Their respiratory rate, however, was twice that of intact filaments. As might be expected, no nitrogenase activity could be detected, although the extreme oxygen-sensitivity of the enzyme might have resulted in loss of activity during preparation of the akinetes. Fay suggested that the increased endogenous respiration is isolated akinetes could indicate the reorganization of enzyme activities prior to germination and indeed it must be stressed that in cultures in which akinete development occurs in a non-synchronous manner, some isolated akinetes may be entering the germination stage. Cultures in which akinete development is well synchronized should therefore be a prerequisite for studies of akinete metabolism.

15.4 Factors controlling akinete development

The factors that control the onset of akinete development are poorly understood, although many nutrients have been implicated. A deficiency of fixed-nitrogen has been reported to induce akinete formation (Harder, 1917; Demeter, 1956) and these observations have been supported by more recent findings which indicated that the addition of nitrogen (nitrate, nitrite or ammonium) had the opposite effect (Singh & Srivastava, 1968; Tyagi, 1974a, 1978c). Canabaeus (1929), however, showed that high concentrations of nitrate induced akinete formation and indeed in both *Anabaena* CA (Adams, unpubl.) and *Nostoc* 7524 (Sutherland *et al.*, 1979), akinete formation occurs in the presence of a fixed-nitrogen concentration sufficient to inhibit heterocyst development. Phosphate has also been considered important, since Wolk (1965b) and Gentile and Maloney (1969) found that a deficiency in this nutrient favoured akinete formation. This might seem an attractive candidate since akinetes do not appear to contain polyphosphate bodies. However, studies by Rother and Fay (1977, 1979a) indicated that a deficiency of phosphate did not stimulate akinete production in a number of planktonic species, although growth was reduced. Sutherland *et al.* (1979) also found that with *Nostoc* 7524, removal of phosphate caused eventual cessation of growth and although cyanophycin accumulated, akinetes were never produced.

Numerous and varied compounds have been tested for their influence on akinete development. Harder (1917) and Sutherland *et al.* (1979) found that akinete production was inhibited by sucrose, while Tyagi (1974a, 1978c) reported a stimulation by glucose. Sodium glutamate (Demeter, 1956), sodium chloride (Canabaeus, 1929) and drying have been shown to induce or increase akinete formation. In *Anabaena cylindrica*, Sinclair and Whitton (1977a) have shown that a deficiency of Fe increased akinete frequency, whereas deficiencies in other trace ions either decreased their frequency (Mg, Ca) or had no effect (Mo, SO_4). The response of *Gloeotrichia ghosei* was quite different in that a deficiency in Fe decreased akinete frequency. Wolk (1965b) showed that akinete formation in *Anabaena cylindrica* was favoured by the addition of 25 mM sodium acetate, the presence of an organic buffer, high carbon concentration and an optimum light intensity, while the addition of calcium glucuronate resulted in the formation of longer chains of akinetes. In *Cylindrospermum licheniforme* Kütz., no such increase in chain length was observed in the presence of calcium glucuronate, although an increase in akinete production was detected (Hirosawa & Wolk, 1979a).

Various amino acids (tryptophan, aspartic acid, phenylalanine, proline and isoleucine) have been shown to increase akinete production in *C. licheniforme* (Hirosawa & Wolk, 1979a). Cyclic nucleotides had little or no effect. In *Anabaena cylindrica* an analogue of arginine, canavanine, increased akinete formation five-fold, arginine itself having no effect (Nichols *et al.*, 1980; see 15.6.1). Another arginine analogue, cyanoalanine and the serine analogue, aminobutyric acid, also increased the number of akinetes but to a lesser extent.

Hirosawa & Wolk (1979a) showed that akinete formation in *Cylindrospermum licheniforme* was completely inhibited by acetylene (1%) and ethylene (0.5%) but was stimulated 2.5-fold by H_2 (12.5–15% v/v) both in air and in $CO_2/O_2/Ar$. This stimulation was not mediated by an effect on nitrogen fixation, since hydrogen did not affect the reduction of acetylene and it also stimulated akinete production in an atmosphere lacking nitrogen. An uptake hydrogenase activity was detected which appeared to be restricted to the heterocysts and the authors suggested that akinete formation might be controlled by some substance (other than nitrogen) which was reduced by this hydrogenase.

In *Nostoc* 7524 a continuous but limited supply of carbon was required for akinete differentiation (Sutherland *et al.*, 1979). Nitrogen starvation stopped growth, cyanophycin production and arrested akinete formation, even in cultures which had commenced differentiation. The addition of inhibitors of protein, RNA and DNA synthesis inhibited or immediately stopped growth and akinetes were not observed.

15.4.1 Growth phase

In a number of species studied in the laboratory in nitrogen-fixing batch cultures, akinete formation has been shown to occur towards the end of, or after the cessation of exponential growth. This was first demonstrated in *Anabaena cylindrica* by Fay (1969b). Using the same species Simon (1977b) showed that akinetes began to develop after the exponential phase of growth and that during the period of akinete formation the optical density at 730 nm increased slowly (this might have been the result, not of growth, but the release of a brown pigment), whereas dry weight and chlorophyll increased substantially but at different rates, indicating unbalanced growth. The development of akinetes at the end of exponential growth in *A. cylindrica* has been clearly demonstrated by Nichols *et al.* (1980; Fig. 15.4).

Gradients of akinete maturity, either towards or away from a heterocyst, have been observed within strings of akinetes, in a number of cyanobacterial

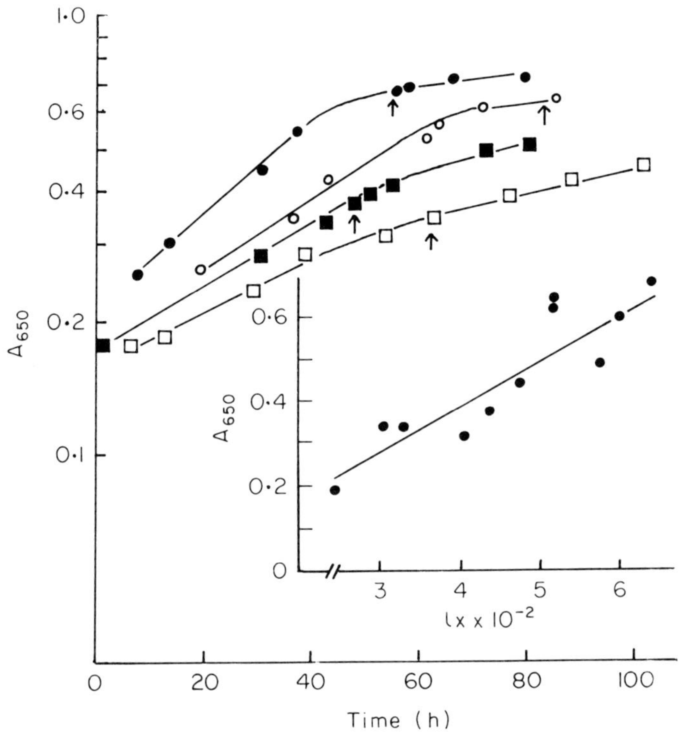

Fig. 15.4. Growth and time of onset of akinete formation (arrows) in *Anabaena cylindrica* as a function of light intensity. *A. cylindrica* was grown on medium lacking fixed-nitrogen at (●) 640 lx, (○) 520 lx. (■) 440 lx and (□) 305 lx. The inset figure shows the relationship between the optical density of the culture at which akinetes were first evident and the light intensity (lx) at which the cultures were grown. (From Nichols *et al.*, 1980)

species (Fritsch, 1945; Wolk, 1965a; Tyagi, 1974a; Simon, 1977b). Wolk (1965a) concluded that in *A. cylindrica*, in which the most mature akinete is adjacent to the heterocyst, these gradients were due to differences in the time required for akinete development. Simon (1977b), however, felt that such gradients were at least partly due to lags in the initiation time of the individual akinetes. This was derived from the observation that at a particular stage of akinete development 83% of cells adjacent to heterocysts had completed, or were in the process of akinete formation, yet at this time no strings of akinetes were present. Such strings (i.e., two or more adjacent akinetes) only formed after this period, but there was no change in the percentage of heterocysts with akinetes attached. Thus, the second akinete of a string must have initiated development at some time after the one adjacent to the heterocyst.

Akinete formation has also been studied in detail in *Nostoc* 7524 by Sutherland *et al.* (1979). In this species a short period (30 h) of exponential growth (generation time 8.5 h) was followed by a 100 h period of non-exponential growth. Akinete differentiation was first observed in the middle of the non-exponential phase. The rate of photosynthetic CO_2 fixation decreased by the end of exponential growth to 60% of its previous level, but maintained this rate during akinete differentiation and maturation. The rate of acetylene reduction, however, slowly decreased during non-exponential growth and akinete differentiation, becoming negligible as akinete maturation continued. This fall in acetylene reduction activity at the end of exponential growth has also been observed in *Anabaena cylindrica* (Nichols *et al.*, 1980). In contrast to these results there have been reports of akinete development in logarithmically growing cultures of *Gloeotrichia echinulata* (Roelofs & Oglesby, 1970) and *Cylindrospermum licheniforme* (Hirosawa & Wolk, 1979a). One should remember, however, that the techniques employed to measure growth, particularly in filamentous cyanobacteria, do not permit the detection of small numbers of cells which may not be growing exponentially. It could be argued that akinetes might develop from such cells during apparently exponential growth. This can be especially true during the late stages of the log phase, when an exact point of transition from exponential to non-exponential growth is extremely difficult to locate. In the absence of conclusive evidence to the contrary, therefore, we must conclude that akinetes develop in batch cultures only at or after the end of exponential growth. One possible exception to this, however, is the stimulation of akinete development in a non-akinete-containing culture, by the filtrate from an akinete-containing culture (Fisher & Wolk, 1976; see 15.4.3).

15.4.2 Light

Light is clearly one of the most important factors governing the growth of an obligate phototroph and one might therefore expect it to have considerable influence on akinete formation. In the laboratory, growth rate and the time of onset of akinete formation can be regulated by light intensity. In both *Nostoc* 7524 (Sutherland *et al.*, 1979) and *Anabaena cylindrica* (Nichols *et al.*, 1980) changes in light intensity alter the cell density at which akinete formation occurs. Decreasing the light intensity results in akinete development at a lower cell density (Fig. 15.4) and vice versa. Under these conditions therefore, akinete differentiation cannot be induced by limitation of inorganic nutrients, since increasing the light intensity would have no effect. In *Nostoc* 7524 addition of exogenous sources of combined nitrogen (NO_3^-, NH_4^+ or glutamine) resulted in more rapid growth and although akinete development occurred after a shorter time than nitrogen-fixing cultures, it did so at the same culture density (Sutherland *et al.*, 1979). However, addition of exogenous sucrose (which could serve as sole carbon source under photoheterotrophic growth conditions in the presence of DCMU), prolonged the exponential growth phase and akinete differentiation commenced at a higher cell density. From field studies which included a number of species of planktonic cyanobacteria, Rother and Fay (1977) concluded that the coincidence of akinete production with dense surface accumulation of algal material was suggestive of light playing a critical role, as there was no deficiency of inorganic nutrients, the orthophosphate concentration being particularly high. While it may be argued that light limitation, caused by dense surface accumulations, could stimulate akinete formation, Rother and Fay (1977) considered the possibility that the quality and high intensity of the light at the surface could have a direct photoinductive effect, or alternatively that akinete production might result from a physiological imbalance between nitrogen fixation and photosynthesis, resulting from inhibition of the latter by the high light intensity. However, when some of these planktonic species were isolated and studied in the laboratory, exposure to extremes of light intensity and to u.v. light of 365 nm reduced and often stopped growth, but did not induce akinete formation (Rother & Fay, 1979a). Limitation of energy supply, whether in the form of light or sucrose, would thus appear to be a major factor in the induction of akinete differentiation. We can therefore explain the development of akinetes at the end of logarithmic growth in terms of light (and hence energy) limitation, which occurs when the culture becomes dense. This does not exclude the possibility that limitation of inorganic nutrients might also be involved in initiating akinete formation and, indeed, in view of the perennating role of akinetes, this would not be surprising, although the results at present are unclear and, on occasions, contradictory.

15.4.3 Extracellular compounds

Cylindrospermum licheniforme produces a substance which stimulates akinete production (Fisher & Wolk, 1976). When a young culture was inoculated into the cell-free supernatant from an akinete-containing culture (conditioned medium) akinete formation occurred almost immediately, but only after a 5-day lag in the control, the final akinete frequency being identical in both cultures. A 1:19 dilution of the conditioned medium retained 50% of its activity. The akinete-stimulating substance has since been isolated and its chemical formula found to be C_7H_5OSN, probably consisting of two fused five-membered rings, one of which is a lactam and the other of which has a thioketone group (Hirosawa & Wolk, 1979b). A concentration as low as 0.3 μM could stimulate the formation of akinetes. An extract from the supernatant of a 2-day-old culture, containing no akinetes, was still able to stimulate akinete formation, although the activity increased as extracts were taken from older cultures. This observation, together with the fact that the activity can also be isolated from the cells themselves, suggests that the stimulating substance is released by healthy cells and is not a product of decomposition (Hirosawa & Wolk, 1979a). The stimulatory substance also increased akinete production in a medium containing phosphate (previous experiments were carried out in a medium devoid of phosphate; Wolk, 1965a), indicating that it is not merely a promoter of a developmental process which occurs in response to phosphate deficiency. While this compound has been detected extracellularly in the culture medium (and indeed, the title of this section reflects this), it is not clear whether its function is inter- or intrafilamentous, or both. Fisher and Wolk (1976) have suggested that it may become concentrated during desiccation or stagnation of a natural body of water and thus stimulate the development of akinetes to survive an impending period of drought.

Rother and Fay (1977) have suggested that breakdown of cells by photooxidative death and bacterial degradation in dense surface accumulations of algal material could release organic compounds which might initiate akinete formation, in a similar manner to the stimulatory factor reported by Fisher and Wolk (1976). However, in laboratory experiments, they found no stimulation of akinete formation by filtrates of cultures containing akinetes. Similarly, when exponentially growing cells of *Nostoc* 7524 were inoculated into cell-free medium from old, akinete-containing cultures, growth ceased after 48 h but no akinetes developed (Sutherland *et al.*, 1979). Following the addition of phosphate, growth resumed and akinetes were observed at the normal cell density, suggesting that the culture had been phosphate limited. Culture filtrates of akinete-containing *Anabaena cylindrica* also seem unable to stimulate akinete formation in young cultures of the same organism (Nichols & Adams, unpubl.). Whether the apparent lack, in other cyanobacteria, of the akinete-stimulating compound found in *Cylindrospermum licheniforme*, is a true reflection of its absence in these organisms, or differences in their ability to excrete or take up the compound, is not known. It might, indeed, be of interest to test cell extracts of akinete-containing cultures of these organisms on non-akinete containing cultures of *C. licheniforme*. It would be surprising if such a compound were restricted to one species. However, if it is present in species such as *Nostoc* 7524 or *Anabaena cylindrica*, then the absence of stimulatory activity in culture filtrates of these organisms would suggest that the compound is functioning as an intrafilamentous, rather than an extracellular agent.

15.5 Germination

Akinetes are propagating bodies which show some resistance to adverse conditions and in this respect resemble bacterial endospores. It follows, therefore, that to fulfil this function, akinetes must be able to germinate and produce vegetative cells (filaments) in response to improved environmental conditions.

Miller and Lang (1968) have studied the morphological changes associated with akinete germination in *Cylindrospermum* sp. Germination, induced by transferring mature akinetes to fresh culture medium, occurred by three different mechanisms. Most frequently the akinete envelope remained intact, except for a single pore which developed at the end, through which the germling emerged after undergoing one or two cell divisions. In some instances the germling began to divide only after emerging from the akinete envelope. The third mechanism, which occurred only occasionally, involved the dissolution of the entire envelope into globular and rod-shaped fragments. In the first two cases, the germlings shrank from the akinete envelope and the inner layers then loosened and dissolved. The germling appeared to be pushed out of the envelope by a layer of mucilage deposited between it and the envelope (Fig. 15.5). Empty akinete envelopes were often observed. Miller and Lang suggested that these different modes of germination might be due to age differences in the akinetes when induced to germinate. The cytoplasm of the germinating akinete differed from that of the mature akinete, being densely packed with glycogen granules and lipid deposits. Cyanophycin was usually absent at germination, indicating that it must have been utilized either by the resting akinete or very early in germination. The thylakoids which had never been completely lost in the akinete, appeared to increase by the fusion of small vesicles. Grilli Caiola and de Vecchi (1980) have also observed three mechanisms of germination of akinetes in *Nostoc* sp., corresponding with those of *Cylindrospermum* sp. Where the germling underwent several divisions inside the akinete envelope, differentiation of some of the cells to heterocysts had occurred before the trichome emerged. Generally, each strain of *Nostoc* sp. showed only one type of germination.

The mode of germination of *Aphanizomenon flos-aquae* akinetes has similarities with *Cylindrospermum* sp. and *Nostoc* sp. in that cell division can occur before emergence of the germlings, which occurs by breakage of the envelope at one or two points at one end (Wildman *et al.*, 1975). The envelope was seen to pull away from the akinete, due either to the expansion of a mucilaginous substance between them, or shrinkage of the germinating akinete. Cyanophycin granules were still observed but stained much less intensely than those in mature akinetes. Gas vesicles increased in number, as did lipid bodies. The intrathylakoid spaces widened and ribosomes were still abundant, while in contrast to *Cylindrospermum* sp. glycogen granules were not observed.

In *Anabaena variabilis*, Braune (1980) found that the undivided akinete was pushed out of its coat which ruptured at only one of the cell poles, cell

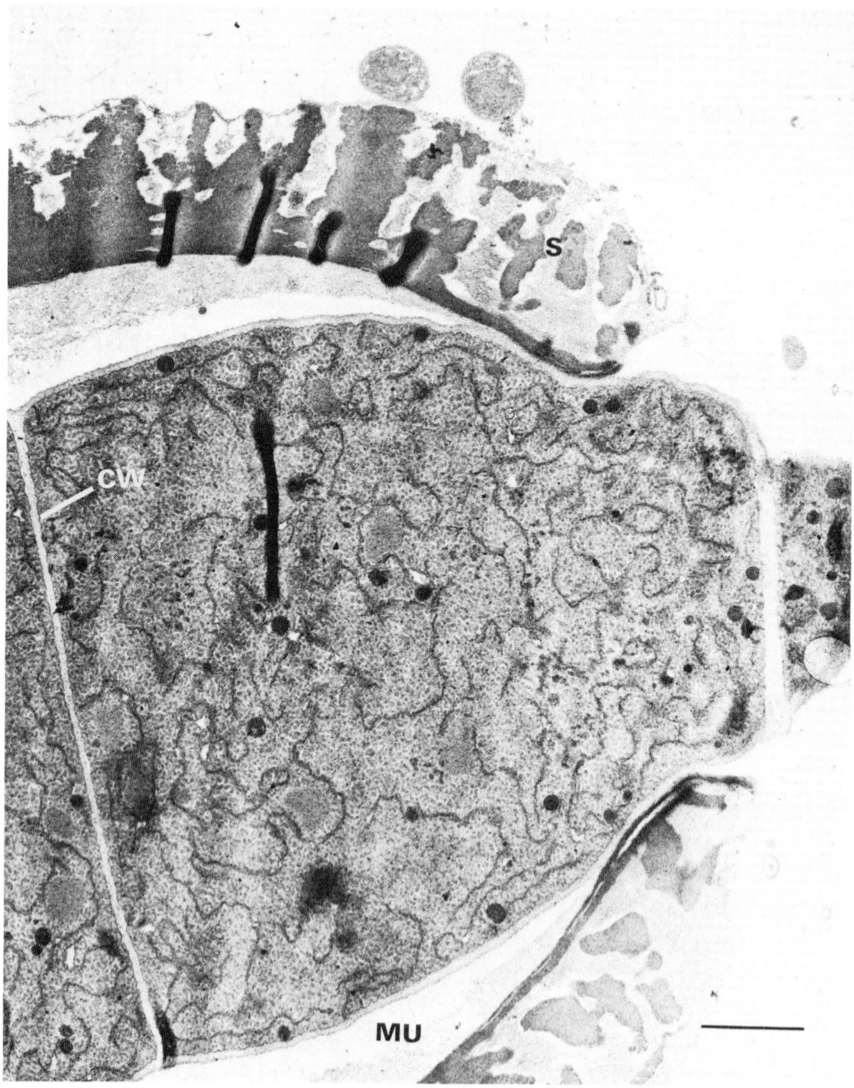

Fig. 15.5. A germinating akinete which has undergone division before emerging from the akinete envelope. Glutaraldehyde–osmium fixation. $Ba(MnO_4)_2$ post stain. CW, cross-wall; MU, mucilaginous deposit; S, sheath. The bar represents 1 μm. Reproduced by permission of N. J. Lang. (From Miller & Lang, 1968)

division occurring later (Fig. 15.6e, f). The akinete coat was formed by condensation and thickening of the fibrillar sheath of the filament. Twelve hours after induction of germination an electron-dense layer (ML) formed between the outer membrane of the cell wall and the akinete coat (Fig. 15.6).

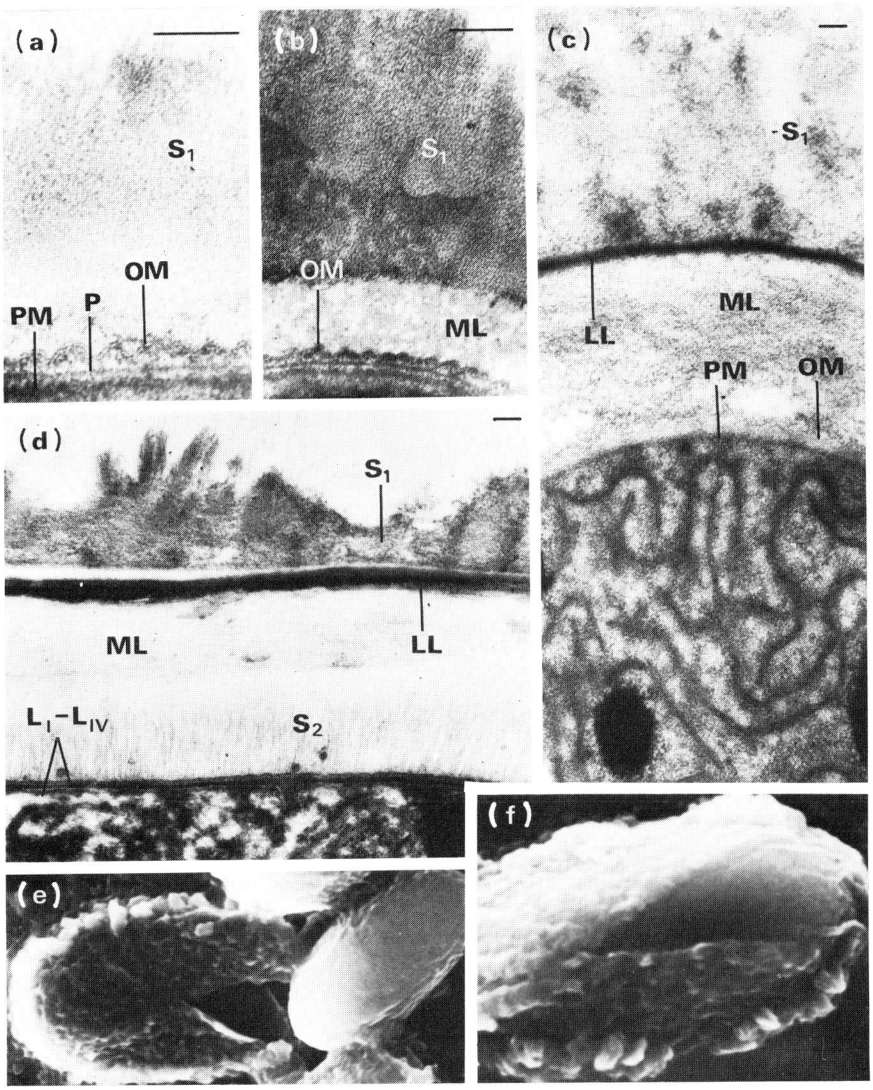

Fig. 15.6. Morphological and ultrastructural changes associated with akinete germination in *Anabaena variabilis*. (a) (b) and (c) Formation of a mucilaginous layer (ML) and laminated layer (LL) between the cell wall layers (OM, P and PM) and condensed sheath of the akinete coat (S_1). (d) An early stage of a germinating akinete showing the coat layers of the akinete together with the remains of the mucilaginous layer (top) and the wall layers, and the newly developed sheath (S_2) of the germinating cell (bottom). L_I–L_{IV} are the cell wall layers. (e) and (f) Scanning electron micrographs showing the rupture of the akinete coat and emergence of an undivided germinating cell. Bar markers represent 0.1 μm. (Reproduced by permission of Braune, 1980)

This may be equivalent to the mucilaginous layers described in *Cylindrospermum* sp. and *Aphanizomenon flos-aquae* (Miller & Lang, 1968; Wildman *et al.*, 1975). Thick lamellar layers (LL) were deposited between this expanding mucilaginous layer and the akinete coat (Fig. 15.6). These became brittle at various points at the first sign of germination and finally broke. In agreement with Miller and Lang (1968), Braune (1980) suggested that the mucilaginous layer acted as a swelling body, causing the expulsion of the germling.

When *Anabaena* CA is grown on medium containing nitrate, filaments consist entirely of vegetative cells and after the end of exponential growth almost total conversion of these into akinetes can occur (Fig. 15.1g). These can be induced to germinate with high frequency by dilution with fresh medium (Adams, unpubl.). If this is done in unshaken culture the akinetes remain in contact and entire filaments can be seen to germinate (Fig. 15.7a). Under such circumstances, the cell wall ruptures to one side of the point of

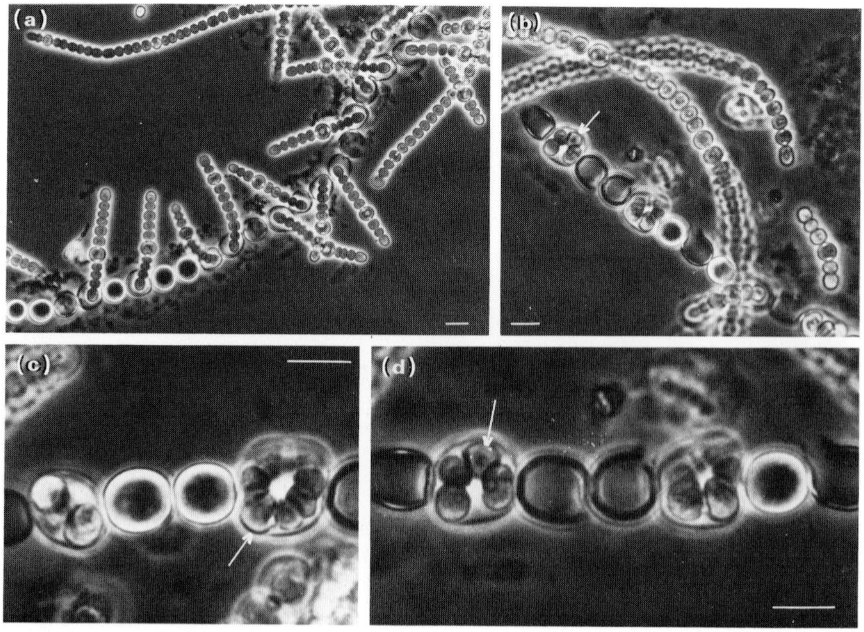

Fig. 15.7. Germination of *Anabaena* CA akinetes. *Anabaena* CA was grown as a standing culture in medium containing nitrate which completely suppressed heterocyst development. After the end of exponential growth, most vegetative cells were transformed into akinetes (see also Fig. 15.1). A small amount of culture was gently resuspended and diluted in medium lacking nitrate, to induce germination. (a) A 'filament' of akinetes runs from bottom left to top right of the field. Short germlings, each containing a heterocyst, can be seen emerging from many of these akinetes. (b), (c) and (d) Sometimes quite long germlings remained within the akinete coat, often containing heterocysts (arrows). Empty akinete coats can also be seen, with the single break where the germling emerged, clearly visible. Bar markers represent 10 μm.

contact with the neighbouring akinete, permitting the germling to emerge (Fig. 15.7a). Sometimes even quite long germlings remain within the akinete coat (Fig. 15.7b–d). When akinetes are diluted in medium free of combined nitrogen, each germling develops a heterocyst when approximately seven cells in length (Fig. 15.7a), and this occurs even when the germling is inside the akinete coat (Fig. 15.7b–d).

Although the morphological changes that occur during akinete germination have been examined in some detail, the factors that induce germination are poorly understood—light, temperature and nutrients all having been implicated. In a laboratory culture of *A. cylindrica* grown at a light intensity of 3500 lx, akinete formation occurred at 9 days and was closely followed by akinete germination, liberated germlings being seen 1 day later (Fay, 1969b). In *Nostoc* 7524, akinetes have been shown to germinate without renewal of the medium, following an increase in light intensity (Sutherland *et al.*, 1979).

While field studies have provided some information on the induction of akinete germination, the results are extremely difficult to interpret because of the simultaneous variation of numerous parameters, which frequently occurs in the natural environment. The use of samples collected in the field can also cause problems. This is amply demonstrated by the work of Wildman *et al.* (1975), who collected akinetes of *Aphanizomenon flos-aquae* from mud sediments in November and attempted to induce germination in the laboratory by variations of light, temperature and nutrient medium. Remarkably, no germination occurred until late May which coincided with germination of akinetes in nature. Reynolds (1972) suggested that germination of akinetes of *Anabaena circinalis* was triggered by external conditions such as temperature, increased light penetration and increased day length, the akinetes having been lifted from mud sediments by turbulent weather. Observations of *Gloeotrichia echinulata* by Roelofs and Oglesby (1970) showed that only an increase in solar radiation could be correlated with the germination of akinetes between May and July, hence the appearance of the alga in the plankton. By contrast, Rother and Fay (1977) were able to detect akinetes in *Anabaena circinalis* and *A. flos-aquae* following the breakdown of blooms, germination occurring soon after akinete maturation.

It is clear from the work described above that unequivocal results can only be obtained in the laboratory, where the factors which stimulate akinete germination can be carefully controlled and monitored. In this way Yamamoto (1976) showed that akinete germination in *A. cylindrica* was dependent on light intensity, since none occurred in the dark or in the presence of 10^{-5} M DCMU. The pH optimum for the process was between 7 and 8 and the optimum temperature (27°C) was the same as for vegetative growth, no germination occurring above 35°C. Germination was about five times greater under aerobic than anaerobic conditions, indicating a requirement for oxygen. Among amino acids, carboxylic acids and other compounds, only acetate (10^{-3}–10^{-4} M) increased germination to a significant extent. Rates of respiration and photosynthesis of both akinetes and vegetative cells were not affected by the addition of acetate and its role in germination was not clear; however, acetate is known to effect germination of bacterial and fungal spores. Heat shock will break the dormancy of bacterial and fungal spores (Sussman & Halvorson, 1966; Gould, 1969) and Yamamoto (1976) has investigated the effect of heating akinetes at various temperatures for 2–5 min. No promotion of germination was observed and viability of akinetes was lost at temperatures higher than 55°C, a clear contrast with bacterial endospores.

The influence of light wavelength on germination of akinetes has been investigated by Kaushik and Kumar (1970) who concluded that the akinetes of *Anabaena doliolum* and *Fischerella mucicola* required non-photosynthetic light for germination, since green, blue, yellow, red and white light all supported germination, while vegetative growth occurred only in photosynthetic (yellow, red or white) light. Akinetes were unable to germinate in the dark, even under heterotrophic conditions. Increasing light intensity had no effect on akinete germination but did stimulate germling growth. Whereas Kaushik and Kumar concluded that akinetes were able to germinate in non-photosynthetic light, Yamamoto (1976) and Reddy *et al.* (1975) were unable to detect germination of various *Anabaena* sp. and *Anabaenopsis arnoldii* under blue or green (non-photosynthetic) light. In addition, Yamamoto (1976) obtained complete inhibition of germination by DCMU, a specific inhibitor of the electron transport system in photosynthesis. Braune (1979) has determined the action spectrum for the light-dependent germination of *Anabaena variabilis* Kütz. akinetes. The most active spectral range was between 620 and 630 nm, which coincided with the maximum of light absorption by C-phycocyanin. Although high light inhibited akinete germination, some akinetes were able to germinate at light intensities which were insufficient to support vegetative growth. Germling growth was also wavelength dependent, but the maximum was shifted to a longer wavelength and blue light also promoted the elongation of filaments (Kaushik & Kumar, 1970). Since DCMU, at a level which reduced carbon incorporation to 5%, only reduced germination by 50% (although the germinated cells died either immediately or after 1–2 cell divisions) it was concluded that the formation of photosynthetic products was not required and indeed, Braune and Sanke (1979) have shown that increased dry matter content of germinating akinetes is not a necessary precondition for outgrowth. Braune (1979) suggested that C-phycocyanin might be involved as a photoreceptor for the germination process in two ways. One was via its role in the energy yielding process of photosystem I (cyclic photophosphorylation), by transferring its absorbed light energy. The other was by a regulative function of the pigment, comparable with phytochrome action in plants, there being a close chemical relationship between the chromophores of plants and the photochromic phycobiliproteins which could be responsible for photoconvertible photomorphoses in cyanobacteria (Bogorad, 1975a). Indeed Reddy *et al.* (1975) reported a red/far-red photoreversal for germination, similar to phytochrome action in comparable activating processes in plants.

The only biochemical studies of akinete germination have been carried out by Singh and Sunita (1974) on *Anabaena doliolum* in which total conversion of vegetative cells into akinetes occurs. To synchronize germination, cultures which were beginning to form akinetes were transferred from the light to the dark for 3–4 days, during which period akinete formation continued. When re-admitted to the light, 90% of the dark-synthesized akinetes germinated as far as the first cell division within 72–80 h. The order of macromolecular synthesis was RNA at 4 h (continuing until 20 h), chlorophyll-*a* at 8 h, phycocyanin at 12 h and DNA at 24 h. If akinetes were incubated in the dark, some RNA synthesis occurred at 8 h, but only reached a level equivalent to 40% of that in the light-incubated akinetes. Synthesis of RNA, chlorophyll, phycocyanin and DNA all showed a requirement for protein synthesis during the 0–4 h period, since addition of chloramphenicol (an inhibitor of protein synthesis) before 4 h, caused complete inhibition. Chlorophyll, phycocyanin and DNA synthesis also required light. Singh and Sunita suggested that since the chloramphenicol-sensitive pigment formation occurred much earlier than that of the photosynthetic pigments, it was unlikely that photosynthetic light participated in the formation of proteins involved in the synthesis of these pigments, unless perhaps it was a photoactivation effect. Photosynthesis may activate DNA replication, however, as this occurred after the formation of chlorophyll-*a* and phycocyanin. This is supported by the reports that DNA replication appeared to stop after transfer of vegetatively grown cells of *Anacystis nidulans* from light to dark (Craig *et al.*, 1969; Asato & Folsome, 1970; Herdman *et al.*, 1970; Herdman & Carr, 1971b).

Although experiments with akinetes from a limited range of species indicate that they are not very resistant to extremes of temperature (e.g. heating akinetes of *Anabaena cylindrica* to temperatures higher than 55°C caused complete loss of viability (Yamamoto, 1976)), they might constitute an over-wintering stock of planktonic cyanobacteria, but as pointed out by Rother and Fay (1977), this is by no means conclusive. They themselves demonstrated, in field studies on populations of cyanobacteria in two separate lakes, that in one, germination of akinetes sedimented into the mud surface may have been an important source of algal material in the spring, while in the other, a large over-wintering planktonic population of vegetative filaments was detected, germination of akinetes therefore being less important. Rother and Fay (1977) proposed that the function of the akinetes was probably to ensure survival of the cyanobacterial population after the period of blooming and subsequent photooxidative death. Other workers provide opposing views. Roelofs and Oglesby (1970) and Reynolds (1972) demonstrated that over-wintering akinetes of *Gloeotrichia echinulata* and *Anabaena circinalis* were the source of the new vegetative material in the spring, whereas in another study of *A. circinalis* at a later date, Reynolds (1975) concluded that germination of over-wintering akinetes contributed very little to the summer maximum, which apparently originated principally from a small vegetative stock which had survived from the previous autumn. A recent study by Livingstone and Jaworski (1980) suggested that akinetes of *Aphanizomenon* and *Anabaena* may ensure long-term survival since those isolated from sediments, 18 and 64 years old respectively, were viable. Whether or not akinetes are important in preserving the cyanobacterial stock during periods of non-vegetative growth is probably a function of the species and the particular environmental conditions.

15.6 Spatial relationship between heterocysts and akinetes

Akinetes normally develop in positions which have a specific spatial relationship to heterocysts (Chapter 14). Thus, they frequently occur adjacent to heterocysts as in *Anabaena cylindrica* (see Fig. 15.1c,d), or at the mid-point of the interheterocyst interval as in *Anabaena* CA (see Fig. 15.1e,f). While this has encouraged investigators from Carter in 1856 to the present day to implicate heterocysts in akinete development, it must be stressed that there are many exceptions to the rule (Burger, 1974; Wildman *et al.*, 1975). There have been reports of akinetes developing in the complete absence of heterocysts (Eberly, 1966; Hill, 1970) and, indeed, this can occur in organisms which also form akinetes adjacent to or between heterocysts (see Adams & Carr, 1981). In *Nostoc* 7524 grown in the presence of ammonium chloride, akinetes develop at random positions within the filament (Sutherland *et al.*, 1979). The presence of heterocysts in nitrogen-fixing cultures, however, imposed a regularity on akinete formation, such that almost half of the first-formed akinetes were at the exact centre of the interheterocyst interval, or within 5% (mean of one cell) of the centre. Thus, although heterocysts were not required for akinete development, their presence exerted some influence on akinete pattern. In a similar way, *Anabaena* CA will form akinetes at the mid-point between heterocysts in a nitrogen-fixing culture (Fig. 15.1e,f), but will also form akinetes in nitrate-grown, heterocyst-free cultures, finally resulting in almost total conversion of vegetative cells to akinetes (Fig. 15.1g,h; Adams, unpubl.). In *A. cylindrica*, akinetes normally develop next to heterocysts. However, in cultures grown in the presence of ammonium chloride, after the end of exponential growth isolated akinetes can sometimes be observed (Nichols *et al.*, 1980). Such cultures contain a low frequency of heterocysts and proheterocysts, and akinetes were also found adjacent to these. The isolated akinetes, however, were usually more than 20 cells from any heterocyst. It is clear then that while the presence of heterocysts is not essential for akinete formation they do impose a regularity on the spatial arrangement of akinetes.

It might be argued that the mere presence of heterocysts imposes a degree of symmetry upon a filament, since heterocysts themselves occur in precisely located positions, whether this be terminally at either end of a filament, basally at one end of a tapered trichome, or as is often the case, at regularly spaced intervals within the filament. Such spacial symmetry could result in biochemical heterogeneity within groups of vegetative cells, because of the nitrogen-fixing ability of the heterocyst which would lead to gradients of fixed nitrogen, highest in the vegetative cells adjacent to the heterocyst and lowest mid-way between. It would be wrong to assume, however, that this explains the placement of akinetes relative to heterocysts, since it is readily shown that nitrogen fixation *per se* is not essential for akinete development or positioning. In cultures grown in the presence of ammonia, akinetes still develop adjacent to non-nitrogen-fixing proheterocysts, which occur at low frequency (see above). In *Nostoc* 7524, akinetes still develop mid-way between heterocysts in cultures grown on Ar/CO_2 in which nitrate (at a concentration which has no effect on heterocyst development) acts as sole nitrogen source (Sutherland *et al.*, 1979). Under such circumstances, when no nitrogen-fixation occurs and one assumes nitrate is assimilated equally by all vegetative cells, no gradients of fixed nitrogen should exist. Nevertheless, the location of akinetes either adjacent to, or mid-way between heterocysts, together with the presence of gradients of akinete maturity (probably caused by differences in the time of initiation of their development—see section 15.4.1) strongly suggests that some form of gradient is involved when heterocysts are present (see Adams & Carr, 1981). One might speculate that akinete development will occur either below a threshold concentration of a compound which inhibits their formation, or above a threshold of a compound which stimulates their formation. If one implicates the heterocyst in the propagation of such a gradient, it must be appreciated that the formation of akinetes adjacent to heterocysts in *Anabaena cylindrica*, but mid-way between in *Nostoc* 7524, becomes a paradox, unless the mechanism involved in these organisms also differs.

15.6.1 Alterations in akinete pattern

No mutants with altered akinete pattern have yet been isolated. However, a number which are unable to form akinetes have been reported (see Chapter

11), some occurring spontaneously (Singh & Sinha, 1965), others induced by u.v. treatment (Singh, 1967) or nitrosoguanidine (Singh, 1978c).

The tryptophan analogue, 7-azatryptophan, is known to induce a remarkable change in heterocyst pattern in *Anabaena*, resulting in both an increase in heterocyst frequency and the formation of adjacent (multiple) heterocysts (Mitchison & Wilcox, 1973). This reduction in the regularity of the spatial pattern of extant and developing heterocysts has also been observed in *Nostoc* 7524 (Sutherland *et al.*, 1979). Interestingly, the analogue had a similar effect on the akinete pattern of this organism, in which akinetes normally develop at the mid-point between heterocysts. Addition of 7-azatryptophan to nitrogen-fixing cultures inhibited nitrogen fixation and consequently akinete development. This problem was overcome, however, by supplying nitrate which could serve as sole nitrogen source without affecting heterocyst development. Indeed, the heterocyst and akinete patterns of cultures grown on Ar/CO_2 + nitrate were almost identical to those of N_2-grown cultures (Fig. 15.8). Addition of 7-azatryptophan to a culture growing exponentially on Ar/CO_2 + nitrate resulted in the formation of multiple heterocysts, newly developing heterocysts occurring further from the centre of the interheterocyst interval (i.e. closer to existing heterocysts) than in a control culture (Fig. 15.8). A very similar effect was observed with akinete development (Fig. 15.8)

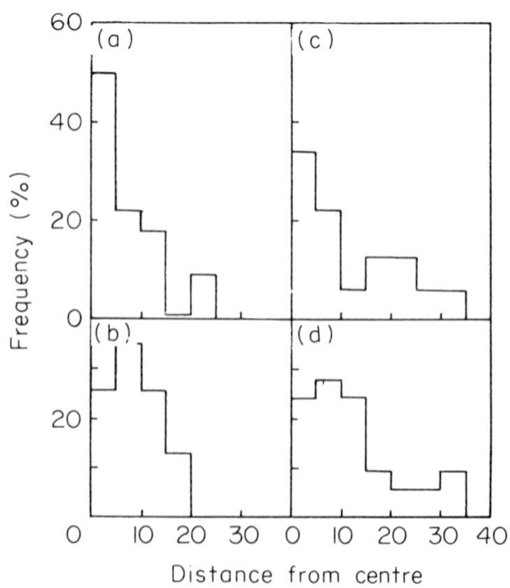

Fig. 15.8. The positions, relative to the centre of existing interheterocyst intervals, of newly differentiated heterocysts in exponentially growing cultures in the presence of Ar/CO_2 plus NO_3^- (a) and 7-azatryptophan plus NO_3^- (c), and the positions of akinetes at the onset of differentiation in Ar/CO_2 plus NO_3^- (b) and 7-azatryptophan plus NO_3^- (d). (Reproduced with permission from Sutherland *et al.*, 1979)

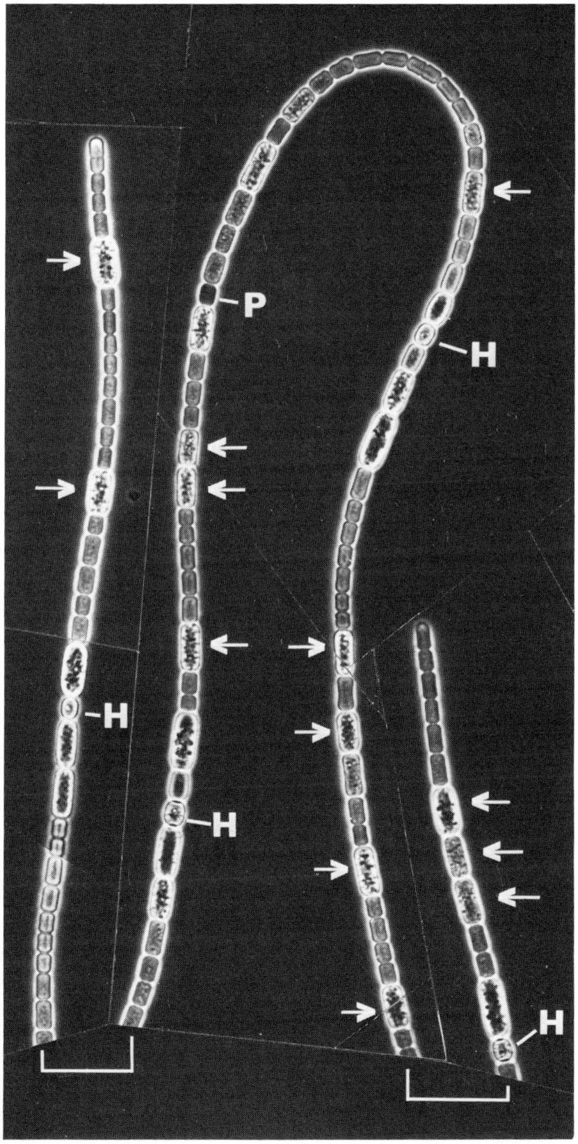

Fig. 15.9. Effect of canavanine on akinete pattern in *Anabaena cylindrica*. A culture of *A. cylindrica* was grown on medium lacking fixed nitrogen and canavanine (40 μM) was added. The photograph was taken 5 days later. Heterocysts (H), a proheterocyst (P) and some of the akinetes in positions not adjacent to the heterocysts are indicated (arrows). The square brackets indicate the continuation of the filament. Phase contrast, ×430.

and the number of akinete sites also increased from 40 to 66 per 1000 vegetative cells. Sutherland *et al.* (1979) concluded from these results that the control of heterocyst and akinete spatial patterns was related in this organism.

The only other compound known to specifically alter the akinete spatial pattern is the arginine analogue, canavanine. The presence of 40 μM canavanine in nitrogen-fixing cultures of *Anabaena cylindrica* at the onset of akinete formation, resulted in the development of akinetes at random positions between heterocysts, in addition to their usual place adjacent to heterocysts (Fig. 15.9; Nichols *et al.*, 1980). An increase in the number of akinetes adjacent to heterocysts was also observed, the total akinete frequency increasing up to five-fold. The effect seemed to be specific for canavanine since, although another arginine analogue, cyanoalanine (4 μM) and a serine analogue, aminobutyric acid (160 μM), caused increases in akinete frequency, no effect on akinete pattern was observed. The authors concluded that since canavanine resulted in the development of akinetes at random positions between heterocysts, it was acting upon vegetative cells rather than heterocysts. If the latter was the case, only an increase in the number of akinetes adjacent to heterocysts would have been expected. They further concluded that during exponential growth, some aspect of the physiology of vegetative cells prevented their development into akinetes. The function of the heterocyst would then be to remove or negate this inhibition in neighbouring cells, rather than directly stimulate akinete formation.

CHAPTER 16
MOTILITY AND TAXES

R. W. CASTENHOLZ

List of abbreviations

Inhibitors

CCCP	carbonyl cyanide m-chlorophenylhydrazone
DBMIB	dibromomethyl-isopropyl-p-benzoquinone
DCCD	dicyclohexyl carbodiimide
DCMU	3-(3,4-dichlorophenyl)-1, 1-dimethylurea
DNP	2,4-dinitrophenol
HOQNO	2-n-heptyl-4-hydroxyquinoline N-oxide
TTFB	4,5,6,7-tetrachloro-2-trifluoromethyl benzimidazole

Miscellaneous

AMP	adenosine-5′-monophosphate
cAMP	adenosine-3′,5′-cyclic monophosphate
BChl	bacteriochlorophyll
Chl	chlorophyll
GMP	guanosine-5′-monophosphate
cGMP	guanosine-3′,5′-cyclic monophosphate
PC	phycocyanin
PE	phycoerythrin

16.1 Introduction

When motile, cyanobacteria move by gliding. There has been only limited progress in understanding the mechanism for this gliding since a similar chapter was published in 1973 (Castenholz, 1973). Several new observations and suggestions on gliding motility have come from studies of other gliding prokaryotes (e.g. flexibacteria and myxobacteria), but it is not clear that the mechanism in these groups is of one type or similar to that of the cyanobacteria. Various tactic and phobic reactions are also known but, among these, only photoresponses have been studied extensively. Considerable progress has occurred in the understanding of the photophobic response in cyanobacteria, but little has been gained in the knowledge of phototaxis *per se* (see Nultsch & Häder, 1979). An enormous literature on bacterial chemotaxis has accumulated in the past several years, but is almost entirely about the flagellated, swimming forms.

The main objective of this chapter, after recapitulating the essential features of gliding behaviour and its basis, is to describe and evaluate the material and concepts presented in the past few years. Taxes will be similarly treated.

16.2 Extent and the nature of gliding

16.2.1 Gliding

Gliding in a prokaryote or eukaryote is a self-propulsion across a solid or semi-solid material without the aid of any visible organ (flagellum) or apparent change in the shape of the organism. However, 'visible' is defined by the limits of the light microscope and by the inability of the electron microscope to utilize living material.

Sliding seems a more appropriate term, but Henrichsen (1972) has used it to mean 'a type of surface translocation produced by the expansive forces in a growing culture in combination with special surface properties of the cells resulting in reduced friction between cell and substrate'. Thus, a sheet of closely packed cells moves or is moved as a unit. *Twitching* is a jerky movement of individual flagellated or non-flagellated cells not necessarily following the long axis of the cell (Henrichsen, 1972). Other unexplained movements of prokaryotes (e.g. darting, swarming) are also described by Henrichsen.

Gliding in cyanobacteria has been described in recent years by Castenholz (1973) and Halfen (1979) and reviewed in prokaryotes generally by Burchard (1981) and Reichenbach (1981); Burkholder (1934) has reviewed thoroughly earlier descriptions and hypotheses. Only some features will be redescribed here. In general, I will try to summarize the most recent information and synthesize an interim conclusion. A rather large amount of information on gliding in other prokaryotes has been presented in recent years. Because of the nature of this information it is now essentially impossible to envision a uniform mechanism for gliding motility in prokaryotes. Thus, gliding motility is probably only superficially a single category of movement. Its structural basis may be diverse and polyphyletic in origin.

Among photosynthetic prokaryotes, gliding occurs in cyanobacteria and the Chloroflexaceae, while among non-photosynthetic prokaryotes it occurs in the fruiting myxobacteria, the flexibacteria (*sensu* Soriano, 1973), the Beggiatoaceae, the spirochaetes, and even in mycoplasmas (Bredt & Radestock, 1977). The phylogenetically distinct 'gliding' forms among the eukaryotes include the raphe-bearing diatoms, the unicellular desmids and related green algae (order: Conjugales), and some of the unicellular red algae and their unicellular disseminules. Eukaryotic gliding motility is discussed briefly by Halfen (1979) and Harper (1977).

16.2.2 Motility system in cyanobacteria

16.2.2.1 Observations

Gliding occurs at nearly all times for many filamentous members of the Oscillatoriaceae. Hormogonial trichomes form and migrate only periodically in most members of the Nostocaceae, Scytonemataceae, Rivulariaceae, and Stigonemataceae. Some unicellular cyanobacteria (e.g. *Synechococcus*) display twitching or pivoting movements as well as gliding and the baeocytes ('endospores') of many members of the Pleurocapsales glide slowly (Waterbury & Stanier, 1978).

Information on the structural basis of motility in the cyanobacteria is little more than existed in 1973 (Castenholz, 1973; Halfen, 1973). Nevertheless, a few observations and experiments have tended to support, or at least not refute, the hypothesis that seemed most adequate at that time (Halfen & Castenholz, 1971a).

Oscillatoria princeps is a large-diametered (30–35 μm) trichome which is quite rigid, being able to glide not only on the surface of, but through, 0.5–4.0% agar. It can emerge from agar and protrude upright for several millimetres into humid air before bending (Halfen, 1970). It is rigid and does not 'flex', a type of movement which in other cyanobacteria is only an illusion that occurs when crooked or bent free ends swing about with a rotation of the gliding trichome. *O. princeps* has a wall that certainly does not undulate, but nevertheless there is presumably a surface translocation phenomenon that pushes this cylinder of cells along at rates as high as about 10 μm s^{-1} (3.6 cm h^{-1}). When it glides it rotates and any surface point traces a right-handed helix of 60° pitch. When it stops and reverses, a right-handed 60° helix is again followed. There is no slippage discernible; when rotation was prevented, forward progress also ceased until the torsion caused the trichome to snap free (Halfen, 1979). It moves through an elastic, but tight-fitting sheath which is deposited by the cells and which sticks to essentially all substrates. If movement continues in one direction a collapsed and sometimes twisted gossamer sheath is left behind as a trail. The trichome may re-enter the collapsed portion if a reversal occurs. As back and forth movement continues and more and more materials are added to the inside of the sheath, it thickens considerably. If the trichome is suspended in a viscous liquid, unable to move, it pushes off its sheath (and any particles attached to it) in a helical manner as before, but instead of trailing off it tends to accumulate and thicken at one end (Fig. 16.1; Halfen & Castenholz, 1971a).

Several past researchers (see Doetsch & Hageage, 1968) have suggested that expulsion of sheath material or the expansive force of its swelling upon hydration may constitute the propulsion mechanism in gliding. The observations that tend to refute this hypothesis are:

(1) In *Oscillatoria princeps* and many other filamentous cyanobacteria there are minute junctional pores (12–14 nm wide or less) circling both sides of each crosswall which might provide an exit for sheath material or mucilage (Halfen & Castenholz, 1971a). But they are lateral and would have no translational effect unless the material were ejected with extreme force (of which there is no evidence). The pores in *O. princeps* are slightly angled in each cell end, but it appears that they do not penetrate the outer lipopolysaccharide envelope (see Halfen, 1979). Even with clear penetration the tight coupling of rotation with forward movement would be most difficult to explain by mucilage expulsion. Lamont (1969b) has suggested that mucilage is not expelled through pores (which are lacking in some mucilage producing cyanobacteria) but that it is shear polymerized on the exterior of the cells after diffusion of small units through membrane, wall and envelope.

(2) Because of the constant speed maintained in gliding (although dependent on temperature, pH and medium viscosity) and the sudden stops and reversals it is difficult to envision a mucilage ejection mechanism. A possible but unlikely way in which it could work is pictured in Fig. 16.2.

(3) Walsby (1968) made the important observation that, in suspended *Anabaena cylindrica*, mucilage 'rings' were moved over the cell surface but tended to hesitate or stop at the only probable site of forceful mucilage ejection, the cross wall area. The rings did not move over the surfaces of heterocysts nor generally over the cells on either side, the incipient akinetes; both types of cells synthesize additional cell walls, masking the original wall. The most logical conclusion is that the propulsion mechanism lies on the main surface area of each vegetative cell. Forceful mucilage expulsion in *A. cylindrica* is probably not even possible, since no pores have been seen.

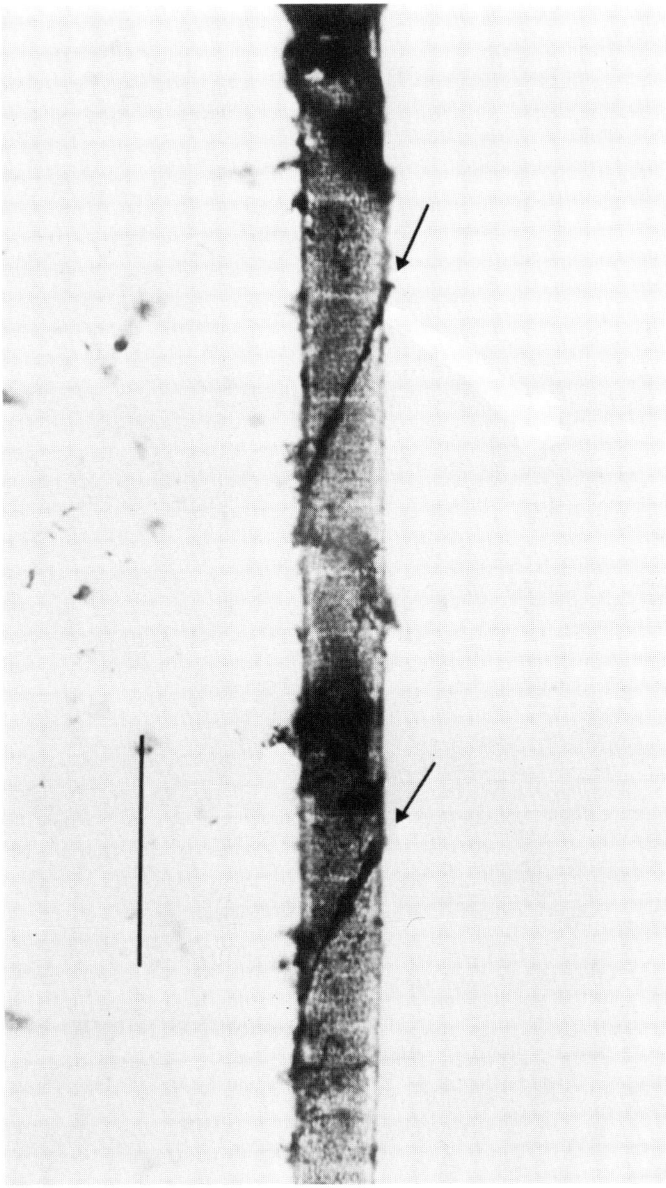

Fig. 16.1. *Oscillatoria princeps.* Bands of mucilage (sheath) that have been transported in a right-handed helical pattern (of 60° pitch) while the trichome is immobilized in suspension. The mucilage is stained with India ink particles. Arrows indicate direction of traverse of the mucilage; bar = 100 μm. (From Halfen & Castenholz, 1971a)

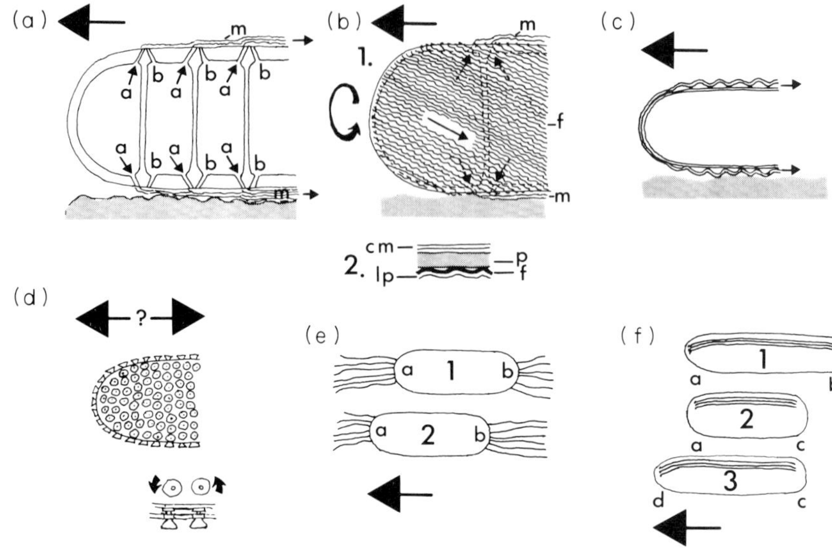

Fig. 16.2. Possible motility mechanisms in various gliding prokaryotes.

(a) Forceful ejection of mucilage method via junctional pores in cyanobacteria. This, apparently untenable, method would require one set of angled pores (i) for movement in one direction, and the other set (ii) for movement in the other direction. Small arrows indicate direction of mucilage movement (m = mucilage).

(b) Periplasmic contractile fibril method as illustrated in *Oscillatoria*. (1) Surface view of trichome showing helical pattern of fibrils with some lateral banding and wave pattern fixed. Small solid arrow indicates direction of wave movement; other arrows indicate mucilage release (f = fibrils, m = mucilage). (2) Details of cell envelope showing apparent location of fibrils (f); cm = cytoplasmic membrane, p = peptidoglycan layer, lp = lipopolysaccharide envelope.

(c) 'Make and Break' rhythmic wave pattern of lipopolysaccharide envelope. Small arrows indicate direction of waves.

(d) Rotary assembly method of surface translocation. Small disc-like structures which may be analogous or homologous to modified flagellar bases. This hypothesis does not easily explain directionality of movement.

(e) Motility by fimbriae. Polar fimbriae are shown in position 1 as extended (a) and contracted (b). In position 2 the cell has moved ahead either by fibrial contraction at pole (a) or by extension at pole (b) or both. Temporary fimbrial adhesion to substrate is assumed.

(f) Movement by extension and contraction of cytoplasmic filaments and by use of Stefan adhesive. Cell in extended position 1. In position 2, adhesion of pole a and slippage of pole b to c with contraction of internal filaments. In position 3, extension of filaments, adhesion of pole at c and slippage of pole from a to d.

(4) There is an observation with the flexibacterium, *Flexibacter* BH3, that slime secretions occurred ahead of advancing cells (Humphrey *et al.*, 1979). The authors emphasized 'that the forceful ejection of slime was not involved . . .'. Although not a cyanobacterium, these observations dictate against a universal propulsion mechanism based on slime ejection.

(5) Calculations by Holton and Freeman (1965 and pers. comm.) suggest that propulsion at moderate cyanobacterial speeds (e.g. 3 μm s^{-1}) by the forceful ejection of slime through the 8–10 nm wide junctional pores would require a cell to empty its total contents many times per second. However, this might not be true for slow movement if the material polymerized and/or swelled after the expulsion of low molecular weight monomers.

Some eukaryotes (desmids) are definitely motile (although slowly) because of the polar extrusion of slime which forces the cell ahead as increments of slime are added to. In the bacterium *Acetobacter xylinum* excretion actually does work as a means of propulsion (Brown *et al.*, 1976). Cellulose microfibrils are extruded from a series of minute pores; these continue to form a flat ribbon that elongates at 0.03–0.04 μm s^{-1} pushing the cell ahead.

16.2.2.2 Viable hypotheses

In *Oscillatoria* there is a well-documented band of fibrils that surrounds the structural wall layers and runs continuously from end-to-end of each trichome (Halfen & Castenholz, 1971a; Halfen, 1973). It is fairly convincingly shown to lie underneath the lipopolysaccharide outer 'membrane'. In *O. princeps* and *O. terebriformis* these 5–8 nm fibrils (apparently with some lateral attachments into bands) run in a right-handed helical pattern around the cylinder of the filament with a pitch of 60° in each case (Fig. 16.2). In *O. animalis* the helix is left-handed. In these species forward movement occurs with rotations that trace right-handed helices in the first two species and a left-handed helix in the case of *O. animalis*. The tight coupling between 'forward' movement and the rotation which describes a specific helix implies strongly that the physical apparatus propelling the filament is arranged in a helix with a 60° pitch in the case of *O. princeps*.

The hypothesis proposed by Halfen and Castenholz (1971a) and amplified by Halfen (1973, 1979) is that the fibrillar array has contractile properties and that rhythmic wave patterns would be generated from one trichome pole to the other. These would also form waves in the outer 'membrane' which overlies the fibrillar layer. These would push in one direction against the tight sheath surrounding the entire trichome (Fig. 16.1). A reversal in gliding direction would result in a reversal of polarity, i.e. the generation of waves in the opposite direction. The location of the fibrillar layer beneath this membrane seems correct in *O. princeps*, at least. The contours of the unraised fibrils are insufficient merely to act as a screw-thread (imprinted through the outer membrane) to cause the filament rotation with linear translation caused by some other mechanism. Although a fibrillar layer has been seen in other cyanobacteria (G. Guglielmi, pers. comm.), its apparent absence in some may also be real.

Fibrils have been seen in approximately the same site in one flexibacterium (Pate & Ordal, 1967). However, there is a definite absence of such fibrils in most of the non-photosynthetic gliding bacteria examined. It has been suggested that the outer membrane itself may serve as the propulsive mechanism in *Flexibacter* BH3. 'A form of rhythmic contraction in *Flexibacter* might be generated by a directional sequence of "make-and-break" interactions between the flexible outer membrane and the more rigid peptidoglycan layer' (Humphrey *et al.*, 1979). A directional sequence of 'make-and-break' interactions between the fibrillar layer and the peptidoglycan wall in the larger and more rapidly gliding cyanobacteria could also be imagined. Wavelike distortions in the fibrillar layer of *Oscillatoria* have been seen by electron microscopy with platinum–carbon replication (Halfen, 1973). Similar wave distortions in the outer 'membrane' have been seen in other cyanobacteria (Halfen & Castenholz, 1971a; Jost, 1965), and in myxobacteria and flexibacteria (Costerton *et al.*, 1961; Follett & Webley, 1965; Humphrey *et al.*, 1979) and in some other Gram-negative bacteria. Whether these are artifacts or possibly motility associated distortions transfixed at the time of fixation is uncertain.

In *O. princeps*, Halfen (1973) and Halfen and Castenholz (1971a) have demonstrated that the fibrillar layer is protein in nature, that there is some lateral linkage among fibrils, and that when the wall is removed and fragmented by ultrasonic disruption the 'elastic' fibrillar layer tightens its helical coil, turning the wall fragment inside out.

Jarosch (1964, 1967), predicting the presence of a contractile layer on (or near) the surface of cyanobacterial filaments, proposed that helically wound protein fibrils rotate. This would form higher order helices (as twisting a taught attached rope) that would move directionally down the length of the fibril, creating a propulsive force.

The cyanobacterium *Spirulina*, particularly those species with tight coils, show jerking movements of the trichome ends when clusters of trichomes are viewed under the microscope (Fig. 16.3). It appears that these twitches are those of unattached but temporarily obstructed ends which need to swing whenever there is gliding motility. The entire coil in *Spirulina* moves as a screw riding on its raised edges. The large number of turns per forward increment in this type of organism result in more dramatic oscillations and jerking movements. A fibrillar layer could completely parallel the coil of the trichome and still cause the type of movement seen.

Abeliovich (1979) measured the jerking of the trichome tips in *Spirulina* photographically in 5 s time exposures as a criterion of motility. He was thus able to easily quantify movement with exposure to a large variety of chemical agents. Some of the reactions, particularly to neuromuscular effectors, strikingly resemble the behaviour of smooth muscle cells. Motility ceased

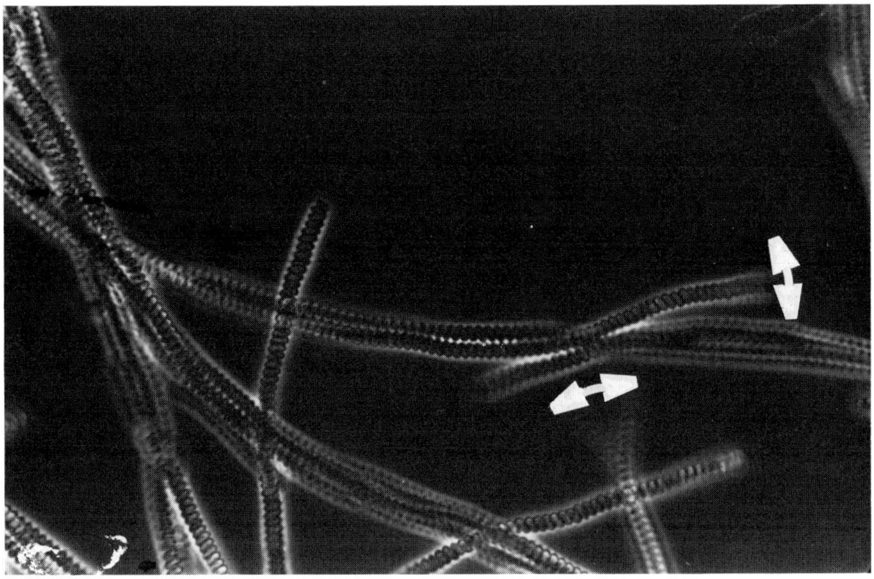

Fig. 16.3. *Spirulina* sp. (OS-76-sp). A portion of an 'interwoven' cluster. Free ends which are not in contact with substrate (e.g. other trichomes) are free to swing, tracing a helical pattern (arrows) as the trichome rotates. The width of the coil is about 4.5 μm.

when the culture was dialysed against distilled water, 0.5 M mannitol or NaCl. Calcium addition (4 mM) specifically restored motility. Calcium denial (chelation by EGTA) had a similar inhibitory effect on motility in *Oscillatoria terebriformis* (Castenholz, 1967). At pH 5.0–6.0 (non-permissible for gliding) motility was restored in *Spirulina* with norepinephrine (noradrenaline), the best 'all-round' stimulator of motility (Abeliovich, 1979). The discovery of actin-like proteins in *Escherichia coli* and *Mycoplasma pneumoniae* (Minkoff & Damadian, 1976; Neimark, 1977) adds to the likelihood of a more general occurrence of contractile proteins in prokaryotes.

16.2.3 Motility systems in other gliding prokaryotes

Beggiatoaceae

Beggiatoa, commonly thought of as an ancient apochlorotic derivative of cyanobacteria because of its striking morphological resemblance to *Oscillatoria*, also glides at rates comparable to *Oscillatoria* (e.g. 2–8 μm s^{-1} with a distinct rotation of the trichome (Nelson & Castenholz, 1982). Because of this speed and behaviour it is presumed to have a mechanism similar to *Oscillatoria*, but no studies have been made.

Chloroflexaceae

The speed of movement in *Chloroflexus aurantiacus* (0.01–0.04 μm s^{-1}) and in an unnamed related organism, 'F-1' (0.1–0.4 μm s^{-1}) is much slower than most Oscillatorian type cyanobacteria (Pierson & Castenholz, 1971, 1974). However, it was recently discovered that at least two pale-green coloured species of *Oscillatoria* are, instead, bacteriochlorophyll- and chlorosome-containing anoxygenic phototrophs (more closely related to *Chloroflexus* than to cyanobacteria). To date two species of this new genus, *Oscillochloris*, have been described (Gorlenko & Korotkov, 1979). *Oscillochloris chrysea* has a superficially very great resemblance to *Oscillatoria* and glides at a similarly rapid rate (Gorlenko & Pivovarova, 1977).

Myxobacteria and flexibacteria

The fruiting myxobacteria are heterotrophic Gram-negative rods that live in many terrestrial and aquatic habitats. They exhibit gliding motility and a swarming aggregative behaviour on solid surfaces; this usually results in 'fruiting bodies' of considerable morphological complexity. The heterogeneous flexibacteria in the sense of Soriano (1973) include heterotrophic Gram-negative rods and filaments common in many aquatic habitats. They utilize a wide variety of organic substrates. Species of *Cytophaga* are commonly lytic and attack cyanobacteria (Stewart & Brown, 1969; Shilo, 1970). Nonlytic filamentous types such as *Herpetosiphon* possess carotenoid pigments very similar to aerobically grown *Chloroflexus* (Kleinig and Reichenbach, 1977; Schmidt, 1976) and have other similarities as well (Fautz *et al.*, 1979). Within these broad and loosely defined groups of gliding bacteria, the motility system of only a few genera and species have been studied. Although gliding motility appears to occur in all, the mechanisms may be very different.

Rhythmic waves of outer membrane

Humphrey *et al.* (1979) have demonstrated in *Flexibacter* BH3 a highly convoluted outer membrane which closely conformed to the microcontours of a thin gold surface on agar. They also established that the water-soluble glycoprotein slime produced by the cells exhibited the characteristics of a Stefan adhesive. This would allow cells to temporarily adhere to solid surfaces yet slide easily with little horizontal drag. As mentioned earlier, these authors suggested that motility was caused by a rhythmical series of 'make-and-break' interactions between the flexible outer membrane and the relatively rigid peptidoglycan layer. These would constitute wave patterns flowing from the anterior to the posterior of the cell (Fig. 16.2). If an oscillation of the outer membrane is indeed a possible basis for gliding motility, then this may also be the basis in some (particularly slow-moving) cyanobacteria, whereas the presence of contractile fibrils under this membrane in others may account for a more rapid motility. Nevertheless, Burchard (1974) was able to abolish gliding in *Myxococcus xanthus* with the proteolytic enzyme pronase. Gliding resumed in about an hour after treatment except in the presence of chloramphenicol, indicating the involvement of surface or periplasmic proteins in the motility of this myxobacterium.

Rotary assemblies of cell wall

Pate and Ordal (1967) revealed the presence of a parallel array of fibrils between the peptidoglycan layer and outer membrane of *Flexibacter columnaris* IR43 (formerly *Chondrococcus columnaris*) and suggested an association with gliding motility. Later, Pate and Chang (1979) discovered numerous small discs covering the cell walls of *F. columnaris* and *Cytophaga johnsonae*. They resemble the basal parts of the rotary flagella of enteric bacteria. The more elaborate goblet-shaped surface structures found in *Flexibacter polymorphus* (Ridgway *et al.*, 1975) may be similar. These units, together with the observation of particle rotation and translocation on the cell surface (Pate & Chang, 1979), suggested to the authors that gliding may be caused by the rotation (spinning) of these discs, thus allying the origins of this form of gliding with the more common swimming of the flagellated bacteria (Fig. 16.2). It is unclear how the rotations of these structures would translocate a cell in one direction rather than the other.

In further experiments, Pate *et al.* (1979) found that all phages active against *Cytophaga johnsonae* would not adsorb to cells in which motility had ceased. They suggested that active movement of the cell surface (i.e. by rotary discs) may be necessary for phages to 'find' their attachment sites.

The role of fimbriae

The cell surface of the fruiting myxobacterium, *Myxococcus xanthus* (strains M36 and M300), has been studied extensively (MacRae & McCurdy, 1976a,b; Dobson & McCurdy, 1979; Dobson *et al.*, 1979). The discovery of polar fimbriae (pili) in this organism and other myxobacteria led to the hypothesis that fimbriae, perhaps extending and retracting, push or pull the cell forward (Fig. 16.2). MacRae *et al.* (1977) have demonstrated polar fimbriae in other myxobacteria, flexibacteria and cyanobacteria. Peritrichous fimbriae occur in *Chloroflexus* (MacRae *et al.*, 1977) and in the cyanobacteria *Synechocystis* CB3 (Lounatmaa *et al.*, 1980) and *Nostoc* sp. (Dick & Stewart, 1980). In the *Nostoc* abundant fimbriae (*c.* 7 nm wide) occurred on the hormogonia. The authors point out a similarity in width between these structures and the periplasmic fibrils of the faster-moving *Oscillatoria* spp. and they suggest a possible relationship between fimbriae and gliding and twitching in cyanobacteria.

Recently, the laboratory group of McCurdy has established that some variants of *Myxococcus* lacking fimbriae also glide, and that anti-fimbrial antiserum does not inhibit gliding motility *per se*, but rather the aggregative swarming which involves cell to cell interactions (Dobson & McCurdy, 1979; Dobson *et al.*, 1979). Thus, the fimbrial hypothesis for gliding motility appears without basis at present.

Intracellular filaments

Burchard *et al.* (1977) have found acentric bundles of 4–5 nm thick filaments running longitudinally under the cytoplasmic membrane of *Myxococcus xanthus*, strain FB2. In the motile forms these bundles showed a herringbone-like periodicity which was lacking in one of the non-motile mutants. It appears that the bands may attach to the cytoplasmic membrane of at least one pole. The authors suggest that motility might be caused by the bundle and cell contracting, then relaxing or extending (Fig. 16.2). This method would also require alternating cell attachment to and release from the substrate—as would be provided by a mucilage with the properties of a Stefan adhesive (as in *Flexibacter* BH3). This 'inch worm' method of movement has not been observed with present microscopic resolution. It is possible that the type of flexing seen in *Flexibacter* FS-1 (Dayrell-Hart & Burchard, 1979) is also related to the activity of intracellular filaments.

The basis of motility in spirochaetes may be somewhat analogous to the system seen by Burchard *et al.* (1977) in *Myxococcus*. Spirochaetes contain acentric, periplasmic filaments and are able to swim in liquid and some at least can also glide on a solid substrate (Canale-Parola, 1978). The periplasmic or 'axial' filaments are chemically and structurally similar to flagella of enteric bacteria and are inserted (complete with proximal hook) only at one cell pole. They are not axial but form a helix usually of somewhat greater pitch than that of the helix of the cell itself. The rapid rotation of the periplasmic 'flagella' is thought by some to generate a swimming motion in which rotation occurs about the helical axis but also around the cell body axis (Bromley & Charon, 1979; Paster & Canale-Parola, 1980). Thus, contractility is probably not involved, but rather a flagellar 'spinning motor'. The basis of the gliding movement has not been investigated.

16.3 Energetics of gliding motility

It has been calculated, after several assumptions, that *Oscillatoria princeps* may expend 0.2–5% of the energy available from oxidative phosphorylation for motility (Halfen & Castenholz, 1971b). Similar and even lower estimates exist for swimming energy expenditure in flagellated bacteria (Winet & Keller, 1976). It was assumed from the knowledge of eukaryotic flagellar movement and muscle-cell contraction that ATP would be directly involved in the activation of motility organelles in prokaryotes. The rotary 'motor' of the flagellar basal assembly in bacteria apparently is driven by a membrane protonmotive force (pmf) generated by a proton gradient (Δ pH) and electrical potential ($\Delta\Psi$) across the cytoplasmic membrane. This electrochemical gradient may be established by an initial outward release of protons during electron transport, or as in anaerobes, by an extrusion of protons coupled to the hydrolysis of ATP. In either case, the pathway for proton re-entry or for dissipation of charge difference may be coupled directly to the turning of the flagellum (and to the synthesis of ATP in aerobic heterotrophs) (Maloney, 1979; Matsuura *et al.*, 1977; Manson *et al.*, 1977; Larsen *et al.*, 1974). The expression for pmf includes both gradients that may provide energy, Δ pH and $\Delta\Psi$ (pmf $= \Delta\Psi - 2.3\,(RT/F)\,\Delta$ pH). R is the gas constant, T is the absolute temperature, F is the faraday, Δ pH is the difference in pH between inside and outside of the cell, and $\Delta\Psi$ is the difference in electrical potential between the inside and outside of the cell. At 24°C, $2.3\,(RT/F) = 59$ mV. There is now increasing evidence that the force driving gliding motility may be similar to that driving flagellar rotation. Thus, a unitary model may be realized in the near future, although the structural aspects of the propulsion mechanism may differ among gliding prokaryotes.

In *Oscillatoria* sp., the use of a number of uncouplers indicated that ATP *per se* was also not the driving agent of motility but that a component of the protonmotive force across the membrane was able to sustain gliding (Gusev, 1979). The addition of the uncoupler TTFB, which destroys pmf by inducing non-physiological H^+ leaks, immediately stopped gliding. Gliding ceased even with possibly large cell pools of ATP which were maintained by inhibition of ATP hydrolysis by DCCD. DCCD by itself has no effect on gliding. Valinomycin, which dissipates $\Delta\Psi$ by making cells permeable to K^+, did not affect motility when added alone, probably because Δ pH was still maintained. Nigericin, however, dissipates Δ pH, and it also had no effect alone. Valinomycin and nigericin together completely stopped gliding, presumably by destroying both $\Delta\Psi$ and Δ pH.

Phormidium uncinatum eventually loses motility in darkness. However, gliding of these de-energized cells was temporarily restored in darkness with the artificial establishment of a sharp pH gradient by lowering the medium pH from 9 to 5 (Gusev, 1979). Other workers have also partially inhibited gliding in cyanobacteria with the phosphorylation uncoupler DNP and have assumed that lack of ATP synthesis was responsible (Halfen & Castenholz, 1971a; Castenholz, 1973). The effect of DNP (or CCCP) as a general uncoupler is by greatly facilitating H^+ movement across the membrane, thus dissipating pH and increasing electrical conductance (Hopfer *et al.*, 1968). The incomplete inhibition in some cases could be a result of the large concentrations of DNP required at physiological pH (7.7–7.8) in contrast to CCCP (Hopfer *et al.*, 1968).

In cyanobacteria the unpigmented cytoplasmic membrane is almost certainly the membrane directly associated with the motility mechanism. Since oxidative electron transport apparently does not occur in this membrane, pmf at this site may be dependent on the hydrolysis of ATP supplied from other sources in the cell. It becomes difficult with some inhibitors (general uncouplers) to distinguish between a direct effect on motility and an effect on the interior suppliers of ATP (photo- or oxidative phosphorylation at the thylakoid site).

In a fast-gliding (*c.* 12 $\mu m\ s^{-1}$) flexibacterium, *Flexibacter polymorphus*, respiration and oxidative phosphorylation are presumably restricted to the cytoplasmic membrane (Ridgway, 1977). Respiration (electron transport) and motility were inhibited in parallel with increasing concentrations of CN^-, azide, or HOQNO, while cellular levels of ATP dropped to about 20% of the control. The generation of Δ pH and $\Delta\Psi$, in this organism would be dependent on respiratory electron transport. The uncoupler CCCP ($>5\ \mu M$), which makes membranes freely permeable to protons, inhibited gliding completely even when intracellular ATP levels and respiratory rates were still relatively high (both about 40% of controls). In contrast, sodium arsenate (10^{-2} M) decreased the cellular ATP pool to very low values (5% of control) but had little effect on gliding rate. Arsenate uncouples phosphorylation by acting competitively with phosphate rather than facilitating transmembrane proton movement. These and other experiments by Ridgway (1977) tend to substantiate the hypothesis that either or both components of pmf, Δ pH and $\Delta\Psi$, are necessary for driving motility, and that ATP is not the immediate energy donor for propulsion.

To explain the anterior–posterior cell polarity Harold (1977) has suggested more localized proton ports, perhaps near one pole of the cell. This would presumably create a potential difference along the long axis in addition to that across the cell membrane. Duxbury *et al.* (1980) suggest that this type of polarization could provide the trigger for a rhythmical series of 'make-and-break' patterns which generate wave patterns in the outer membrane (in *Flexibacter* BH3). Reversals could then result from a switch of the proton port to the other end of the cell.

16.4 Photic motility responses

16.4.1 Terminology

Stimulus-induced movements in microorganisms bear a new uniform terminology recommended by an *ad hoc* committee on behavioural terminology

(Diehn *et al.*, 1979). In *photokinesis* the rate of motility is enhanced or inhibited by changes in light intensity. In a *photophobic* response a rapid temporal change in light intensity elicits a transient change in motility; this is usually a stop followed by a reversal of direction. These responses may be further characterized as step-up or step-down responses, depending on the direction of intensity change. Finally, *phototaxis* is a movement oriented in relation to the light direction. A positive reaction is towards the light, a negative response is away from the light. In cyanobacteria this may be brought about by random movements until orientation occurs or, in some cases, by steering.

Photomovements of prokaryotic and eukaryotic microorganisms have been reviewed recently by Häder (1979b), Nultsch and Häder (1979), Feinleib (1978) and Lenci and Colombetti (1978).

16.4.2 Photokinesis

The increase of gliding speed in some cyanobacteria with increase of light intensity, or simply when in light rather than in darkness, would seem to indicate that an increase in phosphorylating activity is responsible. A direct increase in protonmotive force caused by light would provide an explanation only if the light harvesting and reaction centre pigments were located in a membrane intimately associated with the motility apparatus. Since the cytoplasmic membrane of cyanobacteria is non-pigmented but is probably the membrane most closely associated with motility, it would seem that light-enhancement of motility would occur only if the additional ATP provided by light absorption in the thylakoid membranes were transported to the cytoplasmic membrane where the extrusion of protons was coupled to the hydrolysis of ATP. This is presumably the only manner in which pmf is generated in the cytoplasmic membrane of some anaerobes (Harold, 1977; Maloney, 1979). It probably is the sole manner in cyanobacteria also, since an electron transport chain is restricted to the thylakoids (Paschinger, 1977). Even if not motile, all organisms must drive active transport and maintain ionic balance via the cytoplasmic membrane. In *Synechococcus* AN, DCCD (which inhibits membrane ATPase) dissipated the H^+ gradient but a measurable membrane potential was maintained until ungoverned Na^+ uptake eventually abolished it (Paschinger, 1977). The application of DCCD to a motile cyanobacterium should abolish photokinesis if the above hypothesis is correct. However, a slow gliding rate might remain transiently by reason of the persistent electrical potential. As expected, photokinesis does not occur in 'colourless' *Beggiatoa* (Nelson & Castenholz, 1982), although a photophobic reaction does. Oxidative electron transport presumably occurs only in the cell membrane.

Cyanobacteria that do not show photokinesis or more rapid motility in light than in darkness (e.g. *Oscillatoria princeps*: Halfen & Castenholz, 1971a) may be saturated with ATP produced via oxidative metabolism. *Oscillatoria princeps* contains large stores of polyglucans and consumes O_2 at the rate of 2 μl/mg dry wt/h at 25°C (Halfen & Castenholz, 1971b). *O. terebriformis*, in which motility is highly light-dependent, has little store of polyglucans and presumably is dependent on transport of ATP from the thylakoid source. In some cyanobacteria (e.g. *Phormidium autumnale*) extremely low light intensities (0.02 lx) elicit photokinetic responses (Nultsch, 1962a); in *Anabaena variabilis* the threshold lies at about 10 lx (Nultsch & Hellman, 1972). Only extremely high light intensities ($>5\times10^3$ lx) produce a negative response in *Phormidium* and *Anabaena* (Nultsch & Hellman, 1972).

In two species of *Phormidium* and other Oscillatoriaceae light absorbed by chlorophyll-*a* (and possibly carotenoids) activates photokinesis, but in *Anabaena variabilis* and other Nostocaceae phycobilin absorption may account for the entire effect (Nultsch & Häder, 1979). In many cyanobacteria the maximum efficiency of photosynthesis rests with light absorbed by the phycobilins. The energy seems to be well distributed to both photosystems (via PS II) e.g. Ley and Butler (1976). Thus, it would appear that photokinesis in the Oscillatoriaceae responds only to cyclic photophosphorylation, while that of *A. variabilis* is driven by non-cyclic or pseudocyclic photophosphorylation. The use of inhibitors in all species tend to confirm these conclusions. DCMU, an inhibitor of non-cyclic electron transport, inhibits photokinesis in *A. variabilis* at concentrations as low as 10^{-7}–10^{-6} M (Nultsch & Hellman, 1972). In *Phormidium ambiguum* only concentrations of DCMU over 5×10^{-5} M were effective inhibitors, at which point it probably acts as a general agent. In addition, *Anabaena variabilis*, was more sensitive to DBMIB than *Phormidium ambiguum*, verifying the involvement of non-cyclic electron transport in *Anabaena variabilis*. In either system, the increase in speed with greater light probably reflects an increase in ATP production.

Of the other phototrophic prokaryotes, photokinesis in flagellated *Rhodospirillum rubrum* has been studied and the equivalence of light and ATP has been demonstrated (Nultsch & Throm, 1968; Häder, 1979b).

16.4.3 Photophobic reactions

Schreckbewegung or shock movement in gliding prokaryotes may occur with a shift in light intensity upward or downward. For a photophobic reaction to occur the change in intensity must occur within a few seconds or less. The change must also exceed a certain critical range. In flagellated purple bacteria the ΔI (I=intensity) necessary to invoke a normal phobic response (a stepdown response) remained a constant percentage of initial intensity (I_0), except at very low or very high I_0 (Clayton, 1959). Another possible extreme would be a response to a constant value of I (final intensity) regardless of I_0. Most cyanobacteria appear to respond in a manner intermediate to the two theoretical extremes.

Among prokaryotes, photophobic reactions occur in cyanobacteria, in the flagellated purple bacteria, in the non-pigmented Beggiatoaceae, and in *Halobacterium*. The behaviours of some other non-photosynthetic bacteria have been termed photophobic responses. Brief exposures to high intensity blue light initiates a tumbling behaviour in otherwise normal swimming in *Escherichia coli* and *Salmonella typhimurium*.

Cyanobacteria

The photoperception phenomenon that results in a cessation of gliding and a resumption in the reverse direction has been well reviewed (Nultsch, 1974, 1975; Nultsch & Häder, 1979; Häder, 1979b). Since the action spectrum of the efficiency of response to a step-down in light intensity correlates closely with some of the light-harvesting pigments, it is assumed that photoperception occurs at the site of the thylakoids, and that some message then must be transduced to the motility system. In brief, it appears that light harvested by phycobilins and by chlorophyll-*a* is most effective, although the blue portion of the spectrum seems less effective than the red.

The step-down reaction has been the most studied. Cyanobacteria in general may accumulate in light traps by reason of this response alone. A trichome entering a sharp-bordered bright light field from darkness (or more generally very low light) will continue its traverse across the light field until the first several cells of the anterior (5–10% of the length) have re-entered the dark area. In about 10 s the trichome will have stopped and begun its traverse in the opposite direction. Shading the original posterior of the trichome (up to 50% of the length) does not have the same effect. As a consequence of this response, every trichome once having entered the light field will tend to remain. A step-up in light intensity (of white light) elicits no reaction in most cyanobacteria so far studied except *Oscillatoria jenensis* (see Häder, 1979b) and *Pseudanabaena galeata* (the author, unpubl.). A response to a step-up in light intensity would obviously keep a sharply bordered light field essentially free of trichomes. In light traps (i.e. darkness outside a sharp-bordered light field), a light intensity threshold of 1–20 lx is generally found, although Drews (1959) and Nultsch (1962b) have reported some phobic responses in light traps of about 0.1 lx (white light).

Contrary to earlier hypotheses that change in ATP supply initiated the reversal, a variety of approaches have fairly well confirmed that a sudden change in electron transport is the cause. Uncouplers had to be used at much higher concentrations than were necessary to inhibit photokinesis in order to inhibit the phobic reaction (Häder, 1979a). However, inhibitors of photosynthetic electron transport (DCMU, DBMIB, and others) were extremely effective in inhibiting the photophobic response. This led, after a series of investigations (see Häder, 1978a,b; Nultsch & Häder, 1979; Häder, 1979a,b), to the now well-tested theory that the photophobic step-down response is 'triggered' by the sudden depletion of an 'electron pool' between photosystem (PS) II and PS I. Häder (1976) has used two techniques in which an organism will traverse from a region illuminated with wavelengths primarily absorbed by one photosystem over a sharp border to a region illuminated by wavelengths mainly absorbed by the other photosystem. Phobic responses occurred when *Phormidium* trichomes were beginning to leave a field illuminated with 'PS II + PS I light' and entering a 'PS I' field (Fig. 16.4). Thus, accumulation took place in the 'PS II' field. It appears that the main effect of imbalance favouring PS I light is the rapid drain of an electron pool in reduced electron carriers between PS II and PS I. DCMU has relatively little effect on the traverse to a PS I light field, but DBMIB (preventing plastoquinone reoxidation) does, indicating a 'pool' between the DCMU block and the point of quinone inhibition. The most likely candidate is plastoquinone (PQ) which occurs normally in pool sizes of ten times that of P700 (the reaction centre of PS I). Thus, without inhibitor an imbalanced promotion of PS I would oxidize the PQ, depleting the electron pool. A manipulated decrease in PS II light has a similar effect, presumably by decreasing the reduction rate of PQ. In natural

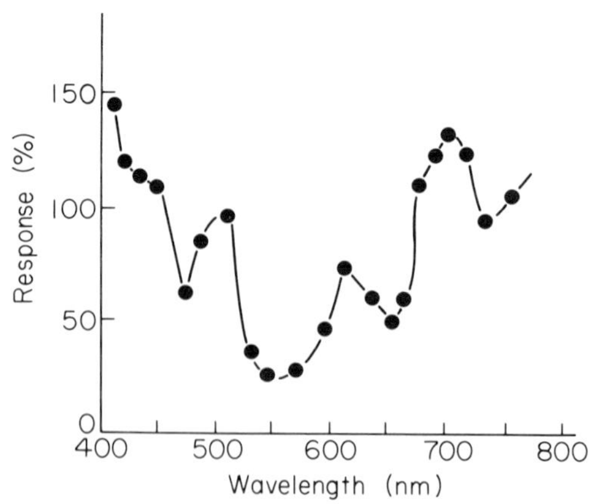

Fig. 16.4. Photophobic reactions of *Phormidium uncinatum* when trichomes enter a light field (λ_2) of variable wavelength from a uniform field (λ_1) of 557 nm light (absorbed mainly by PE associated with PS II). Both fields were adjusted to 5.6×10^{-4} μEinsteins/cm^2/s. The ordinate represents the response (accumulation rate of trichomes in light field λ_1) as a percentage of the accumulation rate in light field λ_1 with reference to an adjacent dark border. (After Häder, 1976)

light (white light) the sudden change from high intensity to low may favour PS I ('Gaffron effect' known in H_2 photoreduction studies), and result in a reversal. The transduction process, i.e. a message from change in electron flow pattern to a change in directional polarity of the motility apparatus, is now thought to be a change in cell electrical potential. The elegant series of experiments by Häder on which this conclusion is based is summarized in 2.8.1.

Purple bacteria

The current evidence indicates that the photophobic swimming response of *Rhodospirillum rubrum*, when grown anaerobically and pigmented, is also due to a change in photo-electron transport after a step-down in light intensity (Häder, 1979b). Inhibitors of electron transport blocked the response, whereas uncouplers had no effect (Throm, 1968). The same uncouplers inhibited photokinesis, however. Both BChl-*a* and carotenoids were involved in absorbing the effective light. Aerobic non-pigmented cells also showed a phobic response at very high intensities but this proved to be the result of light-absorbing ferric ions incorporated into the cells.

Halobacterium

Halobacterium, although photosynthetic in its ability to generate ATP by a light driven H^+ or ion translocation across the bacteriorhodopsin membrane, it is without this pigment under continuously aerobic conditions. With bacteriorhodopsin the flagellated cells show a step-down photophobic reaction (swimming reversal) with an action spectrum matching bacteriorhodopsin absorption. A step-up reaction occurs also, but with an action spectrum in the blue-green, blue and u.v. which indicates a possible flavonoid pigment (Hildebrand & Dencher, 1975).

Beggiatoa

The species of *Beggiatoa* are generally rapidly gliding and rotating trichomes that architecturally resemble a few species of *Oscillatoria*. However, they are ostensibly colourless (appearing white) and usually harbour elemental sulphur granules (within cytoplasmic membrane invaginations). They have frequently been classified as colourless derivatives of cyanobacteria (Pringsheim, 1963), but recently have been placed (for convenience only) with the Flexibacteria (Soriano, 1973) or the Cytophagales (Leadbetter, 1974).

Nevertheless, internal membranes are unknown, and the strong phobic response to a step-up in light intensity must surely involve a photoreceptor in the cytoplasmic membrane (Nelson & Castenholz, 1982). The trichomes of *Beggiatoa leptomitiformis* react negatively by stopping and reversing whenever a short anterior position enters a brighter light field. This response is to blue light only with a maximum at about 430 nm (Nelson, 1979). This maximum corresponds closely to the Soret band absorption of some cytochromes in *Beggiatoa* (Cannon *et al.*, 1979). Electron transport and thus the electrochemical gradient of the cytoplasmic membrane could be directly modulated by a step-up in light intensity. Step-down responses did not occur (Nelson & Castenholz, 1982).

16.4.4 Phototaxis

This, the traditional topic response (phototopotaxis), is the sensing of the direction of light. In order for a pigment to act as a photoreceptor, it must be located at a site where light from one direction would be perceived in a different manner than from the other side. This could be as the elaborate paraflagellar photoreceptor of *Euglena* with its partial shading or filtering device (the stigma) (see Häder, 1979b). In a filamentous cyanobacterium, however, the end of a trichome pointing toward the incoming light would probably realize more light than the end of the trichome farther from the light source, for it (or each cell towards the 'tail') would have a substantial quantity of the directional light absorbed and filtered out. Thus, a light gradient would exist within the trichome. In this so-called 'one-instant' mechanism, two photoreceptors (or photoreceptor regions) may simultaneously detect the internal light intensity, 'comparing' light absorption in the two regions.

Most cyanobacteria, however, 'gravitate' towards light (positive phototaxis) or away from light (negative phototaxis) after apparently random movements have resulted in the long axis lying parallel to the light field. It is only then that phototaxis becomes apparent through the cancellation of the regular stops and reversals of direction (see Castenholz, 1973; Nultsch, 1974; Burkholder, 1934) (Fig. 16.5). The trichome anterior–posterior polarity seems to become locked and gliding towards the light is extended while reversals are shortened or nullified. Many variations in path are possible. For example, trichomes that normally travel in arcing circular paths because of rotation (Castenholz, 1973) may reverse at the end of each 220–250° arc (Fig. 16.5). Since the sense of rotation relative to the existing anterior will remain either right- or left-handed, a series of arcs will bring the trichome closer to the light source (Fig. 16.5; Burkholder, 1934).

Steering, however, occurs in some cyanobacteria, apparently only those with articulated trichomes such as members of the Nostocaceae. Because of the deep constriction at each cross-wall, these trichomes bend easily. Drews (1959) described how 'U'-shaped trichomes of *Anabaena* could steer towards light (blunt end forward) without any unusual requirement except that the two trailing ends of the 'U' are pushing in opposite directions relative to the centre of the trichome (Fig. 16.5). Such a method is also used by the non-articulated

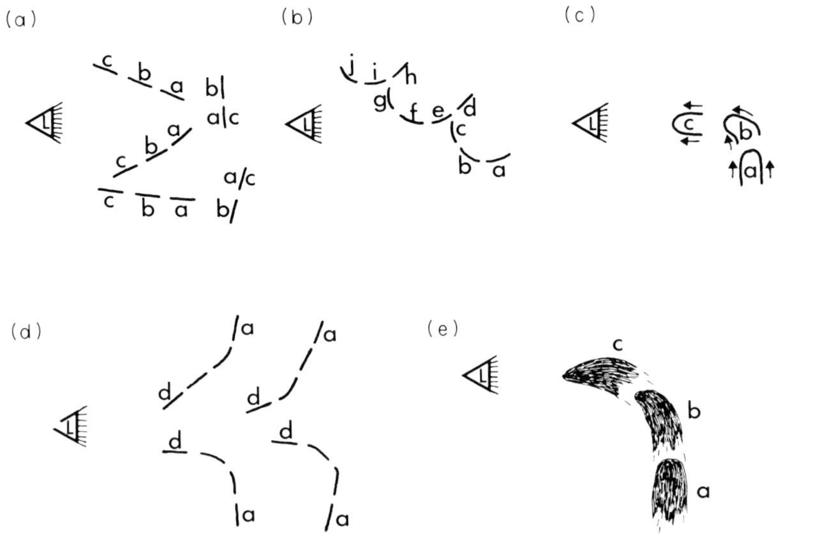

Fig. 16.5. Cyanobacterial methods of approaching a light source (L) (positive phototaxis).

(a) Random movements with reversals followed by nullification of reversals once movement towards light begins. Positions a, b, to c.

(b) Movements in arcs associated (in this case) with a right-handed rotation. Positions a–j. Light effects a reversal at positions d and h.

(c) Steering of a 'U' trichome. Arms of the trichome are moving in opposite directions relative to the whole trichome. The blunt end is being pushed ahead. Turning is probably achieved when more cells near the blunt end of the right arm begin to move positively in response to the light (position b).

(d) True steering by trichomes. One or more of the anterior cells of the articulated chain turn towards the light and are followed by the others.

(e) Turning of a *Pseudanabaena* 'comet'. The swarm or aggregate of hundreds or thousands of trichomes turns towards the light. Positions a–c.

filamentous bacterium *Toxothrix trichogenes* (Krul *et al.*, 1970). However, the simple steering of trichomes by the turning of anterior cells (or anterior cell) towards the light after being originally oriented at normal angles to the light field has been observed in *Anabaena variabilis* (Nultsch *et al.*, 1979) and in *Pseudanabaena galeata* (the author, unpubl.) (Fig. 16.5). *Pseudanabaena*, although perhaps not closely related to the heterocystous Nostocaceae, is 'articulated' in a similar manner. The basis of steering in trichomes is unknown, but features of gliding movements which may help explain it are discussed in 2.8.1.2.

Gliding aggregates of *P. galeata* may consist of a few to several thousands of trichomes. These 'green comets' turn towards the light by curving the anterior portion (Fig. 16.5) (the author, unpubl.). 'U'-trichomes are apparently not involved in the steering method.

There are apparently two classes of action spectra for positive phototaxis in cyanobacteria. Two articulated cell types (Nostocaceae) are responsive to light absorbed by phycobilins and probably Chl-*a*, but Chl-*a* and carotenoid absorption in the blue portion of the spectrum is without effect (Nultsch *et al.*, 1979). In *Anabaena variabilis* the positive response is nullified at about 1000 lx and at intensities above this a negative phototaxis occurs. It appears that photoreceptors absorbing above 700 nm and about 550 nm are also effective in the negative reaction (Nultsch *et al.*, 1979). Although it appears that Chl-*a* absorption in the red is effective in both reactions, the action spectra are crude enough that phototactic photoreception in the 670 nm region by allophycocyanin B instead of Chl-*a* cannot be excluded. The action spectra of two members of the Oscillatoriaceae (*Phormidium autumnale* and *P. uncinatum*) may not be very different (see Häder, 1979b). Only phycobilin absorption (PE and PC) and carotenoids appear to be effective in the positive reaction (Nultsch, 1961, 1962b). Again, the threshold intensity for positive is very low, ranging from 0.05 lx to 5 lx in *Phormidium* and *Anabaena* (Nultsch, 1962b; Nultsch *et al.*, 1979).

Photosynthesis (i.e. electron transport through both photosystems) is not involved in phototaxis of *A. variabilis* since even a concentration of DCMU as high as 5×10^{-5} M did not prevent positive or negative phototaxis *per se*. DBMIB, likewise, had no effect. The reported negative influence of DCMU on phototaxis in *Cylindrospermum* (Tyagi, 1976) appears to be merely the effect of slower gliding, not lack of orientation.

At present, there has been little or no speculation on how differential light absorption by phycobilins in two portions of a trichome could affect the polarity of the motility mechanism by preventing reversals or by effecting a reversal when movement is away from light. Since bulk phycobilins are probably restricted to phycobilisomes, which are attached to the outer surface of the thylakoid membranes, no transmembrane proton or electrical gradients would be affected. It is conceivable, however, that all phycobilin-absorbed energy effective in phototaxis is routed through the 'reaction centre' of the phycobilisomes (e.g. allophycocyanin B) which is in contact with the thylakoid membrane. The transducing mechanism, however, is still a complete puzzle. Several photoinductive phenomena (PE synthesis, filamentation, hormogonial migration) are known to involve one or more special phycobilin pigments which are present in small amounts and are probably photoreversible (Vogelmann & Scheibe, 1978). These may or may not be a part of the phycobilisomes, but the manner of transduction is again unknown.

True phototaxis, although common in photosynthetic eukaryotes (e.g. flagellates, diatoms), is unconfirmed in prokaryotes other than the cyanobacteria. A weak response has been reported in the flagellated purple bacterium *Rhodospirillum rubrum* by Throm (1968). The negative response of *Beggiatoa*

(Schmid, 1923) is almost surely a step-up photophobic response. *Chloroflexus*, the gliding anoxygenic phototrophic bacterium, may show a true phototaxis but this has not been quantified (the author, unpubl.).

16.5 Chemotaxis and aggregative behaviour

Chemotaxis is essentially undescribed in the cyanobacteria. Fechner (1915) has pointed out negative phobic responses to various inorganic and organic acids. Chemotaxis, however, in flagellated bacteria is very well described. A brief introduction is given in 17.5.1 and more detailed reviews have been made by MacNab (1978, 1979), Parkinson (1977), Hazelbauer and Parkinson (1977), Koshland (1977) and Adler (1975). Chemotaxis is also known among the gliding myxobacteria (Ho & McCurdy, 1979). *Myxococcus xanthus* was found to be positively chemotactic for cGMP and AMP when stable gradients were established. GMP, ADP, and cAMP at concentrations of 1–10 mM were not chemo-attractants.

In cyanobacteria, there have been a number of responses observed that are surely chemotactic, but none has been quantified. Negative responses (chemophobic) to soluble sulphide (or oxidation products of sulphide) occur in thermophilic *Oscillatoria terebriformis* (the author, unpubl.). The movement of all motile organisms is modified by many environmental factors. Temperature, viscosity, and various chemical features will influence rate of movement and frequency of reversal while any factors which change rapidly or form a concentration gradient are apt to influence the direction of movement. For a photosynthetic organism, light would be expected to be one of the most influential of factors, but chemical factors must also be important. A large potential area for research has therefore been left untouched.

Aggregation of cells or trichomes through chemotactic attraction and cell to cell interactions have been studied primarily in the myxobacteria, among gliding prokaryotes (Kaiser *et al.*, 1979; Dobson *et al.*, 1979). The movements of the aggregated cells is coordinated in that the cells move together as a swarm. If a cell moves away, it soon returns. The fimbriae of myxobacteria have been implicated in the retention of cell-to-cell contacts in the migratory swarms of *Myxococcus xanthus* (Dobson *et al.*, 1979). Henrichsen (1972) has described other aggregative behaviour in bacteria.

Recently, rather elaborate gliding aggregates have been discovered in the cyanobacterium *Pseudanabaena galeata* (the author, unpubl.). The formation of these 'comets' appears to be a chemotactic swarming of short trichomes (20–30 cells). The 'comets' vary in size but grow from a few trichomes to those of several thousand (measuring $0.5 \times > 1$ mm). The trichomes are also layered from 1 to >10 tiers in height (Fig. 16.6). The comet-like swarms respond

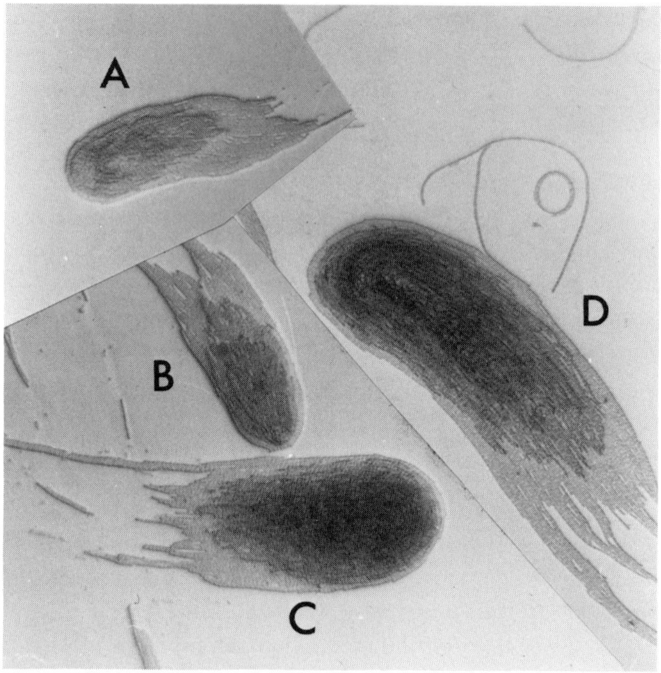

Fig. 16.6. 'Comets' of *Pseudanabaena galeata* on agar. Swarm C is approximately 130 μm in breadth. Swarms A and B are 3 trichome layers thick near the anterior (blunt) end. Swarms C and D are 6–8 trichome layers thick. Swarm C contains over 350 trichomes of 20–30 cells each.

phototactically and the trichomes show a mass reversal of direction in a step-up photophobic response; the 'comets' then reform with movement in the opposite direction. The aggregates will form or maintain their integrity in uniform light or in darkness, but are most rapidly formed in a unidirectional light field (the author, unpubl.). The manner in which trichome-to-trichome communication maintains the same polarity with respect to movement is entirely unknown. Although fimbriae have now been discovered in several cyanobacteria, their presence in the strain of *P. galeata* that forms gliding aggregates has not yet been confirmed. The simple mechanics of turning an entire 'comet' in phototactic or chemotactic response is also not understood at present (Fig. 16.6). Aggregates of rather simpler geometry and characteristics have been described in a few other cyanobacteria (Drews, 1959; Nultsch *et al.*, 1979).

The integrity of large flakes or fascicles of planktonic *Aphanizomenon flos-aquae* may be influenced by chemotactic phenomena. Trichomes glide in contact with each other; those on one side of the flake moving in a direction opposite to those on the other side (Dodd, 1953). When the fascicle is fully stretched out, the trichomes reverse and for a while thicken the fascicle until it is stretched out again. Some subtle thigmophobic reaction could also be involved when the trichomes near the end of their contact. Contact is maintained presumably by some type of trichome–trichome interaction.

The circling of trichomes of many cyanobacteria on agar surfaces has been observed by many. It was noted by Burkholder (1934), Thomas (1970a), Halfen (1970) and Castenholz (1973) that trichomes with a right-handed rotation circle in a clockwise direction, when viewed from above. In *Oscillatoria princeps* this is surely influenced by the slightly bent apex of a few cells which would act as a pry every time it is brought around by a rotation. Circles may consist of many trichomes. They occur also in cyanobacteria that are reported not to rotate (e.g. Nostocaceae). In such cases, at least, it is possible that chemotaxis is involved, together with a steering or bending movement. The following of trails of pioneer trichomes could also be a chemotactic or a mechanical phenomenon. Trichomes will often follow stress or score lines in agar, and when artificially imposed, these can be used for isolation (Vaara *et al.*, 1979).

The mass tightening and consequent contraction of trichome masses of many cyanobacteria described by Castenholz (1967, 1968) and Walsby (1968), and much earlier by Funk (in Haupt, 1959) are caused by trichome adhesions and mass gliding movements, but chemotaxis is probably not a factor. A photokinetic effect (i.e. faster contraction with increased light) is obvious in the case of *O. terebriformis* (Castenholz, 1973).

16.6 Ecological aspects of gliding motility and tactic responses

Gliding, like any type of motility in a microorganism, almost certainly improves survival rates, although motility means some dangers as well. Specifically, moving microorganisms will constantly assure themselves of a steeper gradient of absorbable nutrients from outside to inside than if they had remained in one spot. This is true in microhabitats, such as mud surfaces or interstices, or elsewhere where water movements are slight. Gliding may be thought of as being more restrictive than swimming, not only because of slower speeds, but because of the requisite of a solid surface. However, the ability to move on a solid (or semi-solid) surface is not enjoyed by most flagellated microbes. Solid surfaces in several aquatic situations are the richest sources of organic and inorganic nutrients. Attachment by means of a Stefan-type adhesive while moving also may provide the benefits of not being washed away. For example, in some hot springs motile gliding organisms (e.g. *O. terebriformis*) can remain within their optimal temperature range while washout downstream (to cool temperatures) is generally lethal. Gliding at random (as with swimming) also provides a means of dispersal. It is improbable that it provides protection from the usual predators. It is simply too slow to escape anything except perhaps lytic gliding bacteria, such as *Cytophaga*. The aggregative contractional behaviour of some cyanobacteria, such as the bright day condensation of *Oscillatoria terebriformis* mats in hot springs as a result of photokinesis, may provide increased self-shading and thus protection from mid-day light damage (Castenholz, 1968).

However, it is the photophobic and phototactic responses of many cyanobacteria that are most likely to provide the kind of advantages in the natural world that are readily assessable. Little assessment of this type has been done. Several phycologists have mentioned (unpubl.) the disappearance of cyanobacterial mats or thin coatings on sand or mud in both marine and fresh waters during mid-day periods of bright sun. It is apparent in hot springs where *O. terebriformis* (Castenholz, 1968) and other *Oscillatoria* species occur (the author, unpubl.). When dusk approaches, during mornings, and generally under overcast conditions, the *Oscillatoria* returns to the surface of the sediment. A similar behaviour of *Oscillatoria* was observed in intertidal and subtidal sediments (B. Bohlool, pers. comm.).

Either phototaxis or a photophobic response or both could explain the escape into the sediment. *Beggiatoa leptomitiformis* has only a step-up phobic response and its daytime disappearance from the surface of muds can be explained by this reaction alone (Nelson & Castenholz, 1982). Random gliding movements adhering to sediment particles will eventually allow all trichomes to experience an end pointed in toward darker interstices and the other pointed towards brighter light. The movement into bright light brings on long-lasting reversals which result in a downward migration of several millimetres (Nelson & Castenholz, 1982). Presumably a step-up phobic response to very bright light could explain the downward migration in *Oscillatoria* also, but a negative phototaxis is a simpler explanation, and the re-emergence could reflect positive phototaxis when the light intensity above had passed below a certain threshold value. Critical experiments to explain natural migrations of these types have not been performed.

Thus, either phototaxis or photophobic responses may allow some cyanobacteria to position themselves to an optimal light intensity, to avoid the damage of extremely high intensity, or simply to move to where light is available for photosynthesis.

Chemotactic or chemophobic responses may also be intimately involved in some of the diel vertical migrations observed. One of the most puzzling events is the night-time downward migration of *Oscillatoria terebriformis* into the undermat or sediment in hot springs. During the course of 24 h, the masses of trichomes disappear downward during mid-day under clear summer skies, reappear in late afternoon, and then disappear downward again after dark to reappear again with the first light of dawn. At any time during the night, they can be brought forth by shining a spot light on the sediment. But since the downward migration at night occurs in the absence of any light gradients, another type of tactic response must be involved. This phenomenon is currently under study.

CHAPTER 17
INTERACTIONS WITH BACTERIA

H. W. PAERL

17.1 Introduction

17.1.1 Historical perspective

Since the practical application of the light microscope, biologists have been fascinated by the complexities and ubiquity of microbial associations with non-living particles, other microorganisms and higher plants and animals. It is generally regarded that in nature very few particles or organisms are ever found completely free of microbial epiflora. The cyanobacteria are no exception, as indicated by Whitton's (1973b) previous review. In fact, numerous observations (Guseva, 1951; Zagallo, 1953; Kuentzel, 1969; Paerl, 1976, 1978; Caldwell, 1977) show that cyanobacteria are frequently the sites of intense bacterial and fungal colonization. Colonization occurs in diverse terrestrial and aquatic habitats, during a range of growth conditions, and during various stages of cyanobacterial decomposition. Henrici and McCoy (1938) made routine microscopic observations of bacteria specifically associated with phytoplankton. Their findings were the forerunner of more recent studies, describing bacterial attachment sites as microzones, where chemical and physical properties are determined by microbial exchange processes (Lange, 1967, 1971; Kuentzel, 1969; Paerl & Kellar, 1978; Caldwell & Caldwell, 1978). In such microzones localized oxidative and reductive processes leave the chemical and physical environment distinctly different from ambient conditions.

Attached bacteria have traditionally been considered precursors to subsequent decomposition of the attachment substrate. This fact can be supported with numerous examples, including wood degradation, surface fouling, breakdown of humic substances, decomposing macrophytes, sea and marsh grasses and decaying algal blooms. Frequently, however, bacterial attachment and decomposition of attachment substrates are not contemporaneous events. For example, photosynthetically active terrestrial and aquatic algae are often colonized throughout their entire life cycle. In such associations neither the bacterial epiphyte nor the algal host appear to be negatively affected. More often both groups of interacting microorganisms appear to benefit (in terms of metabolic capabilities and growth rates) from such associations. Hence, traditional concepts of parasitism followed by decomposition are often not the rule, or even the exception. More likely, very close and long-lasting associations between algae and bacteria hint of functional exchange mechanisms, which may be crucial to the survival and growth strategies of host and epiphytic microorganisms. Only recently have such exchange mechanisms been examined in any detail, though a clear example was noted as early as 1881 by Engelmann. This is the close association between the band-like chloroplasts of the green alga *Spirogyra* and bacteria attached to the cell walls bordering the chloroplasts. Engelmann demonstrated that aerotactic bacteria were attracted by the localized formation of minute amounts of dissolved oxygen in the vicinity of the chloroplasts. In addition to illustrating an example of the transfer of a product of algal photosynthesis to a bacterial epiphyte, his experiments also showed that the chloroplast was the site of oxygen evolution during photosynthesis. Presumably, such oxygen-rich microzones would enhance aerobic breakdown of organics, used as energy and growth sources by associated heterotrophic bacteria. Such associations can often last the entire lifetime of an algal filament without marked detrimental effects. We are at the threshold of uncovering numerous other functional aspects of such associations, opening up a vast area of research in the field of microbial ecology.

17.1.2 General occurrence of algal/bacterial interactions

Prior to a more in-depth discussion of cyanobacterial/microbial associations some parallel examples can be shown for eukaryotic algal/microbial associations as well. Freshwater and marine phytoplankton are often heavily populated with microbial epiphytes, even during bloom conditions. Examples are abundant among green algae and diatoms (Golterman, 1973; Paerl *et al.*, 1974). Mucilaginous solitary or colonial green algae virtually always maintain bacterial populations found in their slime coating. The presence of bacteria on a variety of algal hosts does not necessarily dictate algal death or even poor physiological conditions (Paerl *et al.*, 1974; Paerl, 1979). Periphyton communities in streams and rivers are often complex associations of algae, bacteria and protozoans. Autoradiographic examinations of the epiphytized periphyton revealed highly active photosynthetic algal cells supporting actively metabolizing heterotrophic bacteria. Such communities often remained intact for periods up to 3 months, with little, if any, sign of active bacterial degradation of the host diatoms (Perkins, 1974).

Seaweeds (Waite & Mitchell, 1976; Fogg & Burton, 1977) and marine angiosperms (Allen, 1971; Finke & Seeley, 1978) can also support luxuriant growths of bacteria and fungi during optimal growth periods as well as during periods of decomposition. The occurrence of bacteria in colonies of the marine planktonic *Trichodesmium* has been noted by many authors and is discussed in 19.3.4 and 19.5.3.

Soil algae show similar non-degradative associations with bacteria and fungi. Layers of algae are often bordered by layers of bacteria. Physical contact and exchange of metabolites is suspected of being intimate, but decay or bacterial attack of host algae appears absent. Layering of microorganisms leads to specialized functions and exchange processes such as: (a) photosynthetic generation of organic carbon compounds which can be utilized by heterotrophic bacteria and fungi; (b) excretion of growth factors and vitamins by bacteria utilizing algal photosynthate; (c) nitrogen transformation steps restricted to specific oxygen requirements/tolerances (N_2 fixation, nitrification, denitrification, ammonification); (d) solubilization of soil mineral constituents; (e) creation of oxygen-rich or oxygen-poor microzones in order to provide microenvironments suitable for specific redox reactions; (f) assuring CO_2 availability and proper alkalinity conditions for promoting specific metabolic reactions and pH requirements.

17.2 Ultrastructure of cyanobacterial/bacterial associations

Cyanobacterial/bacterial associations have been observed by both scanning (SEM) and transmission (TEM) electron microscopy. Both endophytic and epiphytic associations have been reported, but a vast majority of observations stress epiphytic associations. Recently Wujek (1979) reported the first observations of intracellular rod-shaped bacteria in *Pleurocapsa minor* cells. Intracellular bacteria were present in less than 5% of *Pleurocapsa* cells.

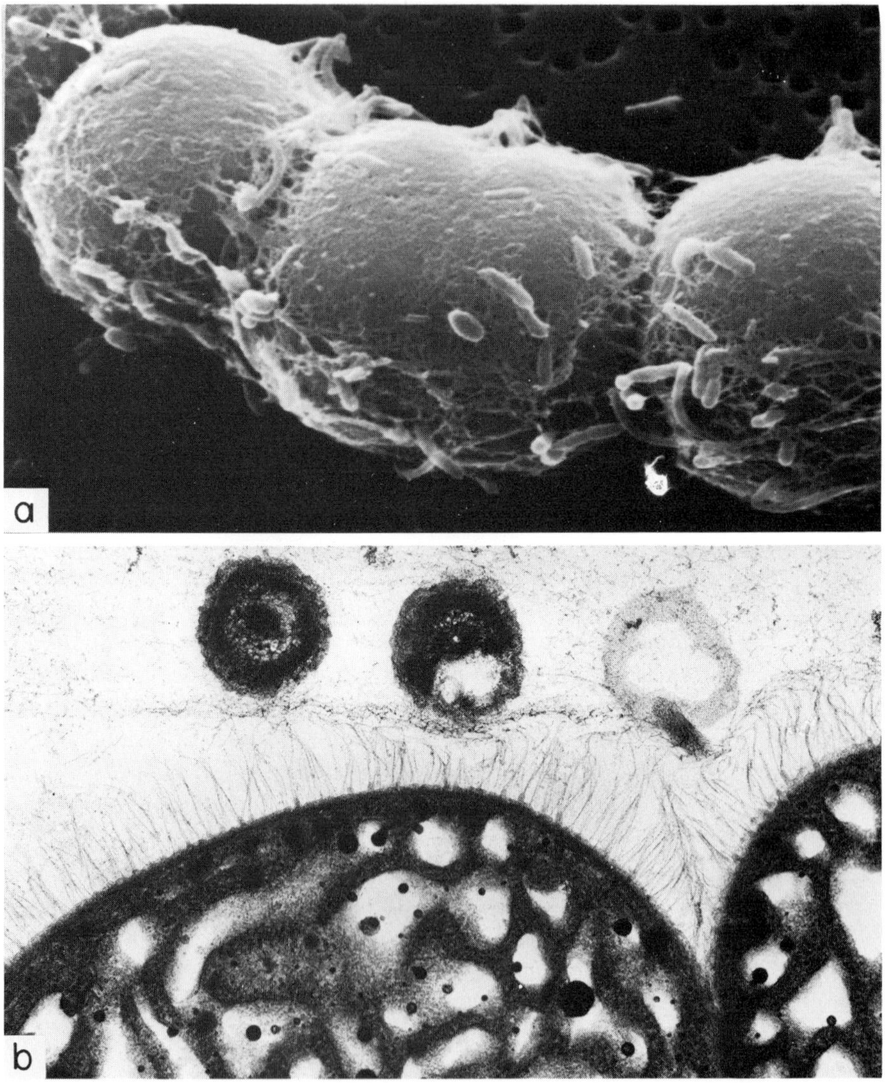

Fig. 17.1. Scanning (SEM) and transmission electron micrographs (TEM) illustrating extracellular fibrils of *Anabaena*. Although SEM does not allow for adequate resolution of individual fibrils (a), TEM reveals the fine structure of such fibrils (b). (a) *A. spiroides* from Lake Okaro, New Zealand (b) *A. oscillarioides*, originally isolated from the Waikato River, New Zealand and grown on nitrogen-free medium. (TEM photo courtesy of Dr A.M. Massalski)

Infected host cells showed similar morphological features to uninfected cells, indicating that cellular damage, including endocytosis, lysis or degradation of photosynthetic lamellar structures, did not occur as a result of endophytic bacteria. Although no physiological data were reported, it was assumed through morphological features that intracellular bacteria were not pathogens.

Epiphytic bacteria are widespread on both naturally occurring and laboratory cultures of cyanobacteria. Morphological features in terms of the ultrastructure of attachment sites have only been described in the few studies which have dealt specifically with bacterial attachment to cyanobacteria (Paerl & Lean, 1976; Leppard *et al.*, 1977; Bitton & Marshall, 1980), while thorough ultrastructural studies have been conducted on either isolated bacterial or algal adhesive appendages and extracellular polymers (Lang & Fay, 1971; Wang & Tischer, 1973; Cagle, 1975; Leppard *et al.*, 1977). Both SEM and TEM observations of cyanobacterial/bacterial associations reveal the presence of electron-opaque non-rigid fibrils of approximately 2–10 nm in diameter present on the surfaces of both groups of microorganisms (Fig. 17.1). Fibrillar lengths may vary dramatically from approximately 10 nm to several microns. Often fibrils appear concentrated between the host cyanobacteria and attached bacteria, forming a 'pad-like' attachment site (Floodgate, 1972), thought to be composed largely of acidic polysaccharides including uronic acid residues (Colvin & Leppard, 1973). Fibrils are apparently easily dislodged from the cell surface, for individual as well as aggregated fibrils are often suspended in lakewater free of host cells (Leppard *et al.*, 1977). A model of interacting fibrils has recently been presented by Costerton *et al.* (1978). Attachment by virtue of cross-linkage of charged residual glycocalyx groups located in polysaccharide chains which form the backbone of fibrils has been proposed. Such cross-linkage would lead to strong binding without the formation of copious amounts of additional cementing material. Indeed, both SEM and TEM observations generally agree that the deposition of fibrillar material at attachment sites is usually not extensive and that the bacterial cell surfaces forming direct contact with host cyanobacteria are often a small percentage ($< 20\%$) of the total cell surface. In fact, only exceptionally are the walls of host and epiphyte in direct contact. There is usually a space, often occupied by a fibrillar 'pad', of approximately 0.1–1.0 μm width separating associated cells (Fig. 17.2). It follows that epiphytic bacteria are rarely seen penetrating cyanobacterial cell walls. Cellular communication in these associations must therefore be through chemical as opposed to physical means, since no direct interaction of protoplasmic material appears evident in the above range of published reports on cyanobacterial/bacterial interactions.

Bacteria are frequently found in polysaccharide-rich organic material surrounding cyanobacterial cells (Fig. 17.3). This material presumably

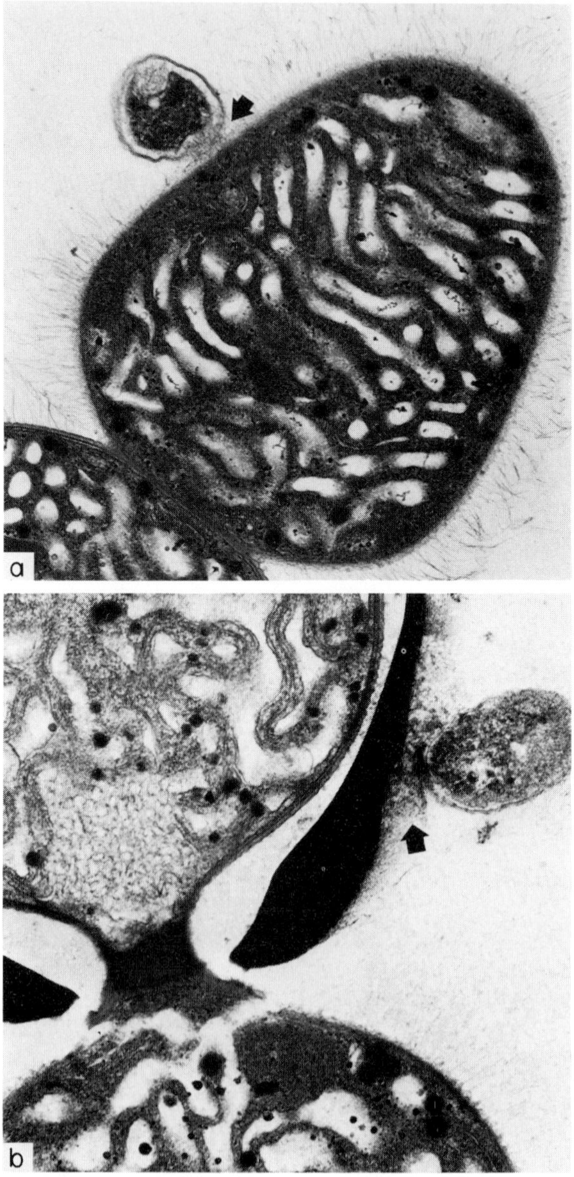

Fig. 17.2. (a) Fibrillar 'pads' (arrow) promoting adhesion of bacteria to vegetative cells and (b) heterocysts of *Anabaena oscillarioides*. Such pads were observed in a wide variety of other *Anabaena*–bacterial associations, including host species *A. circinalis*, *A. flos-aquae* and *A. spiroides*. (Photos courtesy of Dr A.M. Massalski)

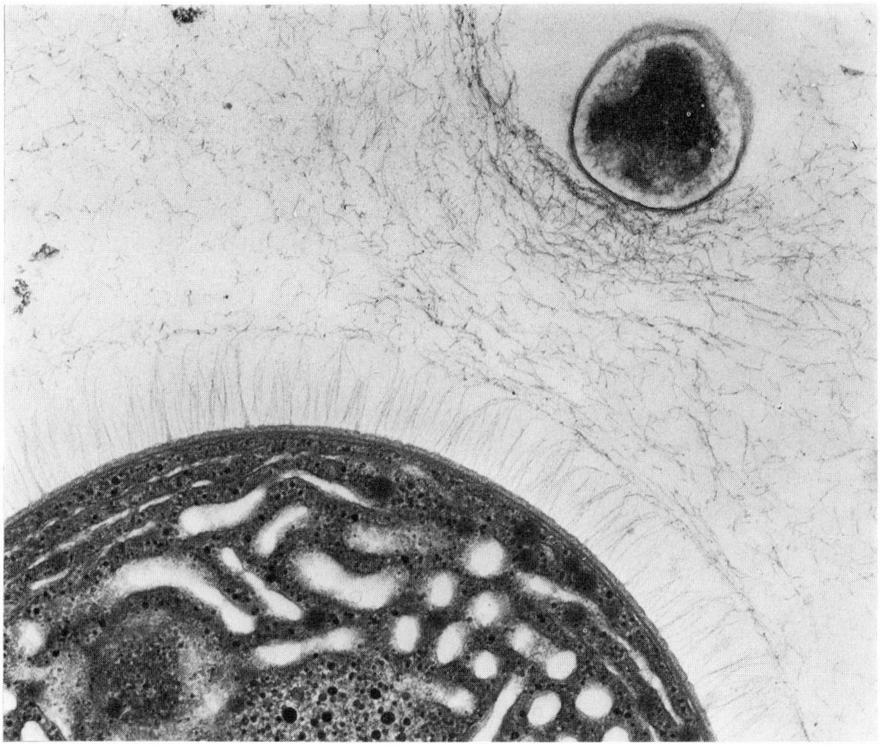

Fig. 17.3. A cross-section of an *Anabaena oscillarioides* vegetative cell, showing extracellular fibrils coating the cell wall as well as more amorphous mucilage fibrils extending several microns away from the inner fibril coat. Associated bacteria often penetrate the mucilage layer, as illustrated here. (Photo courtesy of Dr A.M. Massalski)

includes slime or mucus making up the sheaths which are characteristic of a wide range of cyanobacteria. When located in sheaths, bacterial fibrils seem to act as anchoring webs, mechanically criss-crossing with the fibrillar webbing laid down by the host cyanobacteria. Among the common heterocystous genera, *Anabaena* and *Aphanizomenon*, bacteria appear to frequent the heterocysts more often than vegetative cells (Fig. 17.4; see also Lupton & Marshall, 1981). Geitler (1932) and Schwabe and Mollenhauer (1967) made earlier mention of the high frequency of bacterial attachment to the terminal heterocysts of *Cylindrospermum*. Bacteria are often lodged in the junction of heterocysts and neighbouring vegetative cells, again by means of fibrillar pads and interactions of individual fibrils. The physiological significance of these observations is discussed elsewhere.

Fig. 17.4. A general SEM view of *Aphanizomenon flos-aquae* filaments freshly sampled from Clear Lake, California. Note the distinct preference that associated bacteria have for the heterocysts. (Sample collected by Dr A.J. Horne)

17.3 The concept of microzones

17.3.1 Importance of environmental gradients

In natural ecosystems, the formation and alteration of physical and chemical gradients are of paramount importance in determining community structure and diversity of the biota. On a macro-scale, gradients often lead to distinctions between aerobic and anaerobic, euphotic and aphotic, hypo- and hyperbaric, hypo- and hypersaline, and thermo- and psychrophilic microbial communities. In lacustrine and marine water columns, contrasting turbulence regimes can often exist between the well-mixed epilimnia and less turbulent hypolimnia, leading to the development of dramatically different means of maintaining and regulating flotation among planktonic algae and bacteria.

Ecologists have been aware of such determinative physical features ever since the development of limnology, oceanography and terrestrial ecology as quantitative disciplines. The recent explosion of environmental monitoring equipment, including sophisticated data-logging capabilities, has added immensely to our understanding of the importance of physical and chemical control and regulation of microbial communities.

17.3.2 Identification and measurement of microzones

In contrast to the wealth of information on large-scale environmental gradients, we are only beginning to form an appreciation for the presence of diverse microscopic gradients associated with surfaces of particles in aquatic and terrestrial ecosystems. These gradients, which are often confined to less than 10 μm of a surface, can harbour a highly specialized microbial community which is normally not found in a free-living state and whose growth would be suboptimal in the ambient macroenvironment. Such microenvironments, or 'microzones' (Nikitin, 1973), were observed several decades ago by Waksman and Carey (1935), as well as numerous other investigators who noted that microbial populations on surfaces often exceeded, by orders of magnitude, planktonic or interstitial populations. In understanding the conditions enhancing standing stocks of surface micro-organisms, it would be desirable to measure pertinent physical/chemical gradients in the microzones. Unfortunately, it is extremely difficult to measure physical/chemical gradients in microzones by mechanical devices (O_2, Eh, pH probe analyses), although efforts are underway to design and construct microprobes. At present, semi-micro pH, Eh, and O_2 probes are available which can detect gradients in regions as small as 50 μm across.

Until the development of microzone probes makes this approach practical, there are several alternative means of identifying and detecting microzones on a semi-quantitative scale. Redox indicator salts and dyes such as methylene blue, benzyl and methyl viologen, phenol–indophenol and the tetrazolium salts have proven to be useful in demonstrating the formation and maintenance of microzones, particularly in association with detrital particles and filamentous and colonial algae, including cyanobacteria. It can be shown for example, that the surfaces of particles are often inhabited by actively metabolizing (and hence respiring) bacteria and fungi, which can lower microzone oxygen tension to subsaturated, microaerophilic and anaerobic levels in waters showing ambient oxygen supersaturation. Such reduced microzones can allow highly reductive nutrient-transforming steps (i.e. denitrification, N_2 fixation, methanogenesis) to occur in systems which by macromeasurements would not be expected to yield such processes. In particular, the tetrazolium salts, nitroblue tetrazolium (NBT), triphenyl tetrazolium chloride (TTC) and iodonitrotetrazolium violet (INT) have served as useful indicators of reduced microenvironments surrounding the filamentous cyanobacteria *Anabaena* and *Aphanizomenon* (Paerl & Kellar, 1978). Bacteria present in cyanobacterial mucus sheaths are capable of reducing these tetrazolium salts to coloured formazan crystals. Reduced formazan crystals can be detected by light microscopy in fresh preparations of cyanobacterial/bacterial associations. Since each tetrazolium salt is converted to its respective formazan crystal at a specific redox potential, the magnitude of reduced conditions in microzones can be estimated.

Using both TTC and NBT we (Paerl & Kellar, 1978; Paerl, 1978) have observed that bacteria present on *Anabaena* heterocysts often form a reduced microzone around this N_2 fixing cell, effectively providing protection from ambient oxygen supersaturation by minimizing the diffusion of oxygen into the heterocysts through the heterocyst–vegetative cell junction. The presence of such microzones has been linked to optimal N_2 fixation potentials in *Anabaena*. Likewise, bacterial epiphytes located in *Anabaena* mucilage consume enough oxygen to form reduced microzones. Combined with the fact that CO_2 is evolved through respiration, the presence of bacterial microzones would appear to be of multiple benefit to the host cyanobacteria since CO_2 is replenished and O_2 consumed, both potentially enhancing photosynthesis and N_2 fixation.

Clearly, microzones are dynamic regions which promote gaseous and nutrient exchange processes, and can lead to nutrient transformation steps which would be unfavourable under ambient environmental conditions. A great deal is still to be learned about such microbial habitats, and a host of useful research findings can be expected on this subject in the near future.

17.3.3 Presence of cyanobacteria in microzones

It is noteworthy that numerous cyanobacterial genera often function at optimal rates when associated with surfaces and microzones located on or in host organisms. The N_2 fixing associations between *Anabaena* and *Azolla* and *Nostoc* and *Gunnera* are prime examples. Recently, diverse active N_2-fixing cyanobacteria have also been found associated with surfaces of marsh grasses, seagrasses and seaweeds (Goldner, 1980) as well as higher aquatic (Moeller & Roskoski, 1978) and terrestrial plants. Bacteria are virtually always found closely situated to cyanobacteria in such associations.

17.4 Ecological considerations

17.4.1 Dominant characteristics of naturally occurring cyanobacterial/bacterial associations

The extensive surveys of Bershova *et al.* (1968) of bacterial epiphytes on natural populations of cyanobacteria yielded several species of associated heterotrophic bacteria. *Pseudomonas* and *Caulobacter* proved to be dominant genera. Subsequent studies have generally indicated that diverse bacteria are virtually always present in slime layers and mucilage of cyanobacteria (Kuentzel, 1969; Kessel & Eloff, 1975; Overbeck, 1975; Caldwell & Caldwell, 1978). The types, frequencies and extent of attachment can vary profoundly in nature, but there are some generalizations which can be drawn from these studies. (1) In aquatic systems Gram-negative bacteria appear to predominate as cyanobacterial epiphytes, while terrestrial environments support a large proportion of Gram-positive epiphytic bacteria. (2) Bacterial epiphytes are present during optimal as well as suboptimal growth periods of the host cyanophytes, but epiphytic community diversity and biomass can often contrast sharply between these periods. (3) Bacterial growth on cyanobacteria is often most profound in regions of high primary productivity or N_2 fixation (i.e. the epilimnia of lakes and the near-surface algal layers in soils). (4) The presence of bacterial epiphytes seldom signals the decline of a bloom. In fact, epiphytes are often most abundant during periods of optimal metabolic activities of the host. (5) Cyanobacterial/bacterial associations are both highly specific (in terms of particular attachment sites, chemotactic substrates, antagonism) or general, often involving entire cyanobacterial filaments, colonies and bacterial colonies or clumps.

17.4.2 Antagonism in cyanobacterial/bacterial associations

The importance of antagonism in cyanobacterial/bacterial associations has not been resolved, and much conflicting data can be presented to substantiate this belief. Both the studies of Safferman and Morris (1964) as well as Bershova *et al.* (1968) yielded mixed results. Safferman and Morris examined actinomycete filtrate antagonism on cyanobacteria growth and found that 24.6% of 403 actinomycete strains exerted a negative effect. Bershova *et al.* found that many natural bacterial strains (2166 in all!) taken from lake waters containing cyanobacteria had antagonistic capabilities. In the same study, however, it could be shown that bacteria isolated from cyanobacterial mucilage had few antagonistic tendencies. Daft *et al.* (1975) and Stewart and Daft (1976) have demonstrated the abilities of bacterial strains to lyse cyanobacteria, but at rather high densities of lytic bacteria (10^6 cells ml^{-1}). Fallon and Brock (1979a,b) found lytic bacteria to be present in Lake Mendota at concentrations of 10^3 cells ml^{-1}, leading them to conclude that autolysis as opposed to bacterial lysis may play a more important role in the decline of the lake's *Microcystis* and *Anabaena* blooms. Furthermore, evidence for an extracellular lytic substance has not been found in the above experiments and in earlier experiments with either *Myxobacterium* (Shilo,

1970) or bacterial isolates (Daft & Stewart, 1971). Cyanobactericidal activity has also been recorded among associated *Bacillus* (Reim *et al.*, 1974) and *Bdellovibrio* (Burnham *et al.*, 1976; Wilkinson, 1979) species.

A complication in interpreting cyanobacterial/bacterial antagonism experiments stems from the difficulty in separating direct bacterial attack (lysis) from bacterial competition for nutrients and growth factors as the cause of cyanobacterial death. In both natural waters (Paerl & Lean, 1976; Faust & Correll, 1976) and culture conditions (Rhee, 1972) bacteria have been shown to compete effectively with phytoplankton (including cyanobacteria) for orthophosphate, a nutrient which can often limit microbial growth when depleted. The effectiveness of bacteria to outcompete cyanobacteria for orthophosphate depends on a variety of factors, including concentration, relative abundance to other nutrients and the uptake and assimilative enzyme kinetics of each set of microorganisms. Physical factors, such as light availability and quality, temperature and turbulence also play an important role in determining advantages for either bacteria or cyanobacteria in a competitive situation. In addition to competition for phosphorus, parallel competition for nitrogenous compounds, iron and trace metals is likely to occur in natural ecosystems. Studies have revealed the presence of algal chelators (Price, 1968; Barber & Ryther, 1969; Simpson & Neilands, 1976). Recently Murphy *et al.* (1976) have demonstrated the presence of a powerful Fe-hydroxymate chelating agent in the filtrate of *Anabaena flos-aquae* cultures. Such chelators offered a competitive advantage to *Anabaena* when grown in the presence of the eukaryotic alga *Scenedesmus*. Since it is known that hydroxymate chelators are also excreted by a wide range of heterotrophic bacteria, similar competition between cyanobacteria and bacteria would be expected (see 18.5.2.3).

The excretion of toxic substances by either cyanobacteria or associated bacteria may play an important role in shaping the types and degrees of associations in nature. Decreases in bacterial numbers have been reported in response to phytoplankton (including cyanobacterial) blooms (Chróst, 1975). The suppression of bacterial growth by algae appears related to Gram specificity. Gram-positive bacteria appear more sensitive to growth inhibitors produced by algae (Jørgensen, 1956; Matusiak *et al.*, 1971). Chróst (1975) showed that Gram-positive bacteria were effectively eliminated by cyanobacterial blooms in a eutrophic Mazurian lake in Poland. The reasons for higher sensitivity in Gram-positive bacteria remain unclear. However, caution must be practised in interpreting bacterial population decreases in response to cyanobacterial blooms. For example, cyanobacterial excretion products may promote growth and competitive advantages of specific bacterial strains which appear subdominant during non-cyanobacterial growth periods. Hence, an induced change in bacterial dominance as opposed to selective toxicity may accompany cyanobacterial blooms. Furthermore, some cyanobacterial excretory products may inhibit bacterial growth not by direct toxicity, but rather through the chelation or precipitation of bacterial growth factors and nutrients.

17.4.3 Taxonomic and morphological considerations

Only sketchy taxonomic information exists regarding the bacterial epiphytes of cyanobacteria. Gram-negative bacteria often isolated from aquatic cyanobacteria include *Achromobacter*, *Aerobacter*, *Flavobacterium*, *Pseudomonas* and *Vibrio*. Recently, Caldwell and Caldwell (1978) isolated and described a *Zoogloea* sp. associated with *Anabaena flos-aquae*. The *Zoogloea* was attached to the heterocysts and embedded in the mucilage of *Anabaena* and the mucilage of *Microcystis aeruginosa*. By the use of electron microscopy a density of 7.4×10^5 cells ml^{-1} was found in the planktonic environment, while 2.6×10^{11} cells ml^{-1} were present in the mucilage microenvironment. Kessel and Eloff (1975) obtained similar ratios of planktonic versus epiphytic bacteria in their studies of *Microcystis marginata* colonization and Paerl and Kellar (1978) found 1000–5000 fold higher abundance of a motile rod-shaped Gram-negative bacterium attached to the heterocysts of both *Anabaena oscillarioides* and *Aphanizomenon flos-aquae* than in the water columns of eutrophic lakes supporting these cyanobacteria.

Both Caldwell and Caldwell (1978) and Paerl and Kellar (1978) attribute similar morphological characteristics to epiphytic bacteria. They found that the bacteria appeared to show an obligatory requirement for the mucilaginous microenvironments surrounding cyanobacteria.

When not attached bacteria were planktonic swimmers, showing polar flagellation as they encircled cyanobacterial hosts. Once attached to cyanobacteria the bacterial epiphytes anchor themselves firmly in either mucilage layers or on smooth surfaces. In both Caldwell and Caldwell's (1978) and Paerl and Kellar's (1978) studies the bacteria lost motility once attached. The former study revealed bacteria generally associated with mucilage of the cyanobacterial host while in the latter study bacteria showed a strong preference for the heterocysts of *Anabaena* and *Aphanizomenon*. Once motility was lost the sessile bacteria remained firmly attached, exhibited dividing stages and grew actively, often leaving short filaments of rapidly dividing cells behind.

17.4.4 Exchange processes and beneficial interactions

There are numerous lines of evidence linking nutrient and growth factor exchange to cyanobacterial/bacterial associations. The ubiquity of nutritional exchange via metabolites in marine organisms was initially brought to light by Lucas (1951). Since then voluminous work has solidified the belief that diverse sets of micro- and macroorganisms are normally engaged in metabolite exchange processes. Phycological studies have revealed that numerous marine and freshwater algae have external vitamin requirements (Provasoli & Pintner, 1953), the vitamins being supplied by associated microbes (Robbins *et al.*, 1951). Burkholder (1963) specifically showed that cyanobacteria isolated from marine muds were capable of producing and excreting B vitamins which were utilized by vitamin requiring bacteria.

Antibiotics have been found in algal (including cyanobacterial) excretory products (Pratt *et al.*, 1951; Burkholder, 1963; Sieburth, 1959, 1968). Algae are also known to release bacterial nutrients (Fogg & Westlake, 1955; Lewin, 1956; Hellebust, 1965; Nalewajko, 1978), although the magnitude of release processes in nature is a disputed topic (Sharp, 1977; Harris, 1978). Walsby (1974a,b) and Chróst and Brzeska (1978) found a variety of carbohydrate and nitrogenous organic compounds to be characteristic excretion products of *Anabaena*. The most common excretion products were polysaccharides and polypeptides. Among the polysaccharides, galactose and glucose were dominant sugars. Among excreted polypeptides and shorter peptides (including individual amino acids) serine was shown to be an important constituent.

Bacteria attracted by *Anabaena* showed strong chemotactic responses to many amino acids and sugars (Paerl, 1978). In particular, the amino acid serine showed consistent tendencies to attract these bacteria over a wide range of concentrations (Table 17.1). Autoradiographic analyses of *Anabaena* and associated microorganisms revealed the ability of attached bacteria to incorporate a range of amino acids and sugars in lakewater samples. The fact that attached bacteria readily show chemotaxis indicates that their establishment in mucilaginous and slime microhabitats is promoted by attractive substances which are utilized for growth. Caldwell and Caldwell (1978) determined that mucilaginous bacteria associated with *Anabaena* utilized a low-molecular-weight (1300) component of the mucilage in order to produce their own mucilage while attached. Paerl (1979) showed that some ^{14}C originally fixed as $^{14}CO_2$ by *Anabaena* photosynthesis in nature was eventually utilized by epiphytic bacteria. In this study it was also shown that excretion products streaming from heterocyst–vegetative cell junctions were rapidly colonized and utilized by epiphytic bacteria. Chróst and Brzeska (1978) have also accumulated evidence that extracellular products of cultured *Anabaena* are directly utilized by bacterial microflora associated with this cyanobacterium. In a set of microecosystem experiments, Herbst and Overbeck (1978) demonstrated that the extracellular production of photosynthate by axenic cultures of *Oscillatoria redekei* was linked to the uptake of this photosynthate by introduced bacterial strains. Their results revealed that the

Table 17.1. Chemotactic response of Lake Okaro (New Zealand) bacteria to organic substrates. Extent of attraction judged by relative numbers of motile bacteria in capillaries after 30 min exposure.

Substrate	Concentration (M)	Attraction
Amino acids		
L-alanine	10^{-5}, 10^{-3}	−
L-cysteine (free base)	10^{-3}	−
L-glutamine	10^{-5}, 10^{-3}	−
L-glycine	10^{-6}	−
	10^{-5}	+
L-leucine	10^{-5}	−
L-lysine	10^{-5}	−
	10^{-3}	+
L-proline	10^{-5}	−
L-serine	10^{-6}	+
	10^{-5}	++
	10^{-3}	+
L-threonine	10^{-5}	++
	10^{-3}	+
L-valine	10^{-5}	−
	10^{-3}	+
Sugars		
D-arabinose	10^{-5}, 10^{-3}	−
D-galactose	10^{-5}, 10^{-3}	−
D-glucose	10^{-5}	+
	10^{-3}	++
Organic acids		
Sodium acetate	10^{-5}, 10^{-3}	−
Sodium citrate	10^{-5}, 10^{-3}	−

+ = moderate; ++ = strong; − = none.

rate of cyanobacterial photosynthate production was much lower than the bacterial uptake rate, emphasizing close metabolic coupling, and hence nutrient exchange, between cyanobacteria and associated heterotrophic bacteria.

Evidence that excretion products from non-cyanobacterial taxa are similarly utilized by epiphytic bacteria comes from comparisons of axenic versus bacterial *Chlorella* cultures. Nalewajko *et al.* (1976) compared extracellular release of ^{14}C labelled photosynthate between these cultures and found that algal–bacterial cultures released less ^{14}C label than axenic cultures. Lower release rates were presumably due to incorporation of photosynthate by bacterial epiphytes.

These results all point to a similar conclusion. The sites of microbial attachment support actively metabolizing cells which are directly utilizing host excretion products. Judging by the magnitude of differences in net excretion between axenic and algal-bacterial conditions (Nalewajko *et al.*, 1976), it is concluded that bacterial attachment sites are dynamic regions, where the turnover of organic and inorganic compounds is of quantitative significance. Often associations do not appear detrimental to hosts or epiphytes. In fact, associations are often most profound during optimal growth periods of the host. It is strongly suspected that numerous associations carry on exchange of nutrients, gases, reducing equivalents, and growth factors which can serve a mutual benefit to each organism. Even though the evidence exists for mutualistic and symbiotic associations, there is also equally strong evidence that some associations are clearly antagonistic. Furthermore, competition for similar nutrients, such as phosphate, would also appear antagonistic. Hence, cyanobacterial/bacterial associations appear mixed in terms of benefit/harm. It appears that the growth state of microbial species in such associations is a key factor in determining the type of interaction which will ensue.

17.5 Chemotaxis in cyanobacterial/bacterial associations

17.5.1 Characteristics of chemotaxis

Chemotaxis is the ability of an organism to move toward or away from molecules it can sense. As early as the 1880s, Pfeffer (1884) discovered that bacteria possessed chemotactic abilities. His experimental design was simple, yet the basic procedures are still practised today. If a thin capillary tube containing an attractive substance is placed in a bacterial suspension, soil suspension or water sample, motile bacteria showing positive chemotaxis will accumulate in the capillary. The degree of attraction can be determined by quantifying the number of bacteria entering the capillary. Low-molecular-weight compounds are often excellent substrates for eliciting chemotactic responses. Examples of the wide range of organic molecules capable of attracting bacteria through chemotaxis are given in the reviews of Adler (1975) and Berg (1975). Adler (1969) found that specific chemical sensors were present in bacteria. Sensors for amino acids (Mesibov & Adler, 1972), sugars (Adler *et al.*, 1973), as well as repelling agents (Tsang, Macnab & Koshland, 1973; Tso & Adler, 1974), have been described. Sensor sites for a variety of sugars, including D-galactose, D-ribose and maltose, were found to be present in the periplasmic cavities (Hazelbauer & Adler, 1971). Membrane-bound binding proteins are also induced in recognizing attractive substances (Hazelbauer & Adler, 1971). Most bacteria can be attracted by a variety of substrates, each having optimal concentrations capable of attracting. The study of amino-acid chemotactic responses in *Bacillus subtilis* by van der Drift and de Jong (1974) is an excellent example.

17.5.2 Functional chemotaxis in cyanobacterial/bacterial associations

In natural ecosystems, where organic, inorganic compounds and dissolved gasses can often be very dilute (compared to culture conditions), chemotactic behaviour would seem useful for a variety of motile microorganisms. Paerl and Kellar (1978) and Paerl (1978) hypothesized that chemotaxis might be operative in cyanobacterial/bacterial associations, especially those showing highly specific spatial associations (i.e. specific attachment of bacteria to the heterocysts of *Anabaena* and *Aphanizomenon*). Accordingly, a variety of sugars, amino acids and organic acids were tested for chemotactic potentials in natural lakewater samples, using the capillary technique as described by Adler (1976). These studies revealed a host of attractive substances, operative at specific concentrations, which attracted bacteria similar in morphology and motility to those associated with dominant cyanobacteria (Table 17.1). Bacterial 'sensing' of *Anabaena* cells was observed microscopically. Aquatic bacteria initially show a general attraction to cyanobacterial filaments. As bacteria approach the filaments, motility, in terms of flagellar movement, shows a marked increase over planktonic conditions. Bacteria will often bump into several *Anabaena* cells, scanning the filament for particularly attractive cells. Once intimate contact (remaining in contact for more than 30 s) has been made, bacteria will often show a dramatic increase in flagellar movement (while still attached), imparting a rapid rotational movement, also observed by Adler (1976) and other investigators. Such violent rotation may continue for approximately 5–10 min after attachment. Then, attached bacteria often show a complete loss of motility, dividing and colonizing the host cyanobacteria while in a sessile state. On numerous occasions bacteria initially made contact with vegetative cells of *Anabaena*, but abandoned these cells for heterocysts. Heterocyst–vegetative cell junctions were particularly attractive, often drawing numerous bacterial cells simultaneously. Bacteria were observed wedging themselves into the junctions, maintaining flagellar rotation. Following 20–30 min of such activity, motility ceased. The same bacteria could be seen in various dividing stages a few hours later, while still located in the junction (Fig. 17.5).

As mentioned elsewhere, the heterocyst–vegetative cell junctions often excrete copious amounts of mucilage, which show positive staining with safranin and ruthenium red, indicating the presence of carbohydrates. Fluorescamine assays also indicate the presence of peptides and free amino acids. Both groups of compounds showed strong chemotactic properties when

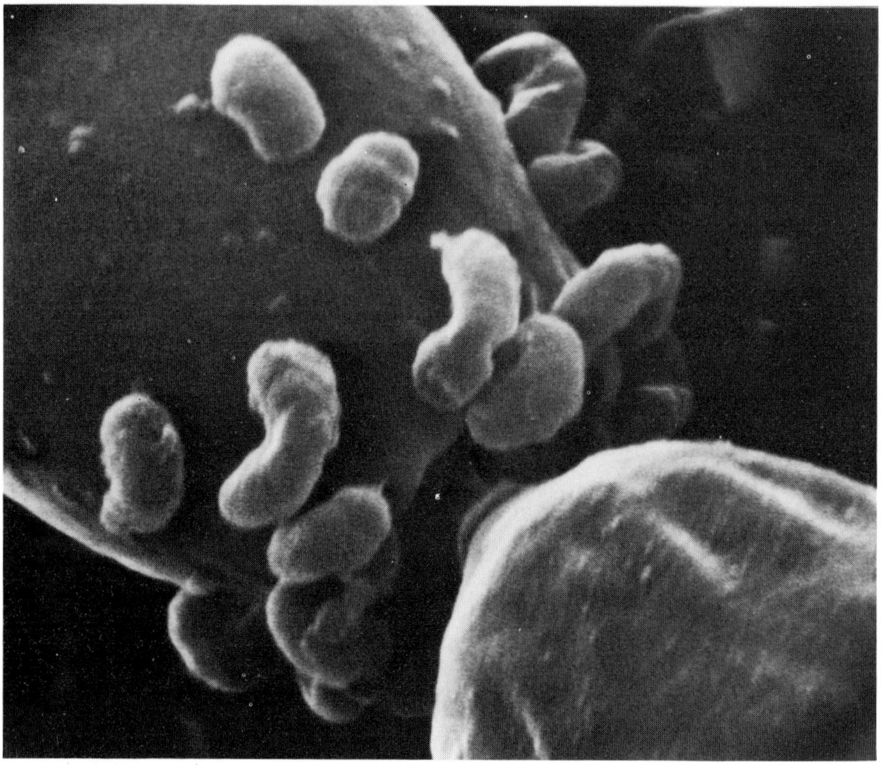

Fig. 17.5. An SEM view of bacteria attached to the heterocyst of *Anabaena oscillarioides* near the heterocyst–vegetative cell junction. Due to some cellular shrinkage resulting from critical point drying, the cellular junction is more exposed than in nature. Accordingly, even though bacteria show a definite preference for heterocysts, they are normally more tightly wedged in the heterocyst–vegetative cell junction.

combined with bacterial isolates from *Anabaena* cultures. In particular, the amino-acid serine showed profound chemotactic characteristics over a 1000-fold concentration range (Paerl & Kellar, 1978) (Table 17.1). Serine is of particular interest since it has been shown to be an abundant amino acid in *Anabaena* filtrate (Walsby, 1974a) and has also been shown to be a product of glycollate metabolism (photorespiration) in a range of prokaryotic and eukaryotic photosynthetic microorganisms (Stewart, 1974a). It would be highly advantageous for epiphytic bacteria to sense and utilize this excretory product during conditions promoting cyanobacterial photorespiration (high O_2 and light levels and low CO_2 levels), which are present during surface bloom periods in aquatic ecosystems.

17.6 Evolutionary significance of cyanobacterial/bacterial associations

Present-day mutualistic cyanobacterial/bacterial associations appear due to adaptational measures in response to long-term changes in biospheric chemical and physical conditions. Most ecological/physiological explanations for the associations emphasize short-term nutrient and gas exchange requirements (Lange, 1967; Kuentzel 1969; Paerl & Kellar, 1978) during daily or weekly periods of CO_2 and nutrient depletion as well as O_2 supersaturation. Presumably such events also played an important long-term role during the evolution and proliferation of cyanobacteria in the biosphere. Prior to the advent of cyanobacteria, autotrophic generation of organic carbon compounds was most likely restricted to the anaerobic (or at least microaerophilic) photosynthetic and chemosynthetic bacteria. These bacteria were restricted to anaerobic environments, such as the hypolimnia of lakes and subsurface soil layers. The advent of O_2-evolving cyanobacterial photosynthesis brought a new dimension to microbial habitats, namely, a need to exclude O_2 from O_2-sensitive metabolism. It is well known that O_2 can be inhibitory to both cyanobacterial CO_2 and N_2 fixation (Warburg, 1929; Stewart & Pearson, 1970; Stewart, 1974a). Endogenous O_2-consuming reactions, including substrate and photorespiration, superoxide dismutase activity (Gregory and Fridovich, 1969) and an oxyhydrogen 'Knallgas' reaction in which H_2 and O_2 are eventually combined to form H_2O (Bothe *et al.*, 1978) may not suffice at times in maintaining low internal O_2 concentrations during high rates of photosynthetic O_2 evolution. Some cyanobacteria have minimized the toxic effects of O_2 evolution by confining a bulk of their N_2-fixing activities to heterocysts, which have photosynthetic energy gathering capabilities, but are missing photosystem II-mediated O_2 evolution (Stewart, 1977). Overcoming toxic side effects of O_2 evolution may have been a serious obstacle to the establishment of N_2-fixing cyanobacteria in the face of ever-increasing biospheric O_2 content. The presence of O_2 consuming heterotrophic bacteria could have played an important role in assuring the early success of O_2 evolving cyanobacteria. The presence of attached aerobic heterotrophic bacteria represents a means of consuming O_2 in the vicinity of cyanobacterial cells. The use of tetrazolium salts as indicators of reduced microzones bordering cyanobacteria substantiates this belief. For example, when bacteria concentrate in the mucilage or on the surfaces of cyanobacterial filaments the O_2 consumption associated with such colonization is high enough to form reduced microzones. Such cyanobacterial/bacterial associations appear to be of functional significance, capable of enhancing the N_2 and CO_2 fixation rates of host cyanobacteria in O_2-rich environments (Paerl & Kellar, 1978).

Lange (1967) also suggested another functional aspect of such associations in nature which has evolutionary relevance. He focussed on the potential problem of CO_2 availability during periods of high photosynthetic CO_2 demand. Laboratory studies on *Microcystis* and *Anabaena* revealed that CO_2 availability could at times be low enough to restrict photosynthesis. The restriction could be alleviated by CO_2 enrichment or a lowering of the pH. Oxidative breakdown of organics by heterotrophic bacteria leads to increased CO_2 availability as well as lower pH values. Accordingly, Lange suggested that bacteria associated with cyanobacteria serve to replenish CO_2, thereby enhancing cyanobacterial photosynthesis.

Following the establishment of cyanobacteria both O_2 and CO_2 levels have changed significantly in the biosphere. Oxygen levels steadily increased until the present equilibrium between photosynthesis and respiration was established. Carbon dioxide levels have decreased following the rapidly rising demands after the establishment of oxygenic photosynthesis. The establishment of cyanobacterial/bacterial associations during various stages of photosynthetic evolution appears to have been an important step in buffering cyanobacteria against environmental constraints, brought about by localized CO_2 depletion and O_2 supersaturation.

It could be asked at this point whether bacteria were actively recruited during the evolution of cyanobacteria or whether associations were simply fortuitous events, eventually serving to benefit both sets of microorganisms. It is clear that the appreciable amounts of extracellular organic matter (mucilage, slime, and associated low-molecular-weight amino acids, sugars and lipids) often characteristic of cyanobacteria (Walsby, 1974b; Fogg, 1971) serve as an attractive source of nutrition and as a physical substrate for attached bacteria. It has recently been shown that products of cyanobacterial CO_2 fixation can be transferred to epiphytic bacteria (Paerl, 1978). In addition, bacteria often concentrate in regions where extracellular organic matter accumulates (Paerl & Kellar, 1978; Caldwell & Caldwell, 1978). One place which has been illustrated is the junction of the heterocysts and the vegetative cells in filamentous genera (Paerl, 1976). Junctions are often the sites of filament breakage and such breakage can lead to cessation of metabolic activities in heterocysts. The presence of bacteria in this junction region, as well as on the broken ends of heterocysts has been shown to promote maximum N_2 fixation in *Anabaena* (Paerl & Kellar, 1978), and serves to prolong metabolic activities in heterocysts. It thus appears that extracellular production of organic matter assures and promotes bacterial associations, which in turn optimize metabolic activities of host cyanobacteria. Since the excretion of organic matter has been shown to benefit cyanobacteria in a variety of manners (some of the organics are powerful nutrient chelators, offer resistance to viral attacks and aid in gliding movement through slime formation), it is difficult to conclude that the excretion of organic matter was solely promoted to attract associated bacteria. Most likely, excretion of organics has a multiple function, and hence the attractive properties to bacteria are likely to be a fortuitous event, which did not prove deleterious during cyanobacterial growth and proliferation, but proved helpful during periods of environmental stress and fluctuations.

It is especially fortuitous that bacteria are attracted to the pore regions of heterocysts, which are the weakest links in filaments and prove susceptable to O_2 inactivation in the event of breakage. It appears that the 'leakiness' and production of copious amounts of extracellular material serves to benefit host cyanobacteria by attracting growth promoting bacteria. During the evolution of O_2 evolving photosynthesis accompanied by decreasing CO_2 supplies it may have been more helpful to excrete photosynthate as a means of direct physical protection and for the attraction of beneficial microorganisms, than to diversify biochemically, by developing novel O_2 scavenging reactions and CO_2 recycling pathways. In other words, the benefits of being 'leaky' in natural ecosystems may outweigh the energy losses incurred in producing extracellular photosynthate.

Acknowledgments

I am grateful to the staff and coworkers of the Ecology Division, Department of Scientific and Industrial Research, New Zealand; Canada Centre for Inland Waters, Burlington, Ontario; Institute of Marine Sciences and Marine Sciences Curriculum, University of North Carolina; Limnologisches Institut, Universität Konstanz. Particular thanks go to Dr A.M. Massalski for joint efforts during electron-microscopic studies.

CHAPTER 18
FRESHWATER PLANKTON

C. E. GIBSON & R. V. SMITH

18.1 Introduction

The enrichment of lakes by man has caused an increase in the prominence of cyanobacteria in the freshwater plankton which suggests that in spite of their presence in the early ages of Earth's history, they are well adapted to modern conditions. Planktonic cyanobacteria become abundant in ponds which have been deliberately fertilized to increase fish production and in waters inadvertently enriched by the inflow of sewage or agricultural effluent. The increase in planktonic cyanobacterial populations has sometimes been dramatic and has brought with it considerable practical problems. Dense populations cause great difficulties to the water-supply industry, interfering with treatment processes and imparting tastes and odours to the water. The production of toxins or the removal of oxygen by the respiration or decomposition of dead cells can have serious consequences, causing death amongst fish, birds and occasionally cattle. Bathing in waters densely populated with *Aphanizomenon* can cause an unpleasant condition of the skin known as 'swimmers' itch' (Schwimmer & Schwimmer, 1964). Even without these unpleasant consequences, when the hitherto clear waters of lakes become clouded with plankton their aesthetic appeal is felt to diminish and the occurrence of surface scum (water-blooms) is still more objectionable. There is a voluminous literature on eutrophication and the practical consequences of abundant planktonic cyanobacteria which will not be reviewed here, but useful entry points to it are Rohlich (1969) and Likens (1972).

The intention of this chapter is to present a brief review of the planktonic cyanobacteria, placing particular emphasis on more recent work. Older assertions and hypotheses are examined to see how far they can be supported by critical studies and whether modern physiological data afford any explanation of the distribution and abundance of these organisms. An attempt has been made to avoid covering the ground of earlier reviews (Lund, 1965; Hutchinson, 1967; Fogg, 1969; Whitton, 1973a; Reynolds & Walsby, 1975).

18.2 Taxonomy

18.2.1 Problems of variation

There is no general agreement about the taxonomy of cyanobacteria and they have been treated traditionally as a plant group with taxonomy based on morphological features visible under the light microscope. Because of the lack of good taxonomic characters, identification rests on a small number of relatively detailed differences and it is not clear whether these differences are meaningful. It is still less clear whether the groups erected on this basis are helpful in an ecological context. The morphological variation possible within one isolate is illustrated by the observations of Whitton and Peat (1969) on *Oscillatoria redekei.* Apart from *O. redekei* closely resembling the original description, a range of other vacuolate and non-vacuolate forms was also present. Similarly, in Lough Neagh (N. Ireland), a large population of vacuolate *O. redekei* was replaced relatively abruptly by a non-vacuolate form resembling *O. limnetica* yet behaving in the lake in every way like *O. redekei*, even to the extent of producing characteristic branched filaments (unpubl. observations). Foy (1980) compared the growth rates of a range of planktonic cyanobacteria under identical conditions and found that whilst the '*O. limnetica*' and *O. redekei* isolates from Lough Neagh grew at similar rates, a strain designated '*O. limnetica*' from (Lake) Windermere grew very much faster. It would therefore have been quite misleading to equate the two '*O. limnetica*' strains ecologically. Similarly, *O. agardhii* var. *isothrix* resembles *O. rubescens*, both morphologically and in its ecological responses, more closely than it does type *O. agardhii*.

18.2.2 Taxonomic range

Notwithstanding the limitations of the existing nomenclature, it is possible to draw up a list of genera represented in the freshwater plankton (Table 18.1);

Table 18.1. Principal genera of cyanobacteria in the freshwater plankton

Synechococcus	*Raphidiopsis*
Aphanothece	*Aphanizomenon*
Rhabdoderma	*Anabaenopsis*
Chroococcus	*Anabaena*
Cyanodictyon	*Nostoc*
Merismopedia	*Romeria*
Microcystis	*Pseudanabaena*
Coelosphaerium	*Oscillatoria*
Marssoniella	*Spirulina*
Gomphosphaeria	*Lyngbya*
Gloeotrichia	*Phormidium*

these twenty-two genera include about ninety species. The types represented belong to Chroococcales and Nostocales and the absence of representatives from the Chamaesiphonales and Dermocarpales is noteworthy.

18.3 Morphology and structure

18.3.1 Morphological range

There is a wide range of form in the plankton from minute simple cells (*Synechococcus* sp.: Bailey-Watts *et al.*, 1968) to simple filaments such as *Oscillatoria* spp. to relatively large rafts (*Aphanizomenon*), coils (*Anabaena*), spherical aggregates (*Coelospharerium*, *Gomphosphaeria*) and indefinite masses (*Microcystis*). With the exception of *Merismopedia*, all planktonic forms divide in one plane, usually across the long axis but sometimes (*Gomphosphaeria*) parallel to it. Apart from *Oscillatoria redekei* which, as mentioned above, may form short branches, there are no branched cyanobacteria in the plankton.

18.3.2 Variations in morphology

The trichomes of species of *Aphanizomenon* and *Gloeotrichia* commonly end in tapering colourless cells ('hairs') which appear dead but may have some active function. Sinclair and Whitton (1977a) found that in *G. echinulata* hairs were produced under phosphorus deficiency and in eutrophic conditions (Gibson *et al.*, 1971) found that strains of *Aphanizomenon* lacking hairs may occur. Members of the Nostocaceae and *Gloeotrichia* produce heterocysts and akinetes, at least under certain conditions. Planktonic *Oscillatoria* spp., unlike their terrestrial and subaerial counterparts, do not in general produce 'necridia' or separation discs and the mechanism of filament replication is not clear. Although the mechanism of replication probably differs between 'Pseudanabaena-types' such as *O. redekei* and 'Oscillatoria-types' such as *O. agardhii* (Stanier & Cohen-Bazire, 1977), in nature, both species show a regular variation in filament length and filaments become shorter as the nutrients are depleted (Gibson, 1975a, Gibson & Stevens, 1979). Most, but not all, planktonic cyanobacteria contain gas vacuoles. These are constructed of cylindrical subunits or vesicles (Walsby, 1978b) which are freely permeable to gases and contain gas of the same composition as that dissolved in the surrounding medium. Both their number and their disposition alters in response to growth rate, to turgor pressure within the cell and to light falling on the cell. The gas vesicles may be situated either throughout the cell or at both ends and they impart a positive buoyancy to the cells; a change in the number of vesicles causes a change in flotation rate (see 18.4.3). Filamentous strains may be motile (*e.g. Oscillatoria agardhii* var. *isothrix*) or non-motile. Under calm conditions *O. agardhii* var. *isothrix* may form a surface scum which then clumps together by the active migration of filaments. All planktonic cyanobacteria are invested with mucilage to a greater or lesser extent. In *Oscillatoria* spp., the mucilage is scant but large quantites of mucilage may be produced by colonial forms. In the mucilage of *Microcystis* and *Aphanizomenon* (Sirenko, 1972; Sirenko *et al.*, 1968) sulphydryl groups have been detected and these may have a reducing function.

18.4 Distribution

18.4.1 Geographical and limnological distribution

Cyanobacteria are widely distributed in the freshwater plankton and occur in lakes at almost every latitude. There is an extensive literature on their occurrence, which will not be reviewed here. Certain genera are widely distributed, such as *Microcystis* and *Anabaena*. *Oscillatoria* species seem more characteristic of temperate zones, whilst *Anabaenopsis* and *Spirulina* occur more frequently at lower latitudes. Similarly, at a given latitude some species show marked habitat preferences. Shallow unstratified lakes commonly contain 'narrow' *Oscillatoria* species such as *O. redekei*, *O. limnetica* and a narrow form of *O. agardhii*, as well as small *Pseudanabaena* spp. and single filaments of *Aphanizomenon flos-aquae* fa *gracile*. *Oscillatoria redekei* was early regarded as being associated with reducing conditions and the production of hydrogen sulphide (Bursche, 1955); this is particularly interesting in view of the discovery of anoxygenic photosynthesis in *O. limnetica* (Chapter 9). While the presence of hydrogen sulphide may apply to the Schleswig-Holstein lakes (Meffert & Krambeck 1977), many lakes where *O. redekei* is abundant are well oxygenated throughout the growing season (Whitton, 1969; Gibson *et al.*, 1971; Bailey-Watts, 1972). Lakes developing a stable temperature stratification exhibit a range of larger forms, which often develop a maximum at the metalimnion. *Oscillatoria agardhii* var. *isothrix* (Baker *et al.*, 1969; Klemer, 1976) and *O. rubescens* (Staub, 1961; Zimmermann, 1969) are most commonly associated with this growth strategy, but subsurface maxima of *Merismopedia trolleri*, *Oscillatoria redekei* (Baker & Brook, 1971), *Anabaena flos-aquae* (Utterback *et al.*, 1942) and *Aphanizomenon flos-aquae* (Eberly, 1959) have also been reported. A list of documented subsurface maxima is given in Reynolds and Walsby (1975). It is not clear whether metalimnetic maxima represent a true growth phase or a survival phase during stratification analogous to overwintering populations such as those formed on the mud by *Microcystis*. It is generally true that planktonic cyanobacteria are more commonly encountered in eutrophic lakes, but they are by no means confined to them. Significant populations may occur in nutrient-poor lakes in both high latitudes (Vallentyne, 1971) and the subtropics (Hart & Hart, 1978). The highly alkaline lakes of East Africa produce massive crops of planktonic cyanobacteria (Lind, 1968; Talling *et al.*,

1973), chiefly *Spirulina platensis* and *Microcystis*. However, similar crops may develop in much less alkaline waters such as Lake Chad and Lake George (Burgis *et al*., 1973), so the high alkalinity of the lakes may be incidental. Instances of planktonic cyanobacteria populations in highly saline or alkaline water have been reported from, among others, Hungary (Kol, 1929), India (Subbaramaiah, 1972) and Turkmeniya (Kogan *et al*., 1972). Resistance of up to 4% sodium chloride solution is possible (Schiewer, 1974) even with laboratory strains from non-saline environments, although this tolerance can be abolished by DDT (Batterton *et al*., 1972) which suggests that an active ion pump is involved (10.3.3).

Cyanobacteria may also occur in association with other planktonic organisms. *Cyanodictyon* spp. have been observed living in the mucilage of *Anabaena* spp. (Geitler, 1932), *Phormidium mucicola* is found in the mucilage of many colonial planktonic cyanobacteria and rotifers and *Lyngbya thornensis* was described growing on the antennae of *Daphnia pulex* (Bunting & Lund, 1956).

18.4.2 Temporal distribution

The earliest studies, at the beginning of this century, discerned a seasonal pattern in the growth of phytoplankton in temperate lakes. Pearsall's (1932) pioneering work established that in the English Lake District the succession began with diatoms, which gave way to green algae which in their turn were succeeded by cyanobacteria. This pattern proved to be typical of many other lakes (Lund, 1965; Hutchinson, 1967) but it is by no means universally true. There are many documented instances of lakes with an almost perpetual crop of cyanobacteria which, whilst it may wax and wain during the seasons, nevertheless forms a large proportion of the plankton throughout the year. Such behaviour is not surprising in tropical lakes where the difference in seasons may be less than diurnal variations. Perpetual crops of *Microcystis* are a common feature of Indian ponds (Ganapati, 1941; George, 1962) and *Microcystis* or *Spirulina* dominate the African lakes referred to above (Talling *et al*., 1973; Ganf, 1975). Temperate waters which have been enriched show an increasing tendency towards perpetual crops, especially in lakes where the winter flushing of water is insufficient to wash out a large percentage of the crop. Unlike tropical waters, however, these 'perpetual' crops are more typically composed of *Oscillatoria* spp. (Lund, 1965). It is also not universally true that blue-green algae form their maxima in the summer. Whilst they are usually preceded by diatom maxima, in Lough Neagh a peak of *O. redekei* occurs in the late spring. Nevertheless, the 'water-bloom' species which consist mainly of larger colonial forms usually produce their maximum crops in late summer and autumn. This led to the supposition, supported by the growth of other cyanobacteria in hot springs, that planktonic cyanobacteria were thermophilic and that their seasonal distribution was due to the increased water temperature. However, Foy *et al.* (1976) showed that for four species of planktonic cyanobacteria optimum temperatures were similar to other planktonic autotrophs. The fact that the larger colonial forms are characteristic of the late summer may be significant in two respects. Firstly, larger forms have a fast flotation rate which may be an advantage in stratified water columns and secondly, size affords some protection from grazing. Cyanobacteria in general are apparently not heavily grazed by zooplankton (Arnold, 1971; Bosselman, 1975) and this must confer some competitive advantage on them at times when there are large crops of herbivores. Since the rate of grazing increases with temperature, the occurrence of water-blooms when water temperatures are maximal may be partly attributable to such an indirect effect.

18.4.3 Lake enrichment

Whatever the reasons for it, the enrichment of lakes ('eutrophication') leads to a striking increase in the abundance of planktonic cyanobacteria. The species encouraged may be heterocystous, for example *Aphanizomenon* (Lund, 1978), *Anabaena* (Schindler & Fee, 1974) or non-heterocystous—such as the notorious *Oscillatoria rubescens* (Staub, 1961; Ravera & Vollenweider, 1968; Pelletier, 1968; Zimmermann, 1969; Edmondson, 1972; Jaag, 1972). Most frequently, the source of enrichment has been sewage effluent and diversion of the effluent has usually reversed the process (Edmondson, 1969). There was initially much speculation as to the mechanism causing the change, but it became clear from studies of experimentally enriched lakes (Schindler & Fee, 1974) or correlative investigations (Uhlmann & Albrecht, 1968; Dillon & Rigler, 1974) that phosphorus supply is the main factor limiting biomass in many lakes. It is not perhaps surprising that the addition of phosphate-phosphorus in clear-water lakes results in crops of nitrogen-fixing cyanobacteria, since the addition of phosphorus may make nitrogen the limiting element. What is more puzzling, however, is the increase in other types of cyanobacteria either accompanying nitrogen-fixing species or, in the case of many *Oscillatoria* species, when nitrogen is apparently not limiting.

18.4.4 Spatial distribution

The most spectacular manifestation of crops of planktonic cyanobacteria are the surface 'water-blooms' caused by the accumulation of buoyant cells at the water surface. This is such a noticeable phenomenon that it forms part of folklore and is recorded in historical sources, often being interpreted as presaging a catastrophe. Early references to water-blooms in the British Isles were collected by Griffiths (1939) and other references are discussed in Lund (1965) and Reynolds and Walsby (1975). In the Zürichsee, surface blooms of *Oscillatoria rubescens* led to the fanciful supposition that it was the blood of the Burgundian warriors, drowned in the lake in an earlier age.

Sustained investigations have been carried out on the small meres of the Cheshire Plain, England (Reynolds, 1971), where the yearly occurrence of surface scums is referred to as the 'breaking of the meres' from a resemblance to the crust formation of yeast in a brewer's vat. It has not always been understood that this was the result of redistribution rather than extremely rapid growth (*e.g.* Mackenthun *et al.*, 1968), but it is clear that blooms can form and disperse with great rapidity when calm weather is followed by a wind (Reynolds, 1975). The biomass of the cells in these blooms is sometimes exaggerated and what appear to be large crops may be rather modest but concentrated in a thin layer at the surface, subsequently perhaps skimmed to a lee shore by light winds. The impoundments of river water in the reservoirs of the Oder–Dnieper cascade led to serious water-blooms on a large scale, and much interesting information is to be found in Topachevskii (1968) and Sirenko (1972).

The occurrence of water-blooms is a product of the gas vesicles which occur in nearly all planktonic cyanobacteria. Cells are able to regulate their buoyancy by changes in the number of gas vesicles and several mechanisms seem to be involved (Walsby, 1978b). Under stratified conditions, diatoms and non-flagellate green algae are unable to maintain their position in the water column and sink below the photic zone, whereas the faculty of buoyancy regulation enables cyanobacteria to stay in the photic zone and even choose an optimum depth. It may be that certain forms undergo diurnal migration (e.g. *Microcystis*: Ganf, 1974a) but this has not been demonstrated unequivocally in nature, since convective water movements brought about by changes in surface water temperature are often sufficient to cause the observed movements. Reynolds and Walsby (1975) have reviewed the whole topic of water-blooms and they conclude that surface water-blooms are an incidental consequence of a mechanism which provides a means of survival in stratified lakes. Surface scums are subject to highly unfavourable conditions and often lyse, so that a white cream of gas vesicles and a 'paint slick' of phycocyanin results.

18.4.5 Annual population cycles

Planktonic members of the Nostocaceae (e.g. *Anabaena* and *Aphanizomenon*) or Rivulariaceae (*Gloeotrichia*) commonly form akinetes, some of which germinate almost immediately (Rother & Fay, 1977), others of which pass the winter on the bottom sediments (Reynolds, 1971; Roelofs & Oglesby, 1970), perhaps to germinate in the following year if they become resuspended in the water column and are hence illuminated (Kaushik & Kumar, 1970). Wolk (1965a) found that germination of *Anabaena cylindrica* akinetes was promoted by high phosphate concentrations in the medium, but since the concentration of dissolved phosphate is likely to be higher in the mud than in the water column, it is debatable whether this factor operates in nature. *Microcystis* does not form akinetes, but none the less colonies sink to the bottom in autumn where they overwinter. It seems that such colonies contain some vacuoles, but are sufficiently dense to sink to the bottom (Reynolds & Rodgers, 1976; Reynolds *et al.*, 1981). It may be that the accumulation of reserve products (e.g. carbohydrate) under nutrient deficiency increases the specific gravity of the cells. Reducing conditions may promote the survival of *Microcystis* cells on the mud (Sirenko *et al.*, 1969) and it is suggested that the formation of mucilage round the colonies is also of benefit in this respect (Sirenko, 1972). It is not clear how the regrowth of the colonies is promoted in the spring, whether it is due simply to physical resuspension or whether changes in the colonies themselves are also important. Members of the Oscillatoriaceae survive principally in the water column. The losses from such populations (e.g. by washout and grazing) may be greater than from populations on the mud, but regrowth in the spring is more certain. The time of the spring maximum of *Oscillatoria redekei* in Lough Neagh depends on the size of the overwintering inoculum (Gibson & Fitzsimons, 1982) and when summer losses are heavy, the attainment of the spring maximum in the next year is delayed.

18.5 Physiology

18.5.1 Culture

18.5.1.1 Isolation

The isolation of axenic cultures of cyanobacteria free of contaminating heterotrophic bacteria has only been achieved with difficulty (Allen, 1952; Gorham, 1964; Staub, 1961; Meffert, 1972). Some axenic cultures have then proved difficult to maintain in an active healthy condition without the presence of the heterotrophic bacteria (Safferman, 1967; Vance, 1966). Experience at our laboratory exemplifies this problem: when both axenic *Oscillatoria redekei* and *O. agardhii* have been isolated, we have obtained consistent growth only with *O. redekei*. The *O. agardhii* always appears to grow better in the presence of contaminant heterotrophic bacteria.

The method we routinely use for isolating axenic cultures of planktonic cyanobacteria involves preliminary filtration through a series of either 5.0 or

Table 18.2. Axenic cultures of planktonic cyanobacteria deposited at the Cambridge Culture Collection

Strain	No.	Isolators	Comments
Anabaena flos-aquae	1403/13d	A.E. Walsby (1976)	No gas-vacuoles
Anabaena flos-aquae	1403/13e	A.E. Walsby (1976)	Helical filaments
Anabaena flos-aquae	1403/13h	J.W.G. Lund (1964) and A.G. Fitzsimons (1977)	'Windermere Strain'
Aphanizomenon flos-aquae	1401/1	S.I. Heaney (1968)	—
Oscillatoria agardhii	1459/11G	A.G. Fitzsimons (1975)	—
Oscillatoria redekei	1459/12G	S.I. Heaney (1968) and A.G. Fitzsimons (1973)	—

10 μm pore size membrane filters depending on the size of the organism, accompanied by careful resuspension and a series of washes to remove as high a proportion of heterotrophic bacteria as possible. The resuspension is finally transferred to a 5 μm membrane where the culture is concentrated. The concentrated cells are then diluted in a replicate log series to 10^{-8} and the cells are then cultured in tubes of FB-1 medium (Foy *et al.*, 1976) and left in an incubator (light intensity of 20.0 μE m^{-2} s^{-1}) at 20°C. A list of axenic cultures deposited in the Cambridge Culture Collection is given in Table 18.2. Carmichael and Gorham (1974) use a similar procedure to that described above for isolating axenic cyanobacteria but add phenol in the preliminary stages. Meffert and Chang (1978) report that the addition of phenol, washing with detergents (McDaniel *et al.*, 1962) and washing individual trichomes (Pringsheim, 1965) all tend to increase lysis of cells and inhibit growth. Recently, Caldwell (1979) has suggested that the problem of isolating axenic cultures of planktonic cyanobacteria arises because the cyanobacteria produce extracellular chelates that are autotoxic by removing available iron from the culture medium. Bacteria effectively metabolize these chelates and allow the cyanobacteria to grow. Caldwell (1979) reports that he was able to prepare axenic cultures by substituting activated carbon for bacterial heterotrophs.

18.5.1.2 Growth requirements

Apart from difficulties in isolating planktonic cyanobacteria another factor which has led to relatively few species being studied in culture has been a poor understanding of their growth requirements. Fortunately the pioneering efforts of Gorham and his co-workers led to the development of a successful

Table 18.3. Composition of ASM-1 growth medium in μmol l^{-1} (Gorham *et al.*, 1964)

$NaNO_3$, 2000	$MgSO_4$, 200	$MgCl_2$, 200	$CaCl_2$, 200	K_2HPO_4, 100
Na_2HPO_4, 100	$FeCl_3$, 4	H_3BO_3, 40	$MnCl_2$, 7	$ZnCl_2$, 3.2
$CoCl_2$, 0.08	$CuCl_2$, 0.0008	Na_2. EDTA, 20		

growth medium termed ASM-1 (Table 18.3) which is now used extensively for the culture of planktonic cyanobacteria (Gorham *et al.*, 1964). One drawback of the ASM-1 medium is its very low buffering capacity. Although this is probably not a critical defect in most growth experiments, the problem can be overcome completely by using N-2-hydroxyethylpiperazine-N′-2-ethanesulphonic acid (HEPES) as a buffer (Smith & Foy, 1974). The chelating capacity in ASM-1 is provided by EDTA which is the only organic constituent. Lange (1970) has observed that improved growth can be obtained by substituting fulvic acid for EDTA. However, these observations were made under suboptimal growth conditions and may be a result of greater stability over several months of the fulvic acid iron complex compared with iron EDTA. The provision of an adequate supply of available iron is probably one of the most critical features of the development of media for planktonic cyanobacteria. Morton and Lee (1974) have shown that iron in concentrations of 0.1–1.0 mg l^{-1} causes a shift in the dominant organisms grown in batch culture from green algae to cyanobacteria. We observed a much higher requirement for iron by *Oscillatoria redekei* compared with the green alga *Selenastrum* when we bioassayed Lough Neagh waters (Parr & Smith, 1976).

The effect of trace elements on the growth of planktonic cyanobacteria has received very little study and the optimization of concentrations of trace elements in ASM-1 medium has necessarily been determined in a somewhat arbitrary fashion. Gorham *et al.* (1964) found that the concentration of manganese that would produce 50% growth inhibition was only slightly higher than the optimum concentrations in ASM-1. Four times the manganese concentration or a deficiency in iron promoted a pronounced filament coiling. The effects of iron and manganese on filament growth appear then to counteract one another. Gerloff and Skoog (1957) employing *Microcystis* found that iron antagonizes or reduces the toxicity of manganese to cyanobacteria. The toxicity of manganese to cyanobacteria has been considered a factor in species succession. Both Guseva (1952) and Patrick *et al.* (1961) have studied the effects of manganese and concluded that high manganese concentrations inhibit cyanobacteria more than green algae or diatoms.

Because of the number of reports of difficulties encountered in culturing planktonic cyanobacteria, it would seem of value to summarize the culture conditions that have been successfully employed at our laboratory (Foy *et al.* 1976; Foy, 1980). For batch culture, 100 ml straight-stemmed flasks without a lip are employed on a shaker. A valuable improvement has been the use of Hakuto silicon sponge bungs instead of the traditional use of cotton wool. The optimum light intensity is usually between 40–80 μE $m^{-2}s^{-1}$ (400–700 nm) and is best provided by a mixture of 'Grolux' and warm white fluorescent tubes. We do not recommend growing temperate species in culture above 20°C, although *Microcystis* species appear to withstand higher temperatures. At low culture densities the carbon dioxide concentration of air appears adequate and higher concentrations can be inhibitory. In contrast at higher culture densities the carbon dioxide content of air may be suboptimal and it may be neccessary to supplement the carbon dioxide supply.

One feature of the freshwater planktonic cyanobacteria that appears to distinguish them from marine species is the almost complete absence of reports that they require vitamins for growth in culture. Recently, Dubois (1977) has obtained some evidence of vitamin B_{12} uptake by *Oscillatoria rubescens* in culture but the requirement does not appear to be an absolute one.

18.5.1.3 Growth rates

Growth rates for a range of algal and cyanobacterial species have been collected together by Hoogenhout and Amesz (1965) and Eppley and Strickland (1968). Whitton (1973a) has tabulated growth rates of planktonic cyanobacteria and more recent measurements are shown in Table 18.4. From the available data it is not possible to state unequivocally that the cyanobacteria have faster or slower growth rates than other algal groups such as diatoms or green algae. The most important single factor governing growth rate appears to be the surface: volume ratio of different species. Foy (1980) cultured twenty-two strains of planktonic cyanobacteria under controlled conditions at 20°C and found that just under half the variation in growth rates could be associated with surface: volume differences. The maximum growth rate observed was $\bar{K}=2.74$ doublings day^{-1} for a culture of *Oscillatoria limnetica* isolated from Windermere. This is close to the theoretical maximum of $\bar{K}=3.06$ doublings day^{-1} predicted by Eppley (1972) for growth at 20°C. One feature of cyanobacterial growth observed by Foy *et al.* (1976) and Ahlgren (1978) is its high Q_{10}. In practical terms this means that temperature control of laboratory cultures must be very precise because a 1°C increase in temperature can increase the growth rate by about 10%.

If growth rate *per se* was the single most important factor governing the species composition of the plankton then one would expect to see small species with large surface: volume ratios and hence higher nutrient uptake rates in a

Table 18.4. Laboratory growth rates of planktonic cyanobacteria grown at 20°C. Determinations of cell diameters by Foy (1980). Figures in parentheses are the 95% confidence limits. (More detailed information on several of these strains is given in Table 18.2.)

Organism and source of isolate	Reference	Light intensity (μE m^{-2} s^{-1})	$\bar{K}$ (doublings day^{-1})	Cell diameter (μm)
Anabaena flos-aquae (Windermere)	Foy *et al.* (1976)	44–92	1.13	5.04 (± 0.14)
Anabaena flos-aquae (Windermere)	Fay & Kulasooriya (1973)	155	0.96	—
Aphanizomenon flos-aquae (Lough Neagh)	Foy *et al.* (1976)	64–92	1.41	4.65 (± 0.15)
Aphanizomenon flos-aquae (Kettlemere)	Rother & Fay (1979a)	58	0.87	—
Oscillatoria agardhii (Loughgall)	Foy *et al.* (1976)	28–64	1.19	3.54 (± 0.04)
Oscillatoria agardhii (Veluwemeer)	van Liere (1979)	48–193	1.25	—
Oscillatoria agardhii var. *isothrix* (White Lough)	Foy (1980)	60	0.54	8.53 (± 0.30)
Oscillatoria limnetica (FBA L100)	Foy (1980)	60	2.72	1.62 (± 0.02)
Oscillatoria redekei (Lough Neagh)	Foy (1980)	60	1.25	2.16 (± 0.07)

dominant position. Although under certain conditions this may be true there are important drawbacks to the smaller size. Foy and Smith (1980) in a study of two planktonic species have shown that under a light:dark regime with a short photoperiod the larger *O. agardhii* species was able to compete successfully with the smaller *O. redekei* species, because the former had the capacity to store more carbohydrate during the light and use this to provide the energy requirement for nitrate uptake and protein synthesis during the subsequent dark period. Another factor favouring larger species is that the zooplankton show evidence of favouring small species as a food supply.

18.5.2 Nutrient uptake

18.5.2.1 Nitrogen

The environmental role of nitrogen-fixing cyanobacteria has been the subject of a recent symposium (Granhall, 1978), and their importance in the nitrogen balance of freshwater lakes is reviewed by Horne (1977). As eukaryotic algae lack the ability to utilize atmospheric nitrogen, it is clear that heterocystous forms will be at an advantage in lakes where sources of other nutrients are adequate.

In the situations prevailing at Lake Erken, Sweden (Granhall & Lundgren, 1971) and Clear Lake, California (Horne & Goldman, 1972) as much as 80% and 40%, respectively, of the annual nitrogen budgets of these lakes is reported to originate from nitrogen fixation by cyanobacteria. In such examples it may be argued that the ability to fix nitrogen is the most important single factor contributing to cyanobacterial dominance of the phytoplankton of these lakes. In other lakes such as Windermere, South Basin, U.K. (Horne & Fogg, 1970) and Lough Neagh, N. Ireland (Riddolls, 1975) where less than 1% of nitrogen originates from cyanobacterial fixation it is clear that although nitrogen fixation is important to nitrogen income of individual species it is not of such pre-emptive importance. However, Hutchinson (1967) showed that nitrogen-fixing ability was the apparent controlling factor in the success of *Anabaena circinalis* over the diatom *Fragilaria crotonensis* in Linsley Pond, U.S.A. The relationships between structure and function in nitrogen-fixing cyanobacteria has been the subject of a comprehensive review by Stewart (1980). It is clear that there is no evidence that non-heterocystous planktonic cyanobacteria can fix nitrogen under normal aerobic conditions in freshwater. However, there have been recent reports of nitrogenase activity by non-heterocystous marine species such as *Trichodesmium* (Carpenter & Price, 1976; Carpenter & Walsby 1979; Bryceson & Fay, 1981) and *Microcoleus chthonoplastes* (Pearson *et al.*, 1979). Unfortunately, these cultures still retain contaminating bacteria and absolute proof of active cyanobacterial nitrogenase awaits the isolation of axenic cultures. There is no doubt about the ability of axenic *Plectonema boryanum* to fix nitrogen, and Rogerson (1980) has constructed a sulphide-stat that facilitates logarithmic, nitrogen-fixing, photoheterotrophic growth dependent on the addition of fructose.

A favourable energy balance may be an important prerequisite for cyanobacterial nitrogen fixation. Lundgren (1978) observed that the peak for fixation did not occur until the afternoon. In this connection Paerl and Kellar (1979) observed that diatom communities that do not fix nitrogen have patterns and efficiencies of carbon-dioxide fixation similar to *Anabaena*; both are similarly inhibited during the afternoon. The ability of *Anabaena* to direct photoreductant to nitrogen-fixation pathways during this inhibited period represented a more efficient use of the total daily energy input. Perhaps even more significant was the observation that 20–30% of the daily nitrogen fixation occurred in the dark, utilizing stored carbohydrate as an energy supply. Fay (1976) has studied the factors influencing dark nitrogen fixation in *Anabaenopsis circularis*. His studies emphasize the importance of light intensity and length of exposure as controlling the magnitude of the dark nitrogenase activity. Perhaps more controversial is his suggestion that direct assimilation of organic substrates may play an important role in supporting nitrogen fixation in both dim light and the dark.

Ward and Wetzel (1980) have shown that the lowest light intensity at which growth could occur with their isolates was determined by nitrogen source. Of the three inorganic nitrogen sources utilized (ammonia, nitrate and nitrogen gas), ammonia always supported the highest growth rates under all light regimes. Growth could not be maintained at low light intensities by *Aphanizomenon flos-aquae* grown solely with gaseous nitrogen. Zevenboom *et al.* (1981) have isolated a non-heterocystous mutant of this species which has a lower energy demand than the wild strain. However, a general conclusion that the high energy demand of heterocystous nitrogen-fixing species mainly arises from heterocyst production cannot be made from their data.

O'Brien (1972) has made a distinction between two types of phytoplankton growth response to increased limiting nutrient concentration: increase in yield (type I) and increase in growth rate (type II). Goldman *et al.* (1979), in a review of phytoplankton growth in the oligotrophic oceans, concluded that although the type I growth response is operating and is determined by the supply of nitrogen, the growth rate of plant cells always remains close to maximum and appears to be independent of a type II response. Part of their evidence centred on measurements of carbon, nitrogen and phosphorus contents of marine phytoplankton which consistently gave ratios of close to 106:15:1. These ratios are only achieved at maximal growth rate in laboratory culture. Such a model may have some applicability to oligotrophic freshwaters but in the case of eutrophic waters it is clear that algae markedly change their C:N:P ratios in response to nutrient deficiency (Healey, 1973b).

Having earlier considered the significance of nitrogen fixation one must also consider how cyanobacteria may show evidence of physiological mechanisms which may confer advantages in competition with other algae for the uptake of combined nitrogen. The results of Zevenboom and Mur (1978) show that in examples of natural populations of *Oscillatoria agardhii* K_s^u (the substrate concentration at which the uptake rate by the whole cell is half of the maximum, V^umax) varied between 26 μM nitrate under conditions of nitrogen starvation to 79 μM under nitrogen sufficiency. Such values are of the order of 100 times higher than the K_s^u for those species examined by Goldman *et al.* (1979), and this concentration of nitrate is higher than normally encountered even in very eutrophic freshwaters. However, although V^umax appears to be a constant which probably reflects the density of uptake sites for nitrate on the surface of the cell, K_s^u does become lower in culture in response to low nitrate concentrations (Zevenboom & Mur, 1978). Clearly the ability of an organism to increase its affinity for nitrate when the latter becomes limiting must be of considerable ecological advantage in competition with other species with higher K_s^u values. One possible mechanism for the efficient utilization of low cellular nitrogen concentrations by cyanobacteria is the GS/GOGAT pathway (see Chapter 4) first reported by Tempest and Meers (1970) in *Aerobacter aerogenes* and shown to be present in *Oscillatoria agardhii* by van Liere *et al.* (1977). This pathway is not found in most eukaryotes, which instead use glutamate dehydrogenase for amino acid synthesis. Indeed, the cyanobacteria show abundant evidence of sophisticated regulatory mechanisms because, despite being grown under nitrate limiting conditions, *O. agardhii* increases its phycocyanin content in response to low-light conditions (Zevenboom & Mur, 1978).

Another example of regulatory control is shown by both *O. redekei* and *O. agardhii* (Foy & Smith, 1980) which markedly increase their rates of carbohydrate synthesis under short light:dark cycles (3L:21D) compared with long light:dark cycles (21L:3D). Stored carbohydrate was used by these species to generate cell protein, DNA and chlorophyll *a* during the dark period. A significant observation was that 39% of the nitrate uptake in an *O. redekei* culture took place in the dark under a 6L:18D cycle. Because of the conditions of high ratios of dark to photic zones in eutrophic lakes, where these species are successful, the dark uptake of nitrogen may be of vital importance. In this regard it is interesting to note that eukaryotic nitrate assimilation as exemplified by wheat leaves is strictly light dependent (Sawhney *et al.*, 1978). Apparently, when light is extinguished mitochondrial respiration becomes significant and NADH is oxidized by way of the respiratory chain: thus NADH is no longer available for nitrate reduction. Cyanobacteria have a different mechanism of nitrate reduction which is dependent on reduced ferredoxin as electron donor to nitrate reductase instead of the NADH diaphorase of eukaryotes (Losada & Guerrero, 1979). It is tempting to speculate that the cyanobacterial ferredoxin-dependent system may be more available for dark reduction by mechanisms as described by Smith *et al.* (1971) for nitrogenase than the comparable eukaryotic system.

18.5.2.2 Carbon

King (1970) has suggested that cyanobacteria are more efficient at obtaining carbon dioxide at low concentrations than green algae and that this is an overriding factor explaining their dominance in eutrophic waters. Shapiro (1973) tested this hypothesis by enclosing lake populations of phytoplankton in polythene bags, varying the pH and adding the additional nutrients nitrogen and phosphorus together with and without injected carbon dioxide. On addition of the carbon dioxide in the presence of nutrients he observed a rapid shift from cyanobacteria to green algae and concluded that King's hypothesis was essentially correct. The implication of this work is that cyanobacteria possess lower K_s^u values for carbon dioxide than green algae and this gives them a competitive advantage at higher pH values. As reported by Goldman (1973) this interpretation of the Shapiro bag experiments can be criticized on the grounds that the species shift was simply an effect of pH *per se* and not carbon dioxide, and that in any case algal growth is not controlled by free carbon dioxide but by the total inorganic carbon concentration. Talling (1976) points out that King's hypothesis is dependent on unpublished data. It completely ignores the possible uptake of bicarbonate and that affinity for carbon dioxide certainly varies with photosynthetic capacity. Schindler *et al.* (1972) concluded that the invasion of carbon dioxide from the atmosphere is sufficient to permit the development of a large biomass in any water body which has an adequate supply of nitrogen and phosphorus. However, the carbon limitation debate does not end here because atmospheric inputs can only oppose the development of low concentrations of free carbon dioxide and do not alter its rate-limiting significance once it is realized (Talling, 1976). Certainly it appears feasible that cells could be carbon limited if the medium adjacent to cells was depleted, although the system as a whole could not be described as carbon limited. However, more unequivocal experimental work is required with a range of species before the cyanobacteria as a group can be declared to possess lower K_s^u values for carbon dioxide than other algal groups. In this regard the low photorespiratory activity of cyanobacteria reported by Cheng and Colman (1974) may be very pertinent.

18.5.2.3 Iron

Fogg and Westlake (1955) demonstrated that cyanobacteria are capable of releasing polypeptides with chelating properties and that the excretion is enhanced in iron-deficient media (Fogg & Pattnaik, 1968). Because of the complexity of iron and organic matter interactions in freshwaters (Perdue *et al.*, 1976) it is not clear how cyanobacteria could achieve an advantage in competition with other groups of algae by this unspecific mechanism. Recently, more attention has been focused on the microbial excretion of siderochromes which are trihydroxymates of low molecular weight that can selectively chelate ferric iron (Emery, 1971). It has been suggested that these chelators act as carrier molecules transporting iron across membranes. It has also been shown that microbes produce antibiotics that are structural homologues of a competitor's siderochromes (Knusel & Nuesch, 1965). However, Caldwell (1979) has shown that cyanobacteria may be inhibited by their own siderochrome production in axenic cultures of limited volume, although Murphy *et al.* (1976) suggest that the excretion of hydroxymate chelators by cyanobacteria appears to be the agent suppressing green algae. Lean *et al.* (1978) made field observations in Lake Ontario which implied that nitrogen fixation by cyanobacteria caused an increased demand for iron and a concomitant production of hydroxymate chelators. Simpson and Neilands (1976) have also identified a hydroxymate produced by an *Anabaena* sp. Although it is not clear how cyanobacteria actually incorporate the iron that is so strongly chelated by the siderochromes, it seems likely that possession of this mechanism might provide them with a potential competitive advantage in iron-limited freshwaters.

18.5.2.4 Phosphorus

It is widely accepted that phosphorus is the critical nutrient that controls the size of phytoplanktonic crops in freshwater lakes (Schindler, 1977). Although widespread attention has also been given to carbon and nitrogen as possible limiting nutrients, it is clear that cyanobacteria are capable of correcting temporary deficiencies in these elements by eventually withdrawing more from the atmosphere. The uptake and metabolism of phosphate by cyanobacteria is described in detail in Chapter 5 and only a few comments will be made here. There is evidence that cyanobacteria can develop low K_s^u values for phosphate uptake and possess efficient mechanisms for phosphate storage and later remobilization of reserves. In relation to the potential of cyanobacteria to produce extracellular alkaline phosphatases and hydrolyse phosphomonoesters to release orthophosphate the evidence is more uncertain. Carr (1973) suggested that cyanobacterial development of specific alkaline phosphatase activity in relation to phosphorus deficiency is much less than that of other bacteria and this observation has been confirmed by us employing the planktonic species *Oscillatoria redekei*. (Compare, however, 5.2.3. and Pettersson, 1980.)

Jones (1972) found that in samples from the English Lakes the origin of alkaline phosphatase activity could not be clearly distinguished between bacteria, cyanobacteria and algae. Stevens and Parr (1977) in their study on Lough Neagh also found that alkaline phosphatase activity originated from sources such as river inputs and zooplankton and hence increased activity in lake water could not be used as a reliable indicator of phosphorus-limited growth. However, the use of cyanobacteria in bioassays (Parr & Smith, 1976) and studies on the phosphorus content of cyanobacterial populations (Gibson & Stevens, 1979) have clearly demonstrated that phosphorus limitation does occur in Lough Neagh.

18.5.3 Respiration

The ratio of photosynthesis:respiration is of obvious importance to growth rate, particularly under short days or in 'optically deep' lakes where the mixed depth greatly exceeds the photic zone. Most measurements of respiration have been made in the dark and it is uncertain how this relates to oxygen uptake in the light. Measurement of respiration rate in the field is especially difficult because the dilution of the organisms makes the changes in oxygen small and there is serious interference from other organisms. Early laboratory studies (Webster & Frenkel, 1952; Kratz & Myers, 1955b) showed that in cyanobacteria, as in *Chlorella pyrenoidosa* (Cramer & Myers, 1949), dark oxygen uptake declined with prolonged incubation, eventually reaching an apparent plateau called the 'endogenous respiration rate'. Later work with *Oscillatoria redekei* (Gibson, 1975b) showed that this was also true of planktonic species and large changes in oxygen uptake rate have been observed in the field (Ganf, 1974b; Gibson, 1975b). In the dark, anthrone-reactive carbohydrate content declines in parallel with oxygen uptake and both the initial respiration rate and the time taken to reach the plateau value depend on the previous light history of the cells (Gibson, 1975b). Under nitrogen limitation, respiration rate is depressed and carbohydrate utilization in the dark is much less (Gibson, 1978). Very few measurements have been published of the dark oxygen uptake of planktonic autotrophs, but those available give little support for the view (Bone, 1971c) that cyanobacteria have lower respiration rates than other autotrophs, since the variation caused by conditions during growth and incubation is greater than observed differences between organisms. However, van Liere and Mur (1979b) collated the published values for specific maintenance constants (μe) and found that those for *O. agardhii* and *Spirulina platensis* were an order of magnitude smaller than for the green algae *Scenedesmus*, *Chlorella* and *Chlamydomonas*, implying that at low light intensities cyanobacteria should have the faster growth rate. Mur *et al.* (1978) have shown experimentally that this can in fact occur.

The ability to reduce respiratory losses in the dark may be a crucial factor in the success of some organisms. In *Anabaena variabilis*, long-term survival in the dark is enhanced by anaerobic conditions (Sentsova *et al.*, 1975), perhaps because the metabolic rate is depressed. Nitrogen or phosphorus deficiency, by lowering respiration rate and causing the accumulation of carbohydrates in the light (Gibson, 1978) may prepare the cell for dark survival. Certain cells must be able to reduce their respiration rate to a very low level and enter a resting state; for example *Microcystis* colonies overwinter on the bottom sediments to regrow the following year (Reynolds & Rogers, 1976). Some over-wintering cells may retain their viability for many years (Stockner & Lund, 1970).

18.5.4 Heterotrophy

Because of the survival of populations under low light and the early correlation between the occurrence of cyanobacteria and organic matter (Pearsall 1932), there have been frequent suggestions that heterotrophic nutrition is involved (see Chapter 3). Despite the plausibility of this hypothesis, a convincing demonstration of it under natural conditions has not yet been achieved. Many cyanobacteria are thought to be obligate phototrophs (Fogg, 1969) and Khoja and Whitton (1971) who screened a range of strains for heterotrophy, found that growth in the dark was slow. Although *Anabaena flos-aquae* and *Aphanizomenon flos-aquae* are unable to assimilate glucose at ecologically meaningful concentrations, *Oscillatoria agardhii* var. *isothrix* and some other *Oscillatoria* species may be able to (Saunders, 1972), so that heterotrophy cannot be ruled out in metalimnetic populations of these organisms. There must, however, be some doubt as to whether the uptake is sufficiently effective to compete with the obligate heterotrophs also found in these situations (Chapter 3).

18.5.5 Photosynthesis

18.5.5.1 Light requirements

In common with most photosynthetic plankton (Harris, 1978), cyanobacteria are low light organisms. The intensity of irradiance saturating photosynthesis varies with light history but is generally between 75–200 μE m^{-2}s^{-1} visible light. It is not clear whether planktonic cyanobacteria as a group are more efficient harvesters of low light than other groups, but Mur *et al.* (1978) found that *Oscillatoria agardhii* grew much faster at low irradiance than *Scenedesmus protuberans* and explained the seasonal succession of the two species on this basis. The fact the *Oscillatoria* species are unusual among cyanobacteria in forming spring maxima (Lund, 1965) may mean that they behave differently to other types in their light response. In our laboratory we have found that large *Anabaena* species need more light to saturate photosynthesis than small *Oscillatoria* species.

18.5.5.2 Adaptation

It is likely that photosynthetic efficiency is increased by low light incubation and van Liere, de Groot and Mur (1979) showed that in nitrate-limited continuous culture the chlorophyll *a* content of *Oscillatoria agardhii* began to increase at irradiances below 25 W m^{-2}. It is clear that, as shown by Beardall and Morris (1976) in a marine diatom, the previous history of a cell has profound influence on its photosynthetic performance. Hence, experiments carried out in continuous light may be a poor guide to photosynthetic characteristics of natural populations experiencing fluctuating irradiances. In our laboratory we have found relatively little adaptation of photosynthesis or pigment composition when a range of planktonic cyanobacteria were grown in 200 $\mu E\ m^{-2}s^{-1}$ and 40 $\mu E\ m^{-2}s^{-1}$ if the photoperiod was 6L:18D, presumably because even at the high-light intensity, the long dark period was of overwhelming importance. The most striking effect of incubation at low light is a reduced capacity to photosynthesize at high irradiance, as is common in other organisms (Beardall & Morris, 1976). Similarly, correlations between position in the water column and light requirement or temperature should be viewed with caution since the situation is often confused by the differential nutrient availability in stratified water columns. Thomas (1950) demonstrated that *O. rubescens* took up a defined position in the water column relative to the water surface and this and other early data are discussed by Klemer (1976). However, on the basis of his experiments with *O. agardhii* var. *isothrix*, Klemer concluded that the alga was 'neither as oligophotic nor as oligothermal as correlations of its population maxima with ambient light and temperature conditions might suggest'.

18.5.5.3 Light interception

The efficiency of light interception varies with the size of the cell, probably passing through an optimum at cell volumes of 100–200 μm^3 (Harris, 1978). In this regard, the very large colonial forms are relatively inefficient light collectors, but presumably this apparent disadvantage is compensated for by other factors such as buoyancy regulation and resistance to grazing. Self-shading is reduced in large colonial populations and in clear lakes where light attenuation from inorganic particles or coloured water is low, species such as *Anabaena* could produce a larger biomass than small *Oscillatoria* species before self-shading limited photosynthesis. It is also possible that such large species are more tolerant of high light intensities, but data are lacking on this point. Laboratory studies are misleading, since strains which form large colonies in nature normally dissociate into single cells or short filaments in culture.

18.5.5.4 Diurnal changes

Photosynthesis in natural populations probably undergoes diurnal changes. In Lake George (Ganf, 1975), *Microcystis* showed diurnal shifts both in photosynthetic efficiency and in maximum assimilation number, tentatively ascribed by Ganf to variations in nutrient supply but, Harris (1978) suggests that conformational changes of the photosynthetic thylakoids may be involved. In Lough Neagh (Gibson, 1978) the pattern of carbohydrate accumulation in unenclosed water samples does not follow the diurnal radiation changes, unlike photosynthesis (Jewson, 1975). Carbohydrate accumulates in the cells at 0.5 m depth towards the end of the afternoon, and at the time of maximum accumulation much of the carbon fixed in photosynthesis appears in the cells as anthrone-reactive carbohydrate.

18.5.5.5 Pigment composition

The presence of phycocyanin as an accessory pigment in cyanobacteria gives a characteristic absorption peak at 625 nm, which has been documented in natural populations of *Oscillatoria redekei* growing in Lough Neagh (Jewson, 1977). In coloured waters which absorb light strongly below 500 nm, the presence of an accessory pigment absorbing at a longer wavelength may give a slight advantage to cyanobacteria since carotenoid pigments characteristic of other groups absorb at shorter wavelengths which are poorly transmitted through brown water. The ability to alter pigment composition in response to incident light quality (chromatic adaptation) is well known (Bogorad, 1975a; Tandeau de Marsac, 1977), but even within one species, strains may differ in this respect. Some strains of *O. redekei* are very variable in their pigment composition forming large quantities of phycoerythrin under blue light but losing it under green or red light, whilst other morphologically similar strains are red even under red light. Meffert (1971) studied pigment shifts in *O. redekei* and found that under nutrient deficiency, *O. redekei* produced more phycoerythrin and some evidence that this may occur in the field is provided by spectroradiometric data from Lough Neagh (Jewson, 1977). Foy & Gibson (1982) have found that the amount of phycocyanin per cell increases in short photoperiods and it is likely that phycocyanin also functions as a nitrogen reserve which can be mobilized during nitrogen starvation (Allen & Smith, 1969) under high light. Under low light conditions there is some evidence that the phycocyanin content is maintained even when nitrogen is limiting (Zevenboom & Mur, 1978).

18.5.5.6 Light inhibition and photo-oxidation

Although field measurements of photosynthesis show surface inhibition at irradiance values above approximately 350 $\mu E\ m^{-2} s^{-1}$ (Jewson, 1976; Harris, 1978), the significance of this to growth is not clear. Such inhibition has a relatively minor effect on integral daily photosynthesis rates (Jewson, 1976) and although there are no *in situ* growth studies with cyanobacteria, long-term lake incubations of diatoms (Talling, 1955) and desmids (Lund, 1971) showed no evidence of growth being inhibited near the surface. Nevertheless, when surface water-blooms occur cells are exposed to full sunlight and these conditions, particularly when CO_2 is depleted (Abeliovich & Shilo, 1972) lead to photo-oxidation and mass death. However, strains differ in their sensitivity to photo-oxidation which is related to their ability to produce carboxydismutase, and a highly resistant strain of *Microcystis* was found in nature (Eloff *et al.*, 1976). In common with the dark repair mechanism of ultra-violet light damage (Bhattacharjee, 1977) the resistance to photo-oxidation is protein mediated and can be abolished by chloramphenicol (Eloff *et al.*, 1976). Some strains of *Oscillatoria redekei* are prone to cell lysis, especially when grown axenically (Meffert, 1973) and it may be that heterotrophs protect the organism by increasing the CO_2 supply. Gas vesicles have sometimes been thought to have a light-shielding function, but a critical study (Shear & Walsby, 1975) showed than in *Anabaena flos-aquae* the presence of gas vacuoles reduces the light falling on the thylakoids by a maximum of 4% and, in this organism at least, gas vesicles offered no protection against ultra-violet light.

18.5.5.7 Anoxygenic photosynthesis

Under the special conditions where light penetrates to water layers rich in hydrogen sulphide, such as the meromictic Solar Lake, Sinai (Cohen, Krumbein & Shilo, 1977), *Oscillatoria limnetica* may perform anoxygenic photosynthesis (Chapter 9), but is able to switch to oxygenic photosynthesis when the water column mixes. The existence of this phenomenon in other planktonic habitats must be strongly suspected from the early observations of cyanobacteria in H_2S lakes (Bursche, 1955).

18.5.6 Extracellular production

Extracellular substances produced by cyanobacteria can be broadly divided into two categories. Firstly, physiologically active compounds which inhibit or stimulate growth at low concentrations and secondly nutritive substances. An obvious example of the former are the siderochromes (18.5.2.3). Keating (1978) made a systematic study of cell-free filtrates from cultures of the dominant cyanobacteria from Linsley Pond, U.S.A., and showed that they were inhibitory to diatoms isolated from the same lake. Unfortunately, no precise chemical identification was made of the active compound, but it is possible that siderochromes were involved. It was suggested that cyanobacterial dominance of eutrophic lakes is attributable to allelopathy.

Toxin production by cyanobacteria leading to the death of cattle and birds is a world-wide phenomenon that is reviewed in detail by Gorham (1964) and Gentile (1971). In the case of toxin from *Microcystis aeruginosa*, Bishop *et al.* (1959) and Ramamurthy and Capindale (1970) have shown it consists of a polypeptide containing D-serine, L-ornithine and some protein L-amino acids. Because the toxin was insensitive to most proteolytic enzymes it was concluded that the peptide was circular. More recently an improved extraction procedure for the toxin has been described by Rabin and Darbre (1975). The complexity of the factors influencing the toxicity and animal susceptibility of toxin production by cyanobacteria is exemplified by the study of Carmichael and Gorham (1977). They demonstrated that two bacteria from the Enterobacteriaceae depressed toxin production in selected axenic toxic clones of *Anabaena flos-aquae* and that colonies from toxic blooms consisted of mixtures of toxic and non-toxic filaments. It was estimated that a surface-concentrated bloom of *A. flos-aquae* having a biomass density of 20 mg ml^{-1} dry weight would cause death of ducks or calves when 20 ml kg^{-1} body weight was consumed, whereas a monogastric animal such as a rat would require four times this dosage. There is no evidence to suggest that this form of toxin production has any limnological significance, although one wonders whether zooplankton could be killed or inhibited by the toxins.

The second type of molecule excreted by cyanobacteria is nutritive substances. Reports of non-planktonic cyanobacterial production of extracellular nitrogen suggest that as much as 40% of the total nitrogen fixed by species such as *Calothrix scopulorum* is excreted (Jones & Stewart, 1969). Maximum rates of extracellular production of carbon by the planktonic *Oscillatoria redekei* reported by Jüttner and Matuschek (1978) was 5.5% of carbon fixed during a 2-h period of active growth in the summer period. Much lower values of 0.3% were found in stagnant populations in late autumn. The main components of the extracellular compounds during July were 58% sugars and oligosaccharides, 20% glycollic acid and only 12% amino acids. Foy and Smith (1980) could not detect significant extracellular production of nitrogenous compounds by axenic *O. redekei* grown in laboratory culture.

18.6 Control

There is a wide range of suggested methods for controlling the growth of nuisance blooms of cyanobacteria in water-supply reservoirs. The most appealing solution is the removal of phosphorus. This may be achieved by diversion of sewage effluent as, for example, in the best-documented case of all, Lake Washington (Edmondson, 1969), or phosphorus removal from effluents by tertiary treatment (De Renzo, 1978). Alternatively, the water body may be directly dosed with alum to precipitate the phosphorus (May, 1974). Apart from the *pro rata* reduction in cyanobacteria by phosphorus removal a secondary effect is an increase in the nitrogen:phosphorus ratio in the lake which should reduce the likelihood of nitrogen-fixing species of cyanobacteria dominating the plankton (Smith, 1977). These species are usually the most troublesome from both water treatment and amenity considerations.

Other methods of control are based on the particular sensitivity of cyanobacteria to biological agents such as phage and fungi or to algicidal agents (Muchmore, 1978). Fitzgerald and Skoog (1954) observed that planktonic cyanobacteria were much more sensitive to 2,3-dichloro-1,4-naphthoquinone than either other cyanobacteria or other algal groups. The precise biochemical explanation for the delimitation of the planktonic cyanobacteria in this regard is still not understood (Whitton & MacArthur, 1967). Similarly, Aaronson and Ardois (1971) found that planktonic cyanobacteria were extremely sensitive to ethionine and, although discounting its practical application because of carcinogenic side-effects, they suggested that other less harmful amino-acid analogues might have some application in bloom control.

Despite these alternatives, the only algicide that is used widely at the present time is copper sulphate. Gibson (1972) showed that *Anabaena flos-aquae* has an enhanced uptake of copper which may explain its particular sensitivity. Horne and Goldman (1974) observed that additions of 1/200th of the copper normally used in algal control was effective in completely inhibiting nitrogen fixation, by cyanobacteria, in Clear Lake, U.S.A. Since in this particular example nitrogen fixation provided half the nitrogen budget of the lakes, economical control by this means appears possible.

The remaining methods of algal control are the physical methods of artificial destratification and sonic disruption of the gas-vesicles. Ultrasonic radiation was first suggested as a potential control technique by Fogg (1969), but field trials by Menday and Buck (1972) led to the death of fish and they do not recommend it as a regular means of control. Artificial destratification appears to have been more successful and the theoretical basis for this approach has been reviewed by Lorenzen and Mitchell (1975). Artificial destratification of Kezar Lake, U.S.A., by continuous aeration caused an increase in the depth of the mixed layer followed by a significant reduction in cell numbers.

18.7 Discussion

The last twenty years have seen a great increase in research on planktonic cyanobacteria, resulting in a wealth of new analytical and observational data that were not available to earlier workers and it is now possible to go some way towards explaining the paradox that cyanobacteria thrive in nutrient-rich lakes at a time when nutrients appear depleted.

A clear analysis of the rates of flux of nutrients has led to a better understanding of the ways nutrient limitation can operate. Nitrogen and carbon may be withdrawn from the atmosphere and are best seen as limiting growth rate whereas phosphorus supply limits yield in a more direct way. It can be misleading simply to study dissolved phosphorus and to neglect the sources of phosphorus within the cell. In Lough Neagh, for example, *Oscillatoria redekei* produces much of its biomass by utilizing internal phosphorus reserves at a time when soluble reactive phosphorus remains undetectable in the water column.

At the beginning of the chapter, a further paradox was posed: although cyanobacteria have an extremely long history, they nevertheless show great success in the modern conditions of enriched lakes. There is room to speculate that part of this success is due to their ability to make appropriate physiological responses to environmental changes. In our laboratory, isolates observed over several years showed gradual morphological and physiological changes which suggest a great ability to adapt to circumstances. New isolates may be difficult to grow at first but in time growth becomes more predictable and this phenomenon could be interpreted as evidence that a strain adapted to laboratory conditions is being selected. If this is so, then it places a severe constraint on experimental work intended to shed light on the characteristics of field populations. It may also suggest that such adaptability is of advantage in the rapidly changing conditions of a lake.

Notwithstanding the difficulties of meaningful experimentation, several attributes of cyanobacteria seem to set them apart from other planktonic autotrophs. Firstly, their possession of gas vacuoles imparts a positive buoyancy which enables them to take up a favourable position in the water column. Secondly, the fixation of atmospheric nitrogen by the heterocystous species is likely to promote their success in nitrogen-limited environments and may be the main reason for their prominence during the late summer. Thirdly, evidence is accumulating of a significant role for extracellular products of high biological activity which may act either by their growth-promoting properties such as the chelation of metals or perhaps by suppressing the growth of other organisms. Allelopathy remains a difficult field of investigation, however, and the remark of Lund (1965) that 'chemical inhibition is as easy to suggest as it is difficult to prove' remains pertinent.

Finally, there are some reasons for supposing that cyanobacteria may be at an advantage in low light environments. However, although planktonic photosynthesis has been intensively studied, dark metabolism has been relatively neglected. This is as curious as it is unfortunate since, at least in eutrophic lakes, the plankton may spend the greater part of their time in darkness even during the longest days of summer. Adaptations to a low light environment can be perceived in the strategy of storing much of the carbon fixed in the light to drive dark metabolism, and in the ability to reduce metabolism to a low rate when starved. This faculty may be extended to enable cells to exist for long periods in the dark on the bottom mud.

Attempts to make quantitative comparisons between taxonomic groups are inevitably frustrated by a lack of information, since there are few physiological studies of other freshwater planktonic groups. However, even if other groups are shown to possess higher growth rates and superior nutrient uptake kinetics, it may finally prove to be the ability of cyanobacteria to reduce rates of loss arising from grazing, from sedimentation and from energy expenditure in the dark that are the pre-eminent factors in their success in eutrophic waters.

Acknowledgment

The authors are grateful to Dr C.S. Reynolds for helpful correspondence.

CHAPTER 19
MARINE PLANKTON

G. E. FOGG

19.1 Introduction

Cyanobacteria or blue-green algae may sometimes occur in striking abundance in warmer seas. It was *Trichodesmium*, a red-pigmented species, which gave the Red Sea its name. The same organism became known to Cook's sailors as sea sawdust and, being reported on by the botanist Banks in his journal of the 1768/71 voyage of the *Endeavour* (Beaglehole, 1963), is the subject of what seems to be the first scientific description of marine phytoplankton. Nevertheless, blue-green algae are usually inconspicuous in the open sea, although, from their ability to withstand saline conditions and their abundance in the freshwater phytoplankton, one would expect them to be widespread and conspicuous in this habitat too. This unexpected scarcity, which may, however, be more apparent than real, is difficult to explain. As photosynthetic organisms blue-green algae contribute to primary production but their main role in the marine ecosystem seems to be as nitrogen-fixers and, again, there is an unresolved problem of why they are not more important in this respect than they appear to be. Since the review by Sournia (1970) much of interest about planktonic blue-green algae has been published but much still remains to be learnt of their role in the sea.

19.2 Occurrence and distribution

Perhaps largely because the taxonomy is unsatisfactory, reports of the occurrence of blue-green algae in the marine plankton are often indefinite. The following account concentrates on the more well-defined forms.

19.2.1 Chroococcales

Lohmann (1911), the discoverer of nannoplankton, described unicellular blue-green algae in marine plankton and since then representatives of thirteen genera, as well as indeterminate forms, belonging to the Chroococcales have been reported as occurring free-floating in the sea (Sournia, 1970). The small size of most of these forms and the consequent difficulty of distinguishing them from bacteria has discouraged detailed investigation. However, using the technique of epifluorescence on fresh membrane-filtered samples, phycoerythrin-containing algae can readily be distinguished under the microscope, by their orange fluorescence, from other phytoplankton, which fluoresce red, and bacteria, which do not fluoresce. Rhodophyceae and Cryptophyceae also contain phycoerythrin but the minute phycoerythrin-containing cells which, apparently, are widespread and often abundant in the photic zone of temperate and tropical seas, have been shown by electron microscopy to have the characteristic cyanobacterial cell structure (Waterbury *et al.*, 1979; Johnson & Sieburth, 1979). Such cells, in concentrations of 10^2 to 10^5 ml^{-1}, were found in the Arabian Sea, off Peru, in slope water north of the Gulf Stream, in Woods Hole harbour, off north-eastern South America (Waterbury *et al.*, 1979), on Georges Bank in the North Atlantic and in the Caribbean (Johnson & Sieburth, 1979). They formed about 20% of the total bacterioplankton *sensu lato* and 6–15% of the total microbial plankton (Johnson & Sieburth, 1979). The greatest number of cells was found in the top 20 m of water but occasional cells were observed as deep as 400 m (Waterbury *et al.*, 1979).

A strain isolated in culture had a typical synechococcoid ultrastructure, chlorophyll *a* as its major photosynthetic pigment, and a predominance of phycoerythrin so that it appeared red in culture. Its size was around 1.1×2.0 μm and it divided in one plane. It did not grow at salinities below 14‰ and seems, therefore, truly marine. Together with similar isolates, ranging in diameter from 0.8 to 2.0 μm, this organism appears referrable to the genus *Synechococcus* (Waterbury *et al.*, 1979).

Johnson and Sieburth (1979) obtained a sheathed *Gloeocapsa*-like form, described as ubiquitous, in culture. Its cells were 0.8×1.0 μm and its thylakoids were arranged peripherally. Two other oceanic *Synechococcus*-like forms, 0.7×0.9 μm and 0.4×0.5 μm in size, the former with peripheral thylakoids and the latter with intracytoplasmic membranes present as a band of three or four in the centre or to one side of the cell, occurred sporadically in larger concentrations but were not obtained in culture.

Other chroococcaleans, such as various species of *Aphanothece*, *Gloeocapsa*, *Merismopedia*, *Dactylococcopsis* and *Gomphosphaeria*, reported from the Baltic (Grøntved & Steemann Nielsen, 1957; Braarud *et al.*, 1958; Pliński, 1975) are perhaps only to be found in coastal and brackish waters.

19.2.2 Chamaesiphonales

Dermocarpa may sometimes be present in coastal waters or epiphytic on other phytoplankton (Norris, 1964, quoted by Sournia, 1970) but otherwise members of this group have not been reported as occurring in marine phytoplankton.

19.2.3 Pleurocapsales

The only member said to be planktonic in the sea is *Johannesbaptistia pellucida* (Umezaki, 1961, quoted by Sournia, 1970).

19.2.4 Nostocales

19.2.4.1 Non-heterocystous forms

Trichodesmium, the most conspicuous type of blue-green alga in the marine plankton, only differs from *Oscillatoria* in showing aggregation of the trichomes by a thin mucilaginous envelope into macroscopic clumps or bundles. *Trichodesmium* shows the movement characteristic of some *Oscillatoria* spp. (Sournia, 1972). Geitler (1932) therefore transferred *Trichodesmium* to *Oscillatoria*, a transfer supported by Sournia (1968). Golubic (1977), putting emphasis on colony formation, a tendency towards cell differentiation

Fig. 19.1. *Trichodesmium erythraeum* (a) a small group of filaments (× 250), (b) colony (× 100). Courtesy of Dr S.Z. Qasim, National Institute of Oceanography, Goa, India.

(19.3.4) and other features, has argued that *Trichodesmium* should retain generic status. Since it is a long-established name with oceanographers and denotes a distinctive life-form which is not shown by the other *Oscillatoria*s which occasionally occur in the marine plankton, *Trichodesmium* will be the preferred name in what follows.

Apart from the characteristics already mentioned, features of *Trichodesmium* to be noted are the presence of gas-vacuoles (19.3.6), a predominance of phycoerythrin which gives the alga a red, purple or brown colour (19.3.2) and curious cylindrical bodies in the core region of the cell (Van Baalen & Brown, 1969). Ohki and Fujita (1982) have grown *Trichodesmium thiebautii* in persistent, but not axenic culture after careful removal of living contaminants, especially larvae of *Macrosetella gracilis*. It appeared to have a vitamin B_{12} requirement and to be inhibited by low concentrations of nitrate or ammonium salts. A doubling time of as little as two days was approached.

Various species have been described, but Sournia (1968) considered that four only are valid: *Oscillatoria thiebautii* (Gomont ex Gomont) Geitler: filaments flexuous, sometimes solitary but usually united into colonies up to 6 mm in size with filaments arranged radially or, less frequently, parallel (see Walsby, 1978a, Fig. 1 a,b). *O. hildebrandtii* (Gomont) Geitler: filaments more or less straight, arranged in a parallel manner to form bundles from 2 to 5 mm in length. *O. erythraea* (Ehr. ex Gomont) Geitler: filaments, constricted at transverse septa, rarely solitary and normally arranged parallel in bundles not greater than 1 mm in length (see Fig. 19.1 and Walsby, 1978a, Fig. 1c,d). *O. contorta* (Wille in Schütt) Sournia nov. comb.: filaments wider than in the other species and constricted at the transverse septa, extremely flexuous or helical, twisted into colonies up to 1 cm (see Walsby, 1978a, Fig. 1e,f).

Drouet (1968) put all *Trichodesmium* forms together under *Oscillatoria erythraea*. Golubic (1977) was inclined to regard colony form as an important taxonomic feature and to distinguish two species according to the radial or parallel arrangement. Until they have been studied in persistent cultures it is perhaps wise not to attach much weight to the differences between 'species'. No really definite differences in ecology or physiology between the various forms are apparent as yet; in what follows they are treated in the aggregate.

Trichodesmium is most characteristic of tropical waters and records of blooms (see Ferguson Wood, 1965; Devassy *et al.*, 1978) lie within a belt 30° on either side of the equator. It has, however, been reported from colder waters, for example by Ferguson Wood (1965) south of New Zealand at 57°S but not in antarctic plankton samples, in the Venice Lagoon (Voltolina, 1975), and off the south coast of Ireland (Farran, 1932; W.D.P. Stewart & G.T. Boalch, pers. comm.). It is not clear whether *Trichodesmium* is typically oceanic or neritic (Sournia, 1970). Blooms are more often noticed in coastal waters (*e.g.* Bowman & Lancaster, 1965) and Dunstan and Hosford (1977) considered *Trichodesmium* to be endemic to the shelf waters of the Georgia Bight rather than originating from the adjacent Gulf Stream. However, Ferguson Wood (1965) recorded windrows of *Trichodesmium* in the Indian Ocean, about 1000 miles from the nearest shore, and a transect across the North Atlantic (McCarthy & Carpenter, 1979) showed the alga to be present throughout.

The relation to hydrographic conditions is no clearer (Jayaraman, 1972). It is usually said that *Trichodesmium* develops under conditions of calm (Ferguson Wood, 1965; Devassy *et al.*, 1978) but it may be that populations grow under turbulent conditions and only become obvious when it becomes calm. Dunstan and Hosford (1977) found it to be most abundant in the Georgia Bight in July under conditions of water column stability but it also produced nearshore blooms in September and December under unstable conditions. No doubt there is some relation of blooms to oceanic fronts—an alga with positive buoyancy such as *Trichodesmium* would be expected to accumulate at convergences, although this does not seem to have been recorded. *Trichodesmium* is frequently reported from oligotrophic waters such as the Sargasso Sea (Carpenter, 1973) and the North Pacific Central Gyre (Mague *et al.*, 1974), but also occurs in waters relatively rich in nutrients (Dunstan & Hosford, 1977; Devassy *et al.*, 1978). It is generally said that *Trichodesmium* occurs when water temperature is above 25°C and salinity near 35‰ but Voltolina (1975) reported it in water of 19.6‰ salinity and Dunstan and Hosford (1977) found it in fairly high concentrations in the Georgia Bight at temperatures between 19 and 23°C and in salinities ranging between 32 and 36‰.

One of the few series of observations on blooms over successive seasons is that of Steven and Glombitza (1972). Sampling at fortnightly intervals for 2 years at a station off Barbados revealed that *Trichodesmium* underwent a regular cycle of development with a period of 105–200 days. This periodicity could not be linked with the solar cycle nor with variations in nutrient concentrations, which were uniformly low, the water being permanently stratified. However, Carpenter and Price (1976) related these data to wind speed and cloud cover as recorded at Barbados airport and found a statistically significant correlation of greater population densities with favourable weather, defined as no days with wind over 15 knots in the week before sampling and cloud cover less than 0.9 on the day before sampling. They attributed this to the inhibitory effect of turbulence on nitrogen fixation (19.3.4) but the possibility of an intrinsic periodicity is not thereby excluded (19.4.2). Devassy *et al.* (1978) studied *Trichodesmium* in the Arabian Sea off Goa and found a well-defined annual rhythm with the bloom appearing from February to May in the pre-monsoon period when the shallow coastal waters were well mixed but undergoing diurnal stratification in calm weather. Similar periodicity was observed by Bryceson (1977) on the east coast of Africa.

The positive buoyancy of *Trichodesmium* results in massive accumulation at the surface in calm conditions. Devassy, Bhattathiri and Qasim (1978) found the blooms they studied to be largely confined to the top few centimetres of water. Even in the more turbulent waters of the Atlantic *Trichodesmium* was found by McCarthy and Carpenter (1979) to be most abundant in the top metre, where it formed up to 6.3% of the total phytoplankton biomass, whereas at 50 m its concentration was never more than 0.04% of the total. Horizontally, *Trichodesmium* blooms form streaks related to Langmuir circulation (Smayda, 1970) or more complex patches, plumes and whorls (Fig. 19.2) which are often very striking, especially in contrast with adjacent impoverished water. These, together with similar accumulations of dinoflagellates, have become known as 'red tides'. Colour pictures of blooms of *Trichodesmium* are given in Devassy *et al.* (1978). Hart

Fig. 19.2. Oblique aerial photograph of a *Trichodesmium* bloom east of the Whitsunday Group of islands, Great Barrier Reef. Note the ship and its wake just above the centre of the photograph. Courtesy of L. Zell, Great Barrier Reef Marine Park Authority, Queensland, Australia.

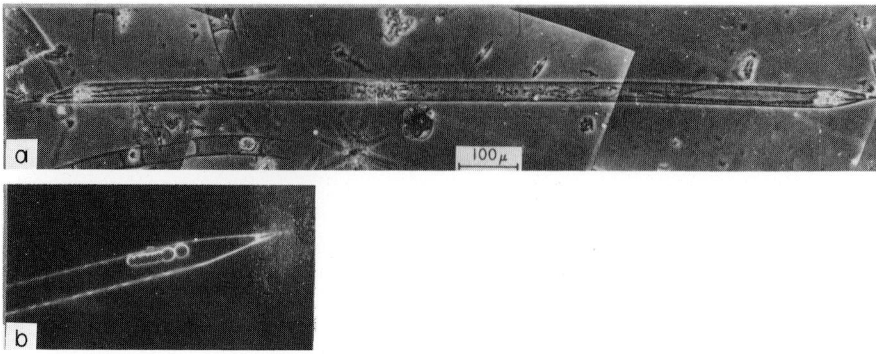

Fig. 19.3. *Richelia.* (a) Natural phytoplankton population, including the diatom *Rhizosolenia* with *Richelia*; (b) *Richelia* filament. (Courtesy of late Dr T.H. Mague)

(1966) estimated that the filament concentration required to give perceptible discolouration of the water as being between 20 and 600 ml^{-1}, but the presence of *Trichodesmium* may be visually evident at concentrations only 1/1000th of this.

Other non-heterocystous blue-green algae, all of minor importance in marine plankton, are listed by Sournia (1970). Some, such as *Oscillatoria* spp. other than those included by him in *Trichodesmium*, *Lyngbya* spp., *Phormidium* spp. and *Spirulina* spp. are probably only to be found in coastal and estuarine waters (see, e.g., Voltolina 1975, 1980). The genera *Katagnymene*, *Haliarachne* and *Pelagothrix* are oceanic and tropical but the first, which has a single trichome in a gelatinous sheath, is little known, and the other two are judged by Sournia (1970) to be 'fantômes bibliographiques'. *Isocystis infusionum*, a species of uncertain affinity, occurs in coastal waters of the northern Adriatic Sea (Voltolina, 1975) and in Liverpool Bay (Voltolina, 1980).

19.2.4.2 Heterocystous forms

Richelia intracellularis Johs. Schmidt has short unbranched filaments with large heterocysts at one, or less commonly, both ends (Fig. 19.3). Although it is occasionally found free-living or epiphytic on the diatoms *Chaetoceros compressum*, *Hemiaulus* spp. and *Rhizosolenia styliformis*, it is usually encountered as an endophyte in various species of *Rhizosolenia* (Sournia, 1970; Mague *et al.*, 1977). In the North Pacific Central Gyre, Venrick (1974) found *R. hebetata* var. *semispina*, *R. cylindrus* and *R. acuminata* with 94–98% of their cells containing *Richelia*. In these species there are usually two or four filaments per cell but the number increases with the size of the host and *R. castracanei* has been found to contain an average of ten and up to thirty-two

(B.M. Sweeney, pers. comm.). The proportion of *Rhizosolenia* cells with *Richelia* varies with season, being low in winter and high in late summer in the North Pacific Gyre (Mague *et al.*, 1977). *Hemiaulus membranaceus* has also been found to contain *Richelia* but this may be a transient phenomenon since the percentage of host cells with the endotype declined from sixty-two to 18.6 in 11 days (Kimor *et al.*, 1978). Mague *et al.* (1974) considered that the *Richelia* is contained in the cytoplasm of its host but Sweeney observed it between the plasmalemma and the frustule. Divisions of the two partners are closely synchronized. The heterocyst is always orientated towards the tip of the diatom but in dividing *Rhizosolenia* a second heterocyst is sometimes observed at the other end of the *Richelia* filament, which then apparently breaks in half, one half migrating into each daughter cell, thus preserving the orientation (B.M. Sweeney, pers. comm.). There is presumably biochemical exchange between diatom and blue-green alga, with the former receiving nitrogen fixed by the latter (19.3.4), but the association is not obligatory and *Richelia* may continue active in photosynthesis after its host has become moribund (Venrick, 1974). *Richelia* is widely distributed in tropical and temperate waters (Sournia, 1970; Venrick, 1974; Gundersen *et al.*, 1976).

The only other heterocystous form reported from oceanic waters is a *Nostoc* which Bernard and Lecal (1960) found in deep water *e.g.* 1000 m, in the tropical Atlantic and Indian Oceans and the Mediterranean. Average population densities were 668 cells ml^{-1} with a maximum of 12 620 ml^{-1}. The existence of such *Nostoc* populations remains unconfirmed.

In coastal and brackish waters, on the other hand, other heterocystous algae may occasionally be extremely abundant. In the Baltic, particularly, freshwater planktonic species such as *Anabaena flos-aquae*, *Aphanizomenon flos-aquae* and *Nodularia spumigena* invade the open sea. Gessner (1959) drew attention to the dominance of blue-green algae in the Gulf of Finland, where the salinity is not more than 6‰, as contrasted with their absence in the North Sea, where the salinity is between 30 and 35‰. Grøntved and Steemann Nielsen (1957) found only occasional filaments of *Nodularia spumigena* off the east coast of Denmark where the salinity was between 12.8 and 17.4‰. A bloom of this species in the Hanö Bight, where the salinity was perhaps around 7‰, was visible in a satellite image (Öström, 1976)—since their gas vacuoles cause them to float at the water surface blue-green algae are well recorded in satellite and aerial photographs in the 730–950 nm infrared (Wrigley & Horne, 1974). Horstmann *et al.* (1978), who have used contrast enhancement and false colour representation to accentuate traces of blue-green blooms in satellite images of the western Baltic (see Fig. 19.4) state that they consist mainly of *Aphanizomenon* and *Nodularia*.

Mention may be made also of *Dichothrix fucicola*, an epiphyte on pelagic *Sargassum* in the western Sargasso Sea, which evidently contributes substan-

Fig. 19.4. Satellite image (MSS band 4) of blue-green bloom in Baltic south of Bornholm Island, Aug. 7 1975 (Courtesy of Nigel Press Associates Ltd).

tially to productivity as a result of its ability to fix nitrogen (Carpenter & Cox, 1974).

19.2.5 Stigonematales

Members of this order, the most advanced morphologically among the blue-green algae, do not appear to occur in marine plankton.

19.3 Physiology

19.3.1 Growth in culture

Some unicellular forms have been grown in pure culture (Waterbury *et al.*, 1979; Johnson & Sieburth, 1979) but so far no detailed studies on such cultures appear to have been reported. Neither *Trichodesmium* nor *Richelia* seems to have been obtained in axenic or even in persistent impure culture. Van Baalen and Brown (1979) failed to get *Trichodesmium* into unialgal culture but Shimura and Fujita (1975) maintained it in bacterized culture in enriched seawater for up to 14 days. Ramamurthy (1972) described the isolation of *Trichodesmium* in axenic culture in a modified Erdschreiber medium containing vitamin B_{12} and streptomycin, stating that such cultures had been maintained in a healthy state for 5 years. However, the growth data presented are ambiguous, the concentration of streptomycin included in the medium (10–12 mg%) was about 100 times that preventing growth of other blue-green algae (Table 4.III in Fogg *et al.*, 1973) and his results have never been confirmed. Among the difficulties in the way of culture of *Trichodesmium* are the great fragility of trichomes, at least when collected from the sea surface (Van Baalen & Brown, 1969; Taylor *et al.*, 1973), which Van Baalen and Brown (1969) attribute to a thinning of the electron-dense layer of the longitudinal wall opposite the transverse septa. There is also remarkable sensitivity to cupric ion. Rueter *et al.* (1979) report that concentrations of 10^{-10}M cupric ion in media with Tris as a chelator or 10^{-8}M total copper added to filtered seawater, inhibited photosynthesis of *Trichodesmium*. The results discussed below were perforce all obtained with natural populations with associated micro-organisms.

19.3.2 Photosynthesis

All forms mentioned in 19.2 are ostensibly phototrophic but there is little detailed knowledge of their photosynthesis. Two strains of *Synechococcus* (from Sargasso Sea and Long Island Sound) have been found to have much greater potential for photorespiration than marine algae belonging to other phyla (Glover & Morris, 1981). Shimura and Fujita (1975) ascertained that the phycoerythrin from *Trichodesmium* has absorption peaks at 495, 547 and 562 nm and showed that light absorbed by this pigment is available for photosynthesis. The alga thus seems to be adapted to absorb the wavelengths which penetrate clear seawater best but McCarthy and Carpenter (1979) found no differences in *in vivo* absorption spectra of material collected from different depths so that this adaptation appears to be genetically fixed. The *Synechococcus* isolated by Waterbury *et al.* (1979) was likewise unable to adapt chromatically. *Richelia*, however, although usually blue-green was found to be red when collected off Halmahera in May 1975 (B.M. Sweeney, pers. comm.), which suggests phenotypic adaptation. McCarthy and Carpenter (1979) determined the relation of rate of photosynthesis of *Trichodesmium* to irradiance. Assuming that in their Fig. 5 the irradiance should be read as μ Einsteins $m^{-2}s^{-1}$, not h^{-1}, the curve shows no unusual features, being similar to that given for the diatom *Asterionella formosa* by Belay and Fogg (1978), with a maximum at about 25% full sunlight and about 50% inhibition at full sunlight. Van Baalen and Brown (1969), who observed that the gas vesicles of *Trichodesmium* are stacked in two ranks at the cell periphery, suggested that this arrangement may serve to protect the photosynthetic apparatus from high irradiance. Carpenter and Walsby (1979) found that photosynthesis was indeed less in *Trichodesmium* in which the gas vesicles had been collapsed, as compared with that in intact material, but it seems possible that disturbance of thylakoids following gas vesicle collapse was the cause of the decrease rather than removal of a light screen. Mague *et al.* (1977) found a maximum carbon assimilation rate for *Trichodesmium* of 0.38 μg C (μg chl. $a)^{-1}h^{-1}$, about a fifth of that reported for other phytoplankton in a comparable situation. This low rate may be attributable to the self-shading and restriction of gaseous diffusion that occurs in its colonies. Release of extracellular products of photosynthesis occurs from healthy *Trichodesmium*, sometimes to the extent of 18% of the total carbon fixed (Pant & Devassy, 1976), or sometimes less than 8% (Shimura *et al.*, 1978), but it may be obscured by the release of phycoerythrin and other soluble components from moribund cells.

19.3.3 Heterotrophy

McCarthy and Carpenter (1979) found that ^{14}C assimilation rates by *Trichodesmium* from glucose, glutamate and aspartate supplied at concentrations ranging from 0.34 to 34 $\mu g\ l^{-1}$, were all less than 0.1% of that at which carbon was assimilated photosynthetically under near surface light conditions. Heterotrophic nutrition thus seems unimportant for this form. Bernard (1963) argued that the *Nostoc* which he believed he had found more abundantly in deep water than in the photic zone, could only exist heterotrophically but he presented no direct evidence in support of this view.

19.3.4 Nitrogen fixation

There is now much evidence, obtained both with ^{15}N as a tracer and with the acetylene reduction technique, associating activity in fixing molecular nitrogen with *Trichodesmium* blooms (Dugdale *et al.*, 1961; Goering *et al.*,

1966; Taylor *et al.*, 1973; Carpenter, 1973; Maruyama, 1975). Nitrogen fixation by *Trichodesmium* has not yet been demonstrated using axenic cultures (see Fogg, 1973) but it seems likely that the alga is the active agent since the rate of fixation is dependent on light intensity (Dugdale *et al.*, 1961; Taylor *et al.*, 1973; Mague *et al.* 1977). Acetylene reduction may occur to some extent (less than 25% of that in the light) in the dark under aerobic conditions but under anaerobic conditions is strictly dependent on light. Organic substrates such as acetate, pyruvate, succinate, glycerol and glucose, stimulated reduction in the dark but depressed it in the light (Taylor *et al.*, 1973). It is conceivable, of course, that nitrogen fixation is carried out by associated heterotrophic bacteria dependent on extracellular products of *Trichodesmium* photosynthesis, although one then has to explain why fixation is never associated with eukaryotic phytoplankton, which presumably liberate similar extracellular products. Bunt *et al.* (1970) and Taylor *et al.* (1973) recorded high acetylene reduction rates by seemingly bacteria-free *Trichodesmium*, but the latter authors point out that even if bacteria constituted only 1/100th of the biomass their rate of fixation need only be 30–100 nmoles mg protein^{-1}min^{-1}, which is within the reported capabilities of *Azotobacter*, in order to account for the observed rate in the natural population. Maruyama (1975) found a high proportion of nitrogen-fixing bacteria associated with colonies of *Trichodesmium* in samples from the western Pacific but, on the other hand, Carpenter and McCarthy (1975) were unable to isolate any nitrogen-fixing bacterium from *Trichodesmium* in the Sargasso Sea, which nevertheless showed high nitrogen-fixing activity. The observation that nitrogenase activity is significantly reduced following collapse of the gas vesicles in *Trichodesmium* is a strong indication that the enzyme is contained in the cells of the alga rather than its associated bacteria, which do not contain gas vacuoles (Carpenter & Walsby, 1979).

A difficulty in accepting *Trichodesmium* as a nitrogen-fixing organism has been that, whereas other blue-green algae possessing nitrogenase avoid its inactivation by oxygen by having heterocysts or fix nitrogen only under near-anaerobic conditions, *Trichodesmium* does not possess heterocysts and is normally found in oxygenated water. Its nitrogen fixation is not, in fact, reduced by oxygen at atmospheric partial pressure (Taylor *et al.*, 1973; Mague *et al.*, 1977). Taylor *et al.* (1973) suggested that the peculiar arrangement of gas vesicles referred to in 19.3.2 might promote nitrogen fixation by facilitating diffusion of oxygen away from the central cytoplasm. Carpenter and Walsby (1979) considered that the decrease in nitrogenase activity following gas vesicle collapse probably arose from disturbance of the cellular architecture rather than from alteration in diffusion rate of oxygen. Fogg (1974) suggested that sufficiently low concentrations of oxygen for nitrogen fixation might exist in the middle of colonies and this idea has been substantiated by Carpenter and Price (1976). A typical colony consists of several hundred trichomes each with perhaps 100 cells, arranged in parallel in a tight bundle (Fig. 19.1). Those trichomes passing through the centre of the bundle were found to have from ten to twenty lightly pigmented cells in a central differentiated region. ^{14}C-radio-autography showed that these central cells did not photosynthesize (whereas the others did) and presumably did not evolve oxygen, thus providing a suitable environment for nitrogen fixation. It follows that disruption of bundles should inhibit nitrogen fixation by allowing access of oxygen to the site of nitrogenase. This is so: shaking sufficient to disrupt colonies reduced the nitrogen-fixing capacity of *Trichodesmium* by 88% as compared with unshaken controls. This effect was not observed if the seawater in which the colonies were shaken was previously purged free from oxygen. Correspondingly, *in situ* nitrogen fixation per cell was greatest at the surface under calm conditions but the maximum sank to 20 m in a force 4 wind. The situation may be more complicated than Carpenter and Price envisage. Bryceson and Fay (1981) observed more bacteria in the central region of bundles than elsewhere and its seems possible that bacterial respiration might be important in reducing oxygen concentration to a level suitable for the functioning of nitrogenase. Herbst and Overbeck (1978) observed a close metabolic coupling between a freshwater *Oscillatoria* and accompanying bacteria. In any case, the hypothesis of Carpenter and Price greatly strengthens the case for *Trichodesmium* being regarded as a nitrogen-fixing organism.

The acetylene reduction technique is now generally used for the estimation of nitrogen-fixation rates by *Trichodesmium*. Mague *et al.* (1974) and Mague *et al.* (1977) have carried out the necessary determination of the C_2H_2 reduction:N_2 fixation ratio using ^{15}N. The former workers found values between 3.3 and 5.6:1 but the latter obtained a value of 1.9:1, lower than the theoretical value of 3:1. They explain this by the surmise that when active synthesis of nitrogenous cell constituents is occurring, reduction of N_2 may be favoured as against that of C_2H_2.

Rates of nitrogen fixation by *Trichodesmium* have been expressed on various bases but some specific rates (μg N mg N^{-1}h^{-1}), recalculated roughly using, where necessary, data given in Table 5 of McCarthy and Carpenter (1979) and a C_2H_2:N_2 ratio of 3.3, are as follows:

0.45 to 1.40	Taylor, *et al.*, 1973
up to 0.64	Mague *et al.*, 1974
about 0.09	Maruyama, 1975
1.85	Mague *et al.*, 1977
a maximum of 5.48	Saino & Hattori, 1978
between 0.05 and 0.72	McCarthy & Carpenter, 1979

The value given by Saino and Hattori (1978) is the highest reported for a marine alga and similar to values for cultured heterocystous algae. Rates vary according to physiological condition. Mague *et al.* (1974) found that fixation by samples from the northern Pacific Ocean was stimulated by addition of phosphate and partially inhibited by surface light intensities. Mague *et al.* (1977), however, found, in the same area, no immediate stimulation by phosphate and no photoinhibition. They also found maximum acetylene reduction at a pO_2 of 0.1 atm. with a more than 75% inhibition at a pO_2 of 0.4 atm. A falling off in rate of acetylene reduction *in vitro* was noted after 2 h by Mague *et al.* (1974). McCarthy and Carpenter (1979) noted a similar falling off, attributing it to sample variation, but Taylor *et al.* (1973) observed that *Trichodesmium* degenerated within 1–2 h of collection. The possible toxic effects of trace amounts of copper introduced during sampling and assay do not seem to have been determined. Taylor *et al.* (1973) detected no seasonal variation in nitrogen-fixing capacity but Saino and Hattori (1978) observed a diel variation of 200-fold in nitrogen-fixing capacity with a peak at mid-day and, since ammonium uptake showed little diel variation, supposed that the variation must be in nitrogenase.

There is good evidence associating nitrogen fixation with *Richelia* (Mague *et al.* 1974), and, since it has heterocysts, this accords with expectation. However, no special information about nitrogen fixation by this species seems to exist. The species occurring in the Baltic, *Aphanizomenon flos-aquae*, *Anabaena flos-aquae* and *Nodularia spumigena* are all heterocystous and nitrogen-fixing (Fogg *et al.*, 1973; Öström, 1976). Since a *Gloeocapsa* of terrestrial origin has been shown to be nitrogen-fixing (Gallon *et al.*, 1975) it is possible that the *Gloeocapsa*-like organism isolated by Johnson and Sieburth (1979) is also.

19.3.5 Phosphate uptake

Mague *et al.* (1977) commented on the relatively low cellular phosphorus and ATP contents of the *Trichodesmium* they collected in the northern Pacific gyre. McCarthy and Carpenter (1979) determined uptake of ^{32}P-labelled orthophosphate by *Trichodesmium* colonies. An initial high rate of uptake was maintained for only 30 min, thereafter the rate declined to about a quarter of its initial value. The rate of uptake versus phosphate concentration curve indicated a surprisingly high half-saturation constant of 9.0 μg-atom PO_4-P l^{-1}, which accords with the impression given by the results of Mague *et al.* (1977) that its ability to compete for phosphate is low. However, *Trichodesmium* was found to have a high alkaline phosphatase activity (170–300 ng-atom P μg chl. $a^{-1}h^{-1}$) and therefore a high potential for using organic phosphates, which diminished after supply of orthophosphate.

19.3.6 Gas vacuolation

Trichodesmium, *Anabaena*, *Aphanizomenon* and *Nodularia* possess gas vacuoles but the chroococcalean forms and *Richelia* evidently do not. Walsby (1978a) demonstrated that the positive buoyancy of *Trichodesmium* depends on the presence of gas vacuoles. A remarkable feature is that the constituent gas vesicles are much stronger than those in freshwater algae (Walsby, 1978a), the mean critical collapse pressure being 12, 34 and 37 bars respectively in the species *T. erythraeum*, *T. contortum* and *T. thiebautii* as compared with 2 to 3 bars after subtracting turgor pressure for those in freshwater *Anabaena flos-aquae*. This greater strength, which appears to depend on stronger wall material rather than on different size or geometry of vesicle, seems to be an adaptation to the greater hydrostatic pressures to which *Trichodesmium* is exposed at depth in the sea (*e.g.* about 20 bars at 200 m). This adaptation is achieved at the expense of the buoyancy control mechanism which in freshwater blue-green algae operates via changes in turgor pressure (Dinsdale & Walsby, 1972)—the vesicles of *Trichodesmium* are too strong to be collapsed by any possible increase in turgor pressure. A slower regulation of buoyancy could still be obtained by the differential growth of gas vesicles and cells. No studies of gas vesicles in brackish water blue-green algae appear to have been carried out as yet.

19.3.7 Physiological differences between colony forms

Taylor *et al.* (1973) observed that both the radial and parallel *Trichodesmium* colonies reduced acetylene. McCarthy and Carpenter (1979) found that the carbon/ATP, carbon/chlorophyll and carbon/nitrogen ratios of these two types were similar, suggesting that there are no major physiological differences between them.

19.3.8 Growth rates

At present, growth rates of *Trichodesmium* can only be estimated from measurements of instantaneous rates of metabolic processes. Assuming a 12-h photosynthetic day, Mague *et al.* (1977) calculated doubling times in the northern Pacific gyre of 43, 14 and 15 to 7 days from total particulate carbon, ATP and nitrogen fixation respectively. McCarthy and Carpenter (1979), again assuming a 12-h photosynthetic day, obtained a value of about 180 days from determinations of carbon fixation rates in the Atlantic, somewhat longer than a previous estimate of 40–110 days obtained in the western Sargasso Sea (Carpenter & McCarthy, 1975). These doubling times are extremely long compared with those of other oceanic phytoplankton. Considering the self-shading and slowness of diffusion within a *Trichodesmium* colony such low rates of growth are not implausible but further determinations ought to be made with material free from any possible inhibition by copper (see 19.3.1).

19.4 Adaptation to the marine planktonic environment

19.4.1 Comparison of freshwater and marine blue-green phytoplankton

Seawater can be an excellent medium for growth of blue-green algae, as is evident from their abundance in littoral habitats (Fogg, 1973) and in certain types of inland water which may resemble seawater in being alkaline or saline, planktonic blue-green algae are conspicuously successful. Nevertheless, although blue-green algae may be more abundant in the marine plankton than was once thought, their success in this environment is limited. Some explanation for this paradox must be sought.

Various factors may contribute to the success of blue-green algae in freshwater (see Chapter 20; Fogg *et al.*, 1973; Reynolds & Walsby, 1975). A capacity for nitrogen fixation and for storage of phosphorus as polyphosphate must be of competitive advantage in some situations but some of the most abundant freshwater planktonic blue-green algae do not fix nitrogen and eukaryotic algae may have high capacities for polyphosphate accumulation (Fogg, 1975). Van Liere (1979) showed that the successful species *Oscillatoria agardhii* is favoured by low irradiance and, when this arises by dense growth of, for example, the green alga *Scenedesmus protuberans*, *O. agardhii* can outgrow this other species. Wymer and Thake (1980) using chemostat cultures have shown that the ability of *Anabaena variabilis* to compete successfully against *Selenastrum gracile* depends on the former's greater ability to take up phosphate from low concentrations. Occasionally, non-gas-vacuolate nannoplankton forms such as *Synechococcus* sp. may be dominant (Bailey-Watts, 1968) but most of the abundant freshwater planktonic blue-green algae are gas-vacuolate. Reynolds and Walsby (1975) have described how gas vacuoles provide a buoyancy control mechanism giving the alga the potentiality of poising itself at the depth at which conditions are most favourable for growth. Species of *Oscillatoria* occurring as single filaments, which because of their great form resistance to movement through the water have a slow response time to changes in buoyancy, are characteristic of the stable metalimnia of stratified lakes. The larger colonial forms, which can sink or float rapidly to the optimum level when turbulence subsides, occur mainly in polymictic lakes and in the unstable epilimnia of stratified lakes. Bloom formation, which appears to serve no biological purpose, occurs in calm periods when most of the population has excess buoyancy as a result of a preceding period of turbulent circulation over a depth exceeding the euphotic zone.

Putting aside the non-gas-vacuolate nannoplankton forms which, in the present state of knowledge, seem to have equivalent status in fresh and salt waters, it appears that buoyancy control is the key to the explanation of our paradox—the open sea is too turbulent for the buoyancy control to be effective and the blue-green algae thereby lose the main advantage which they have in quieter waters. It is significant that they are successful in the Baltic, a shallow sea in which the outflow of freshwater over water of higher salinity stabilizes the water column.

19.4.2 Factors conducive to Trichodesmium *blooms*

If this is accepted, there remains the problem of how *Trichodesmium* is able to be so successful. Margalef (1978) has shown that the over-riding factor for the success of a planktonic species in the sea is adaptation to an unstable and turbulent environment. Firstly, there is the question of whether *Trichodesmium* is holo- or meroplanktonic. Feldmann (1932), who found *Trichodesmium*-like filaments growing attached at 25–30 m in the Mediterranean, suggested that it undergoes its major development as a benthic form, i.e. it is meroplanktonic. Van Baalen and Brown (1969) were unable to find any such attached growth in the Gulf of Mexico but Borstad and Borstad (1977) observed that radial colonies from the Caribbean had characteristic benthic diatoms associated with them; this could however result from preferential attachment rather than a common place of origin. The parallel bundle colonies also contained diatoms but these were exclusively planktonic species. This suggests that the radial colony develops on the sea bottom and the other in the planktonic phase. Since *Trichodesmium* can be oceanic (19.2.4) it seems unlikely that it can be completely dependent on a benthic phase for its continued existence, although, since it is characteristic of gyres it may be that it is transported into these from shallow waters at their margins. Steven and Glombitza (1972) postulated a 'seed population' near the bottom of the photic zone where nutrients might be accumulated by *Trichodesmium*. Mague, Mague and Holm-Hansen (1977) in fact found decreased algal carbon/ATP ratios in plankton dominated by *Trichodesmium* below 40 m, suggesting that the cells there were less nutritionally stressed than in the upper part of the water column. It seems that *Trichodesmium* does not have the competitive advantage of an especially efficient uptake mechanism for inorganic phosphate or adaptation for growth at low light intensities (McCarthy & Carpenter, 1979). However, there is another factor here: forced convection—relative movement of cell and water—is an important means of promoting uptake of scarce nutrients (Smayda, 1970; Fogg, 1975). This may be achieved by swimming or, as with diatoms, by sinking. *Trichodesmium* achieves it by floating upwards. Since its colonies are large, form resistance is low and relatively high velocities are attained. Walsby (1978a) found rates around 0.8 mm s^{-1} whereas most diatoms sink at about a fiftieth of this rate (Smayda, 1970). Absorption of scarce nutrients is presumably correspondingly enhanced. The strengthening of the gas vesicle walls in *Trichodesmium* has resulted in the loss of a means of buoyancy control of little value in the open sea, but has provided a means of achieving buoyancy and high rates of forced convection which is proof against great variation in hydrostatic pressure. As with freshwater blooms, this carries the penalty that in calms much of the population collects at the surface and is killed by exposure to full sunlight. The species will, however, be maintained by non-buoyant filaments remaining at depth; Walsby (1978a) found a wide scatter of floating rates with a high proportion of the colonies showing negative values, i.e. sinking.

The regular cycle of *Trichodesmium* in a stable water column reported by Steven and Glombitza (1972) might thus depend on factors intrinsic to the alga. Filaments grow slowly in nutrient-rich water at depth; those filaments which produce enough gas vesicles begin to rise and as they build up into large colonies in the higher light intensities their upward movement accelerates. This would account for the puzzling insignificance of *Trichodesmium* in the plankton in the pre-bloom phase (Devassy *et al.*, 1978). Moving into lower nutrient concentrations near the surface the capacity for nitrogen fixation becomes important. Colonies accumulating at the surface shade those below so that recruitment stops until the bloom becomes moribund and dissipates, allowing the cycle to begin again from filaments which did not attain positive buoyancy and remained below.

It is remarkable that only *Trichodesmium* of all the blue-green algae has made this adaptation; an investigation of the molecular biology of the gas vesicle wall of this alga should prove of particular interest.

19.4.3 The adaptation of Richelia *to the marine planktonic life*

Richelia appears ill-equipped to compete successfully on its own and presumably relies on the superior nutrient collecting powers of its host diatom, which in turn benefits from nitrogen fixed.

19.5 Role of planktonic blue-green algae in the marine ecosystem

19.5.1 Primary productivity

All the species mentioned presumably contribute to primary production. The indications are that the unicellular nannoplankton forms (19.2.1) may be especially important in this respect. Some are sufficiently small to pass through filters used in the standard radiocarbon method and this may have led to underestimation of primary production (Johnson & Sieburth, 1979). Carpenter and Price (1977) and Dunstan and Hosford (1977), among others, have reported radiocarbon determinations of primary production in *Trichodesmium* blooms.

19.5.2 Nitrogen fixation

Combined nitrogen is commonly the nutrient in shortest supply in the sea and therefore nitrogen-fixing organisms would be expected to have an advantage. However, substrates suitable for heterotrophic fixers are present in seawater in minimal amounts and the only readily available source of energy for fixation is light. The photic zone is normally oxygenated so that nitrogen fixation in it is limited to photosynthetic organisms with mechanisms for avoiding oxygen inactivation of nitrogenase, that is, to blue-green algae. Seawater is an excellent medium for fixation by these micro-organisms—rates averaging 0.18 g N $m^{-2}day^{-1}$ on algal reef flats off the Marshall Islands (Wiebe *et al.* 1975) being among the highest recorded for natural communities. Nevertheless, *Trichodesmium* and *Richelia* appear to be the only micro-organisms of this kind in the plankton; they are limited mainly to warm waters and, unless some of the chroococcalean forms are found to possess the property, nitrogen-fixers are virtually absent from colder oceanic waters. Why this should be so is not clear unless it is that the hypothesis advanced in 19.4.1 is correct and *Trichodesmium* is the only nitrogen-fixer that has been able to adapt gas vacuoles to be of positive advantage in turbulent waters.

Trichodesmium is widespread and often abundant in warmer waters and work carried out in various regions shows that it is always, although to a variable extent, nitrogen-fixing (*e.g.* Dugdale *et al.* 1961; Dugdale *et al.*, 1964; Bunt *et al.*, 1970; Mague *et al.*, 1974; Saino & Hattori, 1978). As expected from its buoyancy and the dependence of fixation on light, fixation is largely confined to the photic zone and the variation of rate with depth is similar to that for photosynthesis. Mague *et al.* (1977) found that in the northern Pacific gyre most fixation occurred in the top 40 m with detectable fixation down to 90 m, which was about the depth at which irradiance was 1% of the surface value. No seasonal variation in nitrogen-fixing capacity was found by Taylor *et al.* (1973) but since denser growths occur in summer in the Atlantic the contribution of fixed nitrogen is greater then, as Mague *et al.* (1977) also found for the northern Pacific gyre. Fixation is reduced if combined nitrogen is available and low rates in the spring in the northern Pacific gyre may be attributable to this. Mague *et al.* (1977) found peaks in *in situ* acetylene reduction in the morning and late afternoon—the existence of a mid-day depression being at variance with the results of their shipboard experiments in which they found no photoinhibition (19.3.4) but time effects may be involved here. Surprisingly, they found high rates, almost approaching day-time values, in the dark, with a maximum at 40 m, in the early part of the night.

A general conclusion is that nitrogen-fixing blue-green algae normally supply only a small fraction of the total combined nitrogen requirements of phytoplankton. Mague *et al.* (1977) found from experiments with ^{15}N-labelled nitrate, ammonium and urea that uptake of these substrates, at a time when their concentrations were appreciable, met 97% of the nitrogen requirements of the phytoplankton, leaving only 3% to be supplied by fixation. However, *Trichodesmium* formed only a small proportion of the total phytoplankton at this time and isolated colonies could supply 100% of their own requirements in this way. Bryceson and Fay (1981) observed, using ^{15}N as a tracer, a transfer of fixed nitrogen from the alga to the nannoplankton and/or bacterial fraction of the phytoplankton. Devassy *et al.* (1978) found a rapid release of ammonium-nitrogen into seawater towards the end of *Trichodesmium* blooms. Gundersen *et al.* (1976) concluded that fixation by *Trichodesmium* and *Richelia* might contribute about 3.8% of the total annual nitrogen requirement of the phytoplankton in the vicinity of the Hawaiian Islands. With phytoplankton in which *Richelia* was abundant Mague *et al.* (1974) found rates suggesting that between 2–5% or perhaps even up to 30% of the daily nitrogen requirement might be provided by fixation.

Fogg (1978) drew up a balance sheet of oceanic nitrogen resources which suggested that annual input of combined nitrogen falls short of annual loss by some 10^8 tonnes. Assuming a balance between input and loss—which is not necessarily so—fixation would be expected to make up the difference and planktonic blue-green algae appear to be the only organisms in a position to make a major contribution. Fogg (1978), assuming an average planktonic rate of 375 μg m^{-2}h^{-1} occurring for 12 h per day for half the year in all waters of less than 200 m depth between the parallels of 25°N and 25°S, calculated a total contribution from *Trichodesmium* of 0.1×10^8 tonnes per year. A revised calculation might be based on the estimate by Mague *et al.* (1977) of 33 μg at N m^{-2}day^{-1} as an average rate for 4 months in the year but takes in all the sea area between the parallels of 30°N and 30°S (188.59×10^6km^2). This gives 0.000105×10^8 tonnes per year. Mague *et al.* (1977) from a similar calculation concluded that *Trichodesmium* fixation could nearly make up the short-fall, between supply of combined nitrogen from other sources and its consumption, in the areas in which it occurs but the value which they take for loss of nitrogen into sediments seems too small (see Fogg, 1978) so that the balance may not be so good as they suppose. It thus seems that planktonic blue-green algae make a minor contribution and that oceanic reserves of combined nitrogen are being depleted. It is more likely that assumptions made in drawing up the balance sheet are wrong, or, possibly, fixation by nannoplanktonic forms is appreciable, or, average rates of fixation by *Trichodesmium* have been underestimated.

In restricted sea areas nitrogen fixation by blue-green algae may contribute substantially to fertility. Öström (1976) calculated that *Nodularia spumigena* fixed 2000 tonnes of nitrogen per year over an area of $2 \times 10^5 km^2$ in the Baltic. Horstmann *et al.* (1978) considered this estimate to be low and quoted an average value of 10^5 tonnes per annum obtained by Rinne and co-workers. This can be compared with the annual total nitrogen load of 10^4 tonnes discharged into the Baltic by Swedish rivers. Horstmann *et al.* (1978) related this activity of blue-green algae to the eutrophication of the Baltic by an annual discharge into it of about 2.2×10^4 tonnes of phosphorus.

19.5.3 Associations of other organisms with Trichodesmium

Trichodesmium blooms often seem unialgal and Devassy *et al.* (1978) found a low diversity index in the phytoplankton at the time of blooms. It has often been said that zooplankton and fish avoid blooms (Ferguson Wood, 1965; Daniel *et al.*, 1978) but it seems that *Trichodesmium* colonies provide a substrate for a characteristic community of micro-organisms and animals. Margalef (1978) went so far as to describe the association of *Trichodesmium* with bacteria and fungi as a symbiosis. Both types of colony contain numerous bacteria but the mucilage characteristic of the radial type may contain at least two fungi, one filamentous and perhaps a phycomycete, and the other a chytrid. Protozoa, dinoflagellates and various diatoms are also frequent (Borstad & Borstad, 1977). An amoeba has been described (Anderson, 1977) which preys on the bacteria growing in the colonies. A wide variety of zooplankton, principally copepods but also chaetognaths, salps and pteropods, are usually present in blooms (Devassy *et al.*, 1978) but additionally there are animals with a specific association with the *Trichodesmium*. Geiselman (1977) described a colonial hydroid found within the colonies in nearly 40% of near-surface samples collected at seven stations in the subtropical Atlantic (see also Borstad & Borstad, 1977). Calef and Grice (1966) observed a relationship between the harpactacoid copepod *Macrosetella gracilis* and *Trichodesmium* off north-eastern South America and the same association was found by Borstad and Borstad (1977) in the Caribbean. Roman (1978) found that *Trichodesmium* is ingested by *Macrosetella* and is necessary for the copepod's growth and development. *Macrosetella* thus appears to be an important agent for secondary production from *Trichodesmium*. Devassy *et al.* (1978) do not consider that *Trichodesmium* is used directly as food by fish. In the Arabian Sea the enrichment following the decay of a bloom results in a rapid succession, beginning with diatoms, followed by cladocerans, then dinoflagellates and green algae, copepods and finally carnivores (Devassy *et al.*, 1979).

19.5.4 Toxicity

Freshwater blue-green algae, among them *Aphanizomenon flos-aquae* and *Anabaena flos-aquae*, are sometimes toxic to higher animals (Fogg *et al.*, 1973) but there does not seem to be any evidence that they are troublesome in this respect in the Baltic (Horstmann *et al.*, 1978). There have been suggestions that *Trichodesmium* blooms cause disease in humans (Sato *et al.*, 1963/4) and Ramamurthy (1970a,b) regarded them as responsible for antibacterial activity detected in the gasto-intestinal tracts of fishes and of a gull. Devassy *et al.* (1978) found no mortality of fish or other animals associated with *Trichodesmium* blooms and noticed no irritation or other ill-effects after being in contact with them. *Trichodesmium* can, of course, cause death of fish by clogging their gills and of fish and other animals when decay leads to depletion of oxygen.

CHAPTER 20
MARINE LITTORAL

B. A. WHITTON & M. POTTS

20.1 Introduction

Benthic cyanobacteria are widespread along maritime coasts, often forming visually conspicuous growths on rocks and sediments. The colouration of zones on limestone shores may, for instance, be due largely to the dominance of cyanobacteria (Schneider, 1976). What are clearly rather similar communities have been recorded from many different countries and examples from salt marshes have been recorded sufficiently often to regard them as cosmopolitan. Some littoral communities dominated by cyanobacteria do however appear to be restricted to warm waters; this is true especially of those which form well-developed stromatolites (Walter, 1976). The present chapter first describes the zonation, geographical distribution and ecology of these communities and then discusses some of the environmental factors influencing them and the processes going on inside them. Although not part of the littoral zone *sensu strictu*, brief accounts are given of the supralittoral and upper sublittoral. Topics dealt with by Fogg (1973) have largely been omitted, but his chapter subheadings have been included to permit easy cross-reference.

A review dealing with coastal cyanobacteria is cursed with various problems of terminology. A variety of different classifications have been suggested to describe the vertical zonation of the littoral system. We use a relatively simple one for general purposes, with three main zones, the supralittoral, eulittoral and upper sublittoral. Schneider (1976, p. 11) provides a useful chart for comparing the zones and names used by various authors. A second problem is introduced by the words sand and silt. Although these terms are used loosely in the great majority of biological papers, nevertheless for geologists they have the strict definitions given by Wentworth (1922). Common sense is needed in reading the literature: no doubt many of the accounts of sandy shores deal with ones which have in fact a high silt component. A third problem lies with the names of the organisms themselves. Many systematic ecological accounts follow the conventions of classical floras such as that of Geitler (1932). However some modern authors, especially from the U.S.A., have adopted all or part of the system of Drouet (Drouet & Daily, 1956; Drouet, 1968). Humm and Wicks (1980), who are among these, give a table for marine species comparing the two sets of names. Species *sensu* Drouet frequently include a range of classical species which differ widely one from another (see Nielsen, 1973; le Campion-Alsumard, 1979), so we have adopted the convention of using quotation marks when giving a Drouet binomial.

20.2 The littoral system

20.2.1 Rocky shores

A diverse flora of epilithic and endolithic cyanobacteria is a common feature of rocky shores throughout the world, particularly those where the substratum is soft and porous (20.3.2). On temperate shores, epilithic communities dominated by species of *Entophysalis*, *Calothrix* and various Oscillatoriaceae are almost always present (Frémy, 1933; Umezaki, 1961; Fogg, 1973), but they tend to be more conspicuous in the upper eulittoral and supralittoral, where the density of grazing molluscs is less (Lewis, 1964). In the uppermost part of the supralittoral, however, free-living cyanobacteria are replaced by a blackish lichen crust. Species of *Rivularia* are frequently present and may be conspicuous. Womersley (1946) concluded that there are few places along the rocky coasts of Australia where one or more of five species (*R. atra*, *R. australia*, *R. firma*, *R. nitida*, *R. polyotis*) are not found. However, at least one species on British coasts, *R. bullata*, can show marked seasonal variation in abundance, occurring along much of the south coast as a summer annual

(Fogg *et al.*, 1973). Marked zonation among communities has been described by many authors. For instance, *Gloeocapsa crepidinum* dominates the upper part of the supralittoral on sandstones near Millport, Scotland, while *Nostoc entophytum* increases towards the splash zone (Little, 1973). Stewart (1975a) recognized three types of *Rivularia* association on British coasts: *R. atra* and *R. biasolettiana* in the supralittoral; *R. atra* and *R. nitida* among and below the *Calothrix* zone on sandstone shores; *R. bullata*. For limestones of the Adriatic coast of Yugoslavia, Schneider (1976) recognized seven zones within the supra- and eulittoral: lichens, halophytes, grey, blue-black, dark-brown, yellow-brown, white. The five colour zones are caused by micro-organisms, among which the cyanobacteria play a dominant role. For instance, *Hormathonema paulocellulare* is especially important in the blue-black zone and *H. luteo-brunneum*, *Hyella tenuior* and *Solentia foveolarum* in the yellow-brown zone. The colours are, however, only partially due to the presence of particular species, since species may develop different colours in the sheath material according to position on the shore, with higher zones having a deeper colour than lower zones. Because the grazing activity of gastropods removes epilithic forms preferentially, the colour of lower zones is due largely to the endoliths (20.4.3). Schneider included a detailed comparison of his own observations with the many earlier ones of Ercegović on the Adriatic coast and also those of le Campion-Alsumard (1969) for the Mediterranean and of his own for Bermuda and Florida; he concluded that the zonation on each of these limestone coasts was, in general, quite similar.

Ecological accounts of rocky shores in the tropics and subtropics are largely restricted to limestone surfaces. On Heron Island, Australia, with its beach, reef flat and seaward reef, the beachrock has three algal bands, the uppermost two of which are dominated by '*Entophysalis deusta*' and *Kyrtuthrix maculans*, respectively (Cribb, 1966). Van den Hoek has suggested that the '*Entophysalis deusta*' subzone is probably world-wide and he used its upper limit to define the limit of the eulittoral on Curaçao coasts (van den Hoek, 1969) and inland bays (van den Hoek *et al.*, 1972). Van den Hoek (1969) refers to *Entophysalis* and *Hyella* as different phases of the same organism (see Prud'homme van Reine and van den Hoek, 1966; le Campion-Alsumard, 1970). In the *Entophysalis* subzone of Curaçao there are occasional *Gloeocapsa*- and *Hormathonema*-like growths, while dark-brown patches of '*Calothrix crustacea*' are nearly always present on blue-green *Hyella*-perforated rocks. The rocky coasts of atolls also support obvious growths of cyanobacteria. Newhouse (1954) commented on the 'slick brown coatings' of *Rivularia polyotis* on intertidal beachrock along the seaward coast of Raroia Atoll in the Pacific. An association of *Entophysalis*, *Calothrix* and *Nostoc* occurs almost everywhere along the coast of this atoll. The same three genera are the dominants of beachrock of the lagoon shore of Onotoa Atoll, Gilbert Islands

Fig. 20.1. Beachrock bounding reef-flat on oceanic shore of Aldabra Atoll.

(Moul, 1957). Zonation on different parts of the Aldabra coast has been described by Whitton and Potts (1980), who concluded that in general the longer rocks are covered by water each tidal cycle, the less conspicuous are the cyanobacteria.

Beachrock itself is a conspicuous feature not only of atolls, but also of many other coasts in the tropics and subtropics, extending further north and south than coral reefs (Krumbein, 1979). Typically it appears as a series of beds, dipping seaward with a slope of 15° and at a strike parallel to or at a slight angle to the trend of the beach (Komar, 1976). It may be composed of quartz and other associated minerals, although it is most common on beaches of calcareous sediment. It occurs near the high-tide level and on high energy shores with pebbles and shingle may provide the only stable substratum for algal colonization. Cyanobacteria are apparently always present as epiliths, chasmoliths and endoliths (20.4.3), including sometimes as a discrete, well-developed cryptoendolith layer (Golubic *et al.*, 1981), such as that described by Moul (1957), with *Oscillatoria nigro-viridis*, *Lyngbya confervoides* and *Phormidium valderianum* at a depth of 6 mm inside the rock. Ever since Darwin's (1841) study at Pernambuco, Brazil, some authors have suspected that the origin of beachrock may be partly biogenic. Krumbein (1979) both reviewed the literature on this subject and reported on his own studies along the shores of the Gulf of Aqaba. As a result of the latter he proposed a model for beachrock development and biogenic calcification, in which cyanobacteria play a major role. This includes a number of stages, starting with the build-up during calm periods on 'high energy' beaches of mats that agglutinate particles. Some carbonates are precipitated by photosynthetic CO_2 uptake, but these do not normally lead to cementation. The more important stages start when reducing conditions are established in the lower zone of the mat, leading to sulphate reduction and sometimes a slight increase in pH. Anaerobic and aerobic decay of the cyanobacteria and other organic compounds yield carbonates that will precipitate from solutions of widely varying Ca and Mg concentrations.

20.2.2 Sandy shores

Visually conspicuous cyanobacterial communities are restricted on sandy shores to areas which remain undisturbed for relatively long periods, although once established they may play an important role in binding the sand. According to Williams (1950), the mat of cyanobacteria and diatoms near Ocracoke, N. Carolina, is sufficiently firm that it was used during the second world war as a landing strip. In recent years there have been many studies on thick intertidal mats and these are described in 20.4.1. The main types of cyanobacterial community of sandy shores are superficial mats, with filaments either predominantly horizontal or both horizontal and vertical, a community of interstitial organisms merging with ones directly associated with the surface of the sand grains and an endolithic community. The type and floristic composition of these communities appears to be determined by a variety of factors. Among these are the chemical composition of the sand, especially whether quartzite or carbonate minerals, nature of the underlying sediments, influence of any water movement through the sand and perhaps also atmospheric humidity and rainfall.

Growths of filamentous cyanobacteria on or near the surface of calcareous sand are mentioned in several accounts of atolls, including growths on sand dunes and true intertidal communities. Moul (1957) reports a community on old dune sands of Onotoa Atoll, Gilbert Islands, with *Microcoleus chthonoplastes* and *Scytonema hofmanni*; the former occurred also on the lagoon beach. A juvenile phase of *Fischerella ambigua* was apparently responsible for the 'somewhat green' colour of a considerable area of beach sand on Rongelap Atoll, Marshall Islands (Taylor, 1950); it did not form a mat binding the sand together. Degener and Degener (1959) refer to various growths colouring marine sand on Canton Island, Southern Pacific, though these are probably all in the upper sublittoral. For instance, *Hydrocoleum glutinosum* imparted a light bluish-green colour to a fine white sand overlying an oceanic reef; this sand was not washed away by wave action, probably being held in place by the *Hydrocoleum*. On Aldabra, old dune sands have a cover of *Microcoleus chthonoplastes*, *Schizothrix arenaria* and *Tolypothrix byssoidea* (authors, unpubl.). Lower down, parts of the supralittoral with coarse sand lack any photosynthetic cover, but the upper fringe of the eulittoral on the lagoon shore has well-developed mats of *Scytonema* sp. Areas of sand from which mat was removed appear to require at least 6 months to re-develop a 'mature' community (Potts & Whitton, 1980). Other cyanobacteria replace *Scytonema* sp. on finer sediments (Potts & Whitton, 1979a).

Many shores lacking a surface mat nevertheless have a well-developed photosynthetic community in the upper few millimetres of sand. Though diatoms were found to be the most abundant group of benthic microalgae in the Hohe Weg shallows of the German Wadden Sea (Colijn & Nienhuis, 1978), *Merismopedia* was also very common, sometimes forming a 'bloom'. It seems probable that this cyanobacterium has increased in abundance in recent years along the shores of Niedersachsen (Michaelis, 1978, cited as '*Agmenellum quadruplicatum*'). In 'sandy-mud' of the Minor River estuary, N.W. Spain, *Merismopedia* was found to reach its maximum density at the end of summer (Niell & Anadon, 1978); in contrast, on the coast of Wangerooge Island, Germany, the dominant *Merismopedia* found in spring was replaced by *Microcoleus* in late summer (M.P., unpubl.). Although cyanobacteria are often an important component of the algal community of sands, it is not clear to what extent they are associated with the surface of particles (epipsammic) or are truly interstitial. However, where the sands are calcareous and coarse, most of the community may be truly endolithic. *Hyella* is often dominant here and sometimes colours sands pale green, as in parts of the lagoon intertidal of Aldabra (Potts & Whitton, 1980). *Mastigocoleus testarum* and *Plectonema terebrans* were found to be the most abundant organisms in experimental studies on colonization of carbonate fragments made by Perkins and Tsentas (1976) and May and Perkins (1979). It seems probable that particle size is an important factor determining species composition, though in the experimental studies availability of inocula may have been the more important. Photosynthetic activity in the surface layer of a calcareous sand on Bermuda brought about a marked diurnal fluctuation in pH (Gnaiger *et al.*, 1978), though details of the organisms were not included.

A colourful zonation of microorganisms in sands and other sediments on sheltered shores was first described by Schulz (1937; Schulz & Meyer, 1939), who gave it the name, Farbstreifen-Sandwatt. Such phenomena are particularly well developed along the German North Sea coast (Hoffmann, 1942; 1949). Here, in its most simple form, it consists of a thin surface layer of bleached, rather white sand, followed by a green layer of cyanobacteria and a lower reddish layer of phototrophic bacteria (e.g. Anagnostidis & Schwabe, 1966). These layers usually comprise the upper few millimetres of the sediment, with much deeper, black anaerobic sediment beneath. A more complex stratification at Sührendorf, Schleswig-Holstein, was studied by Hirsch (1978). A yellowish-brown surface layer 5 mm deep was composed principally of diatoms. A second zone 1–2 mm thick contained six species of cyanobacteria and *Beggiatoa*. A third zone 5 mm thick had dense populations of two *Chromatium* species. The fourth zone, 5 mm thick, was grey to greyish-brown with *Oscillatoria profunda* and *Aphanothece* sp. The *Oscillatoria* had sulphur granules on its surface, contained gas vesicles and was motile. The final black zone contained *Desulfovibrio*. Hoffmann (1942) listed twenty-nine cyanobacteria from this community at Kniepsand, Amrum, including *Microcystis reinboldii*, *Microcoleus chthonoplastes*, *Calothrix scopulorum*, *Oscillatoria nigro-viridis*, *Lyngbya confervoides*, *L. aestuarii* and *L. semiplena*. *Microcystis reinboldii* is one of the binomials most commonly listed for the so-called Sandkörnchen-Flora, but it probably represents an assemblage of diverse unicellular cyanobacteria. Besides light intensity and quality (20.3.4), several other factors, especially O_2, Eh and H_2S, are known to show marked changes on passing through these communities and may perhaps help determine their vertical stratification. On a sand beach in the lagoon of Aldabra, photosynthetic microbial communities undergo marked, cyclical changes in their spatial distribution during each 14-day tidal cycle (Potts & Whitton, 1979b). The communities show many similarities with the vertically zoned ones described for Germany, but because the zonation is predominantly *horizontal*, it is much easier to follow the changes. The horizontal zonation is most clear at a stage in the tidal cycle half way between spring and neap tides. A series of small pools show a pink zone, with *Chromatium* dominant, at the side of the pool closest to the mangrove forest, then a green zone of *Hyella balani* and finally a yellow-brown zone of diatoms at the side nearest the lagoon. This zonation and subsequent changes during the 14-day cycle are thought to be due to the passage of low Eh and probably sulphide-rich water from the mangrove forest draining through the beach. Hirsch (1978) suggested that the regular occurrence of a layer of cyanobacteria below the *Chromatium* layer in the Farbstreifen-Sandwatt at Sührendorf may indicate a capacity for anoxygenic photosynthesis. However, in view of the report by Krumbein *et al.* (1979) of a stromatolitic mat where oxygenic cyanobacteria occur below a layer of purple phototrophic bacteria, the possibility must be considered that a similar phenomenon may occur in beach sands.

20.2.3 Salt marsh and mangrove

Some species of cyanobacteria are more or less cosmopolitan in their occurrence on salt marshes and occur together sufficiently often to be recognized as 'ecological' communities. A general cyanobacterial community with species of *Oscillatoria*, *Lyngbya*, *Phormidium* and *Microcoleus* occurs on bare mud flats and muddy soils between phanerogams and extends up to high intertidal levels; it has been described from Europe, Iceland, eastern N. America, India and New Zealand (Chapman, 1977). A perennial community on the salt marshes of the western Dutch Wadden Sea dominated by *M. chthonoplastes* and *Lyngbya aestuarii* and lacking green algae is sufficiently widespread to have been designated the Microcoleo-Lyngbyetum aestuarii (de Jonge, 1976). A gelatinous community characterized by species of *Nostoc* and *Rivularia* is commonly found on high marshes on both sides of the Atlantic (Chapman, 1977). Brackish soils in southern France which carry the Salicornietum fruticosae association have a surface cover of diatoms and cyanobacteria; among the latter *Calothrix scopulorum*, *Plectonema nostocorum* and *Microcoleus chthonoplastes* are the most important (van den Hoek, 1960).

Salt marsh communities often show obvious seasonal changes, especially those occurring in the upper parts of the marsh. For instance, at Scolt Head, Norfolk, England, a 'marginal' cyanobacterial community develops only between July and October, although the gelatinous community is present throughout the year (Chapman, 1960). De Jonge (1976) and Nienhuis (1978) have confirmed for several Dutch salt marshes many of the earlier observations for Scolt Head of Carter (1933a, 1933b) and Chapman (1939, 1940, 1960), including the development of several communities only in summer. De Jonge reported, however, that the 'Autumn Cyanophyceae Community' of the previous authors was present all the year and considered that the community described at Scolt Head is merely a seasonal aspect of the Microcoleo-Lyngbyetum aestuarii. Webber (1967) identified thirty cyanobacteria from a salt marsh by Cranes Beach, Ipswich, Massachusetts. Of these, twenty-three were non-heterocystous and were collected throughout the year, whereas several of the heterocystous forms such as *Nodularia harveyana* were found only during summer. Another heterocystous species, *Anabaena torulosa*, produced akinetes only during summer.

Light intensity has been suggested by several authors to be a major factor influencing the abundance of cyanobacterial communities in salt marshes, both through the year and in different parts of the marsh. This was, for instance, thought to be the main factor influencing cyanobacterial abundance in a Mississippi salt marsh (Sage & Sullivan, 1978), where the numbers of species recorded in all five different monotypic angiosperm zones was greatest in summer and lowest in winter. In contrast, light conditions in some algal habitats in a Cape Cod salt marsh become marginal as graminoids grow during summer (Blum, 1968). The algal cover under mature *Spartina patens* becomes sparse, with *Calothrix* sp. being the most successful species on moist parts of the dead plant. A continuous algal cover remains under the stunted growths of high marsh *Spartina alterniflora* and Blum suggested that partly as a result of this, dead *S. alterniflora* decays more rapidly than *S. patens*.

Although the importance of light intensity has been mentioned by these and other authors, it is important to remember that many other changes are likely to show some correlation with changes in light intensity in temperate latitudes. Elevated temperatures in summer are likely to favour growths of cyanobacteria. Chapman (1938, 1939, 1940, 1941, 1959) recognized six major non-climatic factors responsible for the appearance and disappearance of salt marsh communities:

(i) frequency, height and duration of tidal floodings;
(ii) precipitation and evaporation;
(iii) salinity of the soil moisture;
(iv) moisture content;
(v) exchangeable calcium;
(vi) soil composition.

Most of these are likely to vary seasonably, with cyanobacteria being favoured selectively in summer. A low soil moisture content eliminated a green algal community at one of the Dutch sites studied by Nienhuis (1978), but still permitted some cyanobacterial cover throughout the year.

Mangrove forests are common along many shores in the tropics and subtropics, particularly where there is protection from severe wave action and where major rivers enter the sea (Chapman, 1977), and the cyanobacterial communities show some similarities with those on salt marshes. The dominants of the principal community of salt marshes, *Oscillatoria*, *Lyngbya*, *Phormidium* and *Microcoleus*, are widespread, as also are heterocystous genera towards the upper tidal limits. Of thirty-seven species of cyanobacteria recorded (Saenger *et al.*, 1977) from mangrove forests in S.-E. Queensland, eleven were heterocystous, including *Calothrix crustacea*, *Anabaena torulosa* and *Fischerella muscicola*. However, judging by the floristic lists of van den Hoek *et al.* (1972), heterocystous species form a smaller component of the mangrove flora of Curaçao. A vertical zonation of heterocystous and non-heterocystous species is sometimes easy to distinguish on pneumatophores or prop roots. For instance, the following zonation sometimes occurs on *Avicennia* pneumatophores along parts of the Sinai coast of the Red Sea where the epiphytic flora is predominantly cyanobacterial (Dor, 1975; Potts, 1979, 1980). The highest level is occupied by a *Scytonema* with thick lamellate yellow sheaths; the living part of the thallus is a layer a few millimetres thick on the periphery of the colony; the inner mass is formed of empty sheaths. Beneath this zone, but above low tide level, there are several species of *Rivularia*, together with *Calothrix* and *Kyrtuthrix*, while in the permanently submerged zone below low tide level there are several non-heterocystous genera (*Lyngbya*, *Hydrocoleum*, *Schizothrix*). Conspicuous horizontal zonation is evident in other parts of the Sinai mangroves (Potts, 1980). *Lyngbya aestuarii* occurs as a cryptoendolith layer in buckled crusts of desert sediments at the extreme upper edge of the supralittoral, whereas lower down the zone where wetting increases, the same species forms thick mats.

There are marked differences in the extent to which the pneumatophores and prop roots of various regions are covered with cyanobacteria or with red and green algae. Eukaryotes such as *Bostrychia*, *Caloglossa*, *Catenella* and *Murrayella* are usually the most evident (Dawson, 1966), while cyanobacteria become dominant only where factors such as high salinity, intense illumination and, presumably, deficiency of combined nitrogen favour their selective advantage. Dor (1975) contrasted two types of mangrove forest in Sinai, the Ras Muhammad with cyanobacteria only, and an area near Nabq with a few green algae present as well. Environmental conditions for the former were regarded as more extreme, due to the greater salinity.

Cyanobacteria are not always conspicuous on mangrove sediments. On Andros Island, Bahamas (B.A.W., unpubl.), extensive cyanobacterial growths (mainly *Scytonema* mats) occur among well-spaced dwarf mangroves, but are sparse in denser forest. Sediments inside well-developed mangrove forest on Aldabra Atoll usually lack an obvious cyanobacterial layer, even where locally the surface is well illuminated (Potts & Whitton, 1980). Nevertheless, in shaded parts of the forest there are almost always visually obvious growths of cyanobacteria on limestone surfaces and often also on prop roots. The authors suggested that high levels of organic material in the sediments may favour heavy grazing by molluscs and direct competition from heterotrophic micro-organisms. It remains uncertain whether light intensity, grazing or the frequency of cover by standing water is the most important factor controlling the development of cyanobacterial mats on mangrove sediments.

20.2.4 Sublittoral

Although there are many records of benthic cyanobacteria in the sublittoral, most of these refer only to scattered tufts of the larger *Lyngbya* spp., epiphytes such as *Dermocarpa* visible only with the microscope or endoliths in carbonate substrata. Records of conspicuous growths come mostly from warmer waters, where films of filamentous species overlying sand in shallow water appear to be relatively common. In fact, Gebelein (1969) reports that superficial mats composed of organic materials and bound sediment grains are absent in shallow subtidal environments around Bermuda only where: (a) rates of surface sediment turnover by grazers and especially by burrowers are very high; and (b) rates of sediment movement over the surface preclude any binding of grains by organic mucilage. These mats can themselves greatly retard sediment erosion (Bathurst, 1967; Neumann *et al.*, 1970). Such mats can include cyanobacteria, green algae, diatoms, red algae and invertebrate mucilage. Mats with photosynthetic organisms have been observed to depths of 50 m, but in general growth is most rapid in depths of less than 10 m (Gebelein, 1976). Sournia (1976) describes a population of *Oscillatoria limosa* at a depth of 20–80 cm in a lagoon at Mooréa Island, which migrated into the sand at periods of intense illumination (20.3.2); if this type of response is widespread, relatively dense populations could easily be overlooked. The Mooréa cyanobacterium was heavily grazed by the holothurian *Halodeima atra*, but large amounts of viable filaments were released among the faecal pellets. Sublittoral stromatolites and oncolites (20.4.3) are perhaps also more widespread than the records would indicate (see Gebelein, 1976). The increasing number of accounts of probable symbiotic associations in the sublittoral of warmer waters (Table 20.1) suggests the likelihood that there are still many more examples to be found.

Mats with organisms which are colourless but otherwise quite similar to cyanobacteria have been described (Jannasch & Wirsen, 1981) at 2500 m depth. Filaments of *Leucothrix* and *Thiothrix* grow as single filaments perpendicular to the substratum, much as *Calothrix* does in the section of a mat illustrated by Golubić (1973).

Many authors have commented that filamentous and fleshy benthic algae are unobtrusive on shallow, but permanently submerged, parts of coral reefs (e.g. Wanders, 1977). The cyanobacteria are no exception, and in the experiments carried out by Wanders, cyanobacteria were not among the species listed which did colonize when grazing was excluded. However, *Calothrix crustacea* was one of three early colonizers on artificial reefs (made of old tyres) in a shallow part of the lagoon on Guam (Tsuda & Kami, 1973). This species and *Microcoleus lyngbyaceus* were the dominants in the 'climax' community which developed within a year. The authors suggested that low light penetration through the silty water and selective browsing by herbivorous fishes were the factors favouring these cyanobacteria. *Lyngbya majuscula* and *Rivularia* were conspicuous among the early colonizers of cleaned glass slides placed at various depths in a lagoon of the Great Barrier Reef (Borowitza *et al.*, 1978). *Calothrix* is frequently present in the periphyton which develops on submerged angiosperms. One such community, forming on leaves of *Posidonia* and *Cymodocea* along the Tunisian coast, is sufficiently abundant to have got the local name 'Muffa' (Feldman & Frémy, 1935). The whitish gelatinous masses, which consist mainly of cyanobacteria and diatoms, frequently become detached due to the presence of oxygen bubbles and float to the surface. According to Golubić (1973), such 'flocculose' mats form in extremely shallow and nutrient-rich waters.

Table 20.1. Examples of close associations between cyanobacteria and other organisms in eulittoral and sublittoral benthos; the literature on *Prochloron* is included, in view of some similarities between this genus and unicellular cyanobacteria. (Most of the relationships are probably symbiotic, but only a few cases have been investigated critically.)

Organism	Associated organism	Morphological relationship	Environment	References
Aphanocapsa spp.	sponge (*Ircinia variabilis*)	extracellular and intracellular	Mediterranean sublittoral	Sara (1965, 1971)
Phormidium (red-coloured)	*Codium bursa*	endophytic	Mediterranean sublittoral	Jacob (1961)
Lyngbya sordida (usually)	shrimp	algal tubes contain animal	sublittoral	Cowles (1913); Taylor (1950); Newhouse (1954); Whitton & Potts (1980)
Synechococcus	sponge (*Verongia aerophoba*)	extracellular, but very close	Mediterranean sublittoral	Vacelet (1971)
unicellular	echiuroid worms	intracellular	eulittoral & sublittoral	Kawaguti (1968, 1971)
unicellular	opistho-branch	intracellular	tropical Pacific, upper sublittoral	Kawaguti *et al.* (1965)
Prochloron (? cyano-bacterial relative)	didemnid ascidians	epizoic and in peribranchial cavities	tropical reef-ridges	Newcomb & Pugh (1975); Lewin (1976); Withers *et al.* (1978a); Thinh (1979); Kott (1980)

20.3 Environmental factors

20.3.1 Substratum

Most of the different types of substratum which occur in the intertidal can support some growth of cyanobacteria. On exposed shores, cyanobacterial communities are most abundant on soft, porous rocks such as sandstone (Little, 1973; Renaut *et al.*, 1975), though if the rocks are extremely soft and friable, wave action and associated abrasion may restrict colonization (Golubic *et al.*, 1975). Harder rocks such as granites and slates usually support a more conspicuous growth of lichens (e.g. Stewart, 1975a,b). Observations of supralittoral and intertidal rockpools along the coast of Oahu, Hawaii, indicate that of the two main substrata present, lava and limestone, the former supports the greater number of species of cyanobacteria (Khan, 1969).

Certain species are restricted to one or the other of the rock types; for instance, '*Brachytrichia quoyi*', '*Hydrocoleum cantharidosum*' and '*H. lyngbyaceum*' are found only on lava.

A hard substratum is not essential for the growth of cyanobacterial mats (Black, 1933). Mats form over other algal surfaces, skeletal sands, aragonite muds and gypsum mushes (Kendall & Skipwith, 1968). They may also bind quartz sands (Schwarz *et al.*, 1975), mangrove muds (Gunatilaka, 1975) or grow within halite crusts (Potts, 1980).

20.3.2 Exposure and desiccation

Wave action, extent of exposure to the atmosphere and degree of desiccation when exposed are all important factors influencing the distribution and growth form of intertidal communities. In the supralittoral of coasts with high energy waves, cyanobacteria appear typically as black to brown, tightly adhering growths, although intense wave action may loosen communities towards the lower supralittoral (Wärmling, 1973). Marine forms may extend well above the limit reached by extreme high water springs where there are large waves breaking against rocks. In the supralittoral of sheltered coasts and lagoons, thick growths of cyanobacteria sometimes cover extensive areas (20.4.1).

Depending on their position on the shore, communities are subject to varying degrees of wetting and subaerial exposure. In general the number of species decreases as extent of exposure increases; changes visible to the naked eye are also often obvious. This can be illustrated with examples from Hamelin Pool, Shark Bay, Australia (Hoffman, 1976). On moving from the supralittoral down through the intertidal, there occur, in sequence: film mat (*Hormathonema violaceo-nigrum*), blister mat (*Microcoleus chthonoplastes*), pustular mat (*Entophysalis major*), tufted mat (*Lyngbya aestuarii*), smooth mat (*Schizothrix helva*), gelatinous mat (*Symploca laete-viridis*), colloform mat (*Microcoleus tenerrimus*). Other examples are given in 20.4.1. On rocky limestone coasts, studies indicate that along a gradient of increasing water supply, average cell size and depth of carbonate penetration tend to increase (Golubic *et al.*, 1975; le Campion-Alsumard, 1970). This usually means that deep-boring forms are restricted to the lower intertidal, whereas shallow-boring and epilithic species are found in the supralittoral (see Schneider, 1976). Extensive boring by *Scytonema endolithicum* does however occur in the upper supralittoral to a depth of 1200 μm on Aldabra (Whitton & Potts, 1980), although this species reaches only 100–200 μm depth on Adriatic coasts (Schneider, 1976).

Critical studies on desiccation and intertidal cyanobacteria are largely lacking (but see 20.4.5); however, it is reasonable to assume that any effects are most pronounced in hot dry regions, especially where the substrata are non-porous rocks.

20.3.3 Temperature

The temperature of coastal waters close to the continents ranges from about $-2°$ to 43°C (Kinne, 1970), but exposed communities in the intertidal may encounter much wider extremes. Most forms living at high latitudes must withstand periodic freezing, but abrasion by sea ice may render colonization of intertidal surfaces impossible (Bunt, 1968). The highest temperatures are presumably reached in the tropics on dark non-porous surfaces; cyanobacteria are certainly widespread on such surfaces, although there have been no studies to investigate whether such high temperatures limit growth in any way. When exposed, mat communities appear to withstand large fluctuations in temperature. On the coast of Øslofjorden, communities of *Calothrix scopulorum* fix nitrogen at temperatures between $-5°$ and 11°C (Wärmling, 1973); an intertidal mat on Aldabra underwent daily changes of as much as 18 C° (Potts & Whitton, 1980). In such mats the surface may be 5°C higher than at a depth of 10 mm (Sorenson & Conover, 1962). Marked temperature fluctuations may also occur in small pools, such as those in the dark brown *Hormathonema luteo-brunneum* and *Hyella caespitosa* zone on the Adriatic coast, where Schneider (1976) recorded a range of 18–35°C.

No *in situ* studies have been carried out on the temperature optima of intertidal cyanobacteria but several have been reported for laboratory isolates. *Anabaena* sp. isolated from a mat on the Texas coast had a growth optimum of 42°C when growing on elemental nitrogen (Stacey *et al.*, 1977). The growth optimum for *Calothrix scopulorum* isolated from a Scottish shore was 30°C, although this form is able to fix nitrogen under field conditions at temperatures as low as 0°C (Fogg *et al.*, 1973).

20.3.4 Light

The distribution of the predominantly phototrophic cyanobacteria is influenced markedly by the quantity and quality of light. Most studies have been concerned with communities associated with sediments, where light penetration varies not only with depth, but also particle size (Taylor & Gebelein, 1966). In one study on wet sand, only 52.2% of the surface incident light was available at 1 mm and 1.56% at 5 mm; in dry sand the values were somewhat less (Hoffman, 1949). The maximum depth of penetration of the visible spectrum ($< 10\ \mu E\ m^{-2}s^{-1}$) in algal communities growing along the Gulf of Aqaba varied between 1.4 and 40 mm (Krumbein & Potts, 1978), but here absorption by algal pigments was the most important factor influencing penetration; when the pigments were extracted by organic solvents, light penetration in the upper layers increased by as much as 1000 times.

Vertical sections of mats usually reveal various groups of prokaryotes in discrete coloured layers, often quite similar to the colour zonation of Farbstreifen-Sandwatt communities (20.2.2), and the colours have been interpreted as various responses to light quantity and quality. Golubić (1973) has described the layering in intertidal mats from the Persian (Arabian) Gulf. Below the brown surface, the layers are successively blue-green, salmon-pink, purple-pink and then black. It was suggested that *Lyngbya aestuarii* in the surface layer was protected from excessive radiation by the brown extracellular sheath pigment 'scytonemin'. The brown colouration apparently takes place only in direct sunlight and is effective in reducing incident light by as much as 95% within the first 0.5–1 mm of algal mats (Birke, 1974). Extracts of sheath material from marine *Lyngbya* mats can apparently also show strong ultra-violet absorption (Rambler *et al.*, 1977). Direct evidence for reduced photosynthetic activity at the surface of a mat was reported by Sorenson and Conover (1962). Activity in mats of *L. confervoides* growing in a Texas lagoon was relatively low at the fully illuminated surface, reached its maximum at depths between 1 and 4 mm below the surface and then declined towards the dark interior of the mat. In many mat communities the bright green layer below the surface is composed of motile filaments capable of responding to fluctuating light conditions (see Potts, 1980). In the case of the population of *Oscillatoria limosa* in a lagoon on Mooréa (Sournia, 1976), the filaments move from the surface into the sediment during the hours of most intense illumination. Phototactic movements may play an important role in the development of stomatolites (20.5.3).

The major spectral component of sparse light in shaded habitats is usually in the green region. Under these light conditions, cyanobacteria capable of chromatic adaptation respond by increased formation of phycoerythrin. *Lyngbya sordida*, for example, is blue-green or brown when growing at surface illumination, but pink or red in deep water (Feldmann, 1959; Whitton & Potts, 1980). Fogg *et al.* (1973) remark that forms found at the surface of salt-marsh sediments are usually brownish to reddish, probably due to their phycoerythrin content, whereas red-pigmented forms such as *Oscillatoria nigro-viridis* are common in shaded parts of tidal pools on Aldabra (Potts & Whitton, 1980). Red-pigmented forms are, however, not necessarily restricted to shaded habitats. A phycoerythrin-rich *Lyngbya* occurs along the coast of El Haroura, Morocco (Renaut *et al.*, 1975), and a red *Oscillatoria* is common on supralittoral beachrock near Elat, Israel (Potts, 1980). In contrast, most endolithic cyanobacteria appear blue-green under the microscope, although violet or pinkish forms have been reported e.g. *Aphanocapsa endolithica* (Ercegović, quoted by Geitler, 1932), *Hyella balani* and *Mastigocoleus testarum* (Potts & Whitton, 1980); there are no reports of endolithic communities forming discrete coloured layers parallel to the surface of the rock. The use of photoacoustic *in vivo* spectra (Schubert *et al.*, 1980) now provides a means for studying critically the pigmentation of such communities.

20.3.5 *Salinity*

Intertidal areas vary greatly in the degree and constancy of the salinity to which they are exposed. In the open oceans, the surface salinity ranges from 32 to 38‰, but with the inclusion of secondary seas, the range extends from near freshwater to 41‰. While oceanic water provides a relatively constant saline environment, rock-pools and the wave-spray zone of coasts are probably the least constant (Golubic, 1980). For instance, the water in small pools along the Adriatic coast with endolithic growth of *Solentia* and *Hormathonema* in the limestone can range from 240‰ to near freshwater after rain (Ercegović, 1930). Lagoons, tidal flats, salt marshes and estuaries are intermediate in the extent to which they are subject to fluctuating salinity. In an account of the types of response shown by cyanobacteria to differing salinity regimes, Golubic (1980) classified them into stenohaline or euryhaline, and within each of these as oligo-, meso- or polyhaline.

Many marine strains have an elevated requirement for Na, not even partially replaceable by K, Cl, Mg or Ca (Stanier & Cohen-Bazire, 1977). Optimum growth of the *Anabaena* sp. isolated from a mat on the Texas coast occurred at 4.6% NaCl, with reduced growth at 0.47% NaCl (Stacey *et al.*, 1977). Only one of five other fast-growing strains of *Anabaena* isolated from similar habitats failed to have a finite requirement for Na (Gotto *et al.*, 1979a). All the isolates described by these authors could, however, grow at salinities well below that of seawater, whereas those from marine environments studied by Batterton and Van Baalen (1971) grew better at lower salinities. The latter authors failed in their attempt to isolate cyanobacteria from brines and concluded that forms found there are not halophilic but merely tolerate high salinities, a view challenged by Miller *et al.* (1976) and Brock (1976), both of whom isolated truly halophilic strains; these strains are discussed further in 10.4.

20.3.6 *Pollution*

Many scattered observations through the literature (e.g. Murray & Littler, 1978) indicate that organic pollution may lead to the increased abundance of marine benthic cyanobacteria, but studies of the Yugoslavian coast summarized by Golubic (1970) provide the only detailed account. Pollution in sheltered places such as harbours leads to the occurrence of mats of cyanobacteria, diatoms and sulphur bacteria; these are developed best in parts of the intertidal subject to the heaviest pollution. The mats show an annual cycle of changes, with maximum development in late winter and disintegration in summer in all except the most polluted harbours. The mats consist of two layers, the lower of which is an epilithic community differing from those in unpolluted areas by the fewer species and shift in dominance; *Xenococcus shousboei* and *Isocystis lithophila* are particularly abundant. The upper layer differs in the lower and upper parts of the intertidal zone; *Hydrocoleum lyngbyaceum* and *Phormidium* spp. dominate in the former and *Lyngbya* spp., particularly *L. confervoides*, in the latter. '*L. confervoides*' has also been noted in the U.S.A. to be tolerant of a variety of pollutants, being especially resistant to creosoted pilings (Humm & Wicks, 1980).

The effect of sewage effluents in the sublittoral of Kaneohe Bay, Hawaii, is complicated, because not only are nutrient levels raised, but also the coral has been killed, thus providing a more favourable substratum for cyanobacteria. A study of nitrogen fixation (Hanson & Gundersen, 1976) by *Calothrix* and *Nostoc* growths showed the highest rates in the middle sector of the Bay, where the influence of terrestrial run-off is least. On Long Island in the Bahamas, the development of thick mats dominated by cyanobacteria has been brought about deliberately in solar salt-producing ponds by fertilizer treatment (Davis, 1978); this has led to increased water colour, salinity and salt precipitation.

In a laboratory study (Pulich *et al.*, 1974) of the toxicity of petroleum oils to several marine algae, a strain of '*Agmenellum quadruplicatum*' was found to be considerably more resistant than one of *Microcoleus chthonoplastes*; the former brought about some emulsification of the oil. Nevertheless, when benthic freshwater marsh communities were subjected to oil pollution (Bott *et al.*, 1979), *Microcoleus* was one of three cyanobacterial genera whose growth was favoured.

20.4 Special topics

20.4.1 Mats

The occurrence of extensive mats of cyanobacteria over coastal sands and silts has already been mentioned (20.2.2, 20.2.3). In recent years it has become clear that these structures increase in abundance and often also complexity on passing from temperate regions to the tropics. Many of the more complex types play an important role in binding and trapping sediments and/or in carbonate precipitation and Golubić (1973) has reviewed these in some detail. Cyanobacterial mats may also trap iron (Skyring & Johns, 1980), though it is not clear whether the main role of the organisms is to fix soluble iron from seawater or to trap submicroscopic, suspended or colloidal iron. The more complex mats sometimes include not only several species of cyanobacteria, but also important contributions from other groups. These may consist almost entirely of prokaryotes (Margulis *et al.*, 1980) or may include eukaryotes, especially red algae. Thicker mats often show obvious vertical zonation (20.3.4). Although not strictly a littoral example, a mat in a small saline lake fed by seawater springs on the western shore of the Gulf of Elat/Aqaba (Krumbein *et al.*, 1979) shows just how complex the structure may be. Eight different layers were recognized within the mat, including alternating ones of cyanobacteria and purple phototrophic bacteria. Oxygen production was observed at a depth of more than 16 mm. At the other end of the scale of complexity, the simplest mats are little more than thin films of interwoven filaments consisting largely of a single species of cyanobacterium.

Nevertheless, although many types of mat have been reported, examples exist from widely scattered locations of structures which are remarkably similar both morphologically and taxonomically. It is reasonable to assume that similar environmental and biological factors occur at such sites, but because so many factors are correlated, it is difficult to establish the importance of any particular one without experimental studies. The extent of flooding and intensity of invertebrate activity are two such variables which have frequently been reported to influence the vertical limits of mats on the shore. Desiccation may limit the upper level of development of mats, as in the Persian (Arabian) Gulf, where they are absent in the supralittoral (Golubić, 1973), although elsewhere they are often present in the lower part of this zone. Thick mats are usually limited to the supralittoral and upper part of the eulittoral, because of the burrowing and reworking of sediment lower down the shore by invertebrates. Above a certain flooding-frequency isograd, activity decreases sufficiently for a surface mat to develop. This isograd varies from place to place. For instance, at Cape Sable, Florida, mats occur where the surface is exposed for at least 75% of the time, whereas on Andros Island this value is 50% (Gebelein, 1976). Ginsberg *et al.* (1970) have used this 'exposure index' to compare structures from different areas on Andros Island. 'Bath-tile' mudcracks and polygons occur in supralittoral levées, where exposure indices are high (95–100%); blistered and curled algal mats in the upper eulittoral have values of 75–100%, while stromatolites of *Scytonema* sp. have values of 50–95%. Although thick mats are usually limited to the upper parts of the shore they can extend to the lower part of the shore in sheltered, moderately hypersaline conditions as on part of the lagoon shore of Aldabra (Potts & Whitton, 1980) and the Guerrero Nego Ojo de Liebre lagoon complex in Baja California (B.J. Javor & R.W. Castenholz, pers. comm.). Kendall and Skipwith (1968) suggested that the distribution of the four main types of mat on the shore of a lagoon in the Persian (Arabian) Gulf was determined by the frequency and duration of subaerial exposure and the salinity of overlying tidal waters. Growth at the edges the lower ('cinder') zone and along ebb channels was limited by wave energy. Another environmental variable responsible for differences in mat morphology is the type and amount of sediment transported by tidal currents and deposited over the mats (Golubić, 1973) and Gunatilaka (1975) related the types of mat in Mannar Lagoon, Sri Lanka, to current velocity. In the Cape Timiris area of Mauritania, with apparently quite similar types of mat to those studied by Kendall and Skipwith (1968) and by Gunatilaka (1975), Schwarz *et al.* (1975) suggested that grazing by fish is important in influencing mat morphology and abundance.

Although the relative importance of various factors in determining the detailed surface morphology of mats is not yet resolved, it is generally accepted that the prevailing environmental conditions have a significant effect on this morphology. However, there is debate about the extent to which the dominant algal species also plays a roles. *Microcoleus chthonoplastes* is the dominant in many of the types of mat described from Andros Island, Mauritania, Arabian/Persian Gulf and Sri Lanka, including the forms termed blister and flat mat by Golubić (1973). On Enewatak Atoll, *Calothrix crustacea* grows as black felt-like mats up to 5 mm thick along the upper eulittoral, yet the same species covers large areas of the reef flat as a yellow-brown film (Wiebe *et al.*, 1975). On the other hand, *Entophysalis major* is characteristic of just one type of mat, the mammillate mat of Golubić (1973) (=cinder mat of Kendall & Skipwith, 1968). At least in the development of one type of stromatolite (20.4.2), there is strong evidence for the dependence of the growth form on the dominant species, *Phormidium hendersonii* (Golubić & Focke, 1978).

20.4.2 Stromatolites

There has been a tremendous increase in interest since the mid 1960s in certain types of mat recognized as Recent stromatolites. Unfortunately, this has led to differing interpretations as to the definition of a stromatolite. Many modern geologists take the broad view that almost any mat with a potential for preservation may be regarded as a stromatolite. For instance, Walter (1976, quoting S.M. Awramik & L. Margulis) defines stromatolites as 'organosedimentary structures produced by trapping, binding and/or precipitation as a result of the growth and metabolic activity of micro-organisms, principally cyanobacteria'. Other authors (especially biologists) follow the older geological literature in restricting the term to laminated structures with predominantly upward convex laminations (Golubić & Focke, 1978). This definition separates them from cryptalgal laminites (Monty, 1976) which are

Fig. 20.2. Stromatolitic mats of *Scytonema* in the upper eulittoral of Freshwater Creek, Andros Island.

laterally continuous stratiform deposits sometimes formed by extensive areas of mat. Oncolites refer to unattached, often subspherical structures. Many examples of supralittoral, eulittoral and sublittoral stromatolites have been described in the volume edited by Walter (1976: organisms—Golubić; Shark Bay, Australia—Hoffman, Playford & Cockbain; Florida, Bahamas and

Fig. 20.3. Close-up of *Scytonema*-dominated stromatolite from Andros Island. One mat (about 10 cm diameter) has been sectioned vertically to show laminations.

Bermuda—Gebelein; Persian Gulf—Kinsman & Park). There are also some striking examples of non-marine stromatolites, but their biology lies outside the scope of this chapter.

Although the development of marine stromatolites appears to be favoured by the same factors as those thought to be important in the formation of mats in general (20.4.1), their occurrence is much more restricted. They are most abundant in lagoons and sheltered coastal localities and some of the best-developed examples such as those in the Arabian/Persian Gulf and at Shark Bay (Fig. 21.1) are associated with waters with elevated salinity. It has often been suggested that this is due to the much reduced grazing in such waters (e.g. Playford & Cockbain, 1976). Because the larger stromatolites take a relatively long time to develop, any disturbance by animals is potentially of even greater importance than for a typical mat community. Monty, however, takes a sceptical view of the theory that grazing and burrowing play a key role in restricting the growth of stromatolites. His arguments were first put forward (Monty, 1973) to oppose Garrett's (1970) suggestion that the decline of stromatolites in the Late Precambrian and Early Palaeozoic was due to animals, but have also (Monty, 1979) been applied to modern observations. He points out that on the one hand recent corals and coral reefs are heavily bored and grazed upon, yet are still highly successful; on the other hand the widespread cyanobacterial stromatolites formed by *Phormidium hendersonii* are (apparently) not grazed by any animal. In spite of Monty's arguments, it seems hard to doubt that grazing is an important factor restricting the distribution of present-day stromatolites, but his doubts do emphasize the need for critical experimental studies. In a comparison of ancient and modern stromatolites, Awramik *et al.* (1978) concluded that the physical and biological factors and processes that shape stromatolite macrostructure are not yet understood. We do know that current and wave action modify columnar and domal shape (Hoffman, 1967), but we do not yet understand the causes of such conspicuous features as branching in ancient columnar stromatolites.

The laminations inside stromatolites may be brought about by combinations of tidal and diel events or by less regular ones. Golubic and Focke (1978) have described the events in a relatively simple stromatolite-forming structure dominated by one species, *Phormidium hendersonii.* The account is based on their own observations on Bonaire and Bermuda and those of Monty (1965) on Andros Island. Under optimal growth conditions in shallow subtidal environments, lamination reflects a diel cycle in cyanobacterial growth, orientation and movement. During the daytime the trichomes move phototactically upwards, supported by their sheaths and bundling of filaments. In the evening, the trichomes resume a prostrate position and glide over the surface of the stromatolite, producing sticky, externally diffluent sheaths. Over the following night there is little size increment in the structure. The thin opaque layer formed is less than one/tenth of the reticulate hyaline layer formed in daylight. Assuming a constant rate of sedimentation, the density of particles accumulated overnight within the prostrate layer will be more than ten times higher than in the hyaline layer. The lamination may however be complicated by currents and tidal influences (Monty, 1979).

20.4.3 Endoliths

The importance of cyanobacteria that live inside rocks, especially limestone, was stressed in 20.2.1. Three groups have been recognized (Golubic *et al.*, 1981). *Euendoliths* actively bore the rock; *chasmoendoliths* grow in fissures and cavities; *cryptoendoliths* occur inside porous rocks as a discrete layer parallel to the surface. The first consists of a sufficiently limited number of species that it is possible to make comments on their biology which appear to be of widespread application; illustrations of the organisms involved can be found in le Campion-Alsumard (1979). Shallow-boring forms (*Hormathonema violaceo-nigrum*, *Hyella balani*) are found in the upper part of the eulittoral, while deeper-boring forms (*H. tenuior*, *Mastigocoleus testarum*) are found lower down the zone (Golubic *et al.*, 1975). Grazing by gastropods is thought to be a key factor influencing this distribution and Golubić and Schneider (1979) report an experimental study where grazing by *Littorina* was apparently necessary for the development of the maximum density of *Hyella balani* thalli. In the lower eulittoral, where grazing is more extreme (20.2.1), epilithic forms are less abundant and thus endoliths are less shaded and as a consequence penetrate deeper into the rock. In the upper eulittoral the compensation depth for boring is much closer to the surface so that endoliths usually have shorter filaments. In *Scytonema endolithicum*, which typically has long filaments, these change to a horizontal mode of boring at the compensation depth (Schneider, 1976). Such a strong dependance of boring on light intensity is reflected in the abundance and distribution in forms from different parts of the littoral (le Campion-Alsumard, 1979; see also 20.3.1). In *Kyrtuthrix* and *Mastigocoleus*, which are relatively deep-boring, the thylakoids occupy most of the cell, but are much less abundant in shallow-boring *Hyella balani*. Even in the same filament there may be marked differences, with thylakoids more abundant in the terminal (innermost) cell than in cells closer to the surface.

Alexandersson (1975) concluded that penetration by *Hormathonema* sp. is essentially an extracellular dissolution process and external structures (1×10 μm) at the apices of perforating filaments were thought to be involved. Similar organelles do not appear to have been described for other endoliths. Nevertheless, le Campion-Alsumard (1979) showed that the form of boring is unique to each species; diagrammatic illustrations were included for sheaths of six species, including *Hormathonema paulocellulare*. She concluded that a loose network of polysaccharide microfibrils, which fills the space between the filament and the bored substratum, probably helps lead to the disintegration of the surrounding limestone. A comparison (Golubic & le Campion-Alsumard, 1973) of the boring pattern morphologies of two heterocystous species, *Kyrtuthrix dalmatica* and *Mastigocoleus testarum*, showed that the course of penetration is predominantly perpendicular to the surface in the former and arched to semi-circular in the latter. Each boring tunnel with *Kyrtuthrix* contains a looped trichome, whereas with *Mastigocoleus* the trichome is single; associated with this, the tunnel cross-section is elongated in the former, circular in the latter. While it is clear that the form of boring may differ from species to species, the substrate itself probably also influences the course of penetration. In a single calcite crystal, the direction of algal growth is likely to follow the planes of crystal twinning, but in compact carbonate rocks, composed of irregularly orientated microcrystals, algal growth and boring is likely to be directed by the distribution of impurities and components of differing solubility. Parts of the sheaths of some endolithic species contain a dark blue pigment (gloeocapsin), whereas other species have a yellow-brown pigment (scytonemin). These are at least partially responsible for the coloured zones noted by Schneider (1976) and many others (20.2.1). Golubić (1969) showed that it was possible to demonstrate the colour differences more clearly under a microscope by acid treatment. *Hormathonema violaceo-nigrum* with gloeocapsin turned red and *Hyella caespitosa* with scytonemin turned green.

The colonization of marine rocks by endoliths can be very rapid. During experimental studies in the Mediterranean, le Campion-Alsumard (1975) found that *Mastigocoleus testarum* began to bore Iceland spar within 8 days and that *Hyella caespitosa* penetrated to a depth of 5 mm by 4 weeks. After 3 months colonization by endolithic algae and fungi had reached its maximum. Recolonization of carbonate substrates placed on the intertidal reef-flat and the sublittoral of St Croix, Virgin Island (Perkins & Tsentas, 1976) showed that $<10\%$ of all grains had been recolonized after 9 days, but 95% grains after 4 months. On Aldabra (Whitton & Potts, 1980), a newly exposed area of beachrock was found to have discernible pale blue patches of *H. balani* within 5 days, contrasting with the pale yellow of uncolonized rock; after 5 weeks the rock was indistinguishable in colour from the steel-blue of undisturbed beachrock. Surfaces exposed on limestone rocks in the middle of the lagoon eulittoral of Aldabra usually become visually indistinguishable again within 3–4 weeks (Potts & Whitton, 1980), making a striking contrast to the (predominantly epilithic) colonization of terrestrial limestones on the same atoll which takes many months. It seems probable that it is the repeated cycle of grazing and subsequent further penetration of endolithic cyanobacteria which is responsible for much of the biological component of intermittently or permanently submerged erosion of limestone (Golubić & Schneider, 1979; Potts & Whitton, 1980).

20.4.4 *Standing crop and productivity*

Quantitative studies have been restricted largely to communities such as periphyton flocs, mats and stromatolites which have relatively high standing crops. Some examples of the relatively few estimates of standing crop are given in Table 20.2; the studies of Bauld *et al.* (1979) show clearly that different types of mat on one shore are characterized by quite different standing crops. Odum (1967) pointed out that the chlorophyll *a* content of a well-developed mat is comparable to that of some terrestrial forests. Under favourable conditions, primary productivity rates of littoral mats may also be high, though apparently not as high as some examples of mats from other types of habitats (see Table in Bauld *et al.*, 1980). In the population of *Oscillatoria limosa* in Mooréa lagoon (Sournia, 1976; Table 20.2) the photosynthesis/respiration ratio was 2, but according to Birke (1974) the ratio probably approaches 1 in most mats. It seems likely that in temperate regions mats show considerable differences in the P/R ratio with season. Bauld *et al.* (1980) found that primary productivity rates in mats at Spencer Gulf, S. Australia, were greater during late winter and spring than during the summer months. Mats

Table 20.2. Examples of standing crop and primary productivity in the literature. (Species composition at Spencer Gulf from J.S. Bauld, pers. comm.)

Author	Species	Location	Lat.	Long.	Environment	Morphology	Standing crop	Primary productivity
Sournia (1976)	*Oscillatoria limosa*	Mooréa			shallow lagoon	green patches	127 μg chl. *a* cm^{-2}	26 $\mu g\ O_2\ cm^{-2}h^{-1}$
Sournia (1976)	*Oscillatoria limosa*	Mooréa			shallow lagoon	average	90 μg chl. *a* cm^{-2}	12.5 $\mu g\ O_2\ cm^{-2}h^{-1}$
Bauld *et al.* (1979)	*Schizothrix helva*	Hamelin Pool	26°S	114°E	eulittoral	smooth mat	12.4 ± 2 μg chl. *a* cm^{-2}	1.7 $\mu g\ C\ cm^{-2}h^{-1}$
	Lyngbya aestuarii	Hamelin Pool	26°S	114°E	eulittoral	tufted mat	71 ± 17 μg chl. *a* cm^{-2}	8.5 $\mu g\ C\ cm^{-2}h^{-1}$
	Microcoleus tenerrimus	Hamelin Pool	26°S	114°E	sublittoral	colloform mat	9.2 ± 3 μg chl. *a* cm^{-2}	11.3 $\mu g\ C\ cm^{-2}h^{-1}$
Potts & Whitton (1980)	*Scytonema* sp.	Aldabra	9°S	46°E	upper eulittoral	mature mat	234 ± 98 μg chl. *a* cm^{-2}	
	Calothrix crustacea + *Microcoleus*	Aldabra	9°S	46°E	mid eulittoral	thin mat	37 ± 26 μg chl. *a* cm^{-2}	
	Hyella balani	Aldabra	9°S	46°E	mid eulittoral	endolithic in sand	28 ± 8 μg chl. *a* cm^{-2}	
Bauld *et al.* (1980)	*Microcoleus*, *Schizothrix* (?*Phormidium*), *Lyngbya*	Spencer Gulf	32°S	137°E	supralittoral + eulittoral	smooth and tufted mat		0.85–12.9 $\mu g\ C\ cm^{-2}h^{-1}$
Johannes *et al.* (1972)	*Schizothrix calcicola*	Eniwetok				'algal pavement'		87 $\mu g\ O_2\ cm^{-2}h^{-1}$

certainly can grow quite fast at times, for Sorensen and Conover (1962) recorded increases as high as 10% dry weight per day. The high primary productivity and photosynthetic efficiency of subtidal (colloform) mat as compared with stratiform intertidal mat in Hamelin Bay, W. Australia, was attributed by Bauld *et al.* (1979) to the continuous submersion of the former. The annual accretion rates of colloform mat were nevertheless considerably less than those of intertidal mat, presumably because colloform mats occupy environments which are mainly erosive (Logan *et al.*, 1974).

20.4.5 Nitrogen fixation

Many *in situ* studies have shown that littoral communities dominated by heterocystous cyanobacteria are capable of fixing nitrogen at rates which are easily detectable using the acetylene reduction assay technique. In the laboratory a marine strain of *Anabaena* from Texas (Stacey *et al.*, 1977) has been shown to be capable of growing faster than any other nitrogen-fixing cyanobacterium so far tested. Only a few authors (e.g. Stewart, 1967; Potts *et al.*, 1978) have used $^{15}N_2$ tracer in littoral environments; Burris (1976) showed experimentally for a range of communities on Lizard Island, Great Barrier Reef, that the average ratio of acetylene to N_2 reduced was 1.9:1, so the estimates of nitrogen-fixation rates given by most authors based on a ratio of 3:1 may eventually need to be corrected. The rates of fixation vary markedly with changes in the environment with, not surprisingly, the extent of desiccation being the most important influence in the supralittoral and upper eulittoral. Renaut *et al.* (1975) followed the changes in rates over 24 h in a *Lichena pygmaea* population on the Moroccan coast. Detectable fixation occurred in the early morning when the lichen was exposed but damp, but the rate dropped as temperature and light intensity increased; activity restarted within 30 min of the lichen being submerged by the tide and subsequently fell off rapidly when the lichen was exposed again. Stewart (1967) reported that a supralittoral population of *Calothrix scopulorum* in Ayrshire, Scotland, showed lower fixation rates in summer than spring or autumn, apparently due to desiccation. In contrast, sublittoral cyanobacteria in a Norwegian fjord showed the highest rates in July and August (Lännergren, 1980); the communities consisted of *Anabaena torulosa* and *Calothrix confervicola* in the vicinity of decaying matter.

Diel studies of nitrogen fixation on the Baltic coast by *Calothrix scopulorum*, *Rivularia* sp. and *Anabaena torulosa* showed a marked dependence of fixation on light (Hübel & Hübel, 1974). Fixation rates were usually six to ten times higher at midday than at night; the difference was however less marked in July and August, presumably due to a greater daytime accumulation of carbohydrate during these months. There was no midday depression in fixation rates as found with Sinai littoral communities (Potts, 1979). Fluctuation in light intensity may influence the liberation of extracellular nitrogenous compounds (Jones & Stewart, 1969) and this may well be an important factor in an environment such as the eulittoral where light conditions can change rapidly.

Some of the highest rates of nitrogen fixation per unit area recorded come from algal-dominated areas of coral reefs, though, as discussed in 20.2.4, cyanobacteria and eukaryotic algae are usually inconspicuous in the coral-dominated stretches of the sublittoral. While phosphorus is apparently conserved extremely efficiently in coral reef ecosystems by recycling, the input of nitrogen by fixation can be sufficiently high that combined nitrogen is actually exported from the reef community (Webb *et al.*, 1975) or can at least contribute a substantial amount of the nitrogen required to maintain this community (Burris, 1976). Wiebe *et al.* (1975) found that nitrogen-fixation rates on the reef surface of Enewetak Atoll, Marshall Islands, are equivalent to an average rate of 1.8 mg $ha^{-1}day^{-1}$. *Calothrix* is the dominant cyanobacterium on this reef, but local areas covered by *Nostoc* show considerably higher rates than those covered by *Calothrix* (Mague & Holm-Hansen, 1975). Goldner (1980) concluded that nitrogen-fixation rates of algae-dominated communities on New Providence Island, Bahamas, are between two and three orders of magnitude lower than those for similar communities on Enewetak; it is not clear whether their lower rates result simply from the sparser cover of cyanobacteria or whether the rates per unit cyanobacterial biomass are actually less. Burris (1976) calculated that 6.8–30.6 kg N_2 $ha^{-1}yr^{-1}$ is fixed on the littoral rocks of Lizard Island, Great Barrier Reef, where *Scytonema* sp. is the dominant.

In addition to those communities dominated by heterocystous cyanobacteria, high nitrogen-fixation rates have been recorded also for a number of communities dominated by non-heterocystous forms; although it seems likely that these cyanobacteria are responsible for much of the observed fixation, few of the organisms involved have been brought into axenic culture. Renaut *et al.* (1975) reported high but variable rates associated consistently with a phycoerythrin-rich *Lyngbya* on littoral rocks in Morocco, while Potts and Whitton (1977) reported high rates for *Hyella balani* endolithic in sand particles on Aldabra. Most other accounts deal with species growing on silts. In cores of sediment taken from the Eden Estuary, Scotland, nitrogen fixation was almost restricted to the surface, where cyanobacteria such as *Oscillatoria laete-virens* and *O. margaretifera* were abundant (Stewart, 1975a). At the surface of Great Sippewissett Saltmarsh, Massachusetts, three of the five nitrogen-fixing mat communities were dominated by non-heterocystous species (Carpenter *et al.*, 1978). Nitrogen fixation on salt marsh sediments at Minas Basin, Nova Scotia, was associated mainly with species of *Oscillatoria* and *Lyngbya* in early and mid-summer (Smith *et al.*, 1979). *Microcoleus chthonoplastes*, but not *Schizothrix* sp., showed high fixation rates on lagoon sediments of Aldabra (Potts & Whitton, 1977: corrected rate for *Microcoleus* = 0.41 nM C_2H_4 μg chl a^{-1}min^{-1}). Pearson *et al.* (1981) confirmed with an axenic culture that aerobic nitrogenase activity can take place in *M. chthonoplastes*; when the organism was grown in a 12:12 light–dark regime; the maximum nitrogenase activity occurred during the dark period, though growth can occur in the absence of combined nitrogen under continuous light. Populations of *M. chthonoplastes* assayed in the Ems Estuary, Netherlands, failed to show detectable activity (V. de Jonge & B.A. Whitton, unpubl.) so it cannot be assumed that the presence of this species necessarily indicates the importance of nitrogen fixation at a site. Nevertheless, non-heterocystous cyanobacterial communities are probably sometimes important contributors to eulittoral nitrogen fixation, as suggested by the observations of Stal and Krumbein (1981). Potts (1979) found for the mangrove forests of the Sinai Peninsula that rates associated with *Lyngbya aestuarii*, *Hydrocoleum* + *Schizothrix* and *Hyella balani* were usually higher than those measured for *Scytonema*, *Calothrix*, *Rivularia* and *Gardnerula corymbosa*.

CHAPTER 21
ORIGIN AND EARLY EVOLUTION OF CYANOBACTERIA: THE GEOLOGICAL EVIDENCE

J. W. SCHOPF & M. R. WALTER

21.1 Historical perspective

Any attempt 20 years ago to delineate the geological history of the cyanobacteria would have been brief indeed, for very little of the now-known fossil record of the group had as yet been discovered. To a considerable extent, this earlier lack of knowledge is a reflection of the fact that 'plant-like' fossils of this type fall within the purview of palaeobotany, a field that traditionally has been concerned largely (indeed, almost exclusively) with the fossil record of megascopic, vascular plants, rather than with that of microscopic 'thallophytes' as cyanobacteria were then regarded. Although fossil cyanobacteria had been detected, and reported, from time to time (for reviews of the earlier literature, see Pia in Hirmer, 1927, and Seward, 1933), prior to recent years they were typically treated more or less as curiosities, minor members of the fossil flora that if mentioned at all in palaeobotanical works tended to be relegated to addenda and included there chiefly for the sake of monographic completeness (e.g. the well-preserved oscillatoriacean *Archaeothrix oscillatoriaeformis* as described by Kidston and Lang in 1921). There have, of course, been notable exceptions (among these, an excellent paper by Croft and George published in 1959 that describes heterocystous nostocaceans and stigonemataceans from Devonian-age cherts of Scotland; for a review of similar reports, see Schopf, 1974a). But for the most part—and until quite recently—the fossil record of this group (like that of prokaryotes generally) has been neglected or ignored. In fact, this would probably still be the case were it not for a series of events that over the past two decades has resulted in a marked upsurge of interest in the early history of life, and the realization that the cyanobacteria played a crucial role in this earliest, Precambrian phase of biotic development.

To appreciate the earlier status of such studies it is appropriate to consider briefly the views of A.C. Seward (1863–1941), the Master of Downing College, Vice-Chancellor and Professor of Botany at Cambridge University, who was certainly among the most influential palaeobotanists of the first half of this century. In his best-known text, *Plant Life Through the Ages* (1st edn, 1931; 2nd edn, 1933), he reviews at some length the then-known fossil evidence of Precambrian life (i.e. that preserved in rocks deposited during the Precambrian Eon, a time segment encompassing the earliest seven-eighths of Earth history that extends from the origin of the Earth, some 4.5 Ga ago ($Ga = 10^9$ years), to the beginning of the Cambrian, about 0.57 Ga ago, the geological period that marks the beginning of the Phanerozoic Eon and the most recent one-eighth of geological time. The Precambrian, in turn, is subdivided into three Eras: the Hadean, extending from the time of origin of the Earth to the age of the oldest known rocks, about 3.8 Ga; the Archaean, extending from 3.8 to 2.5 Ga ago; and the Proterozoic, extending from 2.5 Ga ago to the beginning of the Phanerozoic).

Like other scholars of the day, Seward was aware that large, multilayered, domical limestone structures such as *Cryptozoon*—structures now referred to as 'stromatolites'—had been reported from Precambrian strata and that a number of workers (notably the N. American palaeontologists, Charles Dolittle Walcott and David White, and the Australian geologist Sir Douglas Mawson) attributed their origin to the activities of microscopic blue-green algae. (For an illuminating discussion of Mawson's views regarding these structures, see M.F. Glaessner's Preface in Walter, 1972.) In fact, both Walcott (1915) and White (1928) claimed to have found direct fossil evidence of bacteria or blue-green algae preserved in Precambrian stromatolites. But proof of such claims was scant; indeed, their acceptance was described at the time as being in essence a matter of sheer 'faith that one organism without hard parts, and less than 0.01 mm in diameter, would be preserved in identifiable condition from pre-Cambrian times to the present' (Raymond, 1935, p. 378). They were similarly dismissed by Seward: 'We can hardly expect to find in Pre-Cambrian rocks any actual proof of the existence of bacteria . . . these and similar contributions . . .are by no means convincing' (1933, p. 92). Moreover, since 'one cannot decide whether an alga in a calcareous grain had a share in the precipitation of the lime, or whether it was passively imprisoned in the mineral material deposited through inorganic forces . . . the presence of the plant may be accidental, not causal' (1933, p. 83). Seward averred that even 'if . . . [the] microscopic spherical particles [reported by Walcott] . . . are cells there is no proof that they had any share in the production of the enclosing matrix' (1933, pp. 87, 88). Thus, and based on studies of diffusion phenomena (Liesegang, 1913) indicating that multilayered geological concretions could be formed via solely inorganic processes, Seward concluded that even the laminated, large-scale stromatolites might well be non-biogenic: 'Cyanophyceae or similar primitive algae may have flourished in Pre-Cambrian seas . . . but to regard these hypothetical plants as the creators of reefs of

Cryptozoon and allied structures is to make a demand upon the imagination inconsistent with Wordsworth's definition of that quality as "reason in its most exalted mood" ' (1933, p. 87).

Seward's views were typical of palaeobotanists of the pre-1940's: the Phanerozoic record of cyanobacteria was sparce and unimpressive; their presence in the Precambrian was possible, but by no means firmly established; stromatolites might be, but were very likely not a result of cyanobacterial activity; and there seemed no reason to expect that this bleak picture was likely to change.

But beginning in the early 1930's it did change—first slowly, then with increasing rapidity—as a series of disparate, seemingly unconnected events conspired to make the fossil record of the cyanobacteria a subject worthy of serious investigation. First was the seminal work of Maurice Black (1933) which led to the detailed description of extant stromatolites on Andros Island in the Bahamas, providing what became a very influential model for the interpretation of ancient stromatolites. Then in the early 1950's came the widespread application of radiometric dating techniques to Precambrian strata, a development that for the first time provided a firm indication of the vastness of the Precambrian Eon and a geological time-frame for the palaeontological discoveries that were soon to follow. The startling announcement in 1953 that amino acids could be synthesized via solely abiotic processes under simulated 'primitive Earth conditions' (Miller, 1953) also played a role, for it focused attention on the origin of life and early biological evolution as potentially fruitful areas of scientific inquiry. But in retrospect, perhaps the primary stimulus for the subsequent development of the field came from the publication a year later of a short report by Tyler and Barghoorn (1954) noting their discovery of cellularly preserved microorganisms in stromatolitic cherts of the Gunflint Iron Formation of southern Canada (although, at the time, neither the dominance of cyanobacterium-like microbes in the assemblage, nor the significance of their occurrence in finely laminated stromatolites, was fully appreciated). During this same period, Soviet geologists began to make a concerted effort to extend back into Proterozoic-age strata the techniques of palaeontological dating (biostratigraphy) that had long been successfully applied in the Phanerozoic; the obvious tool for such studies were the stromatolites which—if they were in fact biogenic—were the most abundant evidence of life available in these ancient rocks. Interest thus rapidly grew, both in the Soviet Union and elsewhere, in the genesis, distribution and possible changes through time of these distinctive sedimentary structures (for a recent review of these developments, see Semikhatov, 1976). Shortly thereafter, in the late 1950's, the best-known modern analogues of such fossil stromatolites (occurring at Shark Bay, W. Australia: Fig. 21.1) were studied and first described (Logan, 1961); the

Fig. 21.1. Reefs of columnar carbonate stromatolites, produced by modern cyanobacterium-dominated microbial communities, in Hamelin Pool, a hypersaline lagoon at Shark Bay, about 700 km north-northwest of Perth, W. Australia. The hypersalinity of the lagoon inhibits the growth of grazing metazoans, thus permitting the slow-growing stromatolites to be formed. Geological hammer provides scale.

cyanobacterial component of living stromatolite-forming biocoenoses was thus established, and the morphological similarity between the Shark Bay and Proterozoic structures (Fig. 21.2) suggested, quite strongly, that fossil stromatolites might have been produced by similar microbial communities. Finally, in 1965, Barghoorn and Tyler (1965) and Cloud (1965) first described in some detail the Gunflint microfossils; the prevalence of cyanobacterium-like filaments and coccoids in the assemblage was duly noted and discussed, as was the similarity of their occurrence to modern stromatolitic deposits. It was these reports, more than any other, that ushered in the modern era of Precambrian palaeomicrobiology. Later in that same year, discovery of a second notable Precambrian microbiota was announced (Barghoorn & Schopf, 1965), that of the Bitter Springs Formation of central Australia, subsequent monographic studies of which (Schopf, 1968; Schopf & Blacic, 1971) established the dominance and diversity of cyanobacteria in such Late Proterozoic communities; these studies revealed also the remarkable degree of morphological evolutionary conservatism of the group. Since these germinal reports, nearly 100 such stromatolitic microbial biocoenoses have been discovered in the Precambrian, spanning a time range that extends from 3.5 Ga ago to the Proterozoic–Cambrian boundary (see Schopf, 1975, 1977;

Fig. 21.2. Laminated columnar carbonate stromatolites from the Middle Proterozoic-age (1.3 Ga old) Belt Supergroup of Glacier National Park, Montana, U.S.A.; geological hammer at left provides scale.

Cloud, 1976; Schopf & Prasad, 1978; Hofmann & Schopf, in press; Schopf & Walter, 1982).

From these beginnings, studies of Precambrian cyanobacteria (and other prokaryotes) have flourished in recent years, both because of the inherent fascination of the field—its relevance to understanding the earliest history of life on Earth—and because of its potential economic importance: the possible use of such fossils and of the stromatolites they built for biostratigraphic correlation and basinal palaeoecological analysis, subjects of relevance to the search for economically important mineral resources (especially petroleum, coal and natural gas and iron-bearing and uraniferous deposits). The field has prospered in no small measure due to the stimulation and funding of the U.S. space programme, with its interest in the origin and development of life in the solar system (and now in the universe).

In these introductory paragraphs we have focused largely on the Precambrian, principally because data that have become available in recent years indicate that by the end of the Proterozoic, most of the major families of the group (viz. among coccoid taxa, the Chroococcaceae, Entophysalidaceae and Pleurocapsaceae, and among the Hormogoneae, the Oscillatoriaceae, Nostocaceae and, probably, the Scytonemataceae and Rivulariaceae) had already evolved (Schopf, 1974a); just as the Precambrian can be properly termed 'The Age of Prokaryotes', the latter half of this Eon, the Proterozoic Era, was truly 'The Age of Cyanobacteria'. Despite these notable advances, it must however be emphasized that many aspects of the evolution of the cyanobacteria, such as detailed phylogenetic relations and the time of origin of several of the major families, can be sketched only approximately on the basis of available palaeobiological data. The known fossil record of the group has been reviewed recently by a number of workers (e.g. Schopf, 1974a, 1975, 1977; Cloud, 1976; Walter, 1976; Hofmann & Schopf, 1982; Schopf & Walter, 1982). This chapter is thus restricted largely to an interpretation of that record and a discussion of recent key developments in the field.

21.2 The oldest cyanobacteria

The search for the earliest records of the cyanobacteria takes us back to the Archaean Era of Earth history, since it seems that from the beginning of the Proterozoic (2.5 Ga ago) there is firm, if not yet cogent, evidence that cyanobacterium-like prokaroytes (viz. oxygen-producing photosynthetic microbes) were both present and widespread. We are limited in this search at the extreme end of the time range by the absence on Earth of any rocks known to be older than 3.8 Ga (i.e. rocks of Hadean, rather than of Archaean age). The main limitation, however, is the paucity of unmetamorphosed rock units (i.e. units not subjected to high temperatures and pressures during subsequent geological time) that are known from the Archaean; geologically based knowledge of early evolution can be no better than the rock record allows, and as yet just four key rock sequences are known that have survived to the present in a relatively unmetamorphosed state from the first billion years (from 3.8 to 2.8 Ga ago) of Archaean Earth history: the Warrawoona and George Creek Groups of W. Australia, ~3.5 Ga old; the Onverwacht and associated groups of southern Africa, ~3.5 Ga old; the Pongola Supergroup of Natal, ~3.1 Ga old; and the Fortescue Group of W. Australia, ~2.8 Ga old. Moreover, in interpreting the few fossils known from this record we are faced with the problem that all we can directly observe is the morphology of delicate, and incompletely preserved minute cells and the form and structure of stromatolitic sediments. But the cellular morphology of even the best preserved Archaean microfossils now known provides no information of the sort needed to distinguish primitive cyanobacteria from other structurally undifferentiated prokaryotes (e.g. chloroflexaceans, beggiatoaceans, etc), and although some physiological inferences can be drawn from stromatolite morphology, these are seriously limited by current ignorance of the factors that in extant microbes control the form of stromatolitic mats. In addition, although the carbon isotopic composition (viz. the $^{13}C:^{12}C$ ratio) of the degraded organic matter in ancient sediments (the so-called 'kerogen'), as compared with that exhibited by coexisting carbonate minerals, can provide some indication of whether or not autotrophs played a major role in the Archaean carbon cycle, such measurements cannot be used to discriminate between the several types of CO_2-fixing autotrophy. In particular, they cannot indicate whether Archaean autotrophs were in fact photoautotrophs and, if so, whether they used both Photosystems I and II and thus released molecular oxygen as a byproduct of photosynthesis. This last feature appears to have been the key characteristic that distinguished cyanobacteria from their (presumably photosynthetic, but non-oxygen-producing) evolutionary precursors. In essence, the origin of the cyanobacteria can therefore be equated with the origin of aerobic photosynthesis. But the microfossils, the stromatolites and carbon isotopic data cannot provide unequivocal evidence of the occurrence of this process. Thus, to determine whether or not oxygenic photosynthesis was operative on the Archaean Earth we must turn to indirect indicators, such as the oxidation state of ancient soil horizons, deltaic deposits and iron-rich sedimentary rocks, an approach based on the now widely accepted premise that virtually all free oxygen of the Earth's atmosphere and hydrosphere was generated by oxygenous photosynthesis, and therefore in this ancient context by cyanobacteria (see Holland, 1978). In short, the current approach to Archaean palaeobiology is the postulation of models that account satisfactorily for the relatively few and rather disparate scraps of information that are as yet available.

The established fossil record starts with two rock sequences, each about 3.5 Ga old, the Warrawoona Group of W. Australia and the Onverwacht Group of S. Africa, but there is one still older geological unit that must be mentioned first. This is the suite of rocks from the 3.8 Ga old Isua Supracrustal Belt of western Greenland, a highly metamorphosed and severely altered sequence of originally sedimentary rocks. Objects resembling microfossils have been reported from these metasediments, but they have recently been reinterpreted as being mineralic microstructures of entirely non-biological origin (Bridgwater *et al.*, 1981). The carbon isotopic composition of graphite and carbonate minerals from the Isua sequence was also earlier interpreted as indicating the presence of life, in this case of autotrophic CO_2-fixers, but this interpretation is now regarded as being no more likely than others not involving autotrophy nor, in fact, even requiring the presence of organisms (Chang *et al.*, 1982; Hayes *et al.*, 1982; Schidlowski *et al.*, 1982). Finally, the presence of abundant oxidized iron minerals in some of the rocks of the sequence has been suggested as possibly indicating the existence of oxygenous photosynthesizers in Isua time, but other possible oxidative processes are also known so that this evidence, too, is equivocal (Walker *et al.*, 1982). Life may well have existed 3.8 Ga ago; indeed, its presence would be consistent with the evidence at hand and with evidence from the later geological record, but the data now available are insufficient to prove the point.

By 3.5 Ga ago, a diverse assemblage of micro-organisms was already extant (Schopf & Barghoorn, 1967; Muir & Grant, 1976; Knoll & Barghoorn, 1977; Awramik *et al.*, in press). From rocks of this age are known five morphological varieties of filamentous microbes and perhaps four or more types of coccoid, unicellular micro-organisms (e.g. Figs 21.3–5). All of the filamentous forms, and at least one of the coccoid varieties (Muir & Grant, 1976), were benthic mat formers or mat dwellers; it is possible, although not yet established, that some of the other unicells (e.g. Fig. 21.3 f–i) may have been planktonic. The filamentous forms are known only from the Warrawoona Group, where they occur in stratiform stromatolites, preserved by permineralization in bedded cherts. This is common mode of preservation of such stromatolitic biocoenoses in which the micro-organisms and microbial mats became encased by, and thoroughly embedded within, a siliceous matrix that upon crystallization formed chert, a fine-grained quartz rock; the organisms are thus 'organically preserved'—composed of a residue, now geochemically altered, of the organic matter that originally comprised their cell walls, sheaths, etc—and they are three-dimensional, and in some cases relatively little altered, in form. Some are interpreted as empty, abandoned sheaths (*Siphonophycus antiquus*; Fig. 21.3 a–e); others are septate, unbranched and undifferentiated trichomes (*Eoleptonema australicum* and

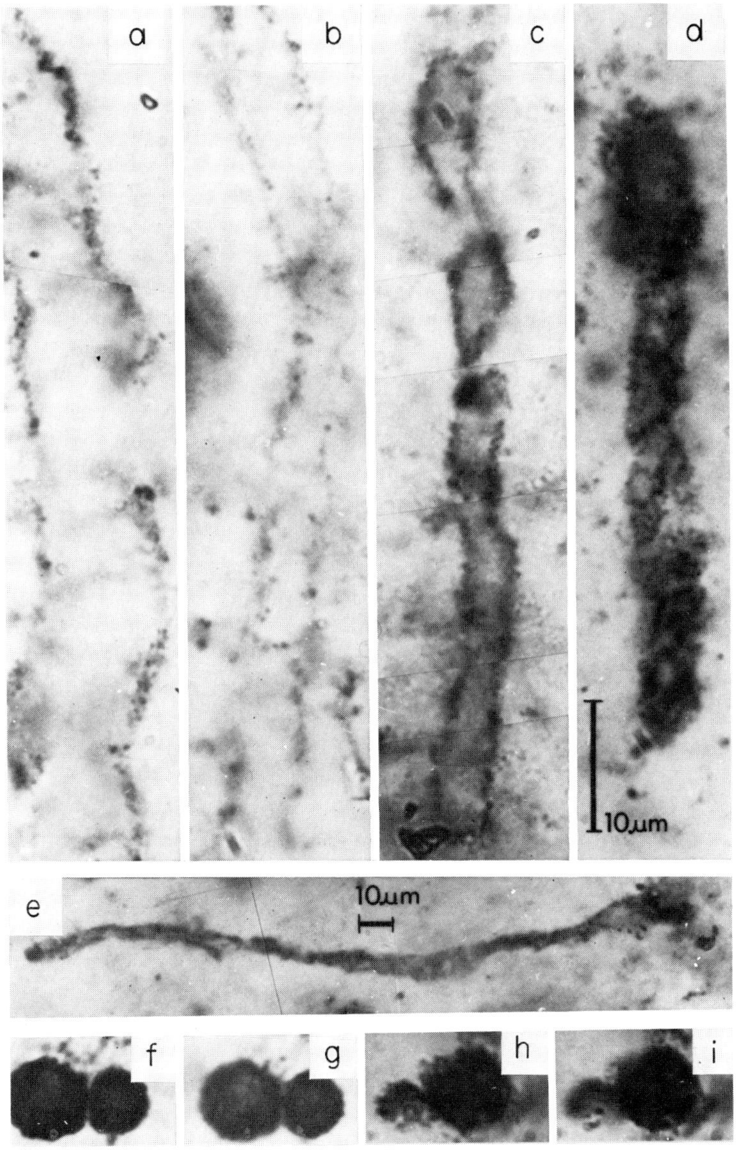

Fig. 21.3. Tubular or partially flattened bacterial or cyanobacterial sheaths (*Siphonophycus antiquus* Schopf, parts a–e) and solitary and paired unicell-like spheroids (*Archaeosphaeroides pilbarensis* Schopf, parts f–i), shown in petrographic thin sections of stromatolitic black chert from the 3.5 Ga old Warrawoona Group of W. Australia. Scale in (d) is for a–d and f–i. (a), (c) and (e) show the microfossils in photomontages, necessitated by their sinuous nature and three-dimensional preservation.

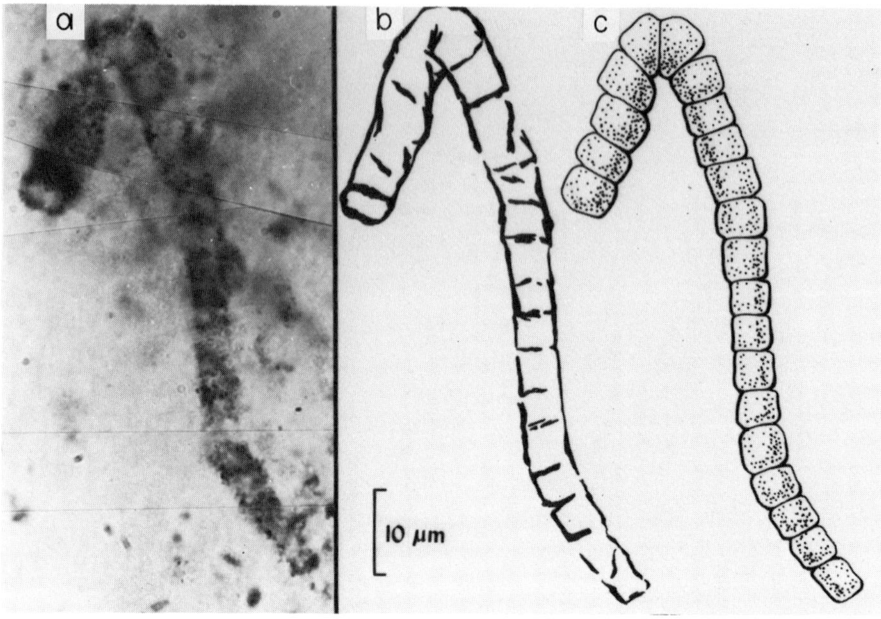

Fig. 21.4. Unbranched and undifferentiated, septate and apparently somewhat tapering bacterial or cyanobacterial trichome (*Primaevifilum septatum* Schopf), shown in (a) a petrographic thin section and (b, c) in schematic reconstructions, from stromatolitic black chert of the 3.5 Ga old Warrawoona Group of W. Australia. Scale for all parts is shown in (b); (a) shows a photomontage, necessitated by the three-dimensional, sinuous nature of the cellularly preserved microfossil.

Primaevifilum septatum; Fig. 21.4); while the last two types are long, narrow (0.3–0.7 μm wide) bacterial threads (*Archaeotrichion contortum*; Fig. 21.5) and rosette-like colonies of fine (0.5–1.0 μm wide) filaments (*Warrawoonella radia*). If this assemblage were preserved in strata of younger geological age (e.g. strata from the Late Proterozoic), the trichomes and possible sheaths very likely would be immediately interpreted as being of cyanobacterial origin. In fact, they could be remnants of cyanobacteria, but based on the data available it seems equally likely that they could be remnants of non-cyanobacterial mat formers such as chloroflexaceans (Schopf & Walter, 1982). The kerogen in these same samples has a carbon isotopic composition ($\delta^{13}C_{PDB} = -35.1‰$ and $-34.7‰$) that is apparently indicative of the former presence of autotrophs, but such analyses are carried out on bulk kerogen and thus cannot be tied to any particular taxon of the fossil assemblage. The mat-forming habit of the preserved micro-organisms is consistent with the possibility that their growth and movement may have been influenced by light, but too little is known about the mat-forming taxa of modern microbial biocoenoses for this possibility to be considered firmly established.

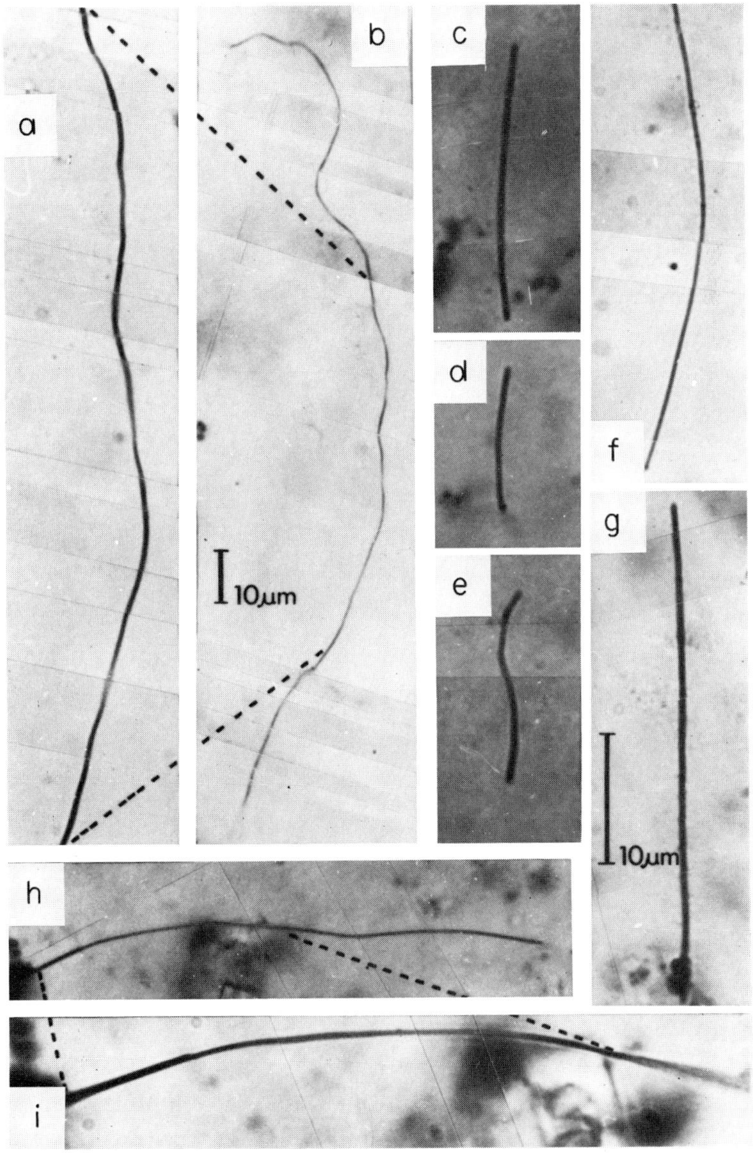

Fig. 21.5. Elongate, rod-shaped, apparently non-septate fossil bacterium (*Archaeotrichion contortum* Schopf), shown in petrographic thin sections of stromatolitic black cherts of the 3.5 Ga old Warrawoona Group of W. Australia. Scale in (b) is for (b) and (h); magnification of all other parts is indicated by the scale in (g). Specimens in (a–c) and (e–i) are shown in photomontages, necessitated by their sinuous nature and three-dimensional preservation. As indicated by the dashed lines, a portion of the specimen in (b) is shown at higher magnification in (a) and a portion of the specimen in (h) is shown at higher magnification in (i). Short filaments shown in (c–e) are interpreted as juvenile stages in the ontogeny of this species.

Relatively large, domical stromatolites are also known from the Warrawoona Group (Lowe, 1980; Walter *et al.*, 1980), structures that are comparable to younger stromatolites that are known or inferred to have been built by cyanobacterium-dominated communities. The Warrawoona structures exhibit, however, none of the distinctive features needed to prove such an interpretation and no microfossils have as yet been detected in them.

An indirect, but important argument *against* the presence of oxygen-producing cyanobacteria 3.5 Ga ago, is that not until the Early Proterozoic is there firm evidence of the presence of appreciable quantities of free oxygen in the Earth's atmosphere. In fact, the presence of easily oxidizable minerals (particularly uraninite, UO_2) in preserved deltaic sediments about 2.3 Ga old and older, and the absence of ferric oxide-pigmented sands and silts from geological units older than about this same age, suggests that there was virtually no free oxygen in the Earth's atmosphere during the Archaean (Walker *et al.*, 1982). Moreover, the first massive appearance of banded iron formations at about 2.5 Ga ago, rock units that contain enormous amounts of oxidized ferric iron that the evidence seems to indicate were deposited over a relatively short period of (geological) time, suggests that copious amounts of photosynthetically produced free oxygen were being released into the upper reaches of the hydrosphere beginning at about that time. In recent attempts to 'model' systems of this sort, Schopf (1980) has shown that given ideal growth conditions and unlimited space, nutrients and light, the progeny of 1 g of cyanobacterium (*Anacystis nidulans*) could generate an amount of oxygen equivalent to that presently in the Earth's atmosphere in just 40 days. Of course this calculation cannot be applied literally to the Earth because before a stable oxygenous atmosphere could have become established, the free oxygen thus generated would first have been used to oxidize ferrous iron in the ocean, unoxidized minerals on land and reduced volcanic gases in the atmosphere (e.g. H_2S and H_2). In addition, nutrients and photosynthetic space would no doubt have been limiting (Schopf, 1980). None the less, attempts to model all of these parameters do lead to the view that it is very unlikely that it would have taken as long as 1.2 Ga to oxygenate the atmosphere, if oxygenous photosynthesis had already been extant 3.5 Ga ago. Thus, it seems probable that cyanobacteria evolved later than 3.5 Ga ago.

In sediments about 3.1 Ga in age there occur stromatolites of a type reasonably interpreted as having been constructed by phototactic, filamentous, phototrophs. These are from the Insuzi Group of S. Africa (Mason & Van Brunn, 1977) and include small, conically laminated structures closely comparable to stromatolites presently forming in hot springs at Yellowstone National Park, U.S.A. The Yellowstone forms have been studied in some detail and seem to owe their distinctive conical form to the filamentous habit,

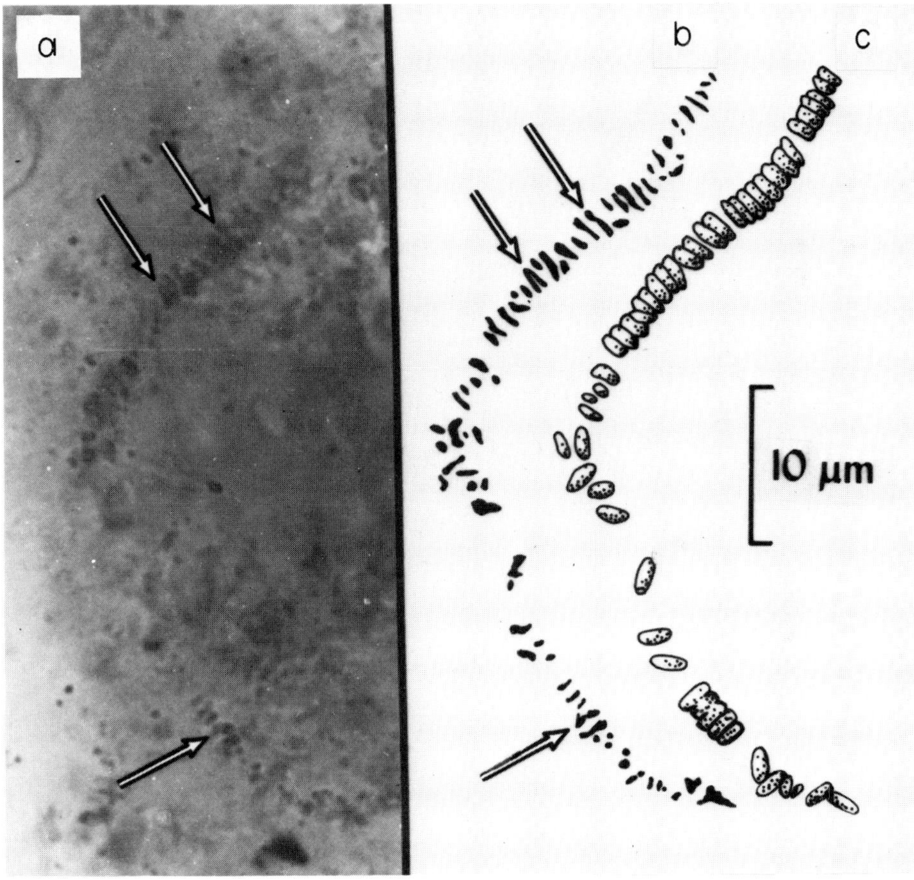

Fig. 21.6. Narrow, relatively poorly preserved, *Oscillatoria*-like filamentous prokaryote, shown in (a) a petrographic thin section and (b, c) in schematic reconstructions, from carbonaceous cherty portions of carbonate stromatolites of the 2.8 Ga old Fortescue Group of W. Australia. Scale for all parts is shown in (c); arrows in (b) point to cellular remnants visible (at arrows) in (a).

phototactic response, and microaerophilic preference of the constructing cyanobacterium, *Phormidium tenue* (Doemel & Brock, 1974; Walter *et al.*, 1976). It seems likely that any micro-organism with these characteristics could build similar stromatolites; there is thus no need to infer that the comparable Insuzi structures were necessarily built by cyanobacteria. The Insuzi stromatolites do not contain preserved microfossils, nor are microfossils of this age known from elsewhere (Schopf & Walter, 1982).

Not until the Late Archaean, about 2.8–2.5 Ga ago, do we begin to find fairly convincing 'presumptive evidence' of the existence of cyanobacteria. The evidence comes primarily from the interpretation of the internal fabric of

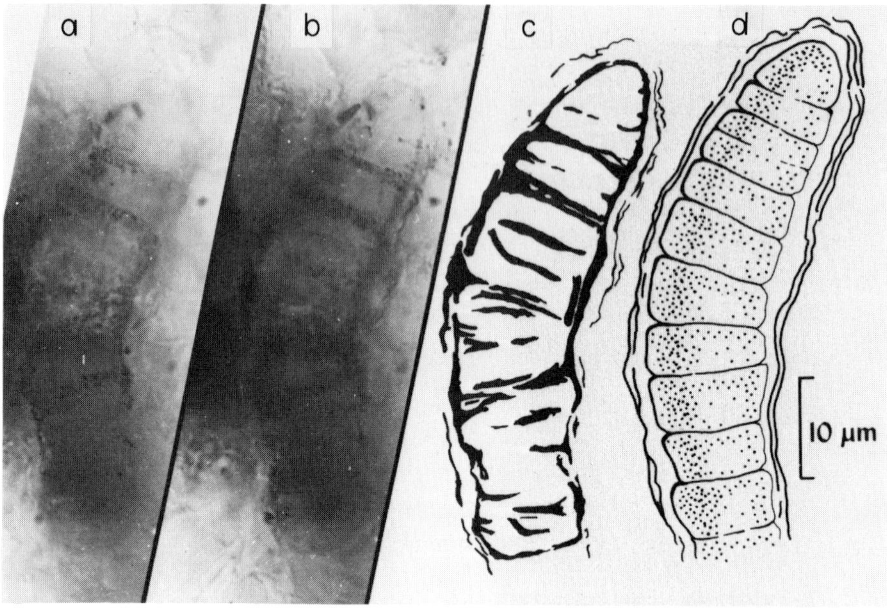

Fig. 21.7. Broad, sheath-enclosed, *Lyngbya*-like filamentous prokaryote, shown in (a, b) a petrographic thin section (at two different focal depths) and (c, d) in schematic reconstructions. from carbonaceous cherty portions of carbonate stromatolites from the 2.8 Ga old Fortescue Group of W. Australia. Scale for all parts is shown in (d).

stromatolites from the Fortescue Group of W. Australia (about 2.8 Ga old), but also from rare filamentous microfossils that have been detected within the same stromatolites. These relatively poorly preserved filaments, the only bona fide microfossils now known from the Late Archaean (Schopf & Walter, 1982), are of two types: narrow (~ 1 μm diameter), unbranched trichomes composed of a uniseriate row of short, disc-shaped cells (Fig. 21.6); and broad (~ 10 μm diameter), unbranched trichomes composed of a series of short, barrel-shaped cells that are capped by a well-defined, subconical terminal cell and that are enclosed by an evident multilamellate sheath (Fig. 21.7). Both forms are comparable to modern oscillatoriaceans, the smaller form to *Oscillatoria* and the larger to *Lyngbya*, but until more examples are found and they can be studied in detail, these similarities can be considered no more than suggestive of possible biological affinities. Nevertheless, as is discussed below, such a cyanobacterial affinity would be consistent with interpretations of the stromatolitic fabrics, with the carbon isotopic composition of the Fortescue kerogen and with indirect, mineralogical evidence from immediately younger Proterozoic sediments.

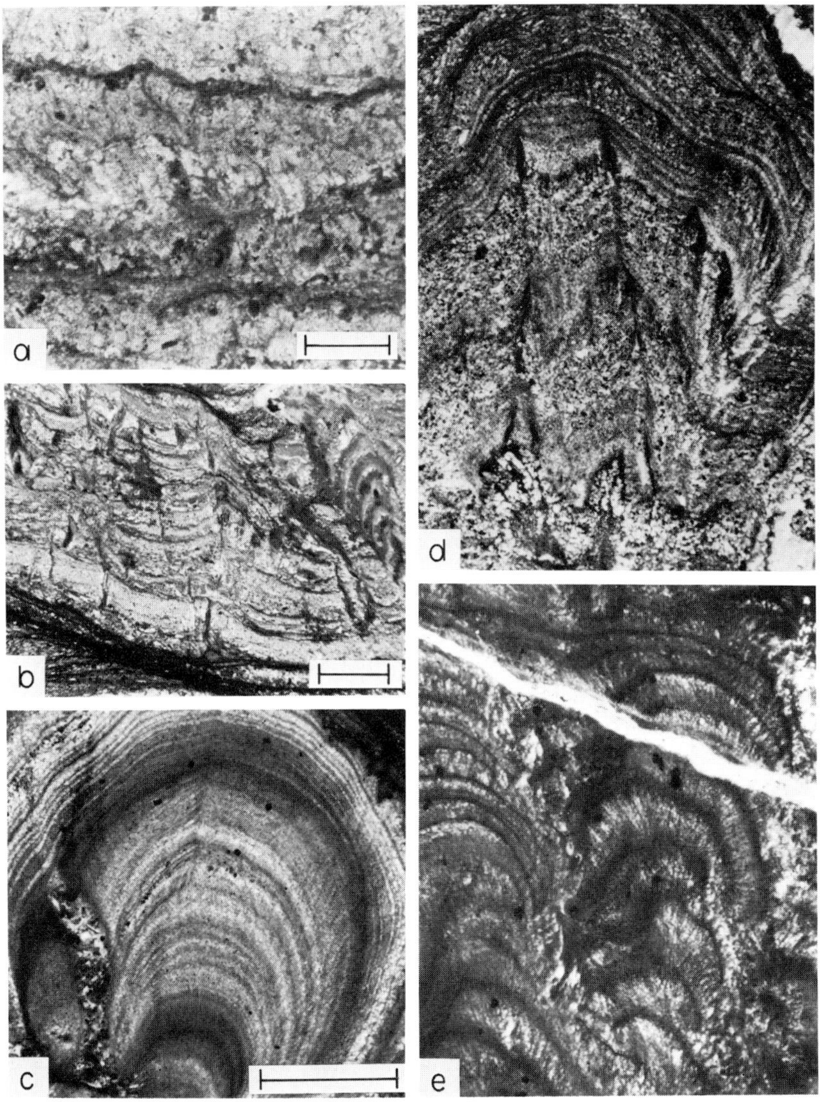

Fig. 21.8. Biologically produced internal fabrics of 2.8 Ga old stromatolites in limestone of the Fortescue Group, W. Australia, as seen in rock thin sections in transmitted light. (a) The horizontal layers mark successive growth stages; in the central layer faint remnants of filaments are arranged more or less vertically, but turn to become parallel to the thin dark lamina in the upper part of the photograph (scale bar 0.5 mm, sample PPRG 026-1-A). (b) The near-horizontal sediment layers are crossed by remnants of tufts of filaments, some so narrow that they may have consisted of single filaments (scale bar 2 mm, sample PPRG 029-1-A). (c) The concentric growth layers (of a formerly gelatinous thallus?) of this small, nodular stromatolite are crossed by faint remnants of filaments with a flabellate arrangement (scale bar 2 mm, applies to d & e also; sample PPRG 025-1-B). (d) Prominant filament remnants in a nodular stromatolite (sample S95c). (e) A flabellate arrangement of filament remnants at an obtuse angle to growth layers (top is to the left; sample PPRG 025-2-2-B).

Of the various intergrading fabrics that occur within the Fortescue stromatolites (Walter, 1982), only the most important end-members will be considered here (Fig. 21.8). One of these is a distinctive 'banded to streaky microstructure' (Walter, 1972) comparable with that presently built by phototactic, fine filamentous oscillatoriaceans. This similarity is particularly evident in those examples that contain remnants of degraded filaments in which it can be seen that the thread-like remnants exhibit the alternately erect and prostrate habit common among modern mat-forming oscillatoriaceans (Gebelein, 1969). In extant cyanobacteria, orientation of this type results from diurnal variation in light intensity, with trichomes migrating toward the light (upward through any overlying sediment cover) during the day, and lying prostrate at night. However, the need for caution with regard to such an interpretation has been clearly demonstrated by Doemel & Brock (1974) who have observed that in the green sulphur bacterium *Chloroflexus aurantiacus* this same filament orientation is due to positive aerotaxis, with the trichomes moving vertically upward at night.

Nodular to bulbous stromatolites, up to a few centimetres wide and high, are abundant in the Fortescue Group sediments; these forms are characterized by a distinctive gradation of fabrics. Among these stromatolites, there is every gradation from individuals consisting of a single filament remnant, 30–40 μm wide (Fig. 21.8b); to individuals that exhibit a banded microstructure but contain no filament remnants; to similar forms that exhibit a prominent radial fabric (Fig. 21.8c); to larger, irregular forms having a thick, streaky microstructure; to similar forms with a sinuous, linear palimpsest microstructure (Fig. 21.8e). It is evident that these stromatolites were built by micro-organisms that grew in erect tufts composed of coarse filaments. In one such form in which the filament remnants range from 60 to 350 μm diameter, the thicker remnants appear to consist of groups of more narrow filaments, about 60 μm wide, some of which contain an axial darker thread that may be the remnant of an internal trichome. Several of these structures rather closely resemble *Frutexites microstroma*, a 2.0 Ga old microfossil described from the Gunflint Iron Formation by Walter and Awramik (1979) that has been interpreted by them as a thick-sheathed scytonematacean. However, these Fortescue structures do not exhibit the combined prostrate and erect growth habits that occur both in *F. microstroma* and in modern scytonemataceans, and they thus seem somewhat more similar to modern rivulariaceans (an attribution which would be consistent with the lacustrine environment inferred for this deposit; Walter, 1982). On the other hand, modern *Lyngbya* commonly forms tufts like those illustrated here (Fig. 21.8), and the broad cellularly preserved trichome noted above (Fig. 21.7) closely resembles this extant cyanobacterium. Although the precise affinities of the micro-organisms involved in the formation of these stromatolites are thus somewhat uncertain, it seems evident that they were built by tuft-forming micro-organisms, probably cyanobacteria. These had trichomes at least 10 μm wide which excreted copious quantities of firm gel to form sheaths up to 60 μm in width, and which commonly formed semiglobose, concentrically laminated masses of gel within which the trichomes exhibited a flabellate arrangement.

The carbon isotopic composition of the Fortescue kerogen and carbonate lends indirect support to the suggested cyanobacterial origin of these stromatolites. In comparison with most ancient organic matter, the stromatolitic kerogen is extraordinarily light ($\delta^{13}C_{PDB} \approx -50‰$); that is, it is enriched in the lighter isotope, ^{12}C, to an unusual degree. The carbonate carbon, in contrast, is essentially 'normal', averaging close to $\delta^{13}C_{PDB} = 0$ (Hayes *et al.*, 1982; Schidlowski *et al.*, 1982). Hayes (1982) has suggested the following scenario as a possible explanation for the abnormally large isotopic difference between the two:

1 fixation of isotopically normal CO_2 by anaerobic, methanogenic bacteria (using hydrogen, presumably bacteriogenic, available in the environment), resulting in generation of isotopically very light methane;

2 incorporation of some of this methane into biomass by aerobic, methane-utilizing (i.e. methanotrophic) bacteria;

3 burial of the methanotrophs to yield the isotopically light kerogen.

Methanogenesis thus serves as the mechanism resulting in the strong enrichment in ^{12}C of the buried kerogen. In this scheme, which at present is regarded as no more than a plausible 'working hypothesis', methane oxidation by aerobes seems likely to be an essential component, since microbial utilization of methane in the modern biota is almost exclusively an aerobic process (Higgins *et al.*, 1980). This, in turn, would suggest that oxygenic photosynthesis—i.e. the cyanobacteria—had already evolved.

Finally, we can draw upon the inorganic geochemical argument mentioned earlier. Easily oxidizable detrital uraninite occurs in sediments older than 2.3 Ga, but is virtually absent from deposits of younger geological age. Conversely, sediments deposited under subaerial conditions later than 2.3 Ga ago commonly contain ferric oxides, minerals that are rare or absent from earlier deposited sediments of this type. If free atmospheric oxygen was involved in such oxidative reactions (rather than, e.g., reactive radicals produced photochemically in the early atmosphere), we may conclude that the atmosphere became oxygenic about 2.3 Ga ago and thus that oxygenic photosynthesis originated at some earlier time. Finally, the evidence from banded iron formations noted above suggests that upper portions of the hydrosphere may have contained appreciable amounts of biologically generated free oxygen as early as 2.5 Ga ago, suggesting that cyanobacteria may have been extant at least this early in Earth history. All of these lines of evidence are thus consistent with the occurrence of cyanobacteria during

Fortescue time, some 2.8 Ga ago, but any attempt to be more precise than this is at present thwarted by the inaccuracy of available radiometric dates of key geological units and the incompleteness of the fossil record as now known.

In sum, the following interpretation is consistent with the geological, geochemical and palaeontological information now available: biological systems originated very early in geological time, possibly earlier than 3.8 Ga ago; by 3.5 Ga ago, life was already abundant and diverse, and multicomponent microbial communities of filamentous and unicellular mat-forming bacteria, which included autotrophs (and possibly, but not necessarily, photoautotrophs), formed laminated, stromatolitic biocoenoses at the sediment–water interface in shallow aqueous environments; sometime between 3.5 and 2.8 Ga ago, oxygenous photosynthesizers (cyanobacteria) evolved. Although the fossil record does not indicate which of the modern classes was first to evolve, by 2.8 Ga ago several lines of evidence point to the existence of mat-forming filamentous cyanobacteria morphologically and ecologically comparable to modern hormogoneans (Oscillatoriaceae and possibly Scytonemataceae and/or Rivulariaceae).

21.3 The Age of Cyanobacteria

The Proterozoic Era is nearly two billion years in duration, spanning the time from the end of the Archaean Era (2.5 Ga ago) to the beginning of the Phanerozoic Eon (0.57 Ga ago). By the end of the Proterozoic, the cyanobacteria were already declining in ecological importance, apparently as a result of 'competitive exclusion' of stromatolite-building microbes by eukaryotic calcareous and non-calcified algae, and of the disruption of stromatolitic habitats by burrowing and grazing metazoans (Garrett, 1970; Monty, 1973; Awramik, 1971). The 'modern' ecosystem, one dominated by multicellular, megascopic metazoans and metaphytes, had become established. Throughout the Proterozoic, cyanobacteria were virtually ubiquitous, at least within the photic zone. They built great carbonate reefs, no less impressive and extensive than the coral–algal reefs of Phanerozoic and modern seas, and their geographic range extended from cold boreal lakes to hot, arid, coastal lagoons, and from the open ocean to the land surface where throughout much of the Proterozoic they were presumably major components of the soil microflora.

This vast 2-billion-year-long span was a time of momentous events in the history of life, developments that as yet can only be dimly perceived due to the inadequacy of available relevant evidence. Within this time span occurred the origin of obligate aerobes, probably about 1.7 Ga ago (Walker *et al.*, 1982), the origin of mitotic eukaryotic unicells, apparently about 1.4 Ga ago (Schopf *et al.*, 1973; Schopf & Oehler, 1976; Schopf, 1978), the origin of meiosis, eukaryotic sexuality and megaplanktonic and megascopic thallophytes, earlier than about 0.8 Ga ago (Hofmann & Aitken, 1979) and possibly earlier than 1.3 Ga ago (Walter *et al.*, 1976b) and the origin of metazoans, sometime prior to about 0.64 Ga ago and possibly substantially earlier. Like the other events, the time of origin of the Metazoa is as yet somewhat uncertain, although it is now well established that by 0·64 Ga ago a diverse fauna of megascopic, but still non-skeletal metazoans, including several tens of species, was globally distributed (Glaessner, 1979); there are hints of still older, smaller metazoans, but this is a field of ongoing research that has yet to yield fully all its secrets. Recognition of early eukaryotic unicells is also difficult, because of the limits imposed by the quality of preservation of such microscopic fossils, but the work of Oehler (1976) strongly suggests the presence of eukaryotic algae in the Bitter Springs microbiota of central Australia about 0.8–0.9 Ga in age, and the occurrence of still older megaplanktonic forms together with other lines of data provides presumptive evidence of the presence of eukaryotes back to about 1.4 Ga ago (see discussions by Knoll & Barghoorn, 1975; Schopf & Oehler, 1976; Schopf, 1978).

The substrate that supported the growing metazoan biomass of the latest Proterozoic was to a considerable extent provided by the benthic cyanobacterium-dominated microbial mats that formed the stromatolites and stromatolitic reefs that are of such common occurrence throughout the Proterozoic. Fig. 21.9 shows a representative sample of fossil cyanobacteria from one such deposit of this age. The record of stromatolites preserved from these times suggests the occurrence of two significant events. Beginning about 0.7 Ga ago, stromatolites became markedly less abundant and diverse, most likely due to the advent of effective metazoan grazing and burrowing (Awramik, 1971). The earlier event has not yet been critically documented, but the evidence suggests that beginning about 1.0 Ga ago, a distinctive and previously abundant type of stromatolite (that characterized by conical laminae and known as *Conophyton*), structures that were typically produced in quiet, subtidal environments, became much less abundant; we can at present only speculate as to the cause of this apparent decrease in abundance, but the appearance of the infaunal grazing habit in a newly evolved meiofauna is a possible explanation.

Although ancient stromatolites (e.g. Figs 21.2, 21.8) are indeed 'true fossils' (in the same sense as are preserved root casts, coprolites or dinosaur footprints, structures that record the presence, if not the detailed anatomy and morphology, of the organisms responsible for their occurrence), they are of course structures built by communities of micro-organisms; they thus differ in this respect from the cellular, microbial, body fossils (e.g. Figs 21.3–21.7, 21.9) preserved in the early rock record. Nevertheless, stromatolites of differing types (e.g. stratiform, domical, conical, columnar, etc.) can be rather easily

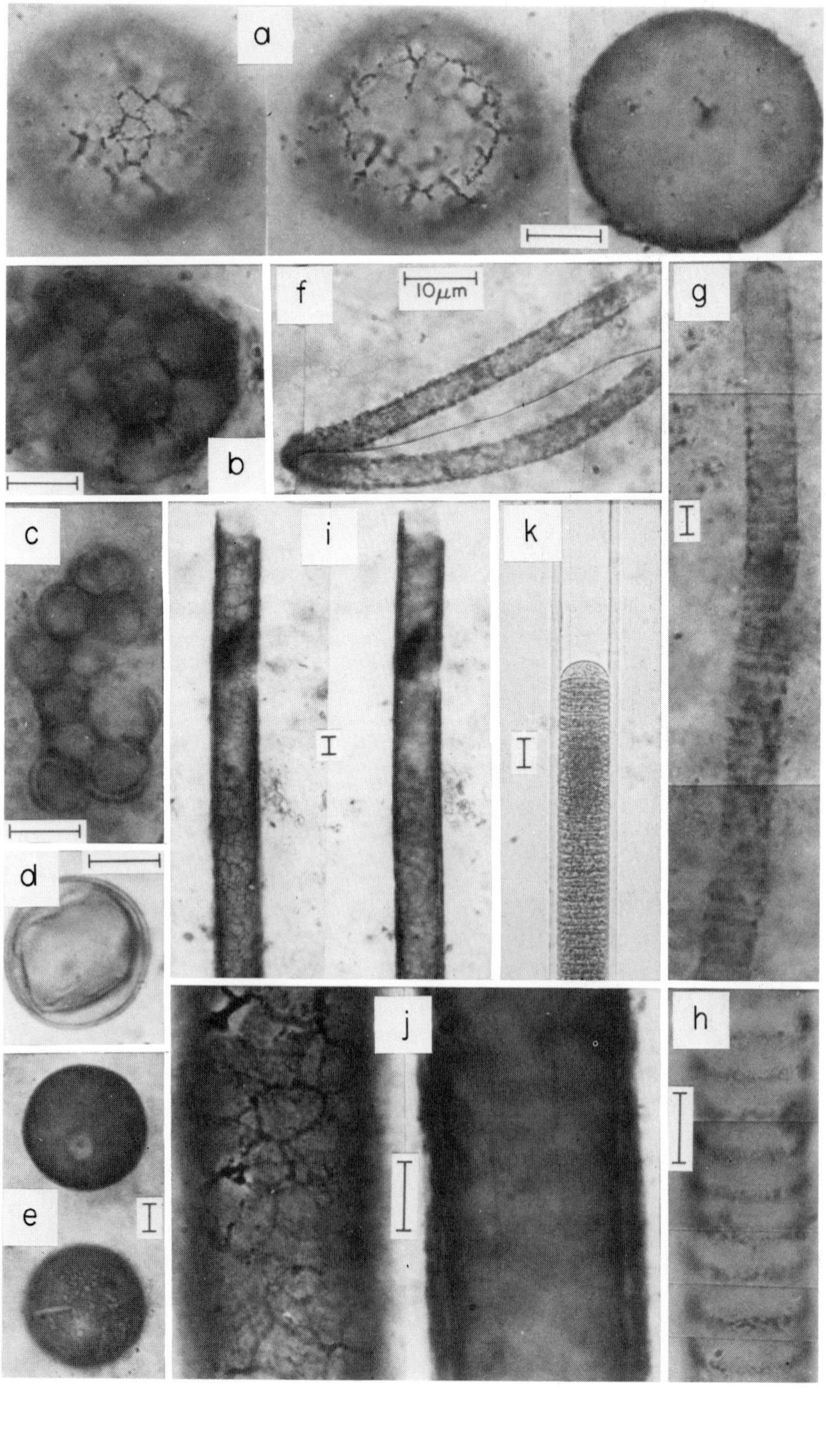
a
b
c
d
e
f
10μm
g
h
i
j
k

distinguished one from the other, and for purposes of reference the various types have thus been given identifying names (e.g. *Stratifera*, *Cryptozoon*, *Conophyton*, etc). Hence, the diversity of stromatolites found in Proterozoic rocks can be readily defined (in terms of the stromatolitic 'taxa' that occur), and the results used as a biostratigraphic tool to determine the comparability of the assemblages of stromatolites that occur in rock sequences of various ages. Such studies appear to show that there are successions of assemblages of stromatolite 'taxa' through the Proterozoic, and that each assemblage is characteristic of a particular time segment. In recent years, this use of stromatolites has become increasingly widespread, although many geologists have yet to be convinced that it can be validly used on a global, rather than a solely intrabasinal scale (for detailed discussions, see Walter 1976, 1977; Bertrand-Sarfati & Walter, 1981). Such a succession could be explained as a result of the evolution of the microbes responsible for stromatolite formation, especially the evolution of the cyanobacteria, and indeed Schopf (1977) has documented broad evolutionary trends in such fossil microbes (as expressed by increases in cell size and in taxonomic diversity) throughout the Proterozoic.

In this short review of the evolutionary history of the cyanobacteria we have chosen to use the classification outlined by Golubic (1976) because of its emphasis on features that can be recognized in cellularly preserved microfossils. Many difficulties are of course encountered in any attempt to compare incompletely preserved fossil micro-organisms with members of specific extant taxa, but such an approach seems at least generally useful for the cyanobacteria because of their extreme, and now well-documented, morphological evolutionary conservativism (see, e.g., Fig. 21.9). Indeed, it seems possible to trace the group at the ordinal level well back into the Proterozoic (viz. for the Chroococcales, Pleurocapsales, Oscillatoriales and Nostocales) and this is apparently true also of the majority of the families (Chroococcaceae, Entophysalidaceae, Pleurocapsaceae, Oscillatoriaceae, Nostocaceae

Fig. 21.9. (a–e) Fossil coccoid and (f–j) filamentous cyanobacteria shown in petrographic thin sections of silicified specimens of the stromatolite *Conophyton* from the Late Proterozoic ($\sim$0.7 Ga old) Chichkan Formation of South Kazakhstan, U.S.S.R. Lines for scale represent 10 μm; (f–h) show photomontages; (a), (e), (i) and (j) show surface textures (produced during preservation and diagenesis) and medial optical sections of individual specimens. Specimens in (a), (d) and (e) are interpreted as solitary unicells; (b) and (c) show colonial coccoid cyanobacteria; (g) and (h) show cellularly preserved oscillatoriacean trichomes; and (f), (i) and (j) show tubular cyanobacterial sheaths. Note the similarity in sheath thickness and morphology and in the shape and size of medial and terminal cells between the broad oscillatoriacean fossil filaments (g–j) and the modern cyanobacterium *Lyngbya majuscula* Harvey ex Gomont shown in (k) (for detailed discussion, see Schopf & Sovietov, 1976).

and probably the Scytonemataceae and Rivulariaceae). Among major types of extant cyanobacteria, only stigonemataceans have as yet to be detected in Precambrian strata (for additional discussion, see Schopf, 1974a). It is to be expected that distinctions between higher taxonomic levels will become blurred as they are traced back toward their origins, but it is a notable feature of the Proterozoic fossil record that it is not until early in the Era that attribution of cyanobacterium-like fossils to higher taxonomic categories becomes difficult to the point of being uncertain (Hofmann & Schopf, 1982). Thus, the roots of the cyanobacterial lineage extend far into the geological past: the group originated, evolved and diversified early in Earth history to become a dominating factor of the world's biota for a major fraction of the total history of life. Indeed, in terms of the longevity of the lineage and its influence on the Earth's early environment, the cyanobacteria must rank as among the most successful and ecologically most significant of the myriad forms of life to have appeared over the long course of biological evolution.

REFERENCES

Quotations in the text to papers written by three or more authors have been abbreviated to the *et al.* format, unless this would lead to ambiguity. Thus .... *et al.* (1978a) and .... *et al.* (1978b) do not necessarily refer to the same combination of authors, so the two references are not necessarily placed adjacent to each other in the list, and may in fact have been quoted in separate chapters. At the end of each reference, on a separate line, is the list of sections in the book in which the reference appears, so that the reference list may also be used as an author index.

The list is composed only of sources quoted in the text; *The Biology of Blue–Green Algae* (Carr & Whitton, 1973) includes many others not quoted here. The following also provide extensive sources of references, some of which are quoted in neither book: Fritsch (1945, pp. 878–98); Desikachary (1959); Geitler (1960, pp. 112–22); Wolk (1973); Roger and Kulasooriya (1980). The *Schweizerische Zeitschrift für Hydrologie* made comprehensive assemblies of the cyanobacterial literature over the period 1960–70: (1964) **26,** 177–87; (1967) **29,** 226–50; (1969) **31,** 81–109; (1970) **32,** 570–90. Safferman and Rohr (1979) compiled *The Practical Directory to Phycovirus Literature* which lists all the literature on cyanophages; this is available from the National Technical Information Services, Springfield, Virginia 22161, USA. A guide to the literature of toxic cyanobacteria is provided in the symposium edited by Carmichael (1981) and to the literature on mass cultivation of *Spirulina* in the volume edited by Shelef and Soeder (1980).

Aaronson S. & Ardois G. (1971) Selective inhibition of blue-green algal growth by ethionine and other amino acid analogs. *J. Phycol.* **7,** 18–20.
18.6

Abad-Zapatero C., Fox J.L. & Hackert M.L. (1977) The quaternary structure of a unique phycobiliprotein: B-phycoerythrin from *Porphyridium cruentum. Biochem. biophys. Res. Commun.* **62,** 266–272.
7.5.1.1, 7.5.1.6

Abeliovich A. (1979) On the mechanism of motility of the trichomes of the filamentous cyanobacterium *Spirulina* sp. In: Nichols J.M. (Ed.) *Abstracts of the Third International Symposium on Photosynthetic Prokaryotes, Oxford.* p. A12. Univ. Liverpool, England.
16.2.2.2

Abeliovich A., Kellenberg D. & Shilo M. (1974) Effect of photooxidative conditions on levels of superoxide dismutase in *Anacystis nidulans. Photochem. Photobiol.* **19,** 379–382.
2.3.1, 8.3

Abeliovich A. & Shilo M. (1972) Photooxidative death in blue-green algae. *J. Bact.* **111,** 682–689.
2.3.1, 6.5.2, 9.5.4, 9.6, 18.5.5.6

Abreu-Grobois F.A., Billyard T.C. & Walton

T.J. (1977) Biosynthesis of heterocyst glycolipids of *Anabaena cylindrica*. *Phytochem.* **16,** 351–354.
14.3.1

Adams D.G. & Carr N.G. (1981) The developmental biology of heterocyst and akinete formation in cyanobacteria. *Crit. Rev. Microbiol.* **9,** 45–100.
15.1, 15.6, 15.6.1

Adams D.G., Phillips D.O., Nichols J.M. & Carr N.G. (1977) The presence and absence of magic spot nucleotide modulation in cyanobacteria undergoing nutritional shift-down. *FEBS Lett.* **81,** 48–52.
14.5

Adams S.M., Kao O.H.W. & Berns D.S. (1979) Psychrophile C-phycocyanin. *Pl. Physiol., Lancaster* **64,** 525–527.
8.8

Adler J. (1969) Chemoreceptors in bacteria. *Science, N.Y.* **166,** 1588–1597.
17.5.1

Adler J. (1975) Chemotaxis in bacteria. *A. Rev. Biochem.* **44,** 341–356.
16.5, 17.5.1

Adler J. (1976) The sensing of chemicals by bacteria. *Scient. Am.* **234,** 40–51.
17.5.2

Adler J., Hazelbauer G.L. & Dahl M.M. (1973) Chemotaxis towards sugars in *Escherichia coli*. *J. Bact.* **115,** 824–847.
17.5.1

Agarwal A. (1979) Blue-green algae to fertilise Indian rice paddies. *Nature, Lond.* **279,** 181.
1

Agarwal M., Jr & Kumar H. (1978) Effect of 7-azatryptophan on heterocyst differentiation in *Anabaena doliolum* Bharadwaja. *Proc. Indian Acad. Sci.* **87B,** 31–39.
14.6.1.1

Ahl T. (1975) Effects of man-induced and natural loading of phosphorus and nitrogen on the large Swedish lakes. *Verh. int. Verein. theor. angew. Limnol.* **19,** 1125–1132.
5.4.3

Ahlgren G. (1971) Limnological studies of Lake Norrviken, a eutrophicated Swedish lake. II. Phytoplankton and its production. *Schweiz. Z. Hydrol.* **32,** 354–396.
2.6.1

Ahlgren G. (1977) Growth of *Oscillatoria agardhii* in chemostat culture. 1. Nitrogen and phosphorus requirements. *Oikos* **29,** 209–224.
5.2.1, 5.2.2, 5.4.1

Ahlgren G. (1978) Growth of *Oscillatoria agardhii* in chemostat culture. 2. Dependence of growth constants on temperature. *Mitt. int. Verein. theor. angew. Limnol.* **21,** 88–102.
5.2.1, 18.5.1.3

Aiba S. & Ogawa T. (1977) Assessment of growth yield of a blue-green alga: *Spirulina platensis*, in axenic and continuous culture. *J. gen. Microbiol.* **102,** 179–182.
2.2.2

Airth R.L. & Blinks L.R. (1956) A new phycoerythrin from *Porphyra naiadum*. *Biol. Bull.* **111,** 321–327.
7.5.1.1

Aitken A. (1975) Prokaryote-eukaryote relationships and the amino acid sequence of plastocyanin from *Anabaena variabilis*. *Biochem. J.* **149,** 675–683.
8.5

Aitken A. (1976) Protein evolution in cyanobacteria. *Nature, Lond.* **263,** 793–796.
8.5, 12.2.2.1

Aitken A. (1977) Purification and primary structure of cytochrome *f* from the cyanobacterium, *Plectonema boryanum*. *Eur. J. Biochem.* **78,** 273–279.
8.5, 12.2.2.1

Aitken A. (1979) Purification and primary structure of cytochrome *c*-552 from the cyanobacterium, *Synechococcus* PCC 6312. *Eur. J. Biochem.* **110,** 297–308.
12.2.2.1

Aitken A. & Rouvière-Janiv J. (1979) Amino and carboxy terminal sequences of the DNA-binding protein HU from the cyanobacterium *Synechocystis* PCC 6701 (ATCC 27170). *Biochem. biophys. Res. Commun.* **91,** 461–467.
11.3.1

Aitken A. & Stanier R.Y. (1979) Characterization of peptidoglycan from the cyanelles of *Cyanophora paradoxa*. *J. gen. Microbiol.* **112,** 219–223.
7.2

Akagi Y. & Taga N. (1980) Uptake of D-glucose and L-proline by oligotrophic and heterotrophic marine bacteria. *Can. J. Microbiol.* **26,** 454–459.
3.4.3

Akinyanju J. & Smith R.J. (1979) Accumulation of ppGpp and pppGpp during nitrogen deprivation of the cyanophyte *Anabaena cylindrica*. *FEBS Lett.* **107,** 173–176.
14.5

Alberte R.J., Tel-Or E., Packer L. & Thornber J.P. (1980) Functional organization of the photosynthetic apparatus in heterocysts of nitrogen-fixing cyanobacteria. *Nature, Lond.* **284,** 481–483.
14.4.3.2.

Alberte R.S. & Thornber J.P. (1978) A rapid procedure for isolating the photosystem 1 reaction center in a highly enriched form. *FEBS Lett.* **91,** 126–130.
8.4

Alexandersson E.T. (1975) Marks of unknown carbonate-decomposing organelles in cyanophyte borings. *Nature, Lond.* **254,** 237–238.
20.4.3

Allen H.L. (1971) Primary productivity, chemoorganotrophy, and nutritional interactions of epiphytic algae and bacteria on macrophytes in the littoral of a lake. *Ecol. Monogr.* **41,** 97–127.
17.1.2

Allen M.B. (1952) The cultivation of Myxophyceae. *Arch. Mikrobiol.* **17,** 34–53.
7.2, 18.5.1.1

Allen M.M. (1968a) Photosynthetic membrane system in *Anacystis nidulans. J. Bact.* **96,** 836–841.
6.2, 6.5.1, 7.8, 7.8.2

Allen M.M. (1968b) Ultrastructure of the cell wall and cell division of unicellular blue-green algae. *J. Bact.* **96,** 842–852.
6.1, 6.2, 13.2.1, 13.3

Allen M.M. (1968c) Simple conditions for the growth of unicellular blue-green algae on plates. *J. Phycol.* **4,** 1–4.
11.1

Allen M.M. (1972) Mesosomes in blue-green algae. *Arch. Mikrobiol.* **84,** 199–206.
6.1

Allen M.M., Hutchinson F. & Weathers P.J. (1980) Cyanophycin granule, polypeptide formation and degradation in the cyanobacterium *Aphanocapsa* 6308. *J. Bact.* **141,** 687–693.
3.2.1, 3.2.2.2

Allen M.M. & Smith A.J. (1969) Nitrogen chlorosis in blue-green algae. *Arch. Mikrobiol.* **69,** 114–120.
3.2.2.2, 7.8, 7.8.1, 7.8.3.2, 11.5.3, 18.5.5.5

Allison E.M. & Walsby A.E. (1981) Potassium in the control of turgor pressure in a gas-vacuolate blue-green alga. *J. exp. Bot.* **32,** 241–249.
2.8.2.2

Allison F.E., Hoover S.R. & Morris H.J. (1937) Physiological studies with the nitrogen-fixing alga *Nostoc muscorum. Bot. Gaz.* **98,** 433–463.
2.6.3

Allsopp A. (1968) Germination of hormocysts of *Scytonema javanicum* and the function of blue-green algal heterocysts. *Nature, Lond.* **220,** 810.
14.1, 15.1

Allsopp A. (1969) Phylogenetic relationships of the Procaryota and the origin of the eukaryotic cell. *New Phytol.* **68,** 591–612.
12.4.1

Almassy R.J. & Dickerson R.E. (1978) *Pseudomonas* cytochrome c_{551} at 2.0 Å resolution: enlargement of the cytochrome *c* family. *Proc. natn. Acad. Sci. U.S.A.* **75,** 2674–2678.
12.1.3, 12.2.2.1

Almon H. & Böhme H. (1980) Components and activity of the photosynthetic electron transport system of intact heterocysts isolated from the blue-green alga *Nostoc muscorum. Biochim. biophys. Acta* **592,** 113–120.
14.4.3.2

Ambler R.P. & Bartsch R.G. (1975) Amino acid sequence similarity between cytochrome *f* from a blue-green bacterium and algal chloroplasts. *Nature, Lond.* **253,** 285–288.
8.5, 12.2.2.1

Ambler R.P., Daniel M., Hermoso J., Meyer T.E., Bartsch R.G. & Kamen M.D. (1979) Cytochrome c_2 sequence variation among the recognized species of purple nonsulphur photosynthetic bacteria. *Nature, Lond.* **278,** 659–660.
12.1.3, 12.2.2.1

Ambler R.P., Meyer T.E. & Kamen M.D. (1979) Anomalies in amino acid sequences of small cytochromes *c′* from two species of purple photosynthetic bacteria. *Nature, Lond.* **278,** 661–662.
12.1.3, 12.2.2.1

Amesz J. & Duysens L.N.M. (1962) Action spectrum, kinetics and quantum requirement of phosphopyridine nucleotide reduction and cytochrome oxidation in the blue-green alga *Anacystis nidulans. Biochim. biophys. Acta* **64,** 261–278.
14.4.3.2

Amla D.V. (1979) Photoreactivation of ultraviolet irradiated blue-green alga *Anacystis nidulans* and cyanophage AS-1. *Arch. Virol.* **59,** 173–179.
11.4.2.1

Anagnostidis K. & Golubić S. (1966) Über die Ökologie einiger *Spirulina*-Arten. *Nova Hedwigia* **11,** 309–335.
9.5.2

Anagnostidis K. & Schwabe G.H. (1966) Über artenreiche Bestände von Cyanophyten und Bacteriophyten in einem Farbstreifensandwatt sowie über das Aufreten Gomontiella-artig deformierter Oscillatoriatrichome. *Nova Hedwigia* **11,** 417–440+4 tables.
20.2.2

Anderson L.E., Price G.B. & Fuller R.C. (1968) Molecular diversity in ribulose-1,5-diphosphate carboxylase from photosynthetic microorganisms. *Science, N.Y.* **161,** 482–484.
3.2.1

Anderson O.R. (1977) Fine structure of a marine amoeba associated with a blue-green alga in the Sargasso Sea. *J. Protozool.* **24,** 370–376.
19.5.3

Andreasen A.A. & Stier T.J.B. (1953) Anaerobic nutrition of *Saccharomyces cerevisiae*. *J. cell. comp. Physiol.* **41,** 23–36.
9.4.1

Andrew P.W., Delaney M.E., Rogers L.J. & Smith A.J. (1975) Isolation and properties of a ferredoxin from *Anabaena flos-aquae*. *Phytochem.* **14,** 931–935.
8.3

Andrews T.J. & Abel K.M. (1981) Kinetics and subunit interactions of ribulose-1,5-bisphosphate carboxylase–oxygenase from the cyanobacterium *Synechococcus* sp. *J. biol. Chem.* **256,** 8445–8451.
3.2.1

Andrews T.J., Abel K.M., Menzel D. & Badger M.R. (1981) Molecular weight and quaternary structure of ribulose-1,5-bisphosphate carboxylase from the cyanobacterium *Synechococcus* sp. *Arch. Microbiol.* **130,** 344–348.
3.2.1

Andrews T.J. & Lorimer G.H. (1978) Photorespiration—still unavoidable? *FEBS Lett.* **90,** 1–9.
3.2.1

Antarikanonda P., Berndt H., Mayer F. & Lorenzen H. (1980) Molecular hydrogen: a new inhibitor of photosynthesis in the blue-green alga (cyanobacterium), *Anabaena* sp. TA 1. *Arch. Microbiol.* **126,** 1–10.
4.6

Antia N.J. (1977) A critical appraisal of Lewin's Prochlorophyta. *Br. phycol. J.* **12,** 271–276.
1

Apte S.K., Rowell P. & Stewart W.D.P. (1978) Electron donation to ferredoxin in heterocysts of the N_2-fixing alga *Anabaena cylindrica*. *Proc. R. Soc. Lond.* **B 200,** 1–25.
4.3, 14.4.1, 14.4.2, 14.4.3.2

Armond P.A. & Staehelin L.A. (1979) Lateral and vertical displacement of integral membrane proteins during lipid phase transition in *Anacystis nidulans*. *Proc. natn. Acad. Sci. U.S.A.* **76,** 1901–1905.
6.3.2, 6.5.3

Armstrong J.E. & Van Baalen C. (1979) Iron transport in microalgae; the isolation and biological activity of hydroxamate siderophore from the blue-green alga *Agmenellum quadruplicatum*. *J. gen. Microbiol.* **111,** 253–262.
2.7.2.1

Arnold D.E. (1971) Ingestion, assimilation, survival and reproduction by *Daphnia pulex* fed seven species of blue-green algae. *Limnol. Oceanogr.* **16,** 906–920.
18.4.2

Arnold W. & Oppenheimer J.R. (1950) Internal conversion in the photosynthetic mechanism of blue-green algae. *J. gen. Physiol.* **33,** 423–435.
7.2

Asato Y. (1971) Photorecovery of γ-irradiated cultures of blue-green alga, *Anacystis nidulans*. *Radiat. Bot.* **11,** 313–316.
11.4.1, 11.4.2.1, 11.5.4

Asato Y. (1972) Isolation and characterization of ultraviolet light-sensitive mutants of the blue-green alga *Anacystis nidulans*. *J. Bact.* **110,** 1058–1064.
11.4.2.1, 11.4.2.2, 11.4.2.3

Asato Y. (1976) Ultraviolet light inactivation and photoreactivation of AS-1 cyanophage in *Anacystis nidulans*. *J. Bact.* **126,** 550–552.
11.4.2.1, 11.4.2.3

Asato Y. & Folsome C.E. (1969) Mutagenesis of *Anacystis nidulans* by *N*-methyl-*N'*-nitro-*N*-nitrosoguanidine and UV irradiation. *Mutation Res.* **8,** 531–536.
11.1, 11.4.2.1, 11.5.4, 11.5.5.1

Asato Y. & Folsome C.E. (1970) Temporal genetic mapping of the blue-green alga *Anacystis nidulans*. *Genetics* **65,** 407–419.
11.4.1, 11.7, 15.5

Astier C. & Espardellier F. (1976) Mise en évidence d'un système de transfert génétique chez une cyanophycée du genre *Aphanocapsa*. *C.r. hebd. Séanc. Acad. Sci., Paris* **282,** D, 795–797.
11.6.3

Astier C. & Joset-Espardellier F. (1981) Characterization and function of a 33 Kd polypeptide in DCMU sensitive and resistant strains of a cyanobacterium, *Aphanocapsa* 6714. *FEBS Lett.* **129,** 47–51.
11.5.4

Astier C., Joset-Espardellier F. & Meyer I. (1979) Conditions for mutagenesis in the cyanobacterium *Aphanocapsa* 6714. Influence of repair phenomena. *Arch. Microbiol.* **120,** 93–96.
11.4.1, 11.4.2.1, 11.4.2.2, 11.5.1

Astier C., Vernotte C., Der-Vartanian M. & Joset-Espardellier F. (1980) Isolation and characterization of two DCMU-resistant

mutants of the blue-green alga *Aphanocapsa* 6714. *Pl. Cell Physiol., Tokyo* **20,** 1501–1510.
11.5.1, 11.5.4

Avissar Y.J. (1980) Biosynthesis of 5-aminolevulinate from glutamate in *Anabaena variabilis. Biochim. biophys. Acta* **613,** 220–228.
3.2.2.1

Awramik S.M. (1971) Precambrian columnar stromatolite diversity: reflection of metazoan appearance. *Science, N.Y.* **174,** 825–827.
21.3

Awramik S.M. (1977) Paleobiology of stromatolites. In: Ponnamperuma C. (Ed.) *Chemical Evolution of the Early Precambrian.* 221 pp., pp. 111–131. Academic Press, New York.
12.3.2

Awramik S.M., Gebelein C.D. & Cloud P. (1979) In: Krumbein W.E. (Ed.) *Environmental Biogeochemistry and Geomicrobiology Volume 1: The Aquatic Environment.* pp. 165–178. Ann Arbor Science, Ann Arbor.
20.4.2

Awramik S.M., Schopf J.W. & Walter M.R. (in press) Filamentous fossil bacteria 3.5 Ga-old from the Archean of Western Australia. *Precambrian Res.*
21.2

Baas Becking L.G.M. & Wood E.J.F. (1955) Biological processes in the estuarine environment. I–II. Ecology of the sulphur cycle. *Proc. K. ned. Akad. Wet. Sect. B* **58,** 160–181.
9.3.3, 9.4.1

Bachmann B.J. & Low K.B. (1980) Linkage map of *Escherichia coli* K-12, Edition 6. *Microbiol. Rev.* **44,** 1–56.
11.2.2

Badger M.R. (1980) Kinetic properties of ribulose-1,5-bisphosphate carboxylase/oxygenase from *Anabaena variabilis. Arch. Biochem. Biophys.* **201,** 247–254.
3.2.1

Bailey-Watts A.E. (1968) Freshwater primary production by a blue-green alga of bacterial size. *Nature, Lond.* **220,** 1344–1345.
19.4.1

Bailey-Watts A.E. (1972) The algal plankton of Loch Leven, Kinross. *Proc. R. Soc. Edinb. (B)* **74,** 135–156.
18.4.1

Bailey-Watts A.E., Bindloss M.E. & Belcher J.H. (1968) Freshwater primary production by a blue-green alga of bacterial size. *Nature, Lond.* **220,** 1344–1345.
6.2, 18.3.1

Bak A.L., Atkins J.F. & Meyer S.A. (1972) Evolution of DNA base compositions in microorganisms. *Science, N.Y.* **175,** 1391–1393.
12.3.4.2

Bak A.L., Black F.T., Christiansen C. & Freundt E.A. (1969) Genome size of mycoplasmal DNA. *Nature, Lond.* **224,** 1209–1210.
11.3.1

Bak A.L., Christiansen C. & Stenderup A. (1970) Bacterial genome sizes determined from DNA renaturation studies. *J. gen. Microbiol.* **64,** 377–380.
11.3.1

Baker A.L. & Brook A.J. (1971) Optical density profiles as an aid to the study of microstratified phytoplankton populations in lakes. *Arch. Hydrobiol.* **69,** 214–233.
2.8.2.1, 18.4.1

Baker A.L., Brook A.J. & Klemer A.R. (1969) Some photosynthetic characteristics of a naturally occurring population of *Oscillatoria agardhii* Gomont. *Limnol. Oceanogr.* **14,** 327–333.
18.4.1

Barber R.T. & Ryther J.H. (1969) Organic chelators: factors of affecting primary production in the Cromwell Current upwelling. *J. exp. mar. Biol. Ecol.* **33,** 191–199.
17.4.2

Barghoorn E.S. & Schopf J.W. (1965) Microorganisms from the Late Precambrian of central Australia. *Science, N.Y.* **150,** 337–339.
21.1

Barghoorn E.S. & Tyler S.A. (1965) Microorganisms from the Gunflint chert. *Science, N.Y.* **147,** 563–577.
21.1

Barlow D.J. & Scott W.E. (1975) Detection of L-aspartate: 2-oxoglutarate aminotransferase in two nitrogen-fixing blue-green algae. *Proc. El. Microsc. Soc. S. Africa* **5,** 57–58.
14.4.1

Barlow D.J. & Scott W.E. (1976) Detection of L-aspartate: 2-oxoglutarate aminotransferase in two nitrogen-fixing Cyanophyta. *S. African J. Sci.* **72,** 240–242.
14.4.1

Bathurst R.G.L. (1967) Subtidal gelatinous mat, sand stabilizer and food, Great Bahama Bank. *J. Geol.* **75,** 736–738.
20.2.4

Batterton J.C., Boush G.M. & Matsumura F. (1972) DDT: inhibition of sodium chloride tolerance by the blue-green alga *Anacystis nidulans*. *Science, N.Y.* **176,** 1141–1143.
18.4.1

Batterton J.C. & Van Baalen C. (1968) Phosphorus deficiency and phosphate uptake in the blue-green alga *Anacystis nidulans*. *Can. J. Microbiol.* **14,** 341–348.
5.1.1, 5.2.2, 5.3.1

Batterton J.C., Jr & Van Baalen C. (1971) Growth responses of blue-green algae to sodium chloride concentrations. *Arch. Mikrobiol.* **76,** 151–165.
20.3.5

Bauld J., Burne R.V., Chambers L.A., Ferguson J. & Skyring G.W. (1980) Sedimentological and geobiological studies of intertidal cyanobacterial mats in North-Eastern Spencer Gulf, South Australia. In: Trudinger P.A., Walter M.R. & Ralph B.J. (Eds) *Biochemistry of Ancient and Modern Environments.* 723 pp., pp. 157–166. Australian Acad. Sci. & Springer-Verlag.
20.4.4

Bauld J., Chambers L.A. & Skyring G.W. (1979) Primary productivity, sulfate reduction and sulfur isotope fractionation in algal mats and sediments of Hamelin Pool, Shark Bay, W.A. *Aust. J. mar. freshwat. Res.* **30,** 753–764.
20.4.4

Baulina O.I., Semenova L.R., Mineyeva L.A. & Gusev M.V. (1978) Cellular ultrastructural organization of the chemoheterophic blue-green alga *Chlorogloea fritschii*. *Mikrobiologia* **47,** 919–923 (Russian).
6.5.2

Baxter R.M., Prosser M.V., Talling J.F. & Wood R.B. (1965) Stratification in tropical African lakes at moderate altitudes (1500 to 2000 m). *Limnol. Oceanogr.* **10,** 510–520.
9.5.4

Bayer M.E. (1968) Areas of adhesion between wall and membrane of *Escherichia coli*. *J. gen. Microbiol.* **53,** 395–404.
6.2

Bayley S.T. (1979) Halobacteria—a problem in biochemical evolution. *Trends biochem. Sci.* **4,** 223–225.
12.2.2.2

Bazin M.J. (1968) Sexuality in a blue-green alga: genetic recombination in *Anacystis nidulans*. *Nature, Lond.* **218,** 282–283.
11.6.1

Bazin M.J. (1969) Quantitative plating of gamma-irradiated cultures of a blue-green alga. *J. Minn. Acad. Sci.* **36,** 45–48.
11.4.1

Bazin M.J. (1971) 5-fluorouracil resistance in *Anacystis nidulans* Drouet. *Br. phycol. J.* **6,** 25–28.
11.5.1

Beaglehole J.C. (1963) *The Endeavour Journal of Joseph Banks 1768–1771.* Vols I & II. The Public Library of New South Wales in association with Angus and Robertson, Sydney.
19.1

Beale S.I., Gough S.P. & Granick S. (1975) Biosynthesis of δ-amino laevulinic acid from the intact carbon skeleton of glutamic acid in greening barley. *Proc. natn. Acad. Sci. U.S.A.* **72,** 2719–2723.
3.2.2.1

Beardall J. & Morris I. (1976) The concept of light intensity adaption in marine phytoplankton. Some experiments with *Phaeodactylum tricornutum*. *Mar. Biol.* **37,** 377–387.
2.4.2, 18.5.5.2

Beauclerk A.A.D. & Smith A.J. (1976) Carbohydrate uptake by cyanobacteria. In: Codd G.A. & Stewart W.D.P. (Eds) *Proceedings of the Second International Symposium on Photosynthetic Prokaryotes.* 258 pp., pp. 213–215. Univ. Dundee, Scotland.
3.4.2

Beauclerk A.A.D. & Smith A.J. (1978) Transport of D-glucose and 3-o-methyl-D-glucose in the cyanobacteria *Aphanocapsa* 6714 and *Nostoc* strain Mac. *Eur. J. Biochem.* **82,** 187–197.
3.4.2, 3.4.3, 3.4.4, 10.3.1, 10.3.4

Beck v. Mannagetta G.R. (1898) Die Sporen von *Microchaete tenera* Thuret und ihre Keimung. *Öst. bot. Z.* **48,** 81–86.
15.1

Belay A. & Fogg G.E. (1978) Photoinhibition of photosynthesis in *Asterionella formosa* (Bacillariophyceae). *J. Phycol.* **14,** 341–347.
19.3.2

Belkin S., Cohen Y. & Padan E. (1979) Sulfide-dependent activities of the cyanobacterium *Oscillatoria limnetica*—further observations. In: Nichols J.M. (Ed.) *Abstracts of the Third International Symposium on Photosynthetic Prokaryotes, Oxford, England.* p. B17. Univ. Liverpool, England.
9.2.2, 9.5.4

Belkin S. & Padan E. (1978) Hydrogen metabolism in the facultative anoxygenic cyanobacteria (blue-green algae) *Oscillatoria limnetica* and *Aphanothece halophytica*. *Arch. Microbiol.* **116,** 109–111.
4.6, 9.3.3, 9.5.4, 12.3.3

Belkin S. & Padan E. (1979) Sulfide-dependent hydrogen evolution in the cyanobacterium

Oscillatoria limnetica. FEBS Lett. **94,** 291–294.
9.2.2, 9.3.3, 9.5.4

Ben-Amotz A. & Avron M. (1973) The role of glycerol in the osmotic regulation of the halophilic alga *Dunaliella parvum. Pl. Physiol., Lancaster* **51,** 875–878.
10.3.1

Bendall D.S. (1978) Oxygen evolution in photosynthesis. *Biochem. Soc. Trans.* **6,** 372–375.
8.6

Benemann J.R. & Weare N.M. (1974a) Nitrogen fixation by *Anabaena cylindrica.* III. Hydrogen-supported nitrogenase activity. *Arch. Microbiol.* **101,** 401–408.
4.3, 4.6, 14.4.3.1, 14.4.3.3

Benemann J.R. & Weare N.M. (1974b) Hydrogen evolution by nitrogen fixing *Anabaena cylindrica* cultures. *Science, N.Y.* **184,** 174–175.
4.6

Bennett A. & Bogorad L. (1971) Properties of subunits and aggregates of blue-green algal biliproteins. *Biochemistry, N.Y.* **10,** 3625–3634.
7.5.1.3, 7.5.1.4, 7.5.1.5, 7.5.1.7, 7.6

Bennett A. & Bogorad L. (1973) Complementary chromatic adaptation in a filamentous blue-green alga. *J. Cell Biol.* **58,** 419–435.
7.2, 7.8.3.1

Bennett A. & Siegelman H.W. (1979) Bile pigments of plants. In: Dolphin D. (Ed.) *The Porphyrins*, Vol. VI, pp. 493–520, Academic Press, New York.
7.5.1.2, 7.5.1.3

Bennett K.J., Silvester W.B. & Brown J.M.A. (1975) The effect of pyruvate on nitrogenase activity in the blue-green alga *Anabaena cylindrica. Arch. Microbiol.* **105,** 61–66.
4.3

Bennett M.E. & Hobbie J.E. (1972) The uptake of glucose by *Chlamydomonas* sp. *J. Phycol.* **8,** 392–398.
3.4.3

Berchthold M. & Bachofen R. (1979) Hydrogen formation by cyanobacteria cultures selected for nitrogen fixation. *Arch. Microbiol.* **123,** 227–232.
4.6

Berg H.C. (1975) Chemotaxis in bacteria. *A. Rev. Biophys. Bioengng* **4,** 119–136.
17.5.1

Berger C. (1975) Occurrence of *Oscillatoria agardhii* Gomont in some shallow eutrophic lakes. *Verh. int. Verein. theor. angew. Limnol.* **19,** 2689–2697.
2.6.1

Bernard F. (1963) Density of flagellates and Myxophyceae in the heterotrophic layers related to environment. In: Oppenheimer C. (Ed.) *Symposium on Marine Microbiology.* 769 pp., pp. 215–228. Charles C. Thomas, Springfield, Illinois.
3.4.3, 19.3.3

Bernard F. & Lecal J. (1960) Plancton unicellulaire recolté dans l'Ocean Indien par le *Charcot* (1950) et le *Norsel* (1955–56). *Bull. Inst. Océanogr., Monaco* **1166,** 1–59.
19.2.4.2

Berner R.A. (1963) An idealised model of dissolved sulphate distribution in recent sediments. *Geochim. cosmochim. Acta* **28,** 1497–1503.
9.5.3

Berns D.S. (1967) Immunochemistry of biliproteins. *Pl. Physiol., Lancaster* **42,** 1569–1587.
7.5.1.7

Berns D.S. & Edwards M.R. (1965) Electron micrographic investigations of C-phycocyanin. *Arch. Biochem. Biophys.* **110,** 511–516.
7.5.1.6

Berns D.S. & Morgenstern A. (1966) Ultracentrifuge investigation of protein aggregation in dilute solutions of C-phycocyanin. *Biochemistry, N.Y.* **5,** 2985–2990.
7.5.1.4

Berns D.S. & Scott E. (1966) Protein aggregation in a thermophilic protein. Phycocyanin from *Synechococcus lividus. Biochemistry, N.Y.* **5,** 1528–1533.
7.5.1.4

Bershova O.I., Kopteva Zh.P. & Tantsyurenko E.V. (1968) The interrelations between blue-green algae—the causative agents of water 'bloom'—and bacteria. In: Topachevsky A.V. (Ed.) *'Tsvetenie' Vodÿ.* 387 pp., pp. 159–171. Naukova Dumka, Kiev, U.S.S.R. (Russian).
3.4.3, 17.4.1, 17.4.2

Bertrand-Sarfati J. & Walter M.R. (1982) Stromatolite biostratigraphy. *Precambrian Res.* **15,** 353–371.
21.3

Beuhler R.J., Pierce R.C., Friedman L. & Siegelman H.W. (1976) Cleavage of phycocyanobilin from C-phycocyanin. Separation and mass spectral identification of the products. *J. biol. Chem.* **251,** 2405–2411.
7.5.1.2

Bhattacharjee S.K. (1977) Unstable protein mediated ultraviolet light resistance in *Anacystis nidulans. Nature, Lond.* **269,** 82–83.
18.5.5.6

Bierman V.J., Jr (1976) Mathematical model of the selective enhancement of blue-green algae by nutrient enrichment. In: Canale R.P. (Ed.) *Modeling Biochemical Processes in Aquatic Ecosystems*. 389 pp., pp. 1–31. Ann Arbor Science, Ann Arbor, Mich.
5.2.1

Biggins J. (1967) Photosynthetic reactions by lysed protoplasts and particle preparations from the blue-green alga, *Phormidium luridum*. *Pl. Physiol., Lancaster* **42,** 1447–1456.
13.4.1

Biggins J. (1969) Respiration in blue-green algae. *J. Bact.* **99,** 570–575.
3.3, 4.4

Binder A. & Bachofen R. (1979) Isolation and characterization of a coupling factor 1 ATPase of the thermophilic blue-green alga (cyanobacterium) *Mastigocladus laminosus*. *FEBS Lett.* **104,** 66–70.
8.7

Binder A., Tel-Or E. & Avron M. (1976) Photosynthetic activities of membrane preparations of the blue-green alga *Phormidium luridum*. *Eur. J. Biochem.* **67,** 187–196.
8.2

Binder A., Wilson K. & Zuber H. (1972) C-phycocyanin from the thermophilic blue-green alga *Mastigocladus laminosus*. Isolation, characterization and subunit composition. *FEBS Lett.* **20,** 111–116.
7.5.1.3

Bindloss M.E., Holden A.V., Bailey-Watts A.E. & Smith I.R. (1972) Phytoplankton production, chemical and physical conditions in Loch Leven. In: Kajak Z. & Hillbricht-Ilkowska A. (Eds) *Productivity Problems of Freshwaters, Proceedings of the UNESCO-IBP Symposium, Warsaw*. 918 pp., pp. 639–659. Polish Scientific Publishers, Warsaw & Krakow.
5.4.2

Birke L. (1974) Marine blue-green algal mats. In: Odum H.T., Copeland B.J. & McMahan E.A. (Eds) *Coastal Ecosystems of the United States. Vol. 1*. 533 pp., pp. 331–345. The Conservation Foundation, Washington D.C., in cooperation with National Oceanic and Atmospheric Administration Office of Coastal Environment.
20.3.4, 20.4.4

Bisalputra T., Oakley B.R., Walker D.C. & Shields C.M. (1975) Microtubular complexes in blue-green algae. *Protoplasma, Wien* **86,** 19–28.
13.3

Bishop C.T., Adams G.A. & Hughes E.O. (1954) A polysaccharide from the blue-green alga, *Anabaena cylindrica*. *Can. J. Chem.* **32,** 999–1004.
13.4.5

Bishop C.T., Anet E.F.L. & Gorham P.R. (1959) Isolation and identification of the fast-death factor in *Microcystis aeruginosa* NRC-1. *Can. J. Biochem. Physiol.* **37,** 453–471.
18.5.6

Biswas B.B. (1956) Chemical nature of nucleic acids in Cyanophyceae. *Nature, Lond.* **177,** 95–96.
11.3.1

Biswas B.B. (1960) Comparative studies on DNA of *Chlorella ellipsoidea* and *Anacystis nidulans*. *Pl. Physiol., Lancaster* **35** (Suppl.), xxx.
11.3.1

Bitton G. & Marshall K.C. (1980) *Adsorption of Microorganisms to Surfaces*. 439 pp. Wiley, N.Y.
17.2

Black M. (1933) The algal sediments of Andros Island, Bahamas. *Phil. Trans. R. Soc. Lond. B.* **222,** 165–192.
20.3.1, 21.1

Blackburn T.H., Kleiber P. & Fenchel T. (1975) Photosynthetic sulfide oxidation in marine sediments. *Oikos* **26,** 103–108.
9.5.3

Blakemore R.P. & Canale-Parola E. (1976) Arginine catabolism by *Treponema denticola*. *J. Bact.* **128,** 616–622.
3.2.2.2

Blaurock A.E. & Walsby A.E. (1976) Crystalline structure of the gas vesicle wall from *Anabaena flos-aquae*. *J. mol. Biol.* **105,** 183–189.
10.2.5.1

Blinks L.R. (1954) The role of accessory pigments in photosynthesis. *Symp. Soc. gen. Microbiol.* **4,** 224–246.
7.2

Bloch K., Barnowsky Y., Goldfine H., Lennarz W.J., Light R., Norris A.T. & Scheuerbrandt G. (1966) Biosynthesis and metabolism of unsaturated fatty acids. *Fed. Proc.* **20,** 921–927.
9.4.2

Blum J.L. (1968) Salt marsh Spartinas and associated algae. *Ecol. Mongr.* **38,** 199–221.
20.2.3

Bogorad L. (1965) Studies of phycobiliproteins. *Rec. Chem. Prog.* **26,** 1–12.
7.5.1.7

Bogorad L. (1975a) Phycobiliproteins and complementary chromatic adaptation. *A. Rev. Pl. Physiol.* **26,** 369–401.
7.5.1.3, 7.5.1.7, 7.8, 7.8.3.1, 15.5, 18.5.5.5

Bogorad L. (1975b) Evolution of organelles and eukaryotic genomes. *Science, N.Y.* **188,** 891–898.
12.4.1

Bogorad L., Mercer F.V. & Mullens R. (1963) Studies with *Cyanidium caldarium*. II. The fine structure of pigment deficient mutants. In: Kok B. & Jagendorf A.T. (Eds) *Photosynthetic Mechanism of Green Plants.* pp. 560–570. Natn. Acad. Sci. Nat. Res. Council 1145, Washington, D.C.
7.2

Bohl M.C. & Binder A. (1980) Photosynthetic activities of a membrane preparation of the thermophilic cyanobacterium *Mastigocladus laminosus*. *Arch. Microbiol.* **124,** 155–160.
8.6

Böhme H. & Böger P. (1978) Reciprocal formation of cytochrome c_{553} and plastocyanin in *Scenedesmus*. *FEBS Lett.* **85,** 337–339.
8.5

Bone D.H. (1971a) Relationship between phosphates and alkaline phosphatase of *Anabaena flos-aquae* in continuous culture. *Arch. Mikrobiol.* **80,** 147–153.
5.2.3

Bone D.H. (1971b) Nitrogenase activity and nitrogen assimilation in *Anabaena flos-aquae* growing in continuous culture. *Arch. Mikrobiol.* **80,** 234–241.
4.4, 5.2.3

Bone D.H. (1971c) Kinetics of synthesis of nitrogenase in batch and continuous culture of *Anabaena flos-aquae*. *Arch. Mikrobiol.* **80,** 242–251.
18.5.3

Bonen L. & Doolittle W.F. (1975) On the prokaryotic nature of red algal chloroplasts. *Proc. natn. Acad. Sci. U.S.A.* **72,** 2310–2314.
12.2.2.4, 12.3.4.3, 12.4.3.2

Bonen L. & Doolittle W.F. (1976) Partial sequences of 16S rRNA and the phylogeny of blue-green algae and chloroplasts. *Nature, Lond.* **261,** 669–673.
12.2.2.4, 12.3.4.3, 12.4.3.2, 12.4.3.3

Bonen L. & Doolittle W.F. (1978) Ribosomal RNA homologies and the evolution of the filamentous blue-green bacteria. *J. mol. Evol.* **10,** 283–292.
12.2.2.4, 12.3.4.3, 12.4.3.2

Bonen L., Cunningham R.S., Gray M.W. & Doolittle W.F. (1977) Wheat embryo mitochondrial 18S ribosomal RNA: evidence for its prokaryotic nature. *Nucl. Acids Res.* **4,** 663–671.
12.4.3.2

Bonen L., Doolittle W.F. & Fox G.E. (1979) Cyanobacterial evolution: results of 16S ribosomal ribonucleic acid sequence analyses. *Can. J. Biochem.* **57,** 879–888.
12.2.2.4, 12.3.4.3, 12.4.3.2

Booker M.J. & Walsby A.E. (1979) The relative form of straight and helical blue-green algal filaments. *Br. phycol. J.* **14,** 141–150.
2.3

Boresch K. (1922) Die komplementäre chromatische Adaptation. *Arch. Protistenk.* **44,** 1–70.
6.5.1, 7.8.3.1

Bornefeld T., Lee-Kaden J. & Simonis W. (1974) Lichtäbhangige Phosphataufnahme und Metabolismus bei *Anacystis nidulans*. *Ber. dt. bot. Ges.* **87,** 493–500.
5.1.1, 5.2.3

Bornefeld T. & Simonis W. (1974) Effects of light, temperature, pH and inhibitors on the ATP level of the blue green alga *Anacystis nidulans*. *Planta, Berl.* **115,** 309–318.
3.3

Borowitzka M.A., Larkum A.W.D. & Borowitzka L.J. (1978) A preliminary study of algal turf communities of a shallow coral reef lagoon using an artificial substratum. *Aquatic Botany* **5,** 365–381.
20.2.4

Borstad G.A. & Borstad L.E. (1977) The *Oscillatoria erythraea* (Cyanophyta) community of associates. In: Steward H.B. (Ed.) *Cooperative Investigations of the Caribbean and Adjacent Regions—II. FAO Fisheries Report No. 200*, pp. 51–57. F.A.O., Rome.
19.4.2, 19.5.3

Borzi A. (1878) Note alla morfologia e biologia delle alghe ficocromacee. *Nuovo G. Bot. Ital.* **10,** 236–288.
14.1

Bosselman S. (1975) Production of *Eudiaptomus gracilis* in Lake Esrom 1970. *Arch. Hydrobiol.* **76,** 43–64.
18.4.2

Bothe H. (1969) The role of phytoflavin in photosynthetic reactions. In: Metzner H. (Ed.) *Progress in Photosynthesis Research, Vol. III.* pp. 1483–1491. University of Tübingen, Germany.
4.3

Bothe H. (1970) Photosynthetische Stickstoffixierung mit einem zellfreien Extrakt aus der Blaualge *Anabaena cylindrica*. *Ber. dt. bot. Ges.* **83,** 421–432.
4.2, 4.3

Bothe H. (1977) Flavodoxin. In: Trebst A. & Avron M. (Eds) *Encyclopedia of Plant Physiology, New Series, Vol. 5.* pp. 217–221. Springer, Berlin.
4.3

Bothe H., Distler E. & Eisbrenner G. (1978) Hydrogen metabolism in blue-green algae. *Biochimie* **60,** 277–289.
4.4, 4.6, 17.6

Bothe H. & Eisbrenner G. (1977) Effect of 7-azatryptophan on nitrogen fixation and heterocyst formation in the blue-green alga *Anabaena cylindrica. Biochem. Physiol. Pflanz.* **171,** 323–332.
4.5, 14.6.1.1

Bothe H., Falkenberg B. & Nolteernsting U. (1974) Properties and function of the pyruvate:ferredoxin oxidoreductase from the blue-green alga *Anabaena cylindrica. Arch. Microbiol.* **96,** 291–304.
4.3

Bothe H. & Floener L. (1980) Nitrogen fixation in *Rhopalodia gibba*, a diatom containing blue-greenish inclusions symbiotically. In: Schwemmler W. & Schenk H.E.A. (Eds) *Endosymbiosis and Cell Biology, Vol. 1.* pp. 541–552. Walter de Gruyter, Berlin.
4.1

Bothe H. & Loos E. (1972) Effect of far red light and inhibitors on nitrogen fixation and photosynthesis in the blue-green alga *Anabaena cylindrica. Arch. Mikrobiol.* **86,** 241–254.
4.3, 14.4.3.2

Bothe H., Neuer G., Kalbe I. & Eisbrenner G. (1980) Electron donors and hydrogenase in nitrogen-fixing microorganisms. In: Stewart W.D.P. & Gallon J.R. (Eds) *Proceedings of an International Symposium on Nitrogen Fixation.* 451 pp., pp. 83–112. Academic Press, N.Y.
4.1, 4.3, 4.6, 14.4.1, 14.4.2

Bothe H. & Nolteernsting U. (1975) Pyruvate dehydrogenase complex, pyruvate:ferredoxin oxidoreductase and lipoic acid content in microorganisms. *Arch. Microbiol.* **102,** 53–57.
4.3

Bothe H., Tennigkeit J. & Eisbrenner G. (1977a) The utilization of molecular hydrogen by the blue-green alga *Anabaena cylindrica. Arch. Microbiol.* **114,** 43–49.
4.6, 14.4.3.1, 14.4.3.3

Bothe H., Tennigkeit J., Eisbrenner G. & Yates M.G. (1977b) The hydrogenase–nitrogenase relationship in the blue-green alga *Anabaena cylindrica. Planta, Berl.* **133,** 237–242.
4.6, 9.3.3

Bothe H., Yates M.G. & Cannon F.C. (1982) Nitrogen fixation. In: Lauchli A. & Bieleski R. (Eds) *Encyclopedia of Plant Physiology, New Series.* Springer, Berlin.
4.1, 4.2, 4.3

Bott T.L., Rogenmuser K. & Thorne P. (1979) Effects of No. 2 fuel oil, Nigerian crude oil, and used crankcase oil on benthic algal communities. *J. environ. Sci.* Ser. A. **13,** 751–779.
20.3.6

Bottomley P.J., Grillo J.F., Van Baalen C. & Tabita F.R. (1979) Synthesis of nitrogenase and heterocysts by *Anabaena* sp. CA in the presence of high levels of ammonia. *J. Bact.* **140,** 938–943.
14.2, 14.5, 14.6.1.1, 14.6.1.3

Bottomley P.J. & Stewart W.D.P. (1976a) ATP pools and transients in the blue-green alga *Anabaena cylindrica. Arch. Microbiol.* **108,** 249–258.
3.3, 3.4.2, 4.3

Bottomley P.J. & Stewart W.D.P. (1976b) The measurement and significance of ATP pools in filamentous blue-green algae. *Br. phycol. J.* **11,** 69–82.
5.2.2, 8.7

Bottomley P.J. & Stewart W.D.P. (1977) ATP and nitrogenase activity in nitrogen-fixing, heterocystous blue-green algae. *New Phytol.* **79,** 625–638.
4.3, 4.5

Bottomley P.J. & Van Baalen C. (1978) Characteristics of heterotrophic growth in the blue-green alga *Nostoc* sp. strain Mac. *J. gen. Microbiol.* **107,** 309–318.
3.4.2

Bottomley P.J., Van Baalen C. & Tabita F.R. (1980) Heterocyst differentiation and tryptophan metabolism in the cyanobacterium *Anabaena* sp. CA. *Arch. Biochem. Biophys.* **203,** 204–213.
14.6.1.3

Bouges-Bocquet B. & Delosme R. (1978) Evidence for a new electron donor to P700 in *Chlorella pyrenoidosa. FEBS Lett.* **94,** 100–104.
8.5

Bourdu R. & Lefort M. (1967) Structure fine, observée en cryodécapage, des lamelles photosynthétiques des Cyanophycées endosymbiotiques: *Glaucocystis nostochinearum* Itzigs. et *Cyanophora paradoxa* Korschikoff. *C.r. hebd. Séanc. Acad. Sci., Paris D.* **265,** 37–40.
6.3.2, 7.2

Boussiba S. & Richmond A.E. (1979) Isolation and characterization of phycocyanins from the blue-green alga *Spirulina platensis. Arch. Microbiol.* **120,** 155–159.
7.5.1.3, 7.5.1.4

Boussiba S. & Richmond A.E. (1980) C-phycocyanin as a storage protein in the blue-

green alga *Spirulina platensis*. *Arch. Microbiol.* **125**, 143–147.
3.2.2.2

Bowen C.C. & Jensen T.E. (1965) Blue-green algae: fine structure of the gas vacuoles. *Science, N.Y.* **147**, 1460–1462.
2.8.2.1., 10.2.5.1

Bowman T.E. & Lancaster L.J. (1965) A bloom of the planktonic blue-green alga *Trichodesmium erythraeum* in the Tonga Islands. *Limnol. Oceanogr.* **10**, 291–293.
19.2.4.1

Boyd C.E. (1973) Summer algal communities and primary productivity in fish ponds. *Hydrobiologia* **41**, 357–390.
5.4.2

Braarud T., Føyn B. & Hasle G.R. (1958) The marine and freshwater phytoplankton of the Dramsfjord and the adjacent part of the Oslofjord March–December 1951. *Hvalbrad. Skr. Oslo.* **No. 43**, pp. 1–102.
19.2.1

Bradley S. & Carr N.G. (1971) The absence of a functional photosystem II in heterocysts of *Anabaena cylindrica*. *J. gen. Microbiol.* **68**, xiii.
14.4.1, 14.4.3.2

Bradley S. & Carr N.G. (1976) Heterocyst and nitrogenase development in *Anabaena cylindrica*. *J. gen. Microbiol.* **96**, 175–184.
4.3, 4.4, 14.4.1, 14.4.3.2, 14.5, 14.6.1.1

Bradley S. & Carr N.G. (1977) Heterocyst development in *Anabaena cylindrica*: the necessity for light as an initial trigger and sequential stages of commitment. *J. gen. Microbiol.* **101**, 291–297.
4.5, 14.5, 14.6.1.1

Brand F. (1901) Bemerkungen über Grenzzellen und über spontan rothe Inhaltskörper der Cyanophyceen. *Ber. dt. bot. Ges.* **19**, 152–159.
14.5

Brand F. (1903) Morphologisch-physiologische Betrachtungen über Cyanophyceen. *Beih. bot. Zbl.* **15**, 31–64.
15.1

Brand J.J. (1977) Spectral changes in *Anacystis nidulans* induced by chilling. *Pl. Physiol., Lancaster* **59**, 970–973.
8.8

Brand J.J. (1979) Spectral changes in membrane fragments and artificial liposomes of *Anacystis* induced by chilling. *Arch. Biochem. Biophys.* **193**, 385–391.
8.2

Brand J.J., Kirchanski, S.J. & Ramierez-Mitchell, R. (1979) Chill-induced morphological alterations in *Anacystis nidulans* as a function of growth temperature. *Planta, Berl.* **145**, 63–68.
8.8

Branton D. (1966) Fracture faces of frozen membranes. *Proc. natn. Acad. Sci. U.S.A.* **55**, 1048–1056.
6.3.1, 7.2

Branton D., Bullivant S., Gilula S., Karnovsky M.J., Moor H., Mühlethaler K., Northcote D.H., Packer L., Satir B., Satir P., Speth V., Staehelin L.A., Steere R.L. & Weinstein R.S. (1975) Freeze etching nomenclature. *Science, N.Y.* **190**, 54–56.
6.3.1

Braun V. & Hantke K. (1977) Bacterial receptors for phages and colicins as constituents of specific transport systems. In: Reissig J.L. (Ed.) *Microbial Interactions: Receptors and Recognition, Series B* **436**, pp. 100–137. Chapman & Hall, London.
13.1

Braune W. (1979) C-phycocyanin—the main photoreceptor in the light dependent germination process of *Anabaena* akinetes. *Arch. Microbiol.* **122**, 289–295.
15.5

Braune W. (1980) Structural aspects of akinete germination in the cyanobacterium *Anabaena variabilis*. *Arch. Microbiol.* **126**, 257–261.
15.2, 15.3, 15.5

Braune W. & Sanke H. (1979) Interferometrische Untersuchungen zur Dynamik von Hydratur und Trockenmassegehalt während der lichtabhärgigen Keimung der Akineten von *Anabaena variabilis*. *Z. allg. Mikrobiol.* **19**, 535–546.
15.5

Bredt W. & Radestock U. (1977) Gliding motility of *Mycoplasma pulmonis*. *J. Bact.* **130**, 937–938.
16.2.1

Bridges B.A. (1972) Evidence for a further dark repair process in bacteria. *Nature, New Biol.* **240**, 52–53.
11.4.2

Bridgwater D., Allaart J.H., Schopf J.W., Klein C., Walter M.R., Barghoorn E.S., Strother P., Knoll A.H. & Gorman B.E. (1981) Microfossil-like objects from the Archaean of Greenland: a cautionary note. *Nature, Lond.* **289**, 51–53.
21.2

Brill W.J. (1975) Regulation and genetics of nitrogen fixation. *A. Rev. Microbiol.* **29**, 109–129.
4.5

Brock T.D. (1973) Evolutionary and ecologi-

cal aspects of the cyanophytes. In: Carr N.G. & Whitton B.A. (Eds) *The Biology of Blue-Green Algae*. 676 pp., pp. 487–500. Blackwell, Oxford.
9.6

Brock T.D. (1975a) Effect of water potential on a *Microcoleus* (Cyanophyceae) from a desert crust. *J. Phycol.* **11,** 316–320.
10.2.4

Brock T.D. (1975b) The effect of water potential on photosynthesis in whole lichens and in their liberated algal components. *Planta, Berl.* **124,** 13–23.

Brock T.D. (1976) Halophilic blue-green algae. *Arch. Microbiol.* **107,** 109–111.
10.4, 20.3.5

Brock T.D. (1978) *Thermophilic Microorganisms and Life at High Temperatures*. 465 pp. Springer, New York.
9.5.2, 9.5.3

Broda E. (1975) *The Evolution of the Bioenergetic Processes.* 221 pp. Pergamon Press, Oxford, New York, Toronto, Sydney, Paris & Braunschweig.
12.1.3, 12.2.1, 12.2.2.2, 12.4.1

Brody M. & Emerson R. (1959) The effect of wavelength and intensity of light on the properties of pigments in *Porphyridium cruentum*. *Am. J. Bot.* **46,** 433–440.
7.8.2

Brody M. & Vatter A.E. (1959) Observations on cellular structures of *Porphyridium cruentum*. *J. biophys. biochem. Cytol.* **5,** 289–294.
7.2

Bromley D.B. & Charon N.W. (1979) Axial filament involvement in the motility of *Leptospira interrogans*. *J. Bact.* **137,** 1406–1412.
16.2.3

Brosius J., Palmer M.L., Kennedy P.J. & Noller H.F. (1978) Complete nucleotide sequence of a 16S ribosomal RNA gene from *Escherichia coli*. *Proc. natn. Acad. Sci. U.S.A.* **75,** 4801–4805.
12.4.3.3

Brown A.S., Foster J.A., Voynow P.V., Franzblau C. & Troxler R.F. (1975) Allophycocyanin from the filamentous cyanophyte, *Phormidium luridum*. *Biochemistry, N.Y.* **14,** 3581–3588.
7.5.1.1

Brown A.S., Offner G.D., Ehrhardt M.M. & Troxler R.F. (1979) Phycobilin-apoprotein linkages in the α and β subunits of phycocyanin from the unicellular rhodophyte, *Cyanidium caldarium*. Amino acid sequences of ^{35}S-labelled chromopeptides. *J. biol Chem.* **254,** 7803–7811.
7.5.1.2, 7.5.1.8

Brown A.S. & Troxler R.F. (1977) Properties and N-terminal sequence of allophycocyanin from the unicellular rhodophyte *Cyanidium caldarium*. *Biochem. J.* **163,** 571–581.
7.5.1.4, 7.5.1.8

Brown C.E. & Romano A.H. (1969) Evidence against necessary phosphorylation during hexose transport in *Aspergillus nidulans*. *J. Bact.* **100,** 1198–1203.
3.4.3

Brown E.J. & Button D.K. (1979) Phosphate-limited growth kinetics of *Selenastrum capricornutum*. *J. Phycol.* **15,** 305–311.
5.1.2, 5.2.1

Brown E.J. & Harris R.F. (1978) Kinetics of algal transient phosphate uptake and the cell quota concept. *Limnol. Oceanogr.* **23,** 35–40.
5.2.3

Brown R.M., Jr, Willison J.H.M. & Richardson C.J. (1976) Cellulose biosynthesis in *Acetobacter xylinum*: visualization of the site of synthesis and direct measurement of the *in vivo* process. *Proc. natn. Acad. Sci. U.S.A.* **73,** 4565–4569.
16.2.2.1

Brunskill G.J. (1973) Rates of supply of nitrogen and phosphorus to Lake Winnipeg, Manitoba, Canada. *Verh. int. Verein. theor. angew. Limnol.* **18,** 1755–1759.
5.4.2

Bryant D.A. (1977) Comparative Studies on Cyanobacterial and Rhodophytan Biliproteins. 422 pp., Ph.D. Thesis, Univ. California, Los Angeles.
7.5.1.7, 7.5.1.8, 7.6

Bryant D.A. (1979) Effects of chromatic illumination on the composition of cyanobacterial phycobilisomes. In: Nichols J.M. (Ed.) *Abstracts of the Third International Symposium on Photosynthetic Prokaryotes, Oxford.* p. D17. Univ. Liverpool, England.
7.5.1.1, 7.5.1.4, 7.5.2, 7.5.3, 7.8.3.1

Bryant D.A. (1981) The photoregulated expression of multiple phycocyanin species. A general mechanism for the control of phycocyanin synthesis in chromatically adapting cyanobacteria. *Eur. J. Biochem.* **119,** 425–429.
7.5.1.1, 7.5.1.4, 7.5.2, 7.5.3, 7.8.3.1

Bryant D.A. & Cohen-Bazire G. (1981) Effects of chromatic illumination on cyanobacterial phycobilisomes. Evidence for the specific induction of a second pair of phycocyanin subunits in *Pseudanabaena* 7409 grown in red light. *Eur. J. Biochem.* **119,** 415–424.
7.5.1.1, 7.5.1.4, 7.5.2, 7.5.3, 7.8.3.1

Bryant D.A., Glazer A.N. & Eiserling F.A. (1976) Characterization and structural properties of the major biliproteins of *Anabaena* sp. *Arch. Microbiol.* **110,** 61–75.
7.5.1.1, 7.5.1.2, 7.5.1.3, 7.5.1.4, 7.5.1.5, 7.5.1.6, 7.5.1.7

Bryant D.A., Guglielmi G., Tandeau de Marsac N., Castets A.M. & Cohen-Bazire G. (1979) The structure of cyanobacterial phycobilisomes: a model. *Arch. Microbiol.* **123,** 113–127.
7.3, 7.4, 7.5.1.4, 7.5.1.6, 7.5.2, 7.5.3, 7.6, 7.7, 7.8.3.1

Bryant D.A., Hixson C.S. & Glazer A.N. (1978) Structural studies on phycobiliproteins. III. Comparison of bilin-containing peptides from the β-subunits of C-phycocyanin, R-phycocyanin and phycoerythrocyanin. *J. biol. Chem.* **253,** 220–225.
7.5.1.2, 7.5.1.3, 7.5.1.8, 12.4.4.2

Bryce T.A., Welti D., Walsby A.E. & Nichols B.W. (1972) Monohexoside derivatives of long-chain polyhydroxy alcohols; a novel class of glycolipid specific to heterocystous algae. *Phytochem.* **11,** 295–302.
14.3.1

Bryceson I. (1977) An Ecological Study of the Phytoplankton of the Coastal Waters of Dar es Salaam. Ph.D. Thesis, Univ. Dar es Salaam.
19.2.4.1

Bryceson I. & Fay P. (1979) Nitrogen fixation associated with surface blooms of *Oscillatoria (Trichodesmium) erythraea* in the coastal waters of Dar Es Salaam. In: Nichols J.M. (Ed.) *Abstracts of the Third International Symposium on Photosynthetic Prokaryotes, Oxford.* p. D2. Univ. Liverpool, England.
4.4

Bryceson I. & Fay P. (1981) Nitrogen fixation in *Oscillatoria (Trichodesmium) erythraea* in relation to bundle formation and trichome differentiation. *Mar. Biol.* **61,** 159–166.
18.5.2.1, 19.3.4, 19.5.2

Buckley C.E. & Houghton J.A. (1975) The repair of U.V. induced damage in the blue-green alga *Gloeocapsa alpicola. Proc. Soc. gen. Microbiol.* **11,** 83.
11.4.2.1

Buckley C.E. & Houghton J.A. (1976) A study of the effects of near UV radiation on the pigmentation of the blue-green alga *Gloeocapsa alpicola. Arch. Microbiol.* **107,** 93–97.
11.4.2.1, 11.4.2.3

Bunt J.S. (1968) Microalgae of the Antarctic pack ice zone. In: Currie R.J. (Ed.) *Antarctic Oceanography*, pp. 198–218. Scott Polar Research Institute, Cambridge.
20.3.3

Bunt J.S. (1969) Observations on photoheterotrophy in a marine diatom. *J. Phycol.* **5,** 37–42.
3.4.3

Bunt J.S., Cooksey K.E., Keeb M.A., Lee C.C. & Taylor B.F. (1970) Assay of algal nitrogen fixation in the marine subtropics by acetylene reduction. *Nature, Lond.* **227,** 1163–1164.
19.3.4, 19.5.2

Bunting W. & Lund J.W.G. (1956) A new blue-green alga epizoic on *Daphnia pulex* L. *Naturalist, Hull* No. 858, pp. 88–90.
18.4.1

Burchard R.P. (1974) Studies on gliding motility in *Myxococcus xanthus. Arch. Microbiol.* **99,** 271–280.
16.2.3

Burchard R.P. (1981) Gliding motility of prokaryotes: ultrastructure, physiology and genetics. *A. Rev. Microbiol.* **35,** 497–529.
16.2.1

Burchard A.C., Burchard R.P. & Kloetzel J.A. (1977) Intracellular periodic structures in the gliding bacterium *Myxococcus xanthus. J. Bact.* **132,** 666–672.
16.2.3

Burger J. (1974) A study of two populations of *Anabaena planctonica* Brunnth. (Cyanophyta) from Minnesota. *Phycologia* **13,** 125–129.
15.6

Burgis M.J., Darlington J.P.E.C., Dunn I.G., Ganf G.G., Gwahaba J.J. & McGowan L.M. (1973) The biomass and distribution of organisms in Lake George, Uganda. *Proc. R. Soc. Lond. B.* **184,** 271–298.
18.4.1

Burkholder P.R. (1934) Movement in the cyanophyceae. *Quart. Rev. Biol.* **9,** 438–459.
2.8.1, 16.2.1, 16.4.4, 16.5

Burkholder P.R. (1963) Some nutritional relationships among microbes of sea sediments and water. In: Oppenheimer C.H. (Ed.) *Symposium on Marine Microbiology.* 769 pp., pp. 133–150. Charles C. Thomas, Springfield, Ill.
17.4.4

Burkholder P.R., Burkholder L.M. & Almodovar L.R. (1960) Antibiotic activity of some marine algae of Puerto Rico. *Bot. Marina* **2,** 149–156.
3.4.3

Burkholder P.R., Repak A. & Sibert J. (1965) Studies on some Long Island Sound littoral

communities of microorganisms and their primary productivity. *Bull. Torrey Bot. Club* **92,** 378–402.
20.4.4

Burnham T.C., Stetak T. & Locher G. (1976) Extracellular lysis of the blue-green alga *Phormidium luridum* by *Bdellovibrio bacteriovorus. J. Phycol.* **12,** 306–313.
17.4.2

Burns R.C. & Hardy R.W.F. (1975) *Nitrogen Fixation in Bacteria and Higher Plants.* 186 pp. Springer, Berlin.
4.6

Burris R.H. (1976) Nitrogen fixation by blue-green algae of the Lizard Island area of the Great Barrier Reef. *Aust. J. Pl. Physiol.* **3,** 41–51.
20.4.5

Bursche E.-M. (1955) Beitrag zur Frages des 'Krautschwundes' in H_2S — Oscillatorien — Seen. *Z. Fisch.* **4,** 53–99.
18.4.1, 18.5.5.7

Bush R.M. & Welch E.B. (1972) Plankton associations and related factors in a hyper-eutrophic lake. *Water, Air Soil Pollut.* **1,** 257–274.
5.4.3

Busson F. (1971) Spirulina platensis *(Gom.) Geitler et* Spirulina geitleri *J. de Toni. Cyanophycées Alimentaires.* 161 pp. Servie de Sante, Parc du Pharo, 13 Marseille 7eme, France.
6.2

Butler R.D. & Allsopp A. (1972) Ultrastructural investigations in the Stigonemataceae. *Arch. Mikrobiol.* **82,** 283–299.
13.2.1, 13.2.2, 13.3

Buttke M. & Ingram L.O. (1975) Comparison of lipopolysaccharides from *Agmenellum quadruplicatum* to *Escherichia coli* and *Salmonella typhimurium* by using thin-layer chromatography. *J. Bact.* **124,** 1566–1573.
13.4.4

Cades J.S., Borden D. & Margoliash E. (1979) The primary structures of ferredoxin and cytochrome *f* of a blue-green alga *Synechococcus lividis.* Abstract no. 03-2-5129. *Abstracts XI Int. Cong. Biochem., Toronto,* p. 172.
8.5

Cagle G.D. (1975) Fine structure and distribution of extra-cellular polymer surrounding selected aerobic bacteria. *Can. J. Microbiol.* **21,** 395–408.
17.2

Caldwell D.E. (1977) The planktonic microflora of lakes. *Crit. Rev. Microbiol.* **5,** 305–370.
17.1.1

Caldwell D.E. (1979) Associations between photosynthetic and heterotrophic prokaryotes in plankton. In: Nichols J.M. (Ed.) *Abstracts of the Third International Symposium on Photosynthetic Prokaryotes, Oxford.* p. A14. Univ. Liverpool, England.
18.5.1.1, 18.5.2.3

Caldwell D.E. & Caldwell S.J. (1978) A *Zoogloea* sp. associated with blooms of *Anabaena flos-aquae. Can. J. Microbiol.* **24,** 922–931.
17.1.1, 17.4.1, 17.4.3, 17.4.4, 17.6

Calef G.W. & Grice G.D. (1966) Relationship between the blue-green alga *Trichodesmium thiebautii* and the copepod *Macrosetella gracilis* in the plankton off northeastern South America. *Ecology* **47,** 855–856.
19.5.3

Cammack R., Luijk L.J., Maguire, J.J., Fry I.V. & Packer L.J. (1979a) EPR spectra of photosystem I and other iron protein components in intact cells of cyanobacteria. *Biochim. biophys. Acta* **548,** 267–275.
8.4

Cammack R., Rao K.K., Bargeron C.P., Hutson K.G., Andrew P.W. & Rogers L.J. (1977) Midpoint potentials of plant and algal ferredoxins. *Biochem. J.* **168,** 205–209.
8.3

Cammack R., Ryan M.D. & Stewart A.C. (1979b) The e.p.r. spectrum of iron-sulfur center B in photosystem I of *Phormidium laminosum. FEBS Lett.* **107,** 422–426.
8.4

Cammack R., Tel-Or E. & Stewart W.D.P. (1976) EPR spectra of photosystem I constituents in heterocyst preparations from *Anabaena cylindrica. FEBS Lett.* **70,** 241–244.
14.4.3.2

Canabaeus L. (1929) Über die Heterocysten und Gasvakuolen der Blaualgen und ihre Beziehungen zueinander. *Pflanzenforschung* **13,** 1–48.
15.4

Canale-Parola E. (1978) Motility and chemotaxis of spirochetes. *A. Rev. Microbiol.* **32,** 69–99.
16.2.3

Cannon G.C., Strohl W.R., Larkin J.M. & Shively J.M. (1979) Cytochromes in *Beggiatoa alba. Current Microbiol.* **2,** 263–266.
16.4.3

Caperon J. & Meyer J. (1972) Nitrogen-limited growth of marine phytoplankton—I.

Changes in population characteristics with steady-state growth rate. *Deep Sea Res.* **19,** 601–618.
5.2.1

Cardemil L. & Wolk C.P. (1976) The polysaccharides from heterocyst and spore envelopes of a blue-green alga. Methylation analysis and structure of the backbones. *J. biol. Chem.* **251,** 2967–2975.
14.3.1, 14.4.2, 15.2

Cardemil L. & Wolk C.P. (1979) The polysaccharides from heterocyst and spore envelopes of a blue-green alga. Structure of the basic repeating unit. *J. biol. Chem.* **254,** 736–741.
14.3.1, 15.2

Cardemil L. & Wolk C.P. (1981a) Isolated heterocysts of *Anabaena variabilis* synthesize envelope polysaccharide. *Biochim. biophys. Acta* **674,** 265–276.
14.3.1

Cardemil L. & Wolk C.P. (1981b) Polysaccharides from the envelopes of heterocysts and spores of the blue-green algae *Anabaena variabilis* and *Cylindrospermum licheniforme*. *J. Phycol.* **17,** 234–240.
14.3.1

Carmichael W.W. (1981) *The Water Environment*. 491 pp. Plenum Press, New York. References.

Carmichael W.W. & Gorham P.R. (1974) An improved method for obtaining axenic clones of planktonic blue-green algae. *J. Phycol.* **10,** 238–240.
18.5.1.1

Carmichael W.W. & Gorham P.R. (1977) Factors influencing the toxicity and animal susceptibility of *Anabaena flos-aquae (Cyanophyta)* blooms. *J. Phycol.* **13,** 97–101.
18.5.6

Carpenter E.J. (1973) Nitrogen fixation by *Oscillatoria (Trichodesmium) thiebautii* in the southwestern Sargasso Sea. *Deep Sea Res.* **20,** 285–288.
19.2.4.1, 19.3.4

Carpenter E.J. & Cox J.L. (1974) Production of pelagic *Sargassum* and a blue-green epiphyte in the Western Sargasso Sea. *Limnol. Oceanogr.* **19,** 429–436.
19.2.4.2

Carpenter E.J. & McCarthy J.J. (1975) Nitrogen fixation and uptake by *Oscillatoria (Trichodesmium) thiebautii* in the western Sargasso Sea. *Limnol. Oceanogr.* **20,** 389–401.
19.3.4, 19.3.8

Carpenter E.J. & Price C.C. IV (1976) Marine *Oscillatoria (Trichodesmium)*: explanation for aerobic nitrogen fixation without heterocysts. *Science, N.Y.* **191,** 1278–1280.
4.4, 18.5.2.1, 19.2.4.1, 19.3.4

Carpenter E.J. & Price C.C. IV (1977) Nitrogen fixation, distribution and production of *Oscillatoria (Trichodesmium)* spp. in the western Sargasso and Caribbean Seas. *Limnol. Oceanogr.* **22,** 60–72.
19.5.1

Carpenter E.J., Van Raalte C.D. & Valiela I. (1978) Nitrogen fixation by algae in a Massachusetts salt marsh. *Limnol. Oceanogr.* **23,** 318–327.
20.4.5

Carpenter E.J. & Walsby A.E. (1979) Gas vacuole collapse in marine *Oscillatoria (Trichodesmium) thiebautii* (Cyanophyta) and the effect on nitrogenase activity and photosynthesis. *J. Phycol.* **15,** 221–223.
18.5.2.1, 19.3.2, 19.3.4

Carr N.G. (1966) The occurrence of poly-β-hydroxybutyrate in the blue-green alga *Chlorogloea fritschii*. *Biochim. biophys. Acta* **120,** 308–310.
3.2.2.2

Carr N.G. (1973) Metabolic control and autotrophic physiology. In: Carr N.G. & Whitton B.A. (Eds) *The Biology of Blue-Green Algae*. 676 pp., pp. 39–65. Blackwell, Oxford.
18.5.2.4

Carr N.G. (1979) Differentiation in filamentous cyanobacteria. In: Parish J.H. (Ed.) *Developmental Biology of Prokaryotes*. pp. 167–201. Blackwell, Oxford.
14.1, 14.3.2

Carr N.G. & Bradley S. (1973) Aspects of development in blue-green algae. In: Ashworth J.A. & Smith J.E. (Eds) *Microbial Differentiation. Symp. Soc. gen. Microbiol.* **23,** 161–188.
14.4.2, 14.4.3.1, 14.4.3.2, 14.6.1.1

Carr N.G. & Whitton B.A. (1973) *The Biology of Blue-Green Algae*. 676 pp. Blackwell, Oxford.

Carter H.J. (1856) Notes on the freshwater infusoria of the island of Bombay. No. 1. Organisation. *Ann. Mag. Nat. Hist.* **18** (2nd ser.), 115–132, 221–249.
15.1, 15.6

Carter N. (1933a, b) A comparative study of the alga flora of two salt marshes. *J. Ecol.* **21,** 128–208; 385–403.
20.2.3

Castenholz R.W. (1967) Aggregation in a thermophilic *Oscillatoria*. *Nature, Lond.* **215,** 1285–1286.
16.2.2.2, 16.5

Castenholz R.W. (1968) The behavior of *Oscil-*

latoria terebriformis in hot springs. *J. Phycol.* **5,** 123–139.
2.8.1, 16.5, 16.6

Castenholz R.W. (1969) Thermophilic blue-green algae and the thermal environment. *Bact. Rev.* **33,** 476–504.
9.5.2

Castenholz R.W. (1973) Movements. In: Carr N.G. & Whitton B.A. (Eds) *The Biology of Blue-Green Algae.* 676 pp., pp. 320–339. Blackwell, Oxford.
16.1, 16.2.1, 16.2.2.1, 16.3, 16.4.4, 16.5

Castenholz R.W. (1976) The effect of sulfide on the blue-green algae of hot springs. I. New Zealand and Iceland. *J. Phycol.* **12,** 54–68.
9.1, 9.3.1, 9.3.2, 9.3.4, 9.5.2, 9.6

Castenholz R.W. (1977) The effect of sulfide on the blue-green algae of hot springs. II. Yellowstone National Park. *Microb. Ecol.* **3,** 79–105.
9.1, 9.3.1, 9.4.2, 9.5.2, 9.6

Catlin B.W. (1956) Extracellular deoxyribonucleic acid of bacteria and a deoxyribonuclease inhibitor. *Science, N.Y.* **124,** 441–442.
11.6.3

Catlin B.W. (1960) Transformation of *Neisseria meningitidis* by deoxyribonucleates from cells and from culture slime. *J. Bact.* **79,** 579–590.
11.6.3

Catlin B.W. & Cunningham L.S. (1958) Studies of extracellular and intracellular bacterial deoxyribonucleic acids. *J. gen. Microbiol.* **19,** 522–539.
11.6.3

Cavalier-Smith T. (1975) The origin of nuclei and of eukaryotic cells. *Nature, Lond.* **256,** 463–468.
12.4.1, 12.4.2

Chadefaud M. (1951) L'evolution morphologique et le systematique des Cyanophycées. *7me Cong. de l'A.F.S.* **4,** 3–8.
12.3.1

Chang, S., Des Marais D., Mack R., Miller S.L. & Strathearn G. (1982) Prebiotic organic synthesis and the origin of life. In: Schopf J.W. (Ed.) *Origin and Evolution of Earth's Earliest Biosphere.* Princeton Univ. Press, Princeton, N.J.
21.2

Chao L. & Bowen C.C. (1971) Purification and properties of glycogen isolated from a blue-green alga *Nostoc muscorum. J. Bact.* **105,** 331–338.
3.2.2.2

Chapman D.J. (1973) Biliproteins and bile pigments. In: Carr N.G. & Whitton B.A. (Eds) *The Biology of Blue-Green Algae.* 676 pp., pp. 162–185. Blackwell, Oxford.
7.5.1.2, 7.10

Chapman D.J., Cole W.J. & Siegelman H.W. (1967) The structure of phycoerythrobilin. *J. Am. chem. Soc.* **89,** 5976–5977.
7.5.1.2

Chapman D.J. & Ragan M.A. (1980) Evolution of biochemical pathways: evidence from comparative biochemistry. *Rev. Plant Physiol.* **31,** 639–678.
12.2.1, 12.3.3

Chapman V.J. (1938) Studies in salt-marsh ecology. I–III. *J. Ecol.* **26,** 144–179.
20.2.3

Chapman V.J. (1939) Studies in salt-marsh ecology. IV and V. *J. Ecol.* **27,** 160–201.
20.2.3

Chapman V.J. (1940) Studies in salt-marsh ecology. VI and VII. *J. Ecol.* **28,** 118–152.
20.2.3

Chapman V.J. (1941) Studies in salt-marsh ecology. VIII. *J. Ecol.* **29,** 69–82.
20.2.3

Chapman V.J. (1959) Studies in salt-marsh ecology. IX. Changes in salt-marsh vegetation at Scolt Head Island. *J. Ecol.* **47,** 619–639.
20.2.3

Chapman V.J. (1960) The plant ecology of Scolt Head Island. In: Steers J.A. (Ed.) *Scolt Head Island.* pp. 85–163. Heffer, Cambridge.
20.2.3

Chapman V.J. (1977) Introduction. In: Chapman V.J. (Ed.) *Ecosystems of the World 1. Wet Coastal Ecosystems.* 428 pp., pp. 1–29. Elsevier, Amsterdam.
20.2.3

Chatton E. (1938) *Titres et Travaux Scientifiques (1906–1937).* 405 pp. Imprimerie E. Sottano, Sète.
Foreword

Chen M. (1974) Kinetics of phosphorus absorption by *Corynebacterium bovis. Microb. Ecol.* **1,** 164–175.
5.1.1, 5.2.2

Chen J.S. & Blanchard D.K. (1978) Isolation and properties of an unidirectional H_2-oxidizing hydrogenase from the strictly anaerobic N_2-fixing bacterium *Clostridium pasteurianum. Biochem. biophys. Res. Commun.* **84,** 1144–1150.
4.6

Cheng K.H. & Colman B. (1974) Measurements of photorespiration in some microscopic algae. *Planta, Berl.* **115,** 207–212.
18.5.2.2

Chernousova V.M., Sirenko L.A. & Arendar-

chuk V.V. (1968) Localization and physiological state of mass species of blue-green algae during late autumn and spring periods. In: Topachevsky A.V. (Ed.) *'Tsvetenie' Vodȳ*. 387 pp., pp. 18–19. Naukova Dumka, Kiev, U.S.S.R. (Russian).
3.4.3

Cheung W.Y. & Gibbs M. (1966) Dark and photometabolism of sugars by a blue-green alga *Tolypothrix tenuis*. *Pl. Physiol., Lancaster* **41**, 731–737.
3.4.2

Chróst R.J. (1975) Inhibitors produced by algae as an ecological factor affecting bacteria in water ecosystems. *Acta microbiol. polon. B.* **7**, 125–133.
17.4.2

Chróst R.J. & Brzeska D. (1978) Extracellular release of organic products and growth of bacteria in *Anabaena cylindrica* (blue-green alga) culture. *Acta microbiol. polon.* **27**, 287–295.
17.4.4

Clark R.L. & Jensen T.E. (1969) Ultrastructure of akinete development in a blue-green alga, *Cylindrospermum* sp. *Cytologica* **34**, 439–448.
15.2

Clarke C.H. (1967) Caffeine- and amino acid-effects upon trp^+ revertant yield in U.V.-irradiated hcr^+ and hcr^- mutants of *E. coli* B/r. *Molec. gen. Genet.* **99**, 97–108.
11.4.2.2

Clayton R.K. (1959) Phototaxis of purple bacteria. In: Bünning E. (Ed.) *Handbuch der Pflanzenphysiologie: 17, Teil 1, Physiologie der Bewegungen*. pp. 371–387. Springer, Berlin.
16.4.3

Clayton R.K. (1977) *Light and Living Matter. A Guide to the Study of Photobiology. Vol. I. The Physical Part*. 158 pp. Krieger, New York.
2.2.1

Clayton R.K. & Sistrom W.R. (1978) *The Photosynthetic Bacteria*. 946 pp. Plenum Press, New York.
12.2.1

Cloern J.E. (1976) Recent limnological changes in southern Kootenay Lake, British Columbia. *Can. J. Zool.* **54**, 1571–1578.
5.4.3

Cloud P. (1965) Significance of the Gunflint (Precambrian) microflora. *Science, N.Y.* **148**, 27–35.
21.1

Cloud P. (1976) Beginnings of biospheric evolution and their biogeochemical consequences. *Paleobiology* **2**, 351–387.
21.1

Codd G.A., Cook C.M. & Stewart W.D.P. (1979) Purification and subunit structure of D-ribulose-1,5-bisphosphate carboxylase from the cyanobacterium *Aphanothece halophytica*. *FEMS Microbiol. Lett.* **6**, 81–86.
3.2.1

Codd G.A. & Cossar J.D. (1978) The site of inhibition of photosystem II by 3-(β,4-dichlorophenyl)-N-N′ dimethylurea in thylakoids of the cyanobacterium *Anabaena cylindrica*. *Biochem. biophys. Res. Commun.* **83**, 342–346.
8.6

Codd G.A., Okabe K. & Stewart W.D.P. (1980) Cellular compartmentation of photosynthetic and photorespiratory enzymes in the heterocystous cyanobacterium *Anabaena cylindrica*. *Arch. Microbiol.* **124**, 149–154.
3.2.1, 14.4.1, 14.4.2

Codd G.A., Rowell P. & Stewart W.D.P. (1974) Pyruvate and nitrogenase activity in cell-free extracts of the blue-green alga *Anabaena cylindrica*. *Biochem. biophys. Res. Commun.* **61**, 424–431.
4.3

Codd G.A. & Stewart W.D.P. (1973) Pathways of glycollate metabolism in the blue-green alga *Anabaena cylindrica*. *Arch. Mikrobiol.* **94**, 11–28.
3.2.1

Codd G.A. & Stewart W.D.P. (1976) Polyhedral bodies and ribulose-1,5-diphosphate carboxylase of the blue-green alga *Anabaena cylindrica*. *Planta, Berl.* **130**, 323–326.
3.2.1

Codd G.A. & Stewart W.D.P. (1977a) D-Ribulose-1,5-diphosphate carboxylase from the blue-green alga *Aphanocapsa* 6308. *Arch. Microbiol.* **113**, 105–110.
3.2.1, 3.3, 14.4.1, 14.4.2

Codd G.A. & Stewart W.D.P. (1977b) Quaternary structure of the D-ribulose-1,5-diphosphate carboxylase from cyanelles of *Cyanophora paradoxa*. *FEMS Microbiol. Lett.* **1**, 35–38.
3.2.1

Codd G.A. & Stewart W.D.P. (1977c) Ribulose-1,5-diphosphate carboxylase in heterocysts and vegetative cells of *Anabaena cylindrica*. *FEMS Microbiol. Lett.* **2**, 247–249.
3.2.1, 14.4.2

Cohen G.N. & Monod J. (1957) Bacterial permeases. *Bact. Rev.* **21**, 169–194.
3.4.4

Cohen Y., Jørgensen B.B., Aizenstat Z. &

Stoler A. (1980) The microbial geochemistry of Solar Lake, Sinai. In: Trudinger P.A. & Walter M. (Eds.) *Environmental Biogeochemistry of Ancient and Recent Stromatolites*. pp. 167–172. Aust. Acad. Sci., Canberra.
9.5.4

Cohen Y., Jørgensen B.B., Padan E. & Shilo M. (1975) Sulphide dependent anoxygenic photosynthesis in the cyanobacterium *Oscillatoria limnetica*. *Nature, Lond.* **257**, 489–492.
3.3, 9.2.1

Cohen Y., Krumbein W.E., Goldberg M. & Shilo M. (1977) Solar Lake (Sinai). 1. Physical and chemical limnology. *Limnol. Oceanogr.* **22**, 597–608.
9.5.4

Cohen Y., Krumbein W.E., & Shilo M. (1977) Solar Lake (Sinai). 2. Distribution of photosynthetic microorganisms and primary production. *Limnol. Oceanogr.* **22**, 609–620.
9.2.1, 9.4.1, 9.5.4, 18.5.5.7

Cohen Y., Padan E. & Shilo M. (1975) Facultative anoxygenic photosynthesis in the cyanobacterium *Oscillatoria limnetica*. *J. Bact.* **123**, 855–861.
9.1, 9.2.1

Cohen-Bazire G., Beguin S., Rimon S., Glazer A.N. & Brown D.M. (1977) Physicochemical and immunological properties of allophycocyanins. *Arch. Microbiol.* **111**, 225–238.
7.5.1.3, 7.5.1.4, 7.5.1.5, 7.5.1.7

Cohen-Bazire G. & Lefort-Tran M. (1968) Structure granulaire du chromatoplasme des cyanophycées. *Fourth European Regional Conference on Electron Microscopy, Rome*. pp. 395–396.
7.2

Cohen-Bazire G., Sistrom W.R. & Stanier R.Y. (1957) Kinetic studies of pigment synthesis by non-sulfur bacteria. *J. cell. comp. Physiol.* **49**, 25–68.
9.4.1

Cohn F. (1875) Untersuchungen über Bakterien. II. *Beitr. Biol. Pfl.* **1**, 141–207.
Foreword

Cole W.J., Chapman D.J. & Siegelman H.W. (1967) The structure of phycocyanobilin. *J. Am. Chem. Soc.* **89**, 3643–3645.
7.5.1.2

Cole W.J., Chapman D.J. & Siegelman H.W. (1968) The structure and properties of phycocyanobilin and related bilatrienes. *Biochemistry, N.Y.* **7**, 2929–2935.
7.5.1.2

Colijn F. & Nienhuis H. (1978) The intertidal microphytobenthos of the 'Hohe Weg' shallows in the German Wadden Sea. *Forsch. Stelle f. Insel. u. Küstenschutz* **29**, 149–174+6 tables, 6 figures.
20.2.2

Collier A.W. (1970) 1. Oceans and coastal waters as life-supporting environments. 1–93 In: Kinne O. (Ed.) *Marine Ecology* Volume 1, part 1. Wiley, New York.

Colman P.M., Freeman H.C., Guss J.M., Murata M., Norris, V.A., Ramshaw J.A. & Venkatappa M.P. (1978) X-ray crystal structure analysis of plastocyanin at 2.7 Å resolution. *Nature, Lond.* **272**, 319–324.
8.5

Colvin J.R. & Leppard G.G. (1973) Fibrillar, modified polygalacturonic acid in, on and between plant cell walls. In: Loewus F. (Ed.) *Biogenesis of Plant Cell Wall Polysaccharides*. 383 pp. Academic Press, New York.
17.2

Conover J.T. (1962) Algal crusts and the formation of lagoon sediments. In: *Symposium on the Environmental Chemistry of Marine Sediments*. Narragansett Marine Laboratory, Graduate School of Oceanography. Univ. Rhode Island Occas. Publ. **1**, 69–76.

Constantopoulos G. & Bloch K. (1967) Isolation and characterization of glycolipids from some photosynthetic bacteria. *J. Bact.* **93**, 1788–1793.
6.4.1

Corry M.J., Payne P.I. & Dyer T.A. (1974) The nucleotide sequence of 5S rRNA from the blue-green alga *Anacystis nidulans*. *FEBS Lett.* **46**, 63–66.
12.4.3.3

Cosner J.C. (1978) Phycobilisomes in spheroplasts of *Anacystis nidulans*. *J. Bact.* **135**, 1137–1140.
7.3

Costerton J.W., Geesey G.G. & Cheng K.J. (1978) How bacteria stick. *Scient. Am.* **238**, 86–95.
17.2

Costerton J.W.F., Murray G.E. & Robinow C.F. (1961) Observations on the motility and structure of *Vitreoscilla*. *Can. J. Microbiol.* **7**, 329–339.
16.2.2.2

Cowles R.P. (1913) The habits of some tropical Crustacea. *Philippine J. Sci.* **8**, 119–124.
20.2.4

Cox R.M. (1966) Physiological studies on nitrogen fixation in the blue-green alga *Anabaena cylindrica*. *Arch. Mikrobiol.* **53**, 263–276.
4.3

Craig S.R. & Budd K. (1979) Chloride uptake

by *Anacystis nidulans* (Cyanophyceae). *J. Phycol.* **15,** 300–304.
10.2.3, 10.3.3.3

Craig I.W., Leach C.K. & Carr N.G. (1969) Studies with deoxyribonucleic acid from blue-green algae. *Arch. Mikrobiol.* **65,** 218–227.
11.3.1, 15.5

Cramer M. & Myers J. (1949) Effects of starvation on the metabolism of *Chlorella. Pl. Physiol., Lancaster* **24,** 255–264.
18.5.3

Crang R.E. & Jensen T.E. (1975) Incorporation of titanium in polyphosphate bodies of *Anacystis nidulans. J. Cell Biol.* **67,** 80a.
5.3.1

Crespi H.L., Boucher L.J., Norman G.D., Katz J.J. & Dougherty R.C. (1967) Structure of phycocyanobilin. *J. Am. Chem. Soc.* **89,** 3642–3643.
7.5.1.2

Crespi H.L., Mandeville S.E. & Katz J.J. (1962) The action of lysozyme on several blue-green algae. *Biochem. biophys. Res. Commun.* **9,** 569–577.
13.4.1

Crespi H.L., Smith U. & Katz J.J. (1968) Phycocyanobilin. Structure and exchange studies by nuclear magnetic resonance and its mode of attachment in phycocyanin. A model for phytochrome. *Biochemistry, N.Y.* **7,** 2232–2242.
7.5.1.2

Cribb A.B. (1966) The algae of Heron Island, Great Barrier Reef, Australia. *University of Queensland Papers, Great Barrier Reef Committee* **1,** 3–23.
20.2.1

Croft W.N. & George E.A. (1959) Blue-green algae from the Middle Devonian of Rhynie (Aberdeenshire). *Bull. Brit. Mus. (Nat. Hist.), Geology* **3** (10), 341–353.
21.1

Cruden D.L. & Stanier R.Y. (1970) The characterization of chlorobium vesicles and membranes isolated from green bacteria. *Arch. Mikrobiol.* **72,** 115–134.
6.4.1

Csatorday K. (1978) Fluorescence from sensitizing phycobilin chromophores in the blue-green alga *Anacystis nidulans. Biochim. biophys. Acta* **504,** 341–343.
8.6

Cséke C., Balogh A. & Farkas G.L. (1981) Redox modulation of glucose-6-phosphate dehydrogenase in *Anacystis nidulans* and its uncoupling by phage infection. *FEBS Lett.* **126,** 85–88.
3.3

Culver D.A. & Brunskill G.J. (1969) Fayetteville Green Lake, New York. 5. Studies of primary production and zooplankton in a meromictic marl lake. *Limnol. Oceanogr.* **14,** 862–873.
9.5.4

Currier T.C., Haury J.F. & Wolk C.P. (1977) Isolation and preliminary characterization of auxotrophs of a filamentous cyanobacterium. *J. Bact.* **129,** 1556–1562.
11.5.3, 14.3.1, 14.6.1.2

Currier T.C. & Wolk C.P. (1978) Regulation of uridylic acid biosynthesis in the cyanobacterium *Anabaena variabilis. J. Bact.* **136,** 682–687.
11.2.2

Currier T.C. & Wolk C.P. (1979) Characteristics of *Anabaena variabilis* influencing plaque formation by cyanophage N-1. *J. Bact.* **139,** 88–92.
11.3.3

Czurda V. (1940–1941) Schwefelwasserstoff als ökologischer factor der Algen. *Zentbl. Bakt. Parasitkde Abt.* **II,** 285–311.
9.5.2

Daday A., Lambert G.R. & Smith G.D. (1979) Measurement *in vivo* of hydrogenase-catalysed hydrogen evolution in the presence of nitrogenase enzyme in cyanobacteria. *Biochem. J.* **177,** 139–144.
14.4.3.3

Daday A., Platz R.A. & Smith G.D. (1977) Anaerobic and aerobic hydrogen gas formation by the blue-green alga *Anabaena cylindrica. Appl. environ. Microbiol.* **34,** 478–483.
4.6, 14.4.1, 14.4.3.2

Daft M.J., McCord S.D. & Stewart W.D.P. (1975) Ecological studies on algal lysing bacteria in freshwaters. *Freshwat. Biol.* **5,** 577–596.
17.4.2

Daft M.J. & Stewart W.D.P. (1971) Bacterial pathogens of freshwater blue-green algae. *New Phytol.* **70,** 819–824.
17.4.2

Dainty J. (1969a) The water relations of plants. In: Wilkins M.B. (Ed.) *The Physiology of Plant Growth and Development.* 695 pp., 419–452. McGraw-Hill, New York.
10.2.1, 10.2.7

Dainty J. (1969b) The ionic relations of plants. In: Wilkins M.B. (Ed.) *The Physiology of Plant Growth and Development.* 695 pp., 453–485. McGraw-Hill, New York.
10.3.3

Dainty J. (1976) Water relations of plant cells.

In: Luttge U. & Pitman M.G. (Eds) *Transport in Plants II. Part A Cells. Encyclopedia of Plant Physiology New Series 2A*. 400 pp., pp. 12–35. Springer-Verlag, Berlin.
10.2.1, 10.2.7

Dale R.E. & Teale F.W.J. (1970) Number and distribution of chromophore types in native phycobiliproteins. *Photochem. Photobiol.* **12,** 99–117.
7.5.1.5

Daniel A., Nagabhushanam A.K. & Krishnamurthy P. (1978) On the occurrence of 'greenish-yellow water' phenomenon caused by the swarming of *Trichodesmium erythraeum* Ehrenburg, in the sea off Madras (India) and its effect on the local marine fauna. *J. Bombay nat. Hist. Soc.* **75,** 88–95.
19.5.3

Darnell J.E., Jr (1978) Implications of RNA:RNA splicing in evolution of eukaryotic cells. *Science, N.Y.* **202,** 1257–1260.
12.4.2

Darwin C. (1841) On a remarkable bar of sandstone off Pernambuco on the coast of Brazil. *Lond. Edinburgh Dublin Phil. Mag. J. Sci.* **19,** 257–261.
20.2.1

Das B. & Singh P.K. (1976) Isolation of gas vacuole-less mutants of the blue-green alga *Anabaenopsis raciborskii. Arch. Microbiol.* **111,** 195–196.
11.5.5.2(a)

David K.A.V. & Thomas J. (1980) Role of amino acids in the regulation of heterocyst production and nitrogenase activity in *Anabaena* L-31. In: Gnanarm A., Krishnaswarmy S. & Kahn J.S. (Eds) *Proc. Int. Symp. Biological Applications of Solar Energy*, pp. 104–108. Macmillan, Madras.
14.6.1.1

Davis B.D. (1950) Non-filterability of the agents of genetic recombination in *E. coli. J. Bact.* **60,** 507–508.
11.6.1

Davis D.J., Krogmann D.W. & San Pietro A. (1980) Electron donation to photosystem 1. *Pl. Physiol., Lancaster*, **65,** 697–702.
8.5

Davis J.S. (1978) Biological communities of a nutrient enriched saline. *Aquatic Botany* **4,** 23–42.
20.3.6

Dawes E.A. & Senior P.J. (1973) The role and regulation of energy reserve polymers in microorganisms. *Adv. microb. Physiol.* **10,** 135–266.
3.2.2.2

Dawson E.Y. (1966) *Marine Botany*. 371 pp. Holt, Rinehart & Winston, New York.
20.2.3

Dayrell-Hart B. & Burchard R.P. (1979) Association of flexing and gliding in *Flexibacter. J. Bact.* **137,** 1417–1420.
16.2.3

de Bary A. (1863) Beitrag zur Kenntnis der Nostocaceen insbesondere der *Rivularien. Flora, Jena* **35,** 553–560.
15.1

Degener O. & Degener I. (1959) Canton Island, South Pacific (Resurvey of 1958). *Atoll Research Bulletin* **64,** 1–23.
20.2.2

Degens E.T., Reuter J.H. & Shaw K.N.F. (1964) Biochemical compounds in off shore California sediments and sea waters. *Geochim. cosmochim. Acta* **28,** 45–66.
3.4.3

de Jonge V.N. (1976) Algal vegetation on salt-marshes along the western Dutch Wadden Sea. *Netherlands J. Sea Res.* **10,** 262–283.
20.2.3

de Kouchkovsky Y. & Sigalat C. (1975) Fractionnement et caracterisation de l'appareil photosynthetique de l'algue bleue unicellulaire *Anacystis nidulans*. II. Analyse fonctionnelle. *Physiol. Vég.* **13,** 831–851.
8.2

Delaney S.F. & Carr N.G. (1975) Temporal genetic mapping in the blue-green alga *Anacystis nidulans* using ethyl methane sulphonate. *J. gen. Microbiol.* **88,** 259–268.
11.3.1, 11.4.1, 11.4.2.2, 11.5.1, 11.5.4, 11.5.5.1, 11.7

Delaney S.F., Dickson A. & Carr N.G. (1973) The control of homoserine-O-transsuccinylase in a methionine-requiring mutant of the blue-green alga *Anacystis nidulans. J. gen. Microbiol.* **79,** 89–94.
11.5.3

Delaney S.F., Herdman M. & Carr N.G. (1976) Genetics of blue-green algae. In: Lewin R.A. (Ed.) *The Genetics of Algae*. 360 pp., pp. 7–28. Blackwell, Oxford.
11.1, 11.5.2, 11.6.1, 11.7

Demerec M., Adelberg A., Clark A.J. & Hartman P.E. (1966) A proposal for a uniform nomenclature in bacterial genetics. *Genetics* **54,** 61–76. Reprinted in *J. gen. Microbiol.* **50,** 61–76 (1968).
11.2.2

Demeter O. (1956) Über Modifikationen bei Cyanophyceen. *Arch. Mikrobiol.* **24,** 105–133.
15.4

Demoulin V. (1979) Protein and nucleic acid

sequence data and phylogeny. *Science, N.Y.* **205,** 1036–1038.
12.1.3, 12.2.2.1

Denoyelles F., Jr & O'Brien W.J. (1978) Phytoplankton succession in nutrient enriched experimental ponds as related to changing carbon, nitrogen and phosphorus conditions. *Arch. Hydrobiol.* **84,** 137–165.
5.4.2, 5.4.3

de Puymaly A. (1957) Les hétérocystes des algues bleues: leur nature et leur rôle. *Le Botaniste* **41,** 209–270.
14.1

De Renzo D.J. (1978) *Nitrogen Control and Phosphorus Removal in Sewage Treatment.* 704 pp. Noyes Data Corporation, New Jersey.
18.6

Desikachary T.V. (1946) Germination of the heterocyst in two members of the Rivulariaceae, *Gloeotrichia raciborskii* Wolosz. and *Rivularia mangini* Fremy. *J. Ind. bot. Soc.* **25,** 11–17.
14.5

Desikachary T.V. (1959) *Cyanophyta,* pp. 623–660. Indian Council of Agricultural Research, New Delhi, India.
References

Desikachary T.V. (1973) Status of classical taxonomy. In: Carr N.G. & Whitton B.A. (Eds) *The Biology of Blue-Green Algae.* 676 pp., pp. 473–481. Blackwell, Oxford.
12.3.1, 12.3.4.2

Desikachary T.V. & Padmaja T.D. (1970) Origin of filamentous condition and phylogeny in the blue-green algae. *Rev. Algol., N.S.* **10,** 8–17.
12.3.1

de Vasconcelos L. & Fay P. (1974) Nitrogen metabolism and ultrastructure in *Anabaena cylindrica.* I. The effect of nitrogen starvation. *Arch. Microbiol.* **96,** 271–279.
14.6.1.1

Devassy V.P., Bhattathiri P.M.A. & Qasim S.Z. (1978) *Trichodesmium* phenomenon. *Indian J. mar. Sci.* **7,** 168–186.
19.2.4.1, 19.4.2, 19.5.2, 19.5.3, 19.5.4

Devassy V.P., Bhattathiri P.M.A. & Qasim S.Z. (1979) Succession of organisms following *Trichodesmium* phenomenon. *Indian J. mar. Sci.* **8,** 89–93.
19.5.3

Devilly C.I. & Houghton J.A. (1977) A study of genetic transformation in *Gloeocapsa alpicola. J. gen. Microbiol.* **98,** 277–280.
11.6.3

de Waard A. & Duyvesteyn M. (1980) Are sequence-specific deoxyribonucleases of value as taxonomic markers of cyanobacterial species? *Arch. Microbiol.* **128,** 242–247.
11.3.3

de Waard A., Korsuize J., van Beveren C.P. & Maat J. (1978) A new sequence-specific endonuclease from *Anabaena cylindrica. FEBS Lett.* **96,** 106–110.
11.3.3

de Waard A., van Beveren C.P., Duyvesteyn M. & van Ormondt H. (1979) Two sequence-specific endonucleases from *Anabaena oscillariodes. FEBS Lett.* **101,** 71–76.
11.3.3

Dewar M.A. & Barber J. (1973) Cation regulation in *Anacystis nidulans. Planta, Berl.* **113,** 143–155.
10.2.3, 10.2.4.1, 10.3.3.2

Dewar M.A. & Barber J. (1974) Chloride transport in *Anacystis nidulans. Planta, Berl.* **117,** 163–172.
10.3.3.3

Dhar B. (1979) Nutritional control of the frequency of MNNG-induced true branching habit in *Anabaena doliolum. Molec. gen. Genet.* **173,** 177–181.
11.5.5.2(a), 11.5.5.2(c)

Dharmawardene M.W.N., Haystead A. & Stewart W.D.P. (1973) Glutamine synthetase of the nitrogen-fixing alga *Anabaena cylindrica. Arch. Mikrobiol.* **90,** 281–295.
14.4.1

Diakoff S. & Scheibe J. (1973) Action spectra for chromatic adaptation in *Tolypothrix tenuis. Pl. Physiol., Lancaster* **51,** 382–385.
7.8.3.2

Dick H. & Stewart W.D.P. (1980) The occurrence of fimbriae on a N_2-fixing cyanobacterium which occurs in lichen symbiosis. *Arch. Microbiol.* **124,** 107–109.
16.2.3

Dickens B.F. & Ingram L.O. (1976) Peptidoglycan synthesis and turnover in cell division mutants of *Agmenellum. J. Bact.* **127,** 334–340.
13.3

Dickerson R.E. (1980a) The cytochromes *c*: an exercise in scientific serendipity. In: Sigman D.S. & Brazier M.A.B. (Eds) *UCLA Forum in Medical Science, Vol. 21.* Academic Press, New York.
12.1.3, 12.2.1, 12.2.2.1

Dickerson R.E. (1980b) Evolution and gene transfer in purple photosynthetic bacteria. *Nature, Lond.* **283,** 210–212.
12.2.2.1

Dicks J.W. & Tempest D.W. (1966) The influence of temperature and growth rate on the quantitative relationship between potassium, magnesium, phosphorus and ribonucleic acid of *Aerobacter aerogenes* grow-

ing in a chemostat. *J. gen. Microbiol.* **45**, 547–557.
5.2.1

Diehn B., Feinleib M.E., Haupt W., Hildebrand E., Lenci F. & Nultsch W. (1979) Terminology of behavioral responses of motile microorganisms *Photochem. Photobiol.* **26**, 559–560.
16.4.1

Dierksheide W.C. & Pfister R.M. (1973) Associated organelles in the blue-green alga, *Anacystis nidulans. Can. J. Microbiol.* **19**, 149–151.
5.3.1, 6.1

Dietrich W.E., Jr & Thornber J.P. (1971) The P 700-Chlorophyll a-protein of a blue-green alga. *Biochim. biophys. Acta* **245**, 482–493.
6.4.2

Dikshit K.L., Verma S.C. & Dikshit R.P. (1979) Effect of nitrogenous compounds on rifampicin-induced heterocyst differentiation in blue-green algae. *Ann. Bot., N.S.* **44**, 633–635.
14.6.1.1., 14.6.1.3

Dillon P.J. & Rigler F.H. (1974) The phosphorus–chlorophyll relationship in lakes. *Limnol. Oceanogr.* **19**, 767–773.
18.4.3

Diner, B.A. (1979) Energy transfer from phycobilisomes to photosystem II reaction centers in wild type *Cyanidium caldarium. Pl. Physiol., Lancaster* **63**, 30–34.
8.6

Diner, B.A. & Wollman, F.A. (1979) Functional comparison of the photosystem II center–antenna complex of a phycocyaninless mutant of *Cyanidium caldarium* with that of *Chlorella pyrenoidosa. Pl. Physiol., Lancaster* **63**, 20–25.
8.6

Dinsdale M.T. & Walsby A.E. (1972) The interrelations of cell turgor pressure, gas vacuolations, and buoyancy in a blue-green alga. *J. exp. Bot.* **23**, 561–570.
2.8.2.2, 10.2.3, 10.2.5.2, 19.3.6

Dmitriev A.P. & Grodzinsky D.M. (1975) Effect of radiosensitizing agents on DNA single-strand breaks and their repair in *Anacystis nidulans. Pl. Sci. Lett.* **4**, 77–83.
11.4.2.2

Dmitrieva E.Y. & Kozlov A.V. (1979) Some physical characteristics of the compact chromosome of *Anabaena variabilis. Biochemistry* **44**, 198–204.
11.3.1

Dobler M., Dover S.D., Laves K., Binder A. & Zuber H. (1972) Crystallization and preliminary crystal data of c-phycocyanin. *J. mol. Biol.* **71**, 785–787.
7.5.1.6

Dobson W.J. & McCurdy H.D. (1979) The function of fimbriae in *Myxococcus xanthus.* I. Purification and properties of *M. xanthus* fimbriae. *Can. J. Microbiol.* **25**, 1152–1160.
16.2.3

Dobson W.J., McCurdy H.D. & MacRae T.H. (1979) The function of fimbriae in *Myxococcus xanthus.* II. The role of fimbriae in cell–cell interactions. *Can. J. Microbiol.* **25**, 1359–1372.
16.2.3, 16.5

Dodd J.D. (1953) A note on the increase in flake size of *Aphanizomenon flos-aquae* (L.) Ralfs. *Proc. Iowa Acad. Sci.* **60**, 117–118.
16.5

Dodson E.O. (1979) Crossing the procaryote-eucaryote border: endosymbiosis or continuous development. *Can. J. Microbiol.* **25**, 652–673.
12.4.1, 12.4.2

Doemel W.N. & Brock T.D. (1974) Bacterial stromatolites: origin of laminations. *Science, N.Y.* **184**, 1083–1085.
12.3.2, 21.2

Doetsch R.N. & Hageage G.J. (1968) Motility in procaryotic organisms: problems, points of view, and perspectives. *Biol. Rev.* **43**, 317–362.
13.2.2, 16.2.2.1

Döhler G. (1974) C_4-Weg der photosynthese in der Blaualge *Anacystis nidulans. Planta, Berl.* **118**, 259–269.
3.2.1

Donze M. (1973) Pigment Systems in Algae. Electron Transport to Nitrogenase in *Anabaena.* Thesis, Rijksuniversiteit, Leiden.
14.2, 14.4.3.1, 14.6.1.1

Donze M., Haveman J. & Schiereck P. (1972) Absence of photosystem 2 in heterocysts of the blue-green alga *Anabaena. Biochim. biophys. Acta* **256**, 157–161.
6.5.2, 14.4.1, 14.4.3.2

Donze M., Raat A.J.P. & van Gorkom H.J. (1974) Supply of ATP and reductant to nitrogenase in the blue-green alga *Anabaena cylindrica. Pl. Sci. Lett.* **3**, 35–41.
14.4.3.1

Doolittle W.F. (1979) The cyanobacterial genome, its expression and the control of that expression. *Adv. microb. Physiol.* **20**, 1–102.
1, 2.6.3, 3.4.4, 11.1, 11.3.1, 11.3.3, 12.1.1

Doolittle W.F. (1980) Revolutionary concepts in evolutionary cell biology. *Trends biochem. Sci.* **5**, 146–149.
12.1.1, 12.4.2

Doolittle W.F. & Bonen L. (1981) Molecular

sequence data indicating an endosymbiotic origin for plastids. *Ann. N.Y. Acad. Sci.* **361,** 248–259.
12.2.2.4, 12.4.3.2

Doolittle W.F. & Singer R.A. (1974) Mutational analysis of dark endogenous metabolism in the blue-green bacterium *Anacystis nidulans. J. Bact.* **119,** 677–683.
3.2.2.2, 3.3, 11.2.2

Doonan B.B. & Jensen T.E. (1977) Ultrastructural localization of alkaline phosphatase in the blue-green bacterium *Plectonema boryanum. J. Bact.* **132,** 967–973.
5.1.3

Dor I. (1975) The blue-green algae of the mangrove forests of Sinaï. *Rapp. Comm. int. Mer Médit.* **23,** 109–110.
20.2.3

Dougherty E.C. (1957) Neologisms needed for structures of primitive organisms. *J. Protozool., Suppl.* **4,** 14.
Foreword

Drawert H. & Metzner I. (1956) Fluoreszenz- u. elektronemikroskopische Beobachtungen an *Cylindrospermum* und einigen anderen Cyanophyceen. *Ber. dt. bot. Ges.* **69,** 291–300.
13.3

Drawert H. & Tischer J. (1956) Über Redox-Vorgänge bei Cyanophyceen unter besonderer Berücksichtigung der Heterocysten. *Naturwissenschaften* **43,** 132.
4.4

Drews G. (1957) Die phototaktischen Reaktionen einiger Cyanophyceen. *Ber. dt. bot. Ges.* **70,** 259–262.
2.8.1

Drews G. (1959) Beiträge zur Kenntnis des phototaktische Reaktionen der Cyanophyceen. *Arch. Protistenk.* **104,** 389–430.
2.8.1, 16.4.3, 16.4.4, 16.5

Drews G. (1973) Fine structure and chemical composition of the cell envelopes. In: Carr N.G. & Whitton B.A. (Eds) *The Biology of Blue-Green Algae.* 676 pp., pp. 96–116. Blackwell, Oxford.
13.2.2, 13.4

Drews G. & Gollwitzer W. (1965) Untersuchungen an der Polysaccharidfraktion der Zellwände von *Anacystis nidulans. Arch. Mikrobiol.* **51,** 179–185.
13.4.2, 13.4.3, 13.4.5

Drews G. & Meyer H. (1964) Untersuchungen zum chemischen Aufbau der Zellwande von *Anacystis nidulans* und *Chlorogloea fritschii. Arch. Mikrobiol.* **48,** 259–267.
13.4.1, 13.4.2, 13.4.3

Drews G. & Niklowitz W. (1956) Beiträge zur Cytologie der Blaualgen. II. Zentroplasma und granuläre Einschlüsse von *Phormidium uncinatum* Gom. *Arch. Mikrobiol.* **24,** 147–162.
3.2.2.2, 6.2

Drews G., Prauser H. & Uhlmann D. (1961) Massenvorkommen von *Synechococcus plancticus* nov. spec., einer solitären, planktischen Cyanophycee, in einem Abwasserteich. *Arch. Mikrobiol.* **39,** 101–115.
6.2

Droop M.R. (1968) Vitamin B_{12} and marine ecology. IV. The kinetics of uptake, growth, and inhibition in *Monochrysis lutheri. J. mar. biol. Ass. U.K.* **48,** 689–733.
5.2.1

Droop M.R. (1974a) The nutrient status of algal cells in continuous culture. *J. mar. biol. Ass. U.K.* **54,** 825–855.
2.5.1, 5.2.1

Droop M.R. (1974b) Heterotrophy of carbon. In: Stewart W.D.P. (Ed.) *Algal Physiology and Biochemistry.* 989 pp., pp. 530–559. Blackwell, Oxford.
3.4.1

Drouet F. (1968) Revision of the classification of the Oscillatoriaceae. *Monogr. Acad. nat. Sci. Philad.* **15,** 334 pp.
19.2.4.1, 20.1

Drouet F. & Daily W.A. (1956) Revision of the coccoid Myxophyceae. *Butler Univ. bot. Stud.* **10,** 1–218.
20.1

Drum R.W. & Pankratz S. (1965) Fine structure of an unusual cytoplasmic inclusion in the diatom genus, *Rhopalodia. Protoplasma, Wien* **60,** 141–149.
4.1

Duane W.C., Hohl M.C. & Krogmann D.W. (1965) Photophosphorylation activity in cell-free preparations of a blue-green alga. *Biochim. biophys. Acta* **109,** 108–116.
8.3

Dubois J.P. (1977) Croissance d'*Oscillatoria rubescens* D.C. en relation avec la vitamine B_{12}. *Annls Hydrobiol.* **8,** 221–230.
18.5.1.2

Dudkin M.S., Peskov V.A. & Areshidze I.V. (1976) Glucan of the blue-green alga *Microcystis aeruginosa. Khim. Prir. Soedin.* **4,** 420–422. (Russian).
3.2.2.2

Dugdale R.C., Goering J.J. & Ryther J.H. (1964) High nitrogen fixation rates in the Sargasso Sea and the Arabian Sea. *Limnol. Oceanogr.* **9,** 507–510.
19.5.2

Dugdale R.C., Menzel D.W. & Ryther J.H.

(1961) Nitrogen fixation in the Sargasso Sea. *Deep Sea Res.* **7,** 298–300.
19.3.4, 19.5.2

Dunn J.H., Simon R.D. & Wolk C.P. (1971) Incorporation of amino sugars into walls during heterocyst differentiation. *Devl Biol.* **26,** 159–164.
14.5

Dunn J.H. & Wolk C.P. (1970) Composition of the cellular envelopes of *Anabaena cylindrica. J. Bact.* **103,** 153–158.
13.4.5, 14.3.1, 14.4.2, 15.2

Dunstan W.M. & Hosford J. (1977) The distribution of planktonic blue-green algae related to the hydrography of the Georgia Bight. *Bull. mar. Sci.* **27,** 824–829.
19.2.4.1, 19.5.1

Duxbury T., Humphrey B.A. & Marshall K.C. (1980) Continuous observations of bacterial gliding motility in a dialysis microchamber: the effects of inhibitors. *Arch. Microbiol.* **124,** 169–175.
16.3

Duysens L.N.M. (1951) Transfer of light energy within the pigment systems present in photosynthesizing cells. *Nature, Lond.* **168,** 548–550.
7.2

Duysens L.N.M. (1952) Transfer of Excitation Energy in Photosynthesis. 96 pp. Ph.D. Thesis, Univ. Utrecht, The Netherlands.
7.2

Duyvesteyn M. & Waard A. de (1980) A new sequence-specific endonuclease from a thermophilic cyanobacterium, *Mastigocladus laminosus. FEBS Lett.* **111,** 423–426.
11.6

Dyer T.A. & Bowman C.M. (1979) Nucleotide sequences of chloroplast 5S ribosomal ribonucleic acid in flowering plants. *Biochem. J.* **183,** 595–604.
12.4.3.3

Eberly W.R. (1959) The metalimnetic oxygen maximum in Myers Lake. *Invest. Indiana Lakes & Streams* **5,** 1–46.
18.4.1

Eberly W.R. (1966) Notes on some new and rare Myxophyceae in laboratory culture. *Trans. Am. microsc. Soc.* **85,** 130–138.
15.6

Echlin P. (1964a) Intra-cytoplasmic membranous inclusions in the blue-green alga, *Anacystis nidulans. Arch. Mikrobiol.* **49,** 267–274.
6.1, 6.2, 6.5.2

Echlin P. (1964b) The fine structure of the blue-green alga, *Anacystis montana* f. *minor* grown in continuous illumination. *Protoplasma, Wien* **58,** 439–457.
6.2, 13.2.2

Echlin P. & Morris I. (1965) The relationship between blue-green algae and bacteria. *Biol. Rev.* **40,** 143–187.
13.4

Edelman M., Swinton D., Schiff J.A., Epstein H.T. & Zeldin B. (1967) Deoxyribonucleic acid of the blue-green algae (Cyanophyta). *Bact. Rev.* **31,** 315–331.
11.3.1

Edmondson W.T. (1969) Eutrophication in North America. In: Rohlich G.A. (Ed.) *Eutrophication: Causes, Consequences, Correctives.* 661 pp., pp. 124–149. National Academy of Sciences, Washington.
5.4.3, 18.4.3, 18.6

Edmondson W.T. (1972) Nutrients and phytoplankton in Lake Washington. In: Likens G.E. (Ed.) *Nutrients and Eutrophication. American Society of Limnology and Oceanography Special Symposia* **1,** 172–193.
18.4.3

Edwards M.R. & Gantt E. (1971) Phycobilisomes of the thermophilic blue-green alga *Synechococcus lividus. J. Cell Biol.* **50,** 896–900.
7.3

Edwards M.R., Berns D.S., Ghiorse W.C. & Holt S.C. (1968) Ultrastructure of the thermophilic blue-green alga, *Synechococcus lividus* Copeland. *J. Phycol.* **4,** 283–289.
6.1, 6.2, 13.2.1

Edwards M.R. & Holt S.C. (1968) The use of freeze-etching and chemical fixation in the electron microscopy study of *Synechococcus lividus. Fourth European Regional Conf. on Electron Microscopy, Rome.* pp. 377–378.
6.2

Eggleston J.R. & Dean W.E. (1976) Freshwater stromatolitic bioherms in Green Lake, New York. In: Walter M.R. (Ed.) *Stromatolites. Developments in Sedimentology* **20.** pp. 479–488. Elsevier, Amsterdam.
9.5.3

Eisbrenner G. & Bothe H. (1979) Modes of electron transfer from molecular hydrogen in *Anabaena cylindrica. Arch. Microbiol.* **123,** 37–45.
4.3, 4.6, 14.4.3.2, 14.4.3.3

Eisbrenner G., Distler, E., Floener L. & Bothe H. (1978) The occurrence of the hydrogenase in some blue-green algae. *Arch. Microbiol.* **118,** 177–184.
4.3, 4.6, 14.4.1, 14.4.3.3

Eisbrenner G., Roos P. & Bothe H. (1981) The

number of hydrogenases in cyanobacteria. *J. gen. Microbiol.* **125,** 383–390.
4.6

Eiserling F.A. & Glazer A.N. (1974) Blue-green algal proteins: assembly forms of C-phycocyanin from *Synechococcus* sp. *J. Ultrastruct. Res.* **47,** 16–25.
7.5.1.4, 7.5.1.6, 7.7

Eley J.H. (1971) The effect of carbon dioxide concentration on pigmentation of the blue-green alga *Anacystis nidulans. Pl. Cell Physiol., Tokyo* **12,** 311–316.
7.8

Ellefson W.L. & Krogmann D.W. (1979) Studies of the multiple forms of ferredoxin-NADP oxidoreductase from spinach. *Arch. Biochem. Biophys.* **194,** 593–599.
8.3

Elliott B.B. & Mortenson L.E. (1977) A molybdenum storage component from *Clostridium pasteurianum.* In: Newton W., Postgate J.R. & Rodriguez-Barrueco C. (Eds) *Proceedings Second International Symposium on Nitrogen Fixation.* 622 pp., pp. 205–218. Academic Press, New York.
4.5

Elo J.E. (1937) Vergleichende Permeabilitatsstudien, besonders an niederen Pflanzen. *Suomal. eläin. ja kasvit seur. van elain. Julk.* **8,** 1–108.
10.3.1

Eloff J.N., Steinitz Y. & Shilo M. (1976) Photooxidation of cyanobacteria in natural conditions. *Appl. environ. Microbiol.* **31,** 119–126.
2.3.1, 6.5.2, 18.5.5.6

Emerson R. & Lewis C.M. (1942) The photosynthetic efficiency of phycocyanin in *Chroococcus*, and the problem of carotenoid participation in photosynthesis. *J. gen. Physiol.* **25,** 579–595.
7.2

Emery T. (1971) Hydroxamic acids of natural origin. *Adv. Enzymol.* **35,** 135–185.
18.5.2.3

Enami I. (1978) Mechanism of the acidophily and thermophily of *Cyanidium caldarium* Geitler. IV. Some physicochemical properties of the acid and heat stable cytochrome c_{553}. *Pl. Cell Physiol., Tokyo* **19,** 901–905.
8.5

Engelmann T.W. (1881) Neue Methode zur Untersuchung der Sauerstoffauscheidung pflanzlicher und thierischer Organismen. *Pflüger's Arch. ges. Physiol.* **25,** 285–292.
17.1.1

Engelmann T.W. (1883). Farbe und Assimilation. *Bot. Ztg* **41,** 1–13.
7.2

Engelmann T.W. (1884) Untersuchungen über die quantitativen Beziehungen zwischen Absorption des Lichtes und Assimilation in Pflanzenzellen. *Bot. Ztg* **42,** 81–93; 97–105.
7.2

Engelmann T.W. (1902) Über experimentelle Erzeugung zweckmässiger Änderungen der Färbung pflanzlicher Chromophylle durch farbiges Licht. *Arch. Anat. Physiol. Physiol. Abt.*, 333–335.
6.5.1

England R.E. & Evans E.H. (1981) A rapid method of extraction of oxygen evolving photosystem 2 preparations from cyanobacteria. *FEBS Lett.* **134,** 175–177.
8.6

Ephrati-Elizur E. (1968) Spontaneous transformation in *Bacillus subtilis. Genet. Res.* **11,** 83–96.
11.7

Eppley R.W. (1972) Temperature and phytoplankton growth in the sea. *Fish. Bull.* **70,** 1063–1085.
18.5.1.3

Eppley R.W. & Strickland J.D.H. (1968) Kinetics of marine phytoplankton growth. In: Droop M.R. & Ferguson Wood E.J. (Eds) *Advances in Microbiology of the Sea.* 239 pp., pp. 23–62. Academic Press, New York.
18.5.1.3

Ercegović A. (1930) Sur la tolérance des Cyanophycées vis-à-vis des variations brusques de la salinité de l'eau de mer. *Acta bot. Inst. bot. Zagreb.* **5,** 48–56.
20.3.5

Eriksson-Quensel I.B. (1938) LXXVI. The molecular weights of phycoerythrin and phycocyanin. I. *Biochem. J.* **32,** 585–589.
7.5.1.4

Ermakova L.R., Nikitina L.A. & Gusev M.V. (1977) Changes in the ultrastructure of the cells of *Anabaena variabilis* with aging. *Mikrobiologiya* **46,** 324–328 (Russian).
6.5.2

Ernst A., Kerfin W., Spiller H. & Böger P. (1979) External factors influencing light-induced hydrogen evolution by the blue-green alga, *Nostoc muscorum. Z. Naturforsch.* **34c,** 820–825.
4.6

Evans E.H., Cammack R. & Evans M.C.W. (1976) Properties of the primary electron acceptor complex of photosystem 1 in the blue-green alga *Chlorogloea fritschii. Biochem. biophys. Res. Commun.* **68,** 1212–1218.
8.4

Evans E.H. & Carr N.G. (1975) Dark-light transitions with a heterotrophic culture of a

blue-green alga. *Biochem. Soc. Trans.* **3,** 373–376.
3.4.1

Evans E.H., Carr N.G. & Evans M.C.W. (1978) Changes in photosynthetic activity in the cyanobacterium *Chlorogloea fritschii* following transition from dark to light growth. *Biochim. biophys. Acta* **501,** 165–173.
6.5.2, 8.6

Evans E.H., Foulds I. & Carr N.G. (1976) Environmental conditions and morphological variation in the blue-green alga, *Chlorogloea fritschii. J. gen. Microbiol.* **92,** 147–155.
13.2.2, 13.4.5, 15.1

Evans E.H., Rush J.D., Johnson C.E. & Evans M.C.W. (1979) Mössbauer spectra of photosystem I reaction centers from the blue-green alga *Chlorogloea fritschii. Biochem. J.* **182,** 861–865.
8.4

Evans E.L. & Allen M.M. (1973) Phycobilisomes in *Anacystis nidulans. J. Bact.* **113,** 403–408.
6.5.1, 7.3

Falkner G., Horner F., Werdan K. & Heldt H.W. (1976) pH changes in the cytoplasm of the blue-green alga *Anacystis nidulans* caused by light dependent proton flux into the thylakoid space. *Pl. Physiol., Lancaster* **58,** 717–718.
8.7, 10.2.3, 10.3.2.1, 10.3.3.3

Falkner G., Werdan K., Horner F. & Heldt H.W. (1974) Energieabhängige Phosphataufnahme der Blaualge *Anacystis nidulans. Ber. dt. bot. Ges.* **87,** 263–266.
5.1.1

Fallon R.D. & Brock T.D. (1979a) Decomposition of blue-green algal (cyanobacterial) blooms in Lake Mendota, Wisconsin. *Appl. environ. Microbiol.* **37,** 820–830.
2.6.3, 17.4.2

Fallon R.D. & Brock T.D. (1979b) Lytic organisms and photo-oxidative effects. Influence on blue-green algae (cyanobacteria) in Lake Mendota, Wisconsin. *Appl. environ. Microbiol.* **38,** 499–505.
17.4.2

Farran G.P. (1932) The occurrence of *Trichodesmium thiebautii* off the south coast of Ireland. *Rapp. P-v. Réun. Cons. perm. int. Explor. Mer.* **775,** 60–64.
19.2.4.1

Faust M.A. & Correll D.L. (1976) Comparison of bacterial and algal utilization of orthophosphate in an estuarine environment. *Mar. Biol.* **34,** 151–162.
17.4.2

Fautz E., Rosenfelder G. & Grotjahn L. (1979) Iso-branched 2- and 3-hydroxy fatty acids as characteristic lipid constituents of some gliding bacteria. *J. Bact.* **140,** 852–858.
16.2.3

Fay P. (1965) Heterotrophy and nitrogen fixation in *Chlorogloea fritschii* Mitra. *J. gen. Microbiol.* **39,** 11–20.
2.6.3, 3.4.1, 3.4.2, 6.5.2

Fay P. (1969a) Cell differentiation and pigment composition in *Anabaena cylindrica. Arch. Mikrobiol.* **67,** 62–70.
6.5.2, 14.3.2, 15.2, 15.3

Fay P. (1969b) Metabolic activities of isolated spores of *Anabaena cylindrica. J. exp. Bot.* **20,** 100–109.
15.3, 15.4.1, 15.5

Fay P. (1970) Photostimulation of nitrogen fixation in *Anabaena cylindrica. Biochim. biophys. Acta* **216,** 353–356.
4.3, 14.4.3.2

Fay P. (1973) The heterocyst. In: Carr N.G. & Whitton B.A. (Eds) *The Biology of Blue-Green Algae.* 676 pp., pp. 238–259. Blackwell, Oxford.
14.1

Fay P. (1976) Factors influencing dark nitrogen fixation in a blue-green alga. *Appl. environ. Microbiol.* **31,** 376–379.
4.3, 18.5.2.1

Fay P. & de Vasconcelos L. (1974) Nitrogen metabolism and ultrastructure in *Anabaena cylindrica.* II. The effect of molybdenum and vanadium. *Arch. Microbiol.* **99,** 221–230.
4.5, 14.6.1.1

Fay P. & Kulasooriya S.A. (1972) Tetrazolium reduction and nitrogenase activity in heterocystous blue-green algae. *Arch. Mikrobiol.* **87,** 341–352.
4.4, 14.2, 14.4.3.2

Fay P. & Kulasooriya S.A. (1973) A simple apparatus for the continuous culture of photosynthetic micro-organisms. *Br. phycol. J.* **8,** 51–57.
18.5.1.3

Fay P. & Lang N.J. (1971) The heterocysts of blue-green algae. I. Ultrastructural integrity after isolation. *Proc. R. Soc. Lond. B* **178,** 185–192.
14.3.2, 14.4.2, 14.5

Fay P., Stewart W.D.P., Walsby A.E. & Fogg G.E. (1968) Is the heterocyst the site of nitrogen fixation in blue-green algae? *Nature, Lond.* **220,** 810–812.
14.2, 14.3.2, 14.4.3.2, 14.6.1.1, 14.6.1.3

Fay P. & Walsby A.E. (1966) Metabolic activities of isolated heterocysts of the blue-

green alga *Anabaena cylindrica*. *Nature, Lond.* **209,** 94–95.
4.4, 14.2

Fechner R. (1915) Die Chemotaxis der Oscillatorien und ihre Bewegungserscheinungen überhaupt. *Z. Bot.* **7,** 289–362.
16.5

Feinleib M.E. (1978) Photomovement of microorganisms. *Photochem. Photobiol.* **27,** 849–854.
16.4.1

Feldman J. (1932) Sur la biologie des *Trichodesmium* Ehr. *Revue Algol.* **6,** 357–358.
19.4.2

Feldmann J. (1959) Sur l'écologie des Rhodophycées et des Cyanophycées marines dans ses rapports avec les problems de leur photosynthèse. *Bull. Soc. fr. Physiol. vég.* **5,** 161–167.
20.3.4

Feldmann J. & Frémy P. (1935) Matériaux pour la flora algologique marine de la Tunisie. II. Contribution à l'étude biologique et systématique de la MUFFA. *Station Océanographique de Salammbo. Notes No. 29*, pp. 1–24.
20.2.4

Fenchel T.M. & Riedl R.J. (1970) The sulfide system: A new biotic community underneath the oxidised layer of marine sand bottoms. *Mar. Biol.* **7,** 225–268.
9.4.1, 9.5.3, 9.6

Fenchel T.M. & Straarup B.J. (1971) Vertical distribution of photosynthetic pigments and the penetration of light in marine sediments. *Oikos* **22,** 172–182.
9.5.3

Ferguson Wood E.J. (1965) *Marine Microbial Ecology*. Chapman & Hall, London.
19.2.4.1, 19.5.3

Feuillade J.B. & Feuillade M.G. (1975) Étude des besoins en azote et en phosphore d'*Oscillatoria rubescens* à l'aide de cultures en chemostats. *Verh. int. Verein. theor. angew. Limnol.* **19,** 2698–2708.
5.2.2

Feuillade J. & Feuillade M. (1981) Le métabolisme photosynthetique d'*Oscillatoria rubescens* D.C. (Cyanophycée). I. La carboxylation initiale dans des conditions favorables à la croissance. *Arch. Hydrobiol.* **90,** 410–426.

Feuillade M. (1972) Croissance d'*Oscillatoria rubescens* et variations quantitatives de la chlorophylle et des différents carotenoïdes en fonction d'éclairement. *Annls Hydrobiol.* **3,** 21–31.
2.3

Fewson C.A., Al-Hafidh M. & Gibbs M. (1962) Role of aldolase in photosynthesis. I. Enzyme studies with photosynthetic organisms with special reference to blue-green algae. *Pl. Physiol., Lancaster* **37,** 402–406.
3.3

Findley D.L., Walne P.L. & Holton R.W. (1970) The effects of light intensity on the ultrastructure of *Chlorogloea fritschii* Mitra grown at high temperatures. *J. Phycol.* **6,** 182–188.
6.5.1

Finke L.R. & Seeley H.W. (1978) Nitrogen fixation (acetylene reduction) by epiphytes of freshwater macrophytes. *Appl. environ. Microbiol.* **36,** 129–138.
17.1.2

Fisher R., Tuli R. & Haselkorn R. (1981) A cloned cyanobacterial gene for glutamine synthetase functions in *Escherichia coli*, but the enzyme is not adenylated. *Proc. natn. Acad. Sci. U.S.A.* **78,** 3393–3397.
11.2.2, 11.8

Fisher R.W. & Wolk C.P. (1976) Substance stimulating the differentiation of spores of the blue-green alga *Cylindrospermum licheniforme*. *Nature, Lond.* **259,** 394–395.
14.6.2, 15.4.1, 15.4.3

Fisher R.G., Woods N.E., Fuchs H.E. & Sweet R.M. (1980) Three-dimensional structure of the phycobiliproteins I: C-phycocyanin and B-phycoerythrin at 5 Å-resolution. *J. biol. Chem.* **255,** 5082–5089.
7.5.1.6

Fitzgerald G.P. (1969) Field and laboratory evaluations of bioassays for nitrogen and phosphorus with algae and aquatic weeds. *Limnol. Oceanogr.* **14,** 206–212.
5.4.1

Fitzgerald M.P., Husain A., Hutber G.N. & Rogers, L.J. (1977) Studies of the flavodoxins from a cyanobacterium and a red alga. *Biochem. Soc. Trans.* **5,** 1505–1506.
8.3

Fitzgerald G.P. & Nelson T.C. (1966) Extractive and enzymatic analyses for limiting or surplus phosphorus in algae. *J. Phycol.* **2,** 32–37.
5.2.2, 5.2.3, 5.4.1

Fitzgerald G.P. & Skoog F. (1954) Control of blue-green algae blooms with 2,3-dichloronaphthoquinone. *Sewage ind. Wastes* **26,** 1136–1140.
18.6

Fleming H. & Haselkorn R. (1973) Differentiation in *Nostoc muscorum*: nitrogenase is

synthesized in heterocysts. *Proc. natn. Acad. Sci. U.S.A.* **70,** 2727–2731.
4.1, 14.2, 14.3.2, 14.5

Fleming H. & Haselkorn R. (1974) The program of protein synthesis during heterocyst differentiation in nitrogen-fixing blue-green algae. *Cell* **3,** 159–170.
14.3.2, 14.5, 14.6.1.3

Floodgate G.D. (1972) The mechanism of bacterial attachment to detritus in aquatic systems. *Memorie Ist ital. Idrobiol., Suppl.* **29,** 309–323.
17.2

Fogg G.E. (1944) Growth and heterocyst production in *Anabaena cylindrica* Lemm. *New Phytol.* **43,** 164–175.
14.5

Fogg G.E. (1949) Growth and heterocyst production in *Anabaena cylindrica* Lemm. II. In relation to carbon and nitrogen metabolism. *Ann. Bot., N.S.* **13,** 241–259.
4.1, 14.2, 14.6.1.1, 14.6.1.3

Fogg G.E. (1951) Growth and heterocyst production in *Anabaena cylindrica* Lemm. III. The cytology of heterocysts. *Ann. Bot., N.S.* **15,** 23–35.
6.5.2, 14.3.2, 14.5

Fogg G.E. (1952) The production of extracellular substances by a blue-green alga. *Proc. R. Soc. B.* **139,** 372–397.
2.7.2.1

Fogg G.E. (1969) The physiology of an algal nuisance. *Proc. R. Soc. Lond. B* **173,** 175–189.
18.1, 18.5.4, 18.6

Fogg G.E. (1971) Extracellular products of algae in freshwater. *Arch. Hydrobiol. Beih. Ergebn. Limnol.* **5,** 1–25.
17.6

Fogg G.E. (1973) Physiology and ecology of marine blue-green algae. In: Carr N.G. & Whitton B.A. (Eds) *The Biology of Blue-Green Algae.* 676 pp. pp. 368–378. Blackwell, Oxford.
19.3.4, 19.4.1, 20.1, 20.2.1

Fogg G.E. (1974) Nitrogen fixation. In: Stewart W.D.P. (Ed.) *Algal Physiology and Biochemistry.* 175 pp. pp. 560–582. Blackwell, Oxford.
19.3.4

Fogg G.E. (1975) *Algal Cultures and Phytoplankton Ecology.* 2nd Edn. 175 pp. University of Wisconsin Press, Madison.
19.4.1, 19.4.2

Fogg G.E. (1978) Nitrogen fixation in the oceans. In: Granhall U. (Ed.) *Environmental role of nitrogen-fixing blue-green algae and asymbiotic bacteria. Ecol. Bull., Stockholm* **26,** 11–19.
19.5.2

Fogg G.E. & Burton N.F. (1977) Utilization of glycolate by bacteria epiphytic on seaweeds. *J. Phycol., Suppl.* **13,** 22.
17.1.2

Fogg G.E. & Pattnaik H. (1968) The release of extracellular nitrogenous products by *Westiellopsis prolifica* Janet. *Phykos.* **5,** 58–67.
18.5.2.3

Fogg G.E., Stewart W.D.P., Fay P. & Walsby A.E. (1973) *The Blue-Green Algae.* 459 pp. Academic Press, London.
9.6, 10.2.2, 12.3.1, 15.1, 19.3.1, 19.3.4, 19.4.1, 19.5.4, 20.2.1, 20.3.3, 20.3,4

Fogg G.E. & Westlake D.F. (1955) The importance of extracellular products of algae in freshwater. *Verh. int. Verein. theor. angew. Limnol.* **12,** 219–232.
17.4.4, 18.5.2.3

Follett E.A.C. & Webley D.M. (1965) An electron microscope study of the cell surface of *Cytophaga johnsonii* and some observations on related organisms. *Antonie van Leeuwenhoek* **31,** 361–382.
16.2.2.2

Ford S.H. & Friedman H.C. (1979) Formation of δ-amino-laevulinic acid from glutamic acid by a partially purified enzyme system. *Biochim. biophys. Acta* **569,** 153–158.
3.2.2.1

Fork D.C., Murata N. & Sato N. (1979) Effect of growth temperature on the lipids and fatty acid composition and the dependence on temperature of light-induced redox reactions of cytochrome f and of light energy redistribution in the thermophilic blue-green algae *Synechococcus lividus. Pl. Physiol., Lancaster* **63,** 524–530.
6.4.1, 6.5.3, 8.8

Foter M.J., Palmer C.M. & Maloney T.E. (1953) Antialgal properties of various antibiotics. *Antibiotics and Chemotherapy* **3,** 505–508.
13.4.1

Foulds I.J. & Carr N.G. (1977) A proteolytic enzyme degrading phycocyanin in the cyanobacterium *Anabaena cylindrica. FEMS Microbiol. Lett.* **2,** 117–119.
7.8.1, 14.5

Fox G.E., Magrum L.J., Balch W.E., Wolfe R.S. & Woese C.R. (1977) Classification of methanogenic bacteria by 16S ribosomal RNA characterization. *Proc. natn. Acad. Sci. U.S.A.* **74,** 4537–4541.
12.2.2.4

Foy R.H. (1980) The influence of surface to

volume ratio on the growth rates of planktonic blue-green algae. *Br. phycol. J.* **15,** 279–289.
18.2.1, 18.5.1.2, 18.5.1.3

Foy R.H. & Gibson C.E. (1982) Photosynthetic characteristics of planktonic blue-green algae: changes in photosynthetic capacity and pigmentation of *Oscillatoria redekei* van Goor under high and low light. *Br. phycol. J.* **17,** 183–193.
18.5.5.5

Foy R.H., Gibson C.E. & Smith R.V. (1976) The influence of daylength, light intensity and temperature on the growth rates of planktonic blue-green algae. *Br. phycol. J.* **11,** 151–163.
2.3, 2.4.1, 2.8.2.2, 18.4.2, 18.5.1.1, 18.5.1.2, 18.5.1.3

Foy R.H. & Smith R.V. (1980) The role of carbohydrate accumulation in the growth of planktonic *Oscillatoria* species. *Br. phycol. J.* **15,** 139–150.
2.6.3, 18.5.1.3, 18.5.2.1, 18.5.6

Frank G., Sidler W., Widmer H. & Zuber H. (1978) The complete amino acid sequence of both subunits of C-phycocyanin from the cyanobacterium *Mastigocladus laminosus. Hoppe–Seyler's Z. physiol. Chem.* **359,** 1491–1507.
7.5.1.2, 7.5.1.3, 7.5.1.8

Frank G., Zuber H. & Lergier W. (1975) How homologous are the α and β subunits of C-phycocyanin? *Experientia* **31,** 23–26.
7.5.1.3, 7.5.1.8

Frank H., Lefort M. & Martin H.H. (1962a) Elektronenoptische und chemische Untersuchungen an Zellwänden der Blaualge *Phormidium uncinatum. Z. Naturf. B.* **7,** 262–268.
13.2.1, 13.3, 13.4.1, 13.4.2

Frank H., Lefort M. & Martin H.H. (1962b) Chemical analysis of a mucopolymer component in cell walls of the blue-green alga, *Phormidium uncinatum. Biochem. biophys. Res. Commun.* **7,** 322–325.
13.4.1, 13.4.2

Fredrick J.F. (Ed.) (1981) *Origins and Evolution of Eukaryotic Intracellular Organelles. Ann. N.Y. Acad. Sci.* **361,** 512 pp.
1

Freidenreich P., Apell G.S. & Glazer A.N. (1978) Structural studies on phycobiliproteins. II. C-phycocyanin: amino acid sequence of the β subunit. Specific cleavage of the α subunit. *J. biol. Chem.* **253,** 212–219.
7.5.1.2, 7.5.1.3, 7.5.1.8

Frémy P. (1933) Les cyanophycées des côtes d'Europe. *Mém. Soc. Nat. Scienc. Nat. Math. Cherbourg* **41,** 1–235.
20.2.1

Frenkel A.W. (1970) Multiplicity of electron transport reactions in bacterial photosynthesis. *Biol. Rev.* **45,** 569–616.
9.4.1

Frey-Wyssling A. & Stecher E. (1954) Über den Feinbau des *Nostoc*-Schleimes. *Z. Zellforsch. mikrosk. Anat.* **39,** 515–519.
13.2.2

Friedberg D. & Seijffers J. (1979) Plasmids in two cyanobacterial strains. *FEBS Lett.* **107,** 165–168.
11.3.2

Fritsch F.E. (1942) The interrelations and classification of Myxophyceae. *New Phytol.* **4,** 134–148.
12.3.1

Fritsch F.E. (1945) Myxophyceae (Cyanophyceae) *The Structure and Reproduction of the Algae*, Vol. II, 939 pp., pp. 780–786. Cambridge Univ. Press, England.
Foreword, 7.8, 15.4.1

Fritsch F.E. (1951) The heterocyst: a botanical enigma. *Proc. Linn. Soc. London* **162,** 194–211.
14.3.2

Fu E., Friedman L. & Siegelman H.W. (1979) Mass-spectral identification and purification of phycoerythrobilin and phycocyanobilin. *Biochem. J.* **179** 1–6.
7.5.1.2

Fuhs G.W. (1958) Enzymatischer Abbau der Membranen von *Oscillatoria amoena* (Kütz.) Gomont mit Lysozym. *Arch. Mikrobiol.* **29,** 51–52.
13.4.1

Fuhs G.W. (1963) Cytochemisch-elektronenmikroskopische Lokalisierung der Ribonukleinsäure und des Assimilats in Cyanophyceen. *Protoplasma, Wien* **56,** 178–187.
11.3.1

Fuhs G.W. (1973) Cytochemical examination of blue-green algae. In: Carr N.G. & Whitton B.A. (Eds) *The Biology of Blue-Green Algae*, 676 pp., pp. 117–143. Blackwell, Oxford.
11.3.1

Fuhs G.W., Demmerle S.D., Canelli E. & Chen M. (1972) Characterization of phosphorus-limited plankton algae. In: Likens G.E. (Ed.) *Nutrients and Eutrophication*. 328 pp., pp. 113–133. American Society of Limnology and Oceanography Special Symposia, Vol. 1.
5.1.1, 5.2.2

Fujita Y. (1975) Further investigations of the

light-induced cytochrome b559 reaction in membrane fragments of the blue-green alga *Anabaena variabilis*. *Pl. Cell Physiol, Tokyo* **16,** 1037–1048.
8.5

Fujita Y. (1976) The C550 photoresponse at room temperature observed in membrane fragments of the blue-green alga *Anabaena variabilis*. *Pl. Cell Physiol., Tokyo* **17,** 187–191.
8.6

Fujita Y. (1978) Carotenoid photobleaching induced by photosystem II action in membrane fragments of the blue-green alga *Anabaena variabilis*. *Pl. Cell Physiol, Tokyo* **19,** 1129–1136.
8.6

Fujita Y. & Hattori A. (1960a) Formation of phycoerythrin in pre-illuminated cells of *Tolypothrix tenuis* with special reference to nitrogen metabolism. *Pl. Cell Physiol., Tokyo* **1,** 281–292.
7.8, 7.8.3.2

Fujita Y. & Hattori A. (1960b) Effect of chromatic lights on phycobilin formation in a blue-green alga *Tolypothrix tenuis*. *Pl. Cell Physiol., Tokyo* **1,** 293–303.
7.8, 7.8.3.1, 7.8.3.2

Fujita Y. & Hattori A. (1962a) Photochemical interconversion between precursors of phycobilin chromoproteids in *Tolypothrix tenuis*. *Pl. Cell Physiol., Tokyo* **3,** 209–220.
7.8.3.2

Fujita Y. & Hattori A. (1962b) Changes in composition of cellular material during formation of phycobilin chromoproteids in a blue-green alga *Tolypothrix tenuis*. *J. Biochem., Tokyo*, **52,** 38–42.
7.8.3.2

Fujita Y. & Shimura S. (1974) Phycoerythrin of the marine blue-green alga *Trichodesmium thiebautii*. *Pl. Cell Physiol., Tokyo* **15,** 939–942.
7.5.1.1

Fulco L., Karfunkel P. & Aaronson S. (1967) Effect of lysozyme (muramidase) on marine and freshwater blue-green algae. *J. Phycol.* **3,** 51–52.
13.4.1

Fuller R.C. (1978) Photosynthetic carbon metabolism in the green and purple bacteria. In: Clayton R.K. & Sistrom W.R. (Eds) *The Photosynthetic Bacteria*. 946 pp., pp. 691–705. Plenum Press, New York.
12.2.1, 12.2.2.3

Furlong C.E. & Preiss J. (1969) Biosynthesis of bacterial glycogen. VII. Purification and properties of the adenosine diphosphoglucose pyrophosphorylase of *Rhodospirillum rubrum*. *J. biol. Chem.* **244,** 2539–2548.
3.2.2.2

Furtado D., Williams W.P., Brain A.P.R. & Quinn P.J. (1979) Phase separations in membranes of *Anacystis nidulans* grown at different temperatures. *Biochim. biophys. Acta* **555,** 352–357.
6.5.3, 8.8

Gaastra P. (1959) Photosynthesis of crop plants as influenced by light, carbon dioxide, temperature and stomatal diffusion resistance. *Med. Landbouw Hogeschool Wageningen*. **59,** 1–68.
2.2.1

Gaffron H. (1944) Photosynthesis, photoreduction and dark reduction of carbon dioxide in certain algae. *Biol. Rev.* **19,** 1–20.
9.4.1

Galanos C., Lüderitz O., Rietschel E.Th. & Westphal O. (1977) New aspects of the chemistry and biology of bacterial lipopolysaccharides, with special reference to their lipid A component. *Int. Rev. Biochem., Biochemistry of Lipids II* **14,** 239–335.
13.4.4

Gallerdo V.A. (1976) On the benthic sulfide system on the continental shelf of north and central Chile. In: *International Symposium on Coastal Upwelling. Proc. Coquimbo, Chile, 1975*. pp. 113–118.
9.5.3

Gallon J.R. (1980) Nitrogen fixation by photoautotrophs. In: Stewart W.D.P. & Gallon J.R. (Eds) *Proceedings of an International Symposium on Nitrogen Fixation*. 451 pp., pp. 197–238. Academic Press, New York.
4.1, 4.3

Gallon J.R. (1981) The oxygen sensitivity of nitrogenase: a problem for biochemists and micro-organism. *Trends biochem. Sci.* **6,** 19–23.

Gallon J.R., Kurz W.G.W. & LaRue T.A. (1975) The physiology of nitrogen fixation by a *Gloeocapsa* sp. In: Stewart W.D.P. (Ed.) *Nitrogen Fixation by Free-living Microorganisms*. pp. 159–173. Cambridge University Press.
19.3.4

Gallon J.R., LaRue T.A. & Kurz W.G.W. (1972) Characteristics of nitrogenase activity in broken cell preparations of the blue-green alga *Gloeocapsa* sp. LB 795. *Can. J. Microbiol.* **18,** 327–332.
4.3

Gallon J.R., Ul-Haque M.I. & Chaplin A.E. (1978) Fluoroacetate metabolism in *Gloeo-*

capsa sp. LB 795 and its relationship to acetylene reduction (nitrogen fixation). *J. gen. Microbiol.* **106,** 329–336.
14.4.3.1

Ganapati S.V. (1941) Ecology of a temple tank containing a permanent bloom of *Microcystis aeruginosa. J. Bombay nat. Hist. Soc.* **42,** 65–77.
18.4.2

Ganf G.G. (1974a) Diurnal mixing and the vertical distribution of phytoplankton in a shallow equatorial lake (Lake George, Uganda). *J. Ecol.* **62,** 641–659.
18.4.4

Ganf G.G. (1974b) Rates of oxygen uptake by the planktonic community of a shallow equatorial lake (Lake George, Uganda). *Oecologia, Berl.* **15,** 17–32.
18.5.3

Ganf G.G. (1975) Photosynthetic production and irradiance—photosynthesis relationships of the phytoplankton from a shallow equatorial lake (Lake George, Uganda). *Oecologia, Berl.* **18,** 165–183.
18.4.2, 18.5.5.4

Ganf G.G. & Horne A.J. (1975) Diurnal stratification, photosynthesis and nitrogen-fixation in a shallow, equatorial lake (Lake George, Uganda). *Freshwat. Biol.* **5,** 13–39.
9.5.4

Ganf G.G. & Viner A.B. (1973) Ecological stability in a shallow equatorial lake (Lake George, Uganda). *Proc. R. Soc. Lond. B.* **184,** 321–346.
9.5.1, 9.5.4

Gantt E. (1969) Properties and ultrastructure of phycoerythrin from *Porphyridium cruentum. Pl. Physiol., Lancaster* **44,** 1629–1638.
7.5.1.6

Gantt E. (1975) Phycobilisomes: light-harvesting pigment complexes. *BioScience* **25,** 781–788.
7.4, 7.6, 7.7, 7.8.2

Gantt E. (1979) Phycobiliproteins of cryptophyceae. In: Hutner H.S. & Levandowski M. (Eds) *Biochemistry and Physiology of Protozoa, 2nd edition.* pp. 121–137. Academic Press, New York.
7.1

Gantt E. (1980) Structure and function of phycobilisomes: light harvesting pigment complexes in red and blue-green algae. *Int. Rev. Cytol.* **66,** 45–80.
7.3, 7.5.3, 7.7

Gantt E. & Conti S.F. (1965) The ultrastructure of *Porphyridium cruentum. J. Cell Biol.* **26,** 365–381.
7.2

Gantt E. & Conti S.F. (1966a) Phycobiliprotein localization in algae. *Brookhaven Symp. Biol.* **19,** 393–405.
7.2

Gantt E. & Conti S.F. (1966b) Granules associated with the chloroplast lamellae of *Porphyridium cruentum. J. Cell Biol.* **29,** 423–434.
6.2, 7.2

Gantt E. & Conti S.F. (1969) Ultrastructure of blue-green algae. *J. Bact.* **97,** 1486–1493.
7.2

Gantt E., Edwards M.R. & Conti S.F. (1968) Ultrastructure of *Porphyridium aerugineum*, a blue-green colored rhodophytan. *J. Phycol.* **4,** 65–71.
7.5.1.1

Gantt E., Edwards M.R. & Provasoli L. (1971) Chloroplast structure of the Cryptophyceae. Evidence for phycobiliproteins within intrathylakoidal spaces. *J. Cell Biol.* **48,** 280–290.
7.1

Gantt E. & Lipschultz C.A. (1972) Phycobilisomes of *Porphyridium cruentum. J. Cell Biol.* **54,** 313–324.
7.4

Gantt E. & Lipschultz C.A. (1973) Energy transfer in phycobilisomes from phycoerythrin to allophycocyanin. *Biochim. biophys. Acta* **292,** 858–861.
7.2, 7.6

Gantt E. & Lipschultz C.A. (1974) Phycobilisomes of *Porphyridium cruentum*: Pigment analysis. *Biochemistry, N.Y.* **13,** 2960–2966.
7.2, 7.5.1.1, 7.5.1.3, 7.5.1.4, 7.5.2, 7.6

Gantt E. & Lipschultz C.A. (1977) Probing phycobilisome structure by immuno-electron microscopy. *J. Phycol.* **13,** 185–192.
7.5.1.7, 7.7

Gantt E., Lipschultz C.A., Grabowski J. & Zimmerman B.K. (1979) Phycobilisomes from blue-green and red algae. *Pl. Physiol., Lancaster* **63,** 615–620.
7.4, 7.6, 8.8

Gantt E., Lipschultz C.A. & Zilinskas B.A. (1976a) Further evidence for a phycobilisome model from selective dissociation, fluorescence emission, immunoprecipitation, and electron microscopy. *Biochim. biophys. Acta* **430,** 375–388.
7.6, 7.7

Gantt E., Lipschultz C.A. & Zilinskas B.A. (1976b) Phycobilisomes in relation to the thylakoid membranes. *Brookhaven Symp. Biol.* **28,** 347–357.
7.6, 7.7

Garlick S., Oren A. & Padan E. (1977) Occur-

rence of facultative anoxygenic photosynthesis among filamentous and unicellular cyanobacteria. *J. Bact.* **129,** 623–629.
9.1, 9.3.1, 9.3.2, 9.3.4, 9.5.3, 9.5.4, 12.3.3, 14.4.2

Garrett P. (1970) Phanerozoic stromatolites: noncompetitive ecologic restriction by grazing and burrowing animals. *Science, N.Y.* **169,** 171–173.
20.4.2, 21.3

Gebelein C.D. (1969) Distribution, morphology, and accretion rate of Recent subtidal algal stromatolites, Bermuda. *J. sed. Petrol.* **39,** 49–69.
20.2.4, 21.2

Gebelein C.D. (1976) Open marine subtidal and intertidal stromatolites (Florida, the Bahamas and Bermuda). In: Walter M.R. (Ed.) *Stromatolites. Developments in Sedimentology* **20.** 790 pp., pp. 381–388. Elsevier, Amsterdam.
9.5.3, 20.2.4, 20.4.1

Geiselman J.A. (1977) Marine colonial hydroid inclusion within planktonic *Oscillatoria* colonies. *Bull. mar. Sci.* **27,** 821–823.
19.5.3

Geitler L. (1921) Versuch einer Lösung des Heterocysten-Problems. *Sitzungsber. Akad. Wiss., Wien., Math.-Naturw. Kl.* **130** (1), 223–245.
14.5

Geitler L. (1925) Synoptische Darstellung der Cyanophyceen in morphologischer und systematischer Hinsicht. *Beih. bot. Cbl.* **41,** 163–294.
14.6.2

Geitler, L. (1932) Cyanophyceae. In: Rabenhorst L. (Ed.) *Kryptogamen-Flora von Deutschland, Österreich und der Schweiz 14.* 1196 pp. Akademische Verlagsgesellschaft, Leipzig.
1, 14.5, 15.1, 17.2, 18.4.1, 19.2.4.1, 20.1, 20.3.4

Geitler L. (1958) Lamelläre Struktur des Chromatoplasmas von Cyanophyceen in mikroskopischen Dimensionen und Baueigentümlichkeiten des Protoplasten von *Chroococcus turgidus. Arch. Mikrobiol.* **29,** 179–188.
6.2

Geitler L. (1960) Schizophyceen. In: Zimmermann W. & Ozenda P. (Eds) *Encyclopedia of Plant Anatomy: Handbuch der Pflanzenanatomie, 2nd edn, pt I.* Borträger, Berlin, Germany.
References

Gelin C. & Ripl W. (1978) Nutrient decrease and response of various phytoplankton size fractions following the restoration of Lake Trummen, Sweden. *Arch. Hydrobiol.* **81,** 339–367.
5.4.3

Gendel S., Ohad I. & Bogorad L. (1979) Control of phycoerythrin synthesis during chromatic adaptation. *Pl. Physiol., Lancaster* **64,** 786–790.
7.8.3.1, 7.8.3.2

Gentile J.H. (1971) Blue-green and green algal toxins. In: Kadis S., Ciegler A. & Ajl S.J. (Eds) *Microbiol Toxins Vol.* **7.** 350 pp., pp. 27–66. Academic Press, New York.
18.5.6

Gentile J.H. & Maloney T.E. (1969) Toxicity and environmental requirements of a strain of *Aphanizomenon flos-aquae* (L.) Ralfs. *Can. J. Microbiol.* **15,** 165–173.
5.2.4, 15.4

George M.G. (1962) Occurrence of a permanent algal bloom in a fish tank at Delhi with special reference to factors responsible for its development. *Proc. Indian Acad. Sci. B* **56,** 354–362.
18.4.2

Georgiyeva V.T. (1976) Effect of light on the uptake of phosphorus by some blue-green algae. *Hydrobiological J.* (English transl.) **12** **12** (3), 67–70.
5.1.1

Gerloff G.C. & Skoog F. (1954) Cell contents of nitrogen and phosphorus as a measure of their availability for growth of *Microcystis aeruginosa. Ecology* **35,** 348–353.
5.2.2, 5.4.1

Gerloff G.C. & Skoog F. (1957) Nitrogen as a limiting factor for the growth of *Microcystis aeruginosa* in southern Wisconsin lakes. *Ecology* **38,** 556–561.
5.4.1, 18.5.1.2

Gessner F. (1959) *Hydrobotanik II. Stoffhaushalt.* VEB Deutscher Verlag der Wissenschaften, Berlin.
19.2.4.2

Gest J. (1972) Energy conversion and generation of reducing power in bacterial photosynthesis. *Adv. microb. Physiol.* **7,** 243–282.
9.4.1

Gibbs M. (1962) Fermentation. In: Lewin R.A. (Ed.) *Physiology and Biochemistry of Algae.* 929 pp., pp. 91–97. Academic Press, London.
9.4.1

Gibbs S.P. (1981) The chloroplasts of some algal groups may have evolved from endosymbiotic eukaryotic algae. *Ann. N.Y. Acad. Sci.* **361,** 193–208.
12.4.4.3

Gibson C.E. (1972) The algicidal effect of copper on a green and a blue-green alga and some ecological implications. *J. appl. Ecol.* **9,** 513–518.
18.6

Gibson C.E. (1975a) Cyclomorphosis in natural populations of *Oscillatoria redekei* van Goor. *Freshwat. Biol.* **5,** 279–286.
18.3.2

Gibson C.E. (1975b) A field and laboratory study of oxygen uptake by planktonic blue-green algae. *J. Ecol.* **63,** 867–880.
18.5.3

Gibson C.E. (1978) Field and laboratory observations on the temporal and spatial variation of carbohydrate content in planktonic blue-green algae in Lough Neagh, Northern Ireland. *J. Ecol.* **66,** 97–115.
18.5.3, 18.5.5.4

Gibson C.E. & Fitzsimons A.G. (1982) Periodicity and morphology of planktonic blue-green algae in an unstratified lake (Lough Neagh, Northern Ireland). *Int. Rev. ges. Hydrobiol.* **67,** 471.
18.4.5

Gibson C.E. & Stevens R.J. (1979) Changes in phytoplankton physiology and morphology in response to dissolved nutrients in Lough Neagh, Northern Ireland. *Freshwat. Biol.* **9,** 105–109.
5.4.1, 18.3.2, 18.5.2.4.

Gibson C.E., Wood R.B., Dickson E.L. & Jewson D.H. (1971) The succession of phytoplankton in Lough Neagh 1968–1970. *Mitt. int. Verein. theor. angew. Limnol.* **19,** 146–160.
18.3.2, 18.4.1

Gibson J.L. & Tabita F.R. (1977) Isolation and characterization of two forms of ribulose 1,5-bisphosphate carboxylase from *Rhodopseudomonas capsulata. J. Bact.* **132,** 818–823.
12.2.2.3

Giddings T.H., Jr & Staehelin L.A. (1978) Plasma membrane architecture of *Anabaena cylindrica*: occurrence of microplasmodesmata and changes associated with heterocyst development and the cell cycle. *Cytobiol.* **16,** 235–249.
6.3.1, 6.5.2, 13.3, 14.3.1, 14.3.2

Giddings T.H., Jr & Staehelin L.A. (1979) Changes in thylakoid structure associated with the differentiation of heterocysts in the cyanobacterium *Anabaena cylindrica. Biochim. biophys. Acta* **546,** 373–382.
6.3.1, 6.3.2, 6.5.2, 7.9, 14.3.2

Giddings T.H., Jr & Wolk C.P. (1981) Factors influencing the stability of nitrogenase activity in isolated cyanobacterial heterocysts. *FEMS Microbiol. Lett.* **10,** 299–302.
14.2

Giddings T.H., Jr, Wolk C.P. & Shomer-Ilan A. (1981) Metabolism of sulfur compounds by whole filaments and heterocysts of *Anabaena cylindrica. J. Bact.* **146,** 1067–1074.
14.2, 14.4.2

Gillham N.W. (1978) *Organelle Heredity.* 602 pp. Raven Press, New York.
12.2.2.3, 12.4.2

Gillis M., De Ley J. & DeCleene M. (1970) The determination of molecular weight of bacterial genome DNA from renaturation rates. *Eur. J. Biochem.* **12,** 143–153.
11.3.1

Gingeras T.R., Milazzo J.P. & Roberts R.J. (1978) A computer assisted method for the determination of restriction enzyme recognition sites. *Nucl. Acids Res.* **5,** 4105–4127.
11.3.3

Ginsburg R.N., Bricker O.P., Wanless H.R. & Garrett P. (1970) Exposure index and sedimentology structures of a Bahama tidal flat. *Geol. Soc. Am., Abstr. Progr.* **2** (7), 744–745.
20.4.1

Giraud G. (1963) La structure, les pigments et les caractéristiques fonctionnelles de l'appareil photosynthétique de diverses algues. *Physiol. Vég.* **1,** 203–255.
7.2

Glaessner M.F. (1979) Precambrian. In: *Treatise on Invertebrate Paleontology. Part A.* 569 pp., pp. A79–A118. Geological Society of America Inc. & University of Kansas.
21.3

Glazer A.N. (1976) Phycocyanins: structure and function. In: Smith D.C. (Ed.) *Photochemical and Photobiological Reviews, Vol. 1.* 612 pp., pp. 71–115. Plenum Press, New York.

Glazer A.N. (1977) Structure and molecular organization of the photosynthetic accessory pigments of cyanobacteria and red algae. *Mol. cell. Biochem.* **18,** 125–141.
7.5.1.2, 12.4.4.2

Glazer A.N. (1981) Photosynthetic accessory proteins with bilin prosthetic groups. In: Hatch M.D. and Boardman N.K. (Eds). *Biochemistry of Plants, Vol. VIII, Photosynthesis.* 500 pp., pp. 51–96. Academic Press, New York.
7.1, 7.5.1.1, 7.5.1.3, 7.5.1.4, 7.5.1.7, 7.5.2, 7.8.2

Glazer A.N. & Apell G.S. (1977) A common evolutionary origin for the biliproteins of

cyanobacteria, rhodophyta and cryptophyta. *FEMS Microbiol. Lett.* **1,** 113–116.
12.4.4.2

Glazer, A.N., Apell G.S., Hixson C.S., Bryant D.A., Rimon S. & Brown D.M. (1976) Biliproteins of cyanobacteria and Rhodophyta: Homologous family of photosynthetic accessory pigments. *Proc. natn. Acad. Sci. U.S.A.* **73,** 428–431.
7.5.1.8

Glazer A.N. & Bryant D.A. (1975) Allophycocyanin B (λ_{max} 671, 618 nm). A new cyanobacterial phycobiliprotein. *Arch. Microbiol.* **104,** 15–22.
7.5.1.1, 7.5.1.2, 7.5.1.4, 7.5.1.5, 7.5.1.7, 7.6, 7.7

Glazer A.N. & Cohen-Bazire G. (1971) Subunit structure of the phycobiliproteins of blue-green algae. *Proc. natn. Acad. Sci. U.S.A.* **68,** 1398–1401.
7.5.1.3, 7.5.1.4

Glazer A.N., Cohen-Bazire G. & Stanier R.Y. (1971) Comparative immunology of algal biliproteins. *Proc. natn. Acad. Sci. U.S.A.* **68,** 3005–3008.
7.5.1.7

Glazer A.N. & Fang S. (1973a) Chromophore content of blue-green algal phycobiliproteins. *J. biol. Chem.* **248,** 659–662.
7.5.1.2, 7.5.1.3, 7.5.1.5

Glazer A.N. & Fang S. (1973b) Formation of hybrid proteins from the α and β subunits of phycocyanins of unicellular and filamentous blue-green algae. *J. biol. Chem.* **248,** 633–671.
7.5.1.3

Glazer A.N., Fang S. & Brown D.M. (1973) Spectroscopic properties of C-phycocyanin and of its α and β subunits. *J. biol. Chem.* **248,** 5679–5685.
7.5.1.3, 7.5.1.4, 7.5.1.5

Glazer A.N. & Hixson C.S. (1975) Characterization of R-phycocyanin. Chromophore content of R-phycocyanin and C-phycoerythrin. *J. biol. Chem.* **250,** 5487–5495.
7.5.1.2, 7.5.1.3, 7.5.1.5, 7.5.1.7

Glazer A.N. & Hixson C.S. (1977) Subunit structure and chromophore composition of rhodophytan phycoerythrins. *Porphyridium cruentum* B-phycoerythrin and b-phycoerythrin. *J. biol. Chem.* **252,** 32–42.
7.5.1.1, 7.5.1.2, 7.5.1.3, 7.5.1.4, 7.5.1.5, 7.5.1.8

Glazer A.N., Hixson C.S. & Delange R.J. (1979a) Determination of the number of thioether-linked cysteine residues in cytochromes c and phycobiliproteins. *Anal. Biochem.* **92,** 489–496.
7.5.1.2

Glazer A.N., Williams R.C., Yamanaka G. & Schachman H.K. (1979b) Characterization of cyanobacterial phycobilisomes in zwitterionic detergents. *Proc. natn. Acad. Sci. U.S.A.* **76,** 6162–6166.
7.4, 7.5.2, 7.6, 7.7

Glazer A.N., Yamanaka G. & Gingrich J.C. (1979c) Phycobiliproteins: Structure and organization. In: Nichols J.M. (Ed.) *Abstracts of the Third International Symposium on Photosynthetic Prokaryotes, Oxford.* p. E4. Univ. Liverpool, England.
7.5.2

Gleason F.K. & Oaka M.P. (1978) Cell cycle and cell wall formation in *Synechococcus* sp. *Cytobiol.* **16,** 224–234.
13.3

Glover H.E. & Morris I. (1981) Photosynthetic characteristics of coccoid marine cyanobacteria. *Arch. Microbiol.* **129,** 42–46.
19.3.2

Gnaiger E., Gluth G. & Wieser W. (1978) pH fluctuations in an intertidal beach in Bermuda. *Limnol. Oceanogr.* **23,** 851–857.
20.2.2

Goad L.H. & Goodwin T.W. (1972) The biosynthesis of plant sterols. *Prog. Phytochem.* **3,** 113–198.
9.4.1

Goedheer J.C. (1976) Spectral properties of the blue-green alga *Anacystis nidulans* grown under different environmental conditions. *Photosynthetica* **10,** 411–422.
7.8, 7.8.2

Goedheer J.C. & Birnie F. (1965) Fluorescence polarization and location of fluorescence maxima of c-phycocyanin. *Biochim. biophys. Acta* **94,** 579–581.
7.5.1.5

Goering J.J., Dugdale R.C. & Menzel D.W. (1966) Estimates of *in situ* rates of nitrogen uptake by *Trichodesmium* sp. in the tropical Atlantic Ocean. *Limnol. Oceanogr.* **11,** 614–620.
19.3.4

Goh S.H. & LéJohn H.B. (1978) Glucose transport in *Achlya* characterization and possible regulatory aspects. *Can. J. Biochem.* **56,** 246–256.
3.4.3

Goldman J.C. (1973) Carbon dioxide and pH: effect on species succession of algae. *Science, N.Y.* **182,** 306–307.
18.5.2.2

Goldman J.C., McCarthy J.J. & Peavey D.G. (1979) Growth rate influence on the chemical composition of phytoplankton in ocean water. *Nature, Lond.* **279,** 210–215.
18.5.2.1

Goldner L.L. (1980) Nitrogen fixation (acetylene reduction) in shallow water Bahamian environments. *Bull. mar. Sci.* **30**, 444–453.
17.3.3, 20.4.5

Golecki J.R. (1977) Studies on ultrastructure and composition of cell walls of the cyanobacterium *Anacystis nidulans*. *Arch. Microbiol.* **114**, 35–41.
6.3.1, 13.2.1, 13.4.1, 13.4.2

Golecki J.R. (1979a) Ultrastructure of cell wall and thylakoid membranes of the thermophilic cyanobacterium *Synechococcus lividus* under the influence of temperature shifts. *Arch. Microbiol.* **120**, 125–133.
6.3.1, 6.3.2, 6.5.3

Golecki J.R. (1979b) Supramolecular organisation of the membrane system in cyanobacteria. In: Nichols J.M. (Ed.) *Abstracts of the Third International Symposium on Photosynthetic Prokaryotes, Oxford.* p. D5. Univ. Liverpool, England.
6.3.1, 6.3.2, 6.5.2

Golecki J.R. & Drews G. (1974) Zur Struktur der Blaualgen-Zellwand. Gefrierätzuntersuchungen an normalen und extrahierten Zellwänden von *Anabaena variabilis*. *Cytobiol.* **8**, 213–227.
6.3.1, 13.2.1, 13.2.2, 13.4.1, 14.3.1

Golecki J.R., Drews G. & Bühler R. (1979) The size and number of intramembrane particles in cells of the photosynthetic bacterium *Rhodopseudomonas capsulata* studied by freeze-fracture electron microscopy. *Cytobiologie* **18**, 381–389.
6.3.2

Golecki J.R. & Oelze J. (1980) Differences in the architecture of cytoplasmic and intracytoplasmic membranes of three chemotrophically and phototrophically grown species of the Rhodospirillaceae. *J. Bact.* **144**, 781–788.
6.3.2

Golterman H.L. (1964) Mineralization of algae under sterile conditions or by bacterial breakdown. *Verh. int. Verein. theor. angew. Limnol.* **15**, 544–548.
17.1.2

Golubic S. (1969) Distribution, taxonomy, and boring patterns of marine endolithic algae. *Am. Zoologist* **9**, 747–751.
20.4.3

Golubic S. (1970) Effect of organic pollution on benthic communities. *Mar. Poll. Bull.* **1** (4), 56–57.
20.3.6

Golubić S. (1973) The relationship between blue-green algae and carbonate deposits. In: Carr N.G. & Whitton B.A. (Eds) *The Biology of Blue-Green Algae.* 676 pp., pp. 434–472. Blackwell, Oxford.
2.7.3.2, 20.2.4, 20.3.4, 20.4.1

Golubic S. (1976) Organisms that build stromatolites. In: Walter M.R. (Ed.) *Stromatolites. Developments in Sedimentology* **20**, 790 pp., pp. 113–126. Elsevier, Amsterdam.
21.3

Golubic S. (1977) Speciation in *Trichodesmium*: occupation of an oceanic niche. In: Zehander A. (Ed.) *Symposium über Fragen der Cyanophytentaxonomic in Lednice (ČSSR), 29 Juli bis 8 August 1976. Schweiz. Z. Hydrol.* **39**, 141–143.
19.2.4.1

Golubic S. (1980) Halophily and halotolerance in cyanophytes. *Origins of Life* **10**, 169–183.
20.3.5

Golubic S. & Barghoorn E.S. (1977) Interpretation of microbial fossils with special reference to the Precambrian. In: Flügel E. (Ed.) *Fossil Algae.* 383 pp., pp. 1–14. Springer-Verlag, Berlin.
12.1.3

Golubic S. & Focke J.A. (1978) *Phormidium hendersonii* Howe: identity and significance of a modern stromatolite building microorganism: *J. sed. Petrol.* **48**, 751–754.
20.4.1, 20.4.2

Golubic S., Friedmann E.I. & Schneider J. (1981) The lithobiontic ecological niche with special reference to microorganisms. *J. sed. Petrol.* **51**, 475–478.
20.2.1, 20.4.3

Golubić S. & le Campion-Alsumard T. (1973) Boring patterns of marine blue-green algae *Mastigocoleus testarum* Lagerheim and *Kyrtuthrix dalmatica* Ercegovic as a taxonomic character. *Schweiz. Z. Hydrol.* **35**, 157–161.
20.4.3

Golubic S., Perkins R.D. & Lukas K.J. (1975) Boring microorganisms and microborings in carbonate substrates. In: Frey R.W. (Ed.) *The Study of Trace Fossils: A Study of Principles, Problems and Procedures in Ichnology.* 562 pp., pp. 229–259. Springer-Verlag, Berlin.
20.3.1, 20.3.2, 20.4.3

Golubic S. & Schneider J. (1979) Carbonate dissolution. In: Trudinger P.A. & Swaine D.J. (Eds) *Biogeochemical Cycling of Mineral-forming Elements. Studies in Environmental Science 3.* 612 pp., pp. 107–129. Elsevier, Amsterdam.
20.4.3

Gomez-Lojero C. & Perez-Gomez B. (1978) Citochromos de *Spirulina maxima*. In: Cha-

goya V., Martuscelli J., Pena A. & Pena E. (Eds) *Temas Bioquimicos de Actualidad.* 518 pp., p. 321–327. UNAM Publicaciones, Mexico.
8.5

Gons H.J. & Mur L.R. (1980) Growth rate and light uptake rate in light-limited continuous cultures of *Scenedesmus protuberans* Fritsch. *Arch. Microbiol.* **125,** 9–17.
2.5.2

Gorham P.R. (1964) Toxic algae. In: Jackson D.F. (Ed.) *Algae and Man,* 434 pp., pp. 307–336. Plenum Press, New York.
18.5.1.1, 18.5.6

Gorham P.R., McLachlan J.R., Hammer V.T. & Kim W.K. (1964) Isolation and culture of toxic strains of *Anabaena flos-aquae* (Lyngb.) de Bréb. *Verh. int. Verein. theor. angew. Limnol.* **15,** 796–804.
18.5.1.2

Gorlenko V.M. & Korotkov S.A. (1979) A new filamentous green bacteria with gas vacuoles, *Oscillochloris trichoides* nov. comb. In: Nichols J.M. (Ed.) *Abstracts of the Third International Symposium on Photosynthetic Prokaryotes, Oxford.* p. A16. Univ. Liverpool, England.
16.2.3

Gorlenko V.M. & Pivovarova T.A. (1977) On the belonging of the blue-green alga, *Oscillatoria coerulescens* Gicklhorn 1921, to a new genus of chlorobacteria, *Oscillochloris* nov. gen. *Izv. Akad. Nauk SSSR, Ser Biol.* 3, 396–409. (in Russian).
16.2.3

Gorlenko V.M., Vainstein M.B. & Kachalkin V.I. (1978) Microbiological characteristic of Lake Mogiluye. *Arch. Hydrobiol.* **81,** 457–492.
9.5.4

Gossauer A. & Hirsch W. (1974) Total synthese des racemischen Phycocyanobilins (Phycobiliverdins) sowie eines 'Homophycobiliverdins'. *Liebigs Ann. Chem.* **1974,** 1496–1513.
7.5.1.2

Gossauer A. & Klahr E. (1979) Total synthese des *racem.* Phycoerythrobilin-dimethylesters. *Chem. Ber.* **112,** 2243–2255.
7.5.1.2

Gotto J.W., Tabita F.R. & Van Baalen C. (1979a) Isolation and characterization of rapidly-growing, marine, nitrogen-fixing strains of blue-green algae. *Arch. Microbiol.* **121,** 155–159.
14.2, 20.3.5

Gotto J.W., Tabita F.R. & Van Baalen C. (1979b) Mutants of *Anabaena* strain CA altered in their ability to grow under nitrogen-fixing conditions. *J. Bact.* **140,** 327–332.
4.4, 11.5.5.2(b)

Gotto J.W., Tabita F.R. & Van Baalen C. (1980) Novel mutant of *Anabaena* sp. strain CA which grows on N_2 but not on combined nitrogen. *J. Bact.* **143,** 1395–1402.
11.5.3

Gould G.W. (1969) Germination. In: Gould G.W. & Hurst A. (Ed.) *The Bacterial Spore.* 747 pp., pp. 397–444. Academic Press, New York.
15.5

Govindjee R., Rabinovitch E. & Govindjee (1968) Maximum quantum yield and action spectrum of photosynthesis and fluorescence in *Chlorella. Biochim. biophys. Acta* **162,** 539–544.
9.2.4

Gozzer C., Zanetti G., Galliano M., Sacchi G.A., Minchiotti L. & Curti B. (1977) Molecular heterogeneity of ferredoxin-NADP reductase from spinach leaves. *Biochim. biophys. Acta* **485,** 278–290.
8.3

Grabowski J. & Gantt E. (1978) Photophysical properties of phycobiliproteins from phycobilisomes: fluorescence lifetimes, quantum yields, and polarization spectra. *Photochem. Photobiol.* **28,** 39–45.
7.5.1.5, 7.6

Granhall U. (1976) The presence of cellulose in heterocyst envelopes of blue-green algae and its role in relation to nitrogen fixation. *Physiol. Plant.* **38,** 208–216.
14.3.1

Granhall U. (1978) (Ed.) Environmental role of nitrogen-fixing blue-green algae and asymbiotic bacteria. *Ecol. Bull., Stockholm* **26,** 1–391.
18.5.2.1

Granhall U. & Lundgren A. (1971) Nitrogen fixation in Lake Erken. *Limnol. Oceanogr.* **16,** 711–719.
18.5.2.1

Grant N.G. & Walsby A.E. (1977) The contribution of photosynthate to turgor pressure rise in the planktonic blue-green alga *Anabaena flos-aquae. J. exp. Bot.* **28,** 409–415.
2.8.2.2., 10.2.3, 10.2.5.2

Gray B.H., Cosner J. & Gantt E. (1976) Phycocyanins with absorption maxima at 637 nm and 623 nm from *Agmenellum quadruplicatum. Photochem. Photobiol.* **24,** 299–302.
7.5.1.1.

Gray B.H. & Gantt E. (1975) Spectral properties of phycobilisomes and phycobilipro-

teins from the blue-green alga *Nostoc* sp. *Photochem. Photobiol.* **21,** 121–128.
7.4, 7.5.2, 7.6

Gray B.H., Lipschultz C.A. & Gantt E. (1973) Phycobilisomes from a blue-green alga *Nostoc* species. *J. Bact.* **116,** 471–478.
7.3, 7.5.2

Greenwood P.H. (1976) Lake George, Uganda. *Phil. Trans. R. Soc. Lond. B.* **274,** 375–391.
9.5.4

Gregory E.M. & Fridovich I. (1969) Oxygen toxicity and the superoxide dismutase. *J. Bact.* **114,** 1193–1197.
17.6

Griffiths B.M. (1939) Early references to water-blooms in the British Isles. *Proc. Linn. Soc. Lond.* **51,** 12–19.
18.4.4

Grilli M. (1964) Infrastrutture di *Anabaena azollae* vivente nelle foglioline di *Azolla caroliniana. Annali Microbiol.* **14,** 69–90.
6.2

Grilli Caiola M. & de Vecchi L. (1980) Akinete ultrastructure of *Nostoc* species isolated from cycad coralloid roots. *Can. J. Bot.* **58,** 2513–2519.
15.2, 15.5

Grillo J.F., Bottomley P.J., Van Baalen C. & Tabita F.R. (1979) A mutant of *Anabaena* sp. CA with oxygen-sensitive nitrogenase activity. *Biochem. biophys. Res. Commun.* **89,** 685–693.
4.4, 11.5.5.2(b), 14.5

Grillo J.F. & Gibson J. (1979) Regulation of phosphate accumulation in the unicellular cyanobacterium *Synechococcus. J. Bact.* **140,** 508–517.
5.1.1, 5.1.2, 5.2.2, 5.2.3, 5.3.1, 5.3.2

Gromet-Elhanan Z. (1977) Electron transport and photophosphorylation in photosynthetic bacteria. In: Trebst A. & Avron M. (Eds) *Encyclopedia of Plant Physiology* **5.** 730 pp., pp. 637–662. Springer-Verlag, Berlin.
9.3.3

Grøntved, J. & Steemann Nielsen E. (1957) Investigations on the phytoplankton in sheltered Danish marine localities. *Meddr. Kommn. Danm. Fisk og Havunders. Serie: Plankton.* **V** (6), 1–52. Copenhagen.
19.2.1, 19.2.4.2

Grossman A. & McGowan R.E. (1975) Regulation of glucose-6-phosphate in blue-green algae. *Pl. Physiol., Lancaster* **55,** 658–662.
3.3, 4.3

Grover I.S. & Puri S. (1979) Alteration of heterocyst pattern in *Cylindrospermum* sp. by rifampicin. *Ind. J. exp. Biol.* **17,** 109–111.
14.6.1.1

Grubb V.M. (1936) Marine algal ecology and the exposure factor at Peveril Point, Dorset. *J. Ecol.* **24,** 392–423.

Gruol D., Rawson J.R.Y. & Haselkorn R. (1975) Ribosomal RNA genes in the nucleus and chloroplast of *Euglena. Biochim. biophys. Acta* **414,** 20–29.
12.4.3.1

Guarino L.A. & Cohen S.S. (1979a) Uptake and accumulation of putrescine and its lethality in *Anacystis nidulans. Proc. natn. Acad. Sci.* **76,** 3184–3188.
10.2.3, 10.3.2

Guarino L.A. & Cohen S.S. (1979b) Mechanism of toxicity of putrescine in *Anacystis nidulans. Proc. natn. Acad. Sci. U.S.A.* **76,** 3660–3664.
10.3.2

Guerts van Kessel A.H.M., Mullahy J.H. & Linjkens H.F. (1977) The relation of heterocysts and hormogonia to N_2 fixation and reproduction in blue-green algae. *Plant Syst. Evol.* **128,** 171–176.
5.2.4

Guglielmi G., Cohen-Bazire G. & Bryant D.A. (1981) The structure of *Gloeobacter violaceus* and its phycobilisomes. *Arch. Microbiol.* **129,** 181–189.
1, 6.2, 7.3, 7.5.1.1, 7.5.1.6, 7.7

Guillard R.R.L., Kilham P. & Jackson T.A. (1973) Kinetics of silicon-limited growth of the marine diatom *Thalassiosira pseudonana* Hasle and Heimdal (*Cyclotella nana* Hustedt). *J. Phycol.* **9,** 233–237.
2.5.1

Gunatilaka A. (1975) Some aspects of the biology and sedimentology of laminated algal mats from Mannar Lagoon, northwest Ceylon. *Sed. Geol.* **14,** 275–300.
20.3.1, 20.4.1

Gundersen K.R., Corbin J.S., Hanson C.L., Hanson M.L., Hanson R.B., Russell D.J., Stollar A. & Yamada O. (1976) Structure and biological dynamics of the oligotrophic ocean photic zone off the Hawaiian Islands. *Pacif. Sci.* **30,** 45–68.
19.2.4.2, 19.5.2

Gunnison D. & Alexander N. (1975) Resistance and susceptibility of algae to decomposition by natural microbial communities. *Limnol. Oceanogr.* **20,** 64–70.
2.6.3

Gupta M. & Carr N.G. (1981) Enzyme activities related to cyanophycin metabolism in heterocysts and vegetative cells of *Anabaena* spp. *J. gen. Microbiol.* **125,** 17–23.
3.2.2.2, 14.3.2

Gupta R.S. & Dashora M.S. (1975) Division of heterocysts in the blue-green alga *Ana-*

baena doliolum Bharadwaja. *Z. allg. Mikrobiol.* **15,** 461–464.
14.3.2

Gupta R.S. & Kumar H.D. (1970) Action of mutagenic chemicals on *Anacystis nidulans*. V. Diethyl sulphate. *Arch. Mikrobiol.* **70,** 313–329.
11.5.5.1

Gusev M.V. (1979) Motility of cyanobacteria. In: Nichols J.M. (Ed.) *Abstracts of the Third International Symposium on Photosynthetic Prokaryotes*, p. A29, Univ. Liverpool, England.
16.3

Gusev M.V., Nikitina K.A. & Korzhenevskaya T.G. (1970) Metabolically active spheroplasts of blue-green algae. *Mikrobiologiya* **39,** 862–868 (Russian).
13.4.1

Guseva K.A. (1951) Interrelation of phytoplankton and saprophyte bacteria in a reservoir. Zoologicheskiy Institute. *Trudy Problemnyleh: Tematicles Kikh Soveschaniy. No. 1.* Russian. Translation: National Technical Information Service AD-650-850. pp. 34–38.
17.1.1

Guseva K.A. (1952) Tsvetenie vodȳ, ego prichinȳ i mery borbȳ snim. *Trudy vses. gidrobiol. Obshch.* **4,** 3.92. (*Russian*: Water blooms, their origin and methods of combating them.)
18.5.1.2

Gysi J. & Zuber H. (1974) Isolation and characterization of allophycocyanin II from the thermophilic blue-green alga *Mastigocladus laminosus* Cohn. *FEBS Lett.* **48,** 209–213.
7.5.1.1, 7.5.1.3

Gysi J. & Zuber H. (1976) Allophycocyanin I—a second cyanobacterial allophycocyanin? Isolation, characterization and comparison with allophycocyanin II from the same alga. *FEBS Lett.* **68,** 49–54.
7.5.1.1, 7.5.1.3

Gysi J. & Zuber H. (1979) Properties of allophycocyanin II and its α- and β-subunits from the thermophilic blue-green alga *Mastigocladus laminosus. Biochem. J.* **181,** 577–583.
7.5.1.4

Haaker H., Laane C. & Veeger C. (1980) Dinitrogen fixation and the proton-motive force. In: Stewart W.D.P. & Gallon J.R. (Eds) *Proceedings of an International Symposium on Nitrogen Fixation.* 451 pp., pp. 113–138. Academic Press, London.
4.3

Hackert M.L., Abad-Zapatero C., Stevens S.E. & Fox J.L. (1977) Crystallization of C-phycocyanin from the marine blue-green alga *Agmenellum quadruplicatum. J. mol. Biol.* **111,** 365–369.
7.5.1.6

Häder D.-P. (1974) Participation of two photosystems in the photo-phobotaxis of *Phormidium uncinatum. Arch. Microbiol.* **96,** 255–266.
2.8.1

Häder D.-P. (1975) The effect of inhibitors on the electron flow triggering photo-phobic reactions in cyanophyceae. *Arch. Microbiol.* **103,** 169–174.
2.8.1

Häder D.-P. (1976) Phobic reactions between two adjacent monochromatic light fields. *Z. Pflanzenphysiol.* **78,** 173–176.
16.4.3

Häder D.-P. (1977) Influence of electric fields on photophobic reactions in blue-green algae. *Arch. Microbiol.* **114,** 83–86.
2.8.1

Häder D.-P. (1978a) Evidence of electrical potential changes in photophobically reacting blue-green algae. *Arch. Microbiol.* **118,** 115–119.
2.8.1, 16.4.3

Häder D.-P. (1978b) Extracellular and intracellular determination of light-induced potential changes during photophobic reactions in blue-green algae. *Arch. Microbiol.* **119,** 75–79.
2.8.1, 10.3.3.1, 16.4.3

Häder D.-P. (1979a) Effect of inhibitors and uncouplers on light-induced potential changes triggering photophobic responses. *Arch. Microbiol.* **120,** 57–60.
16.4.3

Häder D.-P. (1979b) Photomovement. In: Haupt W. & Feinleib M.E. (Eds) *Encycl. Pl. Physiol.* N.S. Vol. 7, *Physiology of Movements*, pp. 268–309. Springer-Verlag, Berlin.
16.4.1, 16.4.2, 16.4.3, 16.4.4

Hagedorn H. (1961) Untersuchungen über die Feinstruktur der Blaualgenzellen. *Z. Naturf.* **16b,** 825–829.
6.2, 13.3

Halfen L.N. (1970) A study of Gliding Motility in the Blue-green Alga *Oscillatoria princeps*. Ph.D. Thesis, Univ. Oregon, Eugene, U.S.A.
16.2.2.1, 16.5

Halfen L.N. (1973) Gliding motility of *Oscilla-*

toria: ultrastructural and chemical characterization of the fibrillar layer. *J. Phycol.* **9,** 248–253.
13.2.2, 16.2.2.1, 16.2.2.2

Halfen L.N. (1979) Gliding movements. In: Haupt W. & Feinleib M.E. (Eds) *Encycl. Pl. Physiol.* N.S. Vol. 7, *Physiology of Movements*, pp. 250–267. Springer-Verlag, Berlin.
16.2.1, 16.2.2.1, 16.2.2.2

Halfen L.N. & Castenholz R.W. (1971a) Gliding motility in the blue-green alga, *Oscillatoria princeps*. *J. Phycol.* **7,** 133–145.
13.1, 13.2.1, 13.2.2, 13.3, 16.2.2.1, 16.2.2.2, 16.3, 16.4.2

Halfen L.N. & Castenholz R.W. (1971b) Energy expenditure for gliding motility in a blue-green alga. *J. Phycol.* **7,** 258–260.
16.3, 16.4.2

Hall D.O., Lumsden J. & Tel-Or E. (1977) Iron-sulphur proteins and superoxide dismutases in the evolution of photosynthetic bacteria and algae. In: Ponnamperuma C. (Ed.) *Chemical Evolution of the Early Precambrian*. 221 pp., pp. 191–210. Academic Press, New York.
12.2.2.2

Hall G.C., Flick M.B. & Jensen R.A. (1980) Approach to recognition of regulatory mutants of cyanobacteria. *J. Bact.* **143,** 981–988.
11.5.1

Hall G.C. & Jensen R.A. (1980) Enzymological basis for growth inhibition by L-phenylalanine in the cyanobacterium *Synechocystis* sp. 29108. *J. Bact.* **144,** 1034–1042.
11.5.1

Halldal P. (1958) Pigment formation and growth in the blue-green algae in crossed gradients of light intensity and temperature. *Physiol. Plant.* **11,** 401–420.
7.8

Halldal P. (1970) The photosynthetic apparatus of microalgae and its adaptation to environmental factors. In: Halldal P. (Ed.) *Photobiology of Microorganisms*. pp. 17–55. Wiley, London.
7.8

Halldal P. & Holmen A.T. (1979) The interrelationship between photosynthetic electron transport, glycolate excretion and amino acid metabolism in the blue-green alga *Anacystis nidulans*. *Pl. Cell Physiol., Tokyo* **20,** 757–763.
8.5

Hallenbeck P.C. & Benemann J.R. (1979) Hydrogen from algae. In: Barber J. (Ed.) *Topics in Photosynthesis, Vol. 3*. pp. 333–364. Elsevier, Amsterdam.
4.6

Hallenbeck P.C., Kostel D.J. & Benemann J.R. (1979) Purification and properties of nitrogenase from the cyanobacterium, *Anabaena cylindrica*. *Eur. J. Biochem.* **98,** 275–284.
4.2

Hamberg M., Samuelsson B., Björkhem I. & Danielsson H. (1974) Oxygenases in fatty acid and steroid metabolism. In: Hayaishi O. (Ed.) *Molecular Mechanism of Oxygen Activation*. 678 pp., pp. 28–85. Academic Press, New York.
9.4.1

Hammer U.T. (1969) Blue-green algal blooms in Saskatchewan Lakes. *Verh. int. Verein. theor. angew. Limnol.* **17,** 116–125.
5.4.3

Hanson R.B. & Gundersen K. (1976) Influence of sewage discharge on nitrogen flux from coral reefs in Kaneohe Bay, Hawaii. *Appl. environ. Microbiol.* **31,** 942–948.
20.3.6

Harder, R. (1917) Ernährungsphysiologische Untersuchungen an Cyanophyceen, hauptsächlich dem endophytischen *Nostoc punctiforme*. *Z. Bot.* **9,** 145–242.
15.4

Harnischfeger G. & Codd G.A. (1978) Factors affecting energy transfer from phycobilisomes to thylakoids in *Anacystis nidulans*. *Biochim. biophys. Acta* **502,** 507–513.
8.6

Harold F.M. (1977) Ion currents and physiological functions in microorganisms. *A. Rev. Microbiol.* **31,** 181–203.
10.3.3, 16.3, 16.4.2

Harold F.M. & Spitz E. (1975) Accumulation of arsenate, phosphate, and aspartate by *Streptococcus faecalis*. *J. Bact.* **122,** 266–277.
5.1.2

Harper M.A. (1977) Movements. In: Werner D. (Ed.) *The Biology of Diatoms*. 498 pp., pp. 224–249. Blackwell, Oxford.
16.2.1

Harris G.P. (1978) Photosynthesis, productivity and growth: the physiological ecology of phytoplankton. *Arch. Hydrobiol. Beih. Ergebn. Limnol.* **10,** 1–171.
17.4.4, 18.5.5.1, 18.5.5.3, 18.5.5.4, 18.5.5.6

Harris J.U. & Berns D.S. (1975) Sequences of the N-terminus portions of biliproteins. *J. mol. Evol.* **5,** 153–163.
7.5.1.8

Hart R.C. & Hart R. (1978) The seasonal cycles of phytoplankton in subtropical Lake Sibaya: a preliminary investigation. *Arch. Hydrobiol.* **80,** 85–107.
18.4.1

Hart T.J. (1966) Some observations on the relative abundance of marine phytoplankton populations in nature. In: Barnes H. (Ed.) *Some Contemporary Studies in Marine Science*. pp. 375–393. George Allen & Unwin, London.
19.2.4.1

Hase T., Wada K. & Matsubara H. (1976a) Amino acid sequence of the major component of *Aphanothece sacrum* ferredoxin. *J. Biochem., Tokyo* **79,** 329–343.
8.3, 12.2.2.2

Hase T., Wada K., Ohmiya M. & Matsubara H. (1976b) Amino acid sequence of the major component of *Nostoc muscorum* ferredoxin. *J. Biochem. Tokyo* **80,** 993–999.
8.3, 12.2.2.2

Hase T., Wakabayashi S., Matsubara H., Kerscher L., Oesterhelt D., Rao K.K. & Hall D.O. (1977) *Halobacterium halobium* ferredoxin: a homologous protein to chloroplast-type ferredoxin. *FEBS Lett.* **77,** 308–310.
12.2.2.2

Hase T., Wakabayashi S., Matsubara H., Kerscher L., Oesterhelt D., Rao K.K. & Hall D.O. (1978a) Complete amino acid sequence of *Halobacterium halobium* ferredoxin containing an N^{ε} acetyllsine residue. *J. Biochem., Tokyo* **83,** 1657–1670.
12.2.2.2

Hase T., Wakabayashi S., Matsubara H., Rao K.K., Hall D.O., Widmer H., Gysi J. & Zuber H. (1978b) The amino acid sequence of ferredoxin from the alga *Mastigocladus laminosus. Phytochem.* **17,** 1863–1867.
8.3

Hase T., Wakabayashi S., Wada K. & Matsubara H. (1978c) The amino acid sequence of *Aphanothece sacrum* ferredoxin II (minor component). Structural characteristics and evolutionary implications. *J. Biochem., Tokyo* **83,** 761–770.
8.3, 12.2.2.2

Haselkorn R. (1978) Heterocysts. *A. Rev. Plant Physiol.* **29,** 319–344.
4.1, 14.1, 14.3.2, 14.5

Haselkorn R. & Rouvière-Yaniv J. (1976) Cyanobacterial DNA-binding protein related to *Escherichia coli* HU. *Proc. natn. Acad. Sci. U.S.A.* **73,** 1917–1920.
11.3.1

Haselkorn R., Mazur B., Orr J., Rice D., Wood N. & Rippka R. (1980) Heterocyst differentiation and nitrogen fixation in cyanobacteria (blue-green algae). In: Newton W.E. & Orme-Johnson W.H. (Eds) *Nitrogen Fixation, Volume II.* 325 pp., pp. 259–278. University Park Press, Baltimore.
14.6.1.3

Hattori A. (1963) Effect of hydrogen on nitrite reduction by *Anabaena cylindrica.* In: *Studies on Microalgae and Photosynthetic Bacteria.* pp. 485–492. Jap. Soc. Plant Cell Physiol., Univ. Tokyo Press.
4.6

Hattori A., Crespi H.L. & Katz J.J. (1965) Association and dissociation of phycocyanin and the effects of deuterium substitution on the processes. *Biochemistry, N.Y.* **4,** 1225–1238.
7.5.1.4, 7.5.1.5

Hattori A. & Fujita Y. (1959) Crystalline phycobilin chromoproteids obtained from a blue-green alga, *Tolypothrix tenuis. J. Biochem., Tokyo* **46,** 633–644.
7.5.1.4

Haupt W. (1959) Die Phototaxis der Algen. In: Bünning E. (Ed.) *Handbuch der Pflanzenphysiologie 17, Teil 1, Physiologie der Bewegungen.* pp. 317–370. Springer, Berlin.
16.5

Haury J.F. & Bogorad L. (1977) Action spectra for phycobiliprotein synthesis in a chromatically adapting cyanophyte *Fremyella diplosiphon. Pl. Physiol., Lancaster* **60,** 835–839.
7.8.3.2

Haury J.F. & Spiller H. (1981) Fructose uptake and influence on growth and nitrogen fixation by *Anabaena variabilis. J. Bact.* **147,** 227–235.
3.4.4

Haury J.F. & Wolk C.P. (1978) Classes of *Anabaena variabilis* mutants with oxygen-sensitive nitrogenase activity. *J. Bact.* **136,** 688–692.
4.4, 11.5.5.2(b), 14.3.1

Haxo F.T. (1960) The wavelength dependence of photosynthesis and the role of accessory pigments. In: Allen M.B. (Ed.) *Comparative Biochemistry of Photoreactive Systems.* pp. 339–360. Academic Press, New York.
7.2

Haxo F.T. & Blinks L.R. (1950) Photosynthetic action spectra of marine algae. *J. gen. Physiol.* **33,** 389–422.
7.2

Haxo F., O'hEocha C. & Norris P. (1955) Comparative studies of chromatographically separated phycoerythrins and phycocyanins. *Arch. Biochem. Biophys.* **54,** 162–173.
7.5.1.1

Hayes J.M. (1982) Geochemical evidence bear-

ing on the origin of aerobiosis, a speculative hypothesis. In: Schopf J.W. (Ed.) *Origin and Evolution of Earth's Earliest Biosphere*. Princeton Univ. Press, Princeton.
21.2

Hayes J.M., Kaplan I.R. & Wedeking K. (1982) Precambrian organic geochemistry, preservation of the record. In: Schopf J.W. (Ed.) *Origin and Evolution of Earth's Earliest Biosphere*. Princeton Univ. Press, Princeton, N.J.
21.2

Hayes W. (1968) *The Genetics of Bacteria and their Viruses*, 2nd edition. 925 pp., pp. 331–336. Blackwell, Oxford.
11.4.2

Haystead A., Robinson R. & Stewart W.D.P. (1970) Nitrogenase activity in extracts of heterocystous and non-heterocystous blue-green algae. *Arch. Mikrobiol.* **74,** 235–243.
4.2

Hazelbauer G.L. & Adler J. (1971) Role of the galactose binding protein in chemotaxis of *Escherichia coli* toward galactose. *Nature, New Biol.* **230,** 101–104.
17.5.1

Hazelbauer G.L. & Parkinson J.S. (1977) Bacterial chemotaxis. In: Reissig J. (Ed.) *Receptors and Recognition: Microbial Interactions*, 436 pp., pp. 60–98. Chapman & Hall, London.
16.5

Healey F.P. (1973a) Characteristics of phosphorus deficiency in *Anabaena*. *J. Phycol.* **9,** 383–394.
5.1.1, 5.1.3, 5.2.3, 5.2.4, 5.3.1, 5.3.2

Healey F.P. (1973b) Inorganic nutrient uptake and deficiency in algae. *CRC Crit. Rev. Microbiol.* **3,** 69–113.
5.1.1, 5.2.2, 5.2.3, 18.5.2.1

Healey F.P. (1975) Physiological indicators of nutrient deficiency in algae. *Fish. Mar. Serv. Res. Dev. Tech. Rep.* **585,** 30 pp.
5.2.2, 5.4.1

Healey F.P. (1979) Short-term responses of nutrient-deficient algae to nutrient addition. *J. Phycol.* **15,** 289–299.
5.3.2

Healey F.P. (1980) The slope of the Monod equation as an indicator of advantage in nutrient competition. *Microb. Ecol.* **5,** 281–286.
5.1.1

Healey F.P. & Hendzel L.L. (1975) Effect of phosphorus deficiency on two algae growing in chemostats. *J. Phycol.* **11,** 303–309.
5.1.1, 5.1.2, 5.1.3, 5.2.1, 5.2.2, 5.2.3, 5.2.4

Healey F.P. & Hendzel L.L. (1976) Physiological changes during the course of blooms of *Aphanizomenon flos-aquae*. *J. Fish. Res. Bd Can.* **33,** 36–41.
5.3.1, 5.4.1, 5.4.3

Healey F.P. & Hendzel L.L. (1979a) Fluorometric measurement of alkaline phosphatase activity in algae. *Freshwater Biol.* **9,** 429–439.
5.1.3, 5.2.3, 5.4.1

Healey F.P. & Hendzel L.L. (1979b) Indicators of phosphorus and nitrogen deficiency in five algae in culture. *J. Fish. Res. Bd Can.* **36,** 1364–1369.
5.2.2, 5.2.3

Healey F.P. & Hendzel L.L. (1980) Physiological indicators of nutrient deficiency in lake phytoplankton. *Can. J. Fish. aquat. Sci.* **37,** 442–453.
5.4.1

Heath R.T. & Cooke G.D. (1975) The significance of alkaline phosphatase in a eutrophic lake. *Verh. int. Verein. theor. angew. Limnol.* **19,** 959–965.
5.1.3, 5.4.1, 5.4.3

Heber U. & Krause G.H. (1980) What is the physiological role of photorespiration? *Trends biochem. Sci.* **5,** 32–34.
2.3.1

Heldt H.W., Werden K., Milovancev M. & Geller G. (1973) Alkalinization of the chloroplast stroma caused by light-dependent proton flux into the thylakoid space. *Biochim. biophys. Acta* **314,** 224–241.
10.3.2, 10.3.2.1

Hellebust J.A. (1965) Excretion of some organic compounds by marine phytoplankton. *Limnol. Oceanogr.* **10,** 192–206.
17.4.4

Hellebust J.A. (1971) Kinetics of glucose transport and growth of *Cyclotella cryptica*. *J. Phycol.* **7,** 1–3.
3.4.3

Henner D.J. & Hoch J.A. (1980) The *Bacillus subtilis* chromosome. *Microbiol. Rev.* **44,** 57–82.
11.2.2

Henning U. (1975) Determination of cell shape in bacteria. *A. Rev. Microbiol.* **29,** 45–60.
13.3

Henrichsen J. (1972) Bacterial surface translocation: a survey and a classification. *Bact. Rev.* **36,** 478–503.
16.2.1, 16.5

Henrici A.T. & McCoy E. (1938) The distribution of heterotrophic bacteria in the bottom deposits of some lakes. *Trans. Wis. Acad. Sci. Arts Lett.* **31,** 323–361.
17.1.1

Henry L.E.A., Gogotov I.N. & Hall D.O. (1978) Superoxide dismutase and catalase in the protection of the proton-donating systems of nitrogen fixation in the blue-green alga *Anabaena cylindrica*. *Biochem. J.* **174,** 373–377.
4.4, 8.3, 14.4.1, 14.4.3.2

Hentschel W.M., Farmer J.L. & Andersen W.R. (1978) Ethionine-resistant mutants of the filamentous blue-green alga *Plectonema boryanum*. *J. Bact.* **133,** 1536–1539.
11.5.1

Herbst V. & Overbeck J. (1978) Metabolic coupling between the alga *Oscillatoria redekei* and accompanying bacteria. *Naturwissenschaften* **65,** 598–599.
17.4.4, 19.3.4

Herdman M. (1973a) Mutations arising during transformation in the blue-green alga *Anacystis nidulans*. *Molec. gen. Genet.* **120,** 369–378.
11.6.1, 11.6.3, 11.7

Herdman M. (1973b) Transformation in the blue-green alga *Anacystis nidulans* and an associated phenomenon of mutation. In: Archer L.J. (Ed.) *Bacterial Transformation*. 413 pp., pp. 369–386. Academic Press, London.
11.6.1, 11.6.3, 11.7

Herdman M. & Carr N.G. (1971a) Recombination in *Anacystis nidulans* mediated by an extracellular DNA/RNA complex. *J. gen. Microbiol.* **68,** xiv.
11.6.3

Herdman M. & Carr N.G. (1971b) Observations on replication and cell division in synchronous cultures of the blue-green alga, *Anacystis nidulans*. *J. Bact.* **107,** 538–584.
15.5

Herdman M. & Carr N.G. (1972) The isolation and characterization of mutant strains of the blue-green alga *Anacystis nidulans*. *J. gen. Microbiol.* **70,** 213–220.
11.5.1, 11.5.3, 11.5.4, 11.5.5.1

Herdman M. & Carr N.G. (1974) Estimation of the genome size of blue-green algae from DNA renaturation rates. *Arch. Microbiol.* **99,** 251–254.
11.3.1

Herdman M., Delaney S.F. & Carr N.G. (1973) A new medium for the isolation and growth of auxotrophic mutants of the blue-green alga *Anacystis nidulans*. *J. gen. Microbiol.* **79,** 233–237.
11.5.3

Herdman M., Delaney S.F. & Carr N.G. (1980) Mutation of the cyanobacterium *Anacystis nidulans* (*Synechococcus* PCC 6301): improved conditions for the isolation of auxotrophs. *Arch. Microbiol.* **124,** 177–184.
11.2.2, 11.4.1, 11.5.3, 11.6.1

Herdman M., Faulkner B.M. & Carr N.G. (1970) Synchronous growth and genome replication in the blue-green alga, *Anacytis nidulans*. *Arch. Mikrobiol.* **73,** 238–249.
15.5

Herdman M., Janvier M., Rippka R. & Stanier R.Y. (1979a) Genome size of cyanobacteria. *J. gen. Microbiol.* **111,** 73–85.
11.3.1

Herdman M., Janvier M., Waterbury J.B., Rippka R., Stanier R.Y. & Mandel M. (1979b) Deoxyribonucleic acid base composition of cyanobacteria. *J. gen. Microbiol.* **111,** 63–71.
11.3.1, 11.3.2, 12.2.1, 12.3.1, 12.3.3, 12.3.4.2

Herdman M., van Liere L. & Mur L. (1982) Cellular macromolecular composition in continuous cultures of the cyanobacterium *Oscillatoria agardhii* under light energy- or inorganic nutrient-limitation. *Arch. Microbiol.* (in press).
11.3.1

Herdman M. & Stanier R.Y. (1977) The cyanelle: chloroplast or endosymbiotic prokaryote? *FEMS Microbiol. Lett.* **1,** 7–12.
7.2

Hertzberg S., Liaaen-Jensen S. & Siegelmann H.W. (1971) The carotenoids of the blue-green algae. *Phytochem.* **10,** 3121–3127.
6.4.2

Higgins I.J., Best D.J. & Hammond R.C. (1980) New findings in methane-utilizing bacteria highlight their importance in the biosphere and their commercial potential. *Nature, Lond.* **286,** 561–564.
21.2

Hildebrand E. & Dencher N. (1975) Two photosystems controlling behavioral responses of *Halobacterium halobium*. *Nature, Lond.* **257,** 46–48.
16.4.3

Hill H. (1970) A new form of *Raphidiopsis mediterranae* Skuja found in Minnesota lakes. *Phycologia* **9,** 73–77.
15.6

Hill S. (1971) Influence of oxygen concentrations on the colony type of *Derxia gummosa* grown on nitrogen-free media. *J. gen. Microbiol.* **67,** 77–83.
4.4

Hirayama O. (1967) Pigments and lipids in blue-green alga *Anacystis nidulans*. *J. Biochem., Tokyo* **61,** 179–185.
6.4.1

Hirosawa T. & Wolk C.P. (1979a) Factors controlling the formation of akinetes adjacent to heterocysts in the cyanobacterium *Cylindrospermum licheniforme* Kütz. *J. gen. Microbiol.* **114**, 423–432.
14.2, 14.4.3.3, 14.6.2, 15.4, 15.4.1, 15.4.3

Hirosawa T. & Wolk C.P. (1979b) Isolation and characterization of a substance which stimulates the formation of akinetes in the cyanobacterium *Cylindrospermum licheniforme* Kütz. *J. gen. Microbiol.* **114**, 433–441.
14.6.2, 15.4.3

Hirsch P. (1978) Microbial stratification in beach sand interstitial water ('Farbstreifen Sandwatt') In: Krumbein W.E. (Ed.) *Environmental Biogeochemistry and Geomicrobiology Volume 3: Methods, Metals and Assessment.* pp. 57–58. Ann Arbor Science, Ann Arbor.
20.2.2

Hixson C.S. (1976) Characterization of Rhodophytan Biliproteins. 222 pp. Ph.D thesis, Univ. California, Los Angeles.
7.5.1.3, 7.5.1.4

Hiyama T., McSwain B.D. & Arnon D.I. (1977) Evidence for two types of P700 in membrane fragments from a blue-green alga. *Biochim. biophys. Acta* **460**, 76–84.
8.4

Hiyama T., Tsujimoto H.Y. & Arnon, D.I. (1979) Photoreduction of membrane-bound paramagnetic component X by water as electron donor. *FEBS Lett.* **98**, 381–385.
8.4

Ho K.K. & Krogmann D.W. (1980) Cytochrome f from spinach and cyanobacteria: purification and characterization. *J. biol. Chem.* **255**, 3855–3861.
8.5

Ho J. & McCurdy H.D. (1979) Demonstration of positive chemotaxis to cyclic GMP and 5'-AMP in *Myxococcus xanthus* by means of a simple apparatus for generating practically stable concentration gradients. *Can. J. Microbiol.* **25**, 1214–1218.
16.5

Ho K.K., Ulrich E.L., Krogmann D.W. & Gomez-Lojero C. (1979) Isolation of photosynthetic catalysts from cyanobacteria. *Biochim. biophys. Acta* **545**, 236–248.
4.3, 6.5.2, 8.3, 8.5

Hoarau J., Remy R. & Leclerk J.C. (1977) Heterogeneite des variations spectrales photoinduites vers 700 nm observers sur les membranes chlorophylliennes et les complexes chlorophylle-proteines isoles de divers organismes photosynthetiques. *Biochim. biophys. Acta* **462**, 659–670.
8.4

Hoare D.S., Ingram L.O., Thurston E.L. & Walkup R. (1971) Dark heterotrophic growth of an endophytic blue-green alga. *Arch. Mikrobiol.* **78**, 310–321.
3.4.1, 3.4.2, 3.4.3, 6.5.2, 12.3.3

Höcht H., Martin H.H. & Kandler O. (1965) Zur Kenntnis der chemischen Zusammensetzung der Zellwand der Blaualgen. *Z. Pflanzenphysiol.* **53**, 39–57.
13.2.1, 13.4.1, 13.4.2, 13.4.4, 13.4.5

Hoffman P.F. (1967) Algal stromatolites: use in stratigraphic correlation and paleocurrent determination. *Science, N.Y.* **157**, 1043–1045.
20.4.2

Hoffman P.F. (1976) Stromatolite morphogenesis in Shark Bay, Western Australia. In: Walter M.R. (Ed.) *Stromatolites. Developments in Sedimentology 20.* 790 pp., pp. 261–271. Elsevier, Amsterdam.
20.3.2

Hoffmann C. (1942) *Beiträge zur Vegetation des Farbstreifen-Sandwattes. Kiele Meeresforschung* **4**, 85–108.
20.2.2

Hoffmann C. (1949) Über die Durchlassigkeit dünner Sandschichten für Licht. *Planta, Berl.* **37**, 48–56.
20.2.2, 20.3.4

Hofmann H.J. & Aitken J.D. (1979) Precambrian biota from the Little Dal Group, Mackenzie Mountains, northwestern Canada. *Can. J. earth Sci.* **16**, 150–166.
21.3

Hofmann H.J. & Schopf J.W. (1982) Early Proterozoic microfossils. In: Schopf J.W. (Ed.) *Origin and Evolution of Earth's Earliest Biosphere.* Princeton Univ. Press, Princeton, N.J.
21.1, 21.3

Holland H.D. (1978) *The Chemistry of the Atmosphere and Oceans.* 351 pp. Wiley, New York.
21.2

Hollerbach M.M. (1928) On the morphology of *Tolypothrix elenkinii* Holl. under natural and laboratory conditions. *Arch. Russk. Protistol.* **7**, 159–178 (Russian).
14.5

Holt T. & Krogmann D.W. (1981) A carotenoid protein complex from cyanobacteria. *Biochim. biophys. Acta* **637**, 408–414.
8.6

Holton R.W., Blecker H.H. & Onore M. (1964) Effect of growth temperature on the fatty acid composition of a blue-green alga. *Phytochem.* **3**, 595–602.
6.5.3

Holton R.W., Blecker H.H. & Stevens T.S. (1968) Fatty acids in blue-green algae: possible relationship to phylogenetic position. *Science, N.Y.* **160,** 545–547.
12.3.3

Holton R.W. & Freeman A.W. (1965) Some theoretical and experimental considerations of the gliding movement of blue-green algae. *Am. J. Bot.* **52,** 640.
16.2.2.1

Honeycutt R.C. & Krogmann D.W. (1972) Further studies on the oxygen-reducing system of *Anabaena variabilis*. *Biochim. biophys. Acta* **256,** 467–476.
8.3

Hood E.E., Armour S., Ownby J.D., Handa A.K. & Bressan R.A. (1979) Effect of nitrogen starvation on the level of adenosine 3′,5′-monophosphate in *Anabaena variabilis*. *Biochim. biophys. Acta* **588,** 193–200.
14.5

Hoogenhout H. & Amesz J. (1965) Growth rates of photosynthetic microorganisms in laboratory cultures. *Arch. Mikrobiol.* **50,** 10–24.
18.5.1.3

Hooper-Reid N.M. & Robinson G.G.C. (1978) Seasonal dynamics of epiphytic algal growth in a marsh pond: composition, metabolism and nutrient availability. *Can. J. Bot.* **56,** 2441–2448.
5.4.1

Hopfer U., Lehninger A.L. & Thompson T.E. (1968) Protonic conductance across phospholipid bilayer membranes induced by uncoupling agents for oxidative phosphorylation. *Proc. natn. Acad. Sci. U.S.A.* **59,** 484–490.
16.3

Hopwood D.A. & Glauert A.M. (1960) The fine structure of the nuclear material of a blue-green alga, *Anabaena cylindrica* Lemm. *J. biophys. biochem. Cytol.* **8,** 813–823.
11.3.1

Hori H. & Osawa S. (1979) Evolutionary change in 5S RNA secondary structure and a phylogenetic tree of 54 5S RNA species. *Proc. natn. Acad. Sci. U.S.A.* **76,** 381–385.
12.1.3

Horne A.J. (1977) Nitrogen fixation—a review of this phenomenon as a polluting process. *Progr. Wat. Technol.* **8,** 359–372.
18.5.2.1

Horne A.J. & Fogg G.E. (1970) Nitrogen fixation in some English Lakes. *Proc. R. Soc. Lond. B.* **175,** 351–366.
18.5.2.1

Horne A.J. & Goldman C.R. (1972) Nitrogen fixation in Clear Lake, California; 1. Seasonal variation and the role of heterocysts. *Limnol. Oceanogr.* **17,** 678–692.
18.5.2.1

Horne A.J. & Goldman C.R. (1974) Suppression of nitrogen fixation by blue-green algae in a eutrophic lake with trace additions of copper. *Science, N.Y.* **183,** 409–411.
18.6

Horstmann U., Ulbricht K.A. & Schmidt D. (1978) Detection of eutrophication processes from air and space. *Proceedings of 12th International ERIM Symposium on Remote Sensing of Environment, Manila, Philippines, April 20–26, 1978.* pp. 1379–1389.
19.2.4.2, 19.5.2, 19.5.4

Hough L., Jones J.K.N. & Wadman W.H. (1952) An investigation of the polysaccharide components of certain freshwater algae. *J. chem. Soc.* 3393–3399.
13.4.5

Houghton J.A. (1971) Attempted isolation of auxotrophic mutants from blue-green algae. *Br. phycol. J.* **6,** 39–40.
11.1

Houghton J.A. (1972) The genetics of blue-green algae. *Sci. Progr.* **60,** 407–412.
11.1

Howarth R.W. & Teal J.M. (1978) Sulfate reduction in a New England salt marsh. *Limnol. Oceanogr.* **24,** 999–1013.
9.5.3

Howsley R. & Pearson H.W. (1979) pH dependent sulphide toxicity to oxygenic photosynthesis in cyanobacteria. *FEMS Microbiol. Lett.* **6,** 287–292.
9.3.4

Hübel von H. & Hübel M. (1974) In-situ-Messungen der diurnalen Stickstoff-Fixierung an Mikrobenthos der Ostseeküste. *Algological Studies* **10,** *Arch. Hydrobiol. Suppl.* **46,** 39–54.
20.4.5

Hughes S.G., Bruce T. & Murray K.H. (1980) The isolation and characterization of a sequence-specific endonuclease from *Anabaena subcylindrica*. *Biochem. J.* **185,** 59–63.
11.3.3

Hughes S.G. & Murray K.H. (1980) The nucleotide sequences recognized by endonucleases *Ava* I and *Ava* II from *Anabaena variabilis*. *Biochem. J.* **185,** 65–75.
11.3.3

Humm H.J. & Wicks S.R. (1980) *Introduction and Guide to the Marine Blue-Green Algae.*

171 pp. Wiley, New York.
20.1, 20.3.6

Humphrey B.A., Dickson M.R. & Marshall K.C. (1979) Physicochemical and *in situ* observations on the adhesion of gliding bacteria to surfaces. *Arch. Microbiol.* **120,** 231–238.
16.2.2.1, 16.2.2.2, 16.2.3

Hunter F. & Thornber J.P. (1976) Further characterization of the P700-Chlorophyll a-protein from blue-green algae. *Brookhaven Symp. Biol.* **28,** 361.
6.4.2

Hurt E. & Hauska G. (1981) A cytochrome f/b6 complex of five polypeptides with plastoquinol-plastocyanin-oxidoreductase activity from spinach chloroplasts. *Eur. J. Biochem.* **117,** 591–599.
8.6

Hutber G.N., Smith A.J. & Rogers L.J. (1978) Isolation of ferredoxin-adenine dinucleotide phosphate (oxidized) reductase from a prokaryote. *Biochem. Soc. Trans.* **6,** 1212–1216.
8.3

Hutchinson G.E. (1967) *A Treatise on Limnology, Vol.* **2,** 1115 pp. Wiley, New York.
9.5.4, 18.1, 18.4.2, 18.5.2.1

Hutson R.G. & Rogers L.J. (1975) Two plant type ferredoxins in light-grown and dark-grown cells of the blue-green bacterium *Nostoc* strain MAC. *Biochem. Soc. Trans.* **3,** 377–378.
4.3

Hutson K.G., Rogers L.J., Haslett B.G., Boulter D. & Cammack R. (1978) Comparative studies on two ferredoxins from the cyanobacterium *Nostoc* strain MAC. *Biochem. J.* **172,** 465–477.
8.3

Ihlenfeldt M.J.A & Gibson J. (1975a) CO_2 fixation and its regulation in *Anacystis nidulans (Synechococcus). Arch. Microbiol.* **102,** 13–21.
3.2.1, 3.3

Ihlenfeldt M.J.A. & Gibson J. (1975b) Phosphate utilization and alkaline phosphatase activity in *Anacystis nidulans (Synechococcus). Arch. Microbiol.* **102,** 23–28.
5.1.3, 5.2.3, 5.3.1, 5.3.2, 7.8

Imafuku H. & Katoh T. (1976) Intracellular ATP level and light induced inhibition of respiration in a blue-green alga, *Anabaena variabilis. Pl. Cell Physiol., Tokyo* **17,** 515–524.
3.3, 8.7

Ingram L.O. & Aldrich H.C. (1974) Cell separation in blue-green bacteria. *J. Bact.* **118,** 708–716.
13.3

Ingram L.O., Calder J.A., Van Baalen C., Plucker F.E. & Parker P.L. (1973) Role of reduced exogenous organic compounds in the physiology of the blue-green bacteria (algae): photoheterotrophic growth of a 'heterotrophic' blue-green bacterium. *J. Bact.* **114,** 695–700.
3.4, 3.4.1, 3.4.2, 3.4.3

Ingram L.O. & Fisher W.D. (1973a) Novel mutant impaired in division: evidence for a positive regulating factor. *J. Bact.* **113,** 999–1005.
11.5.5.1

Ingram L.O. & Fisher W.D. (1973b) Mechanism for the regulation of cell division in *Agmenellum. J. Bact.* **113,** 1006–1014.
11.5.5.1

Ingram L.O., Olson G.L. & Blackwell M.M. (1975) Isolation of a small cell mutant in the blue-green bacterium *Agmenellum quadruplicatum. J. Bact.* **123,** 743–746.
11.5.5.1

Ingram L.O., Pierson D., Kane J.F., Van Baalen C. & Jensen R.A. (1972) Documentation of auxotrophic mutation in blue-green bacteria: characterization of a tryptophan auxotroph in *Agmenellum quadruplicatum. J. Bact.* **111,** 112–118.
11.2.2, 11.5.3

Ingram L.O. & Thurston E.L. (1970) Cell division in morphological mutants of *Agmenellum quadruplicatum* strain BG-1. *Protoplasma, Wien* **71,** 55–75.
6.1, 6.2, 11.5.5.1, 13.3

Ingram L.O. & Van Baalen C. (1970) Characteristics of a stable, filamentous mutant of a coccoid blue-green alga. *J. Bact.* **102,** 784–789.
11.5.5.1

Jaag O. (1972) *Oscillatoria rubescens* D.C. In: Desikachary T.V. (Ed.) *Taxonomy and Biology of Blue-Green Algae.* 591 pp., pp. 296–299. Univ. Madras, India.
18.4.3

Jacob F. (1961) Zur Biologie von *Codium bursa* (L.) AGARDH und seiner endophytischen Cyanophyceen. *Arch. Protistenk.* **105,** 345–406.
20.2.4

Jacobs N.J. (1974) Biosynthesis of heme. In: Neilands J.B. (Ed.) *Microbial Iron Metab-*

olism: A Comprehensive Treatise. 597 pp., pp. 125–148. Academic Press, New York.
3.2.2.1

Jagger D.D. (1967) *Introduction to Research in Ultraviolet Photobiology*. 164 pp. Prentice-Hall, N.J.
2.7.1

Janaki S. & Wolk C.P. (1982) Synthesis of nitrogenase by isolated heterocysts. *Biochim. biophys. Acta* **696,** 187–192.
14.2, 14.3.2

Janczewski E. (1874) Observations sur la reproduction de quelques Nostocacées. *Ann. Sci. Nat. Bot. Ser.* 5 **19,** 119–130.
15.1

Jannasch H.W. & Wirsen C.O. (1981) Morphological survey of microbial mats near deep-sea thermal vents. *Appl. environ. Microbiol.* **41,** 528–538.
20.2.4

Jarosch R. (1964) Gleitbewegung und Torsion von Oscillatorien. *Öst. bot. Z.* **111,** 143–148.
13.2.2, 16.2.2.2

Jarosch R. (1967) Studien zur Bewegungsmechanik der Bakterien und Spirochäten des Hochmoores. *Öst. bot. Z.* **114,** 255–306.
16.2.2.2

Jayaraman R. (1972) On the occurrence of blooms of blue-green algae and the associated oceanographic conditions in the northern Indian Ocean. In: Desikachary T.V. (Ed.) *Taxonomy and Biology of Blue-Green Algae*, pp. 428–432. Univ. Madras, India.
19.2.4.1

Jeanjean R. (1969) Influence de la carence en phosphore sur les vitesses d'absorption du phosphate par les Chlorelles. *Bull. Soc. fr. Physiol. vég.* **15,** 159–171.
5.1.1

Jeanjean R., Blasco F. & Gaudin C. (1970) Études des mécanismes d'absorption de l'ion phosphate par les Chlorelles. *C. r. hebd. Séanc. Acad. Sci., Paris D* **270,** 2946–2949.
5.1.1

Jensen R.A., Stenmark-Cox S. & Ingram L.O. (1974) Mis-regulation of 3-deoxy-D-arabino-heptulosonate-7-phosphate synthetase does not account for growth inhibition by phenylalanine in *Agmenellum quadruplicatum*. *J. Bact.* **120,** 124–1132.
11.5.1

Jensen T.E. (1968) Electron microscopy of polyphosphate bodies in a blue-green alga, *Nostoc pruniforme*. *Arch. Mikrobiol.* **62,** 144–152.
5.3.1

Jensen T.E. (1969) Fine structure of developing polyphosphate bodies in a blue-green alga, *Plectonema boryanum*. *Arch. Mikrobiol.* **67,** 328–338.
5.3.1

Jensen T.E. & Bowen C.C. (1970) Cytology of blue-green algae. II. Unusual inclusions in the cytoplasm. *Cytologia* **35,** 132–152.
6.2

Jensen T.E. & Clark R.L. (1969) Cell wall and coat of the developing akinete of a *Cylindrospermum* species. *J. Bact.* **97,** 1494–1495.
15.2

Jensen T.E. & Sicko L.M. (1971a) Fine structure of poly-β-hydroxybutyric acid granules in a blue-green alga *Chlorogloea fritschii*. *J. Bact.* **106,** 683–686.
3.2.2.2

Jensen T.E. & Sicko L.M. (1971b) The effect of lysozyme on cell wall morphology in a blue-green alga, *Cylindrospermum* sp. *J. gen. Microbiol.* **68,** 71–75.
13.2.1, 13.4.1

Jensen T.E. & Sicko L.M. (1973) The fine structure of *Chlorogloea fritschii* cultured in sodium acetate enriched medium. *Cytologia* **38,** 381–391.
5.3.1.

Jensen T.E. & Sicko L.M. (1974) Phosphate metabolism in blue-green algae. I. Fine structure of the 'polyphosphate overplus' phenomenon in *Plectonema boryanum*. *Can. J. Microbiol.* **9,** 1235–1239.
5.2.4, 5.3.1

Jewson D.H. (1975) The relation of incident radiation to diurnal rates of photosynthesis in Lough Neagh. *Int. Rev. ges. Hydrobiol. Hydrogr.* **60,** 759–767.
18.5.5.4

Jewson D.H. (1976) The interaction of components controlling net phytoplankton photosynthesis in a well-mixed lake (Lough Neagh, Northern Ireland). *Freshwat. Biol.* **6,** 551–576.
18.5.5.6

Jewson D.H. (1977) Light penetration in relation to phytoplankton content of the euphotic zone of Lough Neagh, N. Ireland. *Oikos* **28,** 74–83.
18.5.5.5

Johannes R.E. and 22 others (1972) The metabolism of some coral reef communities: a team study of nutrient and energy flux at Eniwetok. *BioScience* **22,** 541–543.
20.4.4

Johnson P.W. & Sieburth J.McN. (1979) Chroococcoid cyanobacteria in the sea: a

ubiquitous and diverse phototrophic biomass. *Limnol. Oceanogr.* **24,** 928–935.
1, 19.2.1, 19.3.1, 19.3.4, 19.5.1

Jolley E.T. (1976) Aspects of Heterotrophic Nutrition and Interaction of an Epiphytic *Navicula pelliculosa* and an Associated *Flavobacterium.* Ph.D. Thesis, Univ. Wales.
3.4.3

Jolley E.T., Jones A.K. & Hellebust J.A. (1976) A description of glucose uptake in *Navicula pelliculosa* (Bréb.) Hilse including a brief comparison with an associated *Flavobacterium* sp. *Arch. Microbiol.* **109,** 127–133.
3.4.3

Jones D.D. & Jost M. (1970) Isolation and chemical characterization of gas vacuole membranes from *Microcystis aeruginosa* Kuetz. emend. Elenkin. *Arch. Mikrobiol.* **70,** 43–64.
10.2.5.1

Jones J.G. (1972) Studies of freshwater bacteria: association with algae and alkaline phosphatase activity. *J. Ecol.* **60,** 59–75.
18.5.2.4

Jones J.H. & Yopp J.H. (1979) Cell wall constituents of *Aphanothece halophytica. J. Phycol.* **15,** 62–66.
13.4.1, 13.4.2, 13.4.4

Jones K. & Stewart W.D.P. (1969) Nitrogen turnover in marine and brackish habitats. III. The production of extracellular nitrogen by *Calothrix scopulorum. J. mar. biol. Ass. U.K.* **49,** 475–488.
18.5.6, 20.4.5

Jones L.W. & Myers J. (1964) Enhancement in the blue-green alga, *Anacystis nidulans. Pl. Physiol., Lancaster* **39,** 938–946.
9.2.4

Jones L.W. & Myers J. (1965) Pigment variations in *Anacystis nidulans* induced by light of selected wavelengths. *J. Phycol.* **1,** 7–14.
2.4.2, 2.4.3.1, 6.5.1, 7.8.2

Jones O.T.G. (1977a) Electron transport and ATP synthesis in the photosynthetic bacteria. In: Haddock B.A. & Hamilton W.A. (Eds) *Microbial Energetics.* 442 pp., pp. 151–183. *Symp. Soc. gen. Microbiol.* **27,** Cambridge Univ. Press, England.
3.5

Jones R.I. (1977b) Factors controlling phytoplankton production and succession in a highly eutrophic lake (Kinego Bay, Lough Neagh). I. The phytoplankton community and its environment. *J. Ecol.* **65,** 547–549.
2.6.1

Jørgensen B.B. (1977a) The sulphur cycle of a coastal marine sediment (Limfjorden, Denmark). *Limnol. Oceanogr.* **22,** 814–832.
9.5.3

Jørgensen B.B. (1977b) Bacterial sulfate reduction within reduced microniches of oxidised marine sediments. *Mar. Biol.* **41,** 7–17.
9.5.3

Jørgensen B.B. & Cohen Y. (1977) Solar Lake (Sinai). V. The sulfur cycle of the benthic cyanobacterial mats. *Limnol. Oceanogr.* **22,** 657–666.
9.5.3

Jørgensen B.B. & Fenchel T. (1974) The sulfur cycle of a marine sediment model system. *Mar. Biol.* **24,** 189–201.
9.5.3

Jørgensen B.B., Kuenen J.G. & Cohen Y. (1979) Microbial transformations of sulfur compounds in a stratified lake (Solar Lake, Sinai). *Limnol. Oceanogr.* **24,** 799–822.
9.3.4, 9.5.1, 9.5.4

Jørgensen B.B., Revsbech N.P., Blackburn T.H. & Cohen Y. (1979) Diurnal cycle of oxygen and sulfide microgradients and microbial photosynthesis in a cyanobacterial mort sediment. *Appl. environ. Microbiol.* **38,** 46–58.
9.5.3

Jørgensen E.G. (1956) Growth inhibiting substances formed by algae. *Physiol. Plant.* **9,** 712–716.
17.4.2

Joset-Espardellier F., Astier C., Evans E.H. & Carr N.G. (1978) Cyanobacteria grown under photoautotrophic, photoheterotrophic and heterotrophic regimes: sugar metabolism and carbon dioxide fixation. *FEMS Microbiol. Lett.* **4,** 261–264.
3.4.1, 3.4.2, 6.5.2

Jost M. (1965) Die Ultrastruktur von *Oscillatoria rubescens* D.C. *Arch. Mikrobiol.* **50,** 211–245.
3.2.2.2, 6.2, 6.3.1, 6.3.2, 13.2.1, 13.2.2, 16.2.2.2

Jurgenson J.E., Beale S.I. & Troxler R.F. (1976) Biosynthesis of δ-aminolevulinic acid in the unicellular rhodophyte *Cyanidium caldarium. Biochem. biophys. Res. Commun.* **69,** 149–157.
3.2.2.1

Jüttner F. & Carr N.G. (1976) The movement of organic molecules from vegetative cells into the heterocysts of *Anabaena cylindrica.* In: Codd G.A. & Stewart W.D.P. (Eds) *Proceedings of the Second International Symposium on Photosynthetic Prokaryotes.* 258 pp., pp. 121–123. Univ. Dundee, Scotland.
14.4.2

Jüttner F. & Matuschek T. (1978) The release of low molecular weight compounds by the

phytoplankton in an eutrophic lake. *Water Res.* **12,** 251–255.
18.5.6

Kaiser D., Manoil C. & Dworkin M. (1979) Myxobacteria: cell interactions, genetics and development. *A. Rev. Microbiol.* **33,** 595–639.
16.5

Kale S. & Talpasayi E.R.S. (1969) On a spontaneous mutant strain of *Anabaena ambigua* Rao. *Phykos* **8,** 42–45.
11.5.5.2(a)

Kallas T. & Dahlquist F.W. (1981) Phosphorus-31 nuclear magnetic resonance analyses of internal pH during photosynthesis in the cyanobacterium *Synechococcus*. *Biochemistry, N.Y.* **20,** 5900–5907.
10.3.2

Kaney A.R. (1973) Enzymatic characterization of a phenylalanine auxotroph of the blue-green bacterium *Synechococcus cedrorum*. *Arch. Mikrobiol.* **92,** 139–142.
11.2.2, 11.5.3

Kaney A.R. & Dolack M.P. (1972) Temperature-sensitive mutants of *Anacystis nidulans*. *Genetics* **71,** 465–467.
11.1

Kao O.H.W., Berns D.S. & Town W.R. (1973) The characterization of C-phycocyanin from an extremely halo-tolerant blue-green alga, *Coccochloris elabens*. *Biochem. J.* **131,** 39–50.
7.5.1.4, 10.4

Kao O.H.W., Edwards M.R. & Berns D.S. (1975) Physical–chemical properties of C-phycocyanin isolated from an acido-thermophilic eukaryote, *Cyanidium caldarium*. *Biochem. J.* **147,** 63–70.
7.5.1.4

Kappers F.I. (1977) Presence of blue-green algae in sediments of Lake Brielle. *Symposium on Interactions Between Sediments and Fresh Water, Amsterdam, September 6–10, 1976*. pp. 382–386.
3.4.3

Katoh T. & Gantt E. (1979) Photosynthetic vesicles with bound phycobilisomes from *Anabaena variabilis*. *Biochim. biophys. Acta* **546,** 383–393.
8.2

Katoh T. & Ohki K. (1975) Loss of photosystem II induced by nitrate deficiency in photoorganotrophically grown *Anabaena variabilis*. *Pl. Cell Physiol., Tokyo* **16,** 815–828.
8.6

Katoh T. & Ohki K. (1976) Regeneration of photosystem II and phycobilin pigments in photoorganotrophically grown *Anabaena variabilis*. *Pl. Cell Physiol., Tokyo* **17,** 525–536.
8.6

Katz A. (1975) Untersuchungen zur Struktur und den biologischen Eigenschaften des Lipopolysaccharids von *Anacystis nidulans*. Thesis, Freiburg i. Br., Germany.
13.4.1, 13.4.2

Katz A., Weckesser J., Drews G. & Mayer H. (1977) Chemical and biological studies on the lipopolysaccharide (O-antigen) of *Anacystis nidulans*. *Arch. Microbiol.* **113,** 247–256.
13.4.4

Kaushik M. & Kumar H.D. (1970) The effect of light on growth and development of two nitrogen-fixing blue-green algae. *Arch. Mikrobiol.* **74,** 52–57.
15.5, 18.4.5

Kawaguti S. (1968) Electron microscopy on blue-green algae in the body-wall of an echiuroid, *Idedosoma gogoshimense*. *Biol. J. Okayama Univ.* **14,** 67–74.
20.2.4

Kawaguti S. (1971) Blue-green algae in echiuroid worms. In: Cheng T.C. (Ed.) *Aspects of the Biology of Symbiosis*, pp. 265–273. Univ. Park Press, Baltimore, Butterworths, London.
20.2.4

Kawaguti S., Yamamoto M. & Kamishima (1965) Electron microscopy on the symbiosis between blue-green algae and an opisthobranch, *Placobranchus*. *Proc. Japan Acad.* **41,** 614–617.
20.2.4

Kawamura M. & Fujita Y. (1979) A probable lack of function of c type cytochrome in the photosynthetic system of the blue-green alga *Anabaena variabilis*. *Pl. Cell Physiol., Tokyo* **20,** 331–340.
8.5

Kawamura M., Mimuro M. & Fujita Y. (1979) Quantitative relationship between the 2 reaction centers in the photosynthetic system of blue-green algae. *Pl. Cell Physiol., Tokyo* **20,** 697–705.
8.4, 8.6

Kaye A.M., Salomon R. & Fridlender B. (1967) Base composition and presence of methylated bases in DNA from a blue-green alga *Plectonema boryanum*. *J. mol. Biol.* **24,** 479–483.
11.3.1

Keating K.I. (1978) Blue-green algal inhibition

of diatom growth: transition from mesotrophic to eutrophic community structure. *Science, N.Y.* **199,** 971–973.
18.5.6

Keenan J.D. & Auer M.T. (1974) The influence of phosphorus luxury uptake on algal bioassays. *J. Wat. Pollut. Control Fed.* **46,** 532–542.
5.2.3, 5.3.2

Keleti G., Sykora J.L., Lippy E.C. & Shapiro M.A. (1979) Composition and biological properties of lipopolysaccharides isolated from *Schizothrix calcicola* (Ag.) Gomont (cyanobacteria). *Appl. environ. Microbiol.* **38,** 471–477.
13.4.4

Kendall C.G.StC. & Skipwith P.A.d'E. (1968) Recent algal mats of a Persian Gulf lagoon. *J. sed. Petrol.* **38,** 1040–1058.
20.3.1, 20.4.1

Kennedy C. & Postgate J.R. (1977) Expression of *Klebsiella pneumoniae* nitrogen fixation genes in nitrate reductase mutants of *E. coli. J. gen. Microbiol.* **98,** 551–557.
4.5

Kenyon C.N. (1972) Fatty acid composition of unicellular strains of blue-green algae. *J. Bact.* **109,** 827–834.
6.4.1, 6.5.3, 9.4.2, 12.3.3, 13.4.3

Kenyon C.N. (1978) Complex lipids and fatty acids of photosynthetic bacteria. In: Clayton R.K. & Sistrom W.R. (Eds) *The Photosynthetic Bacteria.* 946 pp., pp. 281–313. Plenum Press, New York.
6.4.1, 12.2.1, 12.3.3

Kenyon C.N., Rippka R. & Stanier R.Y. (1972) Fatty acid composition and physiological properties of some filamentous blue-green algae. *Arch. Mikrobiol.* **83,** 216–236.
9.4.2, 12.3.3, 12.3.4.3

Kenyon C.N. & Stanier R.Y. (1970) Possible evolutionary significance of polyunsaturated fatty acid in blue-green algae. *Nature, Lond.* **227,** 1164–1166.
6.4.1, 12.3.3

Kessel M. (1978) A unique crystalline wall layer in the cyanobacterium *Microcystis marginata. J. Ultrastruct. Res.* **62,** 203–212.
13.2.2

Kessel M. & Eloff J.N. (1975) The ultrastructure and development of the colonial sheath of *Microcystis marginata. Arch. Microbiol.* **106,** 209–214.
13.2.2, 17.4.1, 17.4.3

Kessel M., MacColl R., Berns D.S. & Edwards M.R. (1973) Electron microscope and physical chemical characterization of C-phycocyanin from fresh extracts of two blue-green algae. *Can. J. Microbiol.* **19,** 831–836.
7.5.1.4, 7.5.1.5, 7.5.1.6, 7.7

Kessler E. (1974) Hydrogenase, photoreduction and anaerobic growth. In: Stewart W.D.P. (Ed.) *Algal Physiology and Biochemistry.* 989 pp., pp. 456–473. Blackwell, Oxford.
9.4.1

Khan K.R. (1969) Ecology of some littoral blue-green algae of Oahu. *Rev. Algol, N.S.* **3,** 217–229.
20.3.1

Khoja T.M. & Whitton B.A. (1971) Heterotrophic growth of blue-green algae. *Arch. Mikrobiol.* **79,** 280–282.
3.4.1, 18.5.4

Kidston R. & Lang W.H. (1921) On Old Red Sandstone plants showing structure, from the Rhynie chert bed, Aberdeenshire. Part V. The Thallophyta, etc. *Trans. R. Soc. Edinb.* **52,** 855–902.
21.1

Kies L. (1980) Morphology and systematic position of some endocyanomes. In: Schwemmler W. & Schenk H.E.A. (Eds) *Endosymbiosis and Cell Biology Vol. 1.* 1060 pp., pp. 7–19. Walter de Gruyter, Berlin, N.Y.
4.1

Kimor B., Reid F.M.H. & Jordan J.B. (1978) An unusual occurrence of *Hemiaulus membranaceus* Cleve (Bacillariophyceae) with *Richelia intracellularis* Schmidt (Cyanophyceae) off the coast of Southern California in October 1976. *Phycologia* **17,** 162–166.
4.1, 19.2.4.2

King D.L. (1970) The role of carbon in eutrophication. *J. Wat. Pollut. Control Fed.* **42,** 2035–2051.
18.5.2.2

Kinne O. (1970) Temperature. In: Kinne O. (Ed.) *Marine Ecology, A Comprehensive, Integrated Treatise on Life in Oceans and Coastal Waters, Vol. 1.* 681 pp., pp. 321–346. Wiley, London.
20.3.3

Kinsman D.J.J. & Park R.K. (1976) Algal belt and coastal sabkha evolution, Trucial Coast, Persian Gulf. In: Walter M.R. (Ed.) *Stromatolites. Developments in Sedimentology* **20.** 790 pp., pp. 421–434. Elsevier, Amsterdam.
9.5.3

Kipe-Nolt J.A. & Stevens S.E. (1980) Biosynthesis of δ-aminolevulinic acid from glutamate in *Agmnellum quadruplicatum. Pl. Physiol., Lancaster* **65,** 126–128.
3.2.2.1

Kiyohara T., Fujita Y., Hattori A. & Watanabe A. (1960) Heterotrophic culture of a blue-green alga, *Tolypothrix tenuis* I. *J. gen. appl. Microbiol.* **6,** 176–182.
2.6.3

Klein G. & Rüdiger W. (1978) Stereochemie von Modell-Imiden. *Liebigs Ann. Chem.* **1978,** 267–279.
7.5.1.2

Klein G. & Rüdiger W. (1979) Thioether formation of phycocyanobilin: a model reaction of phycocyanin biosynthesis. *Z. Naturforsch.* **34c,** 192–195.
7.5.1.2

Klein S.M. & Vernon L.P. (1977) Composition of a photosystem I chlorophyll protein complex from *Anabaena flos-aquae. Biochim. biophys. Acta* **459,** 364–375.
8.4

Kleinhofs A.M. & Smith J.A. (1976) Effect of excision repair on azide induced mutagenesis. *Mutation Res.* **41,** 233–240.
11.5.5.2(b)

Kleining H. & Reichenbach H. (1977) Carotenoid glucosides and menaquinones from the gliding bacterium *Herpetosiphon giganteus* Hpa2. *Arch. Microbiol.* **112,** 307–310.
16.2.3

Klemer A.R. (1976) The vertical distribution of *Oscillatoria agardhii* var. *isothrix. Arch. Hydrobiol.* **78,** 343–362.
18.4.1, 18.5.5.2

Knaff D.B. (1977) The role of cytochrome b_6 and cytochrome f in cyclic electron flow in a blue-green alga. *Arch. Biochem. Biophys.* **182,** 540–545.
8.5

Knobloch K. (1969) Sulfide oxidation via photosynthesis in green algae. In: Metzner H. (Ed.) Progress in photosynthesis research. Vol. II, pp. 1127., pp. 1032–1034. International Union of Biological Sciences, Tübingen.
9.3.4

Knoll A.H. (1977) Paleomicrobiology. In: Laskins A.I. & Lechevalier H.A. (Eds) *CRC Handbook of Microbiology, 2nd edition, vol. I.* 757 pp., pp. 8–29. CRC Press, Cleveland.
12.1.2, 12.1.3, 12.3.2

Knoll A.H. (1979) Archaean photoautotrophy: some alternatives and limits. *Origins of Life* **9,** 313–327.
12.3.2

Knoll A.H. & Barghoorn E.S. (1975) Precambrian eukaryotic organisms: A reassessment of the evidence. *Science, N.Y.* **190,** 52–54.
12.1.3, 21.3

Knoll A.H. & Barghoorn E.S. (1977) Archean microfossils showing cell division from the Swaziland System of South Africa. *Science, N.Y.* **198,** 396–398.
21.2

Knusel F. & Nuesch J. (1965) Mechanism of action of sideromycins. *Nature, Lond.* **206,** 674–676.
18.5.2.3

Kobayashi T., Degenkolb E.O., Bersohn R., Rentzepis P.M., MacColl R. & Berns D.S. (1979) Energy transfer among the chromophores in phycocyanins measured by picosecond kinetics. *Biochemistry, N.Y.* **18,** 5073–5078.
7.5.1.5

Kobayashi Y., Siegelman H.W. & Hirs C.H.W. (1972) C-phycocyanin from *Phormidium luridum*: isolation of subunits. *Arch. Biochem. Biophys.* **152,** 187–198.
7.5.1.3

Kogan Sh.L., Sadykov Kh.S. & Dubaeva Z.V. (1972) Pervie svedeniya ob algoflore vodoemor Prikarabolgazya i yuzhnogo Ustyurta (severo-zapadnaya Turkmenia). *Bot. Zh.* **57,** 579–585. (Russian: first data on the algal flora of water bodies in Kara-Bogaz-Gol vicinities and South Ustyurt (northwest Turkmeniya).)
18.4.1

Koike H. & Katoh S. (1979) Heat-stabilities of cytochromes and ferredoxin isolated from a thermophilic blue-green alga. *Pl. Cell Physiol., Tokyo* **20,** 1157–1161.
8.3, 8.5

Kol E. (1929) Wasserblüte der Sodateiche auf der Nagy Magyar Alföld. *Arch. Protistenk.* **66,** 515–522.
18.4.1

Koller K.P. & Wehrmeyer W. (1974) Isolierung und Characterisierung der Biliproteide von *Rhodella violacea* (Bangiophycidae). *Arch. Microbiol.* **100,** 253–270.
7.5.1.1, 7.5.1.3

Koller K.P. & Wehrmeyer W. (1975) B-phycoerythrin from *Rhodella violacea. Arch. Microbiol.* **104,** 255–261.
7.5.1.1, 7.5.1.4

Koller K.P., Wehrmeyer W. & Mörschel E. (1978) Biliprotein assembly in the disc-shaped phycobilisomes of *Rhodella violacea*. On the molecular composition of energy-transferring complexes (Tripartite units) forming the periphery of the phycobilisomes. *Eur. J. Biochem.* **91,** 57–63.
7.5.1.1, 7.5.2, 7.5.3, 7.6, 7.7

Koller K.P., Wehrmeyer W. & Schneider H. (1977) Isolation and characterization of disc-shaped phycobilisomes from the red

alga *Rhodella violacea*. *Arch. Microbiol.* **112**, 61–67.
7.5.1.1, 7.6

Komar P.D. (1976) *Beach Processes and Sedimentation*. 429 pp. Prentice-Hall, Englewood Cliffs, New Jersey.
20.2.1

Komárek J. (1975) Cell structure and endospore formation in the blue-green alga *Chroococcidiopsis*. *Arch. Hydrobiol; Suppl.* **46** (*Algological Studies* **12**), 205–223.
6.2

Komor E. & Tanner W. (1971) Characterization of the active hexose transport system of *Chlorella vulgaris*. *Biochim. biophys. Acta* **241**, 170–179.
3.4.3

Konopka A., Brock T.D. & Walsby A.E. (1978) Buoyancy regulation by planktonic blue-green algae in Lake Mendota, Wisconsin. *Arch. Hydrobiol.* **83**, 524–537.
2.8.2.3, 2.8.2.4, 10.2.5.2

Koshland D.E., Jr. (1977) Sensory response in bacteria. In: Agranoff B.W. & Aprison M.H. (Eds) *Advances in Neurochemistry* 2, pp. 277–341. Plenum Press, New York.
16.5

Kost H.-P., Rüdiger W. & Chapman D.J. (1975) Uber die Bindungen zwischen Chromophor und Protein in Biliprotein, I. Abbauversuche und Spektraluntersuchungen an Biliproteiden. *Liebigs Ann. Chem.* **1975**, 1582–1593.
7.5.1.2

Kost-Reyes E., Kost H.-P. & Rüdiger W. (1975) Uber die Bindungen zwischen Chromophor und Protein in Biliproteiden. II. Nachweis von Cystein als bindende Aminosaure in B-phycoerythrin. *Liebigs Ann. Chem.* **1975**, 1594–1600.
7.5.1.2

Kott P. (1980) Algal-bearing didemnid ascidians in the Indo-West-Pacific. *Mem. Queensland Mus.* **20**, 1–48.
1, 20.2.4

Kotyk A. & Höfer M. (1965) Uphill transport of sugars in the yeast *Rhodotorula gracilis*. *Biochim. biophys. Acta* **102**, 410–422.
3.4.3

Kratz W.A. & Myers J. (1955a) Nutrition and growth of several blue-green algae. *Am. J. Bot.* **42**, 282–287.
1, 11.2.1

Kratz W.A. & Myers J. (1955b) Photosynthesis and respiration of three blue-green algae. *Pl. Physiol., Lancaster*, **30**, 275–80.
1, 3.3, 11.2.1, 18.5.3

Krinsky N.I. (1979) Carotenoid pigments: multiple mechanisms for coping with the stress of photosensitized oxidations. In: Shilo M. (Ed.), *Strategies of Microbial Life in Extreme Environments*. 514 pp., pp. 163–177. Berlin, Dahlem Konferenzen.
2.7.2.2

Krogmann D.W. (1973) Photosynthetic reactions and components of thylakoids. In: Carr N.G. & Whitton B.A. (Eds) *The Biology of Blue-Green Algae*. 676 pp., pp. 80–98. Blackwell, Oxford.
8.1

Krogmann D.W. (1977) Blue-green algae. In: Trebst A. & Avron M. (Eds) *Photosynthesis I, Encyclopedia of Plant Physiology, New Series, Vol. 5*. 730 pp., pp. 625–636. Springer, Berlin.
8.1

Krul J.M., Hirsch P. & Staley J.T. (1970) *Toxothrix trichogenes* (Chol.) Beger et Bringmansr: the organism and its biology. *Antonie van Leeuwenhoek* **36**, 409–420.
16.4.4

Krumbein W.E. (1979) Photolithotrophic and chemoorganotrophic activity of bacteria and algae as related to beachrock formation and degradation (Gulf of Aqaba, Sinai). *Geomicrobiol. J.* **1**, 139–203.
9.5.3, 20.2.1

Krumbein W.E., Buchholz H., Franke P., Giani D., Giele C. & Wonneberger K. (1979) O_2 and H_2S coexistence in stromatolites. A model for the origin of mineralogical lamination in stromatolites and banded iron formations. *Naturwissenschaften* **66**, 381–389.
2.7.3.2, 20.2.2, 20.4.1

Krumbein W.E. & Cohen Y. (1974) Biogene, Klastishe und evaporitische Sedimentation im einem mesothermen monomiktischen ufernaharn See (Golf von Aqaba). *Geol. Rundschau* **63**, 1035–1065.
9.5.3

Krumbein W.E., Cohen Y. & Shilo M. (1977) Solar Lake (Sinai). 4. Stromatolitic cyanobacterial mat. *Limnol. Oceanogr.* **22**, 635–656.
9.5.3

Krumbein W.E. & Potts M. (1978) Light penetration, salinity, and other growth regulating factors of four stromatolitic environments along the shores of the Gulf of Aqaba (Sinai). In: Friedman G.M. (Ed.) *Abstracts, 10th International Congress on Sedimentology. Vol. 1*. 401 pp., p. 363. Jerusalem, Israel.
2.7.3.2, 20.3.4

Kuenen G.J. (1975) Colorless sulfur bacteria

and their role in sulfur cycle. *Plant & Soil* **43,** 221–248.
9.5.1

Kuentzel L.E. (1969) Bacteria, carbon dioxide, and algal blooms. *J. Wat. Pollut. Control Fed.* **41,** 1737–1747.
17.1.1, 17.4.1, 17.6

Kuenzler E.J. (1965) Glucose-6-phosphate utilization by marine algae. *J. Phycol.* **1,** 156–164.
5.1.3, 5.2.3

Kuenzler E.J. (1970) Dissolved organic phosphorus excretion by marine phytoplankton. *J. Phycol.* **6,** 7–13.
5.1.2

Kuenzler E.J. & Perras J.P. (1965) Phosphatases of marine algae. *Biol. Bull. Mar. biol. Lab., Woods Hole* **128,** 271–284.
5.1.3, 5.2.3

Kufer W. & Scheer H. (1979) Studies on plant bile pigments, VII. Preparation and characterization of phycobiliproteins with chromophores chemically modified by reduction. *Hoppe Seyler's Z. Physiol. Chem.* **360,** 935–956.
7.5.1.2, 7.5.1.5

Kuhlemeier C.J., Borrias W.E., van den Hondel C.A.M.J.J. & van Arkel G.A. (1981) Vectors for cloning in cyanobacteria: construction and characterization of two recombinant plasmids capable of transformation to *Escherichia coli* K12 and *Anacystis nidulans* R2. *Molec. gen. Genet.* **184,** 249–254.
11.8

Kulasooriya S.A. & Fay P. (1974) On the reversibility of heterocyst differentiation. *Br. phycol. J.* **9,** 97–100.
14.5

Kulasooriya S.A., Lang N.J. & Fay P. (1972) The heterocysts of blue-green algae. III. Differentiation and nitrogenase activity. *Proc. R. Soc. Lond. B* **181,** 199–209.
14.5, 14.6.2

Kullberg R.G. (1977) The effects of some ecological factors on cell size of the hot spring alga *Synechococcus lividus* (Cyanophyta). *J. Phycol.* **13,** 111–115.
6.5.3

Kumar H.D. (1963) Division of heterocyst in *Camptylonema lahorense* Ghose. *Rev. Algol., N.S.* **6,** 330–331.
14.3.2

Kumar A. & Kumar H.D. (1980) Differential effects of amino acid analogs on growth and heterocyst differentiation in two nitrogen-fixing blue-green algae. *Curr. Microbiol.* **3,** 213–218.
14.6.1.1

Kumar H.D., Singh H.N. & Prakash G. (1967) The effect of proflavine on different strains of the blue-green alga *Anacystis nidulans. Pl. Cell Physiol., Tokyo* **8,** 171–179.
11.3.2

Kunisawa R. & Cohen-Bazire G. (1970) Mutations of *Anacystis nidulans* that affect cell division. *Arch. Mikrobiol.* **71,** 49–59.
11.5.5.1, 11.7

Kunisawa R. & Stanier R.Y. (1958) Studies on the role of carotenoid pigments in a chemoheterotrophic bacterium, *Corynebacterium poinsettiae. Arch. Mikrobiol.* **31,** 146–156.
2.7.2.2

Kurz W.G.W. & LaRue T.A. (1971) Nitrogenase in *Anabaena flos-aquae* filaments lacking heterocysts. *Naturwiss.* **58,** 417.
14.2

Kusnezov S.I. (1970) Oppenheimer C.H. (Ed.) *The Microflora of Lakes and its Geochemical Activity*. pp. 424. Univ. Texas Press, Austin.
9.5.1, 9.5.3, 9.5.4

Kützing F.T. (1843) *Phycologia Generalis, oder Antonomie, Physiologie und Systemkunde der Tange*. F.A. Brockhaus, Leipzig.
7.5.1.1

Kylin H. (1910) Über Phykoerythrin, und Phykocyan bei *Ceramium rubrum* (Huds.) Ag. *Hoppe-Seyler's Z. Physiol. Chem.* **69,** 169–239.
7.5.1.1

Laane C., Krone W., Konings W.N., Haaker H. & Veeger C. (1979) The involvement of membrane potential in nitrogen fixation by bacteroids of *Rhizobium leguminosarum. FEBS Lett.* **103,** 328–332.
4.3, 4.5

Laane C., Krone W., Konings W., Haaker H. & Veeger C. (1980) Short-term effect of ammonium chloride on nitrogen fixation by *Azotobacter vinelandii* and by bacteroids of *Rhizobium leguminosarum. Eur. J. Biochem.* **103,** 39–46.
4.5

Lachance M.A. (1979) DNA relatedness among heterocystous cyanobacteria. In: Nichols J.M. (Ed.) *Abstracts of the Third International Symposium on Photosynthetic Prokaryotes, Oxford*, p. E20. Univ. Liverpool, England.
12.3.4.2

Ladha J.K. & Kumar H.D. (1975) Heterocyst division in two blue-green algae. *Arch. Microbiol.* **102,** 171–173.
14.3.2

Ladha J.K. & Kumar H.D. (1976) Isolation

and initial characterization of non nitrogen fixing mutants of *Nostoc linckia*. *Proc. Dunn Dobzh. Symp. Genet.* 361–368. Univ. Mysore, India.
11.5.5.2(b), 14.6.1.1

Ladha J.K. & Kumar H.D. (1977a) Response of wild-type and a non-nitrogen-fixing mutant of *Nostoc linckia* to different carbon sources. *New Phytol.* **79,** 299–308.
11.5.5.2(b), 14.3.2

Ladha J.K. & Kumar H.D. (1977b) Some characteristics of two morphological mutants of *Nostoc linckia* induced by nitrosoguanidine. *Z. allg. Mikrobiol.* **17,** 513–519.
11.5.5.2(b), 14.6.1.1, 14.6.1.2

Ladha J.K. & Kumar H.D. (1978a) Genetics of blue-green algae. *Biol. Rev.* **53,** 355–386.
11.1, 11.4.2.3, 11.5.1, 11.5.2, 11.6.1

Ladha J.K. & Kumar H.D. (1978b) Effects of some amino acid analogues on growth and heterocyst formation in the blue-green alga *Nostoc linckia*. *Proc. Indian Acad. Sci. B* **87** (Exp. Biol.-2), 127–133.
14.6.1.3

Ladha J.K., Rowell P. & Stewart W.D.P. (1978) Effects of 5-hydroxylysine on acetylene reduction and NH_4^+-assimilation in the cyanobacterium *Anabaena cylindrica*. *Biochem. biophys. Res. Commun.* **83,** 688–696.
14.6.1.3

Lagarias J.C., Glazer A.N. & Rapoport H. (1979) Chromopeptides from C-phycocyanin. Structure and linkage of a phycocyanobilin bound to the β subunit. *J. Am. chem. Soc.* **101,** 5030–5037.
7.5.1.2

Lam C.W.Y. (1979) Phosphate requirement of *Anabaena oscillarioides* and its ecological implications. *Hydrobiologia* **67,** 89–96.
5.2.1

Lambein F. & Wolk C.P. (1973) Structural studies on the glycolipids from the envelope of the heterocyst of *Anabaena cylindrica*. *Biochemistry, N.Y.* **12,** 791–798.
4.4, 14.3.1, 14.4.2

Lambert G.R. & Smith G.D. (1981) Hydrogen metabolism by cyanobacteria. *Biol. Rev.* **56,** 589–660.
4.6

Lambert J.A.M., Williams E., O'Brien P.A. & Houghton J.A. (1980) Mutation induction in the cyanobacterium *Gloeocapsa alpicola*. *J. gen. Microbiol.* **121,** 213–219.
11.4.1

Lamont H.C. (1969a) Sacrificial cell death and trichome breakage in an Oscillatoriacean blue-green alga—the role of murein. *Arch. Mikrobiol.* **69,** 237–259.
13.2.1, 13.3

Lamont H.C. (1969b) Shear-oriented microfibrils in the mucilaginous investments of two motile Oscillatoriacean blue-green algae. *J. Bact.* **97,** 350–361.
13.2.1, 13.2.2, 16.2.2.1

Lanaras L. & Codd G.A. (1981a) Ribulose-1,5-bisphosphate carboxylase and polyhedral bodies of *Chlorogloeopsis fritschii*. *Planta, Berl.* **153,** 279–285.
3.2.1

Lanaras L. & Codd G.A. (1981b) Structural and immunoelectrophoretic comparison of soluble and particulate ribulose-1,5-bisphosphate carboxylase from the cyanobacterium *Chlorogloeopsis fritschii*. *Arch. Microbiol.* **130,** 213–217.
3.2.1

Lang N.J. (1965) Electron microscopic study of heterocyst development in *Anabaena azollae* Strasburger. *J. Phycol.* **1,** 127–134.
6.5.2, 14.5

Lang N.J. (1968) The fine structure of blue-green algae. *A. Rev. Microbiol.* **22,** 15–46.
13.4

Lang N.J. (1977) *Starria zimbabweënsis* (Cyanophyceae) gen. nov. et sp. nov.: a filament triradiate in transverse sections. *J. Phycol.* **13,** 288–296.
1, 6.2, 13.2.1

Lang N.J. & Fay P. (1971) The heterocysts of blue-green algae. II. Details of ultrastructure. *Proc. R. Soc. Lond. B* **178,** 193–203.
6.5.2, 14.3.1, 14.3.2, 14.5, 17.2

Lang N.J. & Rae P.M.M. (1967) Structures in a blue-green alga resembling prolamellar bodies. *Protoplasma, Wien* **64,** 67–74.
6.1, 6.2, 6.5.2

Lang N.J., Simon R.D. & Wolk C.P. (1972) Correspondence of cyanophycin granules with structured granules in *Anabaena cylindrica*. *Arch. Mikrobiol.* **83,** 313–320.
3.2.2.2, 14.3.2

Lange W. (1967) Effects of carbohydrates on the symbiotic growth of the planktonic blue-green algae with bacteria. *Nature, Lond.* **215,** 1277–1278.
17.1.1, 17.6

Lange W. (1970) Blue-green algae and humic substances. *Proc. Conf. Great Lakes Res.* **13,** 58–70.
18.5.1.2

Lange W. (1971) Enhancement of algal growth in cyanophyta–bacteria systems by carbonaceous compounds. *Can. J. Microbiol.* **17,** 303–314.
17.1.1

Lännergren C. (1980) Nitrogen fixation by blue-green algae in a Norwegian land-locked fjord, determined by the acetylene reduction technique. *Sarsia* **65,** 73–77.
20.4.5

Larsen S.H., Adler J., Gargus J.J. & Hogg R.W. (1974) Chemomechanical coupling without ATP: the source of energy for motility and chemotaxis in bacteria. *Proc. natn. Acad. Sci. U.S.A.* **71,** 1239–1243.
16.3

Larue B., Cedergren R.J. & Sankoff D. (1977) Nucleic acid sequence evidence for the evolutionary origin of major phyla including plants and algae. In: Bogorad L. & Weil J.H. (Eds) *Acides nucléiques et Synthèse des protéines chez les végétaux. Colloques Internationaux du C.N.R.S.* 714 pp., pp. 219–225. Editions du C.N.R.S., Paris.
12.4.3.4

Lascelles J. (1959) Adaptation to form bacteriochlorophyll in *Rhodopseudomonas spheroides*: change in activity of enzymes concerned in pyrrole synthesis. *Biochem J.* **72,** 508–518.
9.4.1

Lau R.H. & Doolittle W.F. (1979) Covalently closed circular DNAs in closely related unicellular cyanobacteria. *J. Bact.* **137,** 648–652.
11.3.2

Lau R.H. & Doolittle W.F. (1980) *Aqu* I: a more easily purified isoschizomer of *Ava* I. *FEBS Lett.* **121,** 200–202.
11.3.3

Lau R.H., MacKenzie M.M. & Doolittle W.F. (1977) Phycocyanin synthesis and degradation in the blue-green bacterium *Anacystis nidulans. J. Bact.* **132,** 771–778.
7.8, 7.8.1

Lau R.H., Sapienza C. & Doolittle W.F. (1980a) Cyanobacterial plasmids: their widespread occurrence, and the existence of regions of homology between plasmids in the same and different species. *Molec. gen. Genet.* **178,** 203–211.
11.3.2

Lau R.H., Sapienza C. & Doolittle W.F. (1980b) Plasmids in cyanobacteria. In: Stuttard C. & Rozee K.R. (Eds) *Plasmids: Environmental Effects and Maintenance Mechanisms.* pp. 263–273. Academic Press, New York.
11.3.2

Lawrie A.C., Codd G.A. & Stewart W.D.P. (1976) The incorporation of nitrogen into products of recent photosynthesis in *Anabaena cylindrica* Lemm. *Arch. Microbiol.* **107,** 15–24.
14.4.1

Lawry N.H. (1976) Phosphorus utilization in coccoid blue-green algae from large polyphosphate bodies induced by sulfur starvation. *J. Phycol., Suppl.* **12,** 22.
5.3.1

Lawry N.H. & Jensen T.E. (1975) Variation in fine structure of selected coccoid blue-green algae as a reflection of exogenous sulfur source. *J. Phycol., Suppl.* **11,** 8.
5.3.1

Lawry N.H. & Jensen T.E. (1979) Deposition of condensed phosphate as an effect of varying sulfur deficiency in the cyanobacterium *Synechococcus* sp. (*Anacystis nidulans*). *Arch. Microbiol.* **120,** 1–7.
3.2.2.2, 5.1.1, 5.3.1

Lazaroff N. (1973) Photomorphogenesis and nostocacean development. In: Carr N.G. & Whitton B.A. (Eds) *The Biology of Blue-Green Algae.* 676 pp., pp. 279–319. Blackwell, Oxford.
2.1

Leach C.K. & Carr N.G. (1970) Electron transport and oxidative phosphorylation in the blue-green alga, *Anabaena variabilis. J. gen. Microbiol.* **64,** 55–70.
3.3

Leach C.K. & Carr N.G. (1971) Pyruvate: ferredoxin oxidoreductase and its activation by ATP in the blue-green alga *Anabaena variabilis. Biochim. biophys. Acta* **245,** 165–174.
4.3

Leach C.K. & Herdman M. (1973) Structure and function of nucleic acids. In: Carr N.G. & Whitton B.A. (Eds) *The Biology of Blue-Green Algae.* 676 pp., pp. 186–200. Blackwell, Oxford.
11.1, 11.3.1

Leach C.K., Old J.M. & Carr N.G. (1971) Aspects of macromolecular synthesis in the blue-green alga *Anabaena variabilis. J. gen. Microbiol.* **68,** XIV.
15.2

Leadbetter E.R. (1974) Beggiatoaceae Migula 1894. In: Buchanan R.E. & Gibbons N.E. (Eds) *Bergey's Manual of Determinative Bacteriology, 8th edn.*, pp. 112–114. Williams & Wilkins, Baltimore.
16.4.3

Leak L.V. (1965) Electron microscopic autoradiography. Incorporation of ^{3}H-thymidine in a blue-green alga, *Anabaena* sp. *J. Ultrastruct. Res.* **12,** 135–146.
11.3.1, 14.3.2

Leak L.V. (1967) Studies on the preservation and organization of DNA-containing regions in a blue-green alga: a cytochemical and ultrastructural study. *J. Ultrastruct. Res.* **20,** 190–205.
11.3.1

Leak L.V. (1968) Fine structure of the mucilaginous sheath of *Anabaena*. *J. Ultrastruct. Res.* **21,** 61–74.
13.2.2

Leak L.V. & Wilson G.B. (1965) Electron microscopic observations on a blue-green alga, *Anabaena* sp. *Can. J. Genet. Cytol.* **7,** 237–249.
14.3.2, 15.2

Lean D.R.S. (1973) Movements of phosphorus between its biologically important forms in lake water. *J. Fish. Res. Bd Can.* **30,** 1525–1536.
5.1.2, 5.1.3

Lean D.R.S., Charlton M.N., Burnison B.K., Murphy T.P., Millard S.E. & Young K.R. (1975) Phosphorus: changes in ecosystem metabolism from reduced loading. *Verh. int. Verein. theor. angew. Limnol.* **19,** 249–257.
5.4.2

Lean D.R.S., Liao C.F.H., Murphy T.P. & Painter D.S. (1978) The importance of nitrogen fixation in lakes. In: Granhall U. (Ed.) *Environmental Role of Nitrogen-fixing Blue-green Algae and Asymbiotic Bacteria. Ecol. Bull., Stockholm* **26,** 41–51.
18.5.2.3

Lean D.R.S. & Nalewajko C. (1976) Phosphate exchange and organic phosphorus excretion by freshwater algae. *J. Fish. Res. Bd Can.* **33,** 1312–1323.
5.1.2

le Campion-Alsumard T. (1969) Contribution à l'etude des Cyanophycées lithophytes des étages, supralittoral et mediolittoral (Région de Marseille). *Tethys* **1,** 119–172.
20.2.1

le Campion-Alsumard T. (1970) Cyanophycées marines endolithes colonisant les surfaces rocheuses dénudées (Étages supralittoral et médiolittoral de la région de Marseille). *Schweiz. Z. Hydrol.* **32,** 552–558.
20.2.1, 20.3.2

le Campion-Alsumard T. (1975) Étude expérimentale de la colonisation d'éclats de calcite par les cyanophycées endolithes marines. *Cah. Biol. mar.* **16,** 177–185.
20.4.3

le Campion-Alsumard T. (1979) Les Cyanophycées endolithes marines. Systématique, ultrastructure, écologie et biodestruction. *Oceanologica Acta* **2,** 143–156.
20.1, 20.4.3.

Lee J.J. & Berns D.S. (1968) Protein aggregation. Studies of larger aggregates of C-phycocyanin. *Biochem. J.* **110,** 457–464.
7.5.1.4

Lefort M. (1960) Nouvelles recherches sur l'infrastructure du chromatoplasm des Cyanophycées. *C.r. hebd. Séanc. Acad. Sci., Paris* **251,** 3046–3048.
6.2

Lefort M. (1965) Sur le chromatoplasma d'une cyanophycée endosymbiotique: *Glaucocystis nostochinearum* Itzigs. *C.r. hebd. Séanc. Acad. Sci., Paris* **261,** 233–236.
7.2

Lefort-Tran M., Cohen-Bazire G. & Pouphile M. (1973) Les membranes photosynthétiques des algues à biliproteins observées apres cryodécapage. *J. Ultrastruct. Res.* **44,** 199–209.
6.3.2

Lehman J.T., Botkin D.B. & Likens G.E. (1975) The assumptions and rationales of a computer model of phytoplankton population dynamics. *Limnol. Oceanogr.* **20,** 343–364.
5.2.1

Lehmann H. & Jost M. (1971) Kinetics of the assembly of gas vacuoles in the blue-green alga *Microcystis aeruginosa* Kuetz. emend. Elenkin. *Arch. Mikrobiol.* **79,** 59–68.
10.2.5.1

Lehmann M. & Wöber G. (1976) Accumulation, mobilization and turnover of glycogen in the blue-green bacterium *Anacystis nidulans*. *Arch. Microbiol.* **111,** 93–97.
2.6.3, 3.2.2.2, 3.3

Lemasson C., Tandeau de Marsac N. & Cohen-Bazire G. (1973) Role of allophycocyanin as a light-harvesting pigment in cyanobacteria. *Proc. natn. Acad. Sci. U.S.A.* **70,** 3130–3133.
7.2, 7.8.1, 9.2.4

Lemberg R. (1928) Die Chromoproteide der Rotalgen. I. *Justus Liebigs Annln Chem.* **461,** 46–89.
7.5.1.2

Lemberg R. (1930) Die Chromoproteide der Rotalgen. II. Spaltung mit Pepsin und Sauren. Isolierung eines Pyrrolfarbstoffs. *Justus Liebigs Annln Chem.* **477,** 195–245.
7.5.1.2

Lemberg R. & Bader G. (1933) Die Phycobiline der Rotalgen. Überfuhrung in Mesobilirubin und Dehydromesobilirubin. *Justus Liebigs Annln Chem.* **505,** 151–177.
7.5.1.2

Lenci F. & Colombetti G. (1978) Photobeha-

vior of microorganisms: a biophysical approach. *A. Rev. Biophys. Bioeng.* **7,** 341–361.
16.4.1

Leppard G.G., Massalski A. & Lean D.R.S. (1977) Electron opaque microscopic fibrils in lakes: their demonstration, their biological derivation and their potential significance in the redistribution of cations. *Protoplasma, Wien* **92,** 289–309.
17.2

Levi C. & Preiss J. (1976) Regulatory properties of the ADP-glucose pyrophosphorylase of the blue-green bacterium *Synechococcus* 6301. *Pl. Physiol., Lancaster* **58,** 753–756.
3.2.2.2

Levin E., Lennarz W.J. & Bloch K. (1964) Occurrence and localization of α-linolenic acid containing galactolipids in the photosynthetic apparatus of *Anabaena variabilis. Biochim. biophys. Acta* **84,** 471–474.
6.4.1

Lewin R.A. (1956) Extracellular polysaccharides of green algae. *Can. J. Microbiol.* **2,** 665–672.
17.4.4

Lewin R.A. (1976) Prochlorophyta as a proposed new division of algae. *Nature, Lond.* **261,** 697–698.
1, 20.2.4

Lewin R.A. (1977) *Prochloron*, type genus of the Prochlorophyta. *Phycologia* **16,** 217.
1

Lewin R.A. (1979) Formal taxonomic treatment of cyanophytes. *Int. J. syst. Bact.* **29,** 411–412.
12.1.1

Lewin R.A. (1981) *Prochloron* and the theory of symbiogenesis. *Ann. N.Y. Acad. Sci.* **361,** 325–329.
12.3.3

Lewis J.R. (1964) *The Ecology of Rocky Shores*, 323 pp. English Universities Press, London.
20.2.1

Lex M. & Carr N.G. (1974) The metabolism of glucose by heterocysts and vegetative cells of *Anabaena cylindrica. Arch. Microbiol.* **101,** 161–167.
4.3, 14.4.1, 14.4.2

Lex M. & Stewart W.D.P. (1973) Algal nitrogenase, reductant pools and photosystem I activity. *Biochim. biophys. Acta* **292,** 436–443.
4.3, 14.4.3.1

Ley A.C., Babcock G.T. & Sauer K. (1975) Flash kinetics and light intensity dependence of oxygen evolution in the blue-green alga *Anacystis nidulans. Biochim. biophys. Acta* **387,** 379–387.
8.6

Ley A.C. & Butler W.L. (1976) Efficiency of energy transfer from photosystem II to photosystem I in *Porphyridium cruentum. Proc. natn. Acad. Sci. U.S.A.* **73,** 3957–3960.
7.9, 16.4.2

Ley A.C., Butler W.L., Bryant D.A. & Glazer N.A. (1977) Isolation and function of a allophycocyanin-B of *Porphyridium cruentum. Pl. Physiol., Lancaster* **59,** 974–980.
7.5.1.1, 7.5.1.7, 7.6, 7.7

Liao C.F.H. (1977) The effect of nutrient enrichment on nitrogen fixation activity in the Bay of Quinte, Lake Ontario. *Hydrobiologia* **56,** 273–279.
5.4.2

Lichtlé C. (1973) Ultrastructure du plaste de deux rhodophycées: *Rhodochorton purpureum* (Lighft.) Rosenvinge et *Rhodothamniella floridula* (Dillwyn) J. Feldman. *C.r. hebd. Séanc. Acad. Sci., Paris* **277,** 1865–1868.
7.3

Lichtlé C. & Giraud G. (1970) Aspects ultrastructuraux particuliers au plaste du *Batrachospermum virgatum* (Sirodot)-Rhodophycée-Naemalionale. *J. Phycol.* **6,** 281–289.
7.3

Lichtlé C. & Thomas J.C. (1976) Étude ultrastructurale des thylacoïdes des algues à phycobiliproteines, comparaison des résultats obtenus par fixation classique et cryodécapage. *Phycologia* **15,** 393–404.
6.3.2, 7.3, 7.9

Liesegang R.E. (1913) *Geologische Diffusionem.* 180 pp. T. Steinkopff, Dresden, Leipzig.
21.1

Lightbody J.J. & Krogmann D.W. (1966) The role of plastoquinone in the photosynthetic reactions of *Anabaena variabilis. Biochim. biophys. Acta* **120,** 57–64.
8.5

Lightbody J.J. & Krogmann D.W. (1967) Isolation and properties of plastocyanin from *Anabaena variabilis. Biochim. biophys. Acta* **131,** 508–515.
6.5.2, 8.5

Likens G.E. (Ed.) (1972) *Nutrients and Eutrophication.* American Society of Limnology and Oceanography Special Symposia 1.
18.1

Lind E.M. (1968) Notes on the distribution of

phytoplankton in some Kenya waters. *Br. phycol. J.* **3,** 481–493.
18.4.1

Lindsey J.K., Vance B.D., Keeter J.S. & Scholes V.E. (1971) Spheroplast formation and associated ultrastructural changes in a synchronous culture of *Anacystis nidulans* treated with lysozyme. *J. Phycol.* **7,** 65–71.
13.4.1

Little M.G. (1973) The zonation of marine supralittoral blue-green algae. *Br. phycol. J.* **8,** 47–50.
20.2.1, 20.3.1

Livingstone D. & Jaworski G.H.M. (1980) The viability of akinetes of blue-green algae recovered from the sediments of Rostherne Mere. *Br. phycol. J.* **15,** 357–364.
15.5

Lockau W. (1981) Evidence for a dual role of cytochrome C-553 and plastocyanin in photosynthesis and respiration of cyanobacterium *Anabaena variabilis* ATCC 29413. *Arch. Microbiol.* **128,** 336–340.
3.4.4

Lockau W., Peterson R.B., Wolk C.P. & Burris R.H. (1978) Modes of reduction of nitrogenase in heterocysts isolated from *Anabaena* species. *Biochim. biophys. Acta* **502,** 298–308.
4.3, 14.4.3.1, 14.4.3.2

Logan B.W. (1961) *Cyryptozoon* and associate stromatolites from the Recent of Shark Bay, Western Australia. *J. Geol.* **69,** 517–533.
21.1

Logan B.W., Hoffman P. & Gebelein C.D. (1974) Algal mats, cryptalgal fabrics, and structures, Hamelin Pool, Western Australia. *Mem. Am. Ass. petrol. Geol.* **22,** 140–194.
20.4.4

Lohmann H. (1911) Über das Nannoplankton und die Zentrifugierung kleinster Wasserproben zur Gewinnung desselben in lebendem Zustand. *Int. Rev. ges. Hydrobiol. Hydrogr.* **4,** 1–38.
19.2.1

Loogman J.G., Post A.F. & Mur L.-R. (1980) The influence of periodicity in light conditions as determined by the trophic state of the water, on the growth of the green alga *Scenedesmus protuberans* and the cyanobacterium *Oscillatoria agardhii.* In: Barica J. & Mur L.R. (Eds) *Hypertrophic Ecosystems.* 348 pp., pp. 79–82. Junk, Den Haag.
2.4.1

Loogman J.G. & van Liere L. (1978) An improved method for measuring irradiance in algal cultures. *Verh. int. Verein. theor. angew. Limnol.* **20,** 2322–2328.
2.2.1, 2.2.2

Lorch S.K. & Wolk C.P. (1974) Application of gas-liquid chromatography to study of the envelop lipids of heterocysts. *J. Phycol.* **10,** 352–355.
14.3.1

Lord J.M., Codd G.A. & Stewart W.D.P. (1975) Serological comparison of ribulose-1,5-diphosphate carboxylase from *Euglena gracilis*, *Chlorella fusca* and several blue-green algae. *Plant Sci. Lett.* **4,** 377–383.
3.2.1

Lorenzen M.W. & Mitchell R. (1975) An evaluation of artificial destratification for control of algal blooms. *J. Am. Wat. Wks Ass.* **67,** 373–376.
18.6

Lorimer G.H. (1981) The carboxylation and oxygenation of ribulose-1,5-bisphosphate: The primary events in photosynthesis and photorespiration. *A. Rev. Plant Physiol.* **32,** 349–383.
3.2.1

Losada M. & Guerrero M.G. (1979) The photosynthetic reduction of nitrate and its regulation. In: Barber J. (Ed.) *Photosynthesis in Relation to Model Systems.* 434 pp. pp. 365–408. Elsevier, Amsterdam.
18.5.2.1

Lounatmaa K., Vaara T., Österlund K. & Vaara M. (1980) Ultrastructure of the cell wall of a *Synechocystis* strain. *Can. J. Microbiol.* **26,** 204–298.
13.2.2, 16.2.3

Lowe D.R. (1980) Stromatolites 3,400-Myr old from the Archean of Western Australia. *Nature, Lond.* **284,** 441–443.
21.2

Lucas C. (1974) Aspects of the Intermediary Metabolism of Facultatively Heterotrophic Blue-green Algae. Ph.D. Thesis, University of Wales.
3.4.1, 3.4.2

Lucas C.E. (1951) The ecological effects of external metabolities. *Biol. Rev.* **22,** 270–295.
17.4.4

Ludden P.W., Okon Y. & Burris R.H. (1978) Properties of the nitrogenase system from *Spirillum lipoferum. Biochem. J.* **173,** 1001–1003.
4.2

Lumpkin T.A. & Plucknett D.L. (1980) *Azolla*: botany, physiology, and use as a green manure. *Economic Botany* **34,** 111–153.
1

Lumsden J. & Hall D.O. (1975) Superoxide dismutase in photosynthetic organisms provides an evolutionary hypothesis. *Nature, Lond.* **257**, 670–672.
8.3

Lund J.W.G. (1965) The ecology of the freshwater phytoplankton. *Biol. Rev.* **40**, 231–293.
18.1, 18.4.2, 18.4.4, 18.5.5.1, 18.7

Lund J.W.G. (1971) The seasonal periodicity of three planktonic desmids in Windermere. *Mitt. int. Verein. theor. angew. Limnol.* **19**, 3–25.
18.5.5.6

Lund J.W.G. (1978) Changes in the phytoplankton of an English lake, 1945–1977. *Gidrobiol. Zh.* **14**, 10–27. (Russian).
18.4.3

Lundell D.J., Williams R.C. & Glazer A.N. (1981) Molecular architecture of a light-harvesting antenna. *In vitro* assembly of the red substructures of *Synechococcus* 6301 phycobilisomes. *J. biol. Chem.* **256**, 3580–3592.
7.5.2

Lundgren A. (1978) Nitrogen fixation induced by phosphorus fertilisation of a sub-arctic lake. In: Granhall U. (Ed.) *Environmental Role of Nitrogen-fixing Blue-green Algae and Asymbiotic Bacteria. Ecol. Bull. Stockholm* **26**, 52–59.
18.5.2.1

Lupton F.S. & Marshall K.C. (1981) Specific adhesion of bacteria to heterocysts of *Anabaena* spp. and its ecological significance. *Appl. environ. Microbiol.* **42**, 1085–1092.
17.2

Lylis J.C. & Trainor F.R. (1973) The heterotrophic capabilities of *Cyclotella meneghiniana. J. Phycol.* **9**, 365–369.
3.4.3

Lyne R.L. & Stewart W.D.P. (1973a) Emerson enhancement of carbon fixation but not of acetylene reduction (nitrogenase activity) in *Anabaena cylindrica. Planta, Berl.* **109**, 27–38.
4.3, 14.4.3.2

Lyne R.L. & Stewart W.D.P. (1973b) Intermediary metabolism and reductant supply to nitrogenase in *Anabaena cylindrica.* In: *Abstracts of Symposium on Prokaryotic Photosynthetic Organisms.* 228 pp., pp. 103–104. Univ. Freiburg i. Br., Germany.
14.4.3.1

MacColl R. & Berns D.S. (1979) Evolution of the biliproteins. *Trends biochem. Sci.* **4**, 44–47.
7.1, 7.5.1.7

MacColl R., Edwards M.R., Mulks M.H. & Berns D.S. (1974) Comparison of the biliproteins from two strains of the thermophilic cyanophyte *Synechococcus lividus. Biochem. J.* **141**, 419–425.
7.5.1.4

MacColl R., Edwards M.R. & Haaksma C. (1978) Some properties of allophycocyanin from a thermophilic blue-green alga. *Biophys. Chem.* **8**, 369–376.
7.5.1.5

MacColl R., Lee J.J. & Berns D.S. (1971) Protein aggregation in C-phycocyanin. Studies at very low concentrations with the photoelectric scanner of the ultracentrifuge. *Biochem. J.* **122**, 421–426.
7.5.1.4

MacGregor I.M., Truswell J.F. & Eriksson K.A. (1974) Filamentous algae from the 2,300 m.y. old Transvaal Dolomite. *Nature, Lond.* **247**, 538–540.
12.3.2

Mackenthun K.M., Keup L.E. & Stewart R.K. (1968) Nutrients and algae in Lake Sebasticook, Maine. *J. Wat. Pollut. Control Fed.* **40**, R72–81.
18.4.4

MacNab R.M. (1978) Bacterial motility and chemotaxis. *Crit. Rev. Biochem.* **5**, 291–341.
16.5

MacNab R.M. (1979) Bacterial flagella. In: Haupt W. & Feinleib M.E. (Eds) *Encycl. Pl. Physiol. N.S. Vol. 7, Physiol. of Movements.* pp. 207–223. Springer-Verlag, Berlin.
16.5

MacRae T.H. & McCurdy H.D. (1976a) The isolation and characterization of gliding motility mutants of *Myxococcus xanthus. Can. J. Microbiol.* **22**, 1282–1292.
16.2.3

MacRae T.H. & McCurdy H.D. (1976b) Evidence for motility-related fimbriae in the gliding microorganism *Myxococcus xanthus. Can. J. Microbiol.* **22**, 1589–1593.
16.2.3

MacRae T.H., Dobson W.J. & McCurdy H.D. (1977) Fimbriation in gliding bacteria. *Can. J. Microbiol.* **23**, 1096–1108.
16.2.3

Magrum L.J., Luehrsen K.R. & Woese C.R. (1978) Are extreme halophiles actually 'bacteria'? *J. molec. Evol.* **11**, 1–8.
12.2.2.2

Mague T.H. (1977) Ecological aspects of dinitrogen fixation by blue-green algae. In: Hardy R.W.F. & Gibson A.H. (Eds) *A Treatise on Dinitrogen Fixation, Section IV.* pp. 85–140. Wiley, New York.
4.1

Mague T.H. & Holm-Hansen O. (1975) Nitrogen fixation on a coral reef. *Phycologia* **14**, 87–92.
20.4.5

Mague T.H., Mague F.C. & Holm-Hansen O. (1977) Physiology and chemical composition of nitrogen-fixing phytoplankton in the central north Pacific Ocean. *Mar. Biol.* **41**, 213–227.
5.4.1, 19.2.4.2, 19.3.2, 19.3.4, 19.3.5, 19.3.8, 19.4.2, 19.5.2, 20.4.5

Mague T.H., Weare N.M. & Holm-Hansen O. (1974) Nitrogen fixation in the North Pacific Ocean. *Mar. Biol.* **24**, 109–119.
19.2.4.1, 19.2.4.2, 19.3.4, 19.5.2

Makino F. & Tsuzaki J. (1971) Absence of histone in the blue-green alga *Anabaena cylindrica*. *Nature, Lond.* **231**, 446–447.
11.3.1

Makovkina L.E., Mestechkina N.M., Kuznetsov B.A., Mutuskin A.A. & Pshenova K.V. (1978) *Biokhimiya* **43**, 564–567 (Russian = *Biokhimiya, Transl.* **43**, 450–453).
8.5

Malkin R., Bearden A.J., Hunter F.A., Alberte R.S. & Thornber J.P. (1976) Properties of the low temperature photosystem I primary reaction in the P700 chlorophyll a protein. *Biochim. biophys. Acta* **430**, 389–394.
8.4

Maloney P.C. (1979) Membrane H^+ conductance of *Streptococcus lactis*. *J. Bact.* **140**, 197–205.
16.3, 16.4.2

Mann N. & Carr N.G. (1974) Control of macromolecular composition and cell division in the blue-green alga, *Anacystis nidulans*. *J. gen. Microbiol.* **83**, 399–405.
11.3.1, 15.2

Manson M.D., Tedesco P., Berg H.C., Harold F.M. & Van der Drift C. (1977) A protonmotive force drives bacterial flagella. *Proc. natn. Acad. Sci. U.S.A.* **74**, 3060–3064.
16.3

Manzano C., Candau P., Gomez-Moreno C., Relimpio A.M. & Losada M. (1976) Ferredoxin-dependent photosynthetic reduction of nitrate and nitrite by particles of *Anacystis nidulans*. *Molec. Cell. Biochem.* **10**, 161–169.
8.3

Manzano C., Candau P. & Guerrero M.G. (1978) Affinity chromatography of *Anacystis nidulans* ferredoxin-nitrate reductase and NADP reductase on reduced ferredoxin-Sepharose. *Anal. Biochem.* **90**, 408–412.
8.3

Margalef R. (1978) Life-forms of phytoplankton as survival alternatives in an unstable environment. *Oceanologica Acta* **1**, 493–509.
19.4.2, 19.5.3

Margulis L. (1968) Evolutionary criteria in thallophytes: a radical alternative. *Science, N.Y.* **161**, 1020–1022.
9.6

Margulis L. (1970) *Origin of Eukaryotic Cells*. 349 pp. Yale Univ. Press, New Haven.
12.1.3, 12.2.1, 12.4.1, 12.4.2

Margulis L. (1975a) Symbiotic theory of the origin of eukaryotic organelles: criteria for proof. *Symp. Soc. Exp. Biol.* **24**, 21–38.
12.4.2

Margulis L. (1975b) The microbes' contribution to evolution. *Biosystems* **7**, 266–292.
12.4.2

Margulis L. (1981) *Symbiosis in Cell Evolution: Life and its Environment on the Early Earth*. W. H. Freeman & Co., San Francisco.
12.1.3, 12.4.1, 12.4.2

Margulis L., Barghoorn E.S., Ashendorf D., Banerjee S., Chase D., Francis S., Giovannoni S. & Stolz J. (1980) The microbial community in the layered sediments at Laguna Figueroa, Baja California, Mexico: does it have Precambrian analogues? *Precambrian Res.* **11**, 93–123.
20.4.1

Margulis L., Walker J.C.G. & Rambler M. (1976) Reassessment of roles of oxygen and ultraviolet light in Precambrian evolution. *Nature, Lond.* **264**, 620–624.
12.1.2

Markley J.L., Ulrich E.L., Berg S.O. & Krogmann D.W. (1975) Nuclear magnetic resonance studies of the copper binding sites of blue, copper proteins. Oxidized, reduced and apoplastocyanin. *Biochemistry, N.Y.* **14**, 4428–4433.
8.5

Martin T.C. & Wyatt J.T. (1974) Extracellular investments in blue-green algae with particular emphasis on the genus *Nostoc*. *J. Phycol.* **10**, 204–210.
13.2.2

Maruyama Y. (1975) Nitrogen fixation in the open sea. In: Takahashi H. (Ed.) *JIBP Synthesis: Nitrogen Fixation and Nitrogen Cycle*. pp. 79–84. Univ. Tokyo Press.
19.3.4

Masaki R., Wada K. & Matsubara H. (1979) Isolation and characterization of two ferredoxin-NADP reductases from *Spirulina platensis*. *J. Biochem., Tokyo* **86**, 951–962.
8.3

Mason T.R. & Van Brunn V. (1977) 3-Gyr-old stromatolites from South Africa. *Nature, Lond.* **266,** 47–49.
21.2

Matsamoto K. & Nishimura M. (1977) Estimation of internal pH in cells of blue-green algae in the dark and under illumination. *J. Biochem., Tokyo* **82,** 483–487.
10.3.2.1, 10.3.3.3

Matsamoto K. & Nishimura M. (1978) H^+ translocation in lysozyme treated cells of the blue-green alga *Plectonema boryanum. J. Biochem., Tokyo* **82,** 489–493.
8.7,

Matsubara H., Hase T., Wakabayashi S. & Wada K. (1979) Gene duplications during the evolution of chloroplast-type ferredoxins. In: Matsubara H. & Yamanaka T. (Eds) *Proceedings for the Evolution of Protein Molecules.* 561 pp., pp. 209–219. Tokyo Univ. Press.
8.3

Matsuura S., Shioi J. & Imae Y. (1977) Motility in *Bacillus subtilis* driven by an artificial protonmotive force. *FEBS Lett.* **82,** 187–190.
16.3

Matusiak K., Chróst R. & Krzywicka A. (1971) Bacterial microflora in *Chlorella vulgaris* culture. *Acta microbiol. Polon. B.* **3,** 189–193.
17.4.2

Mauzerall D. (1978) Bacteriochlorophyll and photosynthetic evolution. In: Clayton R.K. & Sistrom W.R. (Eds) *The Photosynthetic Bacteria.* 946 pp., pp. 223–231. Plenum Press, New York.
12.2.1, 12.4.1

Mauzy W., Braun W. & Whallon J. (1955) Studies on DNA and DNase in *Brucella* cultures. *Bact. Proc. 1955*, 46.
11.6.3

May J.A. & Perkins R.D. (1979) Endolithic infestation of carbonate substrates below the sediment–water interface. *J. sed. Petrol.* **49,** 357–377.
20.2.2

May V. (1974) Suppression of blue-green algal blooms in Braidwood Lagoon with alum. *J. Aust. Inst. agric. Sci.* **40,** 54–57.
18.6

Mazur B.J., Rice D. & Haselkorn R. (1980) Identification of blue-green algal nitrogen fixation genes by using heterologous DNA hybridization probes. *Proc. natn. Acad. Sci. U.S.A.* **77,** 186–190.
11.2.2, 11.3.2, 11.5.5.2(b), 11.6.1, 11.7, 11.8

McCarthy J.J. & Carpenter E.J. (1979) *Oscillatoria (Trichodesmium) thiebautii* (Cyanophyta) in the central North Atlantic Ocean. *J. Phycol.* **15,** 75–82.
5.1.1, 5.4.1, 19.2.4.1, 19.3.2, 19.3.3, 19.3.4, 19.3.5, 19.3.7, 19.3.8, 19.4.2

McCord J.M. & Fridovich I. (1977) Superoxide dismutases: a history. In: Michelson A.M., McCord J.M. & Fridovich I. (Eds) *Superoxide and Superoxide Dismutases.* 568 pp., pp. 1–10. Academic Press, London.
2.3.1

McCord J.M., Keele B.B., Jr & Fridovich I. (1971) An enzyme based theory of obligate anaerobiosis: the physiological function of superoxide dismutase. *Proc. natn. Acad. Sci. U.S.A.* **68,** 1024–1027.
9.4.1

McDaniel H.R., Middlebrook J.B. & Bowman R.O. (1962) Isolation of pure cultures of algae from contaminated cultures. *Appl. Microbiol.* **10,** 223.
18.5.1.1

McFadden B.A. (1978) Assimilation of one-carbon compounds. In: Ornston L.N. & Sokatch J.R. (Eds) *The Bacteria, Vol. VI.* 603 pp., pp. 219–304. Academic Press, New York.
12.2.2.3

McFadden B.A. & Purohit K. (1978) Chemosynthetic, photosynthetic, and cyanobacterial ribulose bisphosphate carboxylase. In: Siegelman H.W. & Hind G. (Eds) *Basic Life Sciences, Vol. II: Photosynthetic Carbon Assimilation.* 404 pp., pp. 179–207. Plenum Press, New York.
12.2.2.3

McFadden B.A. & Tabita F.R. (1974) D-Ribulose-1,5-diphosphate carboxylase and the evolution of autotrophy. *BioSystems* **6,** 93–112.
3.2.1

McFadden B.A., Tabita F.R. & Kuehn G.D. (1975) Ribulose diphosphate carboxylase from the hydrogen bacteria and *Rhodospirillum rubrum.* In: Colowick S.P. & Kaplan N.O. (Eds) *Methods in Enzymology*, vol. 42. 537 pp., pp. 461–472. Academic Press, London.
3.2.1

McKie J., Lucas C. & Smith A.J. (1981) δ-Aminolaevulinate biosynthesis in the cyanobacterium *Synechococcus* 6301. *Phytochem.* **20,** 1547–1549.
3.2.2.1

McSwain B.D., Tsujimoto H.Y. & Arnon D.I. (1976) Effects of magnesium and chloride

ions on light-induced electron transport in membrane fragments from a blue-green alga. *Biochim. biophys. Acta* **423**, 313–322.
8.2.

Medvecsky N. & Rosenberg H. (1971) Phosphate transport in *Escherichia coli. Biochim. biophys. Acta* **241**, 494–506.
5.1.1

Meeks J.C. (1979) NO_3-assimilation by two *Anabaena* species. In: Nichols J.M. (Ed.) *Abstracts of the Third International Symposium on Photosynthetic Prokaryotes, Oxford*. p. D15. Univ. Liverpool, England.
4.5

Meeks J.C., Wolk C.P., Lockau W., Schilling N., Shaffer P.W. & Chien W.-S. (1978) Pathways of assimilation of $[^{13}N]N_2$ and $^{13}NH_4^+$ by cyanobacteria with and without heterocysts. *J. Bact.* **134**, 125–130.
14.4.1

Meeks J.C., Wolk C.P., Thomas J., Lockau W., Shaffer P.W., Austin S.M., Chien W.-S. & Galonsky A. (1977) The pathways of assimilation of $^{13}NH_4^+$ by the cyanobacterium, *Anabaena cylindrica. J. biol. Chem.* **252**, 7894–7900.
14.4.1

Meffert M.E. (1971) Cultivation and growth of two planktonic *Oscillatoria* species. *Mitt. int. Verein. theor. angew. Limnol.* **19**, 189–205.
18.5.5.5

Meffert M.E. (1972) Zur Isolierung von *Oscillatoria redekei* van Goor. *Arch. Hydrobiol., Suppl.* **41**, 235–241.
18.5.1.1

Meffert M.E. (1973) Kultur und Wachstum von *Oscillatoria redekei* van Goor. Zur Bedeutung der Bakterien. *Verh. int. Verein. theor. angew. Limnol.* **18**, 1359–1366.
,18.5.5.6

Meffert M.E. & Chang T. (1978) The isolation of planktonic blue-green algae (*Oscillatoria*—species). *Arch. Hydrobiol.* **82**, 231–239.
18.5.1.1

Meffert M.-E. & Krambeck H.-J. (1977) Planktonic algae of the *Oscillatoria redekei* group. *Arch. Hydrobiol.* **79**, 149–171.
18.4.1

Megard R.O. & Smith P.D. (1974) Mechanisms that regulate growth rates of phytoplankton in Lake Shagawa, Minnesota. *Limnol. Oceanogr.* **19**, 279–296.
5.4.1, 5.4.3

Mehta V.B. & Vaidya B.S. (1978) Cellular and extracellular polysaccharides of the blue-green alga, *Nostoc. J. exp. Bot.* **29**, 1423–1430.
13.4.5

Menday D.C. & Buck A.A. (1972) A shock-wave technique to collapse the vacuoles of blue-green algae. *Water Res.* **6**, 279–285.
18.6

Menke W. (1961a) Über die Chloroplasten von *Anthoceros punctatus* (5. Mitteilung zur Entwicklungsgeschichte der Plastiden). *Z. Naturforsch.* **16b**, 334–336.
6.1

Menke W. (1961b) Über das Lamellarsystem des Chromatoplasmas von Cyanophyceen. *Z. Naturforsch.* **16b**, 543–545.
6.1

Mereschkowsky C. (1905) Über Natur and Ursprung der Chromatophoren in Pflanzenreiche. *Biol. Centr.* **25**, 593–604.
6.1, 12.4.1

Merrick J.M. (1979) Metabolism of reserve materials. In: Clayton R.K. & Sistrom W.R. (Eds) *The Photosynthetic Bacteria*. 946 pp., pp. 199–219. Plenum Press, New York.
3.2.2.2

Mesibov R. & Adler J. (1972) Chemotaxis toward amino acids in *Escherichia coli. J. Bact.* **112**, 315–326.
17.5.1

Metzner I. (1955) Zur Chemie und zum submikroskopischen Aufbau der Zellwände, Scheiden u. Gallerte von Cyanophyceen. *Arch. Mikrobiol.* **22**, 45–77.
13.2.1, 13.2.2, 13.4

Mevarech M., Rice D. & Haselkorn R. (1980) Nucleotide sequence of a cyanobacterial *nif*H gene coding for nitrogenase reductase. *Proc. natn. Acad. Sci. U.S.A.* **77**, 6476–6480.
11.8

Meyer J., Kelley B.C. & Vignais P.M. (1978) Nitrogen fixation and hydrogen metabolism in photosynthetic bacteria. *Biochimie* **60**, 245–260.
4.5

Michaelis H. (1978) Recent biological phenomena in the German Waddensee. *Rapp. P. v. Réun. Cons. int. Explor. Mer.* 276–277.
20.2.2

Midgley M. & Dawes E.A. (1973) The regulation of transport of glucose and methyl-α-glucoside in *Pseudomonas aeruginosa. Biochem. J.* **132**, 141–154.
3.4.3

Mikheyskaya L.V., Ovodova R.G. & Ovodov Y.S. (1977) Isolation and characterization of lipopolysaccharides from cell walls of blue-green algae of the genus *Phormidium. J. Bact.* **130**, 1–3.
13.4.4

Miller A.G., Chang K.H. & Colman B. (1971) Uptake and oxidation of glycolic acid by blue-green algae. *J. Phycol.* **7**, 97–100.
3.2.1

Miller D.M., Jones J.H., Tindall D.R. & Schmid W.E. (1976) Ion metabolism in a halophilic blue-green alga, *Aphanothece halophytica. Arch. Microbiol.* **111**, 145–149.
10.2.3, 10.4, 20.3.5

Miller J.S. & Allen M.M. (1972) Carbon utilization patterns in the heterotrophic blue-green alga *Chlorogloea fritschii. Arch. Mikrobiol.* **86**, 1–12.
3.4.2

Miller L.S. & Holt S.C. (1977) Effect of carbon dioxide on pigment and membrane content in *Synechococcus lividus. Arch. Microbiol.* **115**, 185–198.
6.5.2, 7.8, 8.6

Miller M.M. & Lang N.J. (1968) The fine structure of akinete formation and germination in *Cylindrospermum. Arch. Mikrobiol.* **60**, 303–313.
3.2.2.2, 15.2, 15.5

Miller M.M. & Lang N.J. (1971) The effect of aging on thylakoid configuration and granular inclusions in *Gloeotrichia.* In: Parker B.C. & Brown R.M., Jr (Eds) *Contributions in Phycology.* 196 pp., pp. 53–58. Allen Press, Lawrence, Kansas, U.S.A.
6.2, 6.5.2

Miller S.L. (1953) A production of amino acids under possible primitive Earth conditions. *Science, N.Y.* **117**, 528–529.
21.1

Miller S.S., Ausubel F.M. & Bogorad L. (1979) Cyanobacterial ribonucleic acid polymerases recognize Lambda promoters. *J. Bact.* **140**, 246–250.
11.3.1

Miller S.S. & Bogorad L. (1978) Purification and characterization of RNA polymerase from *Fremyella diplosiphon. Pl. Physiol. Lancaster* **62**, 995–999.
11.3.1

Millineaux P.M., Gallon J.R. & Chaplin A.E. (1981) Acetylene reduction (nitrogen-fixation) by cyanobacteria grown under alternating light-dark cycles. *FEMS Microbiol. Lett.* **10**, 245–247.
4.3

Mimuro M. & Fujita Y. (1977) Estimation of chlorophyll a distribution in the photosynthetic pigment systems I and II of the blue-green alga *Anabaena variabilis. Biochim. biophys. Acta* **459**, 376–389.
6.4.2, 8.4

Mimuro M. & Fujita Y. (1978) Excitation energy transfer between pigment system II units in blue-green algae. *Biochim. biophys. Acta* **504**, 406–412.
8.6

Minato S. & Werbin H. (1972) Excitation and fluorescence spectra of the chromophore associated with the DNA-photoreactivating enzyme from the blue-green alga *Anacystis nidulans. Photochem. Photobiol.* **15**, 97–101.
11.4.2.1

Minkoff L. & Damadian R. (1976) Actin-like proteins from *Escherichia coli*: concept of cytotonus as the missing link between cell metabolism and the biological ion-exchange resin. *J. Bact.* **125**, 353–365.
16.2.2.2

Mitchell P. (1968) *Chemiosmotic Coupling and Energy Transduction.* Glynn Research Ltd, Bodmin, Cornwall.
10.3.3.1

Mitchison G.J. & Wilcox M. (1972) Rule governing cell division in *Anabaena. Nature, Lond.* **239**, 110–111.
14.6.1.1

Mitchison G.J. & Wilcox M. (1973) Alteration in heterocyst pattern of *Anabaena* produced by 7-azatryptophan. *Nature, New Biol.* **246**, 229–233.
14.6.1.1, 14.6.1.3, 15.6.1

Mitchison G.J., Wilcox M. & Smith R.J. (1976) Measurement of an inhibitory zone. *Science, N.Y.* **191**, 866–868.
14.6.1.2

Mitronova T.N., Shestakov S.V. & Zhevner V.D. (1973) Properties of a radiosensitive filamentous mutant of the blue-green alga *Anacystis nidulans. Mikrobiologiya, Transl.* **42**, 519–524.
11.6.3

Mitsui A. & Kumazawa S. (1977) Hydrogen production by marine photosynthetic organisms as a potential energy source. In: Mitsui A. *et al.* (Eds) *Biological Solar Energy Conversion.* pp. 454, pp. 23–43. Academic Press, New York.
4.6

Mizuno H., Saito T. & Iso N. (1975) Solution properties of phycocyanin. III. Studies of the sedimentation equilibrium of phycocyanin. *Bull. chem. Soc. Japan* **48**, 3496–3499.
7.5.1.4

Moeller R.E. & Roskoski J.P. (1978) Nitrogen-fixation in the littoral benthos of an oligotrophic lake. *Hydrobiologia* **60**, 13–16.
17.3.3

Mohanty P., Mar T. & Govindjee (1971) Action of hydroxylamine in the red alga

Porphyridium cruentum. *Biochim. biophys. Acta* **253**, 213–221.
9.2.3

Monty C.L.V. (1965) Recent algal stromatolites in the Windward Lagoon, Andros Island, Bahamas. *Ann. Soc. géol. Belg.* **88**, 269–276.
20.4.2

Monty C.L.V. (1973) Precambrian background and Phanerozoic history of stromatolite communities, an overview. *Ann. Soc. géol. Belg., Bull.* **96**, 585–624.
20.4.2, 21.3

Monty C.L.V. (1976) The origin and development of cryptalgal fabrics. In: Walter M.R. (Ed.) *Stromatolites; Developments in Sedimentology 20*. 790 pp., pp. 139–249. Elsevier, Amsterdam.
20.4.2

Monty C.L.V. (1979) Scientific reports of the Belgian expedition in the Australian Great Barrier Reefs, 1967. Sedimentology: 2. Monospecific stromatolites from the Great Barrier Reef Tract and their paleontological significance. *Ann. Soc. géol. Belg.* **101** (1978), 163–171.
20.4.2

Moor H. & Mühlethaler K. (1963) Fine structure in frozen-etched yeast cells. *J. Cell Biol.* **17**, 609–628.
6.3.1

Moore B.G. & Tischer R.G. (1964a) An extracellular polysaccharide produced by *Palmella mucosa*. *Arch. Mikrobiol.* **82**, 283–299.
13.4.5

Moore B.G. & Tischer R.G. (1964b) Extracellular polysaccharides of algae: effects on life-support systems. *Science, N.Y.* **145**, 586–587.
13.4.5

Moore B.G. & Tischer R.G. (1965) Biosynthesis of extracellular polysaccharides by the blue-green alga, *Anabaena flos-aquae*. *Can. J. Microbiol.* **11**, 877–885.
13.4.5

More J.E., Williams M.M. & Smith A.J. (1979) Features of organic growth substrate utilization by cyanobacteria. In: Nichols J.M. (Ed.) *Proceedings of the Third International Symposium on Photosynthetic Prokaryotes, Oxford*. p. B49. Univ. Liverpool, England.
3.4.1, 3.4.2

Morel A. & Smith R.C. (1974) Relation between total quanta and total energy for aquatic photosynthesis. *Limnol. Oceanogr.* **19**, 591–600.
2.2.1

Morris I. & Glover H. (1981) Physiology of photosynthesis by marine coccoid cyanobacteria—some ecological implications, *Limnol. Oceanogr.* **26**, 957–961.
1

Mörschel E., Koller K.P. & Wehrmeyer W. (1980) Biliprotein assembly in the disc-shaped phycobilisomes of *Rhodella violacea*. Electron microscopical and biochemical analyses of C-phycocyanin and allophycocyanin aggregates. *Arch. Microbiol.* **125**, 43–51.
7.5.1.3, 7.5.1.6

Mörschel E., Koller K.P., Wehrmeyer W. & Schneider H. (1977) Biliprotein assembly in the disc-shaped phycobilisomes of *Rhodella violacea*. I. Electron microscopy of phycobilisomes *in situ* and analysis of their architecture after isolation and negative staining. *Cytobiologie* **16**, 118–129.
7.7

Mortenson L.E. & Thorneley R.N.F. (1979) Structure and function of nitrogenase. *A. Rev. Biochem.* **48**, 387–418.
4.2

Morton S.D. & Lee T.H. (1974) Algal blooms—possible effects of iron. *Environ. Sci. Technol.* **8**, 673–674.
18.5.1.2

Moul E.T. (1957) Preliminary report on the flora of Onotoa Atoll, Gilbert Islands. *Atoll Res. Bull* **57**, 1–48.
20.2.1, 20.2.2

Muchmore C.B. (1978) Algae control in water supply reservoirs. *J. Am. Wat. Wks Ass.* **78**, 273–279.
18.6

Muckle G., Otto J. & Rüdiger W. (1978) On the linkages between chromophore and protein in biliproteins, VII. Amino acid sequence in the chromophore regions of C-phycoerythrin from *Pseudanabaena* W 1173 and *Phormidium persicinum*. *Hoppe-Seyler's Z. physiol. Chem.* **359**, 345–355.
7.5.1.2, 7.5.1.8

Muckle G. & Rüdiger W. (1977) Chromophore content of C-phycoerythrin from various cyanobacteria. *Z. Naturforsch.* **32c**, 957–962.
7.5.1.2

Mühlradt P.F. & Golecki J.R. (1975) Asymmetrical distribution and artifactual reorientation of lipopolysaccharide in the outer membrane bilayer of *Salmonella typhimurium*. *Eur. J. Biochem.* **51**, 343–352.
6.3.1

Mühlradt P.F., Menzel J., Golecki J.R. & Speth V. (1973) Outer membrane of *Salmonella*. Sites of export of newly synthesised

lipopolysaccharide on the bacterial surface. *Eur. J. Biochem.* **35,** 471–481.
6.2

Muir M.D. & Grant P.R. (1976) Micropalaeontological evidence from the Onverwacht Group, South Africa. In: Windley B.F. (Ed.) *The Early History of the Earth.* pp. 595–604. Wiley, New York.
21.2

Munoz Calvo M.L. & Rodriguez-Lopez M. (1978) Induced synthesis of phosphatases in microalgae by antibiotic inhibitor of protein synthesis: mechanism of action. *Pl. Sci. Lett.* **13,** 275–280.
5.2.3

Mur L.R. & Beijdorff R.O. (1978) A model for succession from green to blue-green algae based on light-limitation. *Verh. int. Verein. theor. angew. Limnol.* **20,** 2314–2321.
2.6.1

Mur L.R., Gons H.J. & van Liere L. (1978) Competition of the green alga *Scenedesmus* and the blue-green *Oscillatoria. Mitt. int. Verein. theor. angew. Limnol.* **21,** 473–479.
2.6.1, 18.5.3, 18.5.5.1

Murai T. & Katoh T. (1975) Photosystem I-dependent oxidation of organic acids in the blue-green alga *Anabaena variabilis. Pl. Cell Physiol., Tokyo* **16,** 759–797.
4.3, 8.5

Murata N. & Fork D.C. (1977) Temperature dependence of the light-induced spectral shift of carotenoids in *Cyanidium caldarium* and higher plant leaves. Evidence for an effect of the physical phase of chloroplast membrane lipids on the permeability of the membrane to ions. *Biochim. biophys. Acta* **461,** 365–378.

Murata N., Sato N., Omata T. & Kuwabara T. (1981) Separation and characterization of thylakoid and cell envelope of the blue-green alga (cyanobacterium) *Anacystis nidulans. Pl. Cell Physiol., Tokyo* **22,** 855–866.
3.4.4

Murata N., Troughton J.H. & Fork D.C. (1975) Relationship between the transition of the physical phase of membrane lipids and photosynthetic parameters in *Anacystis nidulans* and lettuce and spinach chloroplasts. *Pl. Physiol., Lancaster* **56,** 508–517.
6.5.3, 8.8

Murphy R.F. & O'Carra P. (1970) Reversible denaturation of C-phycocyanin. *Biochim. biophys. Acta* **214,** 371–373.
7.5.1.3

Murphy T.P., Lean D.R.S. & Nalewajko C. (1976) Blue-green algae: their excretion of iron-selective chelators enables them to dominate other algae. *Science, N.Y.* **192,** 900–902.
17.4.2, 18.5.2.3

Murray K., Hughes S.G., Brown J.S. & Bruce S.A. (1976) Isolation and characterization of two sequence-specific endonucleases from *Anabaena variabilis. Biochem. J.* **159,** 317–322.
11.3.3

Murray S.N. & Littler M.M. (1978) Patterns of algal succession in a perturbated marine intertidal community. *J. Phycol.* **14,** 506–512.
20.3.6

Murray R.G.E. & Watson S.W. (1965) Structure of *Nitrosocystis oceanus* and comparison with *Nitrosomonas* and *Nitrobacter. J. Bact.* **89,** 1594–1609.
3.2.1

Murray M.A. & Benemann J.R. (1979) Nitrogenase regulation in *Anabaena cylindrica. Pl. Cell Physiol., Tokyo* **20,** 1391–1401.
4.5

Mutuskin A.A., Makovkina L.E., Pshenova K.V. & Vostroknutova G.N. (1976) Effect of metalloproteins on photochemical activity of chloroplasts treated with polyene antibiotics. *Biokhimiya* **42,** 653–658 (Russian): *Biokhimiya, Transl.* **42,** 501–505).
8.5

Myers J., Graham J.R. & Wang R.T. (1978) On spectral control of pigmentation in *Anacystis nidulans* (Cyanophyceae). *J. Phycol.* **14,** 513–518.
8.4

Myers J., Graham J.R. & Wang R.T. (1980) Spontaneous pigment mutants of *Anacystis nidulans* selected by growth under far-red light. *Arch. Microbiol.* **124,** 143–148.
11.5.4

Myers J. & Kratz W.A. (1955) Relations between pigment content and photosynthetic characteristics in a blue-green alga. *J. gen. Physiol.* **39,** 11–22.
7.2, 7.8, 7.8.2

Nagatani H.H. & Haselkorn R. (1978) Molybdenum independence of nitrogenase component synthesis in the non-heterocystous cyanobacterium *Plectonema. J. Bact.* **134,** 597–605.
4.2, 4.5

Nagy L.A. (1974) Transvaal stromatolite: first evidence for the diversification of cells about

2.2×10^9 years ago. *Science, N.Y.* **183**, 514–516.
12.3.2, 14.1

Nalewajko C. (1978) Release of organic substances. In: Hellebust J.A. & Craigie J.S. (Eds) *Handbook of Phycological Methods.* 512 pp., pp. 389–398. Cambridge Univ. Press.
17.4.4

Nalewajko C., Dunstall T.G. & Shear H. (1976) Kinetics of extracellular release in axenic algae and mixed algal-bacterial cultures; significance. *Phycologia* **12**, 1–5.
17.4.4

Nalewajko C. & Lean D.R.S. (1978) Phosphorus kinetics—algal growth relationships in batch cultures. *Mitt. int. Verein. theor. angew. Limnol.* **21**, 184–192.
5.1.1, 5.1.2, 5.2.2, 5.2.3

Neilson A.H. & Nordlund S. (1975) Regulation of nitrogenase synthesis in intact cells of *Rhodospirillum rubrum*: inactivation of nitrogen fixation by ammonia, L-glutamine and L-asparagine. *J. gen. Microbiol.* **91**, 53–62.
4.5

Neilson A., Rippka R. & Kunisawa R. (1971) Heterocyst formation and nitrogenase synthesis in *Anabaena* sp. A kinetic study. *Arch. Mikrobiol.* **76**, 139–150.
14.2, 14.5

Neimark H.C. (1977) Extraction of an actin-like protein from the prokaryote *Mycoplasma pneumoniae*. *Proc. natn. Acad. Sci. U.S.A.* **74**, 4041–4045.
16.2.2.2

Nelson D.C. (1979) The Biology of Beggiatoa–Organic Nutrition, Sulfur Utilization, and Light Responses. Ph.D. Thesis, Univ. Oregon, Eugene, U.S.A.
16.4.3

Nelson D.C. & Castenholz R.W. (1982) Light responses of *Beggiatoa*. *Arch. Microbiol.* **131**, 146–155.
16.2.3, 16.4.2, 16.4.3, 16.6

Neufeld G.J. & Riggs A.F. (1969) Aggregation properties of C-phycocyanin from *Anacystis nidulans*. *Biochim. biophys. Acta* **181**, 234–243.
7.5.1.4, 7.5.1.5

Neumann A.C., Gebelein C.D. & Scoffin T.P. (1970) The composition, structure and erodability of subtidal mats, Abaco, Bahamas. *J. sedim. Petrol.* **40**, 274–297.
20.2.4

Neushul M. (1971) Uniformity of thylakoid structure in red, brown, and two blue-green algae. *J. Ultrastruct. Res.* **37**, 532–543.
6.1, 6.3.2

Newcomb E.B. & Pugh T.D. (1975) Blue-green algae associated with ascidians of the Great Barrier Reef. *Nature, Lond.* **253**, 533–534.
20.2.4.

Newhouse J. (1954) Ecological and floristic notes on the Myxophyta of Raroia. *Atoll Res. Bull.* **33**, 42–54.
20.2.1, 20.2.4

Newman P.J. & Sherman L.A. (1978) Isolation and characterization of photosystem I and II membrane particles from the blue-green alga *Synechococcus cedrorum*. *Biochim. biophys. Acta* **503**, 343–361.
6.4.2, 8.2, 8.6

Nichols B.W. (1973) Lipid composition and metabolism. In: Carr N.G. & Whitton B.A. (Eds) *The Biology of Blue-Green Algae.* 676 pp., pp. 144–161. Blackwell, Oxford.
2.7.2.2

Nichols B.W., Harris R.V. & James A.T. (1965) The lipid metabolism of blue-green algae. *Biochem. biophys. Res. Comm.* **20**, 256–262.
6.4.1, 9.4.2

Nichols B.W. & Wood B.J.B. (1968a) New glycolipid specific to nitrogen-fixing blue-green algae. *Nature, Lond.* **217**, 767–768.
6.4.1, 14.3.1

Nichols B.W. & Wood B.J.B. (1968b) Occurrence and biosynthesis of γ-linolenic acid in a blue-green alga, *Spirulina platensis*. *Lipids* **3**, 46–50.
6.4.1

Nichols J.M., Adams D.G. & Carr N.G. (1980) Effect of canavanine and other amino acid analogues on akinete formation in the cyanobacterium *Anabaena cylindrica*. *Arch. Microbiol.* **127**, 67–75.
15.4, 15.4.1, 15.4.2, 15.6, 15.6.1

Nichols J.M. & Carr N.G. (1978) Akinetes of cyanobacteria. *Spores VII.* pp. 335–343. American Society for Microbiology, Washington D.C.
15.1

Niell F.X. & Anadon R. (1978) Seasonal data on morphology and ecology of *Merismopoedia*-like marine algae. Taxonomical implications of the observed changes. *Bot. mar.* **21**, 39–47.
20.2.2

Nielsen R. (1973) A comparison of the blue-green algae *Gloeocapsa crepidinum* and *Hyella balani* in culture. *Bot. Tidsskr.* **68**, 150–151.
20.1

Niemeyer R. & Richter G. (1969) Schnellmarkierte Polyphosphate und Metaphosphate

bei der Blaualge *Anacystis nidulans*. *Arch. Mikrobiol.* **69,** 54–59.
5.3.1

Nienhuis P.H. (1978) Dynamics of benthic algal vegetation and environment in Dutch estuarine salt marshes studied by means of permanent quadrats. *Vegetatio* **38,** 103–112.
20.2.3

Nikitin D.I. (1973) Direct electron microscopic techniques for the observation of micro-organisms in soil. In: Rosswall T. (Ed.) *Modern Methods in the Study of Microbiol Ecology*. 347 pp., pp. 81–85. Bulletin 17, Swedish Natural Science Research Council.
17.3.2

Nikitina K.A., Ermakova L.R., Khodzaev M.N. & Gusev M.V. (1979) Changes in the ultrastructure of *Anacystis nidulans* cells at different temperatures. *Mikrobiologia* **48,** 476–480 (Russian).
6.5.3

Niklowitz W. & Drews G. (1956) Beiträge zur Cytologie der Blaualgen. I. Mitt.: Untersuchungen zur Substruktur vom *Phormidium uncinatum* Gomont. *Arch. Mikrobiol.* **24,** 134–146.
6.1

Niklowitz W. & Drews G. (1957) Beiträge zur Cytologie der Blaualgen. IV. Mitt.: Vergleichende elektronenmikroskopische Untersuchungen zur Substruktur einiger Hormogonales. *Arch. Mikrobiol.* **27,** 150–165.
6.2

Normore W.M. & Brown J.R. (1970) Guanine plus cytosine (G+C) composition of bacteria. In: Sober H.A. (Ed.) *CRC Handbook of Biochemistry, Selected Data for Molecular Biology*. pp. H-24–H-74. Chemical Rubber Co., Cleveland, Ohio.
12.3.4.2

Norris J.R., Crespi H.L. & Katz J.J. (1972) Characterization of a new photo-ESR signal associated with photosynthesis. *Biochem. biophys. Res. Commun.* **49,** 139–146.
4.3

Norris L., Norris R.E. & Calvin M. (1955) A survey of the rates and products of short-term photosynthesis in plants of nine phyla. *J. exp. Bot.* **6,** 64–74.
3.2.1

Novick R.P., Clowes R.C., Cohen S.N., tiss R., Datta N. & Falkow S. (1976) Uniform nomenclature for bacterial plasmids: a proposal. *Bact. Rev.* **40,** 168–189.
11.2.2

Nultsch W. (1961) Der Einfluss des Lichts auf die Bewegung der Cyanophyceen. I. Mitteilung: Photopotaxis von *Phormidium autumnale*. *Planta, Berl.* **56,** 632–647.
16.4.4

Nultsch W. (1962a) Phototaktische Aktionspektren von Cyanophyceen. *Ber. dt. bot. Ges.* **75,** 443–453.
16.4.2

Nultsch W. (1962b) Der Einfluss des Lichtes auf die Bewegung der Cyanophyceen. II. Photokinesis bei *Phormidium autumnale*. *Planta, Berl.* **57,** 613–623.
2.8.1, 16.4.3, 16.4.4

Nultsch W. (1965) Light reaction systems in Cyanophyceae. *Photochem. Photobiol.* **4,** 613–619.
2.8.1

Nultsch W. (1967) Untersuchungen über den Einfluss von Entkopplern auf die Bewegungsaktivität und das phototaktische Reaktionsverhalten blaugrüner Algen. *Z. Pflanzenphysiol.* **56,** 1–11.
2.8.1

Nultsch W. (1968) Einfluss von Redoxsystemen auf die Bewegungsaktivität und das phototaktische Reaktionsverhalten von *Phormidium uncinatum*. *Arch. Mikrobiol.* **63,** 295–320.
2.8.1

Nultsch W. (1974) Movements. In: Stewart W.D.P. (Ed.) *Algal Physiology and Biochemistry*. 989 pp., pp. 864–893. Blackwell, Oxford.
16.4.3, 16.4.4

Nultsch W. (1975) Phototaxis and photokinesis. In: Carlile M.J. (Ed.) *Primitive Sensory and Communication Systems*. pp. 29–90. Academic Press, London.
16.4.3

Nultsch W. & Häder D.-P. (1979) Photomovement of motile microorganisms. *Photochem. Photobiol.* **29,** 423–437.
16.1, 16.4.1, 16.4.2, 16.4.3

Nultsch W. & Hellman W. (1972) Untersuchungen zur Photokinesis von *Anabaena variabilis* Kützing. *Arch. Mikrobiol.* **82,** 76–90.
2.8.1, 16.4.2

Nultsch W., Schuchart H. & Höhl M. (1979) Investigations on the phototactic orientation of *Anabaena variabilis*. *Arch. Microbiol.* **122,** 85–91.
2.8.1, 16.4.4, 16.5

Nultsch W. & Throm G. (1968) Equivalence of light and ATP in photokinesis of *Rhodospirillum rubrum*. *Nature, Lond.* **218,** 697–699.
16.4.2

Nyholm N. (1976) A mathematical model for microbial growth under limitation by con-

servative substrates. *Biotechnol. Bioengng* **18**, 1043–1056.
5.2.1

Nyholm N. (1977) Kinetics of phosphate-limited algal growth. *Biotechnol. Bioengng* **19**, 467–492.
5.1.1, 5.2.1

O'Brien J.W. (1972) Limiting factors in phytoplankton algae: their meaning and measurement. *Science, N.Y.* **178**, 616–617.
18.5.2.1

O'Carra P. & Killilea S.D. (1971) Subunit structures of C-phycocyanin and C-phycoerythrin. *Biochem. biophys. Res. Commun.* **45**, 1192–1197.
7.5.1.3

O'Carra P. & O'hEocha C. (1966) Bilins released from algae and biliproteins by methanolic extraction. *Phytochem.* **5**, 993–997.
7.5.1.2

O'Carra P. & O'hEocha C. (1976) Algal biliproteins and phycobilins. In: Goodwin T.W. (Ed.) *Chemistry and Biochemistry of Plant Pigments, Vol I, 2nd edn.* 870 pp., pp. 328–376. Academic Press, New York.
7.1, 7.5.1.2

O'Carra P., O'hEocha C. & Carroll D.M. (1964) Spectral properties of the phycobilins. II. Phycoerythrobilin. *Biochemistry, N.Y.* **3**, 1343–1350.
7.5.1.2

Ochiai E.-I. (1978) The evolution of the environment and its influence on the evolution of life. *Origins of Life* **9**, 81–91.
12.1.3, 12.2.1

Odum H.T. (1967) Biological circuits and the marine systems of Texas. In: Olson T.A. & Burgess F.J. (Eds) *Pollution and Marine Ecology*. pp. 99–157. Interscience, New York.
20.4.4

Oehler D.Z. (1976) Transmission electron microscopy of organic microfossils from the late Precambrian Bitter Springs Formation of Australia: techniques and survey of preserved ultrastructure. *J. Paleontol.* **50**, 90–106.
21.3

Oelze J. & Drews G. (1972) Membranes of photosynthetic bacteria. *Biochim. biophys. Acta* **265**, 209–239.
6.2, 6.6

Ogawa K., Tsukihara T., Tahara H., Katsube Y., Matsu-Ura Y., Tanaka N., Kakudo M., Wada K. & Matsubara H. (1977) Location of the iron-sulfur cluster in *Spirulina platensis* ferredoxin by X-ray analysis. *J. Biochem., Tokyo* **81**, 529–531.
8.3

Ogawa R.E. & Carr J.E. (1969) The influence of nitrogen on heterocyst production in blue-green algae. *Limnol. Oceanogr.* **14**, 342–351.
5.2.4

Ogawa T., Sekine T. & Aiba S. (1979) Reappraisal of the so-called light shielding of gas vacuoles in *Microcystis aeruginosa*. *Arch. Microbiol.* **122**, 57–60.
2.2.2, 2.7.3.1

Ogawa T., Vernon L.O. & Mollenhauer H.H. (1969) Properties and structure of fractions prepared from *Anabaena variabilis* by the action of Triton X-100. *Biochim. biophys. Acta* **172**, 216–229.
8.2

O'hEocha C. (1958) Comparative biochemical studies of the phycobilins. *Arch. Biochem. Biophys.* **73**, 207–219.
7.5.1.1, 7.5.1.2

O'hEocha C. (1960) Chemical studies of phycoerythrin and phycocyanins. In: Allen M.B. (Ed.) *Comparative Biochemistry of Photoreactive Systems*. 437 pp., pp. 181–203. Academic Press, London.
7.5.1.2, 7.5.1.5

O'hEocha C. (1963) Spectral properties of the phycobilins. I. Phycocyanobilin. *Biochemistry, N.Y.* **2**, 375–382.
7.5.1.2

O'hEocha C. (1965) Biliproteins of algae. *A. Rev. Plant Physiol.* **16**, 415–434.
7.5.1.2

O'hEocha C. & O'Carra P. (1961) Spectral studies of denatured phycoerythrins. *J. Am. chem. Soc.* **83**, 1091–1093.
7.5.1.2, 7.5.1.5

Ohki K. & Fujita Y. (1978) Photocontrol of phycoerythrin formation in the blue-green alga *Tolypothrix tenuis* growing in the dark. *Pl. Cell Physiol., Tokyo* **19**, 7–15.
7.8.3.2

Ohki K. & Fujita Y. (1982) Laboratory culture of the pelagic blue-green alga *Trichodesmium thiebautii*: condition for unialgal culture. *Mar. Ecol. Prog. Ser.* **1**, 185–190.
19.2.4.1

Ohki K. & Katoh T. (1965) Photoorganotrophic growth of a blue-green alga, *Anabaena variabilis*. *Pl. Cell Physiol., Tokyo* **16**, 53–64.
8.6

Ohmori M. & Hattori A. (1971) Nitrogen fixation and heterocysts in the blue-green alga *Anabaena cylindrica*. *Pl. Cell Physiol., Tokyo* **12**, 961–967.
14.2

Ohmori M. & Hattori A. (1974) Effect of ammonia on nitrogen fixation by the blue-green alga *Anabaena cylindrica. Pl. Cell Physiol., Tokyo* **15,** 131–142.
4.5, 14.3.1, 14.3.2

Ohmori M. & Hattori A. (1978) Transient change in ATP pool of *Anabaena cylindrica* associated with ammonia assimilation. *Arch. Microbiol.* **117,** 17–20.
4.5

Okada S., Kanematsu S. & Asada K. (1979) Intracellular distribution of manganese and ferric superoxide dismutases in blue-green algae. *FEBS Lett.* **103,** 106–113.
8.3

Olson G.J. & Ingram L.O. (1975) Effects of temperature and nutritional changes on the fatty acids of *Agmenellum quadruplicatum. J. Bact.* **124,** 373–379.
6.5.3

Olson J.M. (1970) The evolution of photosynthesis. *Science, N.Y.* **168,** 438–446.
12.1.3, 12.2.1

Olson J.M. (1978) Precambrian evolution of photosynthetic and respiratory organisms. In: Hecht M.K., Steere W.C. & Wallace B. (Eds) *Evolutionary Biology, Vol. II.* 453 pp., pp. 1–37. Plenum Press, New York.
12.1.3, 12.2.1, 12.4.1

Ono T.A. & Murata N. (1977) Temperature dependence of the delayed fluorescence of chlorophyll a in blue-green algae. *Biochim. biophys. Acta* **460,** 220–229.
8.8

Ono T.A. & Murata N. (1978) Photosynthetic electron transport and photophosphorylation reactions in thylakoid membranes from the blue-green alga *Anacystis nidulans. Biochim. biophys. Acta* **502,** 477–485.
8.2

Ono T.A. & Murata N. (1979) Temperature dependence of the photosynthetic activities in thylakoid membranes from the blue-green alga *Anacystis nidulans. Biochim. biophys. Acta* **545,** 69–76.
6.5.3, 8.8

Öquist G. (1971) Changes in pigment composition and photosynthesis induced by iron-deficiency in the blue-green alga *Anacystis nidulans. Physiol. Plant.* **25,** 188–191.
6.5.2

Oren A. & Padan E. (1978) Induction of anaerobic, photoautotrophic growth in the cyanobacterium *Oscillatoria limnetica. J. Bact.* **133,** 558–563.
3.3, 9.2.1, 9.2.3, 9.4.2, 9.5.3

Oren A., Padan E. & Avron M. (1977) Quantum yields for oxygenic and anoxygenic photosynthesis in the cyanobacterium *Oscillatoria limnetica. Proc. natn. Acad. Sci. U.S.A.* **74,** 2152–2156.
4.6, 9.2.1, 9.2.4

Oren A., Padan E. & Malkin S. (1979) Sulfide inhibition of photosystem 2 in cyanobacteria (blue-green algae) and tobacco chloroplasts. *Biochim. biophys. Acta* **546,** 270–279.
8.6, 9.2.3, 9.3.4

Oren A. & Shilo M. (1979) Anaerobic heterotrophic dark metabolism in the cyanobacterium *Oscillatoria limnetica*: sulphur respiration and lactate fermentation. *Arch. Microbiol.* **122,** 77–84.
3.3, 9.4.2, 9.5.4, 12.3.3

Orkwiszewski K.G. & Kaney A.R. (1974) Genetic transformation of the blue-green bacterium, *Anacystis nidulans. Arch. Microbiol.* **98,** 31–37.
11.2.2, 11.5.3, 11.6.3, 11.7

Ortega T., Castillo F. & Cárdenas J. (1976) Photolysis of water coupled to nitrate reduction by *Nostoc muscorum* subcellular particles. *Biochem. biophys. Res. Commun.* **71,** 885–891.
8.3

Ortega T., Castillo F., Cárdenas J. & Losada M. (1977a) Inactivation by ammonia of the photosynthetic reduction of nitrate in *Nostoc muscorum* particles. *Biochem. biophys. Res. Commun.* **75,** 823–931.
8.3

Ortega T., Rivas J., Cárdenas J. & Losada M. (1977b) Metabolic interconversion of ferrodoxin-nitrate reductase and NADP reductase of *Nostoc muscorum. Biochem. biophys. Res. Commun.* **78,** 185–193.
8.3

Öström B. (1976) Fertilization of the Baltic by nitrogen fixation in the blue-green alga *Nodularia spumigena. Remote Sensing of the Environment* **4,** 305–310.
19.2.4.2, 19.3.4, 19.5.2

Ottolenghi E. & Hotchkiss R.D. (1960) Appearance of genetic transforming activity in Pneumococcal cultures. *Science, N.Y.* **132,** 1257–1258.
11.6.3

Overbeck J. (1975) Distribution pattern of uptake kinetic responses in a stratified eutrophic lake. *Verh. int. Verein. theor. angew. Limnol.* **19,** 2600–2615.
17.4.1

Ownby J.D. (1977) Effects of amino acids on methionine-sulfoximine-induced heterocyst

formation in *Anabaena*. *Planta, Berl.* **136,** 277–279.
14.6.1.3

Ownby J.D., Shannahan M. & Hood E. (1979) Protein synthesis and degradation in *Anabaena* during nitrogen starvation. *J. gen. Microbiol.* **110,** 255–261.
14.5, 14.6.1.3

Owners-Narhi L., Robinson S.J., DeRoo C.S. & Yocum C.F. (1979) Reconstitution of cyanobacterial photophosphorylation by a latent Ca^{2+} ATPase. *Biochem. biophys. Res. Commun.* **90,** 1025–1031.
8.7

Paasche E. (1973) Silicon and the ecology of marine plankton diatoms. I. *Thalassiosira pseudonana (Cyclotella nana)* grown in a chemostat with silicate as limiting nutrient. *Mar. Biol.* **19,** 117–126.
5.2.1

Padan E. (1979a) Facultative anoxygenic photosynthesis in cyanobacteria. *A. Rev. Pl. Physiol.* **30,** 27–40.
9.1, 12.3.3

Padan E. (1979b) Impact of facultative anaerobic photoautotrophic metabolism on ecology of cyanobacteria (blue-green algae). *Adv. microbial. Ecol.* **3,** 1–48.
9.1, 9.3.1, 9.3.2, 9.3.4, 9.4.2

Padan E. & Schuldiner S. (1978) Energy transduction in the photosynthetic membranes of the cyanobacterium (blue-green alga) *Plectonema boryanum*. *J. biol. Chem.* **253,** 3281–3286.
3.4.2, 3.5, 8.7, 9.6, 10.3.2.1, 10.3.3.1

Padan E. & Shilo M. (1969) Short-trichome mutant of *Plectonema boryanum*. *J. Bact.* **97,** 975–976.
11.5.5.2(a)

Padhy R.N. & Singh P.K. (1978) Genetical studies on the heterocyst and nitrogen fixation of the blue-green alga *Nostoc muscorum*. *Molec. gen. Genet.* **162,** 203–211.
11.5.1, 11.5.5.2(a), 11.5.5.2(b), 11.6.1, 14.6.1.1

Paerl H.W. (1976) Specific associations of the blue-green algae *Anabaena* and *Aphanizomenon* with bacteria in freshwater blooms. *J. Phycol.* **12,** 431–435.
3.4.3, 17.1.1, 17.6

Paerl H.W. (1978) Role of heterotrophic bacteria in promoting N_2 fixation by *Anabaena* in aquatic habitats. *Microb. Ecol.* **4,** 215–231.
17.1.1, 17.3.2, 17.4.4, 17.5.2, 17.6

Paerl H.W. (1979) Microbial organic carbon recovery in aquatic ecosystems. *Limnol. Oceanogr.* **23,** 927–935.
17.1.2, 17.4.4

Paerl H.W. & Kellar P.E. (1978) Significance of bacterial—*Anabaena* (Cyanophyceae) associations with respect to N_2 fixation in freshwater. *J. Phycol.* **14,** 254–260.
17.1.1, 17.3.2, 17.4.3, 17.5.2, 17.6

Paerl H.W. & Kellar P.E. (1979) Nitrogen-fixing *Anabaena*: physiological adaptations instrumental in maintaining surface blooms. *Science, N.Y.* **204,** 620–622.
18.5.2.1

Paerl H.W. & Lean D.R.S. (1976) Visual observations of phosphorus movement between algae, bacteria and abiotic particles in lake waters. *J. Fish. Res. Bd Can.* **33,** 2805–2813.
17.2, 17.4.2

Paerl H.W., Thomson R. & Goldman C.R. (1974) The ecological significance of detritus formation during a diatom bloom in Lake Tahoe, California-Nevada. *Verh. int. Verein. theor. angew. Limnol.* **19,** 826–834.
17.1.2

Pankratz H.S. & Bowen C.C. (1963) Cytology of blue-green algae. 1. The cells of *Symploca muscorum*. *Am. J. Bot.* **50,** 387–399.
1, 3.2.1, 6.2, 7.2, 11.3.1, 13.3

Pant A. & Devassy V.P. (1976) Release of extracellular matter during photosynthesis by a *Trichodesmium* bloom. *Curr. Sci.* **45,** 487–489.
19.3.2

Papageorgiou G. (1975) Chlorophyll fluorescence: an intrinsic probe of photosynthesis. In: Govindjee (Ed.) *Bioenergetics of Photosynthesis*. 698 pp., pp. 414–492. Academic Press, New York.
9.2.3

Parker B. (1961) Facultative heterotrophy in certain soil algae from the ecological viewpoint. *Ecology* **42,** 381–386.
3.4.3

Parkinson J.S. (1977) Behavioral genetics in bacteria. *A. Rev. Genetics* **11,** 397–414.
16.5

Parr M.P. & Smith R.V. (1976) The identification of phosphorus as a growth-limiting nutrient in Lough Neagh using bioassays. *Water Res.* **10,** 1151–1154.
18.5.1.2, 18.5.2.4

Parrot L.M. & Slater J.H. (1980) The DNA, RNA and protein composition of the cyanobacterium *Anacystis nidulans* grown in light- and carbon dioxide-limited chemostats. *J. gen. Microbiol.* **127,** 53–58.
11.3.1

Paschinger H. (1977) DCCD induced sodium uptake by *Anacystis nidulans. Arch. Microbiol.* **113,** 285–291.
3.4.2, 3.5, 10.2.3, 10.3.3.1, 10.3.3.2, 16.4.2

Paster B.J. & Canale-Parola E. (1980) Involvement of periplasmic fibrils in motility of spirochetes. *J. Bact.* **141,** 359–364.
16.2.3

Pate J.L. & Chang L.E. (1979) Evidence that gliding motility in prokaryotic cells is driven by rotary assemblies in the cell envelopes. *Current Microbiol.* **2,** 59–64.
16.2.3

Pate J.L. & Ordal E.J. (1967) The fine structure of *Chondrococcus columnaris.* III. The surface layers of *Chondrococcus columnaris. J. Cell Biol.* **35,** 37–51.
16.2.2.2, 16.2.3

Pate J.L., Petzold S.J. & Chang L.E. (1979) Phages for the gliding bacterium *Cytophaga johnsonae* that infect only motile cells. *Current Microbiol.* **2,** 257–262.
16.2.3

Patrick R., Crum B. & Coles J. (1961) Temperature and manganese as determining factors in the presence of diatom or blue-green algal floras in streams. *Proc. natn. Acad. Sci. U.S.A.* **64,** 472–478.
18.5.1.2

Pattnaik U. & Singh P.K. (1978) Effect of nitrate nitrogen on the growth, heterocyst differentiation and nitrogen fixation in rice field blue-green algae *Gloeotrichia* sp. *Arch. Hydrobiol., Suppl.* **51,** 318–327.
14.6.1.1

Pearce J. & Carr N.G. (1967) The metabolism of acetate by the blue-green algae *Anabaena variabilis* and *Anacystis nidulans. J. gen. Microbiol.* **49,** 301–313.
3.3

Pearce J. & Carr N.G. (1969) The incorporation and metabolism of glucose by *Anabaena variabilis. J. gen. Microbiol.* **54,** 451–462.
3.3, 3.4.2

Pearsall W.H. (1932) Phytoplankton in English Lakes. II. The composition of the phytoplankton in relation to dissolved substances. *J. Ecol.* **20,** 241–262.
9.5.4, 18.4.2, 18.5.4

Pearson H.W., Howsley R., Kjeldsen C.K. & Walsby A.E. (1979) Aerobic nitrogenase activity associated with a non-heterocystous filamentous cyanobacterium. *FEMS Microbiol. Lett.* **5,** 163–167.
4.1, 18.5.2.1

Pearson H.W., Malin G. & Howsley R. (1981) Physiological studies on *in vitro* nitrogenase activity by axenic cultures of the blue-green alga *Microcoleus chthonoplastes. Br. phycol. J.* **16,** 139.
20.4.5

Peat A. & Whitton B.A. (1967) Environmental effects on the structure of the blue-green alga, *Chlorogloea fritschii. Arch. Mikrobiol.* **57,** 155–180.
6.2, 6.5.1, 6.5.2

Peat A. & Whitton B.A. (1968) Vegetative cell structure in *Anabaenopsis* sp. *Arch. Mikrobiol.* **63,** 170–176.
6.5.2

Pechman K.J. & Woese C.R. (1972) Characterization of the primary structural homology between the 16S ribosomal RNAs of *Escherichia coli* and *Bacillus megaterium* by oligomer cataloging. *J. molec. Evol.* **1,** 230–240.
12.2.2.4

Pelletier J. (1968) Première colonisation du Léman par *Oscillatoria rubescens* D.C. *Rev. Algol., N.S.* **9,** 186–192.
18.4.3

Pelroy R.A. & Bassham J.A. (1972) Photosynthetic and dark carbon metabolism in unicellular blue-green algae. *Arch. Mikrobiol.* **86,** 25–38.
3.2.1, 3.3, 4.3

Pelroy R.A. & Bassham J.A. (1973a) Efficiency of energy conversion of aerobic glucose metabolism in *Aphanocapsa* 6714. *J. Bact.* **115,** 937–942.
3.4.2

Pelroy R.A. & Bassham J.A. (1973b) Kinetics of glucose incorporation by *Aphanocapsa* 6714. *J. Bact.* **115,** 943–948.
3.3, 3.4.2

Pelroy R.A., Kirk M.R. & Bassham J.A. (1976) Photosystem II regulation of macromolecule synthesis in the blue-green alga *Aphanocapsa* 6714. *J. Bact.* **128,** 623–632.
3.3, 3.4.2

Pelroy R.A., Levine G.A. & Bassham J.A. (1976) Kinetics of light-dark CO_2 fixation and glucose assimilation by *Aphanocapsa* 6714. *J. Bact.* **128,** 633–643.
3.3

Pelroy R.A., Rippka R. & Stanier R.Y. (1972) The metabolism of glucose by unicellular blue-green algae. *Arch. Mikrobiol.* **87,** 303–322.
3.3, 3.4.2, 3.4.4

Pemberton J.M. (1974) Size of the chromosome of *Pseudomonas aeruginosa* PAO. *J. Bact.* **119,** 748–752.
11.3.1

Pentecost A. (1978) Blue-green algae and

freshwater carbonate deposits. *Proc. R. Soc. Lond. B.* **200,** 43–61.
2.7.3.2

Perdue E.M., Beck K.C. & Reuter J.H. (1976) Organic complexes of iron and aluminium in natural waters. *Nature, Lond.* **260,** 418–420.
18.5.2.3

Perkins M.S. (1974) *The Dynamics of a Stream Periphyton Community in Ward Creek, California.* 96 pp. Ph.D. Thesis. University of California, Davis.
17.1.2

Perkins R.D. & Tsentas C.I. (1976) Microbial infestation of carbonate substrates planted on the St. Croix Shelf, West Indies. *Geol. Soc. Am. Bull.* **87,** 1615–1628.
20.2.2, 20.4.3

Perry G.J., Gillan F.T. & Johns R.B. (1978) Lipid composition of a prochlorophyte. *J. Phycol.* **14,** 369–371.
12.3.3

Peschek G.A. (1978) Reduced sulfur and nitrogen compounds and molecular hydrogen as electron donors for anaerobic CO_2 photoreduction in *Anacystis nidulans. Arch. Microbiol.* **119,** 313–322.
9.3.1, 12.3.3

Peschek G.A. (1979a) Anaerobic hydrogenase activity in *Anacystis nidulans*: H_2-dependent photoreduction and related reactions. *Biochim. biophys. Acta* **548,** 187–202.
4.6

Peschek G.A. (1979b) Aerobic hydrogenase activity in *Anacystis nidulans.* The oxyhydrogen reaction. *Biochim. biophys. Acta* **548,** 203–215.
4.6

Peschek G.A. (1979c) Evidence for two functionally distinct hydrogenases in *Anacystis nidulans. Arch. Microbiol.* **123,** 81–92.
4.6

Peschek G.A. (1980) Restoration of respiratory electron-transport reactions in quinone depleted particle preparations of *Anacystis nidulans. Biochem. J.* **186,** 515–523.
4.6

Peschek G.A. & Schmetterer G. (1978) Reversible photooxidative loss of pigments and of intracytoplasmic membranes in the blue-green alga *Anacystis nidulans. FEMS Microbiol. Lett.* **3,** 295–297.
6.5.1

Peschek G.A. & Schmetterer G. (1980) Photosynthesis and respiration in blue-green algae: electron-transporting function of thylakoids and plasmamembrane in *Anacystis nidulans. Abst. V. Int. Cong. Photo. Res.* p. 444.
1

Peters G.A. (1975) The *Azolla-Anabaena azollae* relationship. III. Studies on metabolic capacities and a further characterization of the symbiont. *Arch. Microbiol.* **103,** 113–122.
14.4.3.2

Peters G.A. (1977) The *Azolla-Anabaena azollae* symbiosis. In: Hollaender A. (Ed.) *Genetic Engineering for Nitrogen Fixation.* 538 pp., pp. 231–256. Plenum Press, New York.
14.1

Peterson R.B. & Burris R.H. (1976a) Properties of heterocysts isolated with colloidal silica. *Arch. Microbiol.* **108,** 35–40.
4.4, 14.2, 14.4.3.1

Peterson R.B. & Burris R.H. (1976b) Conversion of acetylene reduction rates to nitrogen fixation rates in natural populations of blue-green algae. *Anal. Biochem.* **73,** 404–410.
14.2

Peterson R.B. & Burris R.H. (1978) Hydrogen metabolism in isolated heterocysts of *Anabaena* 7120. *Arch. Microbiol.* **116,** 125–132.
4.3, 4.6, 14.4.3.1, 14.4.3.3

Peterson R.B., Dolan E., Calvert H.E. & Ke B. (1981) Energy transfer from phycobiliproteins to photosystem I in vegetative cells and heterocysts of *Anabaena variabilis. Biochim. biophys. Acta* **634,** 237–248.
14.3.2, 14.4.3.2

Peterson R.B. & Wolk C.P. (1978a) Localization of an uptake hydrogenase in *Anabaena. Pl. Physiol., Lancaster* **61,** 688–691.
4.1, 14.4.1, 14.4.3.3

Peterson R.B. & Wolk C.P. (1978b) High recovery of nitrogenase activity and of ^{55}Fe-labeled nitrogenase in heterocysts isolated from *Anabaena variabilis. Proc. natn. Acad. Sci. U.S.A.* **75,** 6271–6275.
4.1, 4.6, 14.2, 14.4.1, 14.4.3.1, 14.4.3.2, 14.4.3.3

Pettersson K. (1980) Alkaline phosphatase activity and algal surplus phosphorus as phosphorus-deficiency indicators in Lake Erken. *Arch. Hydrobiol.* **89,** 54–87.
5.0.0, 18.5.2.4

Pfeffer W. (1884) Locomotorische Richtungsbewegungen durch chemische Reize. *Unters. bot. Inst. Tübingen* **1,** 363–482.
17.5.1

Pfennig N. (1975) The phototrophic bacteria and their role in the sulfur cycle. *Plant & Soil* **43,** 1–16.
9.3.3, 9.4.1

Pfennig N. (1977) Phototrophic green and purple bacteria: a comparative systematic survey. *A. Rev. Microbiol.* **31,** 275–290.
9.3.3, 9.4.1, 12.2.1

Pfennig N. (1978) General physiology and ecology of photosynthetic bacteria. In: Clayton R.K. & Sistrom W.R. (Eds) *The Photosynthetic Bacteria.* 946 pp., pp. 3–18. Plenum Press, New York.
12.2.1

Phares W. & Chapman L.F. (1975) *Anacystis nidulans* mutants resistant to aromatic amino acid analogues. *J. Bact.* **122,** 943–948.
11.5.1

Phillips D.O. & Carr N.G. (1975) Hybridization of prokaryotic and eukaryotic 5S rRNA to *Euglena gracilis* chloroplast DNA. *FEBS Lett.* **60,** 94–97.
12.4.3.1

Phillips D.O. & Carr N.G. (1977) Nucleic acid analysis and the endosymbiotic hypothesis. *Taxon* **26,** 3–42.
12.4.3.1

Phillips D.O. & Carr N.G. (1981) Molecular approaches to the endosymbiotic hypothesis. *Ann. N.Y. Acad. Sci.* **361,** 298–311.
12.4.3.1

Pia J, (1927) Thallophyta. In: Hirmer M. (Ed.) *Handbuch der Paläeobotanik* 1. pp. 31–136. R. Oldenbourg, Berlin.
21.1

Piccioni R.G. & Mauzerall D.C. (1978a) Calcium and photosynthetic oxygen evolution in cyanobacteria. *Biochim. biophys. Acta* **504,** 384–397.
8.2

Piccioni R.G. & Mauzerall D.C. (1978b) A high potential redox component located within cyanobacterial photosystem II. *Biochim. biophys. Acta* **504,** 398–405.
8.6

Pierson B.K. & Castenholz R.W. (1971) Bacteriochlorophylls in gliding, filamentous prokaryotes from hot springs. *Nature, New Biol.* **233,** 25–27.
16.2.3

Pierson B.K. & Castenholz R.W. (1974) A phototrophic gliding filamentous bacterium of hot springs, *Chloroflexus aurantiacus*, gen. and sp. nov. *Arch. Microbiol.* **100,** 5–24.
9.5.2, 16.2.3

Pigott G.H. & Carr N.G. (1972) Homology between nucleic acids of blue-green algae and chloroplasts of *Euglena gracilis. Science, N.Y.* **175,** 1259–1261.
12.4.3.1

Pipes W.O. & Koutsoyannis S.P. (1962) Light-limited growth of *Chlorella* in continuous cultures. *J. appl. Microbiol.* **10,** 1–5.
2.5.1, 2.5.2

Pirt S.J. (1965) The maintenance energy of bacteria in growing cultures. *Proc. R. Soc. London B.* **163,** 224–231.
2.5.2

Playford P.E. & Cockbain A.E. (1976) Modern algal stromatolites at Hamelin Pool, a hypersaline barred basin in Shark Bay, Western Australia. In: Walter M.R. (Ed.) *Stromatolites. Developments in Sedimentology* **20.** 790 pp., pp. 389–412. Elsevier, Amsterdam.
9.5.3, 20.4.2

Pliński M. (1975) The algae in the surface water of the Bay of Puck (Baltic) in the vegetation period of 1972. *Bot. Mar.* **18,** 183–186.
19.2.1

Pomiluyko V.P. & Stetsenko N.M. (1972) Phosphorus metabolism in common species of blue-green algae. *Hydrobiological J.* (English Translation) **8**(1), 74–76.
5.4.1

Porter G., Tredwell C.J., Searle G.F.W. & Barber J. (1978) Picosecond time-resolved energy transfer in *Porphyridium cruentum.* Part I. In the intact alga. *Biochim. biophys. Acta* **501,** 232–245.
7.6

Porter J. & Jost M. (1976) Physiological effects of the presence and absence of gas vacuoles in the blue-green alga *Microcystis aeruginosa* Kuetz. emend. Elenkin. *Arch. Microbiol.* **110,** 225–231.
2.7.3.1

Postgate J.R. (1978) *Nitrogen fixation. Studies in Biology, No 92.* 68 pp. E. Arnold, London.
4.4

Potts M. (1979) Nitrogen fixation (acetylene reduction) associated with communities of heterocystous and non-heterocystous blue-green algae in mangrove forests of Sinai. *Oecologia, Berl.* **39,** 359–373.
20.2.3, 20.4.5

Potts M. (1980) Blue-green algae (Cyanophyta) in marine coastal environments of the Sinai Peninsula; distribution, zonation, stratification and taxonomic diversity. *Phycologia* **19,** 60–73.
20.2.3, 20.3.1, 20.3.4

Potts M., Krumbein W.E. & Metzger J. (1978) Nitrogen fixation in anaerobic sediments determined by acetylene reduction, a new ^{15}N field assay and simultaneous total N and ^{15}N determinations. In: Krumbein W.E. (Ed.) *Environmental Biochemistry 3: Methods, Metals and Assessment*, pp. 753–769. Ann Arbor Sci., Ann Arbor.
20.4.5

Potts M. & Whitton B.A. (1977) Nitrogen

fixation by blue-green algal communities in the intertidal zone of the lagoon of Aldabra Atoll. *Oecologia, Berl.* **27,** 275–283.
20.4.5

Potts M. & Whitton B.A. (1979a, b) pH and Eh on Aldabra Atoll. 1. Comparison of marine and freshwater environments. 2. Intertidal photosynthetic microbial communities showing zonation. *Hydrobiologia* **67,** 11–17, 99–105.
20.2.2

Potts M. & Whitton B.A. (1980) Vegetation of the intertidal zone of the lagoon of Aldabra, with particular reference to the photosynthetic prokaryotic communities. *Proc. Roy. Soc. Lond. B.* **208,** 13–55.
20.2.2, 20.2.3, 20.3.3, 20.3.4, 20.4.1, 20.4.3, 20.4.4

Prakash G. & Kumar H.D. (1971) Studies on sulfur-selenium antagonism in blue-green algae. I. Sulfur nutrition. *Arch. Mikrobiol.* **77,** 196–202.
6.5.2

Pratt R., Mautner H., Gardner G.M., Sha-Yi-Hsien & Dufrenay J. (1951) Antibiotic activity of seaweed extract. *J. Am. pharm. Ass. (Sci. edn)* **40,** 575–579.
17.4.4

Price C.A. (1968) Iron compounds and plant nutrition. *A. Rev. Pl. Physiol.* **19,** 239–248.
17.4.2

Pringsheim E.G. (1963) *Farblose Algen.* 471 pp. Gustav Fischer, Stuttgart, Germany.
16.4.3

Pringsheim E.G. (1965) *Oscillatoria agardhii* var. *suspensa* nov. var. Kleine Mitteilungen über Algen and Flagellaten X. *Arch. Mikrobiol.* **50,** 401–413.
18.5.1.1

Pringsheim E.G. (1968a) Cyanophyceen-Probleme. *Planta, Berl.* **79,** 1–9.
14.2

Pringsheim E.G. (1968b) Cyanophyceen-Studien. *Arch. Mikrobiol.* **63,** 331–355.
14.2

Pritchard R., Beauclerk A. & Smith A.J. (1975) Glucose dehydrogenase of *Chlorogloeopsis fritschii. Biochem. Soc. Trans.* **3,** 384–385.
3.4.2

Provasoli L. & Pintner I.J. (1953) Ecological implications of *in vitro* nutritional requirements of algal flagellates. *Ann. N.Y. Acad. Sci.* **56,** 839–851.
17.4.4

Prud'homme Van Reine W.F. & van den Hoek C. (1966) Cultural evidence for the morphologic plasticity of *Entophysalis deusta* (MENEGHINI) DROUET and DAILY (Chroococcales, Cyanophyceae). *Blumea* **14,** 277–283.
20.2.1

Pulich W.M. & Van Baalen C. (1973) Pyridine nucleotide dependent glucose dehydrogenase activity in blue-green algae. *J. Bact.* **114,** 28–33.
3.4.2

Pulich W.M., Van Baalen C., Gibson J.L. & Tabita F.R. (1976) Purification and characterization of glucose dehydrogenase from a heterotrophically grown blue-green alga. *Pl. Physiol., Lancaster* **58,** 393–397.
3.4.2

Pulich W.M., Winters K. & Van Baalen C. (1974) The effects of No. 2 fuel oil and two crude oils on the growth and photosynthesis of microalgae. *Mar. Biol.* **28,** 87–94.
20.3.6

Pullin C.A., Brown R.G. & Evans, E.H. (1979) Detection of allophycocyanin in photosystem I preparations from the blue-green alga, *Chlorogloea fritschii. FEBS Lett.* **101,** 110–112.
8.4

Rabenhorst L. (1865) *Flora Europaea Algarum, Vol. 2.* 319 pp. Leipzig.
15.1

Rabin P. & Darbre A. (1975) An improved extraction procedure for the endotoxin from *Microcystis aeruginosa* NRC-1. *Biochem. Soc. Trans.* **3,** 428–430.
18.5.6

Raboy B. & Padan E. (1978) Active transport of glucose and α-methylglucoside in the cyanobacterium *Plectonema boryanum. J. biol. Chem.* **253,** 3287–3291.
3.4.2, 3.4.3, 3.4.4, 8.7, 9.6, 10.2.3, 10.3.1, 10.3.3.1, 10.3.4

Raboy B., Padan E. & Shilo M. (1976) Heterotrophic capacities of *Plectonema boryanum. Arch. Microbiol.* **110,** 77–85.
3.3, 3.4.1, 3.4.2, 3.4.4, 6.5.2, 12.3.3

Ragan M.A. & Chapman D.J. (1978) *A Biochemical Phylogeny of the Protists.* 317 pp. Academic Press, New York.
12.1.3, 12.2.1, 12.3.3, 12.3.4.2, 12.4.1, 12.4.4.3

Rai A.K. (1978) Mutants of the blue-green alga *Anabaena ambigua. Biochem. Physiol. Pflanz.* **172,** 177–180.
11.5.1

Rai A.K., Pandey K.D. & Kashyap A.K. (1978) Heterocyst differentiation and polarity in *Calothrix. New Phytol.* **81,** 647–651,
14.6.1.1

Ramamurthy J. & Capindale J.B. (1970) A new isolation and structure for the endotoxin from *Microcystis aeruginosa* NRC-1. *Can. J. Biochem.* **48**, 508–510.
18.5.6

Ramamurthy V.D. (1970a) Antibacterial activity traceable to the marine blue-green alga *Trichodesmium erythraeum* in the gastro-intestinal contents of two pelagic fishes. *Hydrobiologia* **36**, 159–163.
19.5.4

Ramamurthy V.D. (1970b) Antibacterial activity of the marine blue-green alga *Trichodesium erythraeum* in the gastro-intestinal contents of the sea-gull *Laurus brunicephalus*. *Mar. Biol.* **6**, 74–76.
19.5.4

Ramamurthy V.D. (1972) On the culture of the marine blue-green alga *Trichodesmium erythraeum* Ehr. In: Desikachary T.V. (Ed.) *Taxonomy and Biology of Blue-Green Algae*. 591 pp., pp. 425–427. Univ. Madras, India.
19.3.1

Rambler M., Margulis L. & Barghoorn E.S. (1977) Natural mechanisms of protection of a blue-green alga against ultraviolet light. In: Ponnamperuma C. (Ed.) *Chemical Evolution of the Early Precambrian*. 221 pp., pp. 133–141. Academic Press, New York.
20.3.4

Rao V.S.K., Brand J.J. & Myers J. (1977) Cold shock syndrome in *Anacystis nidulans*. *Pl. Physiol., Lancaster* **59**, 965–969.
8.8

Raven P.H. (1970) A multiple origin for plastids and mitochondria. *Science, N.Y.* **169**, 641–646.
12.4.1, 12.4.4.3

Ravera O. & Vollenweider R.A. (1968) *Oscillatoria rubescens D.C.* as an indicator of Lago Maggiore eutrophication. *Schweiz. Z. Hydrol.* **30**, 374–380.
18.4.3

Raymond P.E. (1935) Pre-Cambrian life. *Geol. Soc. Am. Bull.* **46**, 375–392.
21.1

Reanney D.C. (1978) Coupled evolution: adaptive interactions among the genomes of plasmids, viruses and cells. *Int. Rev. Cytol., Suppl.* **8**, 1–68.
12.1.3

Reaston J., van den Hondel C.A.M.J.J., van der Ende A., van Arkel G.A., Stewart W.D.P. & Herdman M. (1980) Comparison of plasmids from the cyanobacterium *Nostoc* PCC 7524 with two mutant strains unable to form heterocysts. *FEMS Microbiol. Lett.* **9**, 185–188.
11.3.2

Reddy P.M., Rao P.S.N. & Talpasayi E.R.S. (1975) Effect of red and far red illumination on the germination of spores of two blue-green algae. *Curr. Sci.* **44**, 678–679.
15.5

Reddy P.M. & Talpasayi E.R.S. (1974) Heterocyst formation in a blue-green alga, *Cylindrospermum*. *Nature, Lond.* **249**, 493–494.
14.6.1.2

Reed R.H., Rowell P. & Stewart W.D.P. (1981a) Characterization of the transport of potassium ions in the cyanobacterium *Anabaena variabilis* Kütz. *Eur. J. Biochem.* **116**, 323–330.
10.3.3.2

Reed R.H., Rowell P. & Stewart W.D.P. (1981b) Uptake of potassium and rubidium ions by the cyanobacterium *Anabaena variabilis*. *FEMS Microbiol. Lett.* **11**, 233–236.
10.3.3.2

Reichenbach H. (1981) Taxonomy of the gliding bacteria. *A. Rev. Microbiol.* **35**, 339–364.
16.2.1

Reim R.L., Shane M.S. & Cannon R.E. (1974) The characterization of a *Bacillus* capable of blue-green bactericidal activity. *Can. J. Microbiol.* **20**, 981–986.
17.4.2

Reinman S. & Thornber J.P. (1979) The electrophoretic isolation and partial characterization of three chlorophyll-protein complexes from blue-green algae. *Biochim. biophys. Acta* **547**, 188–197.
14.3.2

Renaut J., Sasson A., Pearson H.W. & Stewart W.D.P. (1975) Nitrogen-fixing algae in Morocco. In: Stewart W.D.P. (Ed.) *Nitrogen Fixation by Free-living Microorganisms*. pp. 229–246. Cambridge U.P.
20.3.1, 20.3.4, 20.4.5

Revsbec N.P., Sorensen J., Blackburn T.H. & Lomholt J.P. (1980) Distribution of oxygen in marine sediments measured with microelectrodes. *Limnol. Oceanogr.* **25**, 403–411.
9.5.3

Reynolds C.S. (1971) The ecology of the planktonic blue-green algae in the North Shropshire meres. *Field Studies* **3**, 409–432.
18.4.4, 18.4.5

Reynolds C.S. (1972) Growth, gas vacuolation and buoyancy in a natural population of a planktonic blue-green alga. *Freshwat. Biol.* **2**, 87–106.
5.4.1, 15.5

Reynolds C.S. (1973) Growth and buoyancy of *Microcystis aeruginosa* Kütz. emend. Elenkin in a shallow eutrophic lake. *Proc. R. Soc. Lond. B.* **184**, 29–50.
2.6.3, 2.8.2.3, 5.4.1, 10.2.5.2

Reynolds C.S. (1975) Interrelations of photosynthetic behaviour and buoyancy regulation in a natural population of a blue-green alga. *Freshwat. Biol.* **5**, 323–338.
15.5, 18.4.4

Reynolds C.S. (1976) Succession and vertical distribution of phytoplankton in response to thermal stratification in a lowland mere, with special reference to nutrient availability. *J. Ecol.* **64**, 529–551.
5.4.3

Reynolds C.S., Jaworski G.H.M., Cmiech H.A. & Leedale G.F. (1981) On the annual cycle of the blue-green alga *Microcystis aeruginosa* Kütz. emend. Elenkin. *Phil. Trans. R. Soc.* **293**, 419–477.
18.4.5

Reynolds C.S. & Rogers D.A. (1976) Seasonal variations in the vertical distribution and buoyancy of *Microcystis aeruginosa* Kütz. emend. Elenkin in Rostherne Mere, England. *Hydrobiologia* **48**, 17–23.
18.4.5, 18.5.3

Reynolds C.S. & Walsby A.E. (1975) Waterblooms. *Biol. Rev.* **50**, 437–481.
2.8.2.1, 2.8.2.2, 5.2.4, 18.1, 18.4.1, 18.4.4, 19.4.1

Rhee G-Y. (1972) Competition between an alga and aquatic bacterium for phosphate. *Limnol. Oceanogr.* **17**, 505–514.
17.4.2

Rhee G-Y. (1973) A continuous culture study of phosphate uptake, growth rate and polyphosphate in *Scenedesmus* sp. *J. Phycol.* **9**, 495–506.
5.1.1, 5.2.1, 5.2.3

Rhee G-Y. (1980) Continuous culture in phytoplankton ecology. In: Droop M.R. & Jannasch H.W. (Eds), *Advances in Aquatic Microbiology*. Vol. 2, 356 pp., pp. 150–203. Academic Press, New York.
2.5.1

Rhee G-Y. & Gotham I.J. (1981a) The effects of environmental factors on phytoplankton growth: light and the interactions of light with nitrate limitation. *Limnol. Oceanogr.* **26**, 649–659.
2.5.1

Rhee G-Y. & Gotham I.J. (1981b) The effect of environmental factors on phytoplankton growth: temperature and the interactions of temperature with nutrient limitation. *Limnol. Oceanogr.* **26**, 635–648.
2.5.1

Rhee G-Y., Gotham I.J. & Chisholm S.W. (1981) Use of cyclostat cultures to study phytoplankton ecology. In: Calcott P.C. (Ed.) *Continuous Culture of Cells*. Vol 2, 208 pp., pp. 159–186. CRC Press, Cleveland.
2.5.1, 2.6.2

Richmond M.H. (1979) 'Cells' and 'organisms' as habitats for DNA. *Proc. R. Soc. Lond. B* **204**, 235–250.
12.1.3

Riddolls A. (1975) *Aspects of Nitrogen Fixation by Blue-green Algae in Lough Neagh*. Ph.D. Thesis, New University of Ulster, N. Ireland.
18.5.2.1

Ridgway H.F. (1977) Source of energy for gliding motility in *Flexibacter polymorphus*: effects of metabolic and respiratory inhibitors in gliding movement. *J. Bact.* **131**, 544–556.
16.3

Ridgway H.F., Wagner R.M., Dassey W.T. & Lewin R.A. (1975) Fine structure of the cell envelope layers of *Flexibacter polymorphus*. *Can. J. Microbiol.* **21**, 1733–1750.
16.2.3

Rigby C.H., Craig S.R. & Budd K. (1980) Phosphate uptake by *Synechococcus leopoliensis* (Cyanophyceae): enhancement by calcium ion. *J. Phycol.* **16**, 389–393.

Riley M. & Anilionis A. (1978) Evolution of the bacterial genome. *A. Rev. Microbiol.* **32**, 519–560.
11.3.1

Rippka R. (1972) Photoheterotrophy and chemoheterotrophy among unicellular blue-green algae. *Arch. Mikrobiol.* **87**, 93–98.
3.4, 3.4.1, 3.4.2, 3.4.4, 6.5.2

Rippka R., Deruelles J., Waterbury J.B., Herdman M. & Stanier R.Y. (1979) Generic assignments, strain histories and properties of pure cultures of cyanobacteria. *J gen. Microbiol.* **111**, 1–61.
3.1, 3.4.1, 4.1, 7.5.1.1, 7.5.3, 7.8.3.1, 11.1, 11.2.1, 11.3.1, 11.3.2, 11.3.3, 11.5.5.2(b), 12.3.1, 12.3.3, 12.3.4.2, 12.3.4.3, 13.2.2, 13.3

Rippka R., Neilson A., Kunisawa R. & Cohen-Bazire G. (1971) Nitrogen fixation by unicellular blue-green algae. *Arch. Mikrobiol.* **76**, 341–348.
4.1

Rippka R. & Stanier R.Y. (1978) The effects of anaerobiosis on nitrogenase synthesis and heterocyst development by nostocacean cyanobacteria. *J. gen. Microbiol.* **105**, 83–94.
4.1, 4.5, 11.5.5.2(b), 14.3.1, 14.5

Rippka R. & Stanier R.Y. (1939) Observations on unpublished strain in the Pasteur culture collection of cyanobacteria. In: Nichols J.M. (Ed.) *Abstracts of the Third International Symposium on Photosynthetic Prokar-*

yotes, Oxford. p. A26. Univ. Liverpool, England.

Rippka R. & Waterbury J.B. (1977) The synthesis of nitrogenase by non-heterocystous cyanobacteria. *FEMS. Microbiol. Lett.* **2,** 83–86.
11.5.5.2(b), 12.3.3

Rippka R., Waterbury J.B. & Cohen-Bazire G. (1974) A cyanobacterium which lacks thylakoids. *Arch. Microbiol.* **100,** 419–436.
1, 6.2, 7.3, 7.5.1.1, 10.3.3.1

Ris H. & Singh R.N. (1961) Electron microscope studies on blue-green algae. *J. biophys biochem. Cytol.* **9,** 63–80.
3.2.2.2, 6.2, 7.2, 11.3.1

Robbins S.J., Hervey A. & Stebbins M. (1951) Further observations on *Euglena* and B_{12}. *Bull. Torrey bot. Club* **78,** 363–375.
17.4.4

Roberts R.J. (1978) Restriction and modification enzymes and their recognition sequences. *Gene* **4,** 183–193.
11.3.3

Roberts R.J. (1980) Restriction and modification enzymes and their recognition sequences. *Gene* **8,** 329–343.
11.3.3

Roberts T.M., Klotz L.C. & Loeblich A.R. (1977) Characterisation of a blue-green algal genome. *J. mol. Biol.* **110,** 341–361.
11.3.1

Roberts T.M. & Koths K.E. (1976) The blue-green alga *Agmenellum quadruplicatum* contains covalently closed DNA circles. *Cell* **9,** 551–557.
11.3.2

Roberts T.M., Lauer G.D. & Klotz L.C. (1976) Physical studies on DNA from 'primitive' eucaryotes. *CRC Crit. Rev. Biochem.* **3,** 349–449.
12.4.2

Robertson B.R. & Button D.K. (1979) Phosphate-limited continuous culture of *Rhodotorula rubra*: kinetics of transport, leakage and growth. *J. Bact.* **138,** 884–895
5.1.2, 5.2.1

Roelofs T.D. & Oglesby R.T. (1970) Ecological observations on the planktonic cyanophyte *Gloeotrichia echinulata. Limnol. Oceanogr.* **15,** 224–229.
15.4.1, 15.5, 18.4.5

Roger P.A. & Kulasooriya S.A. (1980) *Blue-green Algae and Rice.* 112 pp. International Rice Research Institute, Los Baños, Laguna, Philippines.
References

Rogerson A.C. (1979) Modifiers of heterocyst repression and spacing and formation of heterocysts without nitrogenase in the cyanobacterium *Anabaena variabilis. J. Bact.* **140,** 213–219.
14.6.1.1, 14.6.1.3

Rogerson A.C. (1980) Nitrogen-fixing growth by non-heterocystous cyanobacterium *Plectonema boryanum. Nature, Lond.* **284,** 563–564.
18.5.2.1

Rogerson A.C., Rowell P. & Stewart W.D.P. (1978) Guanosine 5′-diphosphate-3′-diphosphate in the heterocystous cyanobacterium *Anabaenopsis circularis*: possible lack of involvement in cellular metabolism. *FEMS Microbiol. Lett.* **3,** 299–303.
14.5

Rohatgi A. & Singh S.P. (1979) Isolation and characterization of pigment mutants of the blue-green alga *Aphanothece stagnina. Molec. gen. Genet.* **169,** 59–62.
11.5.4

Rohlich G.A. (Ed.) (1969) *Eutrophication: Causes, Consequences, Correctives.* 661 pp. National Academy of Sciences, Washington.
18.1

Roizes G., Nardeux P.C. & Monier R. (1979) A new specific endonuclease from *Anabaena variabilis. FEBS Lett.* **104,** 39–44.
11.3.3

Roman M.R. (1978) Ingestion of the blue-green alga *Trichodesmium* by the harpactocoid copepod *Macrosetella gracilis. Limnol. Oceanogr.* **23,** 1245–1248.
19.5.3

Romero-Herrera A.E., Lieska N., Goodman M. & Simons E.L. (1979) The use of amino acid sequence analysis in assessing evolution. *Biochimie* **61,** 767–779.
12.1.3

Rother J.A. & Fay P. (1977) Sporulation and the development of planktonic blue-green algae in two Salopian meres. *Proc. R. Soc. Lond. B* **196,** 317–332.
5.2.4, 5.4.1, 15.4, 15.4.2, 15.4.3, 15.5, 18.4.5

Rother J.A. & Fay P. (1979a) Blue-green algal growth and sporulation in response to simulated surface bloom conditions. *Br. phycol. J.* **14,** 59–68.
15.4, 15.4.2, 18.5.1.3

Rother J.A. & Fay P. (1979b) Some physiological-biochemical characteristics of planktonic blue-green algae during bloom formation in three Salopian meres. *Freshwat. Biol.* **9,** 369–379.
5.2.4

Rottenberg H. (1979) The measurement of membrane potential and ΔpH in cells, organelles and vesicles. In: Fleischer S. & Packer L. (Eds) *Methods in Enzymology* Vol.

IV Biomembranes. Part F. Bioenergetics-oxidative phosphorylation. pp. 547–569. Academic Press, London.
10.2.3, 10.3.2.1, 10.3.3.1

Rottenberg, H., Grunwald T. & Avron M. (1972) Determination of ΔpH in chloroplasts. 1. Distribution of [^{14}C] methylamine. *Eur. J. Biochem.* **25**, 54–63.
10.2.3

Rowell P., Enticott S. & Stewart W.D.P. (1977) Glutamine synthetase and nitrogenase activity in the blue-green alga *Anabaena cylindrica. New Phytol.* **79**, 41–54.
4.5

Rubin P.M., Zetooney E. & McGowan R.E. (1977) Uptake and utilization of sugar phosphates by *Anabaena flos-aquae. Pl. Physiol., Lancaster* **60**, 407–411.
5.1.3

Rüdiger W. (1970) Recent chemistry and biochemistry of bile pigments. *Angew. Chem.* **9**, 473–480.
7.5.1.2

Rüdiger W. (1975) Phycobiliproteide. *Ber. dt. bot. Ges.* **88**, 125–139.
7.5.1.2

Rüdiger W. & O'Carra P. (1969) Studies on the structures and apoprotein linkages of the phycobilins. *Eur. J. Biochem.* **7**, 509–516.
7.5.1.2

Rüdiger W., O'Carra P. & O'hEocha C. (1967) Structure of phycoerythrobilin and phycocyanobilin. *Nature, Lond.* **215**, 1477–1478.
7.5.1.2

Rueter J.G., McCarthy J.J. & Carpenter E.J. (1979) The toxic effect of copper on *Oscillatoria (Trichodesmium) thiebautii. Limnol. Oceanogr.* **24**, 558–562.
19.3.1

Rusckowski M. & Zilinskas B.A. (1979) New chlorophyll-protein complexes from the cyanophyte, *Nostoc* sp. Abst. 7th Ann. Meeting, Amer. Soc. Photobiol., Pacific Grove, CA, p. 179.
8.4

Ruvkun G.B. & Ausubel F.M. (1980) Interspecies homology of nitrogenase genes. *Proc. natn. Acad. Sci. U.S.A.* **77**, 191–195.
11.8

Ryther J.H. & Dunstan W.M. (1971) Nitrogen, phosphorus and eutrophication in the coastal marine environment. *Science, N.Y.* **171**, 1008–1013.

Sabatini D.D., Bensch K.G. & Barnett R.J. (1963) Cytochemistry and electron microscopy. The preservation of cellular ultrastructure and enzymatic activity by aldehyde fixation. *J. Cell Biol.* **17**, 19–58.
7.2

Saenger P., Specht M.M., Soecht R. & Chapman V.J. (1977) Mangal and coastal saltmarsh communities in Australia. 293–345. Chapter 15. In: Chapman V.J. (Ed.). *Wet Coastal Ecosystems.* 428 pp. Elsevier, Amsterdam.
20.2.2, 20.2.3

Safferman R.S. (1967) Earthy smelling substance from blue-green algae. *Environ. Sci. Technol.* **1**, 429.
18.5.1.1

Safferman R.S. & Morris M.E. (1964) Growth characteristics of the blue-green algal virus LPP-1. *J. Bact.* **88**, 771–775.
17.4.2

Safferman R.S. & Rohr M.E. (1979) *The Practical Directory to Phycovirus Literature* pp. 1–14. National Technical Information Services, Springfield, Va. 22161, USA.
References

Sage W.W. & Sullivan M.J. (1978) Distribution of bluegreen algae in a Mississippi Gulf Coast salt marsh. *J. Phycol.* **14**, 333–337.
20.2.3

Saino T. & Hattori A. (1978) Diel variation in nitrogen fixation by a marine blue-green alga, *Trichodesmium thiebautii. Deep Sea Res.* **25**, 1259–1263.
19.3.4, 19.5.2

Saito T., Iso N. & Mizuno H. (1974) Solution properties of phycocyanin. I. Studies of dissociation-association by sedimentation measurement. *Bull. chem. Soc. Japan* **47**, 1375–1379.
7.5.1.4

Saito N. & Werbin H. (1970) Purification of a blue-green algal deoxyribonucleic acid photoreactivation enzyme. An enzyme requiring light as a physical cofactor to perform its catalytic function. *Biochemistry, N.Y.* **9**, 2610–2620.
11.4.2.1

Sallal A.-K.J. & Codd G.A. (1977) An immunological and electrophoretic comparison of the thylakoids from vegetative cells and heterocysts of *Anabaena cylindrica. Br. phycol. J.* **12**, 163–169.
14.3.2

Samimi B. & Drews G. (1978) Adsorption of cyanophage AS-1 to unicellular cyanobacteria and isolation of receptor material from *Anacystis nidulans. J. Virol.* **25**, 164–174.
13.1, 13.4.4

Sanchez J.L., Palleroni N.J. & Doudoroff M.

(1975) Lactate dehydrogenases in cyanobacteria. *Arch. Microbiol.* **104,** 57–65.
3.3

Sanders J.G. (1979) The importance of salinity in determining the morphology and composition of algal mats. *Bot. Mar.* **22,** 159–162.

Sanderson K.E. & Hartman P.E. (1978) Linkage map of *Salmonella typhimurium*, Edition V. *Microbiol. Rev.* **42,** 471–519.
11.2.2

Sangar V.K. & Dugan P.R. (1972) Polysaccharide produced by *Anacystis nidulans*: its ecological implication. *Appl. Microbiol.* **24,** 732–734.
13.4.5

Sarà M. (1965) Associations entre éponges et algues unicellulaires dans la Mediterranee. *Rapp. P.-v. Réun. Commn int. Explor. scient. Mer Méditerr.* **18,** 125–127.
20.2.4

Sarà M. (1971) Ultrastructural aspects of the symbiosis between two species of the genus *Aphanocapsa* (Cyanophyceae) and *Ircinia variabilis* (Demospongiae). *Mar. Biol.* **11,** 214–221.
20.2.4

Sarma T.A. & Kantu S. (1979) Biochemical studies on sporulation in blue-green algae. I. Glycogen accumulation. *Z. allg. Mikrobiol.* **19,** 571–575.
15.2

Sato N., Murata N., Miura Y. & Ueta N. (1979) Effects of growth temperature on lipid and fatty acid compositions in the blue-green algae, *Anabaena variabilis* and *Anacystis nidulans. Biochim. biophys. Acta* **572,** 19–28.
6.5.3, 8.8

Sato S., Paranagua M.N. & Eskinake E. (1963/4) On the mechanism of red tide in Recife, North Eastern Brazil, with some considerations of the relation to the human disease 'Tamandare Fever'. *Trab. Inst. Oceanogr. Univ. do Recife* **V** (5/6), 7–49.
19.5.4

Saunders G.W. (1972) Potential heterotrophy in a natural population of *Oscillatoria agardhii* var. *isothrix* Skuja. *Limnol. Oceanogr.* **17,** 704–711.
2.6.3, 3.4.3, 18.5.4

Sawhney S.K., Naik M.S. & Nicholas D.J.D. (1978) Regulation of nitrate reduction by light, ATP and mitochondrial respiration in wheat leaves. *Nature, Lond.* **272,** 647–648.
18.5.2.1

Sawhney S.K. & Nicholas D.J.D. (1978a) Effects of amino acids, adenine nucleotides and inorganic pyrophosphate on glutamine synthetase from *Anabaena cylindrica. Biochim. biophys. Acta* **527,** 485–496.
14.4.1

Sawhney S.K. & Nicholas D.J.D. (1978b) Some properties of glutamine synthetase from *Anabaena cylindrica. Planta, Berl.* **139,** 289–299.
4.5

Schaeffer F. & Stanier R.Y. (1978) Glucose-6-phosphate dehydrogenase of *Anabaena* sp. Kinetic and molecular properties. *Arch. Microbiol.* **116,** 9–19.
3.3, 4.3, 14.4.2

Scheer H. (1976) Studies on plant bile pigments: characterization of a model for the phytochrome Pr chromophore. *Z. Naturforsch.* **31c,** 413–417.
7.5.1.2, 7.5.1.5

Scheer H. (1978) Biliproteins: the chromophore. In: Deutch B., Deutch B.I. & Gyldenholm A.O. (Eds) *Plant Growth and Light Perception.* pp. 25–52. Proc. Ann. Eur. Symp. Photomorphogenesis, Aarhus.
7.5.1.2, 7.5.1.5

Scheer H. & Kufer W. (1977) Conformational studies on C-phycocyanin from *Spirulina platensis. Z. Naturforsch.* **32c,** 513–519.
7.5.1.5

Scheibe J. (1972) Photoreversible pigment: occurrence in a blue-green alga. *Science, N.Y.* 1037–1039.
7.8.3.2

Scheibe J. (1979) A green, red-reversible photomorphogenic system in the cyanobacteria. Abstract. *Annual Meeting of the American Society for Photobiology.* **7,** 180.
7.8.3.2

Scherings G., Haaker H. & Veeger C. (1977) Regulation of nitrogen fixation by Fe-S protein II in *Azotobacter vinelandii. Eur. J. Biochem.* **77,** 631–630.
4.4

Schidlowski M., Hayes J.M. & Kaplan I.R. (1982) Isotopic inferences of ancient biochemistries: carbon, sulfur and other biogenic elements. In: J. W. Schopf (Ed.) *Origin and Evolution of Earth's Earliest Biosphere.* Princeton Univ. Press, Princeton, N.J.
21.2

Schiewer U. (1974) Salztoleranz und der Einfluss steigender NaCl Konzentrationen auf den Stickstoff-, Kohlenhydrat-, Pigmentgehalt und die Kohlenhydratausscheidung bei einigen limnischen Blaualgen. *Arch. Hydrobiol., Suppl.* **46,** 171–184.
18.4.1

Schindler D.W. (1975) Whole-lake eutrophication experiments with phosphorus, nitrogen

and carbon. *Verh. int. Verein. theor. angew. Limnol.* **19,** 3221–3231.
5.4.2, 5.4.3

Schindler D.W. (1976) Biogeochemical evolution of phosphorus limitation in nutrient-enriched lakes of the pre-Cambrian shield. In: Nriagu J.O. (Ed.) *Environmental Biogeochemistry, Vol. 2.* pp. 647–664. Ann Arbor Science Publ. Inc., Mich.
5.4.2

Schindler D.W. (1977) Evolution of phosphorus limitation in lakes. *Science, N.Y.* **195,** 260–262.
5.4.2, 18.5.2.4

Schindler D.W., Brunskill G.J., Emerson S., Broeckner W.S. & Peng T.H. (1972) Atmospheric carbon dioxide: its role in maintaining phytoplankton standing crops. *Science, N.Y.* **177,** 1192–1194.
18.5.2.2

Schindler D.W. & Fee E.J. (1974) Experimental Lakes area: whole lake experiments in eutrophication. *J. Fish. Res. Bd Can.* **31,** 937–953.
5.4.2, 18.4.3

Schleifer K.H. & Kandler O. (1972) Peptidoglycan types of bacterial cell walls and their taxonomical implications. *Bact. Rev.* **36,** 407–477.
13.4.1

Schmid G. (1923) Das Reizverhalten künstlicher Teilstücke, die Kontraktilität und das osmotische Verhalten der *Oscillatoria jenensis. Jb. wiss. Bot.* **62,** 328–419.
16.4.4

Schmid G.H. & Lehmann-Kirk J. (1977) Photooxidation reactions of diphenylcarbazide and their DCMU sensitivity in thylakoids of the blue-green alga *Oscillatoria chalybea. Arch. Microbiol.* **115,** 265–269.
8.6

Schmid G.H., List H. & Radunz A. (1977) Inhibition of photosystem II-reactions in blue-green algae by the antisera to lutein and neoxanthin. *Z. Naturf.* **32C,** 118–124.
8.6

Schmidt K. (1976) Carotenoid glucosides in phototrophic bacteria (Abstract). In: Codd G.A. & Stewart W.D.P. (Eds) *Proceedings of the Second International Symposium on Photosynthetic Prokaryotes.* 258 pp., pp. 58–60. Univ. Dundee, Scotland.
16.2.3

Schmidt K. (1978) Biosynthesis of carotenoids. In: Clayton R.K. & Sistrom W.R. (Eds) *The Photosynthetic Bacteria.* pp. 729–750. Plenum Press, New York.
6.6

Schmidt W., Drews G., Weckesser J., Fromme I. & Borowiak D. (1980a) Characterization of the lipopolysaccharides from eight *Synechococcus* strains. *Arch. Microbiol.* **127,** 209–216.
13.4.4

Schmidt W., Drews G., Weckesser J. & Mayer H. (1980b) Lipopolysaccharides in four strains of the unicellular cyanobacterium *Synechocystis. Arch. Microbiol.* **127,** 217–222.
13.4.4

Schneider J. (1976) Biological and inorganic factors in the destruction of limestone coasts. *Contrib. Sedimentol.* **6,** 1–112.
20.1, 20.2.1, 20.3.2, 20.3.3, 20.4.3

Schneider K.C., Bradbeer C., Singh R.N., Wang L.C., Wilson P.W. & Burris R.H. (1960) Nitrogen fixation by cell-free preparations from microorganisms. *Proc. natn. Acad. Sci. U.S.A.* **46,** 726–733.
4.2

Schneider R.P. & Wiley W.R. (1971) Kinetic characteristics of the two glucose transport systems in *Neurospora crassa. J. Bact.* **106,** 479–486.
3.4.3

Schnepf E. (1964) Zur Feinstruktur von *Geosiphon pyriforme. Arch. Mikrobiol.* **49,** 112–131.
6.2

Scholes P., Mitchell P. & Moyle J. (1969) The polarity of proton translocation in some photosynthetic microorganisms. *Eur. J. Biochem.* **8,** 450–454.
10.3.3.1

Schollhorn R. & Burris R.H. (1967) Reduction of azide by the N_2-fixing enzyme system. *Proc. natn Acad. Sci. U.S.A.* **57,** 1317–1323.
11.5.5.2(b)

Schopf J.W. (1968) Microflora of the Bitter Springs Formation, Late Precambrian, central Australia. *J. Paleontol.* **42,** 651–688.
21.1

Schopf J.W. (1974a) Paleobiology of the Precambrian: the age of blue-green algae. In: Dobzhansky T., Hecht M.K. & Steere W.C. (Eds) *Evolutionary Biology* **7.** pp. 1–43. Plenum Press, New York.
12.1.2, 12.2.1, 12.3.2, 21.1, 21.3

Schopf J.W. (1974b) The development and diversification of Precambrian life. *Origins of Life* **5,** 119–135.
12.1.2, 12.3.2

Schopf J.W. (1975) Precambrian paleobiology: problems and perspectives. *A. Rev. Earth Planet. Sci.* **3,** 213–249.
21.1

Schopf J.W. (1976) Are the oldest 'fossils' fossils? *Origins of Life* **7,** 19–36.
12.3.2

Schopf J.W. (1977) Biostratigraphic usefulness of stromatolitic Precambrian microbiotas: a preliminary analysis. *Precambrian Res.* **5,** 143–173.
21.1, 21.3

Schopf J.W. (1978) The evolution of the earliest cells. *Scient. Am.* **239,** 110–134.
21.3

Schopf J.W. (1980) The origin and Archean evolution of life. In: Hickman M.H. (Ed.) *The Primitive Earth Revisited, A Symposium.* pp. 94–103. Miami Univ. Press, Oxford, Ohio.
12.1.2, 12.3.2, 21.2

Schopf J.W. & Barghoorn E.S. (1967) Alga-like fossils from the Early Precambrian of South Africa. *Science, N.Y.* **156,** 508–512.
12.3.2, 21.2

Schopf J.W. & Blacic J.M. (1971) New microorganisms from the Bitter Springs Formation (late Precambrian) of the north-central Amadeus Basin, Australia. *J. Paleontol.* **45,** 925–960.
21.1

Schopf J.W., Haugh B.N., Molnar R.E. & Satterthwait D.F. (1973) On the development of metaphytes and metazoans. *J. Paleontol.* **47,** 1–9.
21.3

Schopf J.W. & Oehler D.Z. (1976) How old are the eukaryotes? *Science, N.Y.* **193,** 47–49.
12.1.3, 21.3

Schopf J.W. & Prasad K.N. (1978) Microfossils in *Collenia*-like stromatolites from the Proterozoic Vempalle Formation of the Cuddapah Basin, India. *Precambrian Res.* **6,** 347–366.
21.1

Schopf J.W. & Sovietov Yu. K. (1976) Microfossils in *Conophyton* from the Soviet Union and their bearing on Precambrian biostratigraphy. *Science, N.Y.* **193,** 143–146.
21.3

Schopf J.W. & Walter M.R. (1982) Archean microfossils: new evidence of ancient microbes. In: Schopf J.W. (Ed.) *Origin and Evolution of Earth's Earliest Biosphere: An Interdisciplinary Study.* Princeton Univ. Press, Princeton, N.J.
21.1, 21.2

Schrader M., Drews G., Weckesser J. & Mayer H. (1982) Polysaccharide containing 6-o-methyl-D-mannose in *Chlorogloeopsis* PCC 6912. *J. gen. Microbiol.* **128,** 273–277.
13.4.5

Schram B.L. & Kroes H.H. (1971) Structure of phycocyanobilin. *Eur. J. Biochem.* **19,** 581–594.
7.5.1.2

Schreiber U. (1979) Cold induced uncoupling of energy transfer between phycobilins and chlorophyll in *Anacystis nidulans.* Antagonistic effects of monovalent and divalent cations and of high and low pH. *FEBS Lett.* **107,** 4–9.
7.4, 8.8

Schreiber U., Rijgersberg C.P. & Amesz J. (1979) Temperature-dependent reversible changes in phycobilisome-thylakoid membrane attachment in *Anacystis nidulans. FEBS Lett.* **104,** 327–331.
7.4, 8.8

Schubert W., Giani D., Rongen P., Krumbein W.E. & Schmidt W. (1980) Photoacoustic in-vivo spectra of recent stromatolites. *Naturwissenschaften* **67,** 129–132.
20.3.4

Schuldiner S., Rottenberg H. & Avron M. (1972) Determination of ΔpH in chloroplasts. 2. Fluorescent amines as a probe for the determination of ΔpH in chloroplasts. *Eur. J. Biochem.* **25,** 64–70.
10.3.2.1

Schulz E. (1937) Das Farbstreifen-Sandwatt und seine Fauna, eine ökologische-biozönotische Untersuchung an der Nordsee. *Kieler Meeresforschung* **1,** 359–378.
20.2.2

Schulz E. & Meyer H. (1939) Weitere Untersuchungen über das Farbstreifen-Sandwatt. *Kieler Meeresforschung* **3,** 321–336.
20.2.2

Schulz G. (1955) Bewegungsstudien an Cyanophyceen. *Arch. Mikrobiol.* **21,** 335–370.
13.2.2

Schwabe G.H. (1936) Beiträge zur Kenntnis isländischer Thermalbiotope. *Arch. Hydrobiol., Suppl.* **6,** 161–353.
9.5.2

Schwabe G.H. (1960) Über den thermobioten Kosmopoliten *Mastigocladus laminosus* Cohn. Blaualgen und Lebensraum V. *Schweiz Z. Hydrol.* **22,** 757–792.
9.5.2

Schwabe G.H. & Mollenhauer R. (1967) Über drei hormogonale Blaualgen aus indischen Böden. *Nova Hedwigia* **10,** 527–536.
17.2

Schwartz R.M. & Dayhoff M.O. (1978) Origin of prokaryotes, eukaryotes, mitochondria and chloroplasts. *Science, N.Y.* **199,** 395–403.
12.1.3, 12.2.1, 12.2.2.1, 12.2.2.2, 12.4.4.3

Schwartz R.M. & Dayhoff M.O. (1981) Chlor-

oplast origins: inferences from protein and nucleic acid sequences. *Ann. N.Y. Acad. Sci.* **361**, 260–272.
12.1.3, 12.2.1, 12.2.2.1, 12.2.2.2, 12.3.4.1, 12.4.4.3

Schwarz H.-U., Einsele G. & Herm D. (1975) Quartz-sandy, grazing-contoured stromatolites from coastal embayments of Mauritania, West Africa. *Sedimentology* **22**, 539–561.
20.3.1, 20.4.1

Schwarz Zs. & Kössel H. (1980) The primary structure of 16S rDNA from *Zea mays* chloroplast is homologous to *E. coli* 16S rRNA. *Nature, Lond.* **283**, 739–742.
12.4.3.3

Schwimmer D. & Schwimmer M. (1964) Algae and medicine. In: Jackson D.F. (Ed.) *Algae and Man*. 434 pp. Plenum Press, New York.
18.1

Scott E. & Berns D.S. (1965) Protein-protein interaction. The phycocyanin system. *Biochemistry, N.Y.* **4**, 2597–2606.
7.5.1.4, 7.5.1.5

Scott W.E. & Fay P. (1972) Phosphorylation and amination in heterocysts of *Anabaena variabilis*. *Br. phycol. J.* **7**, 283–284.
14.4.1, 14.4.3.1

Searle G.F.W., Barber J., Porter G. & Tredwell C.J. (1978) Pico-second time-resolved energy transfer in *Porphyridium cruentum*. Part II. In the isolated light harvesting complex (phycobilisomes). *Biochim. biophys. Acta* **501**, 246–256.
7.6

Semikhatov M.A. (1976) Experience in stromatolite studies in the U.S.S.R. In: Walter M.R. (Ed.) *Stromatolites. Developments in Sedimentology 20*. 790 pp., pp. 337–357. Elsevier, Amsterdam.
21.1

Senft W.H. (1978) Dependence of light-saturated rates of algal photosynthesis on intracellular concentrations of phosphorus. *Limnol. Oceanogr.* **23**, 709–718.
5.2.2, 5.2.3

Sentsova O. Yu., Nikitina K.A. & Gusev M.B. (1975) Oxygen exchange of the obligate-phototrophic blue-green alga *Anabaena variabilis* in the darkness. *Mikrobiologiya, Transl.* **44**, 250–254.
18.5.3

Serpette M. (1948) Morphologie de *Homoeothrix balearica* (Born. et Fl.) Lemm. et fonction des hétérocystes chez les algues bleues. *Bull. Soc. bot. Fr.* **95**, 295–300.
14.1

Setlow R.B. (1964) Physical changes and mutagenesis. *J. cell. comp. Physiol.* **64** *(Suppl. 1)*, 51–68.
11.4.2.2

Seward A.C. (1933) *Plant Life Through the Ages, 2nd edn*. 603 pp. Cambridge Univ. Press, Cambridge.
21.1

Shaffer P.W., Lockau W. & Wolk C.P. (1978) Isolation of mutants of the cyanobacterium, *Anabaena variabilis*, impaired in photoautotrophy. *Arch. Microbiol.* **117**, 215–219.
11.5.4

Shapiro J. (1973) Blue-green algae: why they become dominant. *Science, N.Y.* **179**, 382–384.
18.5.2.2

Sharma V.K. & Kumar H.D. (1973) Effects of azide, thioglycollic acid and nitrosoguanidine on growth and nitrogen fixation in different clones of the blue-green alga *Cylindrospermum majus*. *Z. allg. Mikrobiol.* **13**, 517–522.
11.5.5.2(b)

Sharma V.K. & Tiagi Y.D. (1975) Isolation of a heterotrophic, non-nitrogen fixing mutant clone of *Anabaena doliolum*. *Z. allg. Mikrobiol.* **15**, 469–471.
11.5.3

Sharp J.H. (1977) Excretion of organic matter by marine phytoplankton. Do healthy cells do it? *Limnol. Oceanogr.* **22**, 381–399.
17.4.4

Shear H. & Walsby A.E. (1975). An investigation into the possible light-shielding role of gas-vacuoles in a planktonic blue-green alga. *Br. phycol. J.* **10**, 241–251.
2.2.2, 2.3, 2.7.3.1, 18.5.5.6

Sheath R.G. & Hellebust J.A. (1974) Glucose transport systems and growth characteristics of *Bracteococcus minor*. *J. Phycol.* **10**, 34–41.
3.4.3

Shelef G. & Soeder C.J. (1980) *Algae Biomass. Cultivation and Use*. 852 pp. Elsevier/North Holland, Amsterdam.
References

Shemin D. & Russell C.S. (1953) δ-Aminolevulinic acid, its role in the biosynthesis of porphyrins and purines. *J. am. Chem. Soc.* **75**, 4873–4874.
3.2.2.1

Shen L. & Preiss J. (1964) The activation and inhibition of bacterial adenosine diphosphoglucose pyrophosphorylase. *Biochem. biophys. Res. Commun.* **17**, 424–429.
3.2.2.2

Sheriff S. & Herriott J.R. (1981) The structure of ferredoxin-$NADP^+$ oxidoreductase and

the location of the NADP binding site: results at 3·7 A° resolution. *J. molec. Biol.* **145,** 441–451.
8.3

Sherman L.A. (1978) Differences in photosynthesis-associated properties of the bluegreen alga *Synechococcus cedrorum* grown at 30 and 40 C. *J. Phycol.* **14,** 427–433.
6.5.3

Sherman L.A. & Cunningham J. (1977) Isolation and characterization of temperature-sensitive, high-fluorescence mutants of the blue-green alga *Synechococcus cedrorum. Pl. Sci. Lett.* **8,** 319–326.
11.5.4

Sherman L.A. & Cunningham J. (1979) Photosynthetic characteristics of temperature-sensitive high fluorescence mutants of the blue-green alga *Synechococcus cedrorum. Pl. Sci. Lett.* **14,** 121–131.
11.5.4

Shestakov S.V. (1972) Mutagenesis and repair systems in unicellular blue-green algae. In: Desikachary T.V. (Ed.) *Taxonomy and Biology of Blue-green Algae.* 591 pp., pp. 262–263. University of Madras, India.
11.4.1

Shestakov S.V. & Jevner V.D. (1968) Study of mutagenesis in blue-green alga *Anacystis nidulans. Proc. XII Int. Congr. Genet.* **1,** 84.
11.4.1, 11.5.3, 11.5.4

Shestakov S.V. & Khyen N.T. (1970) Evidence for genetic transformation in blue-green alga *Anacystis nidulans. Molec. gen. Genet.* **107,** 372–375.
11.5.5.1, 11.6.3

Shestakov S.V., Postnova T.I. & Shakhnabatian L.G. (1975) *In vitro* repair of UV- and X-irradiated bacteriophage T4 DNA by extract from blue-green alga *Anacystis nidulans. Mol. Biol. Reports* **2,** 89–94.
11.4.2.2

Shilo M. (1970) Lysis of blue-green algae by *Myxobacter. J. Bact.* **104,** 453–456.
16.2.3, 17.4.2

Shilo M. (1971) Toxins of Chrysophyceae. In: Kadis S., Ciegler A., & Ajl S.J. (Eds) *Microbial Toxins* Vol. III. *Algal and Fungal Toxins.* pp. 67–103. Academic Press, London, New York.
10.3.2

Shilo M. & Shilo M. (1962) The mechanism of lysis of *Prymnesium parvum* by weak electrolytes. *J. gen. Microbiol.* **29,** 645–658.
10.3.2

Shimke R.T., Berlin C.M., Sweeney E.W. & Carroll W.R. (1966) The generation of energy by the arginine dihydrolase pathway in *Mycoplasma hominis* 07. *J. biol. Chem.* **241,** 2228–2236.
3.2.2.2

Shimura S. & Fujita Y. (1975) Phycoerythrin and photosynthesis of the pelagic blue-green alga *Trichodesmium thiebautii* in the waters of Kuroshio, Japan. *Mar. Biol.* **31,** 121–128.
19.3.1, 19.3.2

Shimura S., Yamaguchi Y. Aruga Y. Fujita Y. & Ichimura S. (1978) Extracellular release of photosynthetic products by a pelagic blue-green alga, *Trichodesmium thiebautii. J. oceanogr. Soc. Japan* **34,** 181–188.
19.3.2

Shin M., Sukenobu M., Oshino R. & Kitazume Y. (1977) Two plant-type ferredoxins from a blue-green alga, *Nostoc verrucosum. Biochim. biophys. Acta* **460,** 85–93.
4.3, 8.3, 12.2.2.2

Shively J.M. (1974) Inclusion bodies of prokaryotes. *A. Rev. Microbiol.* **28,** 167–187.
3.2.1

Shively J.M., Ball F., Brown D.H. & Saunders R.E. (1973) Functional organelles in prokaryotes: polyhedral inclusions (carboxysomes) of *Thiobacillus neapolitanus. Science, N.Y.* **182,** 584–586.
3.2.1

Shukovsky E.S. & Halfen L.N. (1976) Cellular differentiation of terminal regions of trichomes of *Oscillatoria princeps. J. Phycol.* **12,** 336–342.
13.2.1

Sicko-Goad L.M., Crang R.E. & Jensen T.E. (1975) Phosphate metabolism in blue-green algae. IV. In situ analysis of polyphosphate bodies by X-ray energy dispersive analysis. *Cytobiologie* **11,** 430–437.
5.3.1

Sicko-Goad L. & Jensen T.E. (1976) Phosphate metabolism in blue-green algae. II. Changes in phosphate distribution during starvation and the 'polyphosphate overplus' phenomenon in *Plectonema boryanum. Am. J. Bot.* **63,** 183–188.
5.2.2, 5.3.1

Sicko-Goad L., Jensen T.E. & Ayala R.P. (1978) Phosphate metabolism in blue-green algae. V. Factors affecting phosphate uptake in *Plectonema boryanum. Can. J. Microbiol.* **24,** 105–108.
5.1.1, 5.2.3

Sieburth J.M. (1959) Antibacterial activity of Antarctic marine phytoplankton. *Limnol. Oceanogr.* **4,** 419–424.
17.4.4

Sieburth J.M. (1968) The influence of algal antibiosis on the ecology of marine micro-

organisms. In: Droop M.R. & Ferguson Wood E.I. (Eds) *Adv. Microbiol. Sea.* pp. 63–94. Academic Press, New York.
17.4.4

Siegelman H.W., Chapman D.J. & Cole W.J. (1968) The bile pigments of plants. *Biochem. Soc. Symp.* **28,** 107–120.
7.5.1.2

Sigalat C. & de Kouchkovsky Y. (1975) Fractionnement et caractérisation de l'appareil photosynthetique de l'algue bleue unicellulaire *Anacystis nidulans.* I. Obtention de fractions membranaires par 'lyse osmotique' et analyse pigmentaire. *Physiol. Vég.* **13,** 243–258.
8.2

Silvester W.B. (1976) Endophyte adaptation in *Gunnera Nostoc* symbiosis. In: Nutman P.S. (Ed.) *Symbiotic Nitrogen Fixation in Plants.* 584 pp., pp. 521–538. Cambridge Univ. Press, England.
3.4.3

Simon R.D. (1971) Cyanophycin granules from the blue-green alga *Anabaena cylindrica*: a reserve material consisting of copolymers of aspartic acid and arginine. *Proc. natn. Acad. Sci. U.S.A.* **68,** 265–267.
3.2.2.2, 6.5.2

Simon R.D. (1973a) Measurement of the cyanophycin granule polypeptide content in the blue-green alga *Anabaena cylindrica. J. Bact.* **114,** 1213–1216.
3.2.2.2

Simon R.D. (1973b) The effect of chloramphenicol on the production of cyanophycin granule polypeptide in the blue-green alga *Anabaena cylindrica. Arch. Mikrobiol.* **92,** 115–122.
3.2.2.2

Simon R.D. (1976) The biosynthesis of multi-L-arginyl-poly(L-aspartic acid) in the filamentous cyanobacterium *Anabaena cylindrica. Biochim. biophys. Acta* **422,** 407–418.
3.2.2.2

Simon R.D. (1977a) Macromolecular composition of spores from the filamentous cyanobacterium *Anabaena cylindrica. J. Bact.* **129,** 1154–1155.
15.2

Simon R.D. (1977b) Sporulation in the filamentous cyanobacterium *Anabaena cylindrica.* The course of spore formation. *Arch. Microbiol.* **111,** 283–288.
15.4.1

Simon R.D. (1978) Survey of extrachromosomal DNA found in filamentous cyanobacteria. *J. Bact.* **136,** 414–418.
11.3.2

Simon R.D. (1979) Macromolecular composition of heterocysts and spores from *Anabaena variabilis.* In: Nichols J.M. (Ed.) *Abstracts of the Third International Symposium on Photosynthetic Prokaryotes, Oxford.* p. D4. Univ. Liverpool, England.
14.2, 14.3.2

Simon R.D. (1980) DNA content of heterocysts and spores of the filamentous cyanobacterium *Anabaena variabilis. FEMS Microbiol. Lett.* **8,** 241–245.
14.3.2

Simon R.D. & Weathers P. (1976) Determination of the structure of the novel polypeptide containing aspartic acid and arginine which is found in cyanobacteria. *Biochim. biophys. Acta* **420,** 165–176.
3.2.2.2, 14.3.2

Simonis W., Bornefeld T., Lee-Kaden J. & Majumdar K. (1974) Phosphate uptake and photophosphorylation in the blue-green alga *Anacystis nidulans.* In: Zimmerman U. & Dainty J. (Ed.) *Membrane Transport in Plants.* 473 pp., pp. 220–225. Springer, Berlin.
5.1.1

Simonis M. & Urbach W. (1973) Photophosphorylation *in vivo. A. Rev. Pl. Physiol.* **24,** 89–114.
9.4.1

Simpson F.B. & Neilands J.B. (1976) Siderochromes in Cyanophyceae: isolation and characterization of schizokinen from *Anabaena* sp. *J. Phycol.* **12,** 44–48.
2.7.2.1, 17.4.2, 18.5.2.3

Sinclair C. & Whitton B.A. (1977a) Influence of nutrient deficiency on hair formation in the Rivulariaceae. *Br. phycol. J.* **12,** 297–313.
5.2.4, 5.3.1, 14.6.1.1, 15.4, 18.3.2

Sinclair C. & Whitton B.A. (1977b) Influence of nitrogen source on morphology of Rivulariaceae (Cyanophyta). *J. Phycol.* **13,** 335–340.
14.6.1.1

Singer C.E. & Ames B.M. (1970) Sunlight, ultraviolet and bacterial DNA base ratios. *Science, N.Y.* **170,** 822–826.
12.3.4.2

Singer R.A. & Doolittle W.F. (1975) Leucine biosynthesis in the blue-green bacterium *Anacystis nidulans. J. Bact.* **124,** 810–814.
11.2.2, 11.5.3

Singer S.J. (1974) The molecular organization of membranes. *A. Rev. Biochem.* **43,** 805–833.
6.3.1

Singer S.J. & Nicholson G.L. (1972) The fluid

mosaic model of the structure of cell membranes. *Science, N.Y.* **175**, 720–731.
6.3.1

Singh H.N. (1967) Genetic control of sporulation in the blue-green alga *Anabaena doliolum. Planta, Berl.* **75**, 33–38.
11.5.5.2(c), 15.6.1

Singh H.N. (1968) Effect of acriflavine on ultraviolet-sensitivity of normal ultraviolet sensitive and ultraviolet resistant strains of a blue-green alga *Anacystis nidulans. Radiat. Bot.* **8**, 355–361.
11.4.2.2, 11.4.2.3

Singh H.N. & Kumar H.D. (1971) Physiology of heterocyst production in the blue-green alga *Anabaena doliolum*. II. Effects of ultraviolet light. *Z. allg. Mikrobiol.* **11**, 623–626.
11.5.5.2(c)

Singh H.N., Kumar H.D. & Prakash G. (1969) Postirradiation modification of UV-sensitivity of normal and nitrofurazone-resistant strains of the blue-green alga *Anacystis nidulans. Radiat. Bot.* **9**, 105–110.
11.4.2.2

Singh H.N., Ladha J.K. & Kumar H.D. (1977) Genetic control of heterocyst formation in the blue-green algae *Nostoc muscorum* and *Nostoc linckia. Arch. Microbiol.* **114**, 155–159.
11.5.5.2(b), 14.6.1.1, 14.6.1.2, 14.6.1.3

Singh H.N. & Singh H.N. (1978) An azide-resistant mutant of the blue-green alga *Nostoc muscorum* producing heterocyst and nitrogenase in the presence of fixed inorganic nitrogen source. *Arch. Microbiol.* **119**, 197–201.
11.5.1, 11.5.5.2(b), 14.6.1.1

Singh H.N. & Sonie K.C. (1977) Isolation and characterization of chlorate-resistant mutants of the blue-green alga *Nostoc muscorum. Mutation Res.* **43**, 205–212.
11.5.1, 11.5.5.2(b), 14.6.1.2

Singh H.N., Sonie K.C. & Singh H.R. (1977) Nitrate regulation of heterocyst differentiation and nitrogen fixation in a chlorate-resistant mutant of the blue-green alga, *Nostoc muscorum. Mutation Res.* **42**, 447–452.
11.5.1

Singh H.N. & Srivastava B.S. (1968) Studies on morphogenesis in a blue-green alga. I. Effect of inorganic nitrogen sources on developmental morphology of *Anabaena doliolum. Can. J. Microbiol.* **14**, 1341–1346.
15.4

Singh H.N., Vaishampayan A. & Singh P.K. (1978) Evidence for the involvement of a genetic determinant controlling functional specificity of group VIB elements in the metabolism of N_2 and NO_3^- in the blue-green alga *Nostoc muscorum. Biochem. biophys. Res. Commun.* **81**, 67–74.
11.5.1, 11.5.5.2(b)

Singh H.N., Vaishampayan A. & Sonie K.C. (1978) Mutation from molybdenum-dependent growth to tungsten-dependent growth and further evidence for a genetic determinant common to nitrogenase and nitrate reductase in the blue-green alga *Nostoc muscorum. Mutation Res.* **50**, 427–432.
11.5.5.2(b)

Singh P.K. (1973) Nitrogen fixation by the unicellular blue-green alga *Aphanothece. Arch. Mikrobiol.* **92**, 59–62.
4.1

Singh P.K. (1975a) Photoreactivation of UV-irradiated blue-green algae and algal virus LPP-1. *Arch. Microbiol.* **103**, 297–302.
11.4.2.1

Singh P.K. (1975b) Fertilizer. A tolerance of blue-green algae and their effect on heterocyst differentiation. *Phykos* **14**, 81–88.
14.6.2

Singh P.K. (1976a) Growth and nitrogen fixation of mutants defective in heterocyst differentiation in the blue-green alga *Cylindrospermum* sp. *Z. allg. Mikrobiol.* **16**, 453–463.
11.5.5.2(a), 11.5.5.2(b), 11.5.5.2(c), 14.2, 14.5, 14.6.1.1, 14.6.2

Singh P.K. (1976b) Growth and nitrogen fixation of a mutant of blue-green alga *Wollea bharadwajae* defective in heterocyst development. *Biochem. Physiol. Pflanz.* **170**, 237–242.
11.5.5.2(b), 11.5.5.2(c)

Singh P.K. (1977) Multiplication and utilization of fern 'Azolla' containing nitrogen-fixing algal symbiont as green manure in rice cultivation. *Il Riso* **26**, 125–137.
14.1

Singh R.N. (1961) *Role of Blue-green Algae in Nitrogen Economy of Indian Agriculture*. 175 pp. Indian Council of Agricultural Research, New Delhi.
9.5.4, 14.1

Singh R.N. & Dikshit R.P. (1976) Mutagenesis of the blue-green alga, *Anabaena doliolum* Bharadwaja. *Mutation Res.* **35**, 65–78.
11.5.5.2(b), 11.5.5.2(c), 14.3.2

Singh R.N. & Kashyap A.K. (1977) Induction of mutations in the blue-green alga *Plectonema boryanum* Gomont. *Mutation Res.* **43**, 37–44.
11.4.1, 11.4.2.2, 11.5.1

Singh R.N. & Singh H.N. (1964a) Ultra-violet induced mutants of blue-green algae. I.

Anabaena cycadeae. *Arch. Mikrobiol.* **48,** 109–117.
11.5.3

Singh R.N. & Singh H.N. (1964b) Ultra-violet induced mutants of blue-green algae. II. *Anabaena doliolum*. *Arch. Mikrobiol.* **48,** 118–121.
11.5.3

Singh R.N. & Singh P.K. (1972a) UV damage, modification and repair of blue-green algae and their viruses. In: Desikachary T.V. (Ed.) *Taxonomy and Biology of Blue-Green Algae*. 591 pp., pp. 244–257. University of Madras, India.
11.4.2.2

Singh R.N. & Singh P.K. (1972b) Transduction and lysogeny in blue-green algae. In: Desikachary T.V. (Ed.) *Taxonomy and Biology of Blue-Green Algae*. 591 pp., pp. 258–261. University of Madras, India.
11.6.2

Singh R.N., Singh P.K., Kashyap, A.K., Sarma T.A., Dhar B., Chaubey I.J. & Choudhury I.D. (1972a) Isolation and characterization of new cyanophages and mutations of LPP-1 and host alga *Plectonema boryanum*. In: Desikachary T.V. (Ed.) *Taxonomy and Biology of Blue-Green Algae*. 591 pp., pp. 585–591. Univ. Madras, India.
11.5.1

Singh R.N., Singh S.P. & Singh P.K. (1972b) Genetic regulation of nitrogen fixation in blue-green algae. In: Desikachary T.V. (Ed.) *Taxonomy and Biology of Blue-green Algae*. 591 pp., pp. 264–268. Univ. Madras, India.
11.5.1, 11.5.5.2(b), 11.5.5.2(c), 14.6.1.1

Singh R.N., Singh H.N. & Sinha R. (1966) Mutation and recombination in blue-green algae. In: *The Impact of Mendelism on Agriculture, Biology and Medicine. Int. Symp. Indian Soc. Genet. Plant Breeding*. pp. 405–430. Indian Agricultural Research Institute, New Delhi.
11.6.1

Singh R.N. & Sinha R. (1965) Genetic recombination in a blue-green alga, *Cylindrospermum majus* Kütz. *Nature, Lond* **207,** 782–783.
11.5.5.2(c), 15.6.1

Singh R.N. & Sunita K. (1974) A biochemical study of spore germination in the blue-green alga *Anabaena doliolum*. *J. exp. Bot.* **25,** 837–845.
15.5

Singh R.N. & Tiwari D.N. (1969) Induction by ultraviolet irradiation of mutation in the blue-green alga *Nostoc linckia* (Roth) Born. et Flah. *Nature, Lond.* **221,** 62–64.
11.5.5.2(a), 14.6.1.1

Singh R.N., Tiwari D.N. & Singh V.P. (1972) Genetic basis of cellular differentiation in blue-green algae. In: Desikachary T.V. (Ed.) *Taxonomy and Biology of Blue-Green Algae*. 591 pp., pp. 27–37. Univ. Madras, India.
11.5.5.2(c), 14.5, 14.6.1.1, 14.6.1.2, 14.6.2

Singh S.P. (1977) Isolation of sulphate reductaseless mutants of the blue-green alga *Nostoc muscorum*. *Mutation Res.* **43,** 445–447.
11.5.3

Singh S.P. (1978) Isolation of a sulphate reductaseless mutant of the blue-green alga *Nostoc muscorum*. *Z. allg. Mikrobiol.* **18,** 111–114.
11.5.3

Singh V.P. (1978a) Pre-irradiation and post-irradiation treatments with acridine orange monitoring UV killing in the blue-green alga *Anabaena variabilis*. *Microbios* **8,** 63–69.
11.4.2.1, 11.4.2.2

Singh V.P. (1978b) UV dose modifying effects of acriflavine in the blue-green alga *Anabaena variabilis*. *Environ. exp. Bot.* **18,** 193–197.
11.4.2.2

Singh V.P. (1978c) Mutations defective in nitrogen fixation, heterocyst and spore production in the blue-green alga *Anabaena variabilis*. *Biochem. Physiol. Pflanz.* **173,** 365–372.
11.5.5.2(a), 11.5.5.2(b), 11.5.5.2(c), 15.6.1

Sinha B.D. & Kumar H.D. (1973) A mutational study of the nitrogen-fixing blue-green alga *Anabaena doliolum*. *Ann. Bot., N.S.* **37,** 673–679.
11.5.5.2(a), 11.5.5.2(b), 11.5.5.2(c), 14.6.1.3

Sirenko L.A. (1972) *Fiziologicheskie Osnovy Razmnozheniya Sinezelenykh Vodoroslei v Vodokhranilischchakh*. 203 pp. Naukova Dumka, Kiev. (Russian = Physiological principles of the growth of blue-green algae in reservoirs).
18.3.2, 18.4.4, 18.4.5

Sirenko L.A., Chernovsova V.M., Arendarchuk V.V. & Kozitskaya V.N. (1969) Factors of mass development of blue-green algae. *Hydrobiol. J.* **5,** 1–8.
18.4.5

Sirenko L.A., Stetsenko N.M., Arendarchuk V.V. & Kuz'menko M.K. (1968) Role of oxygen conditions in the vital activity of certain blue-green algae. *Mikrobiologiya, Transl.* **37,** 199–202.
18.3.2

Sirevåg R. (1979) Carbon metabolism in

Chloroflexus. In: Nichols J.M. (Ed.) *Abstracts of the Third International Symposium on Photosynthetic Prokaryotes, Oxford*. p. B32. Univ. Liverpool, England.
12.2.1

Sirevåg R. & Castenholz R.W. (1979) Aspects of carbon metabolism in *Chloroflexus*. *Arch. Microbiol.* **120,** 151–153.
12.2.1

Sistrom W.R., Griffiths M. & Stanier R.Y. (1956) The biology of a photosynthetic bacterium which lacks coloured carotenoids. *J. cell. comp. Physiol.* **48,** 473–515.

Skulberg O.M. (1978) Some observations on red-coloured species of *Oscillatoria* (Cyanophyceae) in nutrient-enriched lakes of southern Norway. *Verh. int. Verein. theor. angew. Limnol.* **20,** 776–787.
2.6.1

Skyring G.W. & Johns I.A. (1980) Iron in cyanobacterial mats. *Micron* **11,** 407–408.
20.4.1

Sloan P.R. & Strickland J.D.H. (1966) Heterotrophy of four marine phytoplankters at low substrate concentrations. *J. Phycol.* **2,** 29–32.
3.4.3

Sm̆arda J., Čáslaská J. & Komárek J. (1979) Cell wall structure of *Synechocystis aquatilis* (Cyanophyceae). *Arch. Hydrobiol., Suppl.* **56,** 154–165.
13.2.1, 13.2.2

Smayda T.J. (1970) The suspension and sinking of phytoplankton in the sea. *Oceanogr. mar. Biol. A. Rev.* **8,** 353–414.
19.2.4.1, 19.4.2

Smillie R.M. (1965) Isolation of two proteins with chloroplasts ferredoxin activity from a blue-green alga. *Biochem. biophys. Res. Commun.* **20,** 621–629.
4.3

Smith A.J. (1973) Synthesis of metabolic intermediates. In: Carr N.G. & Whitton B.A. (Eds) *The Biology of Blue-Green Algae*. 676 pp., pp. 1–38. Blackwell, Oxford.
3.1, 3.4

Smith A.J. (1981) Cyanobacterial contributions to the heterotrophic connection. In: Dalton H. (Ed.) *Microbial Growth on C1 Compounds*. 377 pp., pp. 122–130, Heyden, London.
3.4.2

Smith A.J. & Hoare D.S. (1977) Specialist phototrophs, lithotrophs and methylotrophs: a unity among a diversity of prokaryotes? *Microbiol. Rev.* **41,** 419–448.
3.4.4, 12.2.1

Smith A.J. & Lucas C. (1971) Propionate resistance in blue-green algae. *Biochem. J.* **124,** 23p–24p.
11.2.2, 11.5.1

Smith D.L., McLachlan J. & Patriquin D.G. (1979) Nitrogen fixation, as determined by acetylene reduction, in two salt marshes of Minas Basin. *Proc. Nova Scotia Inst. Sci.* **29,** 381–392.
20.4.5

Smith R.V. (1977) Domestic and agricultural contributions to the inputs of phosphorus and nitrogen to Lough Neagh. *Water Res.* **11,** 453–459.
18.6

Smith R.V. & Evans M.C.W. (1970) Soluble nitrogenase from vegetative cells of the blue-green alga *Anabaena cylindrica*. *Nature, Lond.* **225,** 1253–1254.
4.2

Smith R.V. & Evans M.C.W. (1971) Nitrogenase activity in cell-free extracts of the blue-green alga *Anabaena cylindrica*. *J. Bact.* **105,** 913–917.
4.1

Smith R.V. & Foy R.H. (1974) Improved hydrogen ion buffering of media for the culture of freshwater algae. *Br. phycol. J.* **9,** 239–245.
18.5.1.2

Smith R.V., Noy R.J. & Evans M.C.W. (1971) Physiological electron donors to the nitrogenase of the blue-green alga *Anabaena cylindrica*. *Biochim. biophys. Acta* **253,** 104–109.
4.3, 14.4.3.1, 18.5.2.1

Smith R.V. & Peat A. (1967) Comparative structure of the gas-vacuoles of blue-green algae. *Arch. Mikrobiol.* **57,** 111–122.
6.2

Smith R.V., Telfer A. & Evans M.C.W. (1971) Complementary functioning of nitrogenase components from a blue-green alga and a photosynthetic bacterium. *J. Bact.* **107,** 574–575.
4.2

Smith W.W., Ludwig M.L., Pattridge K.A., Tsernogolou D. & Petsko G.A. (1978) Crystallographic studies of flavodoxins: some correlations between structure and redox potential. In: Dutton P.L., Leigh J.S. & Scarpa A. (Eds) *Frontiers of Biological Energetics, Vol. II.* 1561 pp., pp. 957–964. Academic Press, N.Y.
8.3

Sneath P.H.A. & Sokal R.R. (1973) *Numerical Taxonomy: The Principles and Practice of Numerical Classification.* 573 pp. W. H. Freeman & Co., San Francisco.
12.2.1, 12.2.2.4, 12.3.1

Sofrova D., Vilim V. & Leblova S. (1976a) Effect of trypsin on the photochemical reactions of the blue-green alga *Plectonema boryanum*. *Photosynthetica* **10**, 40–46.
8.6

Sofrova D., Vilim V. & Leblova S. (1976b) Non-reactivation of trypsin-inhibited Hill reaction activity in the blue-green alga *Plectonema boryanum*. *Photosynthetica* **10**, 198–200.
8.6

Sorensen L. & Halldal P. (1977) Comparative analyses of action spectra of glycolate excretion ('photorespiration') and photosynthesis in the blue-green alga *Anacystis nidulans*. *Photochem. Photobiol.* **26**, 511–518.
8.5

Sorenson L.O. & Conover J.T. (1962) Algal mat communities of *Lyngbya confervoides* (C. Agardh) Gomont. *Publ. Inst. mar. Sci., Austin* **8**, 61–74.
20.3.3, 20.3.4, 20.4.4

Soriano S. (1973) Flexibacteria. *A. Rev. Microbiol.* **27**, 155–170.
16.2.1, 16.2.3, 16.4.3

Sorokin Yu.I. (1964) On the primary production and bacterial activities in the Black Sea. *J. Explor. Mer.* **29**, 41–60.
9.5.1

Sournia A. (1968) La cyanophycée *Oscillatoria* (= *Trichodesmium*) dans le plancton marin. *Nova Hedwigia* **15**, 1–12.
19.2.4.1

Sournia A. (1970) Les cyanophycées dans le plancton marin. *Annls Biol.* **9**, 63–66.
19.1, 19.2.1, 19.2.2, 19.2.3, 19.2.4.1, 19.2.4.2

Sournia A. (1972) Une période de poussées phytoplanctoniques près de Nosy-Bé (Madagascar) en 1971. I. Espèces rares ou nouvelles du phytoplancton. *Cah. O.R.S.T.O.M., sér Océanogr.* **10**, 151–159.
19.2.4.1

Sournia A. (1976) Ecologie et productivité d'une Cyanophycée en milieu corallien: *Oscillatoria limosa* Agardh. *Phycologia* **15**, 363–366.
20.2.4, 20.3.4, 20.4.4

Sperling J., Grunewald R. & Dovala D. (1974) A solid state pH control system for algal culturing and uptake studies. *Ecology* **55**, 895–898.
5.1.1

Sperry J.F. & Wilkins T.D. (1976) Arginine, a growth limiting factor for *Eubacterium lentum*. *J. Bact.* **127**, 780–784.
3.2.2.2

Spiller H., Ernst A., Kerfin W. & Böger P. (1978) Increase and stabilization of photoproduction of hydrogen in *Nostoc muscorum* by photosynthetic electron transport. *Z. Naturf.* **33c**, 541–547.
4.6

Srivastava B.S. (1969) Ultraviolet-induced mutations to growth factor requirement and penicillin resistance in a blue-green alga. *Arch. Mikrobiol.* **66**, 234–238.
11.5.3

Srivastava B.S. (1970) A simple method of isolating ultraviolet resistant and sensitive strains of a blue-green alga. *Phycologia* **9**, 205–208.
11.4.2.2

Srivastava B.S. & Kumar H.D. (1969) Inactivation and killing of the blue-green alga *Anabaena doliolum* grown on UV-irradiated culture medium. *Radiat. Bot.* **9**, 291–296.
11.4.1

Srivastava B.S., Kumar H.D. & Singh H.D. (1971) The effect of caffeine and light on killing of the blue-green alga *Anabaena doliolum* by ultraviolet irradiation. *Arch. Mikrobiol.* **78**, 139–144.
11.4.2.1, 11.4.2.2

Stacey G., Bottomley P.J., Van Baalen C. & Tabita F.R. (1979) Control of heterocyst and nitrogenase synthesis in cyanobacteria. *J. Bact.* **137**, 321–326.
4.5, 14.5, 14.6.1.1, 14.6.1.3

Stacey G., Van Baalen C. & Tabita F.R. (1977) Isolation and characterization of a marine *Anabaena* sp. capable of rapid growth on molecular nitrogen. *Arch. Microbiol.* **114**, 197–201.
4.6, 20.3.3, 20.3.5, 20.4.5

Stacey G., Van Baalen C. & Tabita F.R. (1979) Nitrogen and ammonia assimilation in the cyanobacteria: regulation of glutamine synthetase. *Arch. Biochem. Biophys.* **194**, 457–467.
4.5, 14.4.1

Stackebrandt E. & Seewaldt E. (1982) Partial sequence of 16S ribosomal RNA and the phylogeny of *Prochloron*. *Nature, Lond.* **295**, 618–620.

Staehelin L.A., Giddings T.H., Badami H. & Krzymowski W.W. (1978) A comparison of the supramolecular architecture of photosynthetic membranes of blue-green, red, and green algae and of higher plants. In: Deamer D.W. (Ed.) *Light Transducing Membranes. Structure, Function and Evolution*. 358 pp., pp. 335–355. Academic Press, New York.
7.9

Staehelin L.A., Golecki J.R., Fuller R.C. & Drews G. (1978) Visualization of the supramolecular architecture of chlorosomes

(chlorobium type vesicles) in freeze-fractured cells of *Chloroflexus aurantiacus*. *Arch. Microbiol.* **119,** 269–277.
12.2.1

Stah L.J. & Krumbein W.E. (1981) Aerobic nitrogen fixation in pure cultures of a benthic marine *Oscillatoria* (cyanobacteria). *FEMS Microbiol. Lett.* **11,** 295–298.
20.4.5

Stam W.T. (1981) Relationships between a number of filamentous blue-green algal strains (Cyanophyta) revealed by DNA-DNA hybridization. *Algological Rev.* **25,** (in press).
12.3.4.2

Stam W.T. & Venema G. (1977) The use of DNA:DNA hybridization for determination of the relationship between some blue-green algae (Cyanophyceae). *Acta bot. neerl.* **26,** 327–342.
12.3.4.2

Stanier R.Y. (1961) La place des bactéries dans le monde vivant. *Annls Inst. Pasteur, Paris* **101,** 297–312.
Foreword

Stanier R.Y. (1970) Some aspects of the biology of cells and their possible evolutionary significance. *Symp. Soc. gen. Microbiol.* **20,** 1–38.
Foreword, 12.4.1

Stanier R.Y. (1973) Autotrophy and heterotrophy in unicellular blue green algae. In: Carr N.G. & Whitton B.A. (Eds) *The Biology of Blue-Green Algae.* 676 pp., pp. 501–518. Blackwell, Oxford.
3.3, 12.3.3

Stanier R.Y. (1974) The origins of photosynthesis in eukaryotes. *Symp. Soc. gen. Microbiol.* **24,** 219–240.
7.1, 7.5.1.7, 12.4.1, 12.4.4.2, 12.4.4.3

Stanier R.Y. (1977) The position of the cyanobacteria in the world of phototrophs. *Carlsberg Res. Commun.* **42,** 77–98.
12.2.1

Stanier R.Y., Adelberg E.A. & Ingraham J. (1976) *The Microbial World, 4th edn.* 871 pp. Prentice-Hall, Engelwood Cliffs, N.J.
12.3.4.2

Stanier R.Y. & Cohen-Bazire G. (1977) Phototrophic prokaryotes: the cyanobacteria. *A. Rev. Microbiol.* **31,** 225–274.
4.1, 4.3, 6.2, 6.6, 7.8, 9.4.2, 9.6, 12.1.1, 12.2.1, 13.1, 13.2.2, 13.4, 18.3.2, 20.3.5

Stanier R.Y., Kunisawa R., Mandel M. & Cohen-Bazire G. (1971) Purification and properties of unicellular blue-green algae (order Chroococcales). *Bact. Rev.* **35,** 171–205.
7.5.1.1, 10.2.4, 11.3.1, 12.3.3

Stanier R.Y., Sistrom W.R., Hansen T.A., Whitton B.A., Castenholz R.W., Pfennig N., Gorlenko V.N., Kondratieva E.N., Eimhjellen K.E., Whittenbury R., Gherna R.L. & Trüper H.G. (1978) Proposal to place the nomenclature of the cyanobacteria (blue-green algae) under the rules of the International Code of Nomenclature of Bacteria. *Int. J. syst. Bact.* **28,** 335–336.
1, 12.1.1

Stanier R.Y. & van Niel C.B. (1941) The main outlines of bacterial classification. *J. Bact.* **42,** 437–466.
Foreword

Stanier R.Y. & van Niel C.B. (1962) The concept of a bacterium. *Arch. Mikrobiol.* **42,** 17–35.
Foreword

Staub R. (1961) Ernährungsphysiologisch — autökologische Untersuchungen an der planktischen Blaualge *Oscillatoria rubescens* D.C. *Schweiz. Z. Hydrol.* **23,** 84–198.
18.4.1, 18.4.3, 18.5.1.1

Steemann Nielsen E. (1976) *Marine Photosynthesis,* 141 pp. Elsevier, Amsterdam.
2.4.2

Steemann Nielsen E. (1978) Principal aspects concerning the batch technique in algal assays. *Mitt. int. Verein. theor. angew Limnol.* **19,** 81–87.
2.5.1

Steere R.L. (1957) Electron microscopy of structural detail in frozen biological specimens. *J. biophys. biochem. Cytol.* **3,** 45–60.
6.3.1

Steinitz Y., Mazor Z. & Shilo M. (1979) A mutant of the cyanobacterium *Plectonema boryanum* resistant to photooxidation. *Plant Sci. Lett.* **16,** 327–335.
2.3.1, 2.7.2.2, 11.5.4

Steinitz Y. & Shilo M. (1976) A study of photodynamic damage in cyanobacteria and mechanisms of their resistance to photooxidative death. In: Codd G.A. & Stewart W.D.P. (Eds). *Proceedings of the Second International Symposium on Photosynthetic Prokaryotes.* 255 pp., pp. 18–20. Univ. Dundee, Scotland.
8.3

Steven D.M. & Glombitza R. (1972) Oscillatory variation of a phytoplankton in a tropical ocean. *Nature, Lond.* **237,** 105–107.
19.2.4.1, 19.4.2

Stevens C.R.L. & Myers J. (1976) Characterization of pigment mutants in a blue-green alga, *Anacystis nidulans. J. Phycol.* **12,** 99–105.
11.5.4

Stevens C.R.L., Stevens S.E. & Myers J. (1975) Isolation and initial characterization of a uracil auxotroph of the blue-green alga *Anacystis nidulans*. *J. Bact.* **124**, 247–251.
11.5.3

Stevens R.J. & Parr M.P. (1977) The significance of alkaline phosphatase activity in Lough Neagh. *Freshwat. Biol.* **7**, 351–355.
18.5.2.4

Stevens S.E. & Porter R.D. (1980) Transformation in *Agmenellum quadruplicatum*. *Proc. natn. Acad. Sci. U.S.A.* **77**, 6052–6056.
11.6.3, 11.7

Stevens S.E. & Van Baalen C. (1969) N-methyl-N′-nitro-N-nitrosoguanidine as a mutagen for blue-green algae: evidence for repair. *J. Phycol.* **5**, 136–139.
11.4.2.2, 11.5.4

Stevens S.E. & Van Baalen C. (1970) Growth characteristics of selected mutants of a coccoid blue-green alga. *Arch. Mikrobiol.* **72**, 1–8.
11.5.3, 11.5.4

Stewart A.C. & Bendall D.S. (1979) Preparation of an active, oxygen evolving photosystem 2 particle from a blue-green alga. *FEBS Lett.* **107**, 308–312.
8.2, 8.6

Stewart A.C. & Bendall D.S. (1980) Photosynthetic electron transport in a cell-free preparation from the thermophilic blue-green alga *Phormidium laminosum*. *Biochem. J.* **188**, 351–361.
8.6

Stewart J.R. & Brown R.M. Jr. (1969) *Cytophaga* that kills or lyses algae. *Science, N.Y.* **164**, 1523–1524.
16.2.3

Stewart R., Auchterlonie C.C. & Codd G.A. (1977) Studies on the subunit structure of ribulose 1,5-diphosphate carboxylase from the blue-green alga *Microcystis aeruginosa*. *Planta, Berl.* **136**, 61–64.
3.2.1

Stewart W.D.P. (1967) Nitrogen turnover in marine and brackish habitats. II. Use of ^{15}N in measuring nitrogen fixation in the field. *Ann. Bot., N.S.* 385–407.
20.4.5

Stewart W.D.P. (1971) Algal metabolism and water pollution in the Tay region. *Proc. R. Soc. Edinb. B* **71**, 209–224.
5.4.1

Stewart W.D.P. (1973a) Nitrogen fixation by photosynthetic microorganisms. *A. Rev. Microbiol.* **27**, 283–316.
4.1

Stewart W.D.P. (1973b) Nitrogen fixation. In: Carr N.G. & Whitton B.A. (Eds) *The Biology of Blue-Green Algae*. 676 pp., pp. 260–278. Blackwell, Oxford.
-9.5.3

Stewart W.D.P. (1974a) *Algal Physiology and Biochemistry*. 989 pp. Blackwell, Oxford.
17.5.2, 17.6

Stewart W.D.P. (1974b) Nitrogen fixation by Myxophyceae from marine environments. *J. gen. Microbiol.* **36**, 415–422.
20.4.5

Stewart W.D.P. (1975a) Biological cycling of nitrogen in intertidal and supralittoral marine environments. In: *Proc. 9th Europ. Mar. Biol. Symp.* pp. 637–660.
20.2.1, 20.4.5

Stewart W.D.P. (Ed.) (1975b) *Nitrogen Fixation by Free-living Microorganisms*. Cambridge Univ. Press, London.
20.3.1

Stewart W.D.P. (1977) A botanical ramble among the blue-green algae. *Br. phycol. J.* **12**, 89–115.
3.2.2.2, 3.4.3, 4.1, 4.3, 4.5, 17.6

Stewart, W.D.P. (1980). Some aspects of structure and function in N_2-fixing cyanobacteria. *A. Rev. Microbiol.* **34**, 497–536.
18.5.2.1

Stewart W.D.P. & Alexander G. (1971) Phosphorus availability and nitrogenase activity in aquatic blue-green algae. *Freshwater Biol.* **1**, 389–401.
3.2.2.2, 5.3.1, 5.3.2

Stewart W.D.P. & Codd G.A. (1975) Polyhedral bodies (carboxysomes) of nitrogen-fixing blue-green algae. *Br. phycol. J.* **10**, 273–278.
3.2.1, 14.4.1, 14.4.2

Stewart W.D.P. & Daft M.J. (1976) Algal lysing agents of freshwater habitats. *Soc. appl. Bact. Symp. Ser.* **4**, 63–90.
17.4.2

Stewart W.D.P., Fitzgerald G.P. & Burris R.H. (1970) Acetylene reduction assay for determination of phosphorus availability in Wisconsin Lakes. *Proc. natn. Acad. Sci. U.S.A.* **66**, 1104–1111.
5.2.3, 5.3.2

Stewart W.D.P., Haystead A. & Dharmawardene M.W.N. (1975) Nitrogen assimilation and metabolism in blue-green algae. In: Stewart W.D.P. (Ed.) *Nitrogen Fixation in*

Free-living Microorganisms. pp. 129–158. Cambridge Univ. Press, Cambridge.
4.5, 14.4.1

Stewart W.D.P., Haystead A. & Pearson H.W. (1969) Nitrogenase activity in heterocysts of blue-green algae. *Nature, Lond.* **224,** 226–228.
14.2, 14.4.2, 14.4.3.2

Stewart W.D.P. & Lex M. (1970) Nitrogenase activity in the blue-green alga *Plectonema boryanum* strain 594. *Arch. Mikrobiol.* **73,** 250–260.
1, 4.1, 14.2

Stewart W.D.P. & Pearson H.W. (1970) Effects of aerobic and anaerobic conditions on growth and metabolism of blue-green algae. *Proc. R. Soc. Lond. B.* **175,** 293–311.
9.6, 17.6

Stewart W.D.P., Pemble M. & Al-Ugaily L. (1978) Nitrogen and phosphorus storage and utilization in blue-green algae. *Mitt. int. Verein. theor. angew. Limnol.* **21,** 224–247.
5.3.1, 5.4.3

Stewart W.D.P. & Rowell P. (1975) Effects of L-methionine-DL-sulphoximine on the assimilation of newly fixed NH_3, acetylene reduction and heterocyst production in *Anabaena cylindrica. Biochem. biophys. Res. Commun.* **65,** 846–856.
4.5, 14.6.1.3

Stewart W.D.P., Rowell P. & Rai A.N. (1980) Symbiotic nitrogen-fixing cyanobacteria. In: Stewart W.D.P. & Gallon J.R. (Eds) *Proceedings of an International Symposium on Nitrogen Fixation*. 451 pp., pp. 239–277.
4.1

Stewart W.D.P. & Singh H.N. (1975) Transfer of nitrogen-fixing (*nif*) genes in the blue-green alga *Nostoc muscorum. Biochem. biophys. Res. Commun.* **62,** 62–69.
11.5.5.2(a), 11.5.5.2(b), 11.6.1

Stewart W.D.P., Tuckwell S.B. & May E. (1975) Eutrophication and algal growths in Scottish fresh water lochs. In: Chadwick M.J. & Goodman G.T. (Eds) *The Ecology of Resource Degradation and Renewal*. 480 pp., pp. 57–80. Blackwell, Oxford.
5.4.2

Stockner J.G. & Lund J.W.G. (1970) Live algae in postglacial lake deposits. *Limnol. Oceanogr.* **15,** 41–58.
18.5.3

Strange J.R., Williamson G.D. & Fletcher D.J. (1975) P-32 uptake in lentic algae. *Environ. Lett.* **9,** 1–5.
5.1.1

Stransky H. & Hager A. (1970) The carotenoid pattern and the occurrence of the light-induced xanthophyll cycle in various classes of algae. IV. Cyanophyceae and Rhodophyceae. *Arch. Mikrobiol.* **72,** 84–96.
6.4.2

Subbaramaiah K. (1972) The biology of the blue-green algae of the Sambhar lake salt works. In: Desikachary T.V. (Ed.) *Taxonomy and Biology of Blue-Green Algae*. 591 pp., pp. 439–441. Univ. Madras, India.
18.4.1

Susor W.A. & Krogmann D.W. (1966) TPN photoreduction with cell-free preparations of *Anabaena variabilis. Biochim. biophys. Acta* **120,** 65–72.
8.2, 8.5

Sussman A.S. & Halvorson H.O. (1966) Spores, their dormancy and germination. 354 pp. Harper & Row, N.Y.
15.5

Sutcliffe J.G. & Church G.M. (1978) The cleavage site of the restriction endonuclease *Ava* II. *Nucl. Acids Res.* **5,** 2313–2319.
11.3.3

Sutherland J.M., Herdman M. & Stewart W.D.P. (1979) Akinetes of the cyanobacterium *Nostoc* PCC 7524: macromolecular composition, structure and control of differentiation. *J. gen. Microbiol.* **115,** 273–287.
11.5.5.2(b), 14.6.1.3, 14.6.2, 15.2, 15.3, 15.4, 15.4.1, 15.4.2, 15.4.3, 15.5, 15.6, 15.6.1

Sutherland J.M., Herdman M. & Stewart W.D.P. (1980) Structure and differentiation of akinetes and their relation to heterocysts of the cyanobacterium *Nostoc* PCC 7524. *Soc. gen. Microbiol. Quarterly* **7,** 105.
11.5.5.2(c)

Suzuki R. & Fujita Y. (1977) Carotenoid photobleaching induced by the action of photosynthetic reaction center. II: DCMU sensitivity. *Pl. Cell Physiol., Tokyo* **18,** 625–631.
8.6

Svedberg T. & Eriksson I.-B. (1932) The molecular weights of phycocyan and of phycoerythrin. III. *J. Am. chem. Soc.* **54,** 3998–4010.
7.5.1.4

Svedberg T. & Katsurai T. (1929) The molecular weights of phycocyan and of phycoerythrin from *Porphyra tenera* and of phycocyan from *Aphanizomenon flos–aquae. J. Am. chem. Soc.* **51,** 3573–3583.
7.5.1.1, 7.5.1.4

Svedberg T. & Lewis N.B. (1928) The molecular weights of phycoerythrin and phycocyan. *J. Am. chem. Soc.* **50,** 525–536.
7.5.1.4

Sweet R.M., Fuchs H.E., Fisher R.G. &

Glazer A.N. (1977) Preliminary crystallographic investigations of two phycobiliproteins. *J. biol. Chem.* **252,** 8258–8260.
7.5.1.6

Sykora J.L., Keleti G., Roche R., Volk D.R., Lippy E.C., Kay G.P., Burgess R.A. & Shapiro M.A. (1980) Endotoxins, algae and *Limulus* amoebocyte lysate test in drinking water. *Water Res.* **14,** 829–839.
13.4.4

Szalonti B. & Csatorday K. (1979) Changes in phycocyanin–carotenoid association during nitrate starvation of *Anacystis nidulans*. *Biochem. biophys. Res. Commun.* **88,** 1294–1300.
8.6

Sze P. (1975) Possible effect of lower phosphorus concentrations on the phytoplankton in Onondaga Lake, New York, U.S.A. *Phycologia* **14,** 197–204.
5.4.2

Tabita F.R., Caruso P. & Whitman W. (1978) Facile assay of enzymes unique to the Calvin cycle in intact cells with special reference to ribulose 1,5-bisphosphate carboxylase. *Anal. Biochem.* **84,** 462–472.
3.2.1

Tabita F.R. & Colletti C. (1979) Carbon dioxide assimilation in cyanobacteria: Regulation of ribulose 1,5-bisphosphate carboxylase. *J. Bact.* **140,** 452–458.
3.2.1, 3.3

Tabita F.R., Gibson J.L., Whitman W.B., Martin M.N. & Robison P.D. (1979) The small subunit of ribulose bisphosphate carboxylase/oxygenase: its role in the molecular regulation of CO_2 fixation and its evolutional significance. In: Nichols J.M. (Ed.) *Abstracts of the Third International Symposium on Photosynthetic Prokaryotes, Oxford*. p. E11. Univ. Liverpool, England.
12.2.2.3

Tabita F.R., Stevens S.E. & Gibson J.L. (1976) Carbon dioxide assimilation in blue-green algae: initial studies on the structure of ribulose 1,5-bisphosphate carboxylase. *J. Bact.* **125,** 531–539.
3.2.1

Tabita F.R., Stevens S.E. & Quijano R. (1974) D-Ribulose 1,5-diphosphate carboxylase from blue-green algae. *Biochem. biophys. Res. Commun.* **61,** 45–52.
3.2.1

Takabe T. (1977) Ribulose 1,5-bisphosphate carboxylase from the blue-green alga *Anabaena cylindrica*. *Agric. Biol. Chem.* **41,** 2255–2260.
3.2.1

Takabe T., Nishimura M. & Akazawa T. (1976) Presence of two subunit types in ribulose 1,5-bisphosphate carboxylase from blue-green algae. *Biochem. biophys. Res. Commun.* **68,** 537–544.
3.2.1

Takahashi I. (1962) Genetic transformation of *Bacillus subtilis* by extracellular DNA. *Biochem. biophys. Res. Commun.* **7,** 467–470.
11.6.3

Takemoto J. & Bogorad L. (1975) Subunits of phycoerythrin from *Fremyella diplosiphon*: chemical and immunochemical characterization. *Biochemistry, N.Y.* **14,** 1211–1216.
7.5.1.7

Takruri I., Haslett B.G., Boulter D., Andrew P.W. & Rogers L.J. (1978) Amino acid sequence of ferredoxin from the red alga. *Porphyra umbilicalis*. *Biochem. J.* **173,** 459–466.
8.3

Talling J.F. (1955) The relative growth rates of three plankton diatoms in relation to underwater radiation. *Ann. Bot., N.S.* **19,** 329–341.
18.5.5.6

Talling J.F. (1976) The depletion of carbon dioxide from lake water by phytoplankton. *J. Ecol.* **64,** 79–121.
18.5.2.2

Talling J.F., Wood R.B., Prosser M.V. & Baxter R.M. (1973) The upper limit of photosynthetic productivity by phytoplankton: evidence from Ethiopian soda lakes. *Freshwat. Biol.* **3,** 53–76.
9.5.4, 18.4.1, 18.4.2

Talpasayi E.R.S. (1962) Acid phosphatase activity of some algae and its inhibition by molybdenum. *Biochim. biophys. Acta* **59,** 710–712.
5.1.3

Talpasayi E.R.S. (1963) Polyphosphate containing particles of blue-green algae. *Cytologia* **28,** 76–80.
5.3.1

Talpasayi E.R.S. (1967) Localisation of ascorbic acid in heterocysts of blue-green algae. *Curr. Sci.* **36,** 190–191.
14.4.3.2

Talpasayi E.R.S. & Bahal M.R. (1967) Cellular differentiation in *Anabaena cylindrica*. *Z. Pflanzenphysiol.* **56,** 100–101.
14.5

Tamiya H., Hase E., Shibata K., Mituya A., Iwamura T., Nikei T. & Sasa T. (1953)

Kinetics of growth of *Chlorella* with special reference to its dependence on quantity of available light and temperature. In: Burlew J.S. (Ed.) *Algal Culture from Laboratory to Pilot Plant*. 357 pp., pp. 204–232. *Carnegie Inst. Wash. Publ.* No. 600.
2.5.2

Tanaka M., Haniu M., Zeitlin S. & Yasunobu K.T. (1975) Amino acid sequence of *Spirulina maxima* ferredoxin—from a prokaryote. *Biochem. biophys. Res. Commun.* **64,** 399–407.
8.3

Tandeau de Marsac N.T. (1977) Occurrence and nature of chromatic adaptation in cyanobacteria. *J. Bact.* **130,** 82–91.
6.5.1, 7.2, 7.7, 7.8, 7.8.3.1, 18.5.5.5

Tandeau de Marsac N. (1978) *Étude sur la Biosynthèse des Phycobiliprotéines chez les Cyanobactéries.* 166 pp. Thèse Doctorat d'Etat, Université Pierre et Marie Curie, Paris VI.
7.8.1

Tandeau de Marsac N., Castets A.-M. & Cohen-Bazire G. (1979) Wavelength modulation of phycoerythrin synthesis in the unicellular cyanobacterium *Synechocystis* 6701. In: Nichols J.M. (Ed.) *Abstracts of Third International Symposium on Photosynthetic Prokaryotes, Oxford.* p. D19. Univ. Liverpool, England.
7.8.3.2

Tandeau de Marsac N. & Cohen-Bazire G. (1977) Molecular composition of cyanobacterial phycobilisomes. *Proc. natn. Acad. Sci. U.S.A.* **74,** 1635–1639.
7.5, 7.5.2, 7.8.3.2

Tang T. & Asato Y. (1978) Ultraviolet light-induction and photoreactivation of thymine dimers in a cyanobacterium, *Anacystis nidulans. Arch. Microbiol.* **118,** 193–197.
11.4.1, 11.4.2.1

Taylor A.R.A., Gebelein C.D. (1966) Plant pigments and light penetration in intertidal sediments. *Helgoländer wiss. Meeresunters.* **13,** 229–237.
20.3.4

Taylor B.F., Lee C.C. & Bunt J.S. (1973) Nitrogen-fixation associated with the marine blue-green alga, *Trichodesmium* as measured by the acetylene-reduction technique. *Arch. Mikrobiol.* **88,** 205–212.
19.3.1, 19.3.4, 19.3.7, 19.5.2

Taylor F.J.R. (1976) Autogenous theories for the origin of eukaryotes. *Taxon* **25,** 377–390.
12.4.1, 12.4.2, 12.4.4

Taylor F.J.R. (1979) Symbionticism revisited: a discussion of the evolutionary impact of intracellular symbioses. *Proc. R. Soc. Lond. B* **204,** 267–286.
12.4.1, 12.4.2

Taylor W.R. (1950) *Plants of Bikini and other Northern Marshall Islands.* 427 pp. Univ. Michigan, Ann Arbor.
20.2.2, 20.2.4

Teale F.W.J. & Dale R.E. (1970) Isolation and spectral characterization of phycobiliproteins. *Biochem. J.* **116,** 161–169.
7.5.1.5

Tel-Or E., Cammack R., Rao K.K., Rogers L.J., Stewart W.D.P. & Hall D.O. (1977a) Comparative immunochemistry of bacterial, algal and plant ferredoxins. *Biochim. biophys. Acta* **490,** 120–131.
8.3, 12.2.2.2, 12.3.4.1

Tel-Or E., Luijk L.W. & Packer L. (1977b) An inducible hydrogenase in cyanobacteria enhances N_2 fixation. *FEBS Lett.* **78,** 49–52.
9.3.3

Tel-Or E., Luijk L.W. & Packer L. (1978) Hydrogenase in N_2-fixing cyanobacteria. *Arch. Biochem. Biophys.* **185,** 185–194.
4.6, 14.4.1, 14.4.3.3

Tel-Or E. & Malkin R. (1977) The photochemical and fluorescence properties of whole cells, spheroplasts and spheroplast particles from the blue-green alga *Phormidium luridum. Biochim. biophys. Acta* **459,** 157–174.
6.4.2, 8.4

Tel-Or E. & Stewart W.D.P. (1975) Manganese and photosynthetic oxygen evolution by algae. *Nature, Lond.* **258,** 715–716.
14.4.3.2

Tel-Or & Stewart W.D.P. (1976) Photosynthetic electron transport, ATP synthesis and nitrogenase activity in isolated heterocysts of *Anabaena cylindrica. Biochim. biophys. Acta* **423,** 189–195.
4.3, 8.7, 14.2, 14.4.3.1, 14.4.3.2

Tel-Or E. & Stewart W.D.P. (1977) Photosynthetic components and activities of nitrogen fixing, isolated heterocysts of *Anabaena cylindrica. Proc. R. Soc. Lond. B* **198,** 61–86.
8.2, 14.4.1, 14.4.3.2

Tempest D.W. (1970) The continuous cultivation of micro-organisms. I. Theory of the chemostat. In: Norris J.R. & Ribbons D.W. (Eds) *Methods in Microbiology, Vol. 2.* 432 pp., pp. 259–277. Academic Press, London, New York.
2.5.1

Tempest D.W. & Meers J.L. (1970) Synthesis of glutamate in *Aerobacter aerogenes* by a hitherto unknown route. *Biochem. J.* **117,** 405–407.
18.5.2.1

Tempest D.W. & Neijssel O.M. (1976) Microbial adaptation to low nutrient environments. In: Dean A.C.R., Ellwood D.C., Evans C.G.T. & Melling J. (Eds) *Continuous Culture, Applications and New Fields, vol 6.* 364 pp., pp. 283–296. Ellis Horwood, Chichester, England.
2.5.2, 3.4.3

Tetley R.M. & Bishop N.I. (1979) The differential action of metronidazole on nitrogen fixation, hydrogen metabolism, photosynthesis and respiration in *Anabaena* and Scenedesmus. *Biochim. biophys. Acta* **546,** 43–53.
4.6, 14.4.3.2

Tharanathan R.N., Mayer H. & Weckesser J. (1978) Location of O-methyl sugars in antigenic (lipo-)polysaccharides of photosynthetic bacteria and cyanobacteria. *Biochem. J.* **171,** 403–408.
13.4.4

Thinh L.V. (1979) *Prochloron* (Prochlorophyta) associated with the ascidian *Tridi-demnium cyclops* Michaelsen. *Phycologia* **18,** 77–82.
20.2.4

Thomas E.A. (1950) Auffallige biologische Folgen von Sprungschichtneigungen im Zürichsee. *Schweiz. Z. Hydrol.* **12,** 1–24.
18.5.5.2

Thomas E.A. (1970a) Beobachtungen über das Wandern von *Phormidium autumnale* und *Oscillatoria princeps* auf Agar. *Schweiz. Z. Hydrol.* **32,** 523–531.
16.5

Thomas J. (1970b) Absence of the pigments of photosystem II of photosynthesis in heterocysts of a blue-green alga. *Nature, Lond.* **228,** 181–183.
4.3, 6.5.2, 14.3.2

Thomas J. (1972) Relationship between age of culture and occurrence of the pigments of photosystem II of photosynthesis in heterocysts of a blue-green alga. *J. Bact.* **110,** 92–95.
14.3.2

Thomas J. & David K.A.V. (1971) Studies on the physiology of heterocyst production in the nitrogen-fixing blue-green alga *Anabaena* sp. L-31 in continuous culture. *J. gen. Microbiol.* **66,** 127–131.
14.6.1.1

Thomas J. & David K.A.V. (1972) Site of nitrogenase activity in the blue-green alga *Anabaena* sp. L-31. *Nature, New Biol.* **238,** 219–221.
14.4.3.2

Thomas J., Meeks J.C., Wolk C.P., Shaffer P.W., Austin S.M. & Chien W.-S. (1977) Formation of glutamine from [^{13}N]-ammonia, [^{13}N]dinitrogen, and [^{14}C] glutamate by heterocysts isolated from *Anabaena cylindrica. J. Bact.* **129,** 1545–1555.
14.2, 14.4.1, 14.6.1.3

Thornber J.P. (1969) Comparison of a chlorophyll a–protein complex isolated from a blue-green alga with chlorophyll–protein complexes obtained from green bacteria and higher plants. *Biochim. biophys. Acta* **172,** 230–241.
6.4.2

Thornber J.P., Alberte R.S., Hunter F.A., Shiozawa J.A. & Kan K.-S. (1976) The organization of chlorophyll in the plant photosynthetic unit. *Brookhaven Symp. Biology* **28,** 132–148.
6.4.2

Thornber J.P., Markwell J.P. & Reinman S. (1979) Plant chlorophyll-protein complexes: recent advances. *Photochem. Photobiol.* **29,** 1205–1216.
8.4

Thorneley R.N.F. & Cornish-Bowden A. (1977) Kinetics of nitrogenase of *Klebsiella pneumoniae. Biochem. J.* **165,** 255–262.
4.5

Throm G. (1968) Untersuchungen zum Reaktionsmechanismus von Phobotaxis und Kinesis an *Rhodospirillum rubrum. Arch. Protistenk.* **110,** 313–371.
16.4.3, 16.4.4

Thurnauer M.C., Bowman M.K. & Norris J.R. (1979) Time resolved electron spin echo spectroscopy applied to the study of photosynthesis. *FEBS Lett.* **100,** 309–312.
8.4

Thurston E.L. & Ingram L.O. (1971) Morphology and fine structure of *Fischerella ambigua. J. Phycol.* **7,** 203–210.
13.2.2, 13.3

Tilden J.E. (1933) A classification of algae based on evolutionary development with special reference to pigmentation. *Bot. Gaz.* **95,** 59–77.
12.3.1

Tilman D. & Kilham S.S. (1976) Phosphate and silicate growth and uptake kinetics of the diatoms *Asterionella formosa* and *Cyclotella meneghiniana* in batch and semi-continuous culture. *J. Phycol.* **12,** 375–383.
5.1.1, 5.2.1

Tischer I. (1957) Untersuchungen über die granulären Einschlüsse und das Reduktions-Oxydations-Vermögen der Cyanophyceen. *Arch. Mikrobiol.* **27,** 400–428.
14.4.3.2

Tischer R.G. (1965) Pure culture of *Anabaena flos-aquae* A-37. *Nature, Lond.* **205,** 419–420.
3.4.3

Tiwari D.N. (1978a) UV-induced developmental abnormalities in the filamentous blue-green alga *Nostoc linckia. Biochem. Physiol. Pflanz.* **172,** 537–540.
11.5.5.2(a)

Tiwari D.N. (1978b) The heterocysts of the blue-green alga *Nostochopsis lobatus*: effects of cultural conditions. *New Phytol.* **81,** 853–856.
14.6.1.1

Tolbert N.E. (1974) Photorespiration. In: Stewart W.D.P. (Ed.) *Algal Physiology and Biochemistry*. 989 pp., pp. 474–504. Blackwell, Oxford.
2.3.1

Tomioka N., Shinozaki K. & Sugiura M. (1981) Molecular cloning and characterization of ribosomal RNA genes from a blue-green alga, *Anacystis nidulans. Molec. gen. Genet.* **184,** 359–363.
11.8

Topachevskii A.V. (Ed.) 1968 *"Tsvetenie" Vody*. 387 pp. Naukova Dumka, Kiev. (Russian = Waterblooms).
18.4.4

Torjesen P.A. & Sletten K. (1972) C-phycocyanin from *Oscillatoria agardhii*. I. Some molecular properties. *Biochim. biophys. Acta* **263,** 258–271.
7.5.1.8

Towe K.M. (1978) Early Precambrian oxygen: a case against photosynthesis. *Nature, Lond.* **274,** 657–661.
12.1.2

Trebst A. & Bothe H. (1966) Zur Rolle des Phytoflavins im photosynthetischen Elektronentransport. *Ber. dt. bot. Ges.* **79,** 44–47.
6.5.2

Troxler R.F. & Brown A.S. (1979) Amino acid sequence of the phycocyanin subunit from the alga, *Cyanidium caldarium. Fed. Proc.* **38,** 410 Abs.
7.5.1.3

Troxler R.F., Brown A.S. & Kost H.-P. (1978) Quantitative degradation of radiolabeled phycobiliproteins. *Eur. J. Biochem.* **87,** 181–189.
7.5.1.2

Troxler R.F., Brown A.S. & Zilinskas B.A. (1979) Amino acid sequence at the amino terminus of allophycocyanin I, II and III from *Nostoc* sp. phycobilisomes. *Pl. Physiol., Lancaster* **63,** 547 Abs.
7.5.1.1., 7.5.1.8

Troxler R.F., Foster J.A., Brown A.S. & Franzblau C. (1975) The α and β subunits of *Cyanidium caldarium* phycocyanin: properties and amino acid sequences at the amino terminus. *Biochemistry, N.Y.* **14,** 268–274.
7.5.1.8

Trüper H.G. (1973) The present state of knowledge of sulfur metabolism in phototrophic bacteria. In: *Abstracts of Symposium on Prokaryotic Photosynthetic Organisms*. 228 pp., pp. 160–166. Univ. Freiburg i. Br., Germany.
9.3.3

Trüper H.G. & Pfennig N. (1978) Taxonomy of the Rhodospirillales. In: Clayton R.K. & Sistrom W.R. (Eds) *The Photosynthetic Bacteria*. 946 pp., pp. 12–27. Plenum Press, New York.
12.2.1, 12.2.2.1

Tsai L.B. & Mortenson L.E. (1978) Interaction of the nitrogenase components of *Anabaena cylindrica* with those of *Clostridium pasteurianum. Biochem. biophys. Res. Commun.* **81,** 280–287.
4.2

Tsang N.G., Macnab R. & Koshland D.E., Jr (1973) Common mechanism for repellents and attractants in bacterial chemotaxis. *Science, N.Y.* **181,** 60–63.
17.5.1

Tso W.W. & Adler J. (1974) Negative chemotaxis in *Escherichia coli. J. Bact.* **118,** 560–576.
17.5.1

Tsuda R.T. & Kami H.T. (1973) Algal succession on artificial reefs in a marine lagoon environment on Guam. *J. Phycol.* **9,** 260–264.
20.2.4

Tsujimoto H.Y., McSwain B.D., Hiyama T. & Arnon D.I. (1976) Effect of $NADP^+$ on light induced cytochrome changes in membrane fragments from a blue-green alga. *Biochim. biophys. Acta* **423,** 303–312.
8.5

Tsukihara T., Fukuyama K., Tahara H., Katsube Y., Matsuura Y., Tanaka N., Kakudo M., Wada K. & Matsubara H. (1978a) Crystallization and a 5Å X-ray diffraction study of *Aphanothece sacrum* ferredoxin. *J. Biochem., Tokyo* **84,** 989–992.
8.3

Tsikihara, T., Fukuyama, K., Tahara, H., Katsube, Y., Matsuura, Y., Tanaka, N., Kakudo, M., Wada, K. & Matsubara, H. (1978b) X-ray analysis of ferredoxin from *Spirulina platensis. J. Biochem., Tokyo* **84,** 1645–1647.
8.3

Tuffery A.A. (1969) Light and electron microscopy of the sheath of a blue-green alga. *J. gen. Microbiol.* **57,** 41–50.
13.2.2

Tuli R. & Thomas J. (1980) Regulation of glutamine synthetase in the blue-green alga *Anabaena* L-31. *Biochim. biophys. Acta* **613,** 526–533.
14.4.1

Turchini M. (1918) Role de l'hétérocyste des nostocées. *Rev. gen. Bot.* **30,** 273–282.
14.1

Tyagi V.V.S. (1974a) Some observations on the pattern of sporulation in a blue-green alga, *Anabaena doliolum. Ann. Bot., N.S.* **38,** 1107–1111.
14.6.2, 15.4, 15.4.1

Tyagi V.V.S. (1974b) Effects of some inhibitors of DNA- and protein-synthesis on heterocyst formation in the blue-green alga *Anabaena doliolum. Beitr. Biol. Pfl.* **50,** 177–183.
14.6.1.1, 14.6.2

Tyagi V.V.S. (1975) The heterocysts of blue-green algae (Myxophyceae). *Biol. Rev.* **50,** 247–284.

Tyagi V.V.S. (1976) Studies on phototopotaxis in the blue-green alga, *Cylindrospermum. Biochem. Physiol. Pflanz.* **170,** 17–22.
16.4.4

Tyagi V.V.S. (1978a) Effect of rifampicin (3-(4-methylpiperazinyl-iminomethyl)-rifamycin) on heterocyst differentiation in the blue-green algae *Anabaena* and *Calothrix. Z. allg. Mikrobiol.* **18,** 495–499.
14.6.1.1

Tyagi V.V.S. (1978b) Pattern formation with N-mediated gradient of cell diameter in the blue-green alga *Calothrix. Ind. J. exp. Biol.* **16,** 275–277.
14.6.1.1

Tyagi V.V.S. (1978c) The pattern of sporulation in *Anabaena circinalis* and comments on the role of heterocysts in sporulation in blue-green algae. *Z. allg. Mikrobiol.* **18,** 281–286.
15.4

Tyler B. (1978) Regulation of the assimilation of nitrogen compounds. *A. Rev. Biochem.* **47,** 1127–1162
4.5

Tyler S.A. & Barghoorn E.S. (1954) Occurrence of structurally preserved plants in pre-Cambrian rocks of the Canadian Shield. *Science, N.Y.* **119,** 606–608.
21.1

Tyndall D.R., Yopp J.H., Miller D.M. & Schmid W.E. (1978) Physicochemical parameters governing the growth of *Aphanothece halophytica* (Chroococcales) in hypersaline media. *Phycologia* **17,** 179–185.

Ueda K. (1971a) Elektronenmikroskopische Struktur der Blaualge *Oscillatoria minima* Gicklhorn unter besonderer Berücksichtigung der Zellteilung. *Biochem. Physiol. Pflanz.* **162,** 345–356.
6.2

Ueda K. (1971b) Die quantitative Bestimmung des DNS-Gehalts in den Zellen von Cyanophyceen durch Fluorochromierung mit Coriphosphin. *Biochem. Physiol. Pflanz.* **162,** 439–449.
15.2

Ueda K. & Sawada M. (1972) DNA-content of akinete cells in *Cylindrospermum. Cytologia* **37,** 519–523.
15.2

Uhlmann D. & Albrecht E. (1968) Biogeochemische Faktoren der Eutrophierung von Trinkwasser–Talsperren. *Limnologica, Berlin.* **6,** 225–245.
18.4.3

Ullrich-Eberius C.I. & Yingchol Y. (1974) Phosphate uptake and its pH-dependence in halophytic and glycophytic algae and higher plants. *Oecologia, Berl.* **17,** 17–26.
5.1.1

Umezaki I. (1961) The Marine Blue-green algae of Japan. *Mem. College Agriculture, Kyoto University* **83,** 1–149.
19.2.3, 20.2.1

Utkilen H.C. (1976) Thiosulfate as electron donor in the blue-green alga *Anacystis nidulans. J. gen. Microbiol.* **95,** 177–180.
9.3.3

Utkilen H. & Castenholz R. (1979) Physiological aspects of adaptation to sulfide in a thermophilic *Oscillatoria*. In: Nichols J.M. (Ed.) *Abstracts of the Third International Symposium on Photosynthetic Prokaryotes, Oxford.* p. D12. Univ. Liverpool, England.
9.3.1, 9.3.4, 9.5.2

Utterback C.L., Phifer L.D. & Robinson R.J. (1942) Some chemical, planktonic and optical characteristics of Crater Lake. *Ecology* **23,** 97–103.
18.4.1

Uzzell T. & Spolsky C. (1974) Mitochondria and plastids as endosymbionts: a revival of special creation? *Am. Sci.* **62,** 334–343.
12.2.1, 12.4.1, 12.4.2

Uzzell T. & Spolsky C. (1981) Two data sets: alternative explanations and interpretations. *Ann. N.Y. Acad. Sci.* **361,** 481–499.
12.1.3, 12.2.2.1, 12.4.2

Vaara T., Varra M. & Niemela S. (1979) Two improved methods for obtaining axenic cultures of cyanobacteria. *J. appl. environ. Microbiol.* **38,** 1011–1014.
16.5

Vaccaro R.F., Hicks S.E., Jannasch H.W. & Carey F.G. (1968) The occurrence and role of glucose in sea water. *Limnol. Oceanogr.* **13,** 356–360.
3.4.3

Vaccaro R.F. & Jannasch H.W. (1966) Studies on heterotrophic activity in sea water based on glucose assimilation. *Limnol. Oceanogr.* **11,** 596–607.
3.4.3

Vacelet J. (1971) Étude en microscopie electronique de l'association entre une Cyanophycée Chroococcale et une éponge du genre *Verongia. J. Microscopie* **12,** 363–380.
20.2.4

Vaillancourt S., Beauchemin-Newhouse N. & Cedergren R.J. (1978) Polyphosphate-deficient mutants of *Anacystis nidulans. Can. J. Microbiol.* **24,** 112–116.
5.3.1, 11.2.2, 11.5.3

Vallentyne J.R. (1971) A description of Hutchinson Lake, Ontario. *Limnol. Oceanogr.* **16,** 473–475.
18.4.1

Van Baalen C. (1965a) Quantitative surface plating of coccoid blue-green algae. *J. Phycol.* **1,** 19–22.
11.1

Van Baalen C. (1965b) Mutation of the blue-green alga *Anacystis nidulans. Science, N.Y.* **149,** 70.
11.5.3, 11.5.4

Van Baalen C. (1968) The effects of UV irradiation on a coccoidal blue-green alga: survival, photosynthesis and photoreactivation. *Pl. Physiol., Lancaster* **43,** 1689–1696.
11.4.2.1

Van Baalen C. (1973) Mutagenesis and genetic recombination. In: Carr N.G. & Whitton B.A. (Eds) *The Biology of Blue-Green Algae*, 676 pp., pp. 201–213. Blackwell, Oxford.
11.1, 11.4.2.1, 11.5.4, 11.6.1

Van Baalen C. & Brown R.M. (1969) The ultrastructure of the marine blue-green alga, *Trichodesmium erythraeum*, with special reference to the cell wall, gas vacuoles, and cylindrical bodies. *Arch. Mikrobiol.* **69,** 79–91.
6.2, 13.3, 19.2.4.1, 19.3.1, 19.3.2, 19.4.2

Van Baalen C., Hoare D.S. & Brandt E. (1971) Heterotrophic growth of blue-green algae in dim light. *J. Bact.* **105,** 685–689.
3.4

Van Baalen C. & O'Donnell R. (1972) Action spectra for ultraviolet killing and photoreactivation in the blue-green alga *Agmenellum quadruplicatum. Photochem. Photobiol.* **15,** 269–274.
11.4.2.1

Vance B.D. (1965) Composition and succession of cyanophycean water blooms. *J. Phycol.* **1,** 81–86.
3.4.3

Vance B.D. (1966) Sensitivity of *Microcystis aeruginosa* and other blue-green algae and associated bacteria to selected antibiotics. *J. Phycol.* **2,** 125–128.
18.5.1.1

van den Hoek C. (1960) Groupements d'algues des étangs saumatres méditerraneéns de la cote française. *Vie et Milieu* **11,** 390–412.
20.2.3

van den Hoek C. (1969) Algal vegetation—types along the open coasts of Curaçao, Netherlands Antilles 1. *Proc. Kon. Ned. Akad. Wetensch.* **C72,** 537–577.
20.2.1

van den Hoek C., Colijn F., Cortel-Breeman A.M. & Wanders J.B.W. (1972) Algal vegetation-types along the shores of inner bays and lagoons of Curacao, and of the Lagoon Lac (Bonaire), Netherlands Antilles. *Verh. Kon. Ned. Akad. Wetensch.* **61** (2), 3–72.
20.2.1, 20.2.3

van den Hondel C.A.M.J.J., Keegstra W., Borrias W.E. & van Arkel G.A. (1979) Homology of plasmids in strains of unicellular cyanobacteria. *Plasmid* **2,** 323–333.
11.3.2

van den Hondel C.A.M.J.J. & van Arkel G.A. (1980) Development of a cloning system in cyanobacteria. *Antonie van Leeuwenhoek* **46,** 228–229.
11.3.2

van den Hondel C.A.M.J.J., Verbeek S., van der Ende A., Weisbeek P.J., Borrias W.E. & van Arkel G.A. (1980) Introduction of the transposon Tn901 into a plasmid of *Anacystis nidulans*: preparation for cloning in cyanobacteria. *Proc. natn. Acad. Sci. U.S.A.* **77,** 1570–1574.
11.3.2, 11.6.3, 11.8

van der Drift C. & de Jong M.H. (1974) Chemotaxis toward amino acids in *Bacillus subtilis. Arch. Microbiol.* **96,** 83–92.
17.5.1

Vanderhoef L.N., Huang C.Y., Musil R. & Williams J. (1974) Nitrogen fixation (acetylene reduction) by phytoplankton in Green Bay, Lake Michigan, in relation to nutrient

concentrations. *Limnol. Oceanogr.* **19,** 119–125.
5.4.2

van der Velde H.H. (1973) The natural occurrence in red algae of two phycoerythrins with different molecular weights and spectral properties. *Biochem. biophys. Acta* **303,** 246–257.
7.5.1.1, 7.5.1.4

Van de Water S. & Simon R.D. (1978) Macromolecular synthesis during heterocyst formation in the filamentous cyanobacterium *Cylindrospermum licheniforme. Abstract, Ann. Mtg Am. Soc. Microbiol.* 1978. 399 pp., p. 100. Am. Soc. Microbiol., Washington, D.C.
14.6.1.2

van Eykelenburg C. (1977) On the morphology and ultrastructure of the cell wall of *Spirulina platensis. Antonie van Leeuwenhoek* **43,** 89–99.
13.2.1

van Eykelenburg C. (1978) A glucan from the cell wall of the cyanobacterium *Spirulina platensis. Antonie van Leeuwenhoek* **44,** 321–327.
13.4.5

van Eykelenburg C. (1979) The ultrastructure of *Spirulina platensis* in relation to temperature and light intensity. *Antonie van Leeuwenhoek* **45,** 369–390.
6.1, 6.5.3, 13.3

van Eykelenburg C., Fichs, A. & Schmidt G.H. (1980) Some theoretical considerations on the *in vitro* shape of the cross walls in *Spirulina* spp. *J. theor. Biol.* **82,** 271–282.
13.3

van Gemerden H. (1974) Coexistence of organisms competing for the same substrate: An example among the purple sulphur bacteria. *Microbial Ecology* **1,** 104–119.
2.6.2

van Gorkom H.J. & Donze M. (1971) Localization of nitrogen fixation in *Anabaena. Nature, Lond.* **234,** 231–232.
14.2

van Liere L. (1979) *On Oscillatoria agardhii* Gomont, *Experimental Ecology of a Nuisance Bloom-forming Cyanobacterium.* 100 pp. Proefschrift Universiteit van Amsterdam. De Nieuwe Schouw Press, Zeist.
2.2.1, 2.2.2, 2.5.1, 3.2.2.2, 18.5.1.3, 19.4.1

van Liere L., de Groot G. & Mur L.R. (1979) Pigment variation with irradiance in *Oscillatoria agardhii* Gomont in nitrogen (nitrate)-limited chemostat cultures. *FEMS Microbiol. Lett.* **6,** 337–340.
2.5.1, 18.5.5.2

van Liere L., Loogman J.G. & Mur L.R. (1978) Measuring light-irradiance in cultures of phototrophic micro-organisms. *FEMS Microbiol. Lett.* **3,** 161–164.
2.5.2

van Liere L. & Mur L.R. (1979a) Decay of *Oscillatoria agardhii. Hydrobiol. Bull.* **13,** 56–60.
2.6.3

van Liere L. & Mur L.R. (1979b) Growth kinetics of *Oscillatoria agardhii* Gomont in continuous culture, limited in its growth by the light energy supply. *J. gen. Microbiol.* **115,** 153–160.
2.5.1, 2.5.2, 2.6.3, 18.5.3

van Liere L. & Mur L.R. (1980) Occurrence of *Oscillatoria agardhii* and some related species, a survey. In: Barica J. & Mur L.R. (Eds) *Hypertrophic Ecosystems,* 348 pp., pp. 67–77. Junk, Den Haag.
2.3

van Liere L., Mur L.R., Gibson C.E. & Herdman M. (1979) Growth and physiology of *Oscillatoria agardhii* Gomont cultivated in continuous culture with a light-dark cycle. *Arch. Microbiol.* **123,** 315–318.
2.5.1, 2.6.3, 3.3, 3.5

van Liere L., Zevenboom W. & Mur L.R. (1977) Nitrogen as a limiting factor for the growth of the blue-green alga *Oscillatoria agardhii. Progr. Wat. Tech.* **8,** 301–312.
3.2.2.2, 18.5.2.1

Vaughan M.H., Jr (1964) *Structural and Comparative Studies of the Algal Protein Phycoerythrin.* 459 pp. Ph.D. Thesis, Massachusetts Institute of Technology, Cambridge.
7.5.1.4, 7.5.1.5, 7.5.1.7

Velthuys B.R. (1978) A third site of proton translocation in green plant photosynthetic electron transport. *Proc. natn. Acad. Sci. U.S.A.* **75,** 6031–6034.
8.7

Venrick E.L. (1974) The distribution and significance of *Richelia intracellularis* Schmidt in the North Pacific Central Gyre. *Limnol. Oceanogr.* **19,** 437–445.
19.2.4.2

Verkleij A.J., Mombers C., Leunissen-Bijvelt J. & Ververgaert P.H.J.T. (1979) Lipid intramembraneous particles. *Nature, Lond.* **279,** 162–163.
13.2.1

Verkleij A.J. & Ververgaert P.H.J.Th. (1978) Freeze-fracture morphology of biological membranes. *Biochim. biophys. Acta* **515,** 303–327.
6.3.1

Vernotte C. (1971) Separation et characterisa-

tion spectroscopique du monomère et des polymères de la phycocyanine. *Photochem. Photobiol.* **14,** 163–173.
7.5.1.5

Verwer W., Ververgaert P.H.J.T., Leunissen-Bijvelt J. & Verkleij A.J. (1978) Particle aggregation in photosynthetic membranes of the blue-green alga *Anacystis nidulans. Biochim. biophys. Acta* **504,** 231–234.
6.5.3, 8.8

Vicente M., de Pedro M.A., de Torrontegui G. & Canovas J.L. (1975) The uptake of glucose and gluconate by *Pseudomonas putida. Mol. Cell Physiol.* **7,** 59–64.
3.4.3

Viner A.B. & Smith I.R. (1973) Geographical, historical and physical aspects of Lake George. *Proc. R. Soc. Lond. B.* **184,** 235–270.
9.5.4

Vogelmann T.C. & Scheibe J. (1978) Action spectra for chromatic adaptation in the blue-green alga *Fremyella diplosiphon. Planta, Berl.* **143,** 233–239.
7.8.3.2, 16.4.4

Voltolina D. (1975) The phytoplankton of the lagoon of Venice: November 1971–November 1972. *Pubbl. Staz. zool. Napoli* **39,** 206–340.
19.2.4.1

Voltolina D. (1980) The phytoplankton of Liverpool Bay. Its taxonomy and quantitative distribution. Ph.D. Thesis, University of Wales.
19.2.4.1

Waaland J.R., Waaland S.D. & Bates G. (1974) Chloroplast structure and pigment composition in the red alga *Griffithsia pacifica*: regulation by light intensity. *J. Phycol.* **10,** 193–199.
7.3, 7.8.2

Waaland J.R., Waaland S.D. & Branton D. (1971) Gas vaculoes. Light shielding in blue-green algae. *J. Cell Biol.* **48,** 212–215.
2.7.3.1, 2.8.2.3

Wada K., Hase T., Tokunaga H. & Matsubara H. (1975) Amino acid sequence of *Spirulina platensis* ferrodoxin: a far divergency of blue-green algal ferredoxins. *FEBS Lett.* **55,** 102–104.
8.3

Wahal C.K., Bhattacharya N.C. & Talpasayi E.R.S. (1973a) Ascorbic acid and heterocyst development in the blue-green alga *Anabaena ambigua. Physiol. Plant.* **28,** 424–429.
14.4.3.2

Wahal C.K., Bhattacharya N.C. & Talpasayi E.R.S. (1974) Lipid synthesis and heterocyst development in *Anabaena ambigua* Rao. *Ind. J. exp. Biol.* **12,** 92–93.
14.5

Wahal C.K., Reddy P.M. & Talpasayi E.R.S. (1973b) Lipid composition of some blue-green algae. *Ind. J. exp. Biol.* **11,** 588–589.
14.3.1

Waite T.D. & Mitchell R. (1976) Some benevolent and antagonistic relationships between *Ulva lactuca* and its microflora. *Aquat. Bot.* **2,** 13–22.
17.1.2

Waksman S.A. & Carey C.L. (1935) Decomposition of organic matter in seawater by bacteria. I. Bacterial multiplication in stored seawater. *J. Bact.* **29,** 531–543.
17.3.2

Walcott C.D. (1915) Discovery of Algonkian bacteria. *Proc. natn. Acad. Sci. U.S.A.* **1,** 256–257.
21.1

Walker J.C.G., Klein C., Schidlowski M., Schopf J.W., Stevenson D.J. & Walter M.R. (1982) Environmental evolution of the Archean-Early Proterozoic Earth. In: Schopf J.W. (Ed.) *Origin and Evolution of Earth's Earliest Biosphere.* Princeton Univ. Press, Princeton, N.J.
21.2, 21.3

Wallace D.C. & Morowitz H.J. (1973) Genome size and evolution. *Chromosoma* **40,** 121–126.
11.3.1

Wallin R., Selset R. & Sletten K. (1978) Characterization of chromophoric peptides from C-phycocyanin. *Biochem. biophys. Res. Commun.* **81,** 1319–1328.
7.5.1.8

Walsby A.E. (1968) Mucilage secretion and the movements of blue-green algae. *Protoplasma, Wien* **65,** 223–238.
2.8.1, 16.2.2.1, 16.5

Walsby A.E. (1969) The permeability of blue-green algal gas-vacuoles' membranes to gas. *Proc. R. Soc. Lond B.* **173,** 233–255.
2.8.2.2, 10.2.5.1, 10.3.1.2

Walsby A.E. (1970) The nuisance alga: curiosities in the biology of planktonic blue-green algae. *Wat. Treat. Exam.,* **19,** 359–373.
2.7.3.1, 2.8.2.2

Walsby A.E. (1971) The pressure relationships of gas vacuoles. *Proc. R. Soc. Lond. B* **178,** 301–326.
2.8.2.2, 10.2.5, 10.2.5.1, 10.2.5.2, 10.3.1.2

Walsby A.E. (1972) Structure and function of gas vacuoles. *Bact. Rev.* **36,** 1–32.
2.7.3.1, 2.8.2.1, 2.8.2.3, 10.2.5.1

Walsby A.E. (1973) A portable apparatus for measuring relative gas vacuolation, the strength of gas vacuoles and turgor pressure in blue-green algae and bacteria. *Limnol. Oceanogr.* **18**, 653–658.
10.2.5.1

Walsby A.E. (1974a) The extracellular products of *Anabaena cylindrica* Lemm. I. Isolation of a macromolecular pigment peptide complex and other components. *Br. phycol. J.* **9**, 371–381.
2.7.2.1, 17.4.4, 17.5.2, 17.6

Walsby A.E. (1974b) The extracellular products of *Anabaena cylindrica* Lemm. II. Fluorescent substances containing serine and threonine, and their role in extracellular pigment formation. *Br. phycol. J.*, **9**, 383–391.
2.7.2.1, 17.4.4, 17.6

Walsby A.E. (1974c) The isolation of gas vesicles from blue-green algae. In: Fleischer S. & Packer L. (Eds) *Methods in Enzymology. XXXI. Biomembranes, Part A.* pp. 678–686. Academic Press, New York.

Walsby A.E. (1977) Absence of gas vacuole protein in a mutant of *Anabaena flos-aquae. Arch. Microbiol.* **114**, 167–170.
2.6.1, 11.5.5.2(a)

Walsby A.E. (1978a) The properties and buoyancy-providing role of gas vacuoles in *Trichodesmium* Ehrenberg. *Br. phycol. J.* **13**, 103–116.
2.8.2.3, 19.2.4.1, 19.3.6, 19.4.2

Walsby A.E. (1978b) The gas vesicles of aquatic prokaryotes. *Symp. Soc. gen. Microbiol.* **28**, 327–358.
2.7.3.1, 10.2.5.1, 10.2.5.2, 18.3.2, 18.4.4

Walsby A.E. (1980) The water relations of gas-vacuolate prokaryotes *Proc. R. Soc. Lond. B.* **208**, 73–102.
10.2.5, 10.2.5.1, 10.2.5.2, 10.2.6, 10.2.7, 10.3.1, 10.3.3.2

Walsby A.B. & Armstrong R.E. (1979) Average thickness of the gas vesicle wall. *J. mol. Biol.* **129**, 279–285.
2.8.2.1

Walsby A.E. & Booker M.J. (1980) Changes in buoyancy of a planktonic blue-green alga in response to light intensity. *Br. phycol. J.* **15**, 311–319.
2.8.2.2

Walsby A.E. & Buckland B. (1969) Isolation and purification of intact gas vesicles from a blue-green alga. *Nature, Lond.* **224**, 716–717.
10.2.5.1

Walsby A.E. & Klemer A.R. (1974) The role of gas vacuoles in the microstratification of a population of *Oscillatoria agardhii* var. *isothrix* in Deming Lake, Minnesota. *Arch. Hydrobiol.* **74**, 375–392.
2.8.2.3, 2.8.2.4, 10.2.5.2

Walsby A.E. & Nichols B.W. (1969) Lipid composition of heterocysts. *Nature, Lond.* **221**, 673–674.
14.3.1, 14.3.2

Walsby A.E. & Xypolyta A. (1977) The form resistance of chitan fibres attached to the cells of *Thalassiosira fluviatilis* Hustedt. *Br. phycol. J.* **12**, 215–223.
10.2.3

Walter M.R. (1972) Stromatolites and the biostratigraphy of the Australian Precambrian and Cambrian. *Palaeontol. Assoc. London, Spec. Pap.* **11**, 190 pp.
21.1, 21.2

Walter M.R. (1976) *Stromatolites. Developments in Sedimentology 20.* 790 pp. Elsevier, Amsterdam.
20.1, 20.4.2, 21.1, 21.3

Walter M.R. (1977) Interpreting stromatolites. *Am. Sci.* **65**, 563–571.
21.3

Walter M.R. (1982) Archaean stromatolites: evidence of the Earth's earliest benthos. In: Schopf J.W. (Ed.) *Origin and Evolution of Earth's Earliest Biosphere.* Princeton Univ. Press, Princeton.
21.2

Walter M.R. & Awramik S.M. (1979) *Frutexites* from stromatolites of the Gunflint Iron Formation of Canada, and its biological affinites. *Precambrian Res.* **9**, 23–33.
21.2

Walter M.R., Bauld J. & Brock T.D. (1976a) Microbiology and morphogenesis of columnar stromatolites (*Conophyton, Vacerrilla*) from hot springs in Yellowstone National Park. In: Walter M.R. (Ed.) *Stromatolites. Developments in Sedimentology 20.* pp. 273–310. Elsevier, Amsterdam.
21.2

Walter M.R., Buick R. & Dunlop J.S.R. (1980) Stromatolies 3,400–3,500 Myr old from the North Pole area, Western Australia. *Nature, Lond.* **284**, 443–445.
21.2

Walter M.R., Oehler J.H. & Oehler D.Z. (1976b) Megascopic algae 1300 million years old from the Belt Supergroup, Montana: a reinterpretation of Walcott's Helminthoidichnites. *J. Paleontol.* **50**, 872–881.
21.3

Wanders J.B.W. (1977) The role of benthic

algae in the shallow reef of Curaçao (Netherlands Antilles) III: The significance of grazing. *Aquatic Botany* **3**, 357–390.
20.2.4

Wang A.W. & Hill A. (1977) Chemical analysis of the phenol-water-extractable materials from *Anabaena flos-aquae*. *J. Bact.* **130**, 558–560.
13.4.4, 13.4.5

Wang R.T. & Myers, J. (1977) Reverse energy transfer from chlorophyll to phycobilin in *Anacystis nidulans*. *Pl. Cell Physiol., Tokyo* **18**, 3–7.
8.6

Wang R.T., Stevens C.L.R. & Myers J. (1977) Action spectra for photoreactions I and II of photosynthesis in the blue-green alga *Anacystis nidulans*. *Photochem. Photobiol.* **25**, 103–108.
7.9, 8.4, 11.5.4

Wang W.S. & Lundgren D.G. (1969) Poly-β-hydroxy butyrate in the chemolithotrophic bacterium, *Ferrobacillus ferrooxidans*. *J. Bact.* **97**, 947–950.
3.2.1

Wang W.S. & Tischer R.G. (1973) Study of the extracellular polysaccharides produced by a blue-green alga, *Anabaena flos-aquae* A-37. *Arch. Microbiol.* **91**, 77–81.
17.2

Warburg O. (1929) Über das Geswindigkeit der photochemischen Kohlensäurezersetzung in lebenden Zellen. *Biochem. Z.* **103**, 188–217.
17.6

Ward A.K. & Wetzel R.G. (1980) Interactions of light and nitrogen source among planktonic blue-green algae. *Arch. Hydrobiol.* **90**, 1–25.
18.5.2.1

Wärmling P. (1973) Nitrogen fixation on rocks in Oslofjord. *Bot. Mar.* **16**, 237–240.
20.3.2, 20.3.3

Watanabe I., Espinas C.R., Berja N.S. & Alimagno B.V. (1977) Utilization of the *Azolla-Anabaena* complex as a nitrogen fertilizer for rice. *IRRI Research Paper Series* **11**, 1–15.
14.1

Watanabe A. & Yamamoto Y. (1967) Heterotrophic nitrogen fixation by the blue-green alga *Anabaenopsis circularis*. *Nature, Lond.* **214**, 738.
3.4.1, 3.4.2

Waterbury J.B. & Stanier R.Y. (1978) Patterns of growth and development in pleurocapsalean cyanobacteria. *Microbiol. Rev.* **42**, 2–44.
13.2.2, 15.1, 16.2.2.1

Waterbury J.B., Watson S.W., Guillard R.R.L. & Brand L.E. (1979) Widespread occurrence of a unicellular marine planktonic cyanobacterium. *Nature, Lond.* **277**, 293–294.
1, 19.2.1, 19.3.1, 19.3.2

Weare N.M. & Benemann J.R. (1973a) Nitrogen fixation by *Anabaena cylindrica*. I. Localization of nitrogen fixation in the heterocysts. *Arch. Mikrobiol.* **90**, 323–332.
4.3, 14.2, 14.3.2, 14.4.3.1

Weare N.M. & Benemann J.R. (1973b) Nitrogen fixation by *Anabaena cylindrica*. II. Nitrogenase activity during induction and aging of batch-cultures. *Arch. Mikrobiol.* **93**, 101–112.
4.3, 4.4

Weare N.M. & Benemann J.R. (1974) Nitrogenase activity and photosynthesis in *Plectonema boryanum*. *J. Bact.* **119**, 258–265.
4.3

Weathers P.J. & Allen M.M. (1978) Variations in short term products of inorganic carbon fixation in exponential and stationary phase cultures of *Aphanocapsa* 6308. *Arch. Microbiol.* **116**, 231–234.
3.2.1, 3.2.2.2

Weathers P.J., Chee H.L. & Allen M.M. (1978) Arginine catabolism in *Aphanocapsa* 6308. *Arch. Microbiol.* **118**, 1–6.
3.2.2.2

Webb K.L., Dupaul W.D., Wiebe W., Sottile W. & Johannes R.E. (1975) Enewetak (Eniwetok) Atoll: aspects of the nitrogen cycle on a coral reef. *Limnol. Oceanogr.* **20**, 198–210.
20.4.5

Webber E.E. (1967) Blue-green algae from a Massachusetts salt marsh. *Bull. Torrey bot. Club* **94**, 98–106.
20.2.3

Weber M. & Wöber G. (1975) The fine structure of the branched α-glucan from the blue-green alga *Anacystis nidulans*. Comparison with other bacterial glycogens. *Carbohydrate Res.* **39**, 295–302.
3.2.2.2

Webster G.C. & Frenkel A.W. (1952) Some respiratory characteristics of the blue-green alga, *Anabaena*. *Pl. Physiol., Lancaster* **28**, 63–69.
3.3, 18.5.3

Weckesser J., Drews G. & Mayer H. (1979) Lipopolysaccharides of photosynthetic prokaryotes. *A. Rev. Microbiol.* **33**, 215–239.
13.1, 13.4.4

Weckesser J., Katz A., Drews G., Mayer H. & Fromme I. (1974) Lipopolysaccharide containing L-acofriose in the filamentous blue-

green alga *Anabaena variabilis*. *J. Bact.* **120,** 672–678.
13.4.4, 13.4.5

Wehrmeyer W. (1971) Elektronmikroskopische Untersuchung zur Feinstruktur von *Porphyridium violaceum* Kornmann mit Bemerkungen über seine taxonomische Stellung. *Arch. Mikrobiol.* **75,** 121–139.
7.3

Wehrmeyer W. & Schneider H. (1975) Elektronmikroskopische Untersuchungen zur reversiblen Veränderung der Chloroplastenfeinstruktur von *Rhodella violacea* bei Stickstoffmangel. *Biochem. Physiol. Pflanz.* **168,** 519–532.
7.3

Weidel W. & Pelzer H. (1964) Bagshaped macromolecules: a new outlook on bacterial cell walls. *Advan. Enzymol.* **26,** 193–232.
13.3

Weise G., Drews G., Jann B. & Jann K. (1970) Identification and analysis of a lipopolysaccharide in cell walls of the blue-green alga, *Anacystis nidulans*. *Arch. Mikrobiol.* **71,** 89–98.
13.4.4

Weisman J.C. & Benemann J.R. (1977) Hydrogen production by nitrogen starved cultures of *Anabaena cylindrica*. *Appl. environ. Microbiol.* **33,** 123–131.
4.6, 9.3.3

Weitzman P.D.J. (1973) Behaviour of enzymes at high concentration: Use of permeabilised cells in the study of enzyme activity and its regulation. *FEBS Lett.* **32,** 247–250.
3.2.1

Weller D., Doemel W. & Brock T.D. (1975) Requirement of low oxidation-reduction potential for photosynthesis in a blue-green alga (*Phormidium* sp.). *Arch. Microbiol.* **104,** 7–13.
9.5.2, 9.6

Wentworth C.K. (1922) A scale of grade and class terms for clastic sediments. *J. Geol.* **30,** 377–392.
20.1

Werbin H. & Rupert C.S. (1968) Photoreactivating enzyme in blue-green alga cells. *Photochem. Photobiol.* **7,** 225–230.
11.4.2.1

Wetzel R.G. (1975) *Limnology 1st edn*, 743 pp. W. B. Saunders Co., Philadelphia.
2.2.1

Whatley J.M. (1977) The fine structure of *Prochloron*. *New Phytol.* **79,** 309–313.
1

Whatley J.M. (1981) Chloroplast evolution—ancient and modern. *Ann. N.Y. Acad. Sci.* **361,** 154–165.
12.4.4.3

Whatley J.M., John P. & Whatley F.R. (1979) From extracellular to intracellular: the establishment of mitochondria and chloroplasts. *Proc. R. Soc. Lond. B.* **204,** 165–187.
12.4.1, 12.4.2, 12.4.4.3

White A.W. (1974a) Growth of two facultatively heterotrophic marine centric diatoms. *J. Phycol.* **10,** 292–300.
3.4.3

White A.W. (1974b) Uptake of organic compounds by two facultatively heterotrophic marine centric diatoms. *J. Phycol.* **10,** 433–438.
3.4.3

White A.W. & Shilo M. (1975) Heterotrophic growth of the filamentous blue-green alga *Plectonema boryanum*. *Arch. Microbiol.* **102,** 123–127.
3.4.1, 3.4.2, 3.4.3

White D. (1928) Algal deposits of Unkar Proterozoic age in the Grand Canyon, Arizona. *Proc. natn. Acad. Sci. U.S.A.* **14,** 597–600.
21.1

Whitfield M. (1969) Eh as an operational parameter in estuarine studies. *Limnol. Oceanogr.* **14,** 547–558.
9.5.3

Whitton B.A. (1965) Extracellular products of blue-green algae. *J. gen. Microbiol.* **40,** 1–11.
2.7.2.1

Whitton B.A. (1967) Phosphate accumulation by colonies of *Nostoc*. *Pl. Cell Physiol., Tokyo* **8,** 293–296.
5.1.1

Whitton B.A. (1969) Seasonal changes in the phytoplankton of St. James's Park Lake, London. *London Naturalist* **48,** 14–39.
18.4.1

Whitton B.A. (1973a) Freshwater plankton. In: Carr N.G. & Whitton B.A. (Eds) *The Biology of Blue-Green Algae*. 676 pp., pp. 353–367. Blackwell, Oxford.
18.1, 18.5.1.3

Whitton B.A. (1973b) Interactions with other organisms. In: Carr N.G. & Whitton B.A. (Eds) *The Biology of Blue-Green Algae*. 676 pp., pp. 415–433. Blackwell, Oxford.
17.1.1

Whitton B.A., Carr N.G. & Craig I. (1971) A comparison of the fine structure and nucleic acid biochemistry of chloroplasts and blue-green algae. *Protoplasma, Wien* **72,** 325–357.
6.2

Whitton B.A. & MacArthur K. (1967) The action of two toxic quinones on *Anacystis nidulans*. *Arch. Mikrobiol.* **57,** 147–154.
18.6

Whitton B.A. & Peat A. (1969) On *Oscillatoria redekei* van Goor. *Arch. Mikrobiol.* **68,** 362–376.
6.1, 18.2.1

Whitton B.A. & Potts M. (1980) Blue-green algae (Cyanobacteria) of the oceanic coast of Aldabra. 8 pp. *Atoll Res. Bull.* No. 238.
20.2.1, 20.2.4, 20.3.2, 20.3.4, 20.4.3

Wiebe W.J., Johannes R.E. & Webb K.L. (1975) Nitrogen fixation in a coral reef community. *Science, N.Y.* **188,** 257–259.
19.5.2, 20.4.1, 20.4.5

Wilcockson, J. (1977) The apparent influence of atmospheric oxygen concentration on nitrogenase activity and slime production in *Klebsiella pneumoniae* grown on a solid medium. *J. gen. Microbiol.* **101,** 311–317.
4.4

Wilcox M. (1970) One-dimensional pattern found in blue-green algae. *Nature, Lond.* **228,** 686–687.
14.6.1.1

Wilcox M., Mitchison G.J. & Smith R.J. (1973a) Pattern formation in the blue-green alga, *Anabaena*. I. Basic mechanisms. *J. Cell Sci.* **12,** 707–723.
14.6.1.2, 14.6.1.3

Wilcox M., Mitchison G.J. & Smith R.J. (1973b) Pattern formation in the blue-green alga, *Anabaena*. II. Controlled proheterocyst regression. *J. Cell Sci.* **13,** 637–649.
14.5, 14.6.1.2

Wilcox M., Mitchison G.J. & Smith R.J. (1975a) Spatial control of differentiation in the blue-green alga *Anabaena*. In: Schlessinger D. (Ed.) *Microbiology—1975*. 521 pp., pp. 453–463. Am. Soc. Microbiol., Washington, D.C.
14.6.1.1, 14.6.1.2

Wilcox M., Mitchison G.J. & Smith R.J. (1975b) Mutants of *Anabaena cylindrica* altered in heterocyst spacing. *Arch. Microbiol.* **103,** 219–223.
14.5, 14.6.1.1, 14.6.1.2

Wildman R.B. & Bowen C.C. (1974) Phycobilisomes in blue-green algae. *J. Bact.* **117,** 866–881.
7.3

Wildman R.B., Loescher J.H. & Winger C.L. (1975) Development and germination of akinetes of *Aphanizomenon flos-aquae*. *J. Phycol.* **11,** 96–104.
15.2, 15.5, 15.6

Wildon D.C. & Mercer F.V. (1963a) The ultrastructure of the heterocyst and akinete of the blue-green algae. *Arch. Mikrobiol.* **47,** 19–31.
14.3.2, 15.2

Wildon D.C. & Ap Rees T. (1965) Metabolism of glucose C^{14} by *Anabaena cylindrica*. *Pl. Physiol., Lancaster* **40,** 332–335.
3.4.2

Wildon D.C. & Mercer F.V. (1963b) The ultrastructure of the vegetative cell of blue-green algae. *Aust. J. biol. Sci.* **16,** 585–596.
7.2, 13.3, 14.3.2

Wilkinson C.R. (1979) *Bdellovibrio*-like parasite of cyanobacteria symbiotic in marine sponges. *Arch. Microbiol.* **123,** 101–103.
17.4.2

Wilkinson, J.F. (1958) Extra-cellular polysaccharides of bacteria. *Bact. Rev.* **22,** 46–73.
4.4

Williams E., Lambert J., O'Brien P. & Houghton J.A. (1979) Evidence for dark repair of far ultraviolet light damage in the blue-green alga *Gloeocapsa alpicola*. *Photochem. Photobiol.* **29,** 543–547.
11.4.2.1

Williams L.G. (1950) The role of algae in stabilizing beach sand. *Bull. Furman Univ. U.S.A.* **33,** 61–63.
20.2.2

Williams R.C., Gingrich J.C. & Glazer A.N. (1980) Cyanobacterial phycobilisomes. Particles from *Synechocystis* 6701 and two pigment mutants. *J. Cell Biol.* **85,** 558–566.
11.5.4

Williams V.P., Freidenreich P. & Glazer A.N. (1974) Homology of amino-terminal regions of C-phycocyanins from a prokaryote and a eukaryote. *Biochem. biophys. Res. Commun.* **59,** 462–466.
7.5.1.8.

Williams V.P. & Glazer A.N. (1978) Structural studies on phycobiliproteins. I. Bilin-containing peptides of C-phycocyanin. *J. biol. Chem.* **253,** 202–211.
7.5.1.2

Wilson A.C., Carlson S.S. & White J.J. (1977) Biochemical evolution. *A. Rev. Biochem.* **46,** 573–639.
12.1.3

Wilson A.T. & Calvin M. (1955) The photosynthetic cycle. CO_2 dependent transients. *J. Am. chem. Soc.* **77,** 5948–5957.
3.2.1

Winet H. & Keller S.R. (1976) Spirillum swimming: theory and observations of propulsion by the flagellar bundle. *J. exp. Biol.* **65,** 577–602.
16.3

Winkenbach F. & Wolk C.P. (1973) Activities of enzymes of the oxidative and the reductive pentose phosphate pathways in heterocysts of a blue-green alga. *Pl. Physiol., Lancaster* **52,** 480–483.
3.2.1, 4.3, 14.4.1, 14.4.2

Winkenbach F., Wolk C.P. & Jost M. (1972) Lipids of membranes and of the cell envelope in heterocysts of a blue-green alga. *Planta, Berl.* **107,** 69–80.
14.3.1, 14.3.2

Withers N.W., Alberte R.S., Lewin R.A., Thornber J.P., Britton G. & Goodwin T.W. (1978a) Photosynthetic unit size, carotenoids, and chlorophyll-protein composition of *Prochloron* sp., a prokaryotic green alga. *Proc. natn. Acad. Sci. U.S.A.* **75,** 2301–2305.
12.2.1, 12.2.3, 12.4.4.3

Withers N.W., Vidaver W. & Lewin R.A. (1978b) Pigment composition, photosynthesis and fine structure of a non-blue-green prokaryotic algal symbiont (*Prochloron* sp.) in a didemnid ascidian from Hawaiian waters. *Phycologia* **17,** 167–171.
20.2.4

Witkin E.M. (1969a) The role of DNA repair and recombination in mutagenesis. *Proc. XII Int. Congr. Genet.* **3,** 225–245.
11.4.2

Witkin E.M. (1969b) Ultraviolet-induced mutation and DNA repair. *A. Rev. Microbiol.* **23,** 487–514.
11.4.2

Woese C.R. & Bleyman M.A. (1972) Genetic code limit organisms, do they exist? *J. molec. Evol.* **1,** 223–229.
12.3.4.2

Woese C.R. & Fox G.E. (1977) Phylogenetic structure of the prokaryotic domain: the primary kingdoms. *Proc. natn. Acad. Sci U.S.A.* **74,** 5088–5090.
12.1.1, 12.1.3, 12.2.2.4, 12.4.2, 12.4.3.2, 12.4.3.3

Woese C.R., Gibson J. & Fox G.E. (1980) Do genealogical patterns in purple photosynthetic bacteria reflect inter-specific gene transfer? *Nature, Lond.* **283,** 212–214.
12.2.2.1, 12.2.2.4

Woese C.R., Magrum L.J. & Fox G.E. (1978) Archaebacteria. *J. molec. Evol.* **11,** 245–252.
12.1.1

Wolk C.P. (1965a) Control of sporulation in a blue-green alga. *Devl Biol.* **12,** 15–35.
14.6.2, 15.1, 15.4.1, 15.4.3, 18.4.5

Wolk C.P. (1965b) Heterocyst germination under defined conditions. *Nature, Lond.* **205,** 201–202.
5.2.4, 14.5, 15.4

Wolk C.P. (1966) Evidence of a role of heterocysts in the sporulation of a blue-green alga. *Am. J. Bot.* **53,** 260–262.
14.6.2

Wolk C.P. (1967) Physiological basis of the pattern of vegetative growth of a blue-green alga. *Proc. natn. Acad. Sci. U.S.A.* **57,** 1246–1251.
14.6.1.2

Wolk C.P. (1968) Movement of carbon from vegetative cells to heterocysts in *Anabaena cylindrica*. *J. Bact.* **96,** 2138–2143.
3.2.1, 14.4.2

Wolk C.P. (1970) Aspects of the development of a blue-green alga. *Ann. N.Y. Acad. Sci.* **175,** 641–647.
14.2, 14.4.3.1, 14.6.1.3

Wolk C.P. (1973) Physiology and cytological chemistry of blue-green algae. *Bact. Rev.* **37,** 32–101.
2.7.2.2, 13.4, 14.3.2, 14.6.1.1, 15.1

Wolk C.P. (1975) Differentiation and pattern formation in filamentous blue-green algae. In: Gerhardt P., Costilow R.N. & Sadoff H.L. (Eds) *Spores VI.* 619 pp., pp. 85–96. *Am. Soc. Microbiol.*, Washington, D.C.
14.1, 14.3.1

Wolk C.P. (1979a) Developmental biology of cyanobacteria: advances and challenges. In: Nichols J.M. (Ed.) *Abst. Third Internat. Symp. Photosyn. Prokaryotes, Oxford.* p. D1. Univ. Liverpool, England.
4.5, 14.1, 14.4.2

Wolk C.P. (1979b) Intercellular interactions and pattern formation in filamentous cyanobacteria. In: Subtelny S. & Konigsberg I.R. (Eds) *Determinants of Spatial Organization, 37th Symp. Soc. Developmental Biology.* 333 pp., pp. 247–266. Academic Press, New York.
4.4, 14.1, 14.2, 14.3.1, 14.6.1.1, 14.6.1.2, 14.6.1.3

Wolk C.P., Austin S.M., Bortins J. & Galonsky A. (1974) Autoradiographic localization of ^{13}N after fixation of ^{13}N-labeled nitrogen gas by a heterocyst-forming blue-green alga. *J. Cell Biol.* **61,** 440–453.
14.4.1

Wolk C.P. & Quine M.P. (1975) Formation of one-dimensional patterns by stochastic processes and by filamentous blue-green algae. *Devl Biol.* **46,** 370–382.
14.6.1.1, 14.6.1.2

Wolk C.P. & Shaffer P.W. (1976) Heterotrophic micro- and macrocultures of a nitrogen-fixing cyanobacterium. *Arch. Microbiol.* **110,** 145–147.
3.4.1, 3.4.2, 3.4.3, 14.4.3.1

Wolk C.P. & Simon R.D. (1969) Pigments and

lipids of heterocysts. *Planta, Berl.* **86,** 92–97.
14.3.1, 14.3.2, 14.4.1, 14.4.3.2, 15.2

Wolk C.P., Thomas J., Shaffer P.W., Austin S.M. & Galonsky A. (1976) Pathway of nitrogen metabolism after fixation of ^{13}N-labeled nitrogen gas by the cyanobacterium, *Anabaena cylindrica. J. biol. Chem.* **251,** 5027–5034.
14.4.1

Wolk C.P. & Wojciuch E. (1971a) Photoreduction of acetylene by heterocysts. *Planta, Berl.* **97,** 126–134.
14.4.3.2, 14.4.3.3

Wolk C.P. & Wojciuch E. (1971b) Biphasic time course of solubilization of nitrogenase during cavitation of aerobically grown *Anabaena cylindrica. J. Phycol.* **7,** 339–344.
14.2

Wolk C.P. & Wojciuch E. (1973) Simple methods for plating single vegetative cells of, and for replica-plating, filamentous blue-green algae. *Arch. Mikrobiol.* **91,** 91–95.
11.5.3, 14.2

Wollman F.A. (1979) Ultrastructure comparison of *Cyanidium caldarium* wild type and III-C mutant lacking phycobilisomes. *Pl. Physiol., Lancaster* **63,** 375–381.
7.9

Womersley H.B.S. (1946) Studies on the marine algae of southern Australia. Introduction and No. 1. The genera *Isactis* and *Rivuleria* (Myxophyceae) *Trans. R. Soc. S. Australia* **70,** 127–136.
20.2.1

Wood B.J.B. (1974) Fatty acids and saponifiable lipids. In: Stewart W.D.P. (Ed.) *Algal Physiology and Biochemistry*. 989 pp., pp. 236–265. Blackwell, Oxford.
9.4.1, 12.3.3

Wood N.B. & Haselkorn R. (1976) Protein degradation during heterocyst development in *Anabaena*. In: Codd G.A. & Stewart W.D.P. (Eds) *Proceedings of the Second International Symposium on Photosyn. Prokaryotes*. 258 pp., pp. 125–127. Univ. Dundee, Scotland.
14.5

Wood N.B. & Haselkorn R. (1977) Protein degradation during heterocyst differentiation in the blue-green alga *Anabaena* 7120. *Fed. Proc. (Abstr.)* **36,** 886.
7.8.1

Wood N.B. & Haselkorn R. (1979) Proteinase activity during heterocyst differentiation in nitrogen-fixing cyanobacteria. In: Cohen G.N. & Holzer H. (Eds) *Limited Proteolysis in Microorganisms*. 255 pp., pp. 159–166. U.S. DHEW Publication No. (NIH) 79–1591, Bethesda, Md.
7.8.1, 14.5

Wood N.B. & Haselkorn R. (1980) Control of phycobiliprotein proteolysis and heterocyst differentiation in *Anabaena. J. Bact.* **141,** 1375–1385.
14.5

Wood P.M. (1977) The roles of c-type cytochromes in algal photosynthesis. Extraction from algae of a cytochrome similar to higher plant cytochrome f. *Eur. J. Biochem.* **72,** 605–612.
8.5

Wood P.M. (1978) Interchangeable copper and iron proteins in algal photosynthesis. Studies on plastocyanin and cytochrome c552 in *Chlamydomonas. Eur. J. Biochem.* **87,** 9–19.
8.5

Wood R.B. & Gibson C.E. (1973) Eutrophication and Lough Neagh. *Wat. Res.* **7,** 173–187.
5.4.3

Work E. & Dewey L.L. (1953) The distribution of α, ε-diamino-pimelic acid among various microorganisms. *J. gen. Microbiol.* **9,** 394–406.
13.4.1

Wright R.T. & Hobbie J.E. (1965) The uptake of organic solutes by planktonic bacteria and algae. *Ocean Sci. Engng* **1,** 116–127.
3.4.3

Wright R.T. & Hobbie J.E. (1966) The use of glucose and acetate by bacteria and algae in aquatic ecosystems. *Ecology* **47,** 447–464.
3.4.3

Wrigley R.C. & Horne A.J. (1974) Remote sensing and lake eutrophication. *Nature, Lond.* **250,** 213–214.
19.2.4.2

Wu U.U., Lewin R.A. & Werbin H. (1967) Photoreactivation of UV-irradiated blue-green algal virus LPP-1. *Virology* **31,** 657–664.
11.4.2.1

Wujek D.E. (1979) Intracellular bacteria in the blue-green alga *Pleurocapsa minor. Trans. Am. microscop. Soc.* **98,** 143–145.
17.2

Wyatt J.T. & Silvey J.K.G. (1969) Nitrogen fixation by *Gloeocapsa. Science, N.Y.* **165,** 908–909.
1, 4.1, 12.3.3

Wymer, P. & Thake B. (1980) The importance of phosphorus in microalgal growth and species competition in mixed populations;

experiments and simulations. *Proc. R. Soc. Lond. B* **209**, 333–353.
19.4.1

Yadava P.K. & Singh S.P. (1979) Histidine sensitive strain of the blue-green alga *Nostoc muscorum*. Possible *Nif-His* interactions. *Molec. gen. Genet.* **173**, 153–158.
11.5.5.2(b)

Yamamoto Y. (1972) The fatty acid composition of akinetes, heterocysts and vegetative cells in *Anabaena cylindrica*. *Pl. Cell Physiol., Tokyo* **13**, 913–915.
14.3.2, 15.2

Yamamoto Y. (1976) Effect of some physical and chemical factors on the germination of akinetes of *Anabaena cylindrica*. *J. gen. appl. Microbiol., Tokyo* **22**, 311–323.
15.5

Yamanaka G. & Glazer A.N. (1980) Dynamic aspects of phycobilisome structure. Phycobilisome turnover during nitrogen starvation in *Synechococcus* sp. *Arch. Microbiol.* **124**, 39–47.
7.8, 7.8.1

Yamanaka G., Glazer A.N. & Williams R.C. (1978a) Cyanobacterial phycobilisomes. Characterization of the phycobilisomes of *Synechococcus* sp. 6301. *J. biol. Chem.* **253**, 8303–8310.
7.4, 7.5, 7.5.1.6, 7.5.2, 7.6

Yamanaka G., Glazer A.N. & Williams R.C. (1980) Molecular architecture of a light-harvesting antenna. Comparison of wild type and mutant *Synechococcus* 6301 phycobilisomes. *J. biol. Chem.* **255**, 11004–11010.
11.5.4

Yamanaka T., Fukumori Y. & Wada K. (1978b) Cytochrome c_{554} derived from the blue-green alga *Spirulina platensis*. *Pl. Cell Physiol., Tokyo* **19**, 117–126.
8.5

Yamaoka T., Satoh K. & Katoh S. (1978) Photosynthetic activities of a thermophilic blue-green alga. *Pl. Cell Physiol., Tokyo* **19**, 943–954.
8.2

Yates M.G. (1980) The biochemistry of nitrogen fixation. In: Miflin B.J. (Ed.) *The Biochemistry of Plants, Vol. 5*. Academic Press, New York.
4.2, 4.4, 4.5

Yopp J.H., Miller D.M. & Tindall D.R. (1979) Regulation of intracellular water potential in the halophilic blue-green alga *Aphanothece halophytica* (Chroococcales). In: Caplan S.R. & Ginzburg M. (Eds) *Energetics and Structure of Halophilic Microorganisms*. pp. 619–624. Elsevier, North-Holland Biomedical Press.
10.4

Yopp J.H., Tindall D.R., Miller D.M. & Schmid W.E. (1978) Isolation, purification and evidence for a halophilic nature of the blue-green alga *Aphanothece halophytica* Frémy (Chroococcales). *Phycologia* **17**, 172–178.
10.4

Yu C.M. & Brand J.J. (1980) Role of divalent cations and ascorbate in photochemical activities of *Anacystis* membranes. *Biochim. biophys. Acta* **591**, 483–487.
8.2

Zablen L.B., Kissil M.S., Woese C.R. & Buetow D.E. (1975) The phylogenetic origin of the chloroplast and the prokaryotic nature of its ribosomal RNA. *Proc. natn. Acad. Sci. U.S.A.* **72**, 2418–2422.
12.4.3.2

Zagallo A.C. (1953) Oxidative metabolism of some bacteria and blue-green algae, alone and in association. *Agronomia lusit.* **15**, 315–345.
17.1.1

Zdrou I. & Tromballa H.W. (1981) Active transport of chloride by *Anacystis nidulans*. *Arch. Microbiol.* **129**, 325–330.
10.3.3

Zepke H.D., Heinz E., Radunz A., Linscheid M. & Pesch R. (1978) Combination and positional distribution of fatty acids from lipids from blue-green algae. *Arch. Microbiol.* **119**, 157–162.
13.4.3

Zevenboom W. & Mur L.R. (1978) Nitrogen-uptake and pigmentation of N-limited chemostat cultures and natural populations of *Oscillatoria agardhii*. *Mitt. int. Verein. theor. angew. Limnol.* **21**, 261–274.
5.4.1, 18.5.2.1, 18.5.5.5

Zevenboom W. & Mur L.R. (1980) N_2-fixing cyanobacteria: why they do not become dominant in Dutch hypertrophic lakes. In: Barica J. & Mur L.R. (Eds). *Hypertrophic Ecosystems*. 348 pp., pp. 123–130. Junk, den Haag.
2.5.2

Zevenboom W., Van der Does J., Bruning K. & Mur L.R. (1981) A non-heterocystous mutant of *Aphanizomenon flos-aquae*, selected by competition in light-limited continuous culture. *FEMS Microbiol. Lett.* **10**, 11–16.
2.6.1, 11.5.5.2(b), 18.5.2.1

Zhevner V.D., Glazer V.M. & Shestakov S.V.

(1973) Mutants of *Anacystis nidulans* with modified cell division. *Mikrobiologiya, Transl.* **42,** 290–297.
11.5.5.1

Zhevner V.D. & Shestakov S.V. (1972) Studies on the ultraviolet-sensitive mutants of blue-green alga *Synechococcus aquatilis* Sanv. *Arch Mikrobiol.* **86,** 349–360.
11.4.2.1, 11.4.2.3

Zilinskas B.A., Greenwald L.S., Bailey C.L. & Kahn P.C. (1980) Spectral analysis of allophycocyanin I, II, III and B from *Nostoc* sp. phycobilisomes. *Biochim. biophys. Acta.* **592,** 267–276.
7.5.1.1

Zilinskas B.A., Zimmerman B.K. & Gantt E. (1978) Allophycocyanin forms isolated from *Nostoc* sp. phycobilisomes. *Photochem. Photobiol.* **27,** 587–595.
7.5.1.1

Zimmermann U. (1969) Ökologische und physiologische Untersuchungen an der planktischen Blaualge *Oscillatoria rubescens* D.C. unter besonderer Berucksichtigung von Licht und Temperatur. *Schweiz. Z. Hydrol.* **31,** 1–58.
18.4.1, 18.4.3

Zimmermann U. (1978) Physics of turgor and osmoregulation. *A. Rev. Pl. Physiol.* **29,** 121–148.
10.2.6, 10.2.7

Zipkas D. & Riley M. (1975) Proposal concerning the evolution of the genome of *Escherichia coli. Proc. natn. Acad. Sci. U.S.A.* **72,** 1354–1358.
11.3.1

Zuber H. (1978) Studies on the structure of the light-harvesting pigment-protein-complexes from cyanobacteria and red algae. *Ber. dt. bot. Ges.* **91,** 459–475.
7.5.1.8

ORGANISM INDEX

This index is composed of genera and species of both cyanobacteria and other organisms. Genera and individual records lacking a specific name are grouped together. The names quoted are those used by the authors; synonyms are not cross-referenced, but see pages 48 and 266.

SUBJECT INDEX